T0210755

Lecture Notes in Computer Science   **9489**

Commenced Publication in 1973
Founding and Former Series Editors:
Gerhard Goos, Juris Hartmanis, and Jan van Leeuwen

Editors
Sabri Arik
University of Istanbul
Istanbul
Turkey

Tingwen Huang
University at Qatar
Doha
Qatar

Weng Kin Lai
Tunku Abdul Rahman University College
Kuala Lumpur
Malaysia

Qingshan Liu
University of Science Technology
Wuhan
China

ISSN 0302-9743   ISSN 1611-3349 (electronic)
Lecture Notes in Computer Science
ISBN 978-3-319-26531-5   ISBN 978-3-319-26532-2 (eBook)
DOI 10.1007/978-3-319-26532-2

Library of Congress Control Number: 2015954339

LNCS Sublibrary: SL1 – Theoretical Computer Science and General Issues

Springer Cham Heidelberg New York Dordrecht London

Printed on acid-free paper

Springer International Publishing AG Switzerland is part of Springer Science+Business Media
(www.springer.com)

Preface

This volume is part of the four-volume proceedings of the 22nd International Conference on Neural Information Processing (ICONIP 2015), which was held in Istanbul, Turkey, during November 9–12, 2015. The ICONIP is an annual conference of the Asia Pacific Neural Network Assembly (APNNA; which was reformed in 2015 as the Asia Pacific Neural Network Society, APNNS). This series of ICONIP conferences has been held annually since 1994 in Seoul and has become one of the leading international conferences in the areas of artificial intelligence and neural networks.

ICONIP 2015 received a total of 432 submissions by scholars coming from 42 countries/regions across six continents. Based on a rigorous peer-review process where each submission was evaluated by an average of two qualified reviewers, a total of 301 high-quality papers were selected for publication in the reputable series of *Lecture Notes in Computer Science* (LNCS). The selected papers cover major topics of theoretical research, empirical study, and applications of neural information processing research. ICONIP 2015 also featured the Cybersecurity Data Mining Competition and Workshop (CDMC 2015), which was jointly held with ICONIP 2015. Nine papers from CDMC 2015 were selected for the conference proceedings.

In addition to the contributed papers, the ICONIP 2015 technical program also featured four invited speakers, Nik Kasabov (Auckland University of Technology, New Zealand), Jun Wang (The Chinese University of Hong Kong), Tom Heskes (Radboud University, Nijmegen, The Netherlands), and Michel Verleysen (Université catholique de Louvain, Belgium).

We would like to sincerely thank to the members of the Advisory Committee and Program Committee, the APNNS Governing Board for their guidance, and the members of the Organizing Committee for all their great efforts and time in organizing such an event. We would also like to take this opportunity to express our deepest gratitude to all the reviewers for their professional review that guaranteed high-quality papers.

We would like to thank Springer for publishing the proceedings in the prestigious series of *Lecture Notes in Computer Science*. Finally, we would like to thank all the speakers, authors, and participants for their contribution and support in making ICONIP 2015 a successful event.

November 2015

Sabri Arik
Tingwen Huang
Weng Kin Lai
Qingshan Liu

Organization

General Chair

Sabri Arik   Istanbul University, Turkey

Honorary Chair

Shun-ichi Amari   Brain Science Institute, RIKEN, Japan

Program Chairs

Tingwen Huang	Texas A&M University at Qatar, Qatar
Weng Kin Lai	School of Technology, Tunku Abdul Rahman College (TARC), Malaysia
Qingshan Liu	Huazhong University of Science Technology, China

Advisory Committee

P. Balasubramaniam	Deemed University, India
Jinde Cao	Southeast University, China
Jonathan Chan	King Mongkut's University of Technology, Thailand
Sung-Bae Cho	Yonsei University, Korea
Tom Gedeon	Australian National University, Australia
Akira Hirose	University of Tokyo, Japan
Tingwen Huang	Texas A&M University at Qatar, Qatar
Nik Kasabov	Auckland University of Technology, New Zealand
Rhee Man Kil	Korea Advanced Institute of Science and Technology (KAIST), Korea
Irwin King	Chinese University of Hong Kong, SAR China
James Kwok	Hong Kong University of Science and Technology, SAR China
Weng Kin Lai	School of Technology, Tunku Abdul Rahman College (TARC), Malaysia
James Lam	The University of Hong Kong, Hong Kong, SAR China
Kittichai Lavangnananda	King Mongkut's University of Technology, Thailand
Minho Lee	Kyungpook National University, Korea
Andrew Chi-Sing Leung	City University of Hong Kong, SAR China
Chee Peng Lim	University Sains Malaysia, Malaysia
Derong Liu	The Institute of Automation of the Chinese Academy of Sciences (CASIA), China

Chu Kiong Loo	University of Malaya, Malaysia
Bao-Liang Lu	Shanghai Jiao Tong University, China
Aamir Saeed Malik	Petronas University of Technology, Malaysia
Seichi Ozawa	Kobe University, Japan
Hyeyoung Park	Kyungpook National University, Korea
Ju. H. Park	Yeungnam University, Republic of Korea
Ko Sakai	University of Tsukuba, Japan
John Sum	National Chung Hsing University, Taiwan
DeLiang Wang	Ohio State University, USA
Jun Wang	Chinese University of Hong Kong, SAR China
Lipo Wang	Nanyang Technological University, Singapore
Zidong Wang	Brunel University, UK
Kevin Wong	Murdoch University, Australia

Program Committee Members

Syed Ali, India
R. Balasubramaniam, India
Tao Ban, Japan
Asim Bhatti, Australia
Jinde Cao, China
Jonathan Chan, Thailand
Tom Godeon, Australia
Denise Gorse, UK
Akira Hirose, Japan
Lu Hongtao, China
Mir Md Jahangir Kabir, Australia
Yonggui Kao, China
Hamid Reza Karimi, Norway
Nik Kasabov, New Zealand
Weng Kin Lai, Malaysia
S. Lakshmanan, India
Minho Lee, Korea
Chi Sing Leung, Hong Kong, SAR China
Cd Li, China

Ke Liao, China
Derong Liu, USA
Yurong Liu, China
Chu Kiong Loo, Malaysia
Seiichi Ozawa, Japan
Serdar Ozoguz, Turkey
Hyeyoung Park, South Korea
Ju Park, North Korea
Ko Sakai, Japan
Sibel Senan, Turkey
Qianqun Song, China
John Sum, Taiwan
Ying Tan, China
Jun Wang, Hong Kong, SAR China
Zidong Wang, UK
Kevin Wong, Australia
Mustak Yalcin, Turkey
Enes Yilmaz, Turkey

Special Sessions Chairs

Zeynep Orman	Istanbul University, Turkey
Neyir Ozcan	Uludag University, Turkey
Ruya Samli	Istanbul University, Turkey

Publication Chair

Selcuk Sevgen   Istanbul University, Turkey

Organizing Committee

Emel Arslan   Istanbul University, Turkey
Muhammed Ali Aydin   Istanbul University, Turkey
Eylem Yucel Demirel   Istanbul University, Turkey
Tolga Ensari   Istanbul University, Turkey
Ozlem Faydasicok   Istanbul University, Turkey
Safak Durukan Odabasi   Istanbul University, Turkey
Sibel Senan   Istanbul University, Turkey
Ozgur Can Turna   Istanbul University, Turkey

Contents – Part I

Learning Algorithms and Classification Systems

Texture Classification with Patch Autocorrelation Features

Radu Tudor Ionescu[1][(✉)], Andreea Lavinia Popescu[2], and Dan Popescu[2]

[1] University of Bucharest, 14 Academiei, Bucharest, Romania
raducu.ionescu@gmail.com
[2] Politehnica University of Bucharest,
313 Splaiul Independentei Street, Bucharest, Romania
andreea.lavinia@ymail.com, dan_popescu_2002@yahoo.com

Abstract. Recently, a novel approach of capturing the autocorrelation of an image termed Patch Autocorrelation Features (PAF) was proposed. The PAF approach was successfully evaluated in a series of handwritten digit recognition experiments on the popular MNIST data set. However, the PAF representation has limited applications, because it is not invariant to affine transformations. In this work, the PAF approach is extended to become invariant to image transformations such as translation and rotation changes. First, several features are extracted from each image patch taken at a regular interval. Based on these features, a vector of similarity values is computed between each pair of patches. Then, the similarity vectors are clustered together such that the spatial offset between the patches of each pair is roughly the same. Finally, the mean and the standard deviation of each similarity value are computed for each group of similarity vectors. These statistics are concatenated in a feature vector called Translation and Rotation Invariant Patch Autocorrelation Features (TRIPAF). The TRIPAF vector essentially records information about the repeating patterns within an image at various spatial offsets. Several texture classification experiments are conducted on the Brodatz data set to evaluate the TRIPAF approach. The empirical results indicate that TRIPAF can improve the performance by up to 10 % over a system that uses the same features, but extracts them from entire images. Furthermore, state of the art accuracy rates are obtained when the TRIPAF approach is combined with a scale invariant model, namely a bag of visual words model based on SIFT features.

Keywords: Patch-based method · Texture classification · Rotation invariance · Translation invariance · Kernel method · Brodatz

1 Introduction

Complex image classification tasks, such as texture classification, require sophisticated methods, naturally because the methods have to take into account several aspects such as translation, rotation, and scale variations, illumination changes,

© Springer International Publishing Switzerland 2015
S. Arik et al. (Eds.): ICONIP 2015, Part I, LNCS 9489, pp. 1–11, 2015.
DOI: 10.1007/978-3-319-26532-2_1

viewpoint changes, and noise. Among the state of the art models used in image classification are elaborate methods such as bag of visual words [5,13], Fisher Vectors [19], and deep learning [14]. Recently, a simple feature representation for images that is based on the autocorrelation of the image with itself was introduced in [11]. In this representation, each feature is determined by the euclidean distance between a pair of patches extracted from the image. To reduce the time necessary to compute the feature representation, patches are extracted by applying a grid over the image. This feature representation is termed Patch Autocorrelation Features (PAF). The authors of [11] have shown that PAF exibits state of the art performance in optical character recognition. However, the PAF approach is affected by affine transformations and it needs to be further extended in order to solve[1] more complex image classification tasks, such as texture classification, for example. The main goal of this work is to provide an extension of PAF that is invariant to image transformations, including but not limited to translation and rotation changes. Naturally, this extension involves more elaborate computations, but the resulted feature vector is actually more compact, since it involves the vector quantization of pairs of patches according to the spatial offset between the patches in each pair. Instead of directly comparing the patches, the extended approach initially extracts a set of features from each patch. The extended feature representation is termed Translation and Rotation Invariant Patch Autocorrelation Features (TRIPAF). Texture classification experiments are conducted to evaluate TRIPAF on a popular data set, namely Brodatz. The empirical results indicate that TRIPAF can significantly improve the performance over a system that uses the same features, but extracts them from entire images. By itself, TRIPAF is invariant to rotation and translation changes, and for this reason, it makes sense to combine it with a scale invariant system in order to further improve the performance. As such, the system based on TRIFAF is combined with a bag of visual words (BOVW) framework [12]. The BOVW framework is based on clustering SIFT descriptors [17] into visual words, which are scale invariant. The performance level of the combined approach is better than two state of the art methods for texture classification.

The paper is organized as follows. Related work on image analysis using autocorrelation and patch-based methods is presented in Sect. 2. The translation and rotation invariant extension of the Patch Autocorrelation Features is presented in Sect. 3. Section 4 describes the texture classification experiments. Finally, the conclusions are drawn in Sect. 5.

2   Related Work

In signal processing, the autocorrelation is used to find repetitive patterns in a signal over time. Images can also be regarded as spatial signals. Thus, it makes sense to measure the spatial autocorrelation of an image. Certainly, the autocorrelation has already been used in image processing [2,10,21]. As many other computer vision techniques [1,4,6], the PAF map considers patches rather than

---

[1] with a reasonable degree of accuracy.

pixels, in order to capture distinctive patterns such as edges, corners, shapes, and so on. In other words, the PAF representation stores information about repeating edges, corners, and other shapes that can be found in the analyzed image. Patches contain contextual information and have advantages in terms of computation and generalization. However, patch-based techniques are still heavy to compute with current machines, as stated in [1]. To reduce the time necessary to compute the PAF representation, patches are compared using a grid over the image. The density of this grid can be adjusted to obtain the desired trade-off between accuracy and speed.

3 Translation and Rotation Invariant Patch Autocorrelation Features

Several modifications are proposed to transform Patch Autocorrelation Features [11] into an approach that takes into account several image variation aspects including translation and rotation changes, illumination changes, and noise. Instead of comparing the patches based on raw pixel values, a set of features is extracted from each image patch. Depending on the kind of patch features, the method can thus become invariant to different types of image variations. In this work, a set of texture-specific features are used since the approach is evaluated on the texture classification task. These features are described in Sect. 3.1. The important remark is that a different set of features can be extracted from patches when PAF is going to be used for a different problem.

The following conventions and mathematical notations are considered throughout this paper to describe TRIPAF. Arrays and matrices are always considered to be indexed starting from position 1, that is $v = (v_1, v_2, ..., v_{|v|})$, where $|v|$ is the number of components of v. The notations v_i or $v(i)$ are alternatively used to identify the i-th component of v. The sequence $1, 2, ..., n$ is denoted by $1 : n$, but if the step is different from the unit, it can be inserted between the lower and the upper bounds. For example, $1 : 2 : 8$ generates the sequence $1, 3, 5, 7$. Moreover, for a vector v and two integers i and j such that $1 \leq i \leq j \leq |v|$, $v_{i:j}$ denotes the sub-array $(v_i, v_{i+1}, ..., v_j)$. In a similar manner, $X_{i:j,k:l}$ denotes a sub-matrix of the matrix X. Since the analyzed images are reduced to gray-scale, the notion of *matrix* and *image* can be used interchangeably, with the same meaning. In this context, a patch corresponds to a sub-matrix.

Rather than computing a single value to represent the similarity between two patches based on the extracted features, the extended PAF approach computes several similarities, one for each feature. More precisely, patches are compared using the Bhattacharyya coefficient between each of their features. Given two feature vectors $f_X, f_Y \in \mathbb{R}^m$ extracted from two image patches X and Y, the vector of similarity values between the two patches $s_{X,Y}$ is computed as follows:

$$s_{X,Y}(i) = \sqrt{f_X(i)} \cdot \sqrt{f_Y(i)}, \forall i \in \{1, 2, ..., m\}. \tag{1}$$

Note that each component of $s_{X,Y}$ is independently computed in the sense that it does not depend on the other features. Therefore, $s_{X,Y}$ can capture different aspects about the similarity between X and Y, since each feature can provide different information. The same features are naturally extracted from each patch, thus $|f_X| = |f_Y| = m$ for every pair of patches (X, Y).

In the basic PAF approach, the next step would be to concatenate the similarity vectors generated by comparing patches two by two. This would be fine, as long as the method relies entirely on the features to achieve invariance to different image transformations. However, further processing can be carried out to ensure that PAF remains invariant to translation and rotation, even if the extracted features are not. Unlike the basic PAF approach, the pairs of patches are vector quantized by the spatial offset between the patches of each pair. Given two patches X and Y having the origins in (x, y) and (u, z), respectively, the spatial offset o between X and Y is measured with the help of the L_2 euclidean distance between their origins:

$$o(X, Y) = \sqrt{(x - u)^2 + (y - z)^2}. \tag{2}$$

In order to cluster pairs of patches together, the spatial offsets are rounded to the nearest integer values. Given two pairs of patches (X, Y) and (U, V), they are clustered together only if $\lfloor o(X, Y) \rceil = \lfloor o(U, V) \rceil$, where $\lfloor x \rceil$ is the rounding function of $x \in \mathbb{R}$, that returns the nearest integer value to x. It is important to note that the similarity vector determined by a pair of patches is included in the cluster, not the patches themselves. Formally, a *cluster* (or a group) of similarity vectors between patches extracted at a given spatial offset k is defined as follows:

$$C_k = \{s_{P_i,P_j} \mid \lfloor o(P_i, P_j) \rceil = k, \forall (P_i, P_j) \in \mathcal{P} \times \mathcal{P}, 1 \leq i < j \leq |\mathcal{P}|\}, \tag{3}$$

where $\mathcal{P} = \{P_1, P_2, ..., P_{|\mathcal{P}|}\}$ is the set of patches extracted from the input image, $o(P_i, P_j)$ is the offset between P_i and P_j determined by Eq. (2), and s_{P_i,P_j} is the similarity vector between patches P_i and P_j computed as in Eq. (1).

In each cluster, the similarity vectors are computed between patches that reside at a certain spatial offset, in all possible directions. The exact position of each patch is simply disregarded in the clustering process. When the image is translated or rotated, these clusters remain mostly unchanged, because the spatial offsets between patches are always the same. Obviously, the patches extracted from an image will not be identical to the patches extracted from the same image after applying a rotation, specifically because the patches are extracted along the vertical and horizontal axes of the image based on a fixed grid, as described in [11]. As such, the clusters may not contain the very same patches when the image is rotated, but in principle, each cluster should capture about the same information, since the distance between patches is always preserved. Therefore, the method can be safely considered as translation and rotation invariant. The final step is to find a representation for each of these clusters. The mean and the standard deviation are computed for each component of the similarity vectors

Algorithm 1. TRIPAF Algorithm

1 **Input:**
2 I - a gray-scale input image of $h \times w$ pixels;
3 p - the size (in pixels) of each square-shaped patch;
4 s - the distance (in pixels) between consecutive patches;
5 d - the number of clusters;
6 $\mathcal{F} = \{F_1, F_2, ..., F_m \mid F_i : \mathbb{R}^p \times \mathbb{R}^p \to \mathbb{R}, \forall i\}$ - a set of m feature extraction functions.

7 **Initialization:**
8 $I \leftarrow$ resized image I such that $\sqrt{(w)^2 + (h)^2} = d$;
9 $(h, w) \leftarrow size(I)$;
10 $n \leftarrow ceil((h - p + 1)/s) \cdot ceil((w - p + 1)/s)$;
11 $\mathcal{P} \leftarrow \emptyset$;
12 $C_i \leftarrow \emptyset$, for $i \in \{1, 2, ..., d\}$;
13 $v_i \leftarrow 0$, for $i \in \{1, 2, ..., m \cdot d\}$;

14 **Computation:**
15 **for** $i = 1 : s : h$ **do**
16   **for** $j = 1 : s : w$ **do**
17    $P \leftarrow I_{i:(i+p-1), j:(j+p-1)}$;
18   **for** $k = 1 : m$ **do**
19     $f_P(k) \leftarrow F_i(P)$;
20    $\mathcal{P} \leftarrow \mathcal{P} \cup f_P$;

21 **for** $i = 1 : n - 1$ **do**
22   **for** $j = i + 1 : n$ **do**
23   **for** $k = 1 : m$ **do**
24     $s_{P_i, P_j}(k) \leftarrow \sqrt{f_{P_i}(k) \cdot f_{P_j}(k)}$;
25    $C_{\lfloor o(P_i, P_j) \rfloor} \leftarrow C_{\lfloor o(P_i, P_j) \rfloor} \cup s_{P_i, P_j}$;

26 $k \leftarrow 1$;
27 **for** $i = 1 : d$ **do**
28   **for** $j = 1 : m$ **do**
29    $v(k) \leftarrow mean(s(j))$, for $s \in C_i$;
30    $v(k + 1) \leftarrow std(s(j))$, for $s \in C_i$;
31    $k \leftarrow k + 2$;

32 **Output:**
33 v - the TRIPAF feature vector with $m \cdot d$ components.

within a group. Finally, the Translation and Rotation Invariant Patch Autocorrelation Features (TRIPAF) are obtained by concatenating these statistics for all the clusters $C_1, C_2, ..., C_d$, in this specific order, where d is a constant integer value that determines the number of clusters. To make sure all the images in a set $\mathcal{I}$ are represented by vectors of the same length, each image needs to be resized such that its diagonal is equal to the constant integer value d:

$$\sqrt{w_I{}^2 + h_I{}^2} = d, \forall I \in \mathcal{I}, \tag{4}$$

where w_I and h_I are the width and the height of image I, respectively. The number of components of the TRIPAF vector is $O(md)$, where m represents the number of features extracted from patches and d is a positive integer value that implicitly controls the number of pairs of patches per cluster.

The TRIPAF representation is computed as described in Algorithm 1. Two predefined functions are used to compute the mean and the standard deviation (as defined in literature), namely $mean$ and std. The TRIPAF algorithm can be divided into three phases. In the first phase (steps 15–20), feature vectors are computed on patches extracted by apply a grid over the image, and then, the resulted feature vectors are stored in the set $\mathcal{P}$. In the second phase (steps 21–25), the similarity vectors are computed and subsequently clustered according to the spatial offsets between patches. In the third phase (steps 26–31), the TRIPAF vector v is generated by computing the mean and the standard deviation of each component of the similarity vectors within each cluster. Algorithm 1 can easily be adapted for a variety of image classification tasks, simply by changing the set of features $\mathcal{F}$. The set of texture-specific features used for the texture classification experiments are presented next.

3.1   Texture Features

The mean and the standard deviation are the first two statistical features extracted from image patches, but the more elaborate features described next are mandatory to adequately discriminate between different textures. One of the most powerful statistical methods for textured image analysis is based on features extracted from the Gray-Level Co-Occurrence Matrix (GLCM) proposed in [8]. Relevant statistical features computed from the GLCM are used inside TRIPAF, namely the contrast, the energy, the homogeneity, and the correlation. Another feature that is relevant for texture analysis is the fractal dimension, which is approximated by an efficient box counting algorithm [20]. The local isotropic phase symmetry measure (LIPSyM) [15] takes the discrete Fourier transform of the input image, and filters this frequency information through a bank of Gabor filters. As stated in [15], local responses of each Gabor filter can be represented in terms of energy and amplitude. Thus, Gabor features, such as the mean-squared energy and the mean amplitude, can be computed through the phase symmetry measure for a bank of Gabor filters with various scales and rotations. The Gabor features are also used in the TRIPAF algorithm.

4   Texture Classification Experiments

4.1   Data Set

Texture classification experiments are presented on a benchmark data set of texture images, namely the Brodatz data set [3]. This data set is probably the best known benchmark used for texture classification, but also one of the most difficult, since it contains 111 classes with only 9 samples per class. The standard procedure in the literature is to obtain samples of 213×213 pixels by cutting them out from larger images of 640×640 pixels using a 3 by 3 grid.

4.2   Learning Methods

All classification systems used in this work rely on kernel methods to learn discriminant patterns. Kernel-based learning algorithms work by embedding the data into a Hilbert space, and searching for linear relations in that space using a learning algorithm. The power of kernel methods lies in the implicit use of a Reproducing Kernel Hilbert Space (RKHS) induced by a positive semi-definite kernel function. For images, many such kernel functions are used in various applications including object recognition, image retrieval, or similar tasks. In this work, two popular kernel functions are chosen, namely the linear kernel and the histogram intersection kernel. After embedding the features with a kernel function, a linear classifier is used to select the most discriminant features. In this work, Support Vector Machines (SVM) and Kernel Discriminant Analysis (KDA) are alternatively used for learning. The KDA method is sometimes able to improve accuracy by avoiding the class masking problem [9].

4.3   Implementation and Evaluation

The TRIPAF representation is compared with a method that extracts texture-specific features from entire images, on the texture classification task. The four GLCM features are averaged on 4 directions (vertical, horizontal and diagonals) using gaps of 1 and 2 pixels. The mean and the standard deviation are also added to the set of texture-specific features. Another feature is given by the box counting dimension. The Gabor features (the mean-squared energy and the mean amplitude) are computed on 3 scales and 6 different rotations, resulting in a total of 36 Gabor features (2 features × 3 scales × 6 directions). There are 43 texture-specific features put together. Alternatively, the two Gabor features are also averaged on 3 scales and 6 different rotations, generating only 4 Gabor features. This produces a reduced set of 9 texture-specific features, which can be more robust to image rotations and scale variations. Results are reported using both representations, one of 43 features and the other of 9 features.

The TRIPAF approach, which is rotation and translation invariant, is combined with the BOVW model described in [12], which is scale invariant (due to the SIFT descriptors), in order to obtain a method that is invariant to all affine transformations. These methods are combined at the kernel level through *multiple kernel learning* (MKL) [7]. The combined representation is alternatively used with the SVM or the KDA. The two classifiers are compared with other state of the art approaches [18, 22].

4.4   Parameter Tuning

A set of preliminary tests on a subset of 40 classes from Brodatz are performed to adjust the parameters of the PAF representation, such as the patch size and the pixel interval used to extract the patches. Patches of 16×16, 32×32 and 64×64 pixels were considered. Better results in terms of accuracy were obtained with either patches of 16×16 or 32×32 pixels, and the rest of the experiments are based on patches of such dimensions. The TRIPAF representations generated by patches of 16×16 or 32×32 pixels are also combined by summing up their kernels to obtain a more robust representation. After setting up the patch sizes, the interest is turned to set the grid density. The grid density was chosen such that the processing time of TRIPAF is less than 5 seconds per image on a machine with Intel Core i7 2.3 GHz processor and 8 GB of RAM using a single Core. As such, patches are extracted at an interval of 8 pixels. Using the same subset of 40 classes from Brodatz, the regularization parameter C of SVM was set to 10^4, while the regularization parameter of KDA was set to 10^{-5}.

4.5 Results on Brodatz Data Set

Typically, the results reported in previous studies [16, 18, 22] on the Brodatz data set are based on randomly selecting 3 training samples per class and using the rest for testing. Likewise, the results presented in this paper are based on the same setup with 3 random samples per class for training. Moreover, the random selection procedure is repeated for 20 times and the resulted accuracy rates are averaged. This helps to reduce the amount of accuracy variation introduced by using a different partition of the data set in each of the 20 trials. To give an idea of the amount of variation in each trial, the standard deviations for the computed average accuracy rates are also reported. The evaluation procedure described so far is identical to the one used in the state of the art approaches [18, 22] that are included in the following comparative study.

Table 1 presents the accuracy rates of the SVM classifier based on several TRIPAF representations, in which the number of texture-specific features and the size of the patches are varied. Several baseline SVM classifiers based on the same texture-specific features used in the TRIPAF representation are included in the evaluation in order to estimate the performance gain offered by TRI-PAF. When the set of 9 texture-specific features is being used, the TRIPAF

Table 1. Accuracy rates of the SVM classifier on the Brodatz data set for the TRIPAF representation versus the standard representation based on texture-specific features. The reported accuracy rates are averaged on 20 trials using 3 random samples per class for training and the other 6 for testing. The best accuracy rate for each set of texture-specific features is highlighted in bold.

Feature map	Texture features	Patches	Kernel	Accuracy
Standard	9	None	Linear	$76.52\% \pm 1.6$
Standard	9	None	Intersection	$77.11\% \pm 1.3$
TRIPAF	9	16×16	Intersection	$87.78\% \pm 1.2$
TRIPAF	9	32×32	Intersection	$91.40\% \pm 0.9$
TRIPAF	9	$16 \times 16 + 32 \times 32$	Intersection	$\mathbf{91.92\% \pm 0.6}$
Standard	43	None	Linear	$89.93\% \pm 1.1$
Standard	43	None	Intersection	$90.42\% \pm 1.2$
TRIPAF	43	16×16	Intersection	$92.11\% \pm 0.8$
TRIPAF	43	32×32	Intersection	$92.28\% \pm 0.9$
TRIPAF	43	$16 \times 16 + 32 \times 32$	Intersection	$\mathbf{92.85\% \pm 0.8}$

Table 2. Accuracy rates of the TRIPAF and BOVW combined representation on the Brodatz data set compared with state of the art methods. The TRIPAF representation is based on 43 texture-specific features extracted from patches of 16 × 16 and 32 × 32 pixels. The BOVW model is based on the PQ kernel. The reported accuracy rates are averaged on 20 trials using 3 random samples per class for training and the other 6 for testing. The best accuracy rate is highlighted in bold.

Model	Accuracy
SVM based BOVW [12]	92.94 % ± 0.8
SVM based TRIPAF	92.85 % ± 0.8
SVM based TRIPAF + BOVW	96.24 % ± 0.6
KDA based TRIPAF + BOVW	**96.51 % ± 0.7**
Best model of [22]	95.90 % ± 0.6
Best model of [18]	96.14 % ± 0.4

approach improves the baseline by more than 10 % in terms of accuracy. On the other hand, the difference is roughly 2 % in favor of TRIPAF when the set of 43 texture-specific features is being used. Both the standard and the TRIPAF representations work better when more texture-specific features are extracted, probably because the Brodatz data set does not contain significant rotation changes within each class of images. Nevertheless, the TRIPAF approach is always able to give better results than the baseline SVM. An interesting remark is that the results of TRIPAF are always better when patches of 16 × 16 pixels are used in conjunction with patches of 32 × 32 pixels, even if the accuracy improvement over using them individually is not considerable (below 1 %). The best accuracy (92.85 %) is obtained by the SVM based on the TRIPAF approach that extracts 43 features from patches of 16 × 16 and 32 × 32 pixels.

The empirical results presented in Table 1 clearly demonstrate the advantage of using the TRIPAF feature vectors. Intuitively, further combining TRIPAF with BOVW should yield even better results. While TRIPAF is rotation and translation invariant, BOVW is scale invariant, and therefore, these two representations complement each other perfectly. Table 2 compares the results of TRIPAF and BOVW combined through MKL with the results of two state of the art methods [18,22]. The intersection kernel used in the case of TRIPAF is summed up with the PQ kernel used in the case of BOVW [12]. The individual results of TRIPAF and BOVW are also listed in Table 2. The two methods obtain fairly similar accuracy rates when used independently, but the accuracy rates are almost 3 % lower than the state of the art methods. However, the kernel combination of TRIPAF and BOVW yields results comparable to the state of the art methods [18,22]. In fact, the best accuracy rate on the Brodatz data set (96.51 %) is given by the KDA based on TRIPAF and BOVW, although the SVM based on the kernel combination is also slightly better than both state of the art models. The kernel sum of TRIPAF and BOVW is much better than using the two representations individually, proving that the idea of combining them up is indeed crucial for obtaining state of the art results.

5   Conclusion

This work proposed a novel representation for texture images termed Translation and Rotation Invariant Patch Autocorrelation Features, which is an extension of the PAF representation that is designed to be invariant to image transformations, including but not limited to translations and rotations. The TRIPAF approach was evaluated in a set of texture classification experiments that require the use of invariant techniques to obtain good performance. The TRIPAF representation improves the accuracy rate over a baseline model based on extracting texture-specific features from entire images. Moreover, the TRIPAF approach was combined with a BOVW model [12] through MKL. The kernel combination of TRIPAF and BOVW yields results that are better than two state of the art texture classification methods [18,22].

Acknowledgments. Dan Popescu has been funded by the National Research Program STAR, project 71/2013: Multisensory robotic system for aerial monitoring of critical infrastructure systems - MUROS. Andreea-Lavinia Popescu has been supported through the Financial Agreement POSDRU/159/1.5/S/134398.

References

1. Barnes, C., Goldman, D.B., Shechtman, E., Finkelstein, A.: The PatchMatch randomized matching algorithm for image manipulation. Commun. ACM **54**(11), 103–110 (2011)
2. Brochard, J., Khoudeir, M., Augereau, B.: Invariant feature extraction for 3D texture analysis using the autocorrelation function. Pattern Recogn. Lett. **22**(6–7), 759–768 (2001)
3. Brodatz, P.: Textures: A Photographic Album For Artists And Designers. Dover Pictorial Archives. Dover Publications, New York (1966)
4. Cho, T.S., Avidan, S., Freeman, W.T.: The patch transform. IEEE Trans. Pattern Anal. Mach. Intell. **32**(8), 1489–1501 (2010)
5. Csurka, G., Dance, C.R., Fan, L., Willamowski, J., Bray, C.: Visual categorization with bags of keypoints. In: Workshop on Statistical Learning in Computer Vision, ECCV, pp. 1–22 (2004)
6. Deselaers, T., Keyser, D., Ney, H.: Discriminative training for object recognition using image patches. In: Proceedings of CVPR, pp. 157–162 (2005)
7. Gonen, M., Alpaydin, E.: Multiple kernel learning algorithms. J. Mach. Learn. Res. **12**, 2211–2268 (2011)
8. Haralick, R.M., Shanmugam, K., Dinstein, I.: Textural features for image classification. IEEE Trans. Syst. Man Cybern. **3**(6), 610–621 (1973)
9. Hastie, T., Tibshirani, R.: The Elements of Statistical Learning. Springer, New York (2003). Corrected edn
10. Horikawa, Y.: Use of autocorrelation kernels in kernel canonical correlation analysis for texture classification. In: Pal, N.R., Kasabov, N., Mudi, R.K., Pal, S., Parui, S.K. (eds.) ICONIP 2004. LNCS, vol. 3316, pp. 1235–1240. Springer, Heidelberg (2004)
11. Ionescu, R.T., Popescu, A.L., Popescu, D.: Patch autocorrelation features for optical character recognition. In: Proceedings of VISAPP, March 2015

12. Ionescu, R.T., Popescu, A.L., Popescu, M.: Texture classification with the PQ kernel. In: Proceedings of WSCG (2014)
13. Ionescu, R.T., Popescu, M., Grozea, C.: Local learning to improve bag of visual words model for facial expression recognition. In: Workshop on Challenges in Representation Learning, ICML (2013)
14. Krizhevsky, A., Sutskever, I., Hinton, G.E.: ImageNet classification with deep convolutional neural networks. In: Proceedings of NIPS, pp. 1106–1114 (2012)
15. Kuse, M., Wang, Y.F., Kalasannavar, V., Khan, M., Rajpoot, N.: Local isotropic phase symmetry measure for detection of beta cells and lymphocytes. J. Pathol. Inform. 2(2), 2 (2011)
16. Lazebnik, S., Schmid, C., Ponce, J.: A sparse texture representation using local affine regions. IEEE Trans. Pattern Anal. Mach. Intell. 27(8), 1265–1278 (2005)
17. Lowe, D.G.: Object recognition from local scale-invariant features. In: Proceedings of ICCV, vol. 2, pp. 1150–1157 (1999)
18. Nguyen, H.G., Fablet, R., Boucher, J.M.: Visual textures as realizations of multivariate log-Gaussian Cox processes. In: Proceedings of CVPR, pp. 2945–2952 (2011)
19. Perronnin, F., Sánchez, J., Mensink, T.: Improving the Fisher kernel for large-scale image classification. In: Daniilidis, K., Maragos, P., Paragios, N. (eds.) ECCV 2010, Part IV. LNCS, vol. 6314, pp. 143–156. Springer, Heidelberg (2010)
20. Popescu, A.L., Popescu, D., Ionescu, R.T., Angelescu, N., Cojocaru, R.: Efficient fractal method for texture classification. In: Proceedings of ICSCS (2013)
21. Toyoda, T., Hasegawa, O.: Extension of higher order local autocorrelation features. Pattern Recogn. 40(5), 1466–1473 (2007)
22. Zhang, J., Marszalek, M., Lazebnik, S., Schmid, C.: Local features and kernels for classification of texture and object categories: a comprehensive study. Int. J. Comput. Vision 73(2), 213–238 (2007)

Novel Architecture for Cellular Neural Network Suitable for High-Density Integration of Electron Devices-Learning of Multiple Logics

Mutsumi Kimura[1,2(✉)], Yusuke Fujita[1], Tomohiro Kasakawa[1], and Tokiyoshi Matsuda[1]

[1] Department of Electronics and Informatics, Ryukoku University, Kyoto, Japan
mutsu@rins.ryukoku.ac.jp
[2] Graduate School of Information Science, Nara Institute of Science and Technology, Ikoma, Japan

Abstract. We will propose a novel architecture for a cellular neural network suitable for high-density integration of electron devices. A neuron consists of only eight transistors, and a synapse consists of just only one transistor. We fabricated a cellular neural network using thin-film devices. Particularly in this time, we confirmed that our neural network can learn multiple logics even in a small-scale neural network. We think that this result indicates that our proposal has a big potential for future electronics using neural networks.

Keywords: Cellular neural network · High-density integration · Electron device · Learning · Multiple logics

1   Introduction

Cellular neural networks are neural networks where a neuron is connected to only neighboring neurons [1], hence suitable for integration of electron devices, and promising for image processing [2], pattern recognition [3], etc. Until now, fundamental theory, working principle, and application potential have been actively investigated using formal models and numerical simulation. However, there exist few reports on actual hardware of cellular neural networks [4], although they are suitable for integration of electron devices as aforementioned. We imagine that this is because the conventional circuits of the neurons and synapses are still complicated, even though the structure of the network is simple.

We are developing neural networks from the viewpoint of device hardware [5, 6]. In this presentation, we will propose a novel architecture for a cellular neural network suitable for high-density integration of electron devices. The main advantage is that the circuits of the neurons and synapses are excellently simple. A neuron consists of only eight transistors, and a synapse surprisingly consists of just only one transistor. As a result, the structure of the cellular neural network and learning principle must be modified. We will explain the device architecture, learning principle, fabrication process, experimental

S. Arik et al. (Eds.): ICONIP 2015, Part I, LNCS 9489, pp. 12–20, 2015.
DOI: 10.1007/978-3-319-26532-2_2

method and result in detail. It should be also noted that we fabricated a cellular neural network using thin-film devices, which are expected as key technologies for micro-giant electronics. Although there will be some repetition of the prior publications, we will explain them again because it is available for the readers in research areas of information technologies who have not yet known our research. Particularly in this time, we confirmed that our neural network can learn multiple logics of AND and OR even in a small-scale neural network of 5×5. Although this result is primitive, we think that it indicates that our proposal has a big potential for future electronics using neural networks.

2   Device Architecture

2.1   Neuron

Figure 1 shows the neuron. We limited the necessary functions of the neuron to that a binary state is maintained by itself and altered by the input signals. In order to realize this simple function, we adopted a latch circuit that circularly connects two inverters with two switches. The firing or non-firing state is maintained using the latch circuit when the switches are turned on, namely, we defined the firing state as a situation when the voltages at node α and node β are high and complementarily low, respectively, whereas we defined the non-firing state as the opposite situation. Although the latch circuit is a well-known circuit for maintaining a binary state, it should be noted that its characteristic is similar to a sigmoid function, a typical function used to provide a favorable soft threshold in neural network models. The binary state is altered after the switches are turned off, the input signals are applied to nodes α and β, and the switches are turned on again. In any case, by employing complementary inverters and switches, we succeeded in making a neuron consist of only eight transistors.

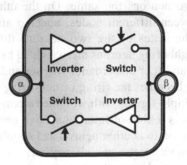

Fig. 1. Neuron.

2.2   Synapse

Figure 2 shows the synapse. We limited the necessary functions of the synapse to that an input signal from a neuron is weighted by its synaptic connection strength and transferred to another neuron, and the synaptic connection strength is adjusted. In order to realize this simple function, we adopted a variable resistor, which will be replaced by a transistor in practical electron devices. An input voltage from a neuron is weighted by the conductance of the variable resistor and transferred to another neuron. The synaptic connection strength corresponds to the conductance of the variable resistor, which is adjusted obeying a modified Hebbian learning as below mentioned. In any case, we succeeded in making a synapse consist of just only one resistor.

Fig. 2. Synapse.

2.3   Network

Figure 3 shows the network. Since the neuron and synapse were dramatically simplified, the structure of the cellular neural network must be modified. We arrayed multiple neurons and connected each neuron to only four up, down, left, and right neighboring neurons through the synapses. In order to compensate the small number of the synapses, we connected a pair of neurons through a pair of synapses, namely, concordant and discordant synapses. The concordant synapse is connected between the same nodes in the two neurons, nodes α and α or nodes β and β, and tends to make the states of the two neurons the same. On the other hand, the discordant synapse is connected between different nodes, nodes α and β, and the discordant synapse tends to make the states of the two neurons different. The eight input voltages from the four neighboring neurons are weighted by the conductances of the four concordant synapses and four discordant synapses and transferred to the target neuron. The target neuron becomes the firing or non-firing state, namely, is subject to the majority rule of multiple signals with weighted strengths. Moreover, it should be noted that this network is a kind of interconnective networks, where a synapse transfer a signal from a neuron to another neuron and simultaneously from the latter neuron to the former neuron vice-versa, which may correspond to the function of two synapses and also compensate the small number of the synapses. In any case, we succeeded in making an interconnective network where we connected each neuron to only four neighboring neurons, witch is exceedingly suitable for integration of electron devices. The detailed information on the structures, sizes, circuits, and characteristics of the neuron, synapse, and networks were also explained in the prior reports [5, 6].

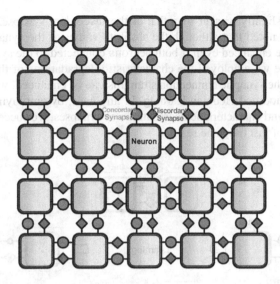

Fig. 3. Network.

3 Learning Principle

Hebbian learning is a typical learning procedure in biological and artificial neural networks [7]. The synaptic connection strength is enhanced when both neurons connected to the synapse are in firing states and impaired otherwise. Since the neuron and synapse were dramatically simplified, the leaning principle must be also modified. Figure 4 shows the modified Hebbian learning. Here, we assume NOT logic as an example. The left and right neurons are assigned to input and output elements, respectively. Initially, at the initial recalling stage, a non-firing state is applied to the input element, and a non-firing state arises from the output elements, and vice versa because the synaptic connection strength of the concordant synapse is accidentally slightly stronger than that of the discordant synapse, which is not NOT logic. Next, at the first learning stage, a non-firing state is applied to the input element, and a firing state is applied to the output element. Since the concordant synapse is connected between the same nodes in the two neurons, and the states at both nodes in two neurons are different, electric current flows through the concordant synapse because of the voltage difference, whereas electric current does not flow through the discordant synapse. Consequently, the characteristic degradation gradually occurs, which is a necessary property of our synapses [8], the conductivity has gradually higher impedance, and only the synaptic connection strength of the concordant synapse becomes gradually weakened. At the second learning stage, a firing state is applied to the input element, and a non-firing state is applied to the output element. Similarly, only the synaptic connection strength of the concordant synapse becomes gradually weakened. Finally, at the final recalling stage, a non-firing state is applied to the input element, and a firing state arises from the output elements, and vice versa because the synaptic connection strength of the concordant synapse becomes slightly weaker than that of the discordant synapse, which is NOT logic. It should be noted that although the absolute values of the synaptic connection strengths cannot be enhanced even if both neurons connected to the synapse are in the firing state because we employed the characteristic degradations of the synapses, the relative values of the synaptic connection strengths can be enhanced, which is a reason that we called it modified Hebbian learning. In any case, by employing the modified Hebbian learning and characteristic degradations of synapses, we succeeded in making a synapse consist of just only one resistor.

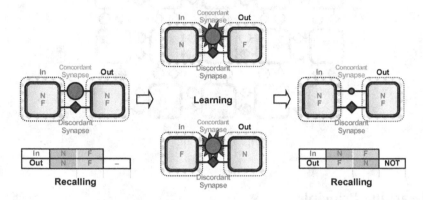

Fig. 4. Modified Hebbian learning.

4   Fabrication Process

We fabricated a cellular neural network using thin-film devices, which are expected as key technologies for micro-giant electronics. Micro-size electron devices can be fabricated on large and inexpensive substrates. Although the device size is in the order of μm in this research, it can be in the order of nm in the most advanced researches [9]. Although they are fabricated on a glass substrate in our research, they can be fabricated on plastic films [10], which can be folded down to compact size as human brains do. Therefore, we believe that thin-film devices are most promising electron devices for cellular neural networks.

We fabricated thin-film devices as follows. First, an amorphous-Si film was deposited using LPCVD of Si_2H_6, crystallized using XeCl excimer pulse laser, and patterned to form poly-Si films [11], whose thickness is 50 nm, which are used as channels for transistors. Next, a SiO_2 film was deposited using PECVD of TEOS to form an insulator film, whose thickness is 75 nm, which is used as gate-insulator films for transistors. Afterward, the first metal film was deposited and patterned, which is used as gate terminals for transistors and simultaneously the first electrode wires. Subsequently, phosphorous and boron were implanted into the poly-Si films and thermally activated to form doping regions, which are used as source and drain regions for transistors. Next, a SiO_2 film was deposited to form an insulator film, which is used as an interlayer-insulator film. After that, and the second metal film was deposited and patterned, which is used as source and drain terminals for transistors and simultaneously the second electrode wires. Finally, water-vapor heat treatment was performed to improve the poly-Si films, the SiO$_2$ film and their interfaces. Consequently, the field effect mobility and threshold voltage of the n-type transistors are 93 cm^2 V^{-1} s^{-1} and 3.6 V, respectively, while those of the p-type transistors are 47 cm^2 V^{-1} s^{-1} and −2.9 V, respectively. These parameters are sufficient for the circuits in the cellular neural network.

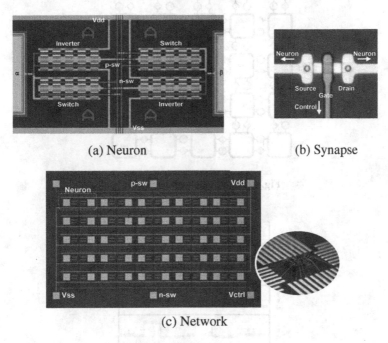

(a) Neuron   (b) Synapse

(c) Network

Fig. 5. Actual devices.

Figure 5 shows the actual devices. We arrayed 5 × 5 neurons and utilized transistors as synapses. The actual chip is die bonded on a printed circuit board and wire bonded to metal contacts.

5   Experimental Result

We tried to make the cellular neural network learn multiple logics. Figure 6 shows the input and output pattern. Four neurons at the corners were assigned to In1 and In2, and two neurons at the inside were assigned to Out1 and Out2, although they can be assigned freely to some extent. The non-firing or firing states were applied to In1 and In2 for each logic pair. At the recalling stage, output voltages generated from the network were measured at Out1 and Out2, whereas at the learning stage, corresponding outputs of AND and OR were applied to Out1 and Out2. Switching pulses were periodically applied to repeat the switch on and off of the switches in the neuron. At the recalling stage, a lower control voltage of 10 V was applied to the gate terminal of the transistor to avoid the characteristic degradations of the synapses, whereas at the learning stage, a higher control voltage of 15 V was applied to induce the characteristic degradations. The recalling and learning stages were repeated several ten times.

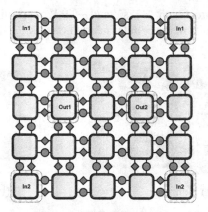

Fig. 6. Input and output pattern.

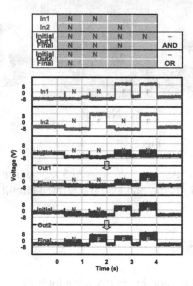

Fig. 7. Input and output pattern.

Figure 7 shows the learning result. It is found that at the first recalling stage, wrong output voltages were generated from Out1 and Out2. On the other hand, at the final recalling stage, correct output voltages corresponding outputs of AND and OR were generated from Out1 and Out2, respectively. Although this is an example, it was checked that the learning was successful in most cases in spite that the number of the times until the correct output voltages were generated was widely distributed. In conclusion, we confirmed that our neural network can learn multiple logics of AND and OR even in a small-scale neural network of 5 × 5.

6   Conclusion

We proposed a novel architecture for a cellular neural network suitable for high-density integration of electron devices. A neuron consists of only eight transistors, and a synapse consists of just only one transistor. We fabricated a cellular neural network using thin-film devices. Particularly in this time, we confirmed that our neural network can learn multiple logics even in a small-scale neural network. We think that this result indicates that our proposal has a big potential for future electronics using neural networks.

Acknowledgments. We thank Prof. Hakaru Tamukoh of Kyushu Institute of Technology and Prof. Yasuhiko Nakashima of Nara Institute of Science and Technology. This research is partially supported by a research project of the Joint Research Center for Science and Technology at Ryukoku University and grant from the High-Tech Research Center Program for private universities from the Ministry of Education, Culture, Sports, Science and Technology (MEXT).

References

1. Chua, L.O., Yang, L.: Cellular neural networks: theory. IEEE Trans. Circ. Syst. **32**, 1257—1272 (1988)
2. Koeppl, H., Chua, L.O.: An adaptive cellular non-linear network and its application. In: 2007 International Symposium on Nonlinear Theory and Its Applications (NOLTA 2007), pp. 15—18 (2007)
3. Crounse, K.R., Chua, L.O., Thiran, P., Setti, G.: Characterization and dynamics of pattern formation in cellular neural networks. Int. J. Bifurcat. Chaos **6**, 1703—1724 (1996)
4. Morie, T., Miyake, M., Nagata, M., Iwata, A.: A 1-D CMOS PWM cellular neural network circuit and resistive-fuse network operation. In: 2001 International Conference on Solid State Devices and Materials (SSDM 2001), pp. 90—91 (2001)
5. Kasakawa, T., Tabata, H., Onodera, R., Kojima, H., Kimura, M., Hara, H., Inoue, S.: An artificial neural network at device level using simplified architecture and thin-film transistors. IEEE Trans. Electron Devices **57**, 2744—2750 (2010)
6. Kimura, M., Miyatani, T., Fujita, Y., Kasakawa, T.: Apoptotic self-organized electronic device using thin-film transistors for artificial neural networks with unsupervised learning functions. Jpn. J. Appl. Phys. **54**, 03CB02 (2015)
7. Hebb, D.O.: The Organization of Behavior. Wiley, London (1949)
8. Kasakawa, T., Tabata, H., Onodera, R., Kojima, H., Kimura, M., Hara, H., Inoue, S.: Degradation evaluation of poly-Si TFTs by comparing normal and reverse characteristics and behavior analysis of hot-carrier degradation. Solid-State Electron. **56**, 207—210 (2011)
9. Kaneda, T., Hirose, D., Miyasako, T., Tue, P.T., Murakami, Y., Kohara, S., Li, J., Mitani, T., Tokumitsu, E., Shimoda, T.: Rheology printing for metal-oxide patterns and devices. J. Mater. Chem. C **2**, 40—49 (2014)

10. Kaltenbrunner, M., Sekitani, T., Reeder, J., Yokota, T., Kuribara, K., Tokuhara, T., Drack, M., Schwoediauer, R., Graz, I., Bauer-Gogonea, S., Bauer, S., Someya, T.: An ultra-lightweight design for imperceptible plastic electronics. Nature **499**, 458——463 (2013)
11. Sameshima, T., Usui, S., Sekiya, M.: XeCl excimer laser annealing used in the fabrication of poly-Si TFT's. IEEE Electron Device Lett. **7**, 276——278 (1986)

Analyzing the Impact of Feature Drifts
in Streaming Learning

Jean Paul Barddal[(✉)], Heitor Murilo Gomes, and Fabrício Enembreck

Graduate Program in Informatics (PPGIa),
Pontifícia Universidade Católica Do Paraná,
R. Imaculada Conceição, 1155, Curitiba, Brazil
jean.barddal@ppgia.pucpr.br

Abstract. Learning from data streams requires efficient algorithms
capable of deriving a model accordingly to the arrival of new instances.
Data streams are by definition unbounded sequences of data that are
possibly non stationary, i.e. they may undergo changes in data distribution, phenomenon named concept drift. Concept drifts force streaming
learning algorithms to detect and adapt to such changes in order to
present feasible accuracy throughout time. Nonetheless, most of works
presented in the literature do not account for a specific kind of drifts:
feature drifts. Feature drifts occur whenever the relevance of an arbitrary attribute changes through time, also impacting the concept to be
learned. In this paper we (i) verify the occurrence of feature drift in a
publicly available dataset, (ii) present a synthetic data stream generator capable of performing feature drifts and (iii) analyze the impact of
this type of drift in stream learning algorithms, enlightening that there
is room and the need for dynamic feature selection strategies for data
streams.

1 Introduction

Mining massives amount of data that arrive at rapid rates, namely data streams,
is a recurring challenge. Extracting useful knowledge from these potentially
unbounded sequences of data requires algorithms capable of acting within limited time, memory space and deal with its peculiarities i.e. concept drifts [9,15]
and evolutions [13]. Concept drifts occur when the data distribution changes
over time and are divided in two types: real and virtual. Real concept drifts
refer to changes in the conditional distribution of the target variable y given
the input (features) $\mathcal{D}$, while its distribution in the data input space $P[x]$ may
stay intact. Conversely, virtual concept drifts occur when the data distribution
$P[x]$ changes, independently of the conditional probability of the output values
$P[y|x]$ [8].

In this paper we review a specific kind of drift that is not commonly addressed
in the literature: feature drifts. Feature drifts occur whenever the relevance of
a feature (dimension) of a data stream grows or shrinks with time, enforcing
the learning algorithm to adapt its model to ignore the irrelevant attributes

© Springer International Publishing Switzerland 2015
S. Arik et al. (Eds.): ICONIP 2015, Part I, LNCS 9489, pp. 21–28, 2015.
DOI: 10.1007/978-3-319-26532-2_3

and account for the newly relevant ones [14]. Several approaches on how to compute the relevance of a feature for the classification task were proposed in the literature, such as Entropy, Information Gain and Gini Index [10].

In order to exemplify a feature drift, we refer to the e-mail spam detection system presented in [12]. This system was a result of a text mining process on an online news dissemination system. Essentially, this work intended on creating an incremental filtering of emails that classifies emails as spam or not and based on this classification, decides whether this email is relevant for dissemination among users. The dataset created contains 9,324 instances and 39,917 features, such that each attribute represents the presence of a single work (feature label) in the instance (e-mail). This dataset is known for containing a concept drift which occurs gradually around the instance of number 1,500 [1,12].

In Fig. 1 we present a plot of the information gain [10] of two specific attributes presented in this problem, namely "directed" and "listinfo", where one can see that the importance of these two attributes exchange gradually around instance 1,500.

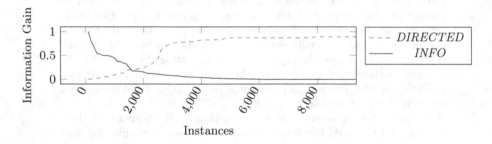

Fig. 1. Information Gain for two specific features of the Spam Corpus dataset.

This paper is divided as follows. The data stream learning and feature drift problems are specified in Sect. 2. In Sect. 3 we present a data stream generator able to simulate feature drifts. In Sect. 4 we empirically show the impact of feature drifts in two algorithms: an updatable naïve bayes algorithm and an incremental decision tree, namely Hoeffding Tree. Finally, in Sect. 5 we state the conclusions of this work and discuss envisioned future works.

2   Problem Statement

Let S be a data stream providing instances $i_t = (\boldsymbol{x}_t, y)$ intermittently, where $\boldsymbol{x}_t$ is a d-dimensional data object arriving at a timestamp t and y is its label. Instances $\boldsymbol{x}_i$ are labeled accordingly to values defined in $\mathcal{Y} = \{y_1, \ldots, y_c\}$. Also, let $\mathcal{D} = \{D_1, D_2, \ldots, D_d\}$ be the set features of a data stream where $d \geq 1$ is the dimensionality of the problem. It is assumed that S is unbounded, i.e. $|S| \to \infty$, thus, it is not feasible to store all instances in memory before processing. This characteristic forces algorithms to either process data in limited size chunks or to incrementally process instances. Firstly, every instance x_i must be processed before an instance x_{i+1} becomes available, otherwise instances start to accumulate and the algorithm may have to discard them. Secondly, there is an inherent temporal aspect associated with a stream process, where the data distribution may change over time, namely concept drift. Therefore, algorithms must also be able to detect and adapt to drifts, updating the algorithm's model.

Definition 1. *Let Eq. 1 denote a concept C, a set of prior probabilities of the classes and class-conditional probability density function [14]. Given a stream S, instances i_t retrieved will be generated by a concept C_t. If during each instant t_i of S we have $C_{t_i} = C_{t_{i-1}}$, it occurs that the concept is stable. Otherwise, if between any two timestamps t_i and t_j occurs that $C_{t_i} \neq C_{t_j}$, we have a concept drift.*

$$C = \{(P[y_1], P[x|y_1]), \ldots, (P[y_c], P[x|y_c])\} \tag{1}$$

Definition 2. *Given a feature space $\mathcal{D}$ at a timestamp t, we are able to select the top discriminative subset $\mathcal{D}_t^* \subseteq \mathcal{D}$. A feature drift occurs if, at any two time instants t_i and t_j, $\mathcal{D}_{t_i}^* \neq \mathcal{D}_{t_j}^*$ betides.*

In this paper we address the feature drift problem, where relevances of features of the data stream vary through time.

Definition 3. *Let $r(D_i, t_j) \in \{0, 1\}$ denote a function which determines the relevance of a feature D_i in a timestamp t_j of the stream. A positive relevance $(r(D_i, t_i) = 1)$ states that $D_i \in \mathcal{D}^*$ in a timestamp t_i and that it impacts the underlying probabilities $P[x|y_i]$ of the concept C_t in S. A feature drift occurs whenever the relevance of an attribute D_α changes in a timespan between t_j and t_k, as stated in Eq. 2*

$$\exists t_j \exists t_k, \; t_j < t_k, \; r(D_\alpha, t_j) \neq r(D_\alpha, t_k) \tag{2}$$

Changes in $r(\cdot, \cdot)$ directly affect the ground-truth decision boundary to be learned by the inductive algorithm. Therefore, feature drifts can be seen as a specific type of real concept drift which can occur with or without changes in the data distribution $P[x]$. As in other concept drifts, changes in $r(\cdot, \cdot)$ may occur during the stream, therefore enforcing algorithms to discard or adapt the model already learned, which is based on features that became irrelevant, which shall be replaced by the most relevant ones [14]. It is important to emphasize that feature drifts differ from concept drifts since concept drifts might occur without changes in attributes relevances but only in the a posteriori probabilities $P[x|y]$.

Additionally, performing dynamic feature selection is desired since it provides a smaller subset of features that gives you as good or better accuracy in the predictive model, while requiring less data. Less attributes (dimensions) is desirable since it reduces the complexity of the model, leading to a smaller chance of overfitting and a model that is simple to understand and explain [5].

In the following section we present a data stream generator able to simulate feature drifts.

3   Simulating Feature Drifts

To verify the impact of feature drifts in existing streaming learning algorithms, we present a data stream generator that extends the SEA generator [16].

The generator here proposed simulates streams with $d > 2$ uniformly distributed features given by the user, where $\forall D_i \in \mathcal{D}, D_i \in [0; 10]$ and only two randomly picked features are relevant to the concept to be learned: D_ω and D_ς. As in [16], the class value y is given accordingly to Eq. 3, where θ is a user-given threshold.

$$y = \begin{cases} 1, \ D_\omega + D_\varsigma \le \theta \\ 0, \ otherwise \end{cases} \tag{3}$$

Additionally, each instance synthesized has a 10 % probability of being generated as noise.

To promote synthetic feature drifts in streams, we adopt the sigmoid framework stated in Eq. 4 and introduced in [4]. This model treats a feature drift as a combination of two pure distributions that characterizes concepts before and after the drift. The variables presented in Eq. 4 are the following: $f(t_i)$ is the probability that an instance x_i belongs to the prior concept, $1 - f(t_i)$ is the probability for the posterior concept, w is the drift window size and t_0 is the drift moment.

$$f(t_i) = \frac{1}{(1 + e^{-w \times (t_i - t_0)})} \tag{4}$$

In [2] authors observe that Eq. 4 has a derivative at time t_0 equal to $f'(t_0) = s/4$ and that $\tan \alpha = f'(t_0)$, thus $\tan \alpha = s/4$. Also, $\tan \alpha = 1/W$ and as $s = 4 \tan \alpha$ then $\alpha = 4/W$, namely t_0 (time of drift), w and α (phase angle). In this sigmoid model there are only two parameters to be specified: t_0 and W.

Nonetheless, it is important to emphasize that any decay function can be applied to simulate feature drifts.

4   Analysis

In this section we evaluate the accuracy of an incremental and updatable Naïve Bayes algorithm and an incremental decision tree, namely Hoeffding Tree [6], in both abrupt and gradual feature drifts. Firstly, we briefly introduce the evaluated algorithms and the experimental protocol adopted. Finally, we discuss the results obtained.

4.1   Evaluated Algorithms

Updatable Naïve Bayes. The updatable Naïve Bayes algorithm is an incremental version of the popular Naïve Bayes algorithm. Both algorithms rely on the assumption that all attributes of the dataset are independent, with the exception of the output value y, which depends on all others $D_1, \ldots, D_d$. Therefore, these algorithms compute the output value for an input instance x_i as stated in Eq. 5, determining the value of y that maximizes the probability $P[x_i|y]$.

$$P[x_i|y] = \frac{P[y|x_i] \times P[x_i]}{P[y]} \tag{5}$$

In order to compute probabilities in a streaming environment, the Updatable Naïve Bayes stores a contingency table, therefore, no windowing process is needed whatsoever.

Hoeffding Tree. Hoeffding Trees algorithms construct decision trees by using constant memory and constant time per sample [6]. These trees are built by recursively replacing leaves with decision nodes, as data arrives. Different heuristic evaluation functions are used to determine whether a split should be performed or not, such as Gain Ratio, Entropy and Gini Coefficient [10]. To do so, Hoeffding Trees assume that the input data meets the Hoeffding bound [11].

Assuming a random variable $r \in \mathbf{R}$ with range R, a number of independent observations n, the mean computed by the latter observations $\bar{n}$; the Hoeffding Inequality states that with probability $1 - \delta$ the true mean of a variable is at least $\bar{r} - \epsilon$, where ϵ is given by Eq. 6 and δ is a user-given confidence bound.

$$\epsilon = \sqrt{\frac{R^2 \ln\left(\frac{1}{\delta}\right)}{2n}} \tag{6}$$

The Hoeffding bound is able to give results regardless the probability distribution that generates data. However, the number of observations needed to reach certain values of δ and ϵ are different across different probability distributions [3]. Generally, with probability $1 - \delta$, one can say that one attribute is superior when compared to others when observed difference of information gain (or any other metric that computes the importance of an attribute) is greater then ϵ.

Finally, all tree's nodes maintain statistics about the data used to derive itself. Periodically, Hoeffdings Trees discard nodes of the tree that are not accessed during traverses and replaces them by new ones accordingly to the Hoeffding bound and the chosen split function.

4.2 Experimental Protocol

Five different scenarios are evaluated in this section. The first scenario is the Spam Corpus dataset presented in [12], while the other four adopt the generator presented in Sect. 3 and were parametrized as follows:

- FD-1: 50,000 instances, $\theta = 7$ and $d = 10$
  - Drift 1: $t_0 = 25,000$, $w = 1$;
- FD-2: 50,000 instances, $\theta = 7$ and $d = 10$
  - Drift 1: $t_0 = 25,000$, $w = 1,000$;

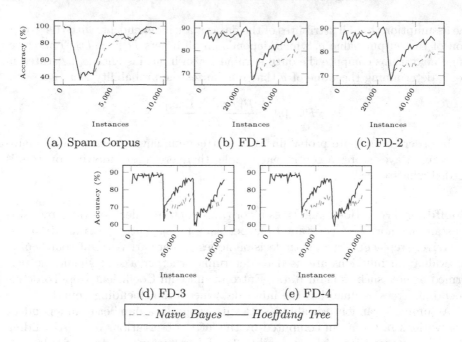

Fig. 2. Accuracy obtained during experiments with feature drifts.

- FD-3: 100,000 instances, $\theta = 9.5$ and $d = 10$
  - Drift 1: $t_0 = 34,000$, $w = 1$;
  - Drift 2: $t_0 = 67,000$, $w = 1$;
- FD-4: 100,000 instances, $\theta = 9.5$ and $d = 10$
  - Drift 1: $t_0 = 34,000$, $w = 1,000$;
  - Drift 2: $t_0 = 67,000$, $w = 1,000$;

In our experiments accuracy is measured using the Prequential test-then-train method. We adopted the Prequential procedure [7] due to the monitoring of the evolution of performance of models over time although it may be pessimistic in comparison to the holdout estimative. Nevertheless, authors in [7] observe that the prequential error converges to an periodic holdout estimative when estimated over a sliding window. Along these lines, we determined an evaluation sliding window of 1,000 instances for these experiments.

Finally, all experiments here presented were implemented and evaluated under the Massive Online Analysis (MOA) framework [4].

4.3 Results Obtained

In Fig. 2a one can see that accuracy drops by 60 % during the known feature drift and slowly recovers after approximately 3,500 instances.

In Figs. 2b through 2e we present the results obtained by the Naïve Bayes and the Hoeffding Tree algorithms in the FD-1, FD-2, FD-3 and FD-4 experiments, respectively.

In Figs. 2b and 2c one can see the impact of one feature drift during the stream. In both cases, it is, abrupt and gradual changes, both algorithms has its accuracy damped in 20 % and the Naïve Bayes fails to completely recover until the end of the stream.

Additionally, in Figs. 2d and 2e one can see that impact of two drifts in accuracy for both algorithms. Again, the mean accuracy drops by 30 %, showing the difficulty of adapting to both abrupt and gradual feature drifts.

The results here presented enable us to argue that existing algorithms do not account for the possibility of feature drifts. Even Hoeffding Trees, which perform feature selection during the stream, fail to quickly adapt to changes in features' relevances, showing that there is room and the need for dynamic feature selection algorithms for data streams.

5   Conclusion

In this paper we analyzed the feature drift problem. Feature drifts differ from conventional concept drifts since they do not occur accordingly to changes in the data distribution, but on the relevance of each attribute in the concept to be learned. Additionally, we presented a data generator capable of synthesizing data streams with this peculiarity. Finally, we benchmarked an incremental and updatable Naïve Bayes classifier and an incremental decision tree on synthetic data streams with feature drifts, showing the impact of feature drifts in their accuracy. We must emphasize that even the Hoeffding Tree fails to quickly adapt to feature drifts, an important trait since it possesses an embedded feature selection algorithm to determine splits in real-time processing, which is however, performed accordingly to user-given parameters and not automatically.

The results here presented highlight the inefficiency of algorithms on tracking which attributes are relevant for classification in data streams. Therefore, dynamic feature selection algorithms are of utmost importance to quickly detect and adapt to feature drifts. In future works we plan to verify the efficiency of state-of-the-art algorithms with the addition of feature selection algorithms using periodical verifications of feature relevances accordingly to a landmark windowing technique. Furthermore, we plan to study the impact of feature evolutions, i.e. appearance and disappearance of features, in streaming learning environments.

References

1. Barddal, J.P., Gomes, H.M., Enembreck, F.: SFNclassifier: a scale-free social network method to handle concept drift. In: Proceedings of the 29th Annual ACM Symposium on Applied Computing (SAC), SAC 2014. ACM March 2014
2. Bifet, A., Holmes, G., Pfahringer, B., Kirkby, R., Gavaldà, R.: New ensemble methods for evolving data streams. In: Proceedings of the 15th ACM SIGKDD International Conference on Knowledge Discovery and Data Mining, pp. 139–148. ACM SIGKDD June 2009

3. Bifet, A., Gavaldà, R.: Learning from time-changing data with adaptive windowing. In: SIAM International Conference on Data Mining (2007)
4. Bifet, A., Holmes, G., Kirkby, R., Pfahringer, B.: MOA: Massive online analysis. J. Mach. Learn. Res. **11**, 1601–1604 (2010)
5. Carvalho, V.R., Cohen, W.W.: Single-pass online learning: performance, voting schemes and online feature selection. In: Proceedings of the 12th ACM SIGKDD International Conference on Knowledge Discovery and Data Mining, KDD 2006, pp. 548–553. ACM, New York (2006)
6. Domingos, P., Hulten, G.: Mining high-speed data streams. In: Proceedings of the Sixth ACM SIGKDD International Conference on Knowledge Discovery and Data Mining, KDD 2000, pp. 71–80. ACM, New York (2000)
7. Gama, J., Rodrigues, P.: Issues in evaluation of stream learning algorithms. In: Proceedings of the 15th ACM SIGKDD International Conference on Knowledge Discovery and Data Mining, pp. 329–338. ACM SIGKDD June 2009
8. Gama, J., Žliobaitė, I., Bifet, A., Pechenizkiy, M., Bouchachia, A.: A survey on concept drift adaptation. ACM Comput. Surv. **46**(4), 1–37 (2014)
9. Gama, J.: Knowledge Discovery from Data Streams. Chapman & Hall/CRC, Boca Raton (2010)
10. Hall, M., Frank, E., Holmes, G., Pfahringer, B., Reutemann, P., Witten, I.H.: The weka data mining software: An update SIGKDD. Explor. Newsl. **11**(1), 10–18 (2009)
11. Hoeffding, W.: Probability inequalities for sums of bounded random variables. J. Am. Stat. Assoc. **58**(301), 13–30 (1963)
12. Katakis, I., Tsoumakas, G., Vlahavas, I.: Dynamic feature space and incremental feature selection for the classification of textual data streams. In: ECML/PKDD-2006 International Workshop on Knowledge Discovery from Data Streams, 2006, pp. 107–116. Springer Verlag, Berlin (2006)
13. Masud, M.M., Gao, J., Khan, L., Han, J., Thuraisingham, B.M.: Classification and novel class detection in concept-drifting data streams under time constraints. IEEE Trans. Knowl. Data Eng. **23**(6), 859–874 (2011)
14. Nguyen, H.-L., Woon, Y.-K., Ng, W.-K., Wan, L.: Heterogeneous ensemble for feature drifts in data streams. In: Tan, P.-N., Chawla, S., Ho, C.K., Bailey, J. (eds.) PAKDD 2012, Part II. LNCS, vol. 7302, pp. 1–12. Springer, Heidelberg (2012)
15. Silva, J.A., Faria, E.R., Barros, R.C., Hruschka, E.R., de Carvalho, A., Gama, J.: Data stream clustering: a survey. ACM Comput. Surv. **46**(1), 1–31 (2013)
16. Street, W.N., Kim, Y.: A streaming ensemble algorithm (sea) for large-classification. In: Proceedings of the seventh ACM SIGKDD International Conference on Knowledge Discovery and Data Mining, pp. 377–382. ACM SIGKDD August 2001

Non-linear Metric Learning Using Metric Tensor

Liangying Yin and Mingtao Pei[✉]

Beijing Laboratory of Intelligent Information Technology,
School of Computer Science, Beijing Institute of Technology,
Beijing 100081, People's Republic of China
{yinliangying,peimt}@bit.edu.cn

Abstract. Manifold based metric learning methods have become increasingly popular in recent years. In almost all these methods, however, the underlying manifold is approximated by a point cloud, and the matric tensor, which is the most basic concept to describe the manifold, is neglected. In this paper, we propose a non-linear metric learning framework based on metric tensor. We construct a Riemannian manifold and its metric tensor on sample space, and replace the Euclidean metric by the learned Riemannian metric. By doing this, the sample space is twisted to a more suitable form for classification, clustering and other applications. The classification and clustering results on several public datasets show that the learned metric is effective and promising.

Keywords: Non-linear metric learning · Manifold · Metric tensor · Gaussian mixture model

1 Introduction

Distance functions or dissimilarity measures are central to many algorithms in machine learning, pattern recognition and computer vision [1–3], which critically determines the performance of these algorithms. Instead of predefining a distance function based on prior knowledge, some more appealing approaches such as metric learning is to learn an appropriate distance function based on information available about the application. Metric learning is a type of algorithm which constructs a metric function in sample space to pull the similar samples close and push the dissimilar samples far, and it can be divided into linear method and non-linear one.

Linear metric learning is a well-studied problem, and many algorithms are proposed, such as the method of Xing et al. [4], large margin nearest neighbor (LMNN) [5], and information-theoretic metric learning (ITML) [6]. The basic idea of these algorithms is to construct a Mahalanobis distance based on different loss functions. The algorithm in [7] learned a cosine similarity for metric samples instead of Mahalanobis distance.

For non-linear metric learning, many algorithms use the kernel approach to learn the metric [8, 9]. GB-LMNN [10] employed a non-linear mapping combined with a traditional Euclidean distance function. By training the non-linear transformation directly in

© Springer International Publishing Switzerland 2015
S. Arik et al. (Eds.): ICONIP 2015, Part I, LNCS 9489, pp. 29–37, 2015.
DOI: 10.1007/978-3-319-26532-2_4

function space as GBRT [11], the resulting algorithm is insensitive to hyper-parameters and robust against over fitting.

Instead of mapping methods, geodesic distance in Riemannian geometry presents a direct and profound method to describe distance function. Recently, the matric tensor, which is the most fundamental concept to describe the manifold, is introduced into metric learning., Some researchers use geodesic metric defined by metric tensor as distance function instead of using homogeneous metric function. However, using metric tensor directly will encounter the problem of combinatorial explosion. Shi et al. [12] learned a sparse combination of locally discriminative metric by assuming that the metric tensor is locally (or globally) constant. Ramanan and Baker [13] use interpolation to integrate the metric tensor along the lines between the test and training points under the assumption that the metric tensor is piecewise constant [14] provides an algorithm for computing geodesics according to the learned metrics, as well as algorithms for computing exponential and logarithmic maps on the Riemannian manifold, which let many Euclidean algorithms take advantage of multi-metric learning.

Different from many current metric tensor algorithms, we propose a new method to approximate and learn the geodesic distance. As Gaussian mixture model can express inhomogeneity of the metric tensor easily and intuitively, we use it to learn the metric tensor on sample space to twist the plane Euclidean space, which can pull the similar samples close, and push the dissimilar samples far.

```
Metric Learning Framework(X,h,B,L)
X: input d*n data matrix
h: Gaussian peak number
B: result of linear algorithm
L: reasonable objective function
Output:{ ñᵢ},all agreements about the peaks
  X'=BX   //Transform samples with matrix B
  for i=1 to h
    ρᵢ⁰=rand() //initialize each ñᵢ
    ρᵢ = argmin L(X' ;∑ᵢ₌₁ⁱ⁻¹ ñₖ + ñᵢ)
         ñₗ
  repeat until L convergence
    for i=1 to h
      ρᵢ⁰=ρᵢ
      ρᵢ = argmin (X' ;ñᵢ + ∑ₖ₌₁ʰ ñₖ)
           ñₗ            k≠i
  end
```

Fig. 1. The optimal framework. n corresponds to vector $(ó, ì, á)$, and L is the loss function.

2   Theoretical Analysis

In the Riemannian geometry, the metric tensor is defined as a second order covariant symmetric tensor field g. We regard it in our method as a symmetric and positive-definition function matrix $[g_{ij}(x)]$ with size $d * d$, where d is the dimension of sample space. For example, the metric tensor is a positive-definition function matrix

$$\begin{bmatrix} g_{11}(x) & g_{12}(x) \\ g_{12}(x) & g_{22}(x) \end{bmatrix}$$ for 2-dimension sample space.

The function matrix indicates the local metric property among and can be interpreted as a Mahalanobis distance on local epsilon neighborhood space among x. Then we can define the arc length L in whole space as:

$$L = \int_a^b \sqrt{dx^t g(x) dx} \tag{1}$$

where both a and b are terminal point of the curve, and path of integral depends on the curve.

With this definition of arc length, we can get a new metric space in which the distance between point a and b is defined as the geodesic distance:

$$d(a,b) = inf \left\{ \int_a^b \sqrt{dx^t g(x) dx} \right\} \tag{2}$$

which is the lower bound in all of the possible integral paths.

From the above mentioned, in our method, we can make a metric tensor in sample space, which twists the space and pull the similar samples close and push the dissimilar samples far.

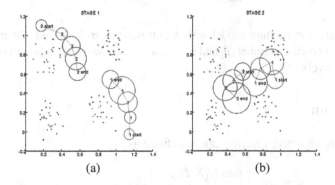

(a)   (b)

Fig. 2. (a) Shows the moves of μ and the changes of σ at stage 1 and (b) is for stage 2. The samples are divided by 2 colors(red and blue) to express their true classifications. We set number of peaks is two. To show the changes of both peaks when running the algorithm, we use '1' and '2' to mark their i respectively and use 'start' and 'end' to indicate the starting and ending i for each peak. Finally, the two radius of circles colored by blue and red indicate changes of σ for the two peaks respectively.

3   Problem Simplification

There are two problems to be solved. One is to compute the $d(x, y)$, which need to solve the variational problem; the other one is to optimize the N-order function matrix. It is very difficult to solve these two problems directly. Therefore, we simplify the problem as follows. We fixed the integral path as straight line:

$$d(a, b) = \int_a^b \sqrt{dx^t g(x) dx} \tag{3}$$

It is a pseudo-distance, which can simplify the variational problem. In general, the distances will be infinite in $d * d$ dimensional function space g, so with a rational parameter space, we can get the solution with small calculation. We define $\rho(x)$ as density function on sample space, and then define:

$$g(x) = B^t \rho(x)^2 B \tag{4}$$

where B can be an arbitrary matrix. We would like to account for the form unrigorously but rationally, that it is equal to transform the space by multiplying matrix B and then define:

$$d(a, b)_\rho = \int_{a'}^{b'} \rho(x) \sqrt{dx^t dx} \tag{5}$$

where $a' = Ba$ and $b' = Bb$. It can be solved by curve integral. This explanation enables us to use the result of linear metric learning method to initialize B, which will be shown in next section.

To describe the inhomogeneity of the metric, we set $\rho(x)$ as:

$$\rho(x) = \sum_{i=1}^h \alpha_i exp(-\frac{\|x - \mu_i\|^2}{\sigma_i^2}) \tag{6}$$

which is a Gaussian mixture model with h components α_i is the weight of Gaussian model, and is a positive number. μ_i and σ_i are the center and the bandwidth of Gaussian model, respectively.

4   Algorithm

Our optimizing objective function can be defined as

$$\min_\rho L\left(X; d(\cdot, \cdot)_\rho\right) \tag{7}$$

where L is the loss functional of distance function $d(\cdot, \cdot)_\rho$, and X is dataset. $d(\cdot, \cdot)_\rho$ is defined by:

$$d\left(a,b\right)_\rho = \int_{a'}^{b'} \sum_{i=1}^{h} \alpha_i \exp\left(-\frac{\|x-\mu_i\|^2}{\sigma_i^2}\right)\sqrt{dx^t dx} = \|a'-b'\| \sum_{i=1}^{h} \alpha_i V_i \qquad (8)$$

where

$$V_i = \left[Ncdf\left(t;\mu_i^*,\sigma_i^*\right)\right]_{t=0}^{t=1} \cdot \frac{\sqrt{\pi}\sigma_i}{\|b'-a'\|} C_i, \; \mu_i^* = -\frac{\langle b'-a', a'-\mu_i\rangle}{\|b'-a'\|^2},$$

$$\mu_i^* = -\frac{\langle b'-a', a'-\mu_i\rangle}{\|b'-a'\|^2}$$

$$C_i = \exp\left(\frac{\langle b'-a', a'-\mu_i\rangle^2 - \|a'-\mu_i\|^2 \|b'-a'\|^2}{\sigma_i^2\|b'-a'\|^2}\right), \sigma_i^* = \frac{\sigma_i}{\sqrt{2}\|b'-a'\|},$$

$Ncdf()$ is short for standard cumulative normal distribution function.
V_i is differentiable with regard to μ_i and σ_i. So $L\left(X;d\left(\cdot,\cdot\right)_\rho\right)$ can be optimized by gradient based methods such as gradient descent or conjugate gradients. We first compute the gradient of V_i. $\frac{\partial V_i}{\partial \mu_i}$ can be computed by:

$$\frac{\partial V_i}{\partial \mu_i} = \frac{(a'-b')}{\sigma_i \|b'-a'\|^2} C_i \left(\sigma_i \left[\exp\left(-\frac{t^2}{2\sigma_i^{*2}}\right)\right]_{t=\mu_i^*}^{t=1-\mu_i^*} + 2\sqrt{\pi}\frac{\langle b'-a', a'-\mu_i\rangle}{\|b'-a'\|}\left[Ncdf\left(t;\mu_i^*,\sigma_i^*\right)\right]_{t=0}^{t=1}\right)$$
$$+ \frac{2(a'-\mu_i)}{\sigma_i^2} V_i \qquad (9)$$

$\frac{\partial V_i}{\partial \sigma_i}$ has no analytic solutions, and difference is used instead. It would not slow down the algorithm because the major ingredient in gradient of V_i is $\frac{\partial V_i}{\partial \mu_i}$. After getting the gradient of V_i, it is easy to compute $\frac{\partial d(a,b)_\rho}{\partial \rho}$ and $\frac{\partial L(X;d(\cdot,\cdot)_\rho)}{\partial \rho}$.

The algorithm framework is shown in Fig. 1. It can be divided into two stages. At the first stage, we initialize ρ with random value and then optimize it. We add optimize $\rho_{i,i=1,2,...,h}$ one by one, that is when ρ_i is optimized, $\rho_{j,j\neq i}$ remains constant. At the second stage, we use the optimized ρ_i as the initial value and continue to optimize ρ_i separately again as in the first stage. Many optimal algorithms can be used in the optimized step in both stages such as steepest descent or conjugate gradients. The algorithm is efficient as it optimizes only one Gaussian peaks in each iteration.

It is worth noting that we can run linear metric learning algorithm firstly and use its result as matrix B to transform the samples, and set each initial α as a positive number near to zero. By doing this, the initial metric tensor will approximate to Euclidean space instead of random space. It means that instead of using our algorithm individually, we can use linear metric learning algorithm to initialize the matrix B and then run our algorithm. The experimental results in Sect. 5 show that using linear metric learning algorithm to initialize matrix B can improve the performance greatly.

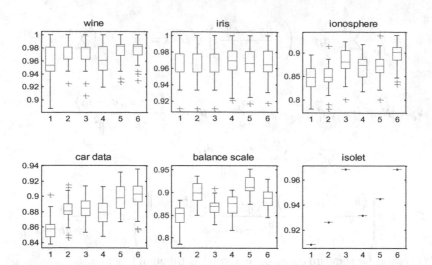

Fig. 3. kNN classification results for 6 UCI data sets. The algorithms are (1) Euclidean kNN, (2) ITML, (3) LMNN, (4) ours, (5) ITML+ ours, (6) LMNN+ ours

Consider the data shown in Fig. 2, which is divided into two classes (shown by the different colors), we use 2 Gaussian peaks, and then run the algorithm. Figure 2 depicts the trajectories of the Gaussian peaks. At the stage one, the algorithm creates the red Gaussian peak first, then the blue peak. At the second stage, these two peaks move in turn. The figure also shows that peaks in both of these two stage do their best to 'avoid' the dense area and have the tendency of moving to position between classes, which show that the algorithm can separate samples with different label exactly.

5 Experiment

5.1 Performance in Supervised Metric Learning

We use the loss function which is similar to LMNN [5] in the supervised metric learning. The loss function of LMNN has two terms, one is for neighboring sample pairs with same label, and the second is for impostor neighbors (i.e., samples with different labels and near to each other). In order to adapt characteristic of our algorithm, we modify the loss function as:

$$L(X;\rho) = \frac{\sum_i \sum_{j \in target\_neighbor(i)} d\left(x_i, x_j\right)_\rho}{\sum_i \sum_{j \in imposter\_neighbor(i)} d\left(x_i, x_j\right)_\rho} \tag{10}$$

The definition of target neighbor and imposter are the same as [10]. This loss function minimizes the distance between an instance and its target neighbors, and maximizes the distance between an instance and its impostor neighbors. The modified loss function can solve two problems effectually. First, because the number of Gaussian increased in each iteration in stage 1, the metric tensor will rise continuously, which means distance between all samples has the tendency to decrease. It makes most of the loss functions fail. Second, if the centers of Gaussian peaks depart from most of samples too much, these peaks cannot improve the classification performance. Fortunately by using the division form, our loss function is changed only by relative distance, which can solve the mentioned problems and pull the centers of Gaussian peaks back.

We evaluate our non-linear metric learning algorithms on several public data sets: Wine, Iris, Ionosphere, car data, Balance scale and ISOLET from the UCI repository. ISOLET predefined testing data and training data. For other data sets, results are averaged over 30 train/test splits (70 %/30 %). ISOLET is reduced the input dimensionality (originally 617) by projecting the data onto its leading 172 principal components which accounts for 95 % of its total variance [5]. We set k = 5 for kNN classification. We compare 6 algorithms: Euclidean kNN, ITML, LMNN, ours, ITML + ours (initialize matrix B by ITML's result and learn the metric by our method) and LMNN + ours. Figure 3 shows the comparison results. We find that initialized by linear algorithms and optimized by our algorithm can get better performances. Using our algorithm individually also can achieve good performance.

5.2 Application in Semi-supervised Clustering

Let S and D be the set of similar and dissimilar pair of the n data points. We define the following performance measure:

$$J_{DS} = \frac{\bar{d}_D}{\bar{d}_S} \tag{11}$$

where $\bar{d}_D = \frac{1}{n_D} \sum_{(x_i,x_j) \in D} d(x_i, x_j)_\rho$ is the mean distance between dissimilar samples with n_D being the number of dissimilar pairs, and $\bar{d}_S = \frac{1}{n_S} \sum_{(x_i,x_j) \in S} d(x_i, x_j)_\rho$. is the mean distance between similar samples with n_s being the number of similar sample pairs.

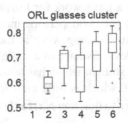

Fig. 4. Semi-supervised clustering results on ORL dataset. The 6 algorithms are (1) Euclidean kmeans, (2) Xing's algorithm, (3) ITML, (4) ours, (5) Xing's + ours, (6)ITML + ours.

We test our algorithm on ORL face data set. The face data set contains 400 grayscale images of 40 individuals in 10 different poses. We down sample the images to 38×31 pixels and use PCA to reduce the dimension to 30 [5]. We want to cluster the data into two classes: with glasses and without glasses. We select 80 sample pairs randomly from all C_{400}^2

pairs which include similar pairs and dissimilar pairs. We compare 6 algorithms: Euclidean k-means, Xing's algorithm [4], ITML [6], ours, Xing's+ ours, and ITML+ ours. We run the experiments for 30 times and the average result is shown in Fig. 4. Figure 5 shows a specific result of one experiment. Like supervised experiment, we can see that each united algorithms has improvement to their ununited ones. On these results, we can see that although result by using our method individually is a bit below ITML, the united one (ITML + ours) can improve ITML greatly. The result of Euclidean kmeans indicates that this problem is so complex that clusters without semi-supervised information is equal to random guess. Finally, the problem is the variance of ours and the united ones are little bigger then original ones, which may be probably caused by initializing the arguments randomly.

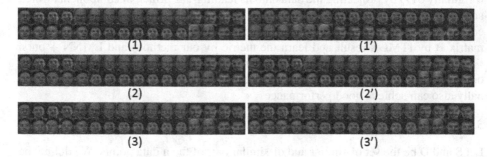

Fig. 5. These subfigures show a specific semi-clustering result on part of ORL data set. The algorithms which are used are (1) Euclidean kmeans, (2) Xing's, (3) ITML, (1') ours individually, (2') Xing's+ ours, (3') ITML+ ours. The faces on first line for each subfigure are with glasses, and others are without glasses. The clustering results are denoted by different color. In these subfigures, the yellow corresponds to whom with glasses, and blue corresponds to whom without glasses.

6   Conclusion

In this paper, we have proposed a framework for learning a distance metric based on metric tensor. Our method can get good performance both using individually and combining with linear algorithm. Ongoing work is focus on following two aspects. Currently, we simplify the Geodesic distance to a straight line integral, which is equal to assume that if point c is at the straight which has terminated point a and b, $d(a, c)$ or $d(c, b)$ must be less than $d(a, b)$, which is little unfit for some real samples. And we simplify the function matrix $[g_{ij}(x)]$ to function $\rho(x)$ for fast computation, which may loss information of the manifold.

References

1. Feng, Z., Jin, R., Jain, A.: Large-scale image annotation by efficient and Robust Kernel metric learning. In: 2013 IEEE International Conference on Computer Vision (ICCV), pp. 1609–1616. IEEE (2013)

2. Lajugie, R., Bach, F., Arlot, S.: Large-margin metric learning for constrained partitioning problems. In: Proceedings of The 31st International Conference on Machine Learning, pp. 297–305 (2014)
3. Cui, Z., Li, W., Xu, D., et al.: Fusing robust face region descriptors via multiple metric learning for face recognition in the wild. In: 2013 IEEE Conference on Computer Vision and Pattern Recognition (CVPR), pp. 3554–3561. IEEE (2013)
4. Xing, E.P., Jordan, M.I., Russell, S., et al.: Distance metric learning with application to clustering with side-information. In: Advances in Neural Information Processing Systems, pp. 505–512 (2002)
5. Weinberger, K.Q., Blitzer, J., Saul, L.K.: Distance metric learning for large margin nearest neighbor classification. In: Advances in Neural Information Processing Systems, 1473–1480 (2005)
6. Davis, J.V., Kulis, B., Jain, P., et al.: Information-theoretic metric learning. In: Proceedings of the 24th International Conference on Machine Learning, pp. 209–216. ACM (2007)
7. Nguyen, H.V., Bai, L.: Cosine similarity metric learning for face verification. In: Kimmel, R., Klette, R., Sugimoto, A. (eds.) ACCV 2010, Part II. LNCS, vol. 6493, pp. 709–720. Springer, Heidelberg (2011)
8. Yeung, D.Y., Chang, H.: A kernel approach for semisupervised metric learning. IEEE Trans. Neural Netw. **18**(1), 141–149 (2007)
9. Wang, F., Zuo, W., Zhang, L., et al.: A kernel classification framework for metric learning. arXiv preprint arXiv:1309.5823 (2013)
10. Kedem, D., Tyree, S., Sha, F., Lanckriet, G.R., Weinberger, K.Q.: Non-linear metric learning. In: Advances in Neural Information Processing Systems, pp. 2573–2581 (2012)
11. Friedman, J.H.: Greedy function approximation: a gradient boosting machine. Ann. Stat. 1189–1232 (2001)
12. Shi, Y., Bellet, A., Sha, F.: Sparse compositional metric learning. arXiv preprint arXiv: 1404.4105 (2014)
13. Ramanan, D., Baker, S.: Local distance functions: A taxonomy, new algorithms, and an evaluation. IEEE Trans. Pattern Anal. Mach. Intell. **33**(4), 794–806 (2011)
14. Hauberg, S., Freifeld, O., Black, M.: A geometric take on metric learning. In: Bartlett, P., Pereira, F., Burges, C., Bottou, L., Weinberger, K. (eds.) Advances in Neural Information Processing Systems (NIPS), vol. 25, pp. 2033–2041. MIT Press, Cambridge (2012)

An Optimized Second Order Stochastic Learning Algorithm for Neural Network Training

Mohamed Khalil-Hani, Shan Sung Liew$^{(\boxtimes)}$, and Rabia Bakhteri

VeCAD Research Laboratory, Faculty of Electrical Engineering,
Universiti Teknologi Malaysia, 81310 Skudai, Johor, Malaysia
{khalil,rabia}@fke.utm.my, ssliew2@live.utm.my

Abstract. The performance of a neural network depends critically on its model structure and the corresponding learning algorithm. This paper proposes bounded stochastic diagonal Levenberg-Marquardt (B-SDLM), an improved second order stochastic learning algorithm for supervised neural network training. The algorithm consists of a single hyperparameter only and requires negligible additional computations compared to conventional stochastic gradient descent (SGD) method while ensuring better learning stability. The experiments have shown very fast convergence and better generalization ability achieved by our proposed algorithm, outperforming several other learning algorithms.

Keywords: Second order method · Fast convergence · Stochastic diagonal Levenberg-Marquardt · Convolutional neural network

1 Introduction

Machine learning mainly comprises of two components: a learnable model (e.g. neural networks) and its corresponding learning algorithm. The structure of the model is often designed towards providing higher nonlinearity behaviour to achieve better learning. Regardless of how well the learning capacity of a model is, the learning performance is still highly dependent on the effectiveness of the learning algorithm. In this paper, we focus on the development of a supervised learning algorithm to train neural networks.

Most learning algorithms are based on iterative methods, which aim to find a set of parameters (weights) that lead to an optimum solution by taking small steps iteratively towards a direction until it reaches a desired solution:

$$W_{t+1} = W_t - \triangle W \tag{1}$$

where W_t are the weights at current iteration t, W_{t+1} denotes the weights for next iteration $t + 1$, and $\triangle W$ represents the step size to be taken at iteration t. In neural networks, $\triangle W$ is usually computed by finding the corresponding error gradients, which can be done using back-propagation algorithm.

© Springer International Publishing Switzerland 2015
S. Arik et al. (Eds.): ICONIP 2015, Part I, LNCS 9489, pp. 38–45, 2015.
DOI: 10.1007/978-3-319-26532-2_5

Gradient descent (GD) is a simple first order optimization algorithm, but often suffers from very slow convergence. Many other methods have been proposed to improve the learning convergence rate, but with the expense of additional computations. This suggests the necessity to develop a learning algorithm that can converge fast while requiring negligible computational overhead.

In this paper, we propose a second order stochastic learning algorithm that is optimized for faster learning and efficient computations. Our proposed algorithm, B-SDLM is inspired by the stochastic diagonal Levenberg-Marquardt (SDLM) algorithm, and utilizes both curvature information and bounding condition for better learning. The paper is organized as follows, Sect. 2 covers common learning algorithms and proposes an optimized algorithm. Section 3 describes the experimental design and discusses the results. The final section concludes the work and suggests possible future work.

2 Proposed Algorithm

2.1 Overview of Learning Algorithms

Conventional GD works by summing and averaging error gradients of all training samples before updating the weights, hence the name batch GD (BGD). BGD is effective and embarrassingly parallel, yet suffers from slow convergence. Meanwhile, stochastic GD (SGD) attempts to avoid local minima by performing noisy updates. Still, SGD can suffer from slow convergence near minimum of ill-conditioned problems.

In general, the convergence rate of a learning algorithm depends on update step sizes, i.e. the learning rate. Many algorithms have been proposed to adapt the global learning rate by having annealing schedule, or rely on the sign of error gradients (e.g. Rprop [6]). Some allow each weight to be tuned using its specific learning rate, mostly based on first derivatives (e.g. AdaGrad [4] and AdaDelta [13]).

Second order algorithms generally make use of both gradient and curvature information for faster optimization, which are mainly based on Newton's method:

$$\triangle W = \eta \left(\frac{d^2 E\left(W\right)}{dW^2} \right)^{-1} \frac{dE\left(W\right)}{dW} \tag{2}$$

Where $\eta \in (0, 1)$ is the global learning rate. Newton's method guarantees the convergence state, but it is prohibitive to compute in practice for large problems. Many alternative methods have been proposed based on approximations to the Hessian (second derivatives) to ease the computations [8]. These include Quasi-Newton [2,11], Gauss-Newton, and Levenberg-Marquardt (LMA) algorithms. Most second order algorithms generally perform better than conventional first order methods, but still can be compute-intensive, plus they only support batch learning mode.

2.2   Stochastic Diagonal Levenberg-Marquardt Algorithm

Realizing the potentials of second-order methods and their computational intensity, algorithms have been developed to reduce computations while still achieving desirable learning performance. One of most well known works was by Becker *et al.* for proposing back-propagation algorithm with "pseudo-Newton step" [1]:

$$\triangle w_{ji} = \frac{\frac{\partial E}{\partial w_{ji}}}{\left| \frac{\partial^2 E}{\partial w_{ji}^2} \right| + \mu} \tag{3}$$

Where μ is regularization parameter to deal with regions of small curvature as suggested in LMA. The "pseudo-Newton step" has later become the underlying basis of stochastic diagonal Levenberg-Marquardt (SDLM) algorithm [7].

SDLM adds a global learning rate which aids to the convergence, and computes running diagonal estimates of Hessian matrix over a subset of training samples:

$$\triangle w_{ji} = \left(\frac{\eta}{\left| \left\langle \frac{\partial^2 E}{\partial w_{ji}^2} \right\rangle \right| + \mu} \right) \frac{\partial E}{\partial w_{ji}} \tag{4}$$

$$\left\langle \frac{\partial^2 E}{\partial w_{ji}^2} \right\rangle_{(t+1)} = (1 - \gamma) \left\langle \frac{\partial^2 E}{\partial w_{ji}^2} \right\rangle_{(t)} + \gamma \frac{\partial^2 E}{\partial w_{ji}^2}_{(t)} \tag{5}$$

Where η is global learning rate, and γ determines the portion of previous estimates to be used for new Hessian estimation [8]. SDLM has been successfully applied to train deep neural networks [7], but there are still some outstanding issues that need to be solved. Firstly, learning stability is highly dependent on the choices of both η and γ. Secondly, more hyperparameters will result in hyperparameter overfitting problem, in which there are endless ways of configuring the learning algorithm. Some approaches have been taken to reduce the total hyperparameters (e.g. layer-specific SDLM algorithm (L-SDLM) [10]), but in most cases adding more computations to the learning algorithm, which is undesirable.

2.3   Bounded SDLM Algorithm

We propose an improved version of SDLM algorithm that requires only a single hyperparameter, fewer computations, while still retaining the second order properties of the method for faster convergence. These improvements are mainly focused on two aspects: hyperparameter overfitting and computational intensity.

Idea 1: Simple averaging of Hessian estimates. Instead of computing running Hessian estimates (Eq. 5), we take simple average of instantaneous diagonal Hessians to omit the memory constant γ while making the computations much simpler:

$$\overline{\frac{\partial^2 E}{\partial w_{ji}^2}} = \frac{\sum_{M_H} \frac{\partial^2 E}{\partial w_{ji}^2}_{(m)}}{M_H} \tag{6}$$

Where M_H is total samples used for Hessian calculation. Then the computation of weight-specific learning rate becomes

$$\eta_{ji} = \frac{\eta}{\left| \frac{\partial^2 E}{\partial w_{ji}^2} \right| + \mu} \tag{7}$$

Where η_{ji} is the learning rate with respect to w_{ji}.

Idea 2: Boundary condition for numerical stability. The main reason of having μ is to prevent the learning rates from blowing up. Our idea is to replace μ with simple boundary condition as in the following equation:

$$\eta_{ji} = \begin{cases} \frac{\eta}{\left| \frac{\partial^2 E}{\partial w_{ji}^2} \right|} & \left| \frac{\partial^2 E}{\partial w_{ji}^2} \right| > 1 \\ \eta & otherwise \end{cases} \tag{8}$$

Where η is now the upper boundary value for the computations of weight-specific learning rates as well. This has three main advantages: the boundary condition serves the same purpose of ensuring learning stability; fewer computations are required; and lastly, the risk of hyperparameter overfitting is significantly reduced by having only a single hyperparameter η to be tuned.

3 Experimental Design

We have evaluated our proposed learning algorithm on handwritten digit classification task based on MNIST dataset. The dataset consists of 60000 training and 10000 testing samples. We performed only z-score normalization on the images [12]. No other preprocessing techniques were applied. Only 1 % of training samples were randomly selected for Hessian calculation, and the training set was shuffled before every new training epoch.

In this work, we conducted the experiments on a convolutional neural network (CNN). All convolutional layers perform convolutions with single step-sized correlation filtering [9]. We used softmax layer and cross-entropy objective function to perform classification and evaluate the fitness of trained model. Outputs of all layers (except max-pooling and softmax) are passed through rectified linear unit (ReLU) activation function due to its superior performance and high computational efficiency.

In terms of free parameters, we have all convolutional layers with shared weights and biases, resulting in fewer parameters. We performed normalized weight initialization on all weights based on each neuron's fan-in and fan-out [5]. The diagonal Hessians were computed through second order back-propagations. Exhaustive search of hyperparameters was conducted to produce best learning results. We ran the experiments using the following global learning rate values: 0.001, 0.0025, 0.005, 0.0075 and 0.01. For SDLM algorithm, we selected from regularization parameter value of 0.01, 0.02 and 0.03. For each experimental

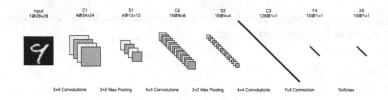

Fig. 1. CNN model for handwritten digit classification

setting, we repeated the training procedure three times, with 50 epochs per repetition. The best results were chosen to be presented in the next section.

All codes were written in C/C++, and were complied in Ubuntu 14.04 64-bit LTS OS with O3 code optimization level. Real-valued data was represented by single-precision floating data type throughout the experiments. We ran all experiments on a PC platform with an overclocked Intel Core i7 4790K 4.5 GHz CPU and 4 GB of RAM.

4   Results and Discussions

We evaluated our proposed algorithm with several other learning algorithms on the same case study using identical neural network model.

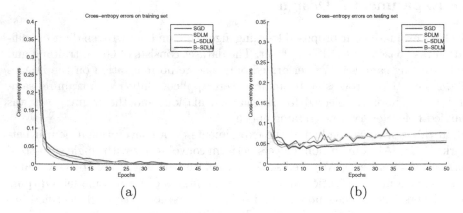

Fig. 2. Average cross-entropy errors for various learning algorithms on training set (a) and testing set (b) (Color figure online)

Figure 2 illustrates the learning curves of various learning algorithms on both training and testing sets. All variants of SDLM algorithms performed better than the conventional SGD algorithm due to utilization of second order information. The modified version of SDLM algorithm proposed by Milakov performed worse than other SDLM variants despite having additional computations of layer-specific hyperparameters. This suggests that these layer-specific hyper-parameters may not work well for all cases as reported in Milakov's work [10]. Meanwhile, our proposed algorithm can perform comparably with a finely-tuned SDLM algorithm, and even outperform it after several epochs as indicated by its cross-entropy error curve in Fig. 2b.

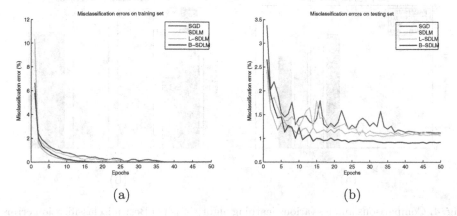

(a)   (b)

Fig. 3. Misclassification errors for various learning algorithms on training set (a) and testing set (b) (Color figure online)

In addition to the learning curves, we examined the learning performances of the algorithms by studying their classification accuracies on both data sets. The misclassification error curves on training set (Fig. 3b) exhibit similar patterns as observed in Fig. 2a. Meanwhile, Fig. 3b gives us a clear indication of how well our algorithm has performed on testing set, even within 20 training epochs only. The steep error curve that is separable from others signifies good generalization performance of the proposed learning algorithm, which is the main objective of any machine learning problems.

Table 1 shows the best errors and accuracies for various learning algorithms within 50 training epochs. The figures in cross-entropy error column suggest that lower cross entropy error on training set does not directly corresponds to

Table 1. Best cross-entropy errors and classification accuracies of various learning algorithms

Learning algorithm	Cross-entropy error		Accuracy (%)	
	Training	Testing	Training	Testing
SGD	6.43×10^{-5}	4.57×10^{-2}	99.99	98.88
SDLM	5.48×10^{-4}	4.09×10^{-2}	99.99	98.92
L-SDLM	5.59×10^{-5}	4.34×10^{-2}	99.99	98.96
Proposed	2.31×10^{-4}	3.76×10^{-2}	99.99	99.10

better classification performance on testing set. A similar pattern is observed in Fig. 4a, where the classification accuracies of the learning algorithms are comparable to each other in training set, with different generalization performances on testing set.

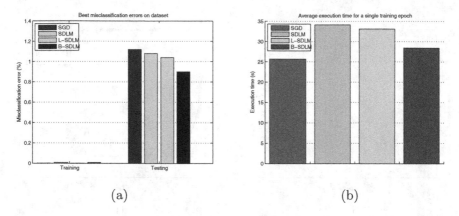

(a)   (b)

Fig. 4. Comparisons among various learning algorithms (a) Best misclassification error on MNIST dataset; and (b) Average execution time of a single training epoch (Color figure online)

We also measured the efficiencies of various learning algorithms in terms of the average execution time per training epoch. By referring to Fig. 4b, it is reasonable to conclude that SGD algorithm consumes least computational time among these learning algorithms. This is due to the fact that it does not involve any additional computations for Hessian approximation. The original SDLM requires longest time to complete a training epoch due to the running estimation of diagonal Hessian, while the layer-specific SDLM algorithm requires less execution time than the former algorithm due to averaging of instantaneous Hessian terms. However, it is still slower than our algorithm, since additional computations are required to determine the layer-specific regularization parameters.

5   Conclusion and Future Works

In this paper, we have presented B-SDLM, an optimized version of SDLM learning algorithm consisting of a single hyperparameter only. The algorithm can achieve very fast convergence, while having negligible computational overhead over SGD method. We evaluated the effectiveness of our learning algorithm on handwritten digit classification problem using CNN model. The results clearly indicate that the proposed algorithm outperforms other learning algorithm significantly, both in convergence rate and generalization performance.

Our future work involves extensive analysis of the proposed algorithm for different neural network models as well as case studies. Mapping the learning algorithm into distributed learning environment can be a good direction of expanding its deep learning capability on big data while maintaining reasonable computational time [3].

Acknowledgements. This work is supported by Universiti Teknologi Malaysia (UTM) and the Ministry of Science, Technology and Innovation of Malaysia (MOSTI) under the ScienceFund Grant No. 4S116.

References

1. Becker, S., Le Cun, Y.: Improving the convergence of back-propagation learning with second order methods. In: Proceedings of the connectionist models summer school, pp. 29–37 (1988)
2. Byrd, R.H., Lu, P., Nocedal, J., Zhu, C.: A limited memory algorithm for bound constrained optimization. SIAM J. Sci. Comput. **16**(5), 1190–1208 (1995)
3. Chen, X.-W., Member, S., Lin, X.: Big data deep learning : challenges and perspectives. IEEE Access 2 (2014)
4. Duchi, J., Hazan, E., Singer, Y.: Adaptive subgradient methods for online learning and stochastic optimization. J. Mach. Learn. Res. **12**, 2121–2159 (2011)
5. Glorot, X., Bengio, Y.: Understanding the difficulty of training deep feedforward neural networks. In: Proceedings of the 13th International Conference on Artificial Intelligence and Statistics (AISTATS), vol. 9, pp. 249–256 (2010)
6. Igel, C., Hüsken, M.: Improving the Rprop learning algorithm. In: Proceedings of the Second International Symposium on Neural Computation (Nc), pp. 115–121 (2000)
7. LeCun, Y., Bottou, L.: Gradient-based learning applied to document recognition. Proc. IEEE **86**(11), 2278–2324 (1998)
8. LeCun, Y.A., Bottou, L., Orr, G.B., Müller, K.-R.: Efficient BackProp. In: Montavon, G., Orr, G.B., Müller, K.-R. (eds.) Neural Networks: Tricks of the Trade, 2nd edn. LNCS, vol. 7700, pp. 9–48. Springer, Heidelberg (2012)
9. Liew, S., Khalil-Hani, M., Syafeeza, A., Bakhteri, R.: Gender classification: a convolutional neural network approach. Turk. J. Elec. Engin. http://journals.tubitak.gov.tr/elektrik/accepted.htm
10. Milakov, M.: Convolutional Neural Networks in Galaxy Zoo Challenge, pp. 1–7 (2014)
11. Shanno, D.F.: Conditioning of Quasi-Newton methods for function minimization. Math. Comput. **24**(111), 647–656 (1970)
12. Syafeeza, A., Khalil-Hani, M., Liew, S., Bakhteri, R.: Convolutional neural network for face recognition with pose and illumination variation. Int. J. Eng. Technol. **6**(1), 44–57 (2014). http://www.enggjournals.com/ijet/vol6issue1.html
13. Zeiler, M.D.: ADADELTA: an adaptive learning rate method. CoRR abs/1212.5701 (2012)

Max-Pooling Dropout for Regularization
of Convolutional Neural Networks

Haibing Wu and Xiaodong Gu[✉]

Department of Electronic Engineering, Fudan University, Shanghai 200433, China
{haibingwu13,xdgu}@fudan.edu.cn

Abstract. Recently, dropout has seen increasing use in deep learning. For deep convolutional neural networks, dropout is known to work well in fully-connected layers. However, its effect in pooling layers is still not clear. This paper demonstrates that *max-pooling dropout* is equivalent to randomly picking activation based on a multinomial distribution at training time. In light of this insight, we advocate employing our proposed probabilistic weighted pooling, instead of commonly used max-pooling, to act as model averaging at test time. Empirical evidence validates the superiority of probabilistic weighted pooling. We also compare max-pooling dropout and stochastic pooling, both of which introduce stochasticity based on multinomial distributions at pooling stage.

Keywords: Deep learning · Convolutional neural network · Max-pooling dropout

1   Introduction

Deep convolutional neural networks (CNNs) have recently been substantially improving on the state of art in computer vision. A standard CNN consists of alternating convolutional and pooling layers, with fully-connected layers on top. Compared to regular feedforward networks with similarly-sized layers, CNNs have much fewer connections and parameters due to the local-connectivity and shared-filter architecture in convolutional layers, so they are far less prone to over-fitting. Another nice property of CNNs is that pooling operation provides a form of translation invariance and benefits generalization. Despite these attractive qualities and despite the fact that CNNs are much easier to train than other regular, deep, feed-forward neural nets, big CNNs with millions or billions of parameters still easily overfit small training data.

Dropout [1] is a recently proposed regularizer to fight against over-fitting. It is a regularization method that stochastically sets to zero the activations of hidden units for each training case at training time. This breaks up co-adaptations of feature detectors since the dropped-out units cannot influence other retained units. Another way to interpret dropout is that it yields a very efficient form of model averaging where the number of trained models is exponential in that of units, and these models share the same parameters. Dropout has also inspired other stochastic model averaging methods such as stochastic pooling [4], drop-connect [5] and maxout networks [3].

© Springer International Publishing Switzerland 2015
S. Arik et al. (Eds.): ICONIP 2015, Part I, LNCS 9489, pp. 46–54, 2015.
DOI: 10.1007/978-3-319-26532-2_6

Although dropout is known to work well in fully-connected layers of convolutional neural nets [1, 5, 6], its effect in pooling layers is, however, not well studied. This paper shows that using max-pooling dropout at training time is equivalent to sampling activation based on a multinomial distribution, and the distribution has a tunable parameter p (the retaining probability). In light of this, probabilistic weighted pooling is proposed and employed at test time to efficiently average all possibly max-pooling dropout trained networks. Our empirical evidence confirms the superiority of probabilistic weighted pooling over max-pooling. Like fully-connected dropout, the number of possible max-pooling dropout models also grows exponentially with the increase of the number of hidden units that are fed into pooling layers, but decreases with the increase of pooling region's size.

As both stochastic pooling [4] and max-pooling dropout randomly sample activation based on multinomial distributions at pooling stage, it becomes interesting to compare their performance. Experimental results show that stochastic pooling performs between max-pooling dropout with different retaining probabilities, yet max-pooling dropout with typical retaining probabilities often outperforms stochastic pooling by large margins.

In this paper, dropout on the input to max-pooling layers is also called *max-pooling dropout* for brevity. Similarly, dropout on the input to fully-connected layers is called *fully-connected dropout*.

2 Related Work

Dropout is a new regularization technique that has been more recently employed in deep learning. Pioneering work by Hinton et al. [1] only applied dropout to fully connected layers. It was the reason they provided that the convolutional shared-filter architecture was a drastic reduction in the number of parameters and thus reduced its possibility to overfit in convolutional layers. Krizhevsky et al. [6] trained a very big convolutional neural net to classify 1.2 million ImageNet images. Two primary methods were used to reduce over-fitting. The first one was data augmentation, an easiest and most commonly used approach to reduce over-fitting on image data. Dropout was exactly the second one. Also, it was only used in fully-connected layers.

Stochastic pooling [4] is a dropout-inspired regularization method. Instead of always capturing the strongest activity within each pooling region as max-pooling does, stochastic pooling randomly picks the activations based on a multinomial distribution.

Maxout network [3] is another model inspired by dropout. Combining with dropout, maxout networks have been shown to achieve best results on five benchmark datasets. However, the authors did not train maxout networks without dropout. Besides, they did not train the rectified counterparts with dropout and directly compare it with maxout networks. Dropout has also motivated other stochastic model averaging methods, such as drop-connect [5] and adaptive dropout [8].

3   Max-Pooling Dropout

We now demonstrate that max-pooling dropout is equivalent to sampling activation according to a multinomial distribution at training time. Basing on this interpretation, we propose to use probabilistic weighted pooling at test time.

3.1   Max-Pooling Dropout at Training Time

Consider a standard CNN composed of alternating convolutional and pooling layers, with fully-connected layers on top. On each presentation of a training example, if layer l is followed by a pooling layer, the forward propagation without dropout can be described as

$$a_j^{(l+1)} = Pool(a_1^{(l)}, \ldots, a_i^{(l)}, \ldots, a_n^{(l)}), \ i \in R_j^{(l)}. \tag{1}$$

Here $R_j^{(l)}$ is pooling region j at layer l and is the activity of each neuron within it. $n = |R_j^{(l)}|$ is the number of units in $R_j^{(l)}$. $Pool()$ denotes the pooling function. Pooling operation provides a form of spatial transformation invariance as well as reduces the computational complexity for upper layers. An ideal pooling method is expected to preserve task-related information while discarding irrelevant image details. Two popular choices are average- and max-pooling. Average-pooling takes all activations in a pooling region into consideration with equal contributions. This may downplay high activations as many low activations are averagely included. Max-pooling only captures the strongest activation, and disregards all other units in the pooling region. We now show that employing dropout in max-pooling layers avoids both disadvantages by introducing stochasticity.

With dropout, the forward propagation becomes

$$\hat{a}^{(l)} \sim m^{(l)} * a^{(l)}, \tag{2}$$

$$a_j^{(l+1)} = Pool(\hat{a}_1^{(l)}, \ldots, \hat{a}_i^{(l)}, \ldots, \hat{a}_n^{(l)}), \ i \in R_j^{(l)}. \tag{3}$$

Here $*$ denotes element wise product and $m^{(l)}$ is a binary mask with each element $m_i^{(l)}$ drawn independently from a Bernoulli distribution. This mask is multiplied with activations $a^{(l)}$ in a pooling region at layer l to produce dropout-modified activations $\hat{a}^{(l)}$. The modified activations are then passed to pooling layers. Figure 1 presents a concrete example to illustrate the effect of dropout in max-pooling layers. Clearly, without dropout, the strongest activation in a pooling regions is always selected as the pooled activation. With dropout, it is not necessary that the strongest activation being the output. Therefore, max-pooling at training time becomes a stochastic procedure. To formulate such stochasticity, suppose the activations $(a_1^{(l)}, a_2^{(l)}, \ldots, a_n^{(l)})$ in each pooling region j are reordered in non-decreasing order, i.e., $0 \leq a_1'^{(l)} \leq a_2'^{(l)} \ldots \leq a_n'^{(l)}$. With dropout, each unit in the pooling region could be possibly set to zero with probability of $q = 1 - p$ is the dropout probability, and p is the retaining probability). As a result,

$a_i'^{(l)}$ will be selected as the pooled activation on condition that (1) $a_{i+1}'^{(l)}, a_{i+2}'^{(l)}, \ldots, a_n'^{(l)}$ are dropped out, and (2) $a_i'^{(l)}$ is retained. This event occurs with probability of p_i according to probability theory:

$$\Pr(a_j^{(l+1)} = a_i'^{(l)}) = p_i = pq^{n-i}, \ (i = 1, 2, \ldots, n). \tag{4}$$

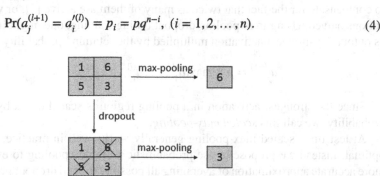

Fig. 1. An illustrating example showing the procedure of max-pooling dropout. The activation in the pooling region is 1, 6, 5 and 3 respectively. Without dropout, the strongest activation 6 is always selected as the output. With dropout, each unit in the pooling region could be possibly dropped out. In this example, only 1 and 3 are retained, then 3 will be the pooled output.

A special event occurring with probability of $p_0 = q^n$ is that all the units in a pooling region is dropped out, and the pooled output becomes $a_0'^{(l)} = 0$. Therefore, performing max-pooling over the dropout-modified pooling region is exactly sampling from the following multinomial distribution to select an index i, then the pooled activation is simply $a_i'^{(l)}$:

$$a_j^{(l+1)} = a_i'^{(l)}, \text{ where } i \sim Multinomial \ (p_0, p_1, p_2, \ldots, p_n). \tag{5}$$

Let s be the size of a feature map at layer l (with r feature maps), and t be the size of pooling regions. The number of pooling region is therefore rs/t for non-overlapping pooling. Each pooling region provides $t + 1$ choices of the indices, then the number of possibly trained models C at layer l is

$$C = (1 + t)^{rs/t} = \left(\sqrt[t]{1 + t}\right)^{rs} = b(t)^{rs}. \tag{6}$$

So the number of possibly max-pooling dropout trained models is exponential in the number of units that are fed pooling max-pooling layers, and the base $b(t)$ $(1 < b(t) = \sqrt[t]{1 + t} \leq 2)$ depends on the size of pooling regions. Obviously, with the increase of the size of pooling regions, the base $b(t)$ decreases, and the number of possibly trained models becomes smaller. Note that the number of possibly fully-connected dropout trained models is also exponential in the number of units that are fed into fully-connected layers, but with 2 as the base.

3.2   Probabilistic Weighted Pooling at Test Time

Using dropout in fully-connected layers during training, the whole network containing all the hidden units should be used at test time, but with their outgoing weights halved to compensate for the fact that twice as many of them are active [1], or with their activations halved. Using max-pooling dropout during training, one might intuitively pick as output the strongest activation multiplied by the retaining probability:

$$a_j^{(l+1)} = p \times \max(a_1^{(l)}, \dots, a_i^{(l)}, \dots, a_n^{(l)}). \tag{7}$$

Since the strongest activation in a pooling region is scaled down by the retaining probability, we call this *scaled max-pooling*.

At test time, scaled max-pooling generally works well in practice, but is not the optimal. Instead we propose to use probabilistic weighted pooling to efficiently get a more accurate approximation of averaging all possibly trained dropout networks. In this pooling scheme, the pooled activity is linear weighted summation over activations in each region:

$$a_j^{(l+1)} = \sum_{i=0}^{n} p_i a_i'^{(l)} = \sum_{i=1}^{n} p_i a_i'^{(l)}. \tag{8}$$

Here p_i is exactly the probability calculated by Eq. (4). This type of probabilistic weighted summation can be interpreted as an efficient form of model averaging where each selection of index i corresponds to a different model. Empirical evidence will confirm that probabilistic weighted pooling is a more accurate approximation of averaging all possible dropout models than scaled max-pooling.

4   Empirical Evaluations

Experiments are conducted on three datasets: MNIST, CIFAR-10 and CIFAR-100. MNIST consists of 28×28 pixel grayscale images, each containing a digit 0 to 9. There are 60,000 training and 10,000 test examples. We do not perform any preprocessing except scaling the pixel values to [0, 1]. The CIFAR-10 dataset [2] consists of ten classes of natural images with 50,000 examples for training and 10,000 for testing. Each example is a 32×32 RGB image taken from the tiny images dataset collected from the web. CIFAR-100 is just like CIFAR-10, but with 100 categories. We also scale to [0, 1] for CIFAR-10 and CIFAR-100 and subtract the mean value of each channel computed over the dataset for each image.

We use rectified linear function [7] for convolutional and fully-connected layers, and softmax activation function for the output layer. More commonly used sigmoidal and tanh nonlinearities are not adopted due to gradient vanishing problem with them. Our models are trained using stochastic mini-batch gradient descent with a batch size of 100, momentum of 0.95, learning rate of 0.1 to minimize the cross entropy loss. The weights in all layers are initialized from a zero-mean Gaussian distribution with 0.1 as standard deviation and the constant 0 as the neuron biases in all layers.

The CNN architecture for MNIST is 1x28x28-20C5-2P2-40C5-2P2-1000N-10N, which represents a CNN with 1 input image of size 28 × 28, a convolutional layer with 20 feature maps and 5 × 5 filters, a pooling layer with pooling region 2 × 2 and stride 2, a convolutional layer with 40 feature maps and 5 × 5 filters, a pooling layer with pooling region 2 × 2 and stride 2, a fully-connected layer with 1000 hidden units, and an output layer with 10 units (one per class). The architecture for CIFAR-10 is 3x32x32-96C5-3P2-128C3-3P2-256C3-3P2-2000N-2000N-10N. The architecture for CIFAR-100 is the same with CIFAR-10 except with 100 output units.

4.1 Probabilistic Weighted Pooling vs. (Scaled) Max-Pooling

We initially validate the superiority of probabilistic weighted pooling over max-pooling and scaled max-pooling using MNIST. The CNNs are trained for 1000 epochs. For max-pooling dropout, CNN models are trained with different retaining probabilities. Figure 2 compares the test performances produced by different pooling methods at test time. Generally, probabilistic weighted pooling performs better than max-pooling and scaled max-pooling with different retaining probabilities. For small p (the retaining probability), max-pooling and scaled max-pooling performs very poorly; probabilistic weighted pooling is considerably better. With the increase of p, the performance gap becomes smaller. This is not surprising as the pooled outputs for different pooling methods are close to each other for large p. An extreme case is that when $p = 1$, scaled max-pooling and probabilistic weighted pooling are exactly the same with max-pooling.

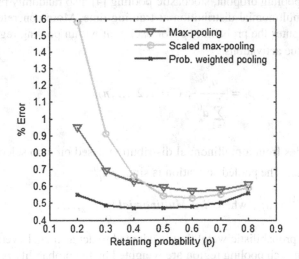

Fig. 2. MNIST test errors for different pooling methods at test time. Max-pooling dropout is used to train CNN models with different retaining probabilities at training time.

We then compares different pooling methods at test time for max-pooling dropout trained models on CIFAR-10 and CIFAR-100. The retaining probability is set to 0.3, 0.5 and 0.7 respectively. At test time, max-pooling, scaled max-pooling and probabilistic weighted pooling are respectively used to act as model averaging. Figure 3 presents the test performance of these pooling methods. Again, for small retaining probability $p = 0.3$, scaled max-pooling and probabilistic weighted pooling perform poorly. Probabilistic weighted pooling is the best performer with different retaining probabilities. The increase of p narrows different pooling methods' performance gap.

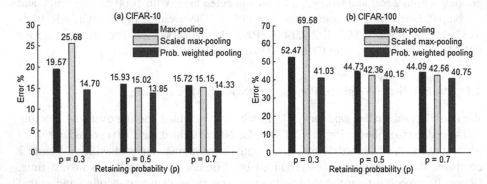

Fig. 3. CIFAR-10 and CIFAR-100 test errors for different pooling methods at test time. Maxpooling dropout is used to train CNNs with different retaining probabilities at training time.

4.2   Max-Pooling Dropout vs. Stochastic Pooling

Similar to max-pooling dropout, stochastic pooling [4] also randomly picks activation according to a multinomial distribution at training time. More concretely, at training time it first computes the probability p_i for each unit within pooling region j at layer l by normalizing the activations:

$$p_i = \frac{a_i^{(l)}}{\sum_{k=1}^{n} a_k^{(l)}}, \ (i = 1, 2, \dots, n).$$

(9)

It then samples from a multinomial distribution based on p_i to select an index i in the pooling region. The pooled activation is simply $a_i^{(l)}$:

$$a_j^{(l+1)} = a_i^{(l)}, \text{ where } i \sim Multinomial\ (p_1, p_2, \dots, p_n).$$

(10)

At test time, probabilistic weighting is adopted to act as model averaging. That is, the activations in each pooling region are weighted by the probability p_i and summed:

$$a_j^{(l+1)} = \sum_{i=1}^{n} p_i a_i^{(l)}.$$

(11)

One may have found that stochastic pooling bears much resemblance to max-pooling dropout, as both involve stochasticity at pooling stage. We are therefore very interested in their performance differences. To compare their performances, we train CNN models with different retaining probabilities on MNIST, CIFAR-10 and CIFAR-100. For max-pooling dropout trained models, only probabilistic weighted pooling is used at test time. Figure 4 compares the test performances of max-pooling dropout with different retaining probabilities against stochastic pooling. The relation between the performance of max-pooling dropout and the retaining probability p is a U-shape. If p is too small or too large, max-pooling dropout performs poorer than stochastic pooling. Yet max-pooling dropout with typical p (around 0.5) outperforms stochastic pooling by a large margin. Therefore, although stochastic pooling is hyper-parameter free and this saves the tuning of retaining probability, its performance is often inferior to max-pooling dropout.

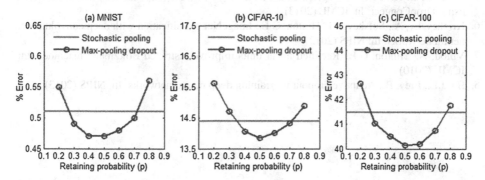

Fig. 4. MNIST, CIFAR-10 and CIFAR-100 test errors for max-pooling dropout with different retaining probabilities against stochastic pooling.

5 Conclusions

This paper mainly addresses the problem of understanding and using dropout on the input to max-pooling layers of convolutional neural nets. At training time, max-pooling dropout is equivalent to randomly picking activation according to a multinomial distribution, and the number of possibly trained networks is exponential in the number of input units to the pooling layers. At test time, a new pooling method, probabilistic weighted pooling, is proposed to act as model averaging. Experimental evidence confirms the benefits of using max-pooling dropout, and validates the superiority of probabilistic weighted pooling over max-pooling and scaled max-pooling. Considering that stochastic pooling is similar to max-pooling dropout, we empirically compare them and show that the performance of stochastic pooling is between those produced by max-pooling dropout with different retaining probabilities.

Acknowledgements. This work was supported in part by National Natural Science Foundation of China under grant 61371148.

References

1. Hinton, G.E., Srivastave, N., Krizhevsky, A., Sutskever, I., Salakhutdinov, R.R.: Improving neural networks by preventing co-adaption of feature detectors. arXiv:1207.0580 (2012)
2. Krizhevsky, A.: Learning multiple layers of features from tiny images. M.S. dissertation, University of Toronto (2009)
3. Goodfellow, I.J., Warde-Farley, D., Mirza, M., Courville, A., Bengio, Y.: Maxout networks. In: ICML (2013)
4. Zeiler, M.D., Fergus R.: Stochastic pooling for regularization of deep convolutional neural networks. In: ICLR (2013)
5. Wan, L., Zeiler, M.D., Zhang, S., LeCun, Y., Fergus, R.: Regularization of neural networks using DropConnect. In: ICML (2013)
6. Krizhevsky, A., Sutskever, I., Hinton, G.E.: ImageNet classification with deep convolutional neural networks. In: NIPS (2012)
7. Vinod, N., Hinton, G.E.: Rectified linear units improve restricted Boltzmann machines. In: ICML (2010)
8. Ba, J.L., Frey, B.: Adaptive dropout for training deep neural networks. In: NIPS (2013)

Predicting Box Office Receipts of Movies with Pruned Random Forest

Zhenyu Guo$^{(\boxtimes)}$, Xin Zhang, and Yuexian Hou$^{(\boxtimes)}$

School of Computer Science and Technology, Tianjin University, Tianjin, China
100662710@qq.com, yxhou@tju.edu.cn

Abstract. Predicting box office receipts of movies in theatres is a difficult and challenging problem on which many theatre managers cogitated. In this study, we use pruned random forest to predict the box office of the first week in Chinese theatres one month before movies' theatrical release. In our model, the prediction problem is converted into a classification problem, where the box office receipt of a movie is discretized into eight categories. Experiments on 68 theatres show that the proposed method outperforms other statistical models. In fact, our model can predict the expected revenue range of a movie, it can be used as a powerful decision aid by theatre managers.

Keywords: Chinese theatres · Box office · Random forest

1 Introduction

Predicting box office receipts of movies in theatres is a difficult and challenging problem on which many theatre managers cogitated. It has an important meaning to control the issue risk and also has a practical significance for the investment decision [4]. Until 2009, there are 37 theatre chains admitted by State Administration of Press, Publication, Radio, Film, and Television of The Peoples Republic of China. Only by predicting the box office receipts of movies correctly will the theatre managers decide the number of release movie theatres, the promotion of the movie cost, the schedule for showing movie and so on. Most theatre chains, e.g., Wanda Media, mainly adopt brainstorming method to predict box office receipts of movies. It may cause low box office receipts for theatres. Thus a good method to predict box office receipts has an important impact for theatres.

Though the work associated with the unpredictable nature of problem domain is difficult, several researches have attempted to develop statistical models to solve it. In [12], Multi Layer Perceptron neural network model is used to classify the box office in nine categories. Bayesian belief network [6] and BP neural network [16] are also constructed to predict box office performance. Google [10] applies a simple linear regression model using film-related search query volume as a predictor of weekend box office performance. In our study, we use pruned random forest to predict the box office performance of movies in

S. Arik et al. (Eds.): ICONIP 2015, Part I, LNCS 9489, pp. 55–62, 2015.
DOI: 10.1007/978-3-319-26532-2_7

Chinese theatres one month before their release. Here, the predict problem is converted into a classification problem, where movies are categorized into eight classes according to box office receipt. Figure 1 shows the diagram of the whole system.

Applications of random forest have been reported in many diverse fields addressing problems in areas such as prediction and classification. Random forest is a combination of several decision tree predictors in machine learning. To classify a new object from an input vector, we put the data into all trees in the forest. Each tree gives a classification result and votes for the predicted class [7]. The forest chooses the class with the most votes. In this paper, we prune random forest through its strength and correlation. And the pruned random forest outperforms conventional random forest.

In predict system, the box office data was supplied by 68 Chinese theatres and the movie information is downloaded from different movie websites. Compares with other box office predict models [6,12,16], out model adds Baidu Index, Trailer searches and Lowest Ticket Price as features. Among them, Baidu Index and Trailer searches could represent the audience's attention of movies. Lowest Ticket Price is the studio's estimate of movies quality.

The remaining of this paper is organized as follows. Section 2 introduces the process of our methodology and how to predict box office in Chinese theatres, followed by experimental results in Sect. 3. Final conclusions are given in Sect. 4.

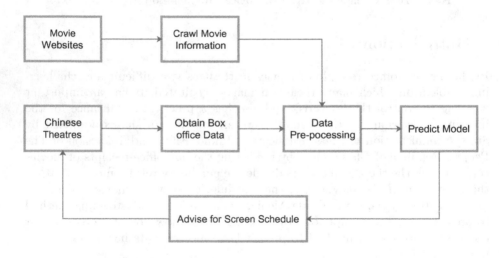

Fig. 1. Framework of the box office prediction system

2   Methodology

The box office data from 2013 June to 2014 June was supplied by 68 Chinese theatres. Each theatre had eight databases, such as Ticket Report, Locations,

Cinemas, Show Details, Ticket, Ticket Bookings, Booking Payments and Distributors. In those databases, we could obtain movies' daily box office, lowest ticket price and other useful information.

2.1 Movie Information Data Collection

It's insufficient to predict box office through the dataset above. In [6,12,16], director, actor, genre, competition and number of screens are used as movie features to predict box office. To improve predict accuracy, we also choose Baidu Index and trailer searches as features based on the Google research [10]. At last, we suppose that studio can also influence the box office. We develop different crawlers for different websites, such as Baidu Index, Youku Index and so on, to crawl the data we need and stored data in MySQL. Next we will detail how to use those data as features. A summary of the above-mentioned and briefly defined decision variables is given in Table 1.

Baidu Index. For most of movies, the Baidu Index of movies concentrates on one month before their release. The larger search volumes are, the higher box offices will be. Movie's Baidu Index increased as the date is closed to its theatrical release, so we use Baidu Index of the first five weeks before theatrical release as features. As movies influence each other at the same time, we also use the Baidu Index of movies which are released closely as features.

Competition. Movie competes against other movies released at the same time, which is still a common phenomenon in theatre. For example, there are two reasons why 'Tiny Times' has a excellent box office performance. One is that it was released at the end of June in the summer schedule. As there are not strong movies released near it, it has a great advantage in schedule. The other is that its schedule is during the summer vacation of students. Sometimes if an ordinary movie has a good schedule, it will have a good box office.

Director and Actor. Some researches [1] show that one of the most important factors that influence the box office is director and actor. As the organizer and leader of a movie, directors determine the movies' quality and artistic style largely. And actors attract audiences attention through their admirable acting skill and grateful characters. Hennig-Thurau [5] use the Hollywood Reporter's Star power and Director power indices as features to predict box office. In our model, we suppose the influences of directors and actors to a movie can embodied through their ever box office performance.

Genre. As different people have different preferences for different genres, the genre of a movie is also important. Genres decide movie content expression, audience basis and movie influence. As different audiences have different cultural background, they have different entertainment consumption demand and spiritual needs. Such phenomenon is more obvious in the movie theatres, because the box office of movies in a movie theatre depends on the people nearby.

Trailer searches. Google research [10] shows that, there is a linear relationship between the box office and trailer-related search volume. We get search and paly volume from Youku Index. Youku Index has authority as the largest video website in China. Like Baidu Index, the higher search and paly volume of trailers have, the more popular of the movie will be. Therefore, through the search and play volume, we can determine the level of box office.

Studio. In China, only Huayi Bros Media Group, Emperor Entertainment Group Limited, Media Asia Group and some other large studios are popular domestic movie studios. A large studio represents good quality of a movie. The studio will also give the lowest ticket price of movies. Through it, we can roughly estimate the box office of a movie.

Number of screens. An intuitive understanding shows that the more number of screens of movies have, the higher box office of movies will be. In our model, we represented the number of screen a movie is scheduled to be shown as its opening with a continuous variable.

Table 1. Summary of independent variables

Variable name	No. of values	Value source
Baidu Index	5	index.baidu.com
Competition	2	box office data
Director and Actor	4	www.cbooo.cn
Genre	1	www.gewara.com
Trailer searches	2	index.youku.com
Studio	2	www.cbooo.cn
Number of screens	1	box office data

2.2 Pruned Random Forest

Random forest has been used widely in many applications as an ensemble method. In this paper, we prune random forest through two parameters which are strength and correlation.

Given an ensemble of random forest classifiers [3] $h_1(x), h_2(x), \ldots, h_k(x)$, and with the training set drawn at random from the distribution of the random vector X, Y, define the margin function for a random forest is

$$mr(X, Y) = \frac{1}{N} \sum_{n=1}^{N} \{I(h(X) = Y) - \max_{j \neq Y} I(h(X) = j)\} \tag{1}$$

where $I(\cdot)$ is the indicator function. The margin measures the extent to which the probability of the votes at X, Y for the right class exceeds the average vote for any other class. The larger the margin is, the more confidence is in the classification. And define the strength function for a random forest as

$$strength = E_{X,Y} mr(X, Y) = \frac{1}{N} \sum_{i=1}^{N} mr(x_i, y_i) \qquad (2)$$

Define the correlation function for a random forest as

$$\overline{\rho} = \frac{var(mr)}{sd(h)^2} \qquad (3)$$

where $sd(h)$ is the standard deviation of random forest.

To improve accuracy, the random forest need to increase the strength while minimize the correlation. In the past few years many researchers have studied the pruning methods for random forest, such as pruning a random forest based on the similarity between tree pairs [15] and margin distance minimization [8,9], which mainly focus on improving the accuracy of the prediction of each tree.

In this paper, we prune random forest through strength and correlation. Changing both of the two parameter needs to compute margin of the random forest. Margin is important for ensemble classifiers, such as Adaboost [11], Mdboost [13], Arc-Gv [2] and so on. Above-mentioned classifiers improve their generalization by changing the margin. In this algorithm, we assume the random forest is H. This method ranks the contribution and importance of a base tree h in the temporary ensemble H by observing the decrease of strength and correlation when removing h from the forest [14]. This is equivalent to evaluating the subensemble $\{H/h\}$. As the redundant tree is pruned, the forest H shrinks to its subset $H' = \{H/h\}$. we compute strength and correlation of forest H first. And then, the strength and correlation of each tree h_k are computed. At last, we eliminate unimportant trees with small strength and large correlation in H. The strength and correlation of each tree on the training set are computed as evaluation metrics $Strength(h_i, H_k, S)$ and $Correlation(h_i, H_k, S)$ as follows,

$$Strength(h_i, H_k, S) = strength(x, y, H) - strength(x, y, H/h_i) \qquad (4)$$

$$Correlation(h_i, H_k, S) = \overline{\rho}(x, y, H) - \overline{\rho}(x, y, H/h_i) \qquad (5)$$

where S is the training set.

2.3   Advice for Screen Schedule

The box office predict model can give a more reliable result than previous methods such as brainstorming method. And as all requisite data could be obtained one month before movie's release, theatre managers have plenty of time to use the predict categories of movies to adjust the movie screen schedule in order to get more profits. Through the results, theatre managers also can give movies which have high box office more attention and drumbeating besides.

3   Results

3.1   The Classification Performance of Pruned Random Forest

Recall that our pruned random forest model aims to categorize a movie in one of the eight categories. In this experiment, a 8-fold cross-validation approach is used to estimate the performance of pruned random forest. And we use average present hit rate (APHR) to measure the predictive performance of our pruned random forest approach [12]. In our case, we have two different hit rates: the exact hit rate (Bingo) which counts the correct classifications to the exact same class and the within 1 class hit rate (1-Away) which reflects the instance that a movie forecasted into its adjacent classes. Algebraically, APHR can be formulated as follows,

$$APHR_{Bingo} = \frac{1}{n} \sum_{i=1}^{c} p_i \tag{6}$$

$$APHR_{1-Away} = \frac{1}{n} \sum_{i=1}^{c} (p_i + p_{i-1} + p_{i+1}) \tag{7}$$

where c is the total number of classes, n is the total number of samples, and p_i is the total number of samples classified as class i.

Table 2 presents our aggregated 8-fold cross-validation pruned random forest results of 68 Chinese theatres in confusion matrix. The columns in a confusion matrix represent the actual classes and the rows represent the predicted classes. The intersection cells of the same classes represent the correct classification of the samples for that class.

Table 2. The total predict results of 68 Chinese theatres

		Actual Categories								Avg.
		1	2	3	4	5	6	7	8	
	1	**1517**	349	58	6	2	0	0	0	
	2	287	**1024**	448	76	12	4	1	0	
Predicted Categories	3	28	378	**885**	417	80	11	0	0	
	4	7	54	344	**942**	398	62	7	0	
	5	0	3	40	368	**919**	375	57	5	
	6	0	2	4	42	359	**975**	349	38	
	7	0	0	0	7	32	347	**1058**	303	
	8	1	1	1	3	6	46	344	**1469**	
	Bingo	82.45	56.54	49.72	50.62	50.83	53.57	58.26	80.94	**60.37**
	1-Away	98.04	96.69	94.21	92.80	92.70	93.24	96.42	97.63	**95.22**

3.2   Comparison with Other Models

Except conventional random forest, we also compare our model's performance with other popular classification methods. Specifically, we use traditional statistical classification methods, such as multiple logistic regression, classification and regression trees (CART), support vector machine (SVM) and neural networks.

Results shows that pruned random forest has a average bingo accuracy rate 60 % and 1-Away accuracy rate 95 % in Table 1. In Fig. 1, we compare our pruned random forest model with conventional random forest (RF) and other popular classification methods, such as decision tree (DT), support vector machine (SVM) and multi layer perceptron (MLP) in Bingo accuracy (Table 3).

Table 3. The Results of Prediction

	PRF	RF	DT	SVM	MLP
Bingo(%)	**60.09**	54.27	52.88	32.15	19.68
1-Away(%)	**95.26**	90.96	92.11	59.62	48.67

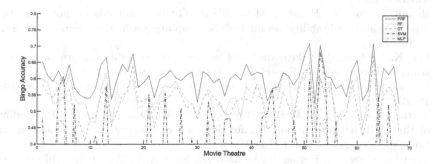

Fig. 2. Comparison Bingo accuracy results of PRF, RF, DT, SVM and MLP in 68 movie theatres (Color figure online)

From the experimental results, we can see that our method achieves a high accuracy and the performance is better than conventional random forest, decision tree, support vector machine and neural networks. We can see both two results are better than conventional random forest, decision tree and support vector machine. And pruned random forest is more stable than support vector machine.

From Fig. 2, we can see that some movie theatres have high predict accuracy and some not. The most important reason is the different scales. First, large movie theatres locate in better place and have a broader audience base. Second, large movie theatres have more movies released.

4   Conclusion

In this paper, we propose a pruned random forest model to predict box office receipts in Chinese movie theatres. The movie information data is crawled from internet and box office data is obtained from movie theatres. After data pre-processing, the pruned random forest is used to classify movie box office. Compared with conventional random forest, our method promote the strength and reduce the correlation of random forest. And the predict results could provide theatres managers a guidance on movie schedule and drumbeating. Experiments on a real word video show our system is able to achieve satisfactory performance in real time.

References

1. Ainslie, A., Drèze, X., Zufryden, F.: Modeling movie life cycles and market share. Mark. Sci. **24**(3), 508–517 (2005)
2. Breiman, L.: Prediction games and arcing algorithms. Neural Comput. **11**(7), 1493–1517 (1999)
3. Breiman, L.: Random forests. Mach. Learn. **45**(1), 5–32 (2001)
4. Eliashberg, J., Elberse, A., Leenders, M.A.: The motion picture industry: critical issues in practice, current research, and new research directions. Mark. Sci. **25**(6), 638–661 (2006)
5. Hennig-Thurau, T., Houston, M.B., Walsh, G.: Determinants of motion picture box office and profitability: an interrelationship approach. Rev. Manag. Sci. **1**(1), 65–92 (2007)
6. Lee, K.J., Chang, W.: Bayesian belief network for box-office performance: a case study on korean movies. Expert Syst. Appl. **36**(1), 280–291 (2009)
7. Liaw, A., Wiener, M.: Classification and regression by randomforest. R News **2**(3), 18–22 (2002)
8. Martınez-Munoz, G., Suárez, A.: Aggregation ordering in bagging. In: Proceedings of the IASTED International Conference on Artificial Intelligence and Applications, pp. 258–263. Citeseer (2004)
9. Martinez-Muoz, G., Hernández-Lobato, D., Suárez, A.: An analysis of ensemble pruning techniques based on ordered aggregation. IEEE Trans. Pattern Anal. Mach. Intell. **31**(2), 245–259 (2009)
10. Panaligan, R., Chen, A.: Quantifying movie magic with google search. Google WhitepaperIndustry Perspectives+ User Insights (2013)
11. Rätsch, G., Onoda, T., Müller, K.R.: Soft margins for adaboost. Mach. Learn. **42**(3), 287–320 (2001)
12. Sharda, R., Delen, D.: Predicting box-office success of motion pictures with neural networks. Expert Syst. Appl. **30**(2), 243–254 (2006)
13. Shen, C., Li, H.: Boosting through optimization of margin distributions. IEEE Trans. Neural Netw. **21**(4), 659–666 (2010)
14. Yang, F., Lu, W.H., Luo, L.K., Li, T.: Margin optimization based pruning for random forest. Neurocomputing **94**, 54–63 (2012)
15. Zhang, H., Wang, M.: Search for the smallest random forest. Stat. Interface **2**(3), 381 (2009)
16. Zhang, L., Luo, J., Yang, S.: Forecasting box office revenue of movies with bp neural network. Expert Syst. Appl. **36**(3), 6580–6587 (2009)

A Novel ℓ^1-graph Based Image Classification Algorithm

Jia-Yue Xu[✉] and Shu-Tao Xia

Graduate School at Shenzhen, Tsinghua University,
Shenzhen 518055, Guangdong, People's Republic of China
xjy13@mails.tsinghua.edu.cn, xiast@sz.tsinghua.edu.cn

Abstract. In original sparse representation based classification algorithms, each training sample belongs to exactly one class, neglecting the association between the training sample and the other classes. However, different classes' features are visually similar and correlated (*e.g.* facial images), which means the association between the training sample and the different classes contain important information, and must be taken into consideration. In this paper, we propose a novel ℓ^1-graph based image classification algorithm (LGC). Our algorithm can automatically calculate associations between training samples and all classes, which are used for future classification. We evaluate our method on some popular visual benchmarks, the experimental results prove the effectiveness of our method.

Keywords: Image classification · Sparse representation · ℓ^1-graph

1 Introduction

Image classification is one of the most attractive and challenging research topics in computer vision and pattern recognition field. Digital images are commonly high-dimensional signals, that it is hard to process using traditional classification algorithms. But in practice these high-dimensional signals usually lie on a special low-dimensional subspace [2]. So, given a set of basis, the signals can be sparsely represented using these basis. The obtained sparse representation is more compact and more useful than the original representation. Sparse representation has already made wide use in image compact [4], image denoising [7] and image super-resolution [18] tasks. Inspired by the success of sparse representation in the image processing field, researchers has applied the sparse representation methods in computer vision and pattern classification tasks. The recently proposed sparse representation based classifier (SRC) [17], ℓ^1-graph methods [6] all achieved state-of-art results.

Among those above sparse representation methods, correctly choosing the basis or dictionary is important to successfully apply sparse representation methods. In classical signal processing applications, the key idea is high fidelity to the original image [3]. We often use some fixed bases (*i.e.*, Fourier, Wavelet

© Springer International Publishing Switzerland 2015
S. Arik et al. (Eds.): ICONIP 2015, Part I, LNCS 9489, pp. 63–70, 2015.
DOI: 10.1007/978-3-319-26532-2_8

[11]), or dictionary learned from the training signals [1]. But in image classification tasks, the image's semantic meaning is more important. SRC [17] algorithm directly uses the training samples as the dictionary, finding the obtained sparse coefficients on the dictionary contains useful information for classification.

After the SRC algorithm has been proposed, recently some researchers have tried to combine SRC algorithm and dictionary learning method to improve the classification performance. Instead of directly using the training samples as the dictionary, these methods try to use dictionary learning methods to learn a more compact and more discriminating dictionary from the training samples. In [10,20], the classification error is incorporated into the K-SVD [1] algorithm's objective function to reduce the dictionary's size. The learned dictionary is shared by all classes, and a linear classifier is simultaneously learned during the dictionary learning process. In [13], to improve the dictionary's discriminating power, an incoherence promoting term is added to the dictionary learning's objective function to encourage class specific sub-dictionaries of different classes as independent as possible. However, some different class's images are highly correlated sometimes. In [15,21], a set of class specific dictionaries are learned for each class that captures the most discriminating features of this class with a common shared dictionary whose atoms are shared by all classes, which only contributes to the representation of the data. In [19], linear Fisher discrimination criterion was combined to the dictionary learning process.

Although those above methods improve dictionary's discriminative power. But the learning process of these algorithms is more complex and time consuming. In practice, sometimes different category's images may be visually similar and correlated [16]. For example, in face recognition tasks, possibly some different people may have similar facial form and hair style, so they may look similar. How to classify these visually similar images is a difficult task. It is a hard problem for human to clarify these images, let alone for computers. The commonly used image classification algorithms also faced this difficulty. In practice, many training samples may have poor quality, such as extreme illumination in facial images, screwed handwritten digits. These poor training samples would make the image classification algorithm degenerate significantly.

To overcome those above difficulties, in this paper, we propose a novel ℓ^1-graph based image classification algorithm (LGC). We still use the original training samples as the dictionary similar to SRC algorithm. Instead of treating each training sample belongs to exactly one class in SRC algorithm. In our new methods, the association between the training samples and all classes are taken into consideration. We use ℓ^1-graph to calculate the association between the training samples and all classes, then use this information for future classification tasks. So, the difficulty of classifying visually similar classes, can be reduced by our algorithm. The experiment results show that LGC algorithm performs better than other classical algorithms in situations in which different classes are visual similar.

2   Background

In this section, we briefly review the sparse representation based classification algorithm (SRC) [17], which is very related to our work. Then we introduce the ℓ^1-graph [6], which is the foundation of our work.

2.1   Sparse Representation Based Classification Algorithm

Given the prior knowledge that all training samples of the same class lie on a low dimensional subspace. Then given plenty training samples for each class, the test image y can be represented as a linear combination of the training samples. Since the classes of the testing samples are unknown, the SRC [17] using whole training samples $D = [D_1, D_2, \ldots, D_k] \in \mathbb{R}^{m \times N}$ (where D_i include training samples from the ith class) as the dictionary. The sparsest representation often selects training samples which have the same class with the testing sample. However, finding the sparsest representation is a NP hard problem. But theory of compressive sensing [5] has pointed out that if the representation is enough sparse, the ℓ^1-minimization results is the sparsest solution. So to get the sparsest solution, we only need to solve the following ℓ^1-minimization problem:

$$\alpha = \operatorname*{argmin}_{\alpha} \| \alpha \|_1 \quad s.t. \quad D\alpha = y \tag{1}$$

Having found the sparsest presentation, we can determine the class of the training sample by finding which classes give the lowest reconstruction error:

$$identity(y) = \operatorname*{argmin}_{i} \| y - D_i \delta_i(\alpha) \| \tag{2}$$

where $\delta_i(\alpha)$ presents the vector that only contains coefficients of the ith class.

The SRC algorithm has not only achieved the state-of-the-art results, but also showed robustness to noise and occlusions. The details of the experiment's results are reported in [17].

2.2 ℓ^1-Graph

Cheng et al. [6] proposes the concept of ℓ^1-graph. This graph uses data samples as vertexes. For each vertex, the algorithms use the remaining vertexes and the noise signals (represented as identity matrix I) as the dictionary, then using ℓ^1-minimization to find a sparse decomposition over this dictionary. The obtained coefficients are used as edge weights in this directed graph, obtaining the ℓ^1-graph. The ℓ^1-graph constructing process can be summarized as follows:

Compared with the other graph models, such as k-nearest-neighbor graph and ϵ-ball graph, the advantage of ℓ^1-graph is the construction process of ℓ^1-graph is a global process. The ℓ^1-minimization process takes all the remaining data samples into consideration, adaptively find the most relevant data samples to assign higher weights.

The ℓ^1-graph has showed better performance in many machine learning tasks such as data clustering, subspace learning, and semi-supervised learning. The details of the experiment results can be found in [6].

Algorithm 1. ℓ^1-graph construction algorithm

Input: Data sample matrix $X = [x_1, x_2, ..., x_N] \in \mathbb{R}^{m \times N}$, where x_i is a m dimensional image sample.

1: Sparse coding process: For each sample x_i, solve following ℓ^1-minimization problem

$$\min_{\alpha^i} \| \alpha^i \|_1, \quad s.t. \quad x_i = D^i \alpha^i, \tag{3}$$

where matrix $D^i = [x_1, ..., x_{i-1}, x_{i+1}, ..., x_N, I] \in \mathbb{R}^{m \times (m+N-1)}$ and $\alpha^i \in \mathbb{R}^{m+N-1}$

2: Graph weights setting: Denote ℓ^1-graph as $G = \{X, W\}$, where X is the graph vertex set, W is the graph's edge weight sets. The vertex set is the sample set. And $w_{ij} = a_j^i$ if $i > j$, and $w_{ij} = \alpha_{j-1}^i$ if $i < j$.

Output: The obtained ℓ^1-graph $G = \{X, W\}$

---

3 ℓ^1-graph Based Image Classification Method

The ℓ^1-graph is originally designed for data clustering algorithms, in this section we will introduce how to apply ℓ^1-graph into classification tasks.

3.1   Relationship Between Training Samples and Classes

In Sect. 2.2 the constructed ℓ^1-graph only considers the edges between different vertices, neglecting the edges that connect a vertex to itself (loops). But in classification tasks, the loops also contain important information, which must be considered. In this section, and the rest part of this paper, the term ℓ^1-graph all means the graph contains loops.

After building the ℓ^1-graph $G = \{X, W\}$, we can use it to construct the association graph $G_a = \{V, E\}$, which demonstrate the association between training samples and classes. From Fig. 1a, we can see that the association graph is bipartite graph. The vertex set V can be divided into two parts. The first part X contains training samples, the second part C represents classes. The edge weight matrix R describes the membership between training samples and classes. The matrix's item r_{ij} represents the numerical measure of the membership between train sample x_i and the jth class, which can be obtained by summing all the weights of edges in the ℓ^1-graph G that connect vertex x_i and the vertexes belongs to the jth class.

3.2   Classification Process

After getting the association graph G_a, we can use it for classification. For every testing sample y, use the training samples as the dictionary, then solve the ℓ^1-minimization problem. The obtained sparse coefficients represent relationship between the testing sample and the training samples. From Fig. 1b, we can see that using this information we can calculate the association between the testing sample and classes easily, and settle the class label of the testing sample. The whole LGC algorithm is summarized in Algorithm 2.

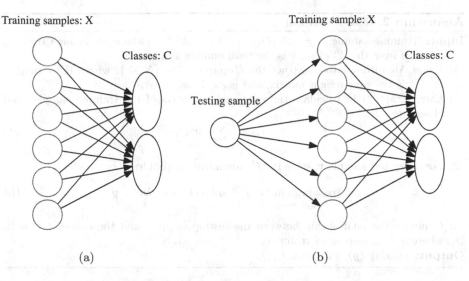

Fig. 1. (a) Association graph G_a, the graph's vertices set include training samples X and classes C. (b) Demonstration of the classification process.

4 Experiment Results

In this section, we evaluate the LGC algorithm in some real image classification tasks such as face recognition and handwritten digit recognition. For comparison purpose, we also test the original SRC [17] algorithm, together with other classical algorithms such as nearest neighbor (NN) algorithm, and the support vector machine (SVM) algorithm. In all the comparing methods, only SVM needs to assign parameters. We use cross-validation to determine SVM's parameter.

4.1 Face Recognition

(a) The Extended Yale B [8] database contains $2,414$ frontal-face images of 38 people, which is taken under various illumination conditions. In our experiment, we use the cropped and normalized 192×168 pixel facial images.In this database, we use EagenFace [14] to extract 300 features from the images. We randomly select half of the images for training (about 32 images per person) and the other half for testing. To remove the effect of different choices of training samples, we take 10 rounds, and calculate the mean recognition rates. The results of LGC, SRC, NN and SVM are listed in Table 1. From the experiment results we can see our LGC algorithm shows significant improvement compared with other methods.

(b) The AR database [12] contains of over $4,000$ colored frontal-face images of 126 people. Each person takes 26 images from two different sessions. Each session contains 13 images, which including expression change, illumination change and partial occlusion. In AR database, we choose a dataset containing 50 male people

---

Algorithm 2. LGC algorithm

Input: Training sample $D = [D_1, D_2, \ldots, D_k] \in \mathbb{R}^{m \times N}$ (where D_i include training samples from the ith class, k is the total number of classes)

1: Using Algorithm 1 to construct the ℓ^1-graph, $G = \{X, W\}$, where $W = [w_{ij}] \in \mathbb{R}^{N \times N}$ is the edge weight matrix, and $w_{ii} = 5$, $w_{ij} = \alpha_j^i$.

2: Using graph G to calculate the relationship matrix of the training samples and classes $R = [r_{ij}] \in \mathbb{R}^{N \times k}$

$$r_{ij} = \sum_{x_k \in D_j} w_{ik} \tag{4}$$

3: For every test sample y, solve the ℓ^1-minimization problem:

$$\alpha = \operatorname*{argmin}_{\alpha} \| \alpha \|_1 \quad subject \quad to \quad D\alpha = y \tag{5}$$

4: Compute the relationship between the testing samples and the classes $t = \alpha' R$, where α' is transpose of vector α

Output: identity $(y) = \operatorname{argmax}_i t_i(y)$.

---

and 50 female people. For each class we select all the first session's images for training and the second session's images for testing. We also use EagenFace [14] to distract 300 dimensional features from the images. The experiment results are listed in Table 2. Again, our algorithm outperforms other competing algorithms.

4.2 Handwritten Digit Recognition

USPS [9] is a handwritten digit database, which contains 7,291 training images and 2,007 testing images. All of the training and testing images are 16×16 resolution images. Because the training and testing images don't have a high resolution, in this database, we directly use the original image for our experiment. For each digit we randomly select 200 samples for training. The experiment results are listed in Table 3. In the handwritten digit recognition tasks our LGC algorithm doesn't perform well, as well the SRC algorithms. Because for each testing sample, our algorithm and the SRC both need to make a sparse representation over the training sample, the data samples must be highly correlated. Unlike face

Table 1. Recognition rates(%) on the Extended Yale B database

NN	SVM	SRC	LGC
88.84	93.67	98.30	**99.22**

Table 2. Recognition rates(%) on the AR database

NN	SVM	SRC	LGC
60.82	82.83	89.15	**89.68**

Table 3. Recognition rates(%) on the USPS database

NN	SVM	SRC	LGC
94.42	94.02	93.52	**92.78**

database, the handwritten digit database doesn't have such a high correlation between the data samples, so both SRC and LGC algorithm don't have a high advantage in this database. But LGC algorithm's recognition accuracy is still acceptable.

5 Conclusion and Future Work

In this paper, we develop a novel ℓ^1-graph based image classification algorithm. Comparing with other algorithm, our algorithm uses the association between the training samples and all classes. We use ℓ^1-graph to evaluate this association, then use it for classification. The experiment results shows this information can help us classify visually similar images. Our algorithm has better performance than the original SRC algorithm, and other classic algorithms such as SVM and NN in face databases.

The main difficulty of our algorithm is to how to evaluate the association between the training samples and the classes. In this paper, we use ℓ^1-graph to calculate it. In the future, we will try to find different methods to get a more precise measure of the association between the training samples and the classes. The experiment results shows that our algorithm performs well in classify visually similar images, but doesn't perform well on other tasks. In the future we will try to modify our algorithm to make it suitable to other image classification tasks, or even make it suitable for other pattern recognition tasks .

Acknowledgments. This research is supported in part by the Major State Basic Research Development Program of China (973 Program, 2012CB315803), the National Natural Science Foundation of China (61371078), and the Research Fund for the Doctoral Program of Higher Education of China (20130002110051).

References

1. Aharon, M., Elad, M., Bruckstein, A.: K-svd: an algorithm for designing overcomplete dictionaries for sparse representation. IEEE Trans. Sig. Proc. **54**(11), 4311–4322 (2006)
2. Basri, R., Jacobs, D.: Lambertian reflectance and linear subspaces. IEEE Trans. Pattern Anal. Mach. Intell. **25**(2), 218–233 (2003)
3. Bruckstein, A.M., Donoho, D.L., Elad, M.: From sparse solutions of systems of equations to sparse modeling of signals and images. SIAM Rev. **51**(1), 34–81 (2009)
4. Bryt, O., Elad, M.: Compression of facial images using the k-SVD algorithm. J. Vis. Commun. Image Represent. **19**(4), 270–282 (2008)

5. Candes, E., Wakin, M.: An introduction to compressive sampling. IEEE Signal Process. Mag. **25**(2), 21–30 (2008)
6. Cheng, B., Yang, J., Yan, S., Fu, Y., Huang, T.S.: Learning with l1-graph for image analysis. IEEE Trans. Image Process. **19**(4), 858–866 (2010)
7. Elad, M., Aharon, M.: Image denoising via sparse and redundant representations over learned dictionaries. IEEE Trans. Image Process. **15**(12), 3736–3745 (2006)
8. Georghiades, A.S., Belhumeur, P.N., Kriegman, D.: From few to many: illumination cone models for face recognition under variable lighting and pose. IEEE Trans. Pattern Anal. Mach. Intell. **23**(6), 643–660 (2001)
9. Hull, J.J.: A database for handwritten text recognition research. IEEE Trans. Pattern Anal. Mach. Intell. **16**(5), 550–554 (1994)
10. Jiang, Z., Lin, Z., Davis, L.: Learning a discriminative dictionary for sparse coding via label consistent k-SVD. In: 2011 IEEE Conference on Computer Vision and Pattern Recognition, pp. 1697–1704 (Jun 2011)
11. Mallat, S.: A wavelet tour of signal processing. Academic press, San Diego, USA (1999)
12. Martinez, A.M.: The AR face database. CVC Technical Report 24 (1998)
13. Ramirez, I., Sprechmann, P., Sapiro, G.: Classification and clustering via dictionary learning with structured incoherence and shared features. In: 2010 IEEE Conference on Computer Vision and Pattern Recognition, pp. 3501–3508 (2010)
14. Turk, M.A., Pentland, A.P.: Face recognition using eigenfaces. In: 1991 IEEE Conference on Computer Vision and Pattern Recognition, pp. 586–591 (1991)
15. Wang, D., Kong, S.: A classification-oriented dictionary learning model: explicitly learning the particularity and commonality across categories. Pattern Recogn. **47**(2), 885–898 (2014)
16. Wright, J., Ma, Y., Mairal, J., Sapiro, G., Huang, T.S., Yan, S.: Sparse representation for computer vision and pattern recognition. Proc. IEEE **98**(6), 1031–1044 (2010)
17. Wright, J., Yang, A.Y., Ganesh, A., Sastry, S.S., Ma, Y.: Robust face recognition via sparse representation. IEEE Trans. Pattern Anal. Mach. Intell. **31**(2), 210–227 (2009)
18. Yang, J., Wright, J., Huang, T., Ma, Y.: Image super-resolution as sparse representation of raw image patches. In: 2008 IEEE Conference on Computer Vision and Pattern Recognition, pp. 1–8 (2008)
19. Yang, M., Zhang, L., Feng, X., Zhang, D.: Sparse representation based fisher discrimination dictionary learning for image classification. Int. J. Comput. Vision **109**(3), 209–232 (2014)
20. Zhang, Q., Li, B.: Discriminative k-SVD for dictionary learning in face recognition. In: 2010 IEEE Conference on Computer Vision and Pattern Recognition, pp. 2691–2698 (2010)
21. Zhou, N., Shen, Y., Peng, J., Fan, J.: Learning inter-related visual dictionary for object recognition. In: 2012 IEEE Conference on Computer Vision and Pattern Recognition, pp. 3490–3497 (2012)

Classification of Keystroke Patterns for User Identification in a Pressure-Based Typing Biometrics System with Particle Swarm Optimization (PSO)

Weng Kin Lai[1(✉)], Beng Ghee Tan[1], Ming Siong Soo[1], and Imran Khan[2]

[1] Tunku Abdul Rahman University College, Kuala Lumpur, Malaysia
laiwk@acd.tarc.edu.my
[2] IIUM, Kuala Lumpur, Malaysia

Abstract. Classification of users' keystroke patterns captured from a typing biometrics system is discussed in this paper. Although the user identification system developed here requires the user to key-in their passwords as they would normally do, the identification of the users will only be based on their keystroke patterns rather than the actual passwords. The keystroke pattern generated is represented by the force applied on a numerical keypad and it is this set of features extracted from a common password that will be submitted to the classifiers to identify the different users. The typing biometrics system had been designed and developed with an 8-bit microcontroller that is based on the AVR enhanced RISC architecture. Classification of these keystroke patterns will be with PSO (particle swarm optimization) and this will be compared with the standard K-Means. The preliminary experimental results showed that the identity of users can be authenticated based solely on their keystroke biometric patterns from a numeric keypad.

Keywords: Biometrics · Keystroke dynamics · PSO · K-Means · Artificial neural networks

1 Introduction

User *identification* and *verification* are two common but different applications of biometric technologies. While verification relates to matching or verifying the patterns against a single user's stored identity, identification on the other hand, involves finding the one unique identity amongst the many stored identities. Essentially, identification seeks to determine the user's identity whereas verification attempts to prove the claimed identity. Although a variety of authentication devices may be used to verify a user's identity, passwords remain the most preferred method especially when the keyboard is the preferred data entry device, due to both the long history of the use of this mechanism to gain access as well as the fact that it is still relatively inexpensive compared to other more sophisticated solutions. However, like most modern technologies, unless it is used correctly, the level of security provided by passwords can be weak. Nevertheless many users still chose to use rather weak passwords and consequently have to bear with some of the associated problems that come with them. When it comes to security, multi-factor

© Springer International Publishing Switzerland 2015
S. Arik et al. (Eds.): ICONIP 2015, Part I, LNCS 9489, pp. 71–78, 2015.
DOI: 10.1007/978-3-319-26532-2_9

approaches may be used to extend and strengthen the security level that passwords provide. Ideally, this reinforcement should be transparent and indiscernible to users, without requiring any additional efforts when gaining entry access. Now in addition to the different and personalized passwords for each user, the users are also known to have developed a unique typing style to enter their important account information. For example, a user may type the characters that constitute the password at different speeds. By leveraging on such differences, one can develop an approach that may be used to enhance the system's security with keystroke biometrics (or in some literature, typing biometrics) to reinforce password-authentication mechanisms. Previous research [1–4] has shown that it is possible to identify a user via his or her typing patterns in the form of keystroke biometrics which attempts to analyze a user's keystroke patterns. It is well known that an individual's keystroke biometrics pattern may be based on any combination of the following features [3, 5],

(a) *the duration each keystroke is pressed, that is the amount of time a user takes to press and release when typing,*
(b) *the latency between consecutive keystrokes,*
(c) *the force exerted on the keys.*

Some prior work has been done on the use of keystroke biometrics as a password hardening technique [6–9]. In addition to the duration between each pair of keystrokes, Obaidat and Sadoun [10] investigated the use of the holding time for each key pressed. The results reported of authenticating users based on just their keystroke have been encouraging. Nonetheless, these works were centered primarily on the common QWERTY computer keyboard. While the QWERTY layout has been used extensively in the past, the simple and inexpensive numeric keypad is gaining popularity especially when used for access entry. A manufacturer of a leading brand of locks believe that a global revolution is taking place in home security as home owners across the globe are looking for smarter, more convenient ways of securing their homes and possessions [11]. Two key advantages stand out for such digital or keyless locks – they cannot be picked nor opened with lock bumping [12].

Nonetheless, the typing style on the numeric keypad would be significantly different than that on the QWERTY due to their dissimilar layout. In this paper, we will be investigating how we may identify an individual's keystroke biometric pattern on a numeric keypad based on the force (or amount of pressure) exerted on each key. The biometric sensors of the system are force sensitive resistors which were used to capture and translate the amount of force exerted on the keys into their equivalent electrical values. The electrical signals will then be digitized into the set of features to represent the user's unique keystroke dynamics. Two main authentication issues are emphasized during the overall design of the system, viz.

(a) the numeric password representing the normal passkey entered by the user, which consists of a combination of numeric keys of the appropriate length created by the user and saved in the system,
(b) the keystroke biometrics pattern associated with the user's password in the form of a *"typing template"*. This is the second factor which will be analyzed by the classifiers.

The remainder of this paper is organised as follows. The next section discusses the hardware design of the keystroke biometrics user identification system which captures the force exerted by users on a numeric keypad. In Sect. 3, we describe the important details of PSO (*Particle Swarm Optimization*) and the *K-Means* that will be used to classify the keystroke patterns. Section 4 discusses the major design issues as well as the important aspects of the experiments and the results obtained. Finally, in Sect. 5 we present some conclusions and potential areas for further work.

2 System Design

2.1 Force Sensor

The force sensitive resistors used here are made from a conductive polymer that changes resistance in a predictable manner following application of force to its surface [13]. Such a force sensitive resistor was placed underneath the numeric keypad to capture the keystroke patterns in the biometrics user identification system. This is shown in Fig. 1.

Fig. 1. The system showing the *FSR* in the centre, and the *Arduino Leonardo* on the right.

2.2 Microprocessor Design with *Arduino*

At the core of the *Arduino Leonardo* micro controller (shown in Fig. 2) is the *ATmega32u4* processor and this was used to digitize the pressure from the keys pressed. The *ATmega32u4* is a low-power CMOS 8-bit microcontroller based on the AVR enhanced RISC (*Reduced Instruction Set Computing*) architecture and has 32 8-bit general purpose working registers. *Leonardo* has 20 digital input-output pins (of which 7 can be used as outputs and 12 as analogue inputs), a 16 MHz crystal oscillator, a micro USB connection, a power jack, an ICSP header, and a reset button. The system would capture the force exerted on each key by the user and generate a biometric pattern that would have the same length as the password entered.

Fig. 2. The Arduino Leonardo

3   Classification

The performance of the PSO was evaluated and compared with that from the standard K-Means.

3.1   Particle Swarm Optimization

Particle Swarm Optimization (PSO) was originally developed by *Kennedy and Eberhart* in 1995, inspired by the social behavior of a bird flock [14]. PSO had been used to solve a range of different problems. Kiran et al. [15] investigated using PSO for Human Posture recognition with good recognition rates for many different postures. At the other end of the spectrum, *Mahamed G. H. Omran* used the PSO to perform image segmentation where the problem was modeled as a data clustering problem [16].

The approach that PSO used here can be summarised into two main stages; a global search stage and a local refining stage. The set of particles move through the search space to find the particle position that will result in the best evaluation of the given fitness function. Each particle will have its own memory that consists of two components – a local best solution (*lbest*) as well as the global best (*gbest*) which is basically the best solution amongst all the *lbest*. This *gbest* defines the best solution which all the other particles will try to emulate on their own individual search for better solutions. The PSO can be represented by the following equations:

$$V_{id}^{new} = W \times V_{id}^{old} + \left(C_1 \times rand_1 \times \left(p_{best} - X_{id}^{old}\right)\right) + \left(C_2 \times rand_2 \times \left(g_{best} - X_{id}^{old}\right)\right) \quad (1)$$

where,

$$X_{id}^{new} = X_{id}^{old} + V_{id} \quad (2)$$

Here, V_{id} represents the velocity that is involved in the updating of the movement and magnitude of the particles, X_{id}, the new position of the particles after updating, W the inertia weight, and C_1 and C_2 a set of acceleration coefficients. Finally, $rand_1$ and $rand_2$ are random values that vary from 0 to 1. These parameters provide the necessary diversity to the particle swarm by changing the momentum of particles to avoid the potential stagnation of particles at the local optima. The values for C_1 and C_2 used here are both set to 1.4 whereas W was set to 0.01. Five particles were used in the experiments here.

3.2 K-Means

The K-Means algorithm [17] and its variants [18] is by far the most commonly used partitional clustering method. The data is partitioned into K clusters C = {cj, $j = 1$ ... K} by repetitive separation of data X = {x_i, $i = 1$ ... n} into clusters with the nearest mean based on the distance measurement between data and the cluster centroids. The standard K-Means algorithm is summarised below:

1. Randomly select a set of K cluster centroids.
2. Compute the Euclidean distance between each data and the K cluster centroids.
3. Assign each data to the closest cluster centroid.
4. Recalculate the cluster centroids.
5. Repeat 2-4 till the stopping criteria is satisfied.

4   Experimental Setup and Results

We had tested this new numerical keypad to authenticate the identity of the users based on their keystroke biometric signatures. Each user's keystroke biometric signature was based on their preferred individual password, each of which consists of 8 digits. In contrast, we are now using the keystroke biometric signatures generated from this numerical keypad based on the *SAME* 8 digit password for all users. Clearly, this will introduce a more challenging set of conditions as the individual identities are now based on nothing more than their keystroke biometric signatures. 6 samples were obtained from each of the 10 users, making a total of 60 sets of data. The users are familiar with the keyboard and have used the keyboard in their daily work but to minimise any inconsistent typing rhythm, they were given sufficient time and opportunities to familiarise themselves with this common password. The data obtained were also pre-processed using a standard sigmoid function (Eq. (3)) to convert the biometric (force) data to the range [0.0, 1.0] i.e.

$$f(x) = \frac{1}{1} + \exp(\lambda - x) \tag{3}$$

where λ, a constant that is set to 10 here. 50 % of the pressure patterns from one class was randomly selected for registration and to train the system at a time. Correct identification of the test samples for the corresponding user contributes towards a *True Positive* (TP) measure. This validation strategy was repeated for a total of 1,000 times for each of the two classifiers, each time with a randomly chosen set of training patterns. Figure 3(a) and (b) shows the average TP and TN values obtained from both the K-Means and PSO respectively. Users #4 and #8 were able to produce consistent typing styles that were easily picked up by the two classifiers. Nevertheless, PSO gave slightly better results and this is evident in correctly identifying users #2 and #6 than K-Means. While the TP classification results for #3 dipped below 30 %, it is probably due to the user rather than the system. Nevertheless, the inconsistent typing style of #3 was significantly different from that of the others for the classifiers to differentiate.

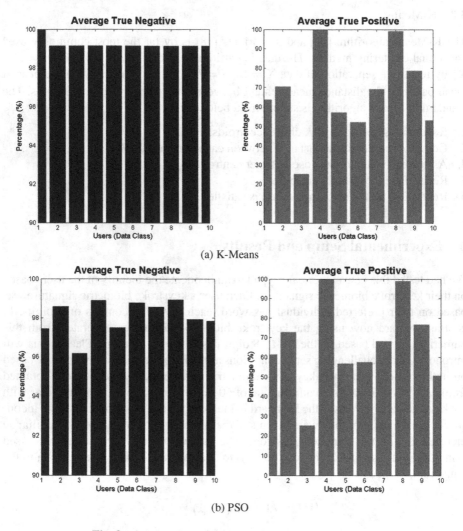

(a) K-Means

(b) PSO

Fig. 3. Average *True Positive* and *True Negative* results

The second set of classification results examines how different are the keystroke dynamics of each user. The typing patterns from the remaining 9 users were presented to the trained classifier that had been trained with the one user. These results then contributed to the *True Negative* (TN) values. The results from Fig. 3(a) and (b) seem to suggest that there is a significant amount of difference in the keystroke dynamics of the users from each other, as looking at the TF rates, both the classifiers were able to correctly identify the incorrect typing patterns at almost 100 %.

They were effective in differentiating the real users from the others just from their typing dynamics. In general, PSO was slightly better in correctly identifying the users than K-Means. This seems to suggest that these users have naturally more consistent typing styles that our system is able to identify. Moreover the results also indicated that the ability of the classifiers to deny access to typing patterns of the other users is consistent and accurate too.

5 Conclusions

Together with the electronic keyboard, passwords have been used for a long time as a simple and convenient means to gain access. However, once the passwords are compromised, there will not be any more protection from unauthorized entry. This is where keystroke biometrics may be used to strengthen this mode of access control. Nonetheless, much of the work in this area has been done with the common QWERTY keyboard using either timing or pressure features. However, the typing style for most people on a numeric keypad is significantly different when compared with that on a QWERTY layout. In this paper we have described the development of a keystroke biometrics with a novel numeric keypad which can generate the keystroke biometric patterns based on the amount of pressure exerted on each key when the users enter their password. This keystroke biometric system was designed with a force sensitive resistor that was integrated into the numeric keypad, and the pressure data from this force sensitive resistor was then acquired via the *Arduino* microcontroller. Preliminary work on this system had produced encouraging results for user authentication - on the set of different passwords [18]. This paper extends the earlier work which described the use of this system for user identification [19] on the CogRAM, a weightless neural net. The experimental results of the keystroke biometrics data, captured from the micro-controller system have shown that the PSO was able to easily identify some aspects of the individual's typing style to either allow or deny access, based on a common password. What this suggests is that through this password hardening, even if the password had been compromised, keystroke biometrics can provide an additional level of authentication for access systems that use the numerical keypad. One key area for further work is to have more data from this keypad system and to investigate if mounting more pressure sensors can help improve the classification accuracies of such a system in the future.

References

1. Joyce, R., Gupta, G.: Identity authentication based on keystroke latencies. Comm. ACM **33**, 168–176 (1990)
2. Monrose, F., Rubin, A.: Authentication via keystroke dynamics. In: Proceedings of the Fourth ACM Conference on Computer and Communications Security, pp. 48–56, Zurich, Switzerland (1997)
3. De Ru, W., Eloff, J.: Enhanced password authentication through fuzzy logic. IEEE Expert **12**, 38–45 (1997)

4. Obaidat, M.S., Sadoun, B.: Verification of computer users using keystroke dynamics. IEEE Trans. Syst. Man Cybern. **27**, 261–269 (1997)
5. Tee, E.R., Selvanathan, N.: Pin signature verification using wavelet transform. Malays. J. Comput. Sci. **9**(2), 71–78 (1996)
6. Maisuria, L.K., Ong, C.S., Lai, W.K.: A comparison of artificial neural networks and cluster analysis for typing biometrics authentication. In: Proceedings of the International Joint Conference on Neural Networks (IJCNN), Washington, DC, 10–16 July 1999
7. Loy, C.C., Lim, C.P., Lai, W.K.: Pressure-based typing biometrics user authentication using the fuzzy ARTMAP neural network. In: Proceedings of the Twelfth International Conference on Neural Information Processing (ICONIP 2005), pp 647–652, Taipei, Taiwan ROC, October 30–November 2, 2005
8. Eltahir, W.E., Salami, M.J.E., Ismail, A.F., Lai, W.K.: Design and evaluation of a pressure-based typing biometric authentication system. EURASIP J. Inf. Sec. **2008** 14 (2008). Article ID 345047 doi:10.1155/2008/345047
9. Yong, S, Lai, W.K., Coghill, G.: Weightless neural networks for typing biometrics authentication. In: Proceedings of the 8th International Conference on Knowledge-Based Intelligent Information and Engineering Systems, KES 2004, vol. II, pp. 284–293, Wellington, New Zealand, 22–24 September 2004
10. Obaidat, M.S., Sadoun, B.: Keystroke dynamics based authentication (Chap. 10). In: Jain, A.K., Bolle, R., Pankanti, S. (eds.) Biometrics: Personal Identification in Networked Society. Springer, US (1996)
11. A global revolution is taking place and YALE Digital Door Lock. http://www.yalelock.com/en/yale/com/Global-Revolution/. Accessed 22 March 2015
12. Lock Bumping. http://e.wikipedia.org/wiki/Lock_bumping. Accessed 22 March 2015
13. Force Sensing resistor. http://en.wikipedia.org/wiki/Force-sensing_resistor. Accessed 18th March 2015
14. Kennedy, J., Eberhart, R.: Particle swarm optimization. In: Proceedings of the IEEE International Joint Conference on Neural Networks, vol. 4, pp. 1942–1948 (1995)
15. Kiran, M., Teng, S.L., Seng, C.C., Kin, L.W.: Human posture classification using hybrid particle swarm optimization. In: Proceedings of the Tenth International Conference on Information Sciences, Signal Processing and their application (ISSPA 2010), Kuala Lumpur, Malaysia 10–13 May 2010
16. Omran, M.G.H.: Particle swarm optimization methods for pattern recognition and image processing. Ph.D. thesis, University of Pretoria (2005)
17. MacQueen, J.: *Some methods for classification and analysis of multivariate observations*. In: The Fifth Berkeley Symposium on Mathematics, Statistics and Probability, pp. 281–297. University of California Press (1967)
18. Jain, A.K.: Data clustering: 50 years beyond k-means. Pattern Recogn. Lett. **31**, 651–666 (2010)
19. Lai, W.K., Tan, B.G., Soo, M.S., Khan, I.: Two-factor user authentication with the CogRAM weightless neural net. In: Proceedings of the 2014 World Congress on Computational Intelligence (WCCI 2014), Beijing, China, 6–11 July 2014

Discriminative Orthonormal Dictionary Learning for Fast Low-Rank Representation

Zhen Dong(✉), Mingtao Pei, and Yunde Jia

Beijing Laboratory of Intelligent Information Technology,
School of Computer Science, Beijing Institute of Technology,
Beijing 100081, People's Republic of China
{dongzhen,peimt,jiayunde}@bit.edu.cn

Abstract. This paper presents a discriminative orthonormal dictionary learning method for low-rank representation. The orthonormal property is beneficial for the representative power of the dictionary by avoiding the dictionary redundancy. To enhance the discriminative power of the dictionary, all the class-specific dictionaries which are encouraged to well represent the samples from the same class are optimized simultaneously. With the learned discriminative orthonormal dictionary, the low-rank representation problem can be solved much faster than traditional methods. Experiments on three public datasets demonstrate the effectiveness and efficiency of our method.

Keywords: Discriminative dictionary learning · Orthonormal · Fast low-rank representation

1 Introduction

Low-rank representation has been widely used in computer vision and pattern recognition due to its strong robustness to the noise of the corrupted data, such as occlusions, lightning variations, and pixel corruptions. The basic idea is to represent the observation data matrix by the addition of a sparse noise matrix and a low-rank data matrix. The low rank data matrix is often described as the linear combinations of the dictionary atoms. Since the quality of the dictionary is quite important to the performance of the low-rank representation, much work focuses on learning a good dictionary from the observation data matrix and presents lots of applications including face recognition [1–7], and image classification [8, 9].

The existing dictionary learning methods can be roughly divided into two categories: unsupervised and supervised. Dictionaries learned by supervised methods [4,9,12,13] are more discriminative and obtain better performances in classification tasks. In most supervised methods, the class-specific dictionary is expected to have the ability of well representing samples from the same class and not be able to represent samples from other classes for better discriminative power of the dictionary. However, the idea is explicitly modeled by imposing constraints on both the class-specific dictionary and the representation [1,3,12,13],

© Springer International Publishing Switzerland 2015
S. Arik et al. (Eds.): ICONIP 2015, Part I, LNCS 9489, pp. 79–89, 2015.
DOI: 10.1007/978-3-319-26532-2_10

and the class-specific dictionaries are optimized one by one, which leads to the loss of the relationship among class-specific dictionaries. The class-specific dictionaries thus usually share same or similar atoms due to the inter-class visual correlations, which makes the learned dictionary redundant. A redundant dictionary with lots of similar atoms will lead to the ambiguity of data representations. For example, it is shown that the dimensionality of the dictionary learned by K-SVD [10] for face images can be reduced by half without too much performance loss [14].

In order to alleviate the problems, we present a discriminative orthonormal dictionary learning method. Unlike previous discriminative dictionary learning methods which impose constraints on both the class-specific dictionaries and the representations, we propose a more concise discriminative regularization term which only restricts the representation. Using the proposed regularization term, all the class-specific dictionaries can be optimized simultaneously instead of being optimized one by one. Furthermore, the dictionary is enforced to be orthonormal, so the class-specific dictionaries are encouraged to be as incoherent as possible, *i.e.*, there is hardly similar atoms in different class-specific dictionaries. The representative power of the compact dictionary is thus improved by avoiding the ambiguity problem of the data representations.

With the learned orthonormal dictionary, a fast low-rank representation method is presented. As proved in [15], the rank of the reconstruction data matrix is upper bounded by the number of non-zero rows of the coefficient matrix when the dictionary is orthonormal. With this theorem, the nuclear norm in the low-rank representation can be replaced by its upper bound. The low-rank representation is thus solved directly by the magnitude shrinkage function instead of the singular value shrinkage function which needs the SVD operation. The solution procedure is much faster than solving the traditional low-rank representation problem. The main contributions of this paper are:

- We present an efficient discriminative orthonormal dictionary learning method which can learn a discriminative and compact dictionary for low-rank representation.
- A fast low-rank representation method over the orthonormal dictionary is presented. With the learned orthonormal dictionary, the low-rank representations can be obtained much fast and achieve comparable classification performances to the state-of-the-art methods.

2   Discriminative Orthonormal Dictionary Learning

2.1   Formulation

Given the observation data matrix $\mathbf{Y} \in \mathbb{R}^{D \times N}$, our goal is to learn the discriminative dictionary $\mathbf{D} \in \mathbb{R}^{D \times K}$, the representation matrix $\mathbf{X} \in \mathbb{R}^{K \times N}$, and the noise matrix $\mathbf{E} \in \mathbb{R}^{D \times N}$ to satisfy $\mathbf{Y} = \mathbf{DX} + \mathbf{E}$. A redundant dictionary have lots of similar atoms, which leads to high computation cost and ambiguity in corresponding representations. In order to alleviate these problems, we impose the orthonormal constraint on the dictionary in our model. The discriminative orthonormal dictionary learning is given by

$$\min_{\mathbf{D},\mathbf{X},\mathbf{E}} \mathcal{LR}(\mathbf{D},\mathbf{X}) + \lambda_1 \mathcal{SN}(\mathbf{E}) + \lambda_2 \mathcal{DC}(\mathbf{X},\mathbf{L})$$
$$s.t.\,\mathbf{Y} = \mathbf{DX} + \mathbf{E},\, \mathbf{D}^\top \mathbf{D} = \mathbf{I}, \tag{1}$$

where the function $\mathcal{LR}(\cdot)$ measures the low-rankness of the reconstruction data matrix $\mathbf{DX}$, the function $\mathcal{SN}(\cdot)$ measures the sparsity of the noise matrix $\mathbf{E}$, the $\mathcal{DC}(\cdot)$ is the discriminative term for improving the discriminative power of the learned dictionary, and λ_1 and λ_2 are parameters to balance the importance of the three terms. The $\mathbf{L}$ is the label matrix of the data $\mathbf{Y}$, and the $\mathbf{I}$ is the identity matrix.

Low-rank Term $\mathcal{LR}(\mathbf{D},\mathbf{X})$: In classification task, training samples from the same class are highly correlated and expected to form a low-dimensionality subspace. The training data without noise represented by $\mathbf{DX}$ should be low-rank, i.e., $\mathcal{LR}(\mathbf{D},\mathbf{X}) = \mathrm{rank}(\mathbf{DX}) = \mathrm{rank}(\mathbf{Z})$, where $rank(\mathbf{Z})$ denotes the rank of the matrix $\mathbf{Z}$. The minimization of $rank(\cdot)$ is an NP-hard problem and difficult to solve due to the discrete nature of the function. Fortunately, Fazel proved that the nuclear norm function $\|\mathbf{Z}\|_*$ (i.e. the sum of the singular values of $\mathbf{Z}$) is the convex envelope of the rank function $rank(\mathbf{Z})$ on the set of $\{\mathbf{Z}|\,\|\mathbf{Z}\|_2 < 1\}$. Furthermore, as demonstrated in [15], $\|\mathbf{DX}\|_*$ is upper bounded by $\|\mathbf{X}^\top\|_{2,1} = \sum_{i=1}^{K} \|\mathbf{x}_i\|_2 = \sum_{i=1}^{K} \sqrt{\sum_{j=1}^{N} \mathbf{x}_{ij}^2}$ under the constraint of $\mathbf{D}^\top \mathbf{D} = \mathbf{I}$, where $\mathbf{x}_i$ represents the i-th row of $\mathbf{X}$, and $\mathbf{x}_{ij}$ denotes the element in the i-th row and j-th column of $\mathbf{X}$. Our low-rank term is thus given by $\mathcal{LR}(\mathbf{D},\mathbf{X}) = \|\mathbf{X}^\top\|_{2,1}$ which can be minimized efficiently.

Sparse Noise Term $\mathcal{SN}(\mathbf{E})$: Real-word data is often noisy or corrupted due to illumination variation, occlusion, and pixel corruption. The classifier trained with these data may overfit and the classification performance may degrade. Motivated by the low-rank recovery, the corrupted data matrix $\mathbf{Y}$ is decomposed into two parts: a low rank component $\mathbf{DX}$ and a sparse noise component $\mathbf{E}$, to alleviate the aforementioned problem. Here, we denote the noisy term as $\mathcal{SN}(\mathbf{E}) = \|\mathbf{E}\|_{2,1}$. The $\|\cdot\|_{2,1}$ is used since we assume that some training samples are corrupted and the others are not. In addition, $\mathbf{E} = \mathbf{Y} - \mathbf{DX}$ measures the reconstruction error, and the minimization of $\mathcal{SN}(\mathbf{E})$ encourages the dictionary $\mathbf{D}$ to well represent the observation data.

Discriminative Term $\mathcal{DC}(\mathbf{X},\mathbf{L})$: In order to enhance the discriminative power of the dictionary $\mathbf{D}$ for classification tasks, we propose the discriminative regularization term, $\mathcal{DC}(\mathbf{X},\mathbf{L}) = \|\mathbf{X} \odot \mathbf{S}\|_F^2$, to the representation $\mathbf{X}$, where $\odot$ means the element-wise multiplication operator, $\|\cdot\|_F$ denotes the Frobenius norm of a matrix. The $\mathbf{S} \in \mathbb{R}^{K \times N}$ is defined as

$$\mathbf{S}(i,j) = \begin{cases} 0, & \text{if } \mathbf{d}_i \text{ and } \mathbf{y}_j \text{ belong to the same class} \\ 1, & \text{otherwise,} \end{cases} \tag{2}$$

where $\mathbf{d}_i$ is the i-th column of the dictionary $\mathbf{D}$, and $\mathbf{y}_j$ is the j-th column of the observation data matrix $\mathbf{Y}$. This term encourages that the class-specific dictionary is able to well reconstruct the samples from the same class and doesn't have the capacity to represent samples from other classes. The $\mathbf{S}$ can be computed from the label matrix $\mathbf{L}$ easily. In [1,13], the idea is explicitly modeled by imposing constraints on $\mathbf{D}$ and $\mathbf{X}$ as $\sum_{i=1}^{C}(\|\mathbf{Y}_i - \mathbf{D}_i\mathbf{X}_i^i\|_F^2 + \sum_{j=1,j\neq i}^{C}\|\mathbf{D}_j\mathbf{X}_i^j\|_F^2)$ which can easily be proved as equivalent as the proposed discriminative regularization when D is orthonormal. By using our discriminative regularization term, all the class-specific dictionary can be optimized simultaneously instead of one by one to obtain a compact and discriminative dictionary.

A similar work is the ideal representation $\mathbf{Q} = \mathbf{U} - \mathbf{S}$ in [8], where $\mathbf{U} \in \mathbb{R}^{K \times N}$ is a matrix whose elements are all 1. In their work, the code-ideal term represented by $\|\mathbf{X} - \mathbf{Q}\|_F^2$ encourages $\mathbf{X}$ to be close to $\mathbf{Q}$. However, it potentially means that all the atoms in a class-specific dictionary provide same contributions to the data, which may bring negative effects to the representations of the data. The malpractice is dislodged by using the element-wise operator in our discriminative regularization term.

2.2   Optimization

In order to optimize Eq. (1), an auxiliary variable $\mathbf{H}$ is introduced, and the optimization problem is rewritten as

$$\min_{\mathbf{D},\mathbf{X},\mathbf{E}} \|\mathbf{X}^\top\|_{2,1} + \lambda_1\|\mathbf{E}\|_{2,1} + \lambda_2\|\mathbf{H} \odot \mathbf{S}\|_F^2$$
$$s.t.\, \mathbf{Y} = \mathbf{DX} + \mathbf{E},\ \mathbf{D}^\top\mathbf{D} = \mathbf{I},\ \mathbf{X} = \mathbf{H}. \tag{3}$$

The augmented Lagrangian function of Eq. (3) is

$$\mathcal{L}(\mathbf{D},\mathbf{X},\mathbf{E},\mathbf{H},\mathbf{A},\mathbf{B})$$
$$= \|\mathbf{X}^\top\|_{2,1} + \lambda_1\|\mathbf{E}\|_{2,1} + \lambda_2\|\mathbf{H} \odot \mathbf{S}\|_F^2 + <\mathbf{A}, \mathbf{Y} - \mathbf{DX} - \mathbf{E}>$$
$$+ <\mathbf{B}, \mathbf{X} - \mathbf{H}> + \frac{\mu}{2}(\|\mathbf{Y} - \mathbf{DX} - \mathbf{E}\|_F^2 + \|\mathbf{X} - \mathbf{H}\|_F^2) \tag{4}$$
$$s.t.\ \mathbf{D}^\top\mathbf{D} = \mathbf{I},$$

where $<\mathbf{A},\mathbf{B}> = tr(\mathbf{AB}^\top)$ denotes the trace of $\mathbf{AB}^\top$, $\mathbf{A}$ and $\mathbf{B}$ are Lagrange multipliers, and μ is the positive penalty parameter. The linearized alternating direction method with adaptive penalty (LADMAP) is used to minimize Eq. (4) with the iterative following steps.

Updating Dictionary D: With fixed $\mathbf{X}$, $\mathbf{H}$, and $\mathbf{E}$, the minimization of Eq. (4) is a quadratic form of $\mathbf{D}$, and the $\mathbf{D}$ is updated as

$$\mathbf{D}^{(i+1)} = (\mathbf{Y} - \mathbf{E}^{(i)} + \mathbf{A}^{(i)}/\mu^{(i)})\mathbf{X}^{(i)\top}(\mathbf{X}^{(i)}\mathbf{X}^{(i)\top})^{-1}. \tag{5}$$

Taking the orthonormal constraint of $\mathbf{D}$ into account, the Gram-Schmidt method is processed on the columns of $\mathbf{D}$ after updating.

Updating X: When updating $\mathbf{X}$, the quadratic term of $\mathbf{X}$ is replaced by its first order Taylor approximation at the previous iteration step $\mathbf{X}^{(i)}$. The representation $\mathbf{X}$ is updated by solving

$$
\mathbf{X}^{(i+1)} = \arg\min_{\mathbf{X}} \|\mathbf{X}^\top\|_{2,1} + \frac{\mu^{(i)}}{2}\left(\|\mathbf{X} - \mathbf{H}^{(i)} + \frac{\mathbf{B}^{(i)}}{\mu^{(i)}}\|_F^2\right.
$$

$$
\left. + \|\mathbf{Y} - \mathbf{D}\mathbf{X} - \mathbf{E}^{(i)} + \frac{\mathbf{A}^{(i)}}{\mu^{(i)}}\|_F^2\right)
$$

$$
= \operatorname*{argmin}_{\mathbf{X}} \|\mathbf{X}^\top\|_{2,1} + \frac{\mu^{(i)}\eta}{2}\|\mathbf{X} - \mathbf{F}\|_F^2, \tag{6}
$$

where $\mathbf{F} = (1 - 2/\eta)\mathbf{X}^{(i)} + (\mathbf{D}^\top(\mathbf{Y} - \mathbf{E}^{(i)} + \mathbf{A}^{(i)}/\mu^{(i)}) + \mathbf{H}^{(i)} - \mathbf{B}^{(i)}/\mu^{(i)})/\eta$, and $\eta > 0$ is a parameter. The $\mathbf{X}$ is updated as $\mathbf{x}_j^{(i+1)} = \mathbb{S}_{1/(\mu^{(i)}\eta)}(\mathbf{f}_j)$, where $\mathbb{S}_\varepsilon(\mathbf{v})$ is a magnitude shrinkage function for vector $\mathbf{v}$ defined as $\mathbb{S}_\varepsilon(\mathbf{v}) = \max(1 - \varepsilon/\|\mathbf{v}\|_2, 0)\mathbf{v}$, and $\mathbf{x}_j$ and $\mathbf{f}_j$ represent the j-th row of $\mathbf{X}$ and $\mathbf{F}$, respectively.

Updating H: Keeping other variables fixed, the auxiliary variable $\mathbf{H}$ is updated by solving

$$
\mathbf{H}^{(i+1)} = \operatorname*{argmin}_{\mathbf{H}} \lambda_2\|\mathbf{H} \odot \mathbf{S}\|_F^2 + \frac{\mu^{(i)}}{2}\|\mathbf{X}^{(i+1)} - \mathbf{H} + \frac{\mathbf{B}^{(i)}}{\mu^{(i)}}\|_F^2 \tag{7}
$$

Setting the derivation of Eq. (7) with respect to $\mathbf{h}_j$ to zero, we have

$$
\mathbf{h}_j^{(i+1)} = (\mu^{(i)}\mathbf{x}_j^{(i+1)} + \mathbf{b}_j^{(i)})\mathbf{M}^{-1},
$$
$$
\mathbf{M} = \lambda_2 diag(\mathbf{s}_j) + \mu^{(i)}\mathbf{I} \tag{8}
$$

where $\mathbf{I}$ is the identity matrix in $\mathbb{R}^{N \times N}$, $diag(\mathbf{s}_j)$ returns a diagonal matrix with $\mathbf{s}_j$ as the main diagonal elements, $\mathbf{s}_j$, $\mathbf{h}_j$, and $\mathbf{b}_j$ represent the j-th row of $\mathbf{S}$, $\mathbf{H}$, and $\mathbf{B}$, respectively.

Updating E: When updating the reconstruction error $\mathbf{E}$, Eq. (3) can be rewritten as

$$
\mathbf{E}^{(i+1)} = \operatorname*{argmin}_{\mathbf{E}} \lambda_1\|\mathbf{E}\|_{2,1} + \frac{\mu^{(i)}}{2}\|\mathbf{E} - \mathbf{N}\|_F^2, \tag{9}
$$

where $\mathbf{N} = \mathbf{Y} - \mathbf{D}^{(i+1)}\mathbf{X}^{(i+1)} + \mathbf{A}^{(i)}/\mu^{(i)}$. Similar to $\mathbf{X}$, the $\mathbf{E}$ is updated as $\mathbf{e}_j^{(i+1)} = \mathbb{S}_{\lambda_1/\mu}(\mathbf{n}_j)$, where $\mathbf{e}_j$ and $\mathbf{n}_j$ are the j-th column of $\mathbf{E}$ and $\mathbf{N}$, respectively.

Updating A, B, and μ: The Lagrange multipliers are updated as $\mathbf{A}^{(i+1)} = \mathbf{A}^{(i)} + \mu^{(i)}(\mathbf{Y} - \mathbf{D}^{(i+1)}\mathbf{X}^{(i+1)} - \mathbf{E}^{(i+1)})$ and $\mathbf{B}^{(i+1)} = \mathbf{B}^{(i)} + \mu^{(i)}(\mathbf{X}^{(i+1)} - \mathbf{H})$, and the penalty parameter μ is updated as $\mu^{(i+1)} = \gamma\mu^{(i)}$, where $\gamma > 1$ is the magnified factor.

The K-SVD method is used to initialize each class-specific dictionary $\mathbf{D}_i$ over the training samples of the i-th class. The whole dictionary is obtained by combing all the class-specific dictionaries as $\mathbf{D}^{(0)} = [\mathbf{D}_1, \mathbf{D}_2, ..., \mathbf{D}_C]$. After the orthonormalization of the initialized dictionary $\mathbf{D}^{(0)}$, the representations $\mathbf{X}^{(0)}$ is initialized by the orthogonal matching pursuit. The error $\mathbf{E}^{(0)}$ and the auxiliary variable $\mathbf{H}^{(0)}$ are initialized as zeros.

3   Fast Low-Rank Representation

With the learned orthonormal discriminative dictionary $\mathbf{D}$, the representations of training and testing samples are computed by

$$\min_{\mathbf{X},\mathbf{E}} \|\mathbf{X}^\top\|_{2,1} + \lambda_1 \|\mathbf{E}\|_{2,1}$$
$$s.t. \mathbf{Y} = \mathbf{DX} + \mathbf{E}, \tag{10}$$

where λ_1 is the same as in Eq. (1). The alternating direction method (ADM) is used to optimize Eq. (10), and the augmented Lagrange function is

$$\tilde{\mathcal{L}} = \|\mathbf{X}^\top\|_{2,1} + \lambda_1 \|\mathbf{E}\|_{2,1} + \frac{\theta}{2}\|\mathbf{Y} - \mathbf{DX} - \mathbf{E} + \frac{\mathbf{J}}{\theta}\|_F^2, \tag{11}$$

where $\mathbf{J}$ is the Lagrange multiplier, and θ is the positive penalty parameter. Compared with the low-rank representation (LRR) method, the nuclear norm of $\mathbf{S}$ is replaced by its upper bound. Similar to updating the noise matrix in Sect. 2.2, both of the $\mathbf{X}$ and $\mathbf{E}$ can be updated easily with the help of the magnitude shrinkage function $\mathbb{S}_\epsilon(v)$ instead of the singular value shrinkage function which needs the SVD operation in LRR.

The ridge regression model is used to obtain a linear classifier from the training representations $\mathbf{X}$:

$$\min_{\mathbf{W}} \|\mathbf{L} - \mathbf{WX}\|_F^2 + \zeta \|\mathbf{W}\|_F^2, \tag{12}$$

where ζ is a parameter, $\mathbf{L} \in \mathbb{R}^{C \times N}$ is the label matrix of $\mathbf{X}$, and each column $[0, ..., 0, 1, 0, ...0]^\top$ describes the label of a sample. The optimal solution of Eq. (12) is $\mathbf{W}^* = \mathbf{LX}^\top(\mathbf{LX}^\top + \zeta\mathbf{I})^{-1}$. With the optimized $\mathbf{W}^*$, a testing sample $\mathbf{x}$ can be predicted by picking the index of the maximum element of $\mathbf{W}^*\mathbf{x}$.

4   Experiments

We evaluate the proposed discriminative orthonormal dictionary learning and low-rank representation methods on three public datasets, two face datasets: Extended Yale B, AR, and an object recognition dataset: Caltech-101.

4.1   Extended Yale B Dataset

The Extended Yale B dataset contains 2,414 frontal-face images from 38 people with 59 to 64 images per person. The images are captured under various controlled lighting conditions. The resolution of the image is 192×168 pixels. Since previous work test their methods on this dataset with different experimental settings, we do various experiments following their settings for a fair comparison.

In [13], the face images are resized to 54×48 pixels. For each people, 20 images are randomly selected for training with the remaining images for testing. We test our method under the same settings and compare with the results reported in [13]. Table 1 shows that our method outperforms other methods. To compare with more recent work, we follow the same experiment setting of [17]. The images are projected to a 504-dimensional feature by using a random matrix of zero-mean normal distribution. For each person, half images are used for training, and the rest half for testing. The comparison results are displayed in Table 2. Our method achieves comparable results as other methods.

Table 1. Comparison results on Extended Yale B dataset under the setting of [13].

Methods	SRC [16]	NN	SVM	D K-SVD [5]	DLSI [12]	FDDL [13]	Our method
Accuracy(%)	90.0	61.7	88.8	75.3	89.0	91.9	**94.4**

Table 2. Comparison results on Extended Yale B dataset under the setting of [17].

Methods	K-SVD [10]	D K-SVD [5]	CIDL [17]	Our method
Accuracy(%)	93.10	94.10	95.72	**95.85**

We also evaluate the computation time of our fast low-rank representation (FLRR) method with the traditional low-rank representation problem solved by linear alternating direction method with adaptive penalty (LRR-LADMAP). Both of the FLRR and LRR-LADMP run on the computer with a 3.40GHz Intel(R) Core(TM) CPU. The FLRR computes all the representations of 2,414 images by 4 iterations, and the average processing time per image is 0.0007 second, while the LRR-LADMAP represents all the images by 185 iterations with each image taking about 0.058 second. Furthermore, with the same learned dictionary, the FLRR achieves accuracy of 94.44%, and the LRR-LADMAP achieves accuracy of 94.80%. It means that our FLRR is much faster on getting low-rank representations than traditional low-rank representation with similar recognition accuracy.

4.2 AR Dataset

The AR dataset consists of more than 4,000 color frontal images from 126 people with 26 images per person. The 26 images are taken in two separated sessions, and each session contains 13 images. In each session, 3 images are obscured by scarves, 6 are obscured by sunglasses, and the remaining faces are with different facial expressions or illuminations conditions and regarded as unobscured images. Following the protocol in [2,8], we conduct experiments under three scenarios:

Sunglass: The sunglass give rise to about 20 % occlusions of the faces. In this scenario, 7 unobscured images and 1 randomly selected image with sunglass from session 1 are used for training, and the remaining unobscured images from session 2 and the images with sunglass from both session 1 and session 2 are used for testing.

Scarf: Compared with sunglass, the scarf occludes more regions of faces. The scarf covers around 40 % of the faces. Similar to the sunglass scenario, 7 unobscured images and 1 randomly selected image with scarf from session 1 are used for training, and the remaining unobscured images and images with scarf are used for testing.

Mixed (Sunglass + Scarf): In this case, the images with sunglass and scarf are mixed. The training samples are 7 unobscured images, 1 image with sunglass, and 1 image with scarf from session 1, and the rest images are used for testing.

For each scenario, the experiments are conducted three times and the average results are reported. The comparison results are displayed in Table 3. As shown in the table, our method outperforms other methods.

Table 3. Comparison results of face recognition under different scenarios on AR dataset.

Methods	Sunglass	Scarf	Mixed
LR [2]	84.9	75.8	78.9
SRC [16]	86.8	83.2	79.2
LLC [18]	65.3	59.2	59.9
SLRR [8]	87.3	83.4	82.4
Our method	**89.6**	**88.1**	**87.8**

On this dataset, we also do the experiments of gender classification. As in [13, 19], an unobscured subset of 50 male and 50 female individuals with 14 images per individual is chosen. Images of the first 25 male and 25 female individuals are used for training, and other 25 male and 25 female individuals are used for testing. Following the protocol in previous works, PCA is used to reduce the dimension for each image to 300. The number of class-specific dictionary atoms (#(C-S DA)) are set as 250 and 25, respectively, and the comparison results with other dictionary learning methods are displayed in Table 4. As shown in the figure, our method outperforms other methods.

4.3   Caltech 101 Dataset

The Caltech-101 dataset contains 101 object categories, such as cameras, chairs, flowers, and vehicles. All the categories are with significant variances in shape and cluttered backgrounds. This dataset has 9, 144 images in all, and the image

Table 4. Comparison results of gender classification on AR dataset.

Methods	#(C-S DA)=250	#(C-S DA)=25
SRC [16]	93.0	–
D K-SVD [5]	86.1	–
LC K-SVD [9]	86.8	–
DLSI [12]	94.0	93.7
COPAR [20]	93.4	93.0
JDL [11]	92.6	91.0
FDDL-LC [13]	94.3	93.7
FDDL-GC [13]	94.3	92.1
LDL-LC [19]	95.3	95.0
LDL-GC [19]	94.8	92.4
Our method	**98.3**	**98.0**

Table 5. Comparison results on Caltech 101 dataset.

Methods	#(TrS)=15	#(TrS)=30
LR [2]	58.3	65.7
SRC [16]	64.9	70.7
LLC [18]	65.4	73.4
SLRR [8]	66.1	73.6
CCLR-Sc$^+$SPM [21]	70.9	76.6
Our method	**71.2**	**77.0**

number varies from 31 to 800 per category. Moreover, it is individually added to an extra "background" category, *i.e.*, BACKGROUND_Google.

Same with the setting in [8], the spatial pyramid features are used to test our method with 15 and 30 randomly selected images for training per class (#(TrS)). The comparison results are displayed in Table 5. Our method achieves the highest classification accuracy.

5   Conclusions

We have presented a discriminative orthonormal dictionary learning method for fast low-rank representation in this paper. The discriminative regularization term encourages the class-specific dictionary to well represent the samples from the same class and cannot represent samples from other classes for the discriminative power of the learned dictionary. The orthonormal property ensures the representative power of the dictionary by avoiding the redundancy. With the learned discriminative orthonormal dictionary, a fast low-rank representation method is proposed.

References

1. Li, L., Li, S., Fu, Y.: Discriminative dictionary learning with low-rank regularization for face recognition. In: International Conference on Automatic Face and Gesture Recognition, pp. 1–6. IEEE (2013)
2. Chen, C.-F., Wei, C.-P., Wang, Y.-C.: Low-rank matrix recovery with structural incoherence for robust face recognition. In: IEEE Conference on Computer Vision and Pattern Recognition, pp. 2618–2625 (2012)
3. Ma, L., Wang, C., Xiao, B., Zhou, W.: Sparse representation for face recognition based on discriminative low-rank dictionary learning. In: IEEE Conference on Computer Vision and Pattern Recognition, pp. 2586–2593 (2012)
4. Yang, M., Zhang, L., Feng, X., Zhang, D.: Sparse representation based fisher discrimination dictionary learning for image classification. Int. J. Comput. Vision 109(3), 209–234 (2014)
5. Zhang, Q., Li, B.: Discriminative K-SVD for dictionary learning in face recognition. In: IEEE Conference on Computer Vision and Pattern Recognition, pp. 2691–2698 (2010)
6. Lu, J., Wang, G., Deng, W., Moulin, P.: Simultaneous feature dictionary learning for image set based face recognition. In: European Conference on Computer Vision, pp. 265–280 (2014)
7. Wei, C., Wang, Y.: Undersampled face recognition via robust auxiliary dictionary learning. IEEE Trans. Image Process. 24(6), 1722–1734 (2015)
8. Zhang, Y., Jiang, Z., Davis, L.S.: Learning structured low-rank representations for image classification. In: IEEE Conference on Computer Vision and Pattern Recognition, pp. 676–683 (2013)
9. Jiang, Z., Lin, Z., Davis, L.S.: Label consistent K-SVD: learning a discriminative dictionary for recognition. IEEE Trans. Pattern Anal. Mach. Intell. 35(11), 2651–2664 (2013)
10. Aharon, M., Elad, M., Bruckstein, A.: K-SVD: an algorithm for designing overcomplete dictionaries for sparse representation. IEEE Trans. Signal Process. 54(11), 4311–4322 (2006)
11. Zhou, N., Shen, Y., Peng, J., Fan, J.: Learning inter-related visual dictionary for object recognition. In: IEEE Conference on Computer Vision and Pattern Recognition, pp. 3490–3497 (2012)
12. Ramirez, I., Sprechmann, P., Sapiro, G.: Classification and clustering via dictionary learning with structured incoherence and shared features. In: IEEE Conference on Computer Vision and Pattern Recognition, pp. 3501–3508 (2010)
13. Yang, M., Zhang, D., Feng, X.: Fisher discrimination dictionary learning for sparse representation. In: International Conference on Computer Vision, pp. 543–550. IEEE (2011)
14. Mazhar, R., Gader, P.D.: EK-SVD: optimized dictionary design for sparse representations. In: International Conference on Pattern Recognition, pp. 1–4. IEEE (2008)
15. Shu, X., Porikli, F., Ahuja, N.: Robust orthonormal subspace learning: efficient recovery of corrupted low-rank matrices. In: IEEE Conference on Computer Vision and Pattern Recognition, pp. 3874–3881 (2014)
16. Wright, J., Yang, A.Y., Ganesh, A., Sastry, S.S., Ma, Y.: Robust face recognition via sparse representation. IEEE Trans. Pattern Anal. Mach. Intell. 31(2), 210–227 (2009)

17. Bao, C., Quan, Y., Ji, H.: A convergent incoherent dictionary learning algorithm for sparse coding. In: European Conference on Computer Vision, pp. 302–316 (2014)
18. Wang, J., Yang, J., Yu, K., Lv, F., Huang, T., Gong, Y.: Locality-constrained linear coding for image classification. In: IEEE Conference on Computer Vision and Pattern Recognition, pp. 3360–3367 (2010)
19. Yang, M., Dai, D., Shen, L., Gool, L.V.: Latent dictionary learning for sparse representation based classification. In: IEEE Conference on Computer Vision and Pattern Recognition, pp. 4138–4145 (2014)
20. Kong, S., Wang, D.: A dictionary learning approach for classification: separating the particularity and the commonality. In: European Conference on Computer Vision, pp. 186–199 (2012)
21. Zhang, C., Liu, J., Liang, C., Xue, Z., Pang, J., Huang, Q.: Image classification by non-negative sparse coding, correlation constrained low-rank and sparse decomposition. Comput. Vis. Image Underst. **123**, 14–22 (2014)

Supervised Topic Classification for Modeling a Hierarchical Conference Structure

Mikhail Kuznetsov[1]([⊠]), Marianne Clausel[2], Massih-Reza Amini[3],
Eric Gaussier[3], and Vadim Strijov[1]

[1] Moscow Institute of Physics and Technology,
Institutskiy Lane 9, Dolgoprudny, Moscow 141700, Russia
{mikhail.kuznecov,strijov}@phystech.edu
[2] Laboratoire Jean Kuntzmann, Université de Grenoble Alpes,
CNRS, 38041 Grenoble Cedex 9, France
marianne.clausel@imag.fr
[3] Laboratoire d'Informatique de Grenoble, Université de Grenoble Alpes,
CNRS, 38041 Grenoble Cedex 9, France
{Massih-Reza.Amini,eric.gaussier}@imag.fr

Abstract. In this paper we investigate the problem of supervised latent modeling for extracting topic hierarchies from data. The supervised part is given in the form of expert information over document-topic correspondence. To exploit the expert information we use a regularization term that penalizes the difference between a predicted and an expert-given model. We hence add the regularization term to the log-likelihood function and use a stochastic EM based algorithm for parameter estimation. The proposed method is used to construct a topic hierarchy over the proceedings of the European Conference on Operational Research and helps to automatize the abstract submission system.

Keywords: Hierarchical topic model · Labeled classification · Probabilistic latent semantic analysis · EM approach

1 Introduction

Probabilistic topic models are generally unsupervised generative models that describe document content in large document collections. These models assume that each document is associated with a set of hidden variables, called topics, that indicate how the words within the document are generated. Formally, a topic is a probability distribution over terms in a vocabulary. The two most popular topic models are the Probabilistic Latent Semantic Indexing (PLSI) [6] and the latent Dirichlet allocation (LDA) model [2] and their variants. The LDA model consists of two types of probability distributions: (*a*) distributions of topics over documents and (*b*) distributions of words over topics. After estimating the model parameters over a training corpus, the obtained distributions of words over topics can then be used to infer per-document topic distributions on unseen documents. LDA has found applications in many areas ranging from document

© Springer International Publishing Switzerland 2015
S. Arik et al. (Eds.): ICONIP 2015, Part I, LNCS 9489, pp. 90–97, 2015.
DOI: 10.1007/978-3-319-26532-2_11

clustering, text categorization, ad-hoc information retrieval, to signal analyzes. Several attempts have been made to extent PLSI and LDA to unsupervised hierarchical topic modeling. In [5], Dirichlet processes are used to model different levels of an hierarchy, while in [3] an extension of the PLSI model is proposed by introducing additional probabilities corresponding to different levels of the hierarchy.

In this paper, we address the problem of hierarchical topic modeling using an expert information over the document-topic correspondence in the form of a labeled document collection with a predefined hierarchical-structured topics. The problem is hence to predict a topic model for a new document collection using such past labeled information.

The application that we consider is the construction of an hierarchical topic model for the "European Conference on Operational Research" (EURO) containing over 3000 abstracts. The structure of the conference papers is shown in Fig. 1. At the upper level there are 26 main areas, each of which contains about 10 streams. Each stream then contains about 10 sessions, and each session is formed by

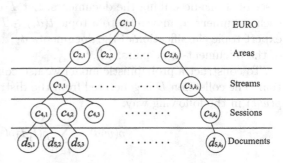

Fig. 1. Hierarchical structure of the conference

four abstracts. The main areas correspond to the broad topics of the operational research field like *Non-smooth optimization, timetabling, logistics,* etc. Every year the program committee, constituted by groups of experts, constructs by hand such an hierarchy for the submitted papers [7]. Each group is responsible for the organization of a stream or a set of streams. After the abstract submission deadline each group of experts starts to fill a stream with unassigned abstracts and to form sessions within a stream. The practical goal of our research is to construct an efficient structure from the supervised expert information of the previous years using the topic modeling methodology. For that, we consider the additive regularization of topic models (ARTM) [9]. In general, this method finds topic-document and word-topic probabilities by optimizing a log-likelihood quality measure with an additional regularization term. Here we propose a regularizer term that penalizes the difference between the predicted and the expert-given topic models.

Compared to [8], where the prior probabilities are modified with respect to the projections of document-topic vectors on the set of identified topics, here we propose a unified formalism to measure the distance between the hierarchy trees by introducing a set of hyperparameters that describes the hierarchy and summarizes penalizations on different hierarchy levels. The optimization of the regularized likelihood is then carried out using a stochastic version of the Expectation-Maximization algorithm [6, 10]. The algorithm has a modified

M-step that takes into account the regularization term and samples a current topic from the conditional distribution on a given word-document pair.

The structure of the paper is as follows. In Sect. 2, we present our framework and a Bayesian interpretation of the proposed supervised topic model. Section 3 presents its hierarchy extension and empirical results are shown in Sect. 4. Finally, in Sect. 5 we discuss the outcomes of this study and give some pointers to further research.

2   Supervised Classification, Flat Case

Let D denote a collection of documents, $d_i \in D$, and W denote a vocabulary, a set of terms describing the documents. Let T denote a set of topics such that each document d_i may refer to a topic $t(d_i) \in T$. Let $t_1, ..., t_n$ denote an initial expert topic classification of the documents $d_1, ..., d_n$. The given sample consists of the document-topic pairs, $\{d_i, t_i\}_{i=1}^n$.

To construct a probabilistic model we use conditional independence assumption. The collection D is generated from the distributions $\theta_{td} = p(t|d)$ and $\phi_{wt} = p(w|t)$ in the following way:

$$p(w|d) = \sum_{t \in T} p(t|d)p(w|t).$$

To estimate the probabilities $(\theta_{td})_{t \in T, d \in D}$ and $(\phi_{wt})_{w \in W, t \in T}$, we consider the PLSI approach [6], where the optimization problem consists in maximizing the log-likelihood $L(\boldsymbol{\Phi}, \boldsymbol{\Theta})$ under non-negativity and normalization conditions:

$$\boldsymbol{\Phi}^*, \boldsymbol{\Theta}^* = \operatorname*{argmax}_{\boldsymbol{\Phi}, \boldsymbol{\Theta}} L(\boldsymbol{\Phi}, \boldsymbol{\Theta}) = \sum_{d \in D} \sum_{w \in d} n_{dw} \ln \sum_{t \in T} \phi_{wt} \theta_{td},$$
$$\text{u.c. } \phi_{wt} \geqslant 0, \quad \theta_{td} \geqslant 0, \text{ and } \sum_{w \in W} \phi_{wt} = 1, \quad \sum_{t \in T} \theta_{td} = 1. \tag{1}$$

The PLSI model (1) does not take into account the initial topic classification $t_1, ..., t_n$, we tackle the problem by introducing the expert-given topic labels using a regularization term $R(\mathbf{t}, \hat{\mathbf{t}})$ that measures the similarity between the predicted and the expert-given topic vectors, $\mathbf{t}$ and $\hat{\mathbf{t}}$:

$$\boldsymbol{\Phi}^*, \boldsymbol{\Theta}^* = \operatorname*{argmax}_{\boldsymbol{\Phi}, \boldsymbol{\Theta}} L(\boldsymbol{\Phi}, \boldsymbol{\Theta}) + \lambda R(\mathbf{t}, \hat{\mathbf{t}}),$$
$$\text{u.c. } \phi_{wt} \geqslant 0, \quad \theta_{td} \geqslant 0, \text{ and } \sum_{w \in W} \phi_{wt} = 1, \quad \sum_{t \in T} \theta_{td} = 1. \tag{2}$$

where is λ the regularization parameter.

Bayesian Interpretation of the Regularized PLSI. As stated in [4], a penalized approach can be interpreted within the Bayesian framework. According to such an interpretation, the penalized likelihood function corresponds to the *a posteriori* density whereas the penalty is the density of the prior. The solution of the maximization of the penalized likelihood of the model is then a maximum *a posteriori* estimate of the parameters of interest. In our setting, adding a regularization to the PLSI model means that we are setting the following prior for the latent variables (θ, ϕ)

$$\pi(\theta, \phi) = C\, exp\left(\lambda R(\phi, \theta)\right) \tag{3}$$

where $C > 0$ is a normalizing constant.

Our corpus can then be assumed to be generated as follows :

- Step 1: Generate the whole set of the topic and of the matrix word–topic $(\theta, \phi) \sim \pi$ where π is the distribution defined in 3.
- Step 2: for each document d and each word of the document
  - Draw the nth topic $t_n^w \sim mult(\theta_{td})$.
  - Draw the nth word w_n with probability ϕ_{w_n, t_n^w}.

Labeled Classification. Let $\mathbf{Z} = \|z_{td}\|$ be a document-topic correspondence matrix of size $D \times T$ such that

$$z_{td} = \mathbb{1}_{\hat{t}_d = t}.$$

where $\mathbb{1}_\pi$ is the indicator function, equal to 1 if the predicate π holds and 0 otherwise. We define similarity $R(\mathbf{t}, \hat{\mathbf{t}})$ as a matrix norm of difference between matrices Θ and $\mathbf{Z}$:

$$R(\mathbf{t}, \hat{\mathbf{t}}) = -\|\Theta - \mathbf{Z}\|_1.$$

This form of regularization leads us to the following optimization problem:

$$\Phi^*, \Theta^* = \operatorname*{argmax}_{\Phi, \Theta} \sum_{d \in D} \sum_{w \in d} n_{dw} \ln \sum_{t \in T} \phi_{wt}\theta_{td} + \lambda \left(\sum_{d \in D} \sum_{t \in T} \theta_{td}(2z_{td} - 1)\right),$$

$$\phi_{wt} \geqslant 0, \quad \theta_{td} \geqslant 0, \quad \sum_{w \in W} \phi_{wt} = 1, \quad \sum_{t \in T} \theta_{td} = 1. \tag{4}$$

Parameter Optimization: EM Approach. To solve the optimization problem (4) we use the Expectation-Maximization algorithm. To derive the explicit expectation and maximization formulas we use theorem 1 from [9] that gives properties of the local optimum of the general expression of (Eq. 2). Following this result if $R(t, \hat{t})$ is continuously differentiable then at the local maximum of R we have:

$$\phi_{wt} \propto \left(n_{wt} + \phi_{wt}\frac{\lambda \partial R}{\partial \phi_{wt}}\right)_+ , \quad \theta_{td} \propto \left(n_{td} + \theta_{td}\frac{\lambda \partial R}{\partial \theta_{td}}\right)_+ . \tag{5}$$

Note that in our problem the function R depends only of θ_{td} variables, therefore we will use only a second equation for θ.

For the problem (4) we hence obtain the following formula for the M-step:

$$\theta_{td} = \frac{\eta_{td}}{\sum\limits_{t \in T} \eta_{td}}, \quad \eta_{td} = \left[\sum_{w \in d} n_{dw} \frac{\phi_{wt}\theta_{td}}{\sum\limits_{t \in T} \phi_{wt}\theta_{td}} + \lambda\theta_{td}\left(2z_{td} - 1\right)\right]_+ . \tag{6}$$

Stochastic EM. To speed up the proposed EM algorithm we rather use its stochastic version that is similar to the Gibbs sampling method for LDA [2]. The
approach consists in sampling a topic t from the estimated distribution $p(t|d, w)$,
where the distribution of a topic t given w, d is given by a formula

$$p(t|d, w) \propto \left(\frac{\hat{n}_{wt}}{\hat{n}_t} \frac{\hat{n}_{dt} + \lambda \hat{n}_{dt}(2z_{td} - 1)}{n_d + \lambda \sum_{t \in T} \hat{n}_{dt}(2z_{td} - 1)} \right)_+,$$

where

$$\hat{n}_{dt} = \sum_{w \in d} n_{dw} \frac{\phi_{wt}\theta_{td}}{\sum_{t \in T} \phi_{wt}\theta_{td}}, \quad \hat{n}_{wt} = \sum_{d \in D} n_{dw} \frac{\phi_{wt}\theta_{td}}{\sum_{t \in T} \phi_{wt}\theta_{td}}, \quad \hat{n}_t = \sum_{w \in d} \hat{n}_{wt}.$$

3   Topics Hierarchy

We extend the model by taking into account the expert-given hierarchy defined
on the set of topics. To model the hierarchical structure we introduce the following notations. Let us denote by $T = T = T_0 \sqcup ... \sqcup T_L$ a set of topics, or
a set of vertices in a hierarchical tree, where the sets $T_0, ..., .T_L$ denote disjoint
sets of topics at different levels of hierarchy. For further reading we consider a
two-level hierarchical structure. However, the proposed method can be used for
any number of levels.

For further convenience we introduce parent $p(t)$ and children $s(t)$ operators
defined as follows:

$$p(t) \in T_{l-1} \quad \text{for} \quad t \in T_l, \quad l = 1, ..., L,$$

$$s(t) \subset T_{l+1} \quad \text{for} \quad t \in T_l, \quad l = 0, ..., L - 1.$$

To define the loss function $R(t, \hat{t})$ between topics we propose to measure a summary loss over the hierarchy levels:

$$R(t, \hat{t}) = \sum_{l=0}^{L-1} r(p^l(t), p^l(\hat{t})).$$

Here the vertex t belongs to the lowest level of hierarchy, $t \in T_L$, and $p^l(t)$ is the
l-th predecessor of the vertex t.

To measure the value of a single loss $r(t_i, \hat{t}_i)$ on a document d_i we expand
Eq. (4) to the different hierarchy levels:

$$r(t, \hat{t}) = |\mathbb{1}_{\hat{t}=t} - \theta'_{td}|,$$

where θ'_{td} is defined for an arbitrary hierarchy level as follows:

$$\theta'_{td} = \begin{cases} \theta_{td}, & t \in T_l, \\ \frac{1}{\#s(t)} \sum_{s \in s(t)} \theta'_{sd}, & \text{otherwise.} \end{cases}$$

According to the introduced hierarchy addition we obtain the following modification of the M-step formula (6):

$$\eta_{td} = \left[\sum_{w \in d} n_{dw} \frac{\phi_{wt}\theta_{td}}{\sum\limits_{t \in T} \phi_{wt}\theta_{td}} + \lambda_1 \theta_{td} \left(2z_{td} - 1\right) + \lambda_2 \theta'_{p(t)d} \left(2z_{p(t)d} - 1\right) \right]_+ . \quad (7)$$

4 Empirical Results

We use the proposed method to construct a topic model for the European Conference on Operational Research. We use the collection of abstracts for the 2012 year. Each abstract contains less than 600 symbols, the collection contains 1342 abstracts, and vocabulary contains 1675 words after preprocessing. The preprocessing stage includes removing stop words and lemmatization. Together with the collection we used an initial expert-given conference structure as described in introduction [1].

To show the hierarchical results we first need to choose the hyperparameters λ_1 and λ_2 (Eq. 7). To do this we perform the following steps.

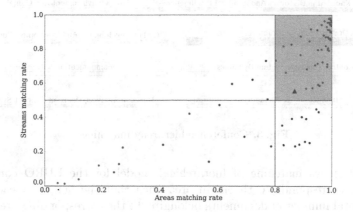

Fig. 2. Hierarchical model matching rates for different regularization values

1. Estimate model parameters for different sets of parameters λ_1, λ_2. We took about 100 different parameter sets from the range $\lambda_1, \lambda_2 \in [0, 1]$.
2. For each set of parameters we obtain three values measuring the quality of a hierarchical model: (1) the normalized number of documents matched with the expert model within the areas $na \in [0, 1]$, (2) the same for the streams, $ns \in [0, 1]$, (3) the value of perplexity.
3. We chose those regularization values λ_1, λ_2 that minimize the perplexity for the values $na > 0.8$, $ns > 0.5$.

Figure 2 illustrates the mentioned steps. x- and y-axis correspond to the values na and ns, respectively. Each point corresponds to the different set of parameters λ_1, λ_2. The color of each point indicates the values of perplexity: the darker the color, the higher the perplexity. The optimal point (of minimum perplexity with $na > 0.8$ and $ns > 0.5$) indicated by the triangle. The regularization values for this point are $\lambda_1 = 0.15$, $\lambda_2 = 0.2$.

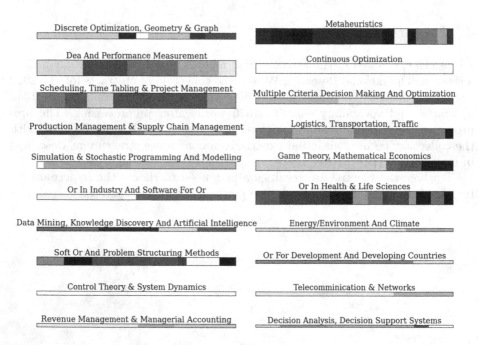

Fig. 3. Conference hierarchy matching

Figure 3 shows matching of hierarchical model for the EURO conference. Each block corresponds to the main area such that the height of each block indicates total number of documents belonging to the corresponding area due to the expert-given model. Each block consists of the subblocks corresponding to the streams; the length of the subblock indicates the size of the stream. The color of each subblock indicates rate of documents ns matched with the expert-given model: the more white is subblock, the better is the matching (ns is closer to 1). According to our method we can specify stable and non-stable areas. We see that good matched areas (mostly white-colored) are "Continuous optimization", "Control theory" and "Revenue management", whereas bad-matched are, e.g., "Metaheuristics" and "OR in health".

5   Conclusion

We proposed a supervised hierarchical topic model, where the expert knowledge is encompassed into a regularization term measuring the distance between the predicted and the expert-given topic models. The optimization of the regularized likelihood is then carried out using a stochastic version of the Expectation-Maximization algorithm where the modified M-step takes into account the regularization term and samples a current topic from the conditional distribution on a given word-document pair. Our experiments on the EURO proceedings showed that the proposed topic model is able to find expert-given topics, with high perplexity.

References

1. EURO conference abstracts and data. http://sourceforge.net/p/mlalgorithms/code/HEAD/tree/EURO_data/. Accessed 14 May 2015
2. Blei, D.M., Ng, A.Y., Jordan, M.I.: Latent Dirichlet allocation. J. Mach. Learn. Res. **3**, 993–1022 (2003)
3. Gaussier, É., Goutte, C., Popat, K., Chen, F.: A hierarchical model for clustering and categorising documents. In: Crestani, F., Girolami, M., van Rijsbergen, C.J. (eds.) ECIR 2002. LNCS, vol. 2291, p. 229. Springer, Heidelberg (2002)
4. Good, I.J., Gaskins, R.A.: Nonparametric roughness penalties for probability densities. Biometrika **58**(2), 255–277 (1971)
5. Griffiths, T.L., Jordan, M.I., Tenenbaum, J.B., Blei, D.M.: Hierarchical topic models and the nested Chinese restaurant process. In: Thrun, S., Saul, L.K., Schölkopf, B. (eds.) Advances in Neural Information Processing Systems 16, pp. 17–24 (2004)
6. Hofmann, T.: Probabilistic latent semantic indexing. In: Proceedings of the 22nd Annual International ACM SIGIR Conference on Research and Development in Information Retrieval, pp. 50–57. ACM (1999)
7. Kuzmin, A.A., Strijov, V.V.: Validation of the thematic models for document collections. Inf. Technol. **4**, 16–20 (2013)
8. Ramage, D., Hall, D., Nallapati, R., Manning, C.D.: Labeled LDA: a supervised topic model for credit attribution in multi-labeled corpora. In: Proceedings of the 2009 Conference on Empirical Methods in Natural Language Processing: Volume 1, pp. 248–256. Association for Computational Linguistics (2009)
9. Vorontsov, K.V., Potapenko, A.A.: Additive regularization of topic models. Mach. Learn. J., Special Issue "Data Analysis and Intelligent Optimization" **101**, 303–323 (2015)
10. Wallach, H.M.: Topic modeling: beyond bag-of-words. In: Proceedings of the 23rd International Conference on Machine Learning, pp. 977–984. ACM (2006)

A Framework for Online Inter-subjects Classification in Endogenous Brain-Computer Interfaces

Sami Dalhoumi[1(✉)], Gérard Dray[1], Jacky Montmain[1],
and Stéphane Perrey[2]

[1] Laboratoire d'Informatique et d'Ingénierie de Production (LGI2P),
Ecole des Mines d'Alès, Parc Scientifique G. Besse, 30035 Nîmes, France
{sami.dalhoumi, gerard.dray,
jacky.montmain}@mines-ales.fr
[2] Movement to Health (M2H), Montpellier University, Euromov,
700 Avenue du Pic Saint-Loup, 34090 Montpellier, France
stephane.perrey@univ-montpl.fr

Abstract. Inter-subjects classification and online adaptation techniques have been actively explored in the brain-computer interfaces (BCIs) research community during the last years. However, few works tried to conceive classification models that take advantage of both techniques. In this paper we propose an online inter-subjects classification framework for endogenous BCIs. Inter-subjects classification is performed using a weighted average ensemble in which base classifiers are learned using data recorded from different subjects and weighted according to their accuracies in classifying brain signals of current BCI user. Online adaptation is performed by updating base classifiers' weights in a semi-supervised way based on ensemble predictions reinforced by interaction error-related potentials (iErrPs). The effectiveness of our approach is demonstrated using two electroencephalography (EEG) data sets and a previously proposed procedure for simulating interaction error potentials.

Keywords: Brain-computer interfaces · Inter-subjects classification · Online adaptation · Weighted average ensembles

1 Introduction

A brain-computer interface (BCI) is a communication and control technology that allows translating brain's electrical or hemodynamic activity patterns into commands for an external device [1]. This technology was originally meant to allow patients with severe neuromuscular disabilities to autonomously interact with their environment. Depending on the modality of interaction, BCIs can be classified as either exogenous or endogenous [2]. Exogenous BCIs rely on brain activity patterns that are elicited spontaneously in response to external stimuli such as visual evoked potentials (VEPs), while endogenous BCIs are based on the voluntary induction of different brain states by the user such as sensorimotor rhythms-based BCIs. Endogenous BCIs can offer a natural way of interaction for the user but they are difficult to set-up because

© Springer International Publishing Switzerland 2015
S. Arik et al. (Eds.): ICONIP 2015, Part I, LNCS 9489, pp. 98–107, 2015.
DOI: 10.1007/978-3-319-26532-2_12

self-regulation of brain rhythms is not a straightforward task. For this reason, a long calibration phase is needful for user-system co-adaptation before every use of the BCI [3]. During this phase, the user interacts with the BCI in a cue-based mode which allows him to learn self-regulating his brain rhythms and the system to create a "robust" classification model. The accuracy of the system depends on the capacity of the classification model to decode brain activity patterns of the user during a feedback phase (self-paced interaction mode).

In order to bring endogenous BCIs out of the lab, many research groups have focused on conceiving new machine learning approaches that allow reducing calibration time without decreasing classification accuracy of the system. Among these approaches, inter-subjects classification has been actively explored during the last years [3–5]. It consists of incorporating labeled data recorded from different BCI users in the learning process of current BCI user. When performed correctly, inter-subjects classification allows capturing information that generalize across users and extend to new users. One way to do that is to use a weighted average ensemble technique in which base classifiers are learned using data from different BCI users and weighted according to their accuracy in classifying signals recorded from current BCI user [4]. In the absence of true class labels during feedback phase of current BCI user, these weights can be estimated in two ways: statically using a small calibration set or dynamically by recalculating these weights for each incoming sample based on its position in the feature space [4, 5]. The first approach may perform poorly because brain activity patterns of the BCI user vary between calibration and online phases and during online phase (non-stationarity). The second approach may not perform well because it does not take into consideration the stochastic dependence between time-contingent feature vectors.

In a preliminary work [6], we found that a static classifiers weighting approach using a small calibration set outperforms dynamic classifiers weighting approaches and we showed that online adaptation of base classifiers' weights using ensemble predictions during feedback phase may increase classification accuracy in comparison to the static approach. However, the update coefficient used for adjusting adaptation speed was subject-dependent which presented a limitation to the proposed approach. In this work, we propose to use interaction error-related potentials (iErrPs) as an additional source of information to reduce uncertainty about ensemble predictions during feedback phase. iErrPs are a type of event-related potentials that occur immediately after the user perceives that the feedback provided by the BCI is in contradiction to his intent [7]. The physiological background of iErrPs has been well established [7] and they have been successfully used to improve BCIs accuracy [8, 9]. However, iErrPs are also subject to a degree of uncertainty because their detection is not perfect. Ferrez and Del R. Millán [7] reported an average recognition rate of iErrPs with 16.5 % of false positives (i.e., the correct predictions are considered erroneous) and 20.8 % of false negatives (i.e., the erroneous predictions are considered correct).

This paper is organized as follows: in Sect. 2, we describe our adaptive weighted average ensemble method. In Sect. 3, we present material used to evaluate this method and the experimental results. Section 4 concludes the paper and gives future directions of this work.

2 Methods

In this section, we describe different steps of our adaptive weighted average ensemble method for binary classification tasks. Base classifiers trained on brain signals recorded from different subjects are weighted according to their accuracies in classifying a small calibration set from current BCI user. These weights are updated during feedback phase using ensemble predictions reinforced by iErrPs. In the absence of an iErrP, the label predicted by the ensemble is considered correct and base classifiers' weights are updated based on their disagreement with the ensemble. When an iErrP is detected, the prediction of the ensemble is considered to be wrong and base classifiers' weights are updated using the opposite label.

2.1 Base Classifiers' Weights Initialization

Let $\{h^1, h^2, \ldots, h^K\}$ be K classification models learned using data from different BCI users (many classification models may be learned using data recorded from the same user and preprocessed in different ways). For each incoming feature vector x and each class label y, the classifier $h^k, k = 1 \ldots K$ outputs the value $h_y^k(x) \in [0 1]$ which is an estimation of the posterior probability $p(y/x)$. Given a small calibration set $L = \{(x_t, y_t), x_t \in \mathbb{R}^d, y_t \in \{0, 1\}, t = 1 \ldots T\}$ recorded from current BCI user, each classifier is assigned a weight w^k inversely proportional to its error in classifying this labeled set:

$$w^k = \max\left(0, MSE^r - MSE^k\right), k = 1..K \tag{1}$$

where, MSE^r is the mean squared error of a random classifier and MSE^k is the mean squared error of the classifier h^k given below.

$$MSE^r = \sum_y p(y).(1 - p(y))^2 \tag{2}$$

$$MSE^k = \frac{1}{T} \cdot \sum_{t=1}^{T} \left(1 - h_{y_t}^k(x_t)\right)^2 \tag{3}$$

For binary classification with equal class priors, $MSE^r = 0.25$.

This weighting scheme allows removing classifiers performing less or equal than a random classifier from the ensemble and assigning weights to the rest of classifiers inversely proportional to their error in classifying calibration data.

2.2 Base Classifiers' Weights Adaptation Using Ensemble Predictions

Given a new labeled sample (x_{t+1}, y_{t+1}), the mean squared errors of base classifiers up to the time step $(t + 1)$ can be updated in the following way:

$$MSE^k(t+1) = \frac{1}{t+1} \cdot \left[t.MSE^k(t) + \left(1 - h^k_{y_{t+1}}(x_{t+1})\right)^2 \right], k = 1...K \qquad (4)$$

where, $MSE^k(t)$ is the mean squared error of the classifier h^k up to the time step t.
Base classifiers' weights can then be updated using the adaptive version of Eq. (1):

$$w^k(t+1) = \max\left(0, MSE^r - MSE^k(t+1)\right), k = 1..K \qquad (5)$$

In order to take into consideration different types of data shift, we add an update coefficient $UC \in [01]$ to Eq. (4) that becomes:

$$MSE^k(t+1) = \frac{1}{(1 - UC).t + UC} \times$$

$$\left[(1 - UC).t.MSE^k(t) + UC.\left(1 - h^k_{y_{t+1}}(x_{t+1})\right)^2\right], k = 1...K \qquad (6)$$

For $UC = 0$, there is no update, for $UC = 1$, only the new data sample is used for calculating error and when $UC = 0.5$, we retrieve exactly the update Eq. (4).

In self-paced interaction mode, the true class labels are unknown for the classification model. One way to alleviate this problem is to use ensemble predictions for online adaptation. For each incoming feature vector, the label predicted by the ensemble is considered to be the true class label and each base classifier's weight is updated according to its disagreement with the ensemble. So, formula (6) becomes:

$$MSE^k(t+1) = \frac{1}{(1 - UC).t + UC} \times$$

$$\left[(1 - UC).t.MSE^k(t) + UC.\left(1 - h^k_{\tilde{y}_{t+1}}(x_{t+1})\right)^2\right], k = 1...K \qquad (7)$$

Where $\tilde{y}_{t+1}$ is the label predicted by the ensemble:

$$\tilde{y}_{t+1} = argmax_y \left(\sum_{k=1}^{K} w^k(t).h^k_y(x_{t+1}) \right) \qquad (8)$$

As ensemble's decisions are subject to a high degree of uncertainty, using them for adaptation may lead to error accumulation and consequently degrades the accuracy of the BCI. Thus, we should use and additional information to minimize uncertainty. In BCIs, such information could come from interaction error-related potentials.

2.3 Base Classifiers' Weights Adaptation Using Ensemble Predictions Reinforced by Interaction Error-Related Potentials

Let $E \in \{0, 1\}$ be the true absence or presence of an iErrP following the output of the BCI. $E = 0$, when the decision of the ensemble $\tilde{y}_{t+1}$ corresponds to the intent of the user y_{t+1} and $E = 1$, in the opposite case. The iErrPs classifier outputs a value $\tilde{E} \in \{0, 1\}$ which is a prediction of E. The predicted value $\tilde{E}$ may or may not correspond to the real value E depending on the accuracy of the iErrPs classifier. This iErrPs classifier can be used to assess the reliability of the predicted labels as follows:

$$
MSE^k(t+1) = \frac{1}{(1 - UC).t + UC} \times
$$
$$
\left[(1 - UC).t.MSE^k(t) + UC.\left((1 - \tilde{E}) - h^k_{\tilde{y}_{t+1}}(x_{t+1}) \right)^2 \right], k = 1 \ldots K
$$

(9)

When $\tilde{E} = 0$, the predicted label is considered correct and the update is the same as in Eq. (7). When $\tilde{E} = 1$, the opposite class label is used for update because $\left(h^k_{\tilde{y}_{t+1}}(x_{t+1}) \right)^2 = \left(1 - h^k_{(1-\tilde{y}_{t+1})}(x_{t+1}) \right)^2$.

3 Experiments

In this section we evaluate our adaptive ensemble approach using two EEG data sets and the procedure for simulating iErrPs used in [8, 9].

3.1 EEG Data Sets

Data set 2A in BCI Competition IV. This data set comprises electroencephalography (EEG) signals recorded from 9 subjects using 22 Ag/AgCl electrodes at 250 Hz sampling rate [10]. Subjects performed left hand, right hand, foot and tongue motor imagery tasks. For the purpose of this study, only EEG signals corresponding to the left hand and right hand motor imagery tasks were used. For each subject, two sessions on different days, each of which comprises 72 trials of duration 7 s, were collected. At the beginning of each trial, a fixating point appeared on a computer screen. After two seconds, a cue appeared informing the subject which motor imagery task to perform until the cue disappeared.

EEG measurements were band-pass filtered using a 5th order Butterworth filter in the frequency bands of 4 Hz width ranging from 8 Hz to 30 Hz with step size of 2 Hz and an additional wide band from 8 Hz to 30 Hz. Time segments 3–5 s after the beginning of each trial were extracted. The common spatial pattern (CSP) algorithm and the logarithmic variance features are used for spatial filtering and feature extraction (the three most discriminative CSP filters for each class are used) [11].

Two Class Motor Imagery Data Set from BNCI Horizon 2020 Project. This data set was provided by the Graz group [12]. 14 subjects performed sustained kinesthetic motor imagery of the right hand and feet. 5 subjects had previously performed BCI experiments and 9 subjects were naïve to the task. Each subject performed a training phase composed of 50 trials per class and a validation phase composed of 30 trials per class. EEG signals were recorded using 15 Ag/AgCl electrodes at 512 Hz sampling rate. Time segments of length 3 s, starting at 3 s after the beginning of each trial were preprocessed in the same way as in the previous data set. CSP algorithm and logarithmic variance features were used to extract relevant features from this data set.

3.2 Procedure for Simulating IErrPs

Llera et al. [8] proposed a simple procedure for simulating iErrPs that allows understanding the relation between the accuracy of the iErrPs classifier and the accuracy of the task classifier. Below we describe it in case of our adaptive ensemble method.

Let α_1 and α_2 be the false positive and false negative rates of the iErrPs classifier, respectively. Given the output of the ensemble classifier $\tilde{y}_t$ and the true class label y_t at time step t, the procedure is performed as follows:

- If $\tilde{y}_t = y_t$, we draw $\tilde{E} = 1$ with probability α_1 and $\tilde{E} = 0$ with probability $1 - \alpha_1$ and apply Eqs. (5) and (9).
- If $\tilde{y}_t \neq y_t$, we draw $\tilde{E} = 1$ with probability $1 - \alpha_2$ and $\tilde{E} = 0$ with probability α_2 and apply Eqs. (5) and (9).

3.3 Results

Evaluation was performed offline using leave-one-subject-out cross-validation. In each step, training data from N-1 subjects (from now called source subjects) were used for learning spatial filters and base classifiers, the calibration set extracted from training data of the Nth subject (from now called target subject) was used to initialize base classifiers' weights and test data of the same subject was used for evaluation (N = 9 in the first data set and N = 14 in the second data set). During training phase, CSP filters and corresponding linear discriminant analysis (LDA) classifiers are learned using EEG signals recorded from each subject and filtered in different frequency bands, resulting in 88 base classifiers in the first data set and 143 base classifiers in the second data set. Calibration set of the target subject is filtered in different frequency bands and projected into the subspaces spanned by the previously learned CSP filter banks. The initial mean-squared error of each base classifier is calculated using the corresponding projection. For evaluation, each trial in the test set of target subject is filtered in different frequency bands and projected into the subspaces spanned by the CSP filter banks.

Figure 1 illustrates the average classification accuracies of a static accuracy-weighted ensemble (AWE) learned using data from source subjects and a baseline LDA classifier learned using only calibration data of target subject filtered in the 8–30 Hz frequency band (traditional approach) for the first data set. As we can see, learning from other users allows increasing classification accuracy when the size of calibration set is small because the subject-independent information captured from large data sets is more robust than subject-specific information learned from a small data set. As the size of calibration set increases, the accuracy of the baseline classifier increases while the accuracy of the inter-subject classification model remains relatively constant. This shows that, in contrary to the traditional classification approach, the performance of the inter-subjects classification approach is not much dependent on the size of calibration set.

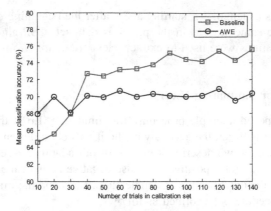

Fig. 1. Average classification accuracy of the standard classification approach (baseline) and the static inter-subjects classification approach (AWE) for different sizes of calibration set in the first data set

In order to assess whether online adaptation of base classifiers' weights allows increasing performance of our inter-subjects classification approach, we performed a comparison between the static accuracy-weighted ensemble and the adaptive accuracy-weighted ensemble. To do so, we evaluated three scenarios for online adaptation of base classifiers' weights:

- Guided: adaptation is performed using only ensemble predictions.
- Realistic iErrPs detection: adaptation is performed using ensemble predictions reinforced by an iErrPs classifier with false positive rate α_1 of 16.5 % and false negative rate α_2 of 20.8 % as found in [7].
- Perfect iErrPs detection: adaptation is performed using ensemble's predictions reinforced by a perfect iErrPs classifier ($\alpha_1 = \alpha_2 = 0$).

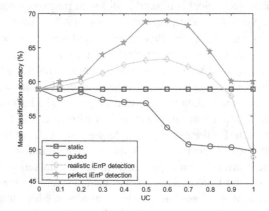

Fig. 2. Average classification accuracies of the static weighted average ensemble and different scenarios of the adaptive weighted average ensemble when the size of calibration set is equal to 10 trials in the second data set

Figure 2 illustrates the comparative results for the second data set when the size of calibration set is equal to 10 trials. The x-axis corresponds to different values of the update coefficient UC and the y-axis to the average classification accuracy over all subjects. The results of the adaptive ensemble method using realistic iErrPs classifier are the average over 100 tests for each subject and each value of UC. This figure shows that using iErrPs for assessing the reliability of the ensemble predictions allows preventing error accumulation and increases classification accuracy especially for values of the update coefficient between 0.5 and 0.7.

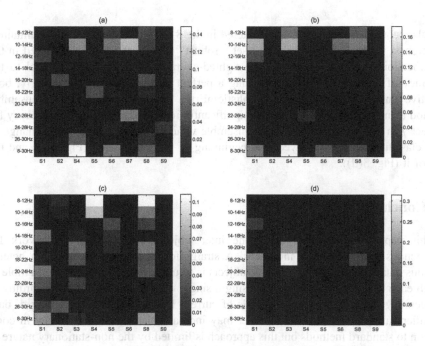

Fig. 3. The evolution of base classifiers' weights during the test session for two different subjects in data set 1. (a) and (b) correspond to base classifiers' weights at the beginning and the end of test session of subject 3. (c) and (d) correspond to base classifiers' weights at the beginning and the end of test session of subject 7

We performed the same comparison for the first data set with predefined update coefficient $UC = 0.5$. Table 1 shows the accuracies of the static ensemble method and the three adaptive methods for different subjects. For most of subjects, the adaptive ensemble method using a realistic iErrPs classifier allows increasing classification accuracy compared to the static method which shows its applicability in online settings.

For further investigation of the behavior of our adaptive ensemble method, Fig. 3 shows an illustration of the evolution of base classifiers' weights between the beginning and the end of test session for two different cases in data set 1. Figure 3(a) and (b) show the normalized weights of the base classifiers for subject 3 at the beginning and the end of test session, respectively. For this subject, the base classifier learned using EEG

Table 1. Classification accuracy of the static weighted average ensemble and different scenarios of the adaptive weighted average ensemble when the size of calibration set is equal to 10 trials in the first data set and the update coefficient UC is equal to 0.5

	S1	S2	S3	S4	S5	S6	S7	S8	S9	Mean	Std.
Static	81.2	51.4	95.8	65.0	48.6	58.3	64.6	89.6	56.3	67.9	17.0
Guided	75.7	50.0	95.8	49.7	50.7	50.7	50.0	90.3	68.1	64.5	18.7
Realistic iErrP	81.9	60.2	94.0	55.5	51.7	54.7	73.7	90.2	71.2	70.3	15.9
Perfect iErrP	82.6	59.7	95.8	60.8	52.7	57.6	83.3	90.3	75.7	73.2	15.8

signals recorded from subject 4 and filtered in the 8–30 Hz frequency band maintained the highest weight during all the test set ("robust" classifier) which is reflected in the classification accuracy of the static weighted-average ensemble that is equal to the accuracy of the adaptive ensemble using a perfect iErrPs classifier. Oppositely, both adaptive ensemble method using realistic iErrPs classifier and adaptive ensemble method using perfect iErrPs classifier significantly increased classification accuracy for subject 7 in comparison to the static ensemble which is related to the huge change of base classifiers' weights between the beginning of the test session (Fig. 3(c)) and the end of it (Fig. 3(d)).

4   Conclusion

In this paper we presented an online inter-subjects classification framework for endogenous brain-computer interfaces. A straightforward way to learn from heterogeneous data recorded from different subjects is to use a weighted average ensemble in which each base classifier is trained using a single data set and weighted according to its accuracy in classifying brain signals of current BCI user. Static weighting of base classifiers using a small calibration set may increase classification accuracy in comparison to standard methods but this approach is limited by the non-stationary nature of brain signals. In the absence of true class labels during feedback phase, we proposed a new online adaptation approach of base classifiers' weights based on ensemble predictions reinforced by interaction error-related potentials (iErrPs). Results on two EEG data sets showed that our adaptive ensemble method based on a realistic iErrPs classifier allows increasing classification accuracy in comparison to the static method and preventing error accumulation compared to the adaptive method based only on ensemble predictions.

The proposed online adaptation method was limited to binary classification tasks. In future work we will extend it to multi-class classification and evaluate it in online experimental settings. Beyond the scope of BCI applications, our approach can be extended to other applications in which online transfer learning is needful and information about user's assessment of the system is accessible such as spam filtering application.

References

1. McFarland, D.J., Wolpaw, J.R.: Brain-computer interfaces for communication and control. Commun. ACM **54**(5), 60–66 (2011)
2. Nicolas-Alonso, L.F., Gomez-Gil, J.: Brain computer interfaces, a review. Sensors **12**, 1211–1279 (2012)
3. Lotte, F., Guan, C.: Learning from other subjects helps reducing brain-computer interface calibration time. In: International Conference on Audio Speech and Signal Processing (ICASSP), pp. 614–617 (2010)
4. Tu, W., Sun, S.: A subject transfer framework for EEG classification. Neurocomputing **82**, 109–116 (2011)
5. Liyanage, S.R., Guan, C., Zhan, H., Ang, K.K., Xu, J., Lee, T.H.: Dynamically weighted ensemble classification for non-stationary EEG processing. J. Neural Eng. **10**(3), 036007 (2013)
6. Dalhoumi, S., Dray, G., Montmain, J., Derosière, G., Perrey, S.: An adaptive accuracy-weighted ensemble for inter-subjects classification in brain-computer interfacing. In: 7th International IEEE EMBS Neural Engineering Conference (2015)
7. Ferrez, P.W., Del R Millan, J.: Error-related EEG potentials generated during simulated brain–computer interaction. IEEE Trans. Biomed. Eng. **55**(3), 923–929 (2008)
8. Llera, A., Van Gerven, M.A.J., Gomez, V., Jensen, O., Kappen, H.J.: On the use of interaction error potentials for adaptive brain computer interfaces. Neural Netw. **24**(10), 1120–1127 (2011)
9. Zeyl, T.J., Chau, T.: A case study of linear classifiers adapted using imperfect labels derived from human event-related potentials. Pattern Recogn. Lett. **37**, 54–62 (2014)
10. Blankertz, B.: BCI Competition IV. http://www.bbci.de/competition/iv (2008)
11. Blankertz, B., Tomioka, R., Lemm, S., Kawanabe, M., Muller, K.R.: Optimizing spatial filters for robust EEG single-trial analysis. IEEE Sig. Process. Mag. **25**(1), 41–56 (2008)
12. Steyrl, D., Scherer, R., Forstner, O., Muller-Putz, G.R.: Motor imagery brain-computer interfaces: random forests vs regularized LDA – non-linear beats linear. In: 6th International Brain-Computer Interface Conference (2014)

A Bayesian Sarsa Learning Algorithm
with Bandit-Based Method

Shuhua You, Quan Liu$^{(\boxtimes)}$, Qiming Fu, Shan Zhong, and Fei Zhu

School of Computer Science and Technology,
Soochow University, Suzhou 215000, China
jstzysh@126.com, quanliu@suda.edu.cn

Abstract. We propose an efficient algorithm called Bayesian Sarsa (BS)
on the consideration of balancing the tradeoff between exploration and
exploitation in reinforcement learning. We adopt probability distributions to estimate Q-values and compute posterior distributions about
Q-values by Bayesian Inference. It can improve the accuracy of Q-
values function estimation. In the process of algorithm learning, we
use a Bandit-based method to solve the exploration/exploitation problem. It chooses actions according to the current mean estimate of Q-
values plus an additional reward bonus for state-action pairs that have
been observed relatively little. We demonstrate that Bayesian Sarsa performs quite favorably compared to state-of-the-art reinforcement learning
approaches.

Keywords: Reinforcement learning · Probability distribution ·
Bayesian Inference · Bandit-based method · Exploration/exploitation

1 Introduction

In general, reinforcement learning (RL) [1] is considered as a kind of Machine
Learning method, which refers to that agent interacts with the unknown environment for trying to obtain maximum accumulative rewards. The RL methods
are classified as on-policy and off-policy on the view of policy. When the policy learned about is the same as the one used to select actions, we call it the
on-policy learning method, otherwise, it is off-policy. Sarsa is one of typical on-
policy algorithms. When we exploit Sarsa to learn an optimal policy, we always
encounter the central problem that is balancing exploration of untested actions
against exploitation of actions. At first, researchers attempted to solve the problem by means of the methods of ϵ-greedy [2] and Boltzmann [3,4]. ϵ-greedy is
the expansion of greedy, which behaves greedily most of the time, but every
once in a while, adopts actions at random with small probability ϵ. Boltzmann
exploration is a more sophisticated approach in which the probabilities of exe-
cuting actions are related to Q-values. In this approach, the parameter t is used
to control exploration over time, but it is difficult to set the appropriate value
for the rate of it decreasing. As a result, designing a more efficient algorithm

© Springer International Publishing Switzerland 2015
S. Arik et al. (Eds.): ICONIP 2015, Part I, LNCS 9489, pp. 108–116, 2015.
DOI: 10.1007/978-3-319-26532-2_13

to solve the problem of balancing exploration and exploitation will be of great importance in RL.

Then, researchers set their sights to Bayesian theory, and so Bayesian RL has been begun to be developed. In 1998, Dearden firstly proposed an algorithm called Bayesian Q-learning by combining Bayesian Inference with RL, while Bayesian Q-learning would cost more time to compute value of perfect information (VPI) [5-7]. Ishii presented a method which adopts Dirichlet distributions to estimate the state-transition probability of the environment, but the performance of their algorithm is significantly dependent on the inver-temperature meta-parameter so that it is not very stable [8]. Brochu utilized the predictive distribution of Gaussian process to balance exploitation and exploration, however, if the parameter ξ would encourage exploration early, then exploitation does not work well, in hence, the value of ξ is hard to choose [9]. Kolter and Ng attempted to use Bayesian Exploration Bonus (BEB) algorithm [10] to drive the agent towards the Bayesian optimal policy by using exploration rewards similar to R-MAX [11], but its drawback is that it may fail to find the optimal policy for some certain MDPs. Strehl developed an algorithm called Delayed Q-learning, unlike traditional Q-learning, which maintains Q-values estimates and waits for m samples to update Q-values, but it reduces the rate of convergence when most of samples are valid [12].

In this paper, We still combine Bayesian Inference with Sarsa and take advantage of normal-gamma distribution to model for Q-value functions, in which we calculate the posterior distribution over the prior distribution and samples. We consider a bandit-based method to select actions that we will consider actions which have been observed relatively little, and then make full use of optimal actions in the long run. This paper is organized as follows: Sect. 2 describes how to model for Q-values by normal-gamma distribution and the formula of updating parameters of probability distribution, and then we demonstrate the method of actions selection more efficient and present BS algorithms in detail. Section 3 describes the results of experiments by comparing our algorithm against other methods. In Sect. 4, we make an final conclusion.

2   Bayesian Sarsa

2.1   Q-values Distribution

In the Bayesian framework, we need to consider prior distributions over Q-values, and then update these priors based on the agent's experience. Formally, let $R_{s,a}$ be a random variable that denotes the total discounted reward received when action a is executed in state s and an optimal policy is followed thereafter. What we are initially uncertain about is how $R_{s,a}$ is distributed, and then the following five assumption are given first.

Assumption 1. $R_{s,a}$ has a normal distribution.

Assumption 2. The prior distribution over $\mu_{s,a}$ and $\tau_{s,a}$ is independent of the prior distribution over $\mu_{s',a'}$ and $\tau_{s',a'}$ for $s \neq s'$ and $a \neq a'$.

Assumption 3. *The prior $P(\mu_{s,a}, \tau_{s,a})$ is a normal-gamma distribution.*

Assumption 4. *The posterior distribution over $\mu_{s,a}$ and $\tau_{s,a}$ is independent of the posterior distribution over $\mu_{s',a'}$ and $\tau_{s',a'}$ for $s \neq s'$ and $a \neq a'$.*

Assumption 5. *The posterior $P(\mu_{s',a'}, \tau_{s',a'})$ is also a normal-gamma distribution.*

In these assumptions, $\mu_{s,a}$ is the mean value of normal distribution. Since the normal distribution is about $R_{s,a}$, $\mu_{s,a}$ can be approximately considered as the value of $Q(s, a)$. $\tau_{s,a}$ is the precision of normal distribution, and its value is the inverse of the variance of normal distribution, $\tau_{s,a} = \frac{1}{\sigma_{s,a}^2}$. As it turns out, it is more convenient to represent the uncertainty about Q-values over the precision than over the variance. $P(\mu_{s,a}, \tau_{s,a})$ is the probability distribution of $R_{s,a}$, and our ultimate aim is to compute the value of $Q^*(s, a)$, where $Q^*(s, a) = E[R_{s,a}]$. In an MDP setting, these assumptions are likely violated. Dearden [5] describes these five assumptions in detail in their paper.

2.2 Updating Q-Values

We can use a collection of hyperparameters $\rho = (\mu_0, \lambda, \alpha, \beta)$ or the normal-gamma posterior for the mean and precision of each $R_{s,a}$, which is said that $P(\mu, \tau) \sim NG(\mu_0, \lambda, \alpha, \beta)$, if

$$P(\mu, \tau) \propto \tau^{\frac{1}{2}} e^{-\frac{1}{2}(\lambda\tau(\mu-\mu_0)^2)} \tau^{\alpha-1} e^{\beta\tau} = \tau^{\alpha-\frac{1}{2}} exp(-\frac{1}{2}(\lambda\tau(\mu - \mu_0)^2) + \beta\tau). \quad (1)$$

Theorem 1. *Let $P(\mu, \tau) \sim NG(\mu_0, \lambda, \alpha, \beta)$ be a prior distribution over the unknown parameter R, and R is a normally distributed variable. Let $r_1, r_2, \ldots, r_n$ be n independent samples of R, $C_1 = \frac{1}{n}\sum_i r_i$ and $C_2 = \frac{1}{n}\sum_i r_i^2$. From the assumption 5, posterior distribution is $P(\mu, \tau|r_1, r_2, \ldots, r_n) \sim NG(\mu_0', \lambda', \alpha', \beta')$, where $\mu_0' = \frac{\lambda\mu_0 + nC_1}{\lambda+n}$, $\lambda' = \lambda + n$, $\alpha = \alpha + \frac{1}{2}n$, and $\beta' = \beta + \frac{1}{2}n(C_2 - C_1^2) + \frac{n\lambda(C_1-\mu_0)^2}{2(\lambda+n)}$.*

The updating equation is obtained based on [13]. However, The difficulty of computing posterior distribution is that how to calculate the values of C_1 and C_2. Suppose that the agent is in state s, executes action a, receives reward r, and lands up in state s'. We would like to know the complete sequence of rewards received from s' onwards, but this is not available. Let $R_{s'}$ be a random variable denoting the discounted accumulative rewards form s'. If we assume that the agent will follow the apparently optimal policy, then $R_{s'}$ is distributed as $R_{s',a'}$, where a' is the action with the highest expected value at s'. Therefore, we can randomly sample values $R_{s'}^1, R_{s'}^2, \ldots, R_{s'}^n$ from $R_{s',a'}$, and then update $P(R_{s,a})$ with the samples $r + \gamma R_{s'}^1, r + \gamma R_{s'}^2, \ldots, r + \gamma R_{s'}^n$, where we take each sample to have weight $\frac{1}{n}$ and r is the immediate reward, γ is the discount factor. Theorem 1 implies that we only need the first two moments of samples to update our distribution. It assumes that n tends to infinity, these two moments are:

$$C_1 = E[r + \gamma R_{s'}] = r + \gamma E[R_{s'}], \quad (2)$$

$$C_2 = E[(r + \gamma R_{s'})^2] = r_2 + 2\gamma r E[R_{s'}] + \gamma_2 E[R_{s'}^2]. \tag{3}$$

Because R is a normally distributed variable with unknown mean μ and unknown precision τ, and $P(\mu, \tau)$ about R is $NG(\mu_0, \lambda, \alpha, \beta)$, then $E[R] = \mu$ and $E[R_2] = \frac{\lambda+1}{\lambda} \cdot \frac{\beta}{\alpha-1} + \mu_0^2$.

Now, we can calculate the C_1 and C_2. Then we will update the parameters of posterior distribution as though we have seen a collection of examples with total weight 1, mean C_1, and second moment C_2. As the agent learns the policy, the precision $\tau_{s,a}$ of each $R_{s,a}$ will all converge to the threshold and this moment, $\mu_{s,a}$ is the value of $Q_{s,a}$.

Algorithm 1. Bayesian Sarsa

1: Input S, A
2: Initialize $P_{s,a}(\mu, \tau) \sim NG(\mu_0, \lambda, \alpha, \beta)$ for each (s, a), ψ, c, δ, step, $N(s)$, $N(s, a)$
3: **repeat**
4:     $s \leftarrow$ initial (nonterminal) state
5:     $a \leftarrow \arg\max_a(\mu(s, a) + c\sqrt{\frac{\ln N(s)}{1+N(s,a)}})$
6:     $N(s) \leftarrow N(s) + 1$
7:     $N(s, a) \leftarrow N(s, a) + 1$
8:     step $\leftarrow 0$
9:     **repeat**
10:        execute a, observe reward r, next state s'
11:        $a' \leftarrow \arg\max_{a'}(\mu(s', a') + c\sqrt{\frac{\ln N(s')}{1+N(s',a')}})$
12:        $N(s') \leftarrow N(s') + 1$
13:        generate n independent samples of $P_{s',a'}(\mu, \tau)$, e.g. $R_{s',a'}^1, R_{s',a'}^s, \ldots, R_{s',a'}^n$
14:        $C_1 \leftarrow r + \gamma E[R_{s',a'}]$
15:        $C_2 \leftarrow r^2 + 2r\gamma E[R_{s',a'}] + \gamma^2(\frac{\lambda_{s',a'}+1}{\lambda_{s',a'}} \cdot \frac{\beta_{s',a'}}{\alpha_{s',a'}-1} + (E[R_{s',a'}])^2)$
16:        $\mu_{0,s,a} \leftarrow \mu_{0,s,a} + \psi(\frac{\lambda_{s,a}\mu_{s,a}+nC_1}{\lambda_{s,a}+n})$
17:        $\lambda_{s,a} \leftarrow \lambda_{s,a} + \psi(\lambda_{s,a})$
18:        $\alpha_{s,a} \leftarrow \alpha_{s,a} + \psi(\frac{1}{2}n)$
19:        $\beta_{s,a} \leftarrow \beta_{s,a} + \psi(\frac{1}{2}n(C_2 - C_1^2) + \frac{n\lambda_{s,a}(C_1-\mu_{0,s,a})^2}{2(\lambda_{s,a}+n)})$
20:        $s \leftarrow s'$
21:        $a \leftarrow a'$
22:        step++
23:     **until** s is terminal
24:     $\mu \leftarrow \mu_0$
25:     $\tau \leftarrow \frac{\lambda \times (\alpha-1)}{\beta}$
26:     Output step
27: **until** time out
28: Output μ, τ of all distributions

2.3 Actions Selection

In every iteration of the Bayesian Sarsa algorithm we need to selection an action to execute, and then we consider a reward bonus for actions selection which is controlled by the count of actions selection. Specifically, Bayesian Sarsa algorithm, at each time step, chooses actions based on the multi-armed bandit algorithm [14].

$$a = \arg\max_a (Q(s,a) + c\sqrt{\frac{\ln N(s)}{1 + N(s,a)}}) \tag{4}$$

where $N(s)$ is the times of state being observed, and $N(s,a)$ is the count of action a being selected in s, $N(s) = \sum_a N(s,a)$. c a bias parameter which defines the proportion of exploitation and exploration. If $c = 0$, this policy becomes a greedy policy. This is different from $\beta/(1 + N(s,a))$ which is proposed by Kolter [10]. Although the method Kolter put forward allows an agent acting in an MDP to perform ϵ-close to the (intractable) optimal Bayesian policy after a polynomial number of time steps, it brings about that algorithms may fail to find the optimal policy for some certain MDPs. Whereas the method of action selection we proposed avoid this problem, since the reward term $Q(s,a)$ encourages the exploitation of higher-reward choices, and the right hand term $c\sqrt{\frac{\ln N(s)}{1+N(s,a)}}$ encourages the exploration of less-chose actions. In hence, we can present an algorithm that combines elements from both the Bayesian Q-learning and UCT algorithms and it shows to perform "nearly as well" as them. we call it Bayesian Sarsa (BS) which is shown as Algorithm 1.

3 Experimental Results

3.1 Gridworld

Figure 1 uses a rectangular grid to illustrate BS for a simple finite MDP. The cells of the grid correspond to the states of the environment. At each cell, four actions are possible: north, south, east, and west, which deterministically cause the agent to move one cell in the corresponding direction on the grid. Actions that would take the agent off the grid leave its location unchanged, but also result in a reward of -10. Other actions also result in a reward of -10, except those that make the agent to arrive at the goal state, resulting in a reward of 400. In this problem, the starting state is (4, 1), and the goal state is (1, 5). The discount factor $\gamma = 0.95$, and the learning rate $\psi = 0.2$.

Figure 2 shows the curve of variance about the distribution of the Q-value which agent executes action east in state (4,2). From the analysis of Fig. 2, we find that the variance of Q-value of one state-action pair reduces with the increasing episodes, and the posterior distribution of its Q-value will be close to the true Q-value in the long run. When episodes tend to infinity, the variance converges to 0, so as others. In hence, it can demonstrate that BS can converge to optimal Q-values.

We mainly take Sarsa, Bayesian Q-learning, BEB, Rmax and BS compared with each other. For Sarsa with softmax actions selection, the temperature $t = 10$, and the value of decay rate b is respectively 0.003, 0.005, 0.03, 0.09. For

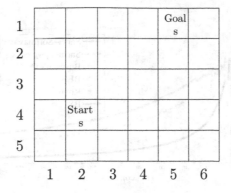

Fig. 1. 5 × 6 Gridworld

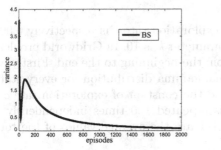

(a) The variance of Q-value

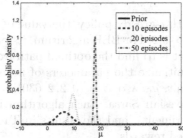

(b) The prior distribution and posterior distribution of Q-value

Fig. 2. Proof of converging to true Q-values

Table 1. The total rewards of Sarsa, Bayesian-Q, BEB, Rmax and BS

Name	t	b	k	ϵ	Total rewards	
					1500 steps	3000 steps
Sarsa-Softmax	10	0.003	/	/	10798	96992
		0.005			14978	104046
		0.03			20552	95738
		0.09			19496	−172032
Sarsa-ϵ-greedy	/	/	/	0.005	50546	136754
				0.01	50976	138086
				0.05	43834	127888
				0.2	29132	84756
Bayesian-Q	/	/	/	/	72878	154054
BEB	/	/	10	/	−5920	−21020
Rmax	/	/	/	/	66570	151170
BS	/	/	/	/	69010	160220

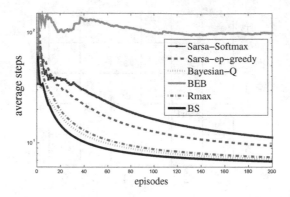

Fig. 3. The average steps of Sarsa, Bayesian-Q, BEB, Rmax and BS

Sarsa with ϵ-greedy policy, the value of exploration rate ϵ is respectively 0.005, 0.01, 0.05, 0.2. For BEB algorithm, its parameter k is 10. In Gridworld problem, our target is to find the optimal path from the beginning to the end. Firstly, we should initialize the parameters of normal-gamma distribution for every state-action pair, $\mu_0, \lambda, \alpha, \beta = (0, 2, 2, 630)$, and the constant of exploration rate $c = 5$ for Bayesian Sarsa. Each algorithm is repeated 100 times independently in the experiment , and the results in Table 1 are the average values of the total discounted rewards.

In Table 1, we can know that BS obtains the more rewards than the other algorithms over a long period of time, even though Bayesian Q-learning gains more rewards than BS in the initial 1500 steps. Because at the beginning of algorithms running, BS attaches more importance to exploration strategy than Bayesian Q-learning, so Bayesian Q-learning will receive more rewards than BS during initial period. However, when BS finds optimal actions and makes full use of them, the total rewards agent receives will be the most. For BEB algorithm, it is unfortunate that BEB can't find optimal policy in this certain MDP due to its own drawback. For Rmax, it also receive less rewards than BS in the process of learning. We run Sarsa algorithms with each method of actions selection, Softmax and ϵ-greedy, and results show that BS is more efficient than Sarsa with both methods for all the parameters. Table 1 shows that BS has the best performance among all the algorithms, and Fig. 3 reflects the speed of convergence of all the algorithms with $t = 10$, $b = 0.05$, $\epsilon = 0.01$, which are the best parameter values in Table 1. As expected, BS is competitive or superior to state of the art exploration techniques such as Bayesian Q-learning, Rmax, and BEB.

From Fig. 4, we can know that prior with larger variances usually lead to better performance, like $(0, 2, 2, 630)$ and $(0, 2, 2, 63)$, and c being different values seems to affect the results of our algorithm, too.

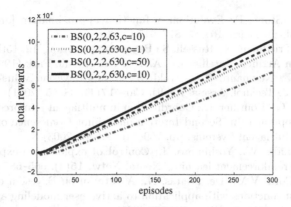

Fig. 4. Total rewards of Bayesian Sarsa with different parameter values

4   Conclusion

We present a model-free approach to exploration/exploitation called Bayesian
Sarsa (BS) that combines Bayesian Inference with Sarsa algorithm that estimates
Q-values with normal-gamma distributions and chooses actions with bandit-
based method that converges quickly. We compared BS with several state-of-the-
art exploration methods and demonstrate it better than some other algorithms.

We are also investigating alternative actions selection schemes, and approximations that can be used to reduce the computational requirements of BS.
Finally, it should be possible to use function approximator to extend this work
to problems with large or continuous state space.

Acknowledgements. This work was funded by National Natural Science Foundation
(61272005, 61303108, 61373094, 61502323, 61272005, 61303108, 61373094, 61472262).
We would also like to thank he reviewers for their helpful comments. Natural Science Foundation of Jiangsu (BK2012616), High School Natural Foundation of Jiangsu
(13KJB520020), Key Laboratory of Symbolic Computation and Knowledge Engineering of Ministry of Education, Jilin University (93K172014K04), Suzhou Industrial
application of basic research program part (SYG201422).

References

1. Sutton, R.S., Barto, A.G.: Reinforcement Learning: An Introduction. MIT Press,
   Cambridge (1998)
2. Kalidindi, K., Bowman, H.: Using ϵ-greedy reinforcement learning methods to further understand ventromedial prefrontal patients' deficits on the Iowa Gambling
   Task. Neural Netw. **20**(6), 676–689 (2007)
3. Coggan, M.: Exploration and exploitation in reinforcement learning. In: 4th International Conference on Computational Intelligence and Multimedia Applications.
   IEEE Press, Japan (2001)

4. Whiteson, S., Stone, P.: Evolutionary function approximation for reinforcement learning. J. Mach. Learn. Res. **7**, 877–917 (2006)
5. Dearden, R., Friedman, N., Russell, S.: Bayesian Q-learning. In: 15th International Conference on Artificial Intelligence. AAAI Press, Menlo Park (1998)
6. WClaxton, K., Neumann, P.J., Araki, S., et al.: Bayesian value-of-information analysis. Int. J. Technol. Assess. Health Care **17**(1), 38–55 (2001)
7. Chalkiadakis, G., Boutilier, C.: Coordination in multiagent reinforcement learning: a Bayesian approach. In: Second International Joint Conference on Autonomous Agents and Multiagent Systems, pp. 709–716. ACM (2003)
8. Ishii, S., Yoshida, W., Yoshimoto, J.: Control of exploitation-exploration meta-parameter in reinforcement learning. Neural Netw. **15**(4), 665–687 (2002)
9. Brochu, E., Cora, V.M., De, Freitas, N.: A tutorial on Bayesian optimization of expensive cost functions with application to active user modeling and hierarchical reinforcement learning. arXiv:1012.2599 (2010)
10. Kolter, J.Z., Ng, A.: Near-Bayesian exploration in polynomial time. In: 26th International Conference on Machine Learning, pp. 513–520 (2009)
11. Brafman, R.I., Tennenholtz, M.: R-max: a general polynomial time algorithm for near-optimal reinforcement learning. J. Mach. Learn. Res. **3**, 213–231 (2003)
12. Strehl, A.L., Li, L., Wiewiora, E., et al.: PAC model-free reinforcement learning. In: 23rd International Conference on Machine Learning, pp. 881–888. ACM (2006)
13. Degroot, M., Schervish, M.: Probability and Statistics, 4th edn. Pearson Education, Inc., New York (2010)
14. Auer, P., Cesa-Bianchi, N., Fischer, P.: Finite-time analysis of the multi-armed bandit problem. Mach. Learn. **47**(2–3), 235–256 (2002)
15. Cox, C., Chu, H., Schneider, M.F., et al.: Parametric survival analysis and taxonomy of hazard functions for the generalized gamma distribution. Stat. Med. **26**(23), 4352–4374 (2007)
16. Shin, J.W., Chang, J.H., Kim, N.S.: Statistical modeling of speech signals based on generalized gamma distribution. IEEE Sig. Process. Lett. **12**(3), 258–261 (2005)

Incrementally Built Dictionary Learning for Sparse Representation

Ludovic Trottier[✉], Brahim Chaib-draa, and Philippe Giguère

Department of Computer Science and Software Engineering,
Université Laval, Québec, QC G1V 0A6, Canada
ludovic.trottier.1@ulaval.ca,
{chaib,philippe.giguere}@ift.ulaval.ca
http://www.damas.ift.ulaval.ca

Abstract. Extracting sparse representations with Dictionary Learning (DL) methods has led to interesting image and speech recognition results. DL has recently been extended to supervised learning (SDL) by using the dictionary for feature extraction and classification. One challenge with SDL is imposing diversity for extracting more discriminative features. To this end, we propose Incrementally Built Dictionary Learning (IBDL), a supervised multi-dictionary learning approach. Unlike existing methods, IBDL maximizes diversity by optimizing the between-class residual error distance. It can be easily parallelized since it learns the class-specific parameters independently. Moreover, we propose an incremental learning rule that improves the convergence guarantees of stochastic gradient descent under sparsity constraints. We evaluated our approach on benchmark digit and face recognition tasks, and obtained comparable performances to existing sparse representation and DL approaches.

Keywords: Supervised dictionary learning · Sparse representation · Digit recognition · Face recognition

1 Introduction

Feature extraction is a crucial step for improving the performance of machine learning algorithms. Various engineering methods were proposed in past decades for extracting the most useful information from raw observations [1]. However, recent developments in *representation learning* (RL) showed that feature engineering has limitations [2]. In particular, engineered features do not generalize to a large amount of problems due to their task-specific design, and the approaches are theoretically cumbersome to analyze and improve.

On the contrary, RL approaches do not suffer from these limitations because they aim to learn the relevant features instead of relying on expert knowledge for creating the pipeline of preprocessing transformations. Their goal is to construct in an unsupervised fashion a high-level representation from unlabeled data and use it for feature extraction. An important class of RL approaches, known as

© Springer International Publishing Switzerland 2015
S. Arik et al. (Eds.): ICONIP 2015, Part I, LNCS 9489, pp. 117–126, 2015.
DOI: 10.1007/978-3-319-26532-2_14

dictionary learning (DL), uses sparse modeling to construct efficient data representations by linearly combining a small number of typical patterns (atoms) learned from data. Significant practical and theoretical contributions for learning a collection of such patterns (called a dictionary) have lead to state-of-the-art results in many signal processing and vision-related tasks [3,4].

Recently, DL has been extended to *supervised dictionary learning* (SDL) by taking into account the label information instead of only relying on unlabeled data [5]. Among the different ways to extend a DL approach to supervised learning, one is to learn class-specific dictionaries. For instance, Ramirez et al. [6] proposed *dictionary learning with structured incoherence* (DLSI), a multi-dictionary approach minimizing the correlation between the class-specific dictionaries. Also, Yang et al. [7] incorporated a Fisher discrimination penalty in their *Fisher discrimination dictionary learning* (FDDL) approach to make the dictionaries more discriminative. Another way is to learn a joint dictionary with class-specific atoms. For example, Zhang et al. [8] proposed *discriminative KSVD* and Jiang et al. [9] proposed *label consistent KSVD*, both extending the well-known KSVD algorithm [10] to supervised learning. Finally, Wright et al. [11] proposed using the whole dataset as the dictionary in their *sparse representation-based classification* (SRC) approach and applied it on face recognition tasks.

The critical part when using multiple dictionaries in SDL is encouraging dictionary diversity for learning discriminative patterns [12]. The most common way to achieve this goal is by incorporating a discriminative term to the learning framework. For instance, DLSI uses the correlation between the class-specific dictionaries while FDDL uses a Fisher discrimination criterion. However, these new terms greatly complexify the learning phase. In this paper, we propose a framework for learning class-specific dictionaries under a diversity constraint without adding new discriminative terms. Moreover, we propose a learning algorithm that simultaneously treats the dictionaries in parallel. We finally propose an incremental learning rule that improves the convergence guarantees of stochastic gradient descent under sparsity constraints.

This paper is organized as follows. We introduce the reader to dictionary learning in Sect. 2 and present the proposed approach in Sect. 3. We show the experimentations in Sect. 4 and conclude in Sect. 5.

2   Background on Dictionary Learning

Let us define $\mathbf{X} = [\mathbf{x}_1 \ldots \mathbf{x}_N] \in \mathbb{R}^{M \times N}$ as the data matrix containing N M-dimensional observations and $\mathbf{D} = [\mathbf{d}_1 \ldots \mathbf{d}_K] \in \mathbb{R}^{M \times K}$ as the dictionary with K atoms. Dictionary learning aims to represent each observation $\mathbf{x}_n$ as a sparse linear composition $\mathbf{w}_n$ of the dictionary atoms $\mathbf{d}_k$ with minimal residual error:

$$\min_{\mathbf{D}, \mathbf{W}} \quad \|\mathbf{X} - \mathbf{D}\mathbf{W}\|_F^2 + \lambda \|\mathbf{W}\|_1 \quad s.t. \quad \|\mathbf{d}_k\|_2 = 1 \ \ \forall k, \qquad (1)$$

where $\mathbf{W} = [\mathbf{w}_1 \ldots \mathbf{w}_N] \in \mathbb{R}^{K \times N}$ is the weight matrix, $\| \cdot \|_F^2$ and $\| \cdot \|_1$ are respectively the squared Frobenius and entry-wise ℓ_1 norms. We define the residual error as $\|\mathbf{X} - \mathbf{DW}\|_F^2$ and the sparsity constraint as $\lambda\|\mathbf{W}\|_1$. The hyper-parameter $\lambda > 0$ governs the sparsity of the weight vectors $\mathbf{w}_n$ and is usually chosen by cross-validation. Constraining the dictionary atoms with $\|\mathbf{d}_k\|_2 = 1$ is needed for removing the trivial solution where $\mathbf{W} \approx 0$ and $\mathbf{D} \approx \infty$.

Minimizing Eq. 1 for both the dictionary $\mathbf{D}$ and the weight matrix $\mathbf{W}$ is however computationally too expensive, due to the coupling between $\mathbf{D}$ and $\mathbf{W}$ and the large amount of data. Approximating the optimum with an *iterative-alternative optimization* scheme is the usual choice for finding a suitable solution. The procedure works as follows. First, initialize randomly the dictionary $\mathbf{D}$. Second, minimize Eq. 1 w.r.t. the weight matrix $\mathbf{W}$, considering $\mathbf{D}$ fixed. We call this step *sparse coding*. Then, minimize Eq. 1 w.r.t. the dictionary $\mathbf{D}$, considering $\mathbf{W}$ fixed. We refer to this step as *dictionary learning*. Finally, alternate the two last steps until convergence.

Sparse coding is generally reduced to the LASSO [13] problem and has been solved by many approaches such as *orthogonal matching pursuit* (OMP) [14], *lest angle regression* (LARS) [15] and *marginal regression* (MR) [16]. On the other hand, dictionary learning is usually viewed as a constrained least squares problem. *Method of optimal direction* (MOD) [17], *online dictionary learning* [18] and KSVD [10] are examples of well-known methods for solving it.

3   Incrementally Built Dictionary Learning

In this section, we describe the proposed approach for imposing dictionary diversity. We first develop the optimization framework that will be used for learning the parameters and elaborate on its convergence guaranty. The section ends with a discussion on sparse-coding based features.

3.1   Approach Description

Let $\mathcal{D}$ be the unknown data distribution that generated the dataset $\{(\mathbf{x}_n, y_n)\}_{n=1}^N$, where observation $\mathbf{x}_n \in \mathbb{R}^M$ has class label $y_n \in \{1 \ldots C\}$. Let us define the following residual error-based model:

$$f_c(\mathbf{x}) = \min_{\mathbf{w}} \|\mathbf{x} - \mathbf{D}_c\mathbf{w}\|_2^2 + \lambda\|\mathbf{w}\|_1 , \tag{2}$$

where $\mathbf{D}_c$ are class-specific dictionaries. Our goal is then to learn the dictionaries minimizing the expected classification error:

$$h = \arg\min_h \mathbb{E}_{(\mathbf{x},y) \sim \mathcal{D}} [\ell(h(\mathbf{x}), y)] \qquad \text{s.t.} \qquad h(\mathbf{x}) = \arg\min_c f_c(\mathbf{x}) , \tag{3}$$

where ℓ is the 0–1 loss. The classification rule $h(\mathbf{x})$ assigns class label c to observation $\mathbf{x}$ when the combined residual error and sparsity penalty using dictionary $\mathbf{D}_c$ is the smallest. Since Eq. 3 is a multi-dictionary approach, the main concern is imposing dictionary diversity for extracting discriminative features. We propose to maximize the distance between $f_c(\mathbf{x}_i)$ and $f_c(\mathbf{x}_j)$ where observations

$\mathbf{x}_i$ are labeled c ($y_i = c$) and observations $\mathbf{x}_j$ are not labeled c ($y_j \neq c$). This learning principle is contrary to the one generally used in a SDL approach where class-dictionary $\mathbf{D}_c$ must achieved the smallest residual error on observations from class c. Here, we rather encourage that the residual $f_c(\mathbf{x}_i)$ is far enough from the residual $f_c(\mathbf{x}_j)$. In a sense, we want the dictionaries to learn features maximizing the residual error margin to improve the separation of the classes. We believe that better classification performances could be achieved by tuning the class-c dictionary to increase the residual for observations not from class c rather than to reduce the residual for observations from class c. Therefore, we define the following optimization problem, $\forall c \in \{1 \dots C\}$:

$$\mathbf{D}_c = \arg\min_{\mathbf{D}_c} \sum_{\substack{\mathbf{x}_i \\ s.t.\ y_i = c}} \sum_{\substack{\mathbf{x}_j \\ s.t.\ y_j \neq c}} \mathcal{L}(\gamma f_c(\mathbf{x}_i) - f_c(\mathbf{x}_j)) \quad s.t. \quad \|\mathbf{d}_{ck}\|_2 = 1 \quad \forall k, \quad (4)$$

where $\mathcal{L}(x) = \max\{x, 0\}$ is a Hinge-based loss. The hyper-parameter $\gamma \geq 1$ governs the importance of minimizing the residual for $\mathbf{x}_i$ and will be inferred by cross validation. We emphasis that our framework is easily parallelized by solving Eq. 4 simultaneously for each class c. Minimizing Eq. 4 is done with projected stochastic gradient descent and the gradient of the loss function $\mathcal{L}(\gamma f_c(\mathbf{x}_i) - f_c(\mathbf{x}_j))$ is computed as follows:

$$\nabla_{\mathbf{D}_c}\mathcal{L} = \begin{cases} \mathbf{D}_c \left(\gamma \tilde{\mathbf{w}}_i \tilde{\mathbf{w}}_i^\top - \tilde{\mathbf{w}}_j \tilde{\mathbf{w}}_j^\top\right) + \mathbf{x}_j \tilde{\mathbf{w}}_j^\top - \gamma \mathbf{x}_i \tilde{\mathbf{w}}_i^\top & \text{if } \gamma f_c(\mathbf{x}_i) > f_c(\mathbf{x}_j) \\ 0 & \text{otherwise} \end{cases}, \quad (5)$$

where $\tilde{\mathbf{w}}_i = \arg\min_{\mathbf{w}} \|\mathbf{x}_i - \mathbf{D}_c\mathbf{w}\|_2^2 + \lambda\|\mathbf{w}\|_1$ is the sparse coding solution. The final update rule is then:

$$\mathbf{D}_c \leftarrow \Pi_{\mathbb{D}} \left(\mathbf{D}_c - \alpha \nabla_{\mathbf{D}_c}\mathcal{L}(\gamma f_c(\mathbf{x}_i) - f_c(\mathbf{x}_j))\right) \quad (6)$$

where $\alpha > 0$ is the learning rate and $\Pi_{\mathbb{D}}$ is a projection onto the subspace $\mathbb{D}$ of dictionaries having normalized columns.

3.2   Incremental Learning Rule

Our approach does not have a convergence guarantee. Due to sparsity, many dictionary atoms are never updated and only a small subset truly represent the latent structure. Consequently, we observe a *rich get richer* phenomenon where the same atoms get activated over and over again resulting in suboptimal dictionaries. To prevent this, we propose the following incremental learning rule. We initialize the dictionary $\mathbf{D}_c$ with $K_0 < K$ atoms and perform gradient descent. At each iteration, we keep track of the usage statistics by incrementing $\mathbf{u}_k$ by 1 when atom $\mathbf{d}_k$ is active ($\tilde{\mathbf{w}}_{ik} \neq 0$), $\forall k \in \{1 \dots K\}$. After T_0 iterations, we reconsider the dictionary in two ways. Let $u^* = \arg\min_k \mathbf{u}_k$ denotes the least used dictionary atom[1] and $\mathbf{p} = \mathbf{u}/\|\mathbf{u}\|_2$ denotes the probability distribution computed

---

[1] In the case where there are several atoms, randomly select one of them.

Algorithm 1. Incrementally Built Dictionary Learning

Initialize $\mathbf{D}_c, \forall c \in \{1 \ldots C\}$, each with K_0 random and normalized atoms.

for $c = 1 \ldots C$ **do**

    Initialize the usage statistics: $\mathbf{u} \leftarrow \mathbf{0}$

    for $t = 1 \ldots T$ **do**

        Sample $\mathbf{x}_i \sim \mathcal{D}$, $\mathbf{x}_j \sim \mathcal{D}$ such that $y_i = c$ and $y_j \neq c$

        Compute $\tilde{\mathbf{w}}_i$ and $\tilde{\mathbf{w}}_j$ using, for example, LARS.

        $\mathbf{D}_c \leftarrow \mathbf{D}_c - \alpha \nabla_{D_c} \mathcal{L}(\gamma f_c(\mathbf{x}_i) - f_c(\mathbf{x}_j))$ (Eq. 5)

        $\mathbf{d}_{ck} \leftarrow \mathbf{d}_{ck}/\|\mathbf{d}_{ck}\|_2, \forall k \in \{1 \ldots K\}$

        $\mathbf{u} \leftarrow \mathbf{u} + \mathbf{1}_{\tilde{\mathbf{w}}_i \neq 0}$ (element-wise indicator function)

        if $0 \equiv t \mod T_0$ **and** $K_0 < K$ **then**

            $u^* = \arg\min_k \mathbf{u}_k$

            $i, j \sim \mathcal{GB}(\mathbf{p})$, where $\mathbf{p} = \mathbf{u}/\|\mathbf{u}\|_2$

            $\mathbf{d}_{cu^*} \sim \mathcal{N}(\mathbf{d}_{ci}, \sigma^2 \mathbf{1}), \quad \mathbf{d}_+ \sim \mathcal{N}(\mathbf{d}_{cj}, \sigma^2 \mathbf{1})$

            $\mathbf{D}_c \leftarrow \mathbf{D}_c \cup \{\mathbf{d}_+\}, \quad K_0 \leftarrow K_0 + 1, \quad \mathbf{u} \leftarrow \mathbf{0}$

        end if

    end for

end for

---

from the usage statistics. We first resample the least used dictionary atom $\mathbf{d}_{u^*}$ and second add a new atom to the dictionary according to the following scheme:

$$i, j \sim \mathcal{GB}(\mathbf{p}), \qquad \mathbf{d}_{u^*} \sim \mathcal{N}(\mathbf{d}_i, \sigma^2 \mathbf{1}) \qquad \mathbf{d}_+ \sim \mathcal{N}(\mathbf{d}_j, \sigma^2 \mathbf{1}), \qquad (7)$$

where $\mathcal{GB}$ is the generalized Bernoulli distribution and $\mathcal{N}$ is the Gaussian distribution with variance parameter σ^2. Based on our experimentations, we found that $\sigma^2 = 1/M$ and $T_0 = T/2K$ achieved good performances, where T is the total number of iterations. The resulting algorithm is presented in Algorithm 1 and we refer to our approach as *incrementally build dictionary learning* (IBDL).

There are two aspects motivating this incremental learning. First, unused dictionary atoms are either useless or specialized. At the beginning of the gradient descent, unused atoms are necessarily useless for reasons given earlier. However, at the end of the descent, unused atoms might represent a relevant structure even though they are rarely used for sparse coding. This explains our choice for T_0 because no atom resampling is permitted after $T/2$ iterations to allow learning specialized atom. Second, overused atoms need specialization. To understand this principle, let us study the following limit case. Suppose that a dictionary atom $\mathbf{d}_p$ has discovered such a complex structure that it always gets activated for encoding. However, most observations only uses subregions of $\mathbf{d}_p$ (corresponding to simpler structures) due to its overly complex nature. As a consequence, the gradient descent updates only target subregions of $\mathbf{d}_p$ (those found in the observation used for the gradient descent) and the content of $\mathbf{d}_p$ never settles to a stationary point. Sampling $\mathbf{d}_{u^*}$ and $\mathbf{d}_+$ according to the usage statistic allows them to share the complex structure found by $\mathbf{d}_p$.

Table 1. Error rate results (%) of our approaches (IBDL-E and IBDL-C) and the state-of-the-art on the MNIST and USPS digit recognition tasks.

	IBDL-E IBDL-C	SDL-G SDL-D	REC L REC BL	ℓ_2-KNN	SVM-Gauss	FDDL	DLSI	SRSC
MNIST	2.88 2.21	3.56 1.05	4.33 3.41	5.00	1.4	-	1.26	-
USPS	4.63 3.99	6.67 3.54	6.83 4.38	5.2	4.2	3.69	3.98	6.05

Table 2. Optimal hyper-parameter values of our approaches on the MNIST and USPS digit recognition tasks.

	MNIST				USPS			
	Encoder	K	λ	γ	α	Encoder	K	λ	γ	α
IBDL-E	LARS	75	0.66	11.36	0.00009	LARS	50	0.13	2.86	0.00006
IBDL-C	MR 25 0.06 10.60 0.0003 5NN classifier					MR 50 11.18 1.84 0.0001 SVM, $gam = 0.00005$, $C = 11851.15$			

3.3 Sparse Coding-Based Feature Extraction

Another alternative to using the residual error for classification is to train a classifier on sparse coding-based features. We therefore define $\mathbf{e} = [f_1(\mathbf{x}) \dots f_C(\mathbf{x})]$ the error vector and $\mathbf{r} = [\mathbf{w}_1 \dots \mathbf{w}_C]$ the representation vector of observation $\mathbf{x}$ computed from the class-specific dictionaries. We construct a feature vector by concatenating the normalized vectors $\mathbf{e}$ and $\mathbf{r}$ to form the vector $\phi = [\frac{\mathbf{e}}{\|\mathbf{e}\|_2}, \frac{\mathbf{r}}{\|\mathbf{r}\|_2}]$ and use it as input for the classifier.

4 Experimentations

We tested IBDL on digit recognition using MNIST and USPS datasets and on face recognition using the Extended Yale B dataset. We evaluated the two proposed types of classification: (1) minimal residual error (IBDL-E), as defined by Eq. 3, and (2) classification with our sparse coding-based features (IBDL-C), as defined in Sect. 3.3. For all tasks, we cross-validated (3-fold) the hyper-parameters $\Theta = \{\lambda, \gamma, \alpha\}$ using Bayesian optimization [19] and tested the OMP, MR and LARS sparse coding algorithms with $K \in \{25, 50, 75\}$ and $T \in \{10000, 25000, 50000\}$. We report the test score of the approach achieving the best validation score. For IBDL-C, we evaluated the KNN and RBF-SVM classifiers. We cross-validated their hyper-parameters using Bayesian optimization [19] and report the test score of the best approach.

4.1 Digits Recognition

The USPS dataset contains 7,291 training and 2,007 testing 16×16 images and the MNIST contains 60,000 training and 10,000 testing 28×28 images. We report in Table 1 the performances of our approach in comparison to others in the literature: REC-L, REC-BL, SDL-G and SDL-D [5], KNN, SVM and DLSI [6], FDLL [7] and SRSC [20]. The optimal hyper-parameters are also reported in Table 2.

Discussion. Our approach IBDL-C has competitive performances on the USPS dataset and comparable performances on MNIST with the state-of-the-art. The main difference between these two tasks is the number of observations and the image size. Since our approach samples one observation from class c and another one from class $\neg c$, it appears that more iterations are needed when dealing with larger datasets. Due to the curse of dimensionality, the image size may also affect the accuracy. Even though IBDL-E does not work well on either tasks, our results show that the KNN classifier has better accuracy with our sparse coding-based features. We believe that using both the structure of the weight vectors $\mathbf{w}_c$ and the error values $\mathbf{e}$ for constructing the feature vector ϕ makes it more discriminative. Further investigations for explaining why IBDL-C achieve better accuracy than IBDL-E is needed.

4.2   Face Recognition

The Extended Yale B dataset contains 2,414 frontal-face images of 38 individuals (approximatively 64 images per subject). We used the experimental setup of [11] and compared against SRC, NN and SVM from [11] for downsampled images and Eigenfaces. The accuracy results are reported in Fig. 1. We tested for $K = 25$ and $T = 10000$. IBDL-E achieved a maximum accuracy of 95.39 % with the OMP encoding and $\lambda = 10$, $\gamma = 14.66$ and $\alpha = 0.000009$.

Discussion. The IBDL-E approach has state-of-the-art performances on face recognition. Interestingly, our approach achieved its best accuracy with down-sampled images. This was unexpected since eigenfaces are usually known to outperform them (it is indeed the case for SRC, NN and SVM). We believe that the eigenface transformation removes important information during the orthog-onalization affecting the latent structure learning. Using downsampled images is advantageous because no training data is required for extracting the features. Also, IBDL-E has better accuracy than SRC with low-dimensional features but does not outperform it with high-dimensional features. We believe that this is due to the curse of dimensionality. However, since our approach learns the dictio-naries independently, the 38 class-specific models were trained in parallel. This parallel linear speed up is important for recognition task with many classes, such as face recognition which requires on class per individual.

4.3   The Effects of Incremental Learning

In this section, we demonstrate the beneficial effects of the proposed incremental learning rule. We trained two IBDL models on the USPS dataset by optimiz-ing Eq. 4, one with incremental learning and the other without, using a LARS encoding with hyper-parameters $\lambda = 3$, $\alpha = 0.0001$, $\gamma = 15$, $T = 5000$ and

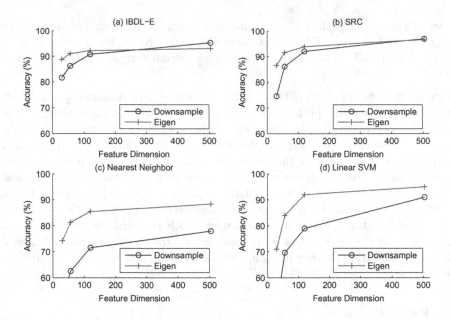

Fig. 1. Face recognition results on the Extended Yale B dataset.

$K = 25$. Figure 2 shows the 25 learned dictionary atoms of both dictionaries for the class digit 0. As explained in Sect. 3, the dictionary trained without incremental learning is suboptimal containing atoms unrelated to the structure of a 0. This can clearly be seen in the top row of Fig. 2 where many entries (e.g. 4th) are just noise. Those atoms were never updated during the gradient descent (their $\mathbf{u}_k$ were 0). On the contrary, the dictionary learned with incremental learning (bottom row of Fig. 2) used all atoms to represent the latent structure. The same phenomenon appeared for all class-specific dictionaries. Therefore, this shows empirically that incremental learning improves the convergence guarantee of the gradient descent under a sparsity constraint for finding a more representative dictionary.

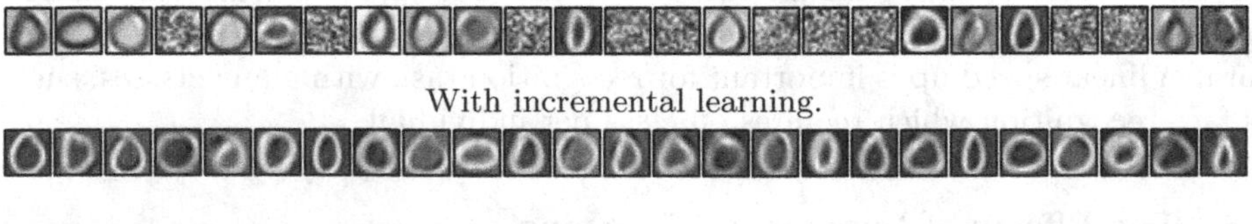

Without incremental learning.

With incremental learning.

Fig. 2. The effects of incremental learning on the USPS dataset for class digit 0. Each image corresponds to a dictionary atom $\mathbf{d}_k$. The dictionary learned without incremental learning (top row) contains uninformative features in comparison to the dictionary learned with incremental learning (bottom row).

5 Conclusion

In this paper, we proposed *incrementally built dictionary learning* (IBDL), a supervised multi-dictionary approach for classification. The IBDL aims to learn class-specific dictionaries with high diversity by optimizing the between-class residual error distance. We proposed a parallel optimization framework based on stochastic gradient descent that allows learning the dictionaries simultaneously. The preliminary experimental results on digit and face recognition show that IBDL achieves good accuracy on face recognition and improved the KNN classifier performances with the proposed sparse coding-based features. As future work, we will extend the notion of diversity by using probability simplexes as dictionary atoms and by considering a Kullback-Leibler based distance between the dictionaries.

References

1. Guyon, I., Gunn, S., Nikravesh, M., Zadeh, L.: Feature Extraction: Foundations and Applications. Springer, New York (2006)
2. Bengio, Y., Courville, A., Vincent, P.: Representation learning: a review and new perspectives. IEEE Trans. Pattern Anal. Mach. Intell. **35**(8), 1798–1828 (2013)
3. Kreutz-Delgado, K., Murray, J.F., Rao, B.D., Engan, K., Lee, T.W., Sejnowski, T.J.: Dictionary learning algorithms for sparse representation. Neural Comput. **15**(2), 349–396 (2003)
4. Coates, A., Ng, A.Y.: The importance of encoding versus training with sparse coding and vector quantization. In: Proceedings of the 28th International Conference on Machine Learning (ICML 2011), pp. 921–928 (2011)
5. Mairal, J., Ponce, J., Sapiro, G., Zisserman, A., Bach, F.R.: Supervised dictionary learning. In: Advances in Neural Information Processing Systems, pp. 1033–1040 (2009)
6. Ramirez, I., Sprechmann, P., Sapiro, G.: Classification and clustering via dictionary learning with structured incoherence and shared features. In: 2010 IEEE Conference on Computer Vision and Pattern Recognition (CVPR), pp. 3501–3508 (2010)
7. Yang, M, Zhang, D, Feng, X: Fisher discrimination dictionary learning for sparse representation. In: 2011 IEEE International Conference on Computer Vision (ICCV), pp. 543–550 (2011)
8. Zhang, Q., Li, B.: Discriminative K-SVD for dictionary learning in face recognition. In: 2010 IEEE Conference on Computer Vision and Pattern Recognition (CVPR), pp. 2691–2698 (2010)
9. Jiang, Z., Lin, Z., Davis, L.S.: Learning a discriminative dictionary for sparse coding via label consistent K-SVD. In: 2011 IEEE Conference on Computer Vision and Pattern Recognition (CVPR), pp. 1697–1704 (2011)
10. Aharon, M., Elad, M., Bruckstein, A.: K-SVD: an algorithm for designing overcomplete dictionaries for sparse representation. IEEE Trans. Signal Process. **54**(11), 4311–4322 (2006)
11. Wright, J., Yang, A.Y., Ganesh, A., Sastry, S.S., Ma, Y.: Robust face recognition via sparse representation. IEEE Trans. Pattern Anal. Mach. Intell. **31**(2), 210–227 (2009)

12. Donoho, D.L., Elad, M.: Optimally sparse representation in general (nonorthogonal) dictionaries via ℓ_1 minimization. Proc. Natl. Acad. Sci. **100**(5), 2197–2202 (2003)
13. Tibshirani, R.: Regression shrinkage and selection via the lasso. J. Roy. Stat. Soc. Ser. B (Methodol.) **58**, 267–288 (1996)
14. Pati, Y.C., Rezaiifar, R., Krishnaprasad, P.: Orthogonal matching pursuit: recursive function approximation with applications to wavelet decomposition. In: 1993 Conference Record of The Twenty-Seventh Asilomar Conference on Signals, Systems and Computers, pp. 40–44 (1993)
15. Efron, B., Hastie, T., Johnstone, I., Tibshirani, R.: Least angle regression. Ann. Stat. **32**(2), 407–499 (2004)
16. Donoho, D.L., Johnstone, I.M.: Adapting to unknown smoothness via wavelet shrinkage. J. Am. Stat. Assoc. **90**(432), 1200–1224 (1995)
17. Engan, K., Aase, S.O., Husoy, J.H.: Method of optimal directions for frame design. In: 1999 IEEE International Conference on Acoustics, Speech, and Signal Processing, vol. 5, pp. 2443–2446 (1999)
18. Mairal, J., Bach, F., Ponce, J., Sapiro, G.: Online learning for matrix factorization and sparse coding. J. Mach. Learn. Res. **11**, 19–60 (2010)
19. Snoek, J., Larochelle, H., Adams, R.P.: Practical Bayesian optimization of machine learning algorithms. In: Advances in Neural Information Processing Systems, pp. 2951–2959 (2012)
20. Huang, K., Aviyente, S.: Sparse representation for signal classification. In: Advances in Neural Information Processing Systems, pp. 609–616 (2006)

Learning to Reconstruct 3D Structure from Object Motion

Wentao Liu[✉], Haobin Dou, and Xihong Wu

Key Lab of Machine Perception (MOE), Speech and Hearing Research Center,
Peking University, Beijing, People's Republic of China
{liuwt,douhb,wxh}@cis.pku.edu.cn

Abstract. In this paper, we propose a new approach for reconstructing 3D structure from motion parallax. Instead of obtaining 3D structure from multi-view geometry or factorization, a Deep Neural Network (DNN) based method is proposed without assuming the camera model explicitly. In the proposed method, the targets are first split into connected 3D corners, and then the DNN regressor is trained to estimate the relative 3D structure of each corner from the target rotation. Finally, a temporal integration is performed to further improve the reconstruction accuracy. The effectiveness of the method is proved by a typical experiment of the Kinetic Depth Effect (KDE) in human visual system, in which the DNN regressor reconstructs the structure of a rotating 3D bent wire. The proposed method is also applied to reconstruct another two real targets. Experimental results on both synthetic and real images show that the proposed method is accurate and effective.

Keywords: 3D Reconstruction · Structure from Motion · Deep Neural Network · Kinetic Depth Effect

1 Introduction

2D image features like edges and lines are important for visual perception as they convey compact information about objects in the world. However they must be interpreted into 3D structures to make the inferential leap from image to environment [4]. 3D reconstruction is one of the basic problem in computer vision, and has important applications in scene understanding and augmented reality. Recovering 3D structure from 2D images is an inverse problem, however, the lost depth information can also be recovered from various visual cues, such as binocular disparity, motion parallax, accommodation, shading and shadows, *etc.* Among these, motion parallax is one of the most important as motion information figures importantly in both spatial perception and organization of visual perception.

In computer vision, recovering 3D structure from images taken from different views has been widely studied, known as Structure From Motion (SFM). Recent works on SFM concern the geometric differences and assume that the geometric

© Springer International Publishing Switzerland 2015
S. Arik et al. (Eds.): ICONIP 2015, Part I, LNCS 9489, pp. 127–137, 2015.
DOI: 10.1007/978-3-319-26532-2_15

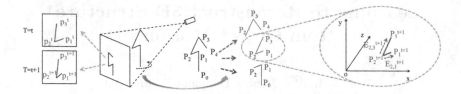

Fig. 1. A bent wire is rotating around the vertical axis with its shadow back-projected onto a translucent screen. Corners are selected to be the reconstruction units.

camera model is accurate. However, these methods are computationally expensive and may not be able to reconstruct correct 3D structures due to local optima. In this paper, a Deep Neural Network (DNN) based 3D reconstruction method, which learns the 2D motion patterns of different 3D structures and then inferences 3D structures from the object rotation, is proposed. As the motion parallax provides relative and quantitative depth information, the proposed method successfully recovers the relative 3D structure of the observed object. Furthermore, a typical experiment of the Kinetic Depth Effect (KDE) [1] is suggested to evaluate the proposed method. The KDE is the first phenomenon that demonstrates the ability of human to perceive depth from the object motion, as is shown in Fig. 1. The back-projected shadow of a wire is presented to the observers. As soon as the wire begins rotating, a 3D rotating wire is perceived. Although reconstructing 3D information from the wire rotation is geometrically underdetermined, human observers have no difficulty to perceive 3D structure from only two views. They obtain the possible interpretation with the gained experience.

In the experiments, the objects are first split into several reconstruction units and the learned DNN model is applied to estimate the 3D structure of each unit. The whole objects are reconstructed by jointing each units. After that, a temporal integration method is performed to integrate the 3D structures and estimate the 3D motion.

The remainder of the paper is organized as follows. Section 2 reviews related work in SFM and learning based 3D reconstruction. The proposed DNN based 3D structure reconstruction method is presented in Sect. 3, and Sect. 4 shows the experimental results on both synthetic and real images. Finally Sect. 5 concludes the paper.

2   Related Work

Structure from motion has been actively studied over decades to estimate 3D structure and camera motion [10,11,20,22]. The bundle adjustment [6] strategy was applied to estimate the 3D structure. During the optimization procedure, nonlinear least squares was often applied to measure the projection errors. The factorization method was also proposed to reconstruct 3D model from image correspondence [2]. It provided a linear method to initialize iterative techniques, *e.g.* bundle adjustment. The shape and motion were able to be reconstructed simultaneously under orthography or related models. The factorization approaches were extended to handle non-rigidity by Bregler *et al.* [5]. However, the camera model was assumed to be orthographic projection.

Many researchers focused on learning based 3D reconstruction methods. They denoted to the problem as training models via supervised learning to infer the 3D structure. Saxena *et al.* [12] presented an algorithm for predicting 3D structure from single still image with a trained Markov Random Field (MRF) model. Their method produced visually-pleasing 3D depth maps but was not accurate enough. The binocular disparity and the monocular cues were incorporated to obtain more accurate depth estimation in [13]. Hedau *et al.* [16] and Xiao *et al.* [17] presented methods to detect individual boxy volumetric structures from single image using 3D cuboid detectors. Edges and corners were modeled when the spatial consistency were enforced to a 3D cuboid model. Fouhey *et al.* [19] aimed at discovering a vocabulary of 3D primitives that are visually discriminative and geometrically informative, and then dense 3D interpretation of single image was created by the primitives. Li *et al.* [21] predicted the depth of a scene from single monocular image by combining deep convolutional neural network and conditional random fields. However, those methods focused on reconstructing 3D structure from single images based on static visual cues, such as texture variations and gradients, edge interpretation, *etc*. Several researchers had interpreted the SFM as a probabilistic generative model which permitted the inclusion of priors on the motion sequence [7,15], but they assumed that the object shape at each time was drawn from a specific distribution.

3 DNN Based 3D Reconstruction Method

In this section, the proposed DNN based 3D structure reconstruction method is detailed. The representation of the reconstruction unit is presented firstly, followed by a description of the proposed model, and the temporal integration is presented at last.

3.1 Reconstruction Unit

In the proposed method, three pairwise point-correspondences from two views are selected as the reconstruction unit as is illustrated in Fig. 1. Three consecutive points form a corner which contains a corner point and two edges in KDE. Thus, the bent wire is first split into reconstruction units, and then the 3D structure of each unit is estimated from two images taken at time t and $t + 1$. The 3D structure of the whole wire is obtained by jointing every parts.

Figure 1 shows the representation of the reconstruction unit. The 3D position of the kth corner point at time t is represented by P_k^t. S_k refers to the 3D reconstruction unit with the corner point P_k and another two points P_{k-1} and P_{k+1}. The vectors $E_{k,k-1}^{t+1} = P_{k-1}^{t+1} - P_k^{t+1}$ and $E_{k,k+1}^{t+1} = P_{k+1}^{t+1} - P_k^{t+1}$ indicate the relative 3D structure of S_k at time $t + 1$. p_k^t represents the position of the kth corner point in image taken at time t, and $v_k^t = p_k^{t+1} - p_k^t \in \mathbb{R}^2$ represents the 2D motion of the corner point p_k from time t to $t + 1$. $e_{k,k-1}^t = p_{k-1}^t - p_k^t$ and $e_{k,k+1}^t = p_{k+1}^t - p_k^t$ indicate the 2D structure of the corner at time t.

3.2 Deep Neural Network for 3D Reconstruction

DNN has emerged as an important research area of machine learning [9,14]. During the past several years, DNN has been impacting a wide range of pattern recognition tasks. In this work, 3D structure reconstruction from image motion is regarded as a nonlinear regression problem since the targets are rotating around different axis at various speed. DNN provides a simple framework to learn such regression models owe to the multiple levels of nonlinear operations [14].

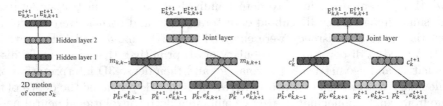

Fig. 2. DNN architecture for 3D reconstruction, the left one is the structureless model. The middle structured architecture is the motion-first model and the corner-first model is on the right.

After obtaining the point correspondences, DNN models are trained to regress image motions to 3D structures. Three neural networks of different architecture are trained in our experiment. The first neural network, as depicted in the left of Fig. 2, which uses positions p_{k-1}, p_k and p_{k+1} at time t and $t+1$ as input followed by 2 layers of hidden units and then output the 3D structure of S_k^{t+1}, i.e. $E_{k,k-1}^{t+1}$ and $E_{k,k+1}^{t+1}$.

Good internal representations are supposed to be hierarchical. Pixels are assembled into edges, and edges are then assembled into parts (like the reconstruction unit applied in this paper), and parts into objects (e.g. a wire) [18]. Instead of taking the unorganized 2D information as input, the second model has a structured architecture, as is shown in the middle of Fig. 2. There are four inputs in this model. $(p_k^t, e_{k,k-1}^t)$ represents the position and 2D structure of edge $e_{k,k-1}$ at time t and $(p_k^{t+1}, e_{k,k-1}^{t+1})$ represents the same information of edge $e_{k,k-1}$ at time $t+1$. The representation of $e_{k,k+1}$ is the same as $e_{k,k-1}$. Layers $m_{k,k-1}$ and $m_{k,k+1}$ extract motion information of $e_{k,k-1}$ and $e_{k,k+1}$ from t to $t+1$. As the extraction of motion information at each time is consistent, the parameters of $m_{k,k-1}$ and $m_{k,k+1}$ are shared. After modeling the 2D motion of edges, a joint layer is applied to integrate all the 2D information of corner S_k. Then the joint layer is full connected to the output layer which indicates the 3D structure of corner S_k^{t+1}. To prove that the motion should be processed in the first layer, another model shown in the right of Fig. 2 is proposed. Edges are assembled into corners firstly, i.e. c_k^t and c_k^{t+1}, and then a joint layer integrates the 2D information.

After each reconstruction unit is reconstructed individually, $\hat{P}_k^{t+1}$, which represents the position of P_k^{t+1} relative to P_0^{t+1}, is obtained by assuming that the first corner point P_0^{t+1} is located at the origin O of coordinate.

$$\hat{P}_k^{t+1} = \begin{cases} O, & \text{if } k = 0; \\ \hat{P}_0^{t+1} - E_{1,0}^{t+1}, & \text{if } k = 1, \\ \hat{P}_{k-1}^{t+1} + \frac{E_{k-1,k}^{t+1} - E_{k,k-1}^{t+1}}{2}, & \text{if } k > 1. \end{cases} \tag{1}$$

3.3   Temporal Integration

The visual system considers motion perception in at least two steps: an early process of 2D image motion and a later process perceiving the 3D objects motion. In the experiment of KDE, the wire is supposed to be rotated around the first part $P_0 P_1$. 3D motion is firstly estimated from the 3D structure obtained from the image sequence, while the 3D rotation is assumed to be constant for a short time. After that, the 3D structure of the wire at each time is integrated to improve the reconstruction accuracy.

The relationship between the relative 3D position of kth point at time $t - 1$ and t, i.e. $\hat{P}_k^{t-1}$ and $\hat{P}_k^t$, is defined as $\hat{P}_k^t = R^t \hat{P}_k^{t-1}$. R^t represents the rotation of the wire from $t - 1$ to t, which is assumed to be fixed for a short time, *i.e.* $R^t = R^{t+1}$. As the first point P_0 is located at the origin and $P_0 P_1$ is the rotation axis. The rotation R^t can be computed by the least square method. Then the 3D structure is extracted at time n by $\overset{*}{P}_k^n = (R^t)^{(t-n)} \hat{P}_k^n$. Instead of considering all reconstructed 3D structures, the 3D structures from $n = t - T$ to t are integrated, $\bar{P}_k^t = \frac{1}{T} \sum_{n=t-T}^t \overset{*}{P}_k^n$. T represents the length of the time window of 3D perception. $\bar{P}_k^t$ indicates the relative position of P_k after temporal integration.

4   Experiments

4.1   Data Generation

The synthetic datasets, which contain the ground truth 3D structures and images of different views, are generated using computer graphics. Human observers place the target of interest on the fovea which has the highest visual acuity, and track the target under movements caused by the object or the observer's head [4]. So more attentions should be paid to the restricted area around the rotation axis when recovering 3D structures. In the experiment, the synthetic wires are rotating near the middle of view. The rotation axises which parallel to y-axis are generated within 20 cm around the y-axis randomly. All corners of the bent wires are limited to certain range around the rotation axis, which is set to 15 cm in the experiment. Besides, the range of possible rotational speed is also restricted as observers could only percept the object motion which is neither too slow nor too fast [4]. Thus, the rotational speed is restricted from 5 to 10 degrees per frame. The images of the wires, as is shown in Fig. 3, are captured from a fixed virtual camera which is 100 cm from the origin. Only the motion parallax extracted from the images is used to recover the 3D structure. The camera parameters is calibrated from a Logitech HD Webcam C270 camera which is also used to capture real images with a resolution of 640*480 pixels.

Fig. 3. Examples of synthetic data generated by computer graphic. The images in the same column are captured from the same rotating corner. The white line is the rotation axis, the yellow point represents the position of a corner and the red lines represent the 2D structure of the corner (Color figure online).

The DNN models are trained on synthetic data and evaluated on both synthetic and real images. 160,000 reconstruction units are generated randomly for training and other 40,000 units for testing. 100 wires each contains 5 reconstruction units are generated to test the model. For each wire, 20 images are generated to evaluate the temporal integration method. Real images are captured from the Logitech HD Webcam C270 camera when the targets are rotating.

4.2   Reconstruction on Synthetic Images

To prove the effectiveness and accuracy of the proposed method, it is first tested on the synthetic dataset. The average distance between reconstructed corner points and the ground truth, *i.e.* the reconstruction error, is suggested as metrics. As the reconstructed 3D structure is relative, the absolute position of point P_k^{t+1} is provided to align the estimated reconstruction units and the ground truth. The first model (Structureless Network, SN) has 12 visible units which represent 2D information of the reconstruction unit, followed by 2 layers of 256 hidden units. The second model (Motion First, MF) and the third model (Corner First, CF) both contain 4*4 visible units. 128 hidden units are applied for each path way in the first level hidden layers and 256 hidden units for the joint layer. The parameters of the first level hidden layers are shared. All networks are trained by the stochastic gradient descent (SGD) with the sum square loss function. The weights are initialized from a zero-mean Gaussian distribution with standard deviation 0.001 and the biases with the constant 0. For all layers, the rectified linear unit (ReLU) is employed as activation function. The initial learning rate is 0.001 and decays during training progress. The training process is terminated when the training error convergences.

Table 1. Quantitative comparison on synthetic images.

Model	Average distance (cm)
CF-unshared	1.52
MF-unshared	1.52
SN	1.49
CF-shared	1.51
MF-shared	1.48

Comparative result is given in Table 1. The following models are evaluated: (a) CF-unshared: combining the motion information after the corner has been processed without the parameters sharing. (b) MF-unshared: combining the motion information before formating corner but not sharing parameters.(c) SN: the two layers structureless network. (d) CF-shared: the architecture is the same as CF-unshared but the parameters in the first layer are shared. (e) MF-shared: MF model with the parameters sharing. The results show that, the DNN based method is able to reconstruct 3D structure from three points of two distinct views accurately. The quantitative comparison shows that the average distance of CF model is larger than both MF and SN models. The results suggest that image motion should be processed in low level of network. Table 1 also shows that the MF model with parameters sharing (MF-shared) outperforms all the other models in reconstruction accuracy with the least trained parameters. The results prove that sharing low level parameters is more reasonable and accurate.

Although 3D structures of wires could be reconstructed from only two views, temporal information should also be performed to decrease reconstruction error. The mean average distance of temporal integration is shown in Fig. 4. The result shows that the integration method improves the reconstruction accuracy effectively when more images are available. The reconstruction error decreases faster at the beginning, and slows down when more images are available just as the human observers do. The results after temporal integration of 20 views are shown in Fig. 5. The higher corners appear to present larger reconstruction error due to error accumulation as the proposed model reconstructs the relative structure.

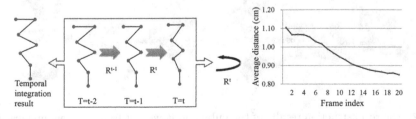

Fig. 4. The temporal integration is represented on the left, in which the rotation R^t and integrated 3D structure are estimated. The mean average distance after temporal integration is on the right.

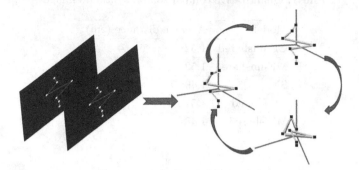

Fig. 5. Reconstruction results on synthetic images. The input images are on the left and the reconstruction result of different views is shown on the right. The red lines represent the ground truth of the wires and the yellow points represent corner points of the ground truth. The reconstructed structure of each wire is represented in green and the reconstructed corners are shown as black (Color figure online).

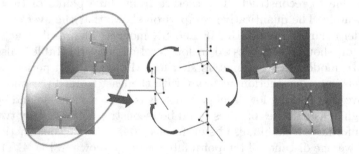

Fig. 6. Reconstruction result of rotation wire on real images. The 3D structure of the wire is reconstructed from the first two images. The reconstruction result (the green lines) of different views is shown with the corresponding real images (Color figure online).

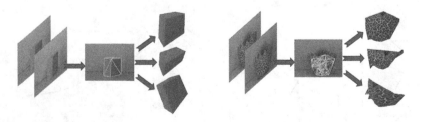

Fig. 7. The reconstruction results of the cube and flowerpot from two real images. The virtual reconstruction units are represented as green and the results of different views are also presented (Color figure online).

4.3 Reconstruction on Real Images

The reconstruction model is also tested on real images. The corners are first detected by Shi-Tomasi corner detector [3] followed by a non-maximal suppression. After that, the correspondence is obtained by tracking corners through each view. The MF-shared model is then performed to reconstruct the 3D structure. The results on real images are shown in Fig. 6. The 3D wires are reconstructed from the correspondence between the first two views. The reconstruction result (the green lines) are shown with the corresponding real images of different views. The results show that the proposed model is reasonable for reconstruction on real images. To further evaluate the proposed method, it is employed to more complex targets. SURF extractor [8] followed by the RANSAC method is applied to extract point-correspondences. Reconstruction unit is assumed to be existent between every three corner points and the whole 3D structure is obtained after estimating the relative 3D structure of each units, and then the texture is mapped to the 3D results. The reconstruction results are shown in Fig. 7. As the 3D structures are reconstructed from two views, only the visible parts of the targets are recovered. The complete 3D structure of each target could be recovered by integrating all the parts when more images are captured, which can be further considered.

5 Conclusions

In this paper, a DNN based method is presented for reconstructing 3D structure from object rotation without predefining the camera model. Since the speed of object motion is various, the 3D reconstruction is treated as a nonlinear regression problem. DNN is adopted to model the relation between 2D motions and 3D structures. Instead of obtaining absolute 3D position, the proposed method reconstructs the relative 3D structure of the observed object. Moreover, the temporal integration estimates 3D structures from multiple frames and further improves the reconstruction accuracy.

The proposed approach is evaluated on both synthetic and real images. The experimental results on synthetic images present that the proposed method reconstructs accurate 3D structures with an average reconstruction error of 1.48 cm. In the experiments on real images, a bent wire and another two real targets are successfully reconstructed. The experimental results demonstrate that the proposed method is both accurate and effective.

Acknowledgement. The work was supported in part by the National Basic Research Program of China (2013CB329304), the "Twelfth Five-Year" National Science&Technology Support Program of China (No.2012BAI12B01), the Major Project of National Social Science Foundation of China (No.12&ZD119), the research special fund for public welfare industry of health (201202001) and National Natural Science Foundation of China (No.81170906).

References

1. Wallach, H., O'Connell, D.N.: The kinetic depth effect. J. Exp. Psychol. **45**, 205–217 (1953)
2. Tomasi, C., Kanade, T.: Shape and motion from image streams under orthography: a factorization method. Int. J. Comput. Vision **9**, 137–154 (1992)
3. Shi, J., Tomasi, C.: Good features to track. In: IEEE Computer Society Conference on Computer Vision and Pattern Recognition, Seattle, pp. 593–600 (1994)
4. Palmer, S.E.: Vision Science: Photons to Phenomenology. MIT Press, Cambridge (1999)
5. Bregler, C., Hertzmann, A., Biermann, H.: Recovering non-rigid 3D shape from image streams. In: IEEE Computer Society Conference on Computer Vision and Pattern Recognition, Hilton Head Island, pp. 690–696 (2000)
6. Triggs, B., McLauchlan, P.F., Hartley, R.I., Fitzgibbon, A.W.: Bundle adjustment – a modern synthesis. In: Triggs, B., Zisserman, A., Szeliski, R. (eds.) ICCV-WS 1999. LNCS, vol. 1883, p. 298. Springer, Heidelberg (2000)
7. Gruber, A., Weiss, Y.: Multibody factorization with uncertainty and missing data using the EM algorithm. In: IEEE Computer Society Conference on Computer Vision and Pattern Recognition, Washington, DC, pp. 707–714 (2004)
8. Bay, H., Tuytelaars, T., Van Gool, L.: SURF: speeded up robust features. In: Leonardis, A., Bischof, H., Pinz, A. (eds.) ECCV 2006, Part I. LNCS, vol. 3951, pp. 404–417. Springer, Heidelberg (2006)
9. Hinton, G.E., Osindero, S., Teh, Y.W.: A fast learning algorithm for deep belief nets. Neural Comput. **18**, 1527–1554 (2006)
10. Snavely, N., Seitz, S.M., Szeliski, R.: Photo tourism: exploring photo collections in 3D. ACM Trans. Graph. **5**, 835–846 (2006)
11. Klein, G., Murray, D.: Parallel tracking and mapping for small AR workspaces. In: 6th IEEE and ACM International Symposium on Mixed and Augmented Reality, Nara, pp. 225–234 (2007)
12. Saxena, A., Sun, M., Ng, A.Y.: Learning 3-D scene structure from a single still image. In: International Conference on Computer Vision Workshop, Rio de Janeiro, pp. 1–8 (2007)
13. Saxena, A., Schulte, J., Ng, A.Y.: Depth estimation using monocular and stereo cues. In: International Joint Conference on Artificial Intelligence, Hyderabad, pp, 2197–2203 (2007)
14. Bengio, Y.: Learning deep architectures for AI. Found. Trends Mach. Learn. **2**, 1–127 (2009)
15. Ross, D.A., Tarlow, D., Zemel, R.S.: Learning articulated structure and motion. Int. J. Comput. Vision **88**, 214–237 (2010)
16. Hedau, V., Hoiem, D., Forsyth, D.: Recovering free space of indoor scenes from a single image. In: IEEE Computer Society Conference on Computer Vision and Pattern Recognition, Rhode Island, pp. 2807–2814 (2012)
17. Xiao, J., Russell, B.C., Torralba, A.: Localizing 3D cuboids in single-view images. In: Advances in Neural Information Processing Systems, Lake Tahoe, pp. 746–754 (2012)
18. Farabet, C., Couprie, C., Najman, L., LeCun, Y.: Learning hierarchical features for scene labeling. Pattern Anal. Mach. Intell. **35**, 1915–1929 (2013)
19. Fouhey, D.F., Gupta, A., Hebert, M.: Data-driven 3D primitives for single image understanding. In: International Conference on Computer Vision, Sydney, pp. 3392–3399 (2013)

20. Tanskanen, P., Kolev, K., Meier, L., Camposeco, F., Saurer, O., Pollefeys, M.: Live metric 3D reconstruction on mobile phones. In: International Conference on Computer Vision, Sydney, pp. 65–72 (2013)
21. Li, B., Shen, C., Dai, Y., Hengel, A., He, M.: Depth and surface normal estimation from monocular images using regression on deep features and hierarchical CRFs. In: Proceedings of the IEEE Conference on Computer Vision and Pattern Recognition, Boston, pp. 1119–1127 (2015)
22. Resch, B., Lensch, H.P.A., Wang, O., Pollefeys, M., Sorkine-Hornung, A.: Scalable Structure from Motion for Densely Sampled Videos. In: Proceedings of the IEEE Conference on Computer Vision and Pattern Recognition, Boston, pp. 3936–3944 (2015)

Convolutional Networks Based Edge Detector Learned via Contrast Sensitivity Function

Haobin Dou[✉], Wentao Liu, Junnan Zhang, and Xihong Wu

Key Lab of Machine Perception (MOE), Speech and Hearing Research Center,
Peking University, Beijing 100871, China
{douhb,liuwt,zhangjn,wxh}@cis.pku.edu.cn

Abstract. Edge detection extracts rich geometric structures of the image and largely reduces the amount of data to be processed, providing essential input to many visual tasks. Traditional algorithms consist of three steps: smoothing, filtering and locating, in which the filters are usually designed manually and thresholds are selected without strictly theoretical support. In this paper, convolutional networks (ConvNets) are trained to detect edges by learning a group of filters and classifiers simultaneously. In addition, the contrast sensitivity function (CSF) in visual psychology is adopted to determine whether an edge is visible to human visual system (HVS). Edge samples of various appearance are synthesised, and then labelled via CSF for model training. Multichannel ConvNets are trained to perceive edges of different frequencies and composed at last. Compared with classical algorithms, ConvNets-CSF model is more robust to contrast variation and more biologically plausible. Evaluated on USF edge detection dataset, it achieves comparable performance as Canny edge detector and outperforms other classical algorithms.

Keywords: Edge detection · Convolutional networks · Contrast sensitivity function

1   Introduction

Edge is defined as the point at which intensity changes abruptly in images. It may be produced from nonuniform light, surface material variation, object border, occlusion etc. Edge detection is an essential pre-processing stage in both biological and machine visual systems and provides important feature cues for image segmentation, stereo matching, object detection and tracking. Therefore over a long period of time, edge detection has been a fundamental research topic in computer vision.

Nowadays lots of algorithms are available for edge detection. They mainly consist of three steps: smoothing, filtering and locating. Edge operators like Sobel [1], Prewitt [2] and Roberts [3] are designed to compute image's intensity gradient, then on which threshold is applied to locate edge positions. These algorithms are simple but also crude. To mimic the receptive field of simple cells in primary

© Springer International Publishing Switzerland 2015
S. Arik et al. (Eds.): ICONIP 2015, Part I, LNCS 9489, pp. 138–146, 2015.
DOI: 10.1007/978-3-319-26532-2_16

visual cortex (V1), (Marr and Hildreth 1980) [4] proposed an algorithm by finding zero-crossings of the Laplacian operator applied to a Gaussian-smoothed image. (Canny 1986) introduced an optimal smoothing filter to satisfy the three mathematical criteria for edge detection [5]. Strengthened with techniques of non-maximum suppression and edge tracking by hysteresis, Canny edge detector has long been the state-of-the-art edge detector.

Although most commonly used by the computer vision community, gradient-based edge detectors suffer from a number of problems: (1) It's hard to choose appropriate threshold values to preserve weak edges while eliminate noise. (2) The differential operators need to be designed elaborately to perceive all the morphological properties of edge. (3) The size of operators (and hence the degree of smoothing) also affects the estimation of edge position.

To solve the first problem, we borrow ideas from the findings of psychophysics. Contrast sensitivity function (CSF) [6] depicts the contrast sensitivity of human visual system (HVS) to sinusoidal gratings of different frequencies, and this can be served as criterion to discriminate edge from non-edge in perceptual sense, as all edges can be decomposed into a series of sinusoidal gratings.

For the second problem, we use a learning method to obtain operators directly from raw data, which has become a trend in edge detection [7,8]. The model used by us is convolutional networks (ConvNets) [9], which have been widely used in many tasks of computer vision and become new state of the art. With ConvNets, differential filtering and edge locating are combined in a unified frame by learning operators and classifiers simultaneously.

Based on CSF and ConvNets, we build a multi-channel edge detector to perceive both coarse and fine edges. Training samples are synthesised with various frequencies, orientations and contrasts, and then labelled automatically as edge or non-edge via CSF. For each frequency channel, a ConvNet is trained with its filter size adaptive to the frequency. Outputs of all the ConvNets are composed and thinned to produce the final edge map.

The proposed model is evaluated qualitatively on some standard images. Compared with classical edge detectors, it is more robust to contrast variation and more biologically plausible. Quantitative evaluation is also made on the USF edge detection dataset and comparable result is achieved as Canny detector.

2   The Model Architecture

Our ConvNets based edge detector has a multi-channel architecture, i.e. one ConvNet is established for each frequency channel. Then outputs of all the channels are composed to give the final edge map, as shown in Fig. 1. This is inspired by the theory of spatial frequency channel in visual psychology.

2.1   Convolutional Networks

Convolutional Networks (ConvNets) were first introduced by (Lecun et al. 1989) [9]. They are hierarchical feature learning neural networks whose structure is biologically inspired. Compared with traditional fully-connected neural networks,

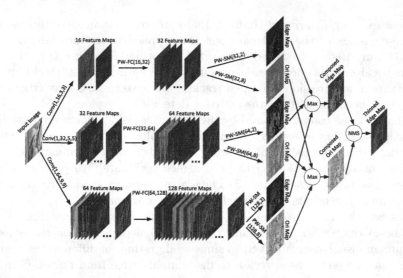

Fig. 1. The architecture of our multi-channel ConvNets based edge detector. The orientation map is drawn in color mode and each color depicts pixels of one orientation.

ConvNets adopt convolution mechanism to tie the weights across spatial domain which largely reduces the parameter amount to be learned.

Formally, let's denote the k-th feature map at layer l as H^{lk}, whose filters are determined by the weights W^{lk} and bias b^{lk}, then the feature map H^{lk} is obtained as follows:

$$H_{ij}^{lk} = f((W^{lk} * H^{l-1})_{ij} + b^{lk}) \tag{1}$$

where $(*)$ is the convolution operation and $f(\cdot)$ is the activation function, here, ReLU is used. After that are the response normalization layers and no pooling layers are used in the model to keep spatial resolution unchanged.

As shown in Fig. 1, for each frequency channel, a convolutional layer is firstly built, denoted as $\mathrm{Conv}(m, n, sx, sy)$, m is the number of input channels, n is the number of filters, (sx, sy) is the size of filters. That means given one input image, a set of n feature maps are obtained after the convolutional layer. Then the n-length feature vector of each pixel inputs into a pixel-wise fully-connected layer, denoted as PW-FC(m, n), that is equivalent to $\mathrm{Conv}(m, n, 1, 1)$. After that, two separate pixel-wise softmax layers are established to output the labels of each pixel, denoted as PW-SM(m, n). One of them performs a binary classification to discriminate edge from non-edge, the other is a multi-category classification to predicate the edge orientation, for which, we quantify its possible values of $[0°, 360°)$ uniformly into 8 categories. Actually our ConvNets can be also regarded as special patch-based neural networks.

2.2 Multi-channel Structure

In order to perceive both smoothly and sharply changed edges, we establish a ConvNet for each frequency channel. Their convolutional filter sizes are frequency dependent, i.e., lower frequency channel with larger filter size, while higher frequency channel with smaller filter size. Three channels are established with filter sizes of 3×3, 5×5, and 9×9. Each channel produces two maps, i.e. edge map and orientation map. A tuple of (*class, likelihood*) is assigned to every pixel of the maps, here *likelihood* refers to the output of softmax function.

To compose the outputs of multi-frequency channels, a MAX operation is used, i.e., the composed output is determined by the channels with the maximal predicated likelihood. On the composed edge map and orientation map, non-maximum suppression is applied to produce final thinned edge map.

3   Training Data Generation and Annotation

A large number of training samples are required to train our ConvNets, but manually building a dataset for edge detection is costly, and moreover, the annotation criterion is hard to determine, which usually results in confliction of annotations from different labelers. To guarantee both quantity and quality, we attempt to find a more efficient and reliable method to generate training samples.

3.1   Training Data Generation

Edges are basic image elements, which can be expressed and analyzed mathematically. Therefore, methods of computer graphics can be used to synthesise edge images conveniently. When synthesising training images, we control four variables: intensity mean (I_{mean}), intensity variation (I_{var}), orientation (o) and sample size (s). Their values are listed in Table 1.

Table 1. Variables controlled to synthesise training samples

Variable	Range	Step
I_{mean}	[0, 255]	16
I_{var}	[0, $\min(I_{mean}, 255 - I_{mean})$]	1
o	[0°, 360°)	45°
s	{3pixel, 5pixel, 9pixel}	–

We firstly generate edge images with large enough fixed size, which we call the base images used to produce training samples of different sizes. Figure 2 shows some examples of the base images. They are half-divided images with a dividing line passing through their centre. Except the size is fixed, their appearance are changed according to Table 1. Then we crop training samples from these base images along and around the dividing line with sample size s varying as Table 1. However, the label of each sample can't be determined completely by its cropping positions, i.e. samples off the dividing line should all be labelled as negative, while samples on the dividing line are not all positive, as they may be invisible to HVS with contrast below HVS's perceiving threshold.

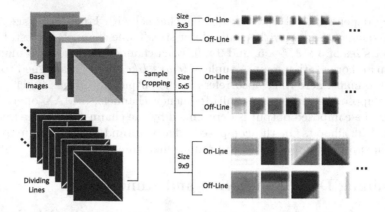

Fig. 2. The process of generating edge image samples.

3.2 Training Data Annotation

Contrast Sensitivity Function. Contrast sensitivity function (CSF) depicts an important character of the HVS. It is defined as the standard measurement of how sensitive observers are to gratings at different frequencies.

Since the early work of (Schade 1956), more complete models have been built for CSF [10,11], which have been applied to many tasks of image processing. The ModelFest dataset was created by (Watson et al. 2005) [12] to provide a public source of data to test and calibrate models of foveal spatial contrast detection. They fitted these data with a variety of simple models and found that simple models with particular parameters were powerful enough to account for the visibility of a wide variety of spatial stimuli.

We use the HPmH model proposed in [12] to model our CSF and its output is the reciprocal of the visible contrast threshold:

$$1/c_{\text{thr}} = G \cdot S_{\text{HPmH}}(f; f_0, f_1, a, p) = G \cdot \{\text{sech}[(f/f_0)^p] - a\text{sech}[f/f_1]\} \quad (2)$$

Including gain, it has five parameters and we use the fitting result of standard A as: $G = 373.08, f_0 = 4.1726, f_1 = 1.3625, a = 0.8493, p = 0.7786$. Here, f is the input frequency measured in cycles per degree (cpd) and the contrast is defined as Michelson-contrast, that is $c = I_{var}/I_{mean}$. The contrast sensitivity curve according to HPmH model is plotted in Fig. 3(a).

Labeling Samples via CSF. As the spatial frequency in CSF is measured in cpd, when applying it to edge detection, the observation model need to be considered. Let us assume for the moment, that the edges shall be detected in an image with a given resolution r measured in pixels (dots) per inch (ppi) and a viewing distance v measured in meters. The sampling frequency, f_S in pixels per degree is then given by $f_S = 2v\tan(0.5°)r/0.0254$. This transition process is intuitively shown in Fig. 3(b).

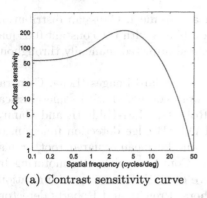

(a) Contrast sensitivity curve

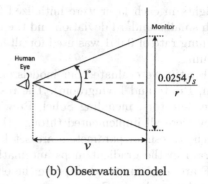

(b) Observation model

Fig. 3. Contrast sensitivity curve of the HPmH model and the transition of spatial frequency to image domain via observation model.

To get the exact f_S, we have made some reasonable assumptions according to the common viewing conditions and monitor settings. We assume that the viewing distance $v = 0.65$ metre and the monitor's resolution $r = 144$ ppi, then $f_S \approx 64$ pixels per degree.

Then for a training sample on the dividing line with size of s pixels, it can be approximately regarded as a one-cycle square-wave grating and its frequency (cpd) f can be computed as $f = f_S/s = 64/s$.

As the CSF is mainly measured for sinusoidal gratings, some transfer rules should be considered for square-wave gratings, which has been well studied in (Campbell and Robson 1968) [13]. They find out that the CSF of square-wave grating is mainly determined by its fundamental component when the frequency is over $0.8cpd$ and the square/sine ratio is measured as $4/\pi \approx 1.273$. So for each sample $X(c, o, f, i, j)$ with contrast c, orientation o, frequency f and the cropped position (i,j) in the base image, its edge predication label $Y_{edge}(X(c, o, f, i, j))$ is determined as follows:

$$Y_{edge}(X(c, o, f, i, j)) = \begin{cases} 1, & i\sin(o) + j\cos(o) = 0 \text{ and} \\ & c > (1.273 \cdot G \cdot S_{\text{HPmH}}(f))^{-1}; \\ 0, & \text{else.} \end{cases} \tag{3}$$

If $Y_{edge}(X(c, o, f, i, j)) = 1$, $X(c, o, f, i, j)$ will also be used to train the orientation predication layer, here, its label is simply $Y_{ori}(X(c, o, f, i, j)) = \lfloor o/45 \rfloor$.

Finally, the amount of training samples for edge predication in three channels is: $N_{pos}^{s=3} = 19,944$, $N_{neg}^{s=3} = 58,392$; $N_{pos}^{s=5} = 41,760$, $N_{neg}^{s=5} = 175,840$; $N_{pos}^{s=9} = 76,824$, $N_{neg}^{s=9} = 314,100$. Raw gray value ranging in [0,255] is used directly.

4   Experiments

We employed Theano [14] to build and train our models. Stochastic gradient descent (SGD) was used during training with a batch size of 500 examples.

Weights in each layer were initialized from a zero-mean Gaussian distribution with small standard deviation and the neuron biases with the constant 0. Equal learning rate of 0.001 was used for all layers and adjusted manually throughout training.

Firstly, we evaluated our models on some standard images (Lena, Cameraman, Pirate and Livingroom). The results were compared with some classical edge detectors, including Sobel, Prewitt, Roberts, Marr-Hildreth and Canny. They were all implemented through the MATLAB edge detection function, of which, threshold parameters are set based on the mean squared root or histogram of the gradient maps automatically. ConvNets without annotating by CSF are also compared to prove the effective of CSF. Due to space limitation, only the results on Cameraman are given here. Prewitt and Roberts detectors are also absent, as their results are very similar to Sobel.

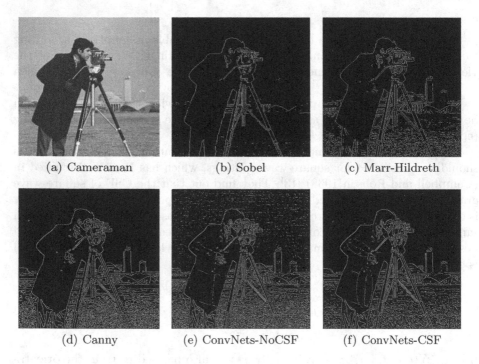

(a) Cameraman   (b) Sobel   (c) Marr-Hildreth

(d) Canny   (e) ConvNets-NoCSF   (f) ConvNets-CSF

Fig. 4. Comparison of our model to classical algorithms on Cameraman image. Red dash lines in (f) show the edges with low contrast detected by our model (Color figure online).

Then the proposed model was evaluated quantitatively on the USF dataset [15], a public edge detection dataset consisting of 60 gray images with manually labelled ground truth for each image. Precision (P), recall (R) and F-measure (F) were used to compare the performance of detectors, here $F = 2PR/(P + R)$ is the balanced mean between precision and recall.

Table 2. Comparison of different detectors on USF dataset

Methods	Precision	Recall	F-measure
Sobel	0.99	0.55	0.71
Prewitt	0.99	0.54	0.70
Roberts	0.99	0.46	0.62
Marr-Hildreth	0.93	0.72	0.81
Canny	0.85	0.82	0.83
ConvNets-NoCSF	0.61	0.87	0.72
ConvNets	0.87	0.80	0.83

As shown in Fig. 4 and Table 2, on the whole, Sobel, Prewitt and Roberts achieve extremely high precision, but their recalls are much low, as they use much simple operators and brutal thresholds, only strong enough edges are detected. Marr-Hildreth is better than them, as finding zero-crossings is more reliable than finding maximum-points. But single threshold is also used in Marr-Hildreth for the final output and lots of weak edges get missed. Canny achieves very excellent performance, as it applies edge tracking by hysteresis with double thresholds to filter out edges caused by noise and bright variation while preserving true weak edges. ConvNets based edge detector without CSF obtains very high recall, but lots of spurious edges are also detected. After adopting CSF to annotate the training samples, the model is trained to learn HVS's perceptual limits. Most of the spurious edges, distributing mainly in high frequencies, are eliminated, as their contrast is under the visible threshold, while the edges strong enough to cause perception are preserved. As edge tracking is not used in our model, continuity of the detected edges is not as good as Canny. While benefiting from CSF, our model can detect edges of very low contrast, such as edges marked out by red dash lines in Fig. 4(f). Finally, our ConvNets-CSF model achieves same value of F-measure as Canny detector and outperforms other methods.

5 Conclusion

In this paper, firstly, multi-channel ConvNets are built and trained for edge detection. They follow the multi-channel processing of HVS and combine filtering and locating into one unified leaning architecture. Edge maps and orientation maps are generated as two classification tasks of the model and final thinned edge maps are produced using non-maximal suppression. Secondly, to obtain enough training data, we proposed an approach to generate and annotate edge samples automatically, i.e. synthetising edge images of various appearance and annotating them via CSF. With these two proposals, our ConvNets-CSF model is more robust to contrast variation and achieves comparable performance to Canny edge detector on the USF edge detection dataset.

Acknowledgment. The work was supported in part by the National Basic Research Program of China (2013CB329304), the "Twelfth Five-Year" National Science & Technology Support Program of China (No. 2012BAI12B01), the Major Project of National Social Science Foundation of China (No.12&ZD119), the research special fund for public welfare industry of health (201202001) and National Natural Science Foundation of China (No. 81170906).

References

1. Sobel, I.: An isotropic 3x3 image gradient operator. Talk at the Stanford, Artificial Intelligence Project (SAIP) (1968)
2. Prewitt, J.M.S.: Object enhancement and extraction. In: Lipkin, B.S., Rosenfeld, A. (eds.) Picture Processing and Psychopictorics, pp. 75–149. Academic Press, New York (1970)
3. Roberts, L.G.: Machine perception of three-dimensional solids. In: Tippett, T., et al. (eds.) Optical and Electro-Optical Information Processing, pp. 159–197. MIT Press, Cambridge (1965)
4. Marr, D., Hildreth, E.: Theory of edge detection. Proc. Roy. Soc. Lond. Ser. B Biol. Sci. **207**(1167), 187–217 (1980)
5. Canny, J.: A computational approach to edge detection. IEEE Trans. Pattern Anal. Mach. Intell. **8**(6), 679–698 (1986)
6. Schade, O.H.: Optical and photoelectric analog of the eye. J. Opt. Soc. Am. **46**, 721–739 (1956)
7. Konishi, S.M., Yuille, A.L., Coughlan, J.M., Zhu, S.C.: Statistical edge detection: learning and evaluating edge cues. IEEE Trans. Pattern Anal. Mach. Intell. **25**(1), 57–74 (2003)
8. Pablo, A., Michael, M., Charless, F., Jitendra, M.: Contour detection and hierarchical image segmentation. IEEE Trans. Pattern Anal. Mach. Intell. **33**(5), 57–74 (2011)
9. LeCun, Y., Bottou, L., Bengio, Y., Haffner, P.: Gradient-based learning applied to document recognition. Proc. IEEE **86**(11), 2278–2324 (1998). IEEE Press, New York
10. Mannos, J.L., Sakrison, D.J.: The effects of a visual fidelity criterion on the encoding of images. IEEE Trans. Inf. Theor. **4**, 525–536 (1974)
11. Barten, P.G.J.: Contrast Sensitivity Function and Its Effect on Image Quality. SPIE Press, Bellingham (1999)
12. Watson, A.B., Ahumada Jr, A.J.: A standard model for foveal detection of spatial contrast. J. Vison **5**, 717–740 (2005)
13. Campbell, F.W., Robson, J.G.: Application of fourier analysis to the visibility of gratings. J. Physiol. **197**, 551–566 (1968)
14. Bastien, F., Lamblin, P., et al.: Theano: new features and speed improvements. In: Deep Learning Workshop, NIPS (2012)
15. Bowyer, K., Kranenburg, C., Dougherty, S.: Edge detector evaluation using empirical ROC curves. Comput. Vis. Image Understand. **1**(84), 77–103 (2001)

Learning Algorithms and Frame Signatures for Video Similarity Ranking

Teruki Horie, Akihiro Shikano, Hiromichi Iwase, and Yasuo Matsuyama[✉]

Department of Computer Science and Engineering,
Waseda University, Tokyo 169-8555, Japan
{t.horie,a.shikano,ihi}@wiz.cs.waseda.ac.jp,
yasuo2@waseda.jp
http://www.wiz.cs.waseda.ac.jp

Abstract. Learning algorithms that harmonize standardized video similarity tools and an integrated system are presented. The learning algorithms extract exemplars reflecting time courses of video frames. There were five types of such clustering methods. Among them, this paper chooses a method called time-partition pairwise nearest-neighbor because of its reduced complexity. On the similarity comparison among videos whose lengths vary, the M-distance that can absorb the difference of the exemplar cardinalities is utilized both for global and local matching. Given the order-aware clustering and the M-distance comparison, system designers can build a basic similar-video retrieval system. This paper promotes further enhancement on the exemplar similarity that matches the video signature tools for the multimedia content description interface by ISO/IEC. This development showed the ability of the similarity ranking together with the detection of plagiarism of video scenes. Precision-recall curves showed a high performance in this experiment.

Keywords: Video similarity ranking · Exemplar · Frame signature · Numerical label · M-distance

1 Introduction

Learning algorithms for mechanical decision have made rapid progress toward various types of practical problems. This development has two motives:

(a) Unexpectedly fast accumulation of raw data exceeded human power.
(b) Fast and inexpensive ICT devices became available.

In this paper, we address item (a) through the problem of video similarity ranking. We present a class of clustering algorithms that finds an order-aware exemplar set for each video. For experiments, we will design a practical system that is indebted to item (b).

Y. Matsuyama—This work was supported by the Grant-in-Aid for Scientific Research 26330286, and Waseda University Special Research Projects 2015K-161.

S. Arik et al. (Eds.): ICONIP 2015, Part I, LNCS 9489, pp. 147–157, 2015.
DOI: 10.1007/978-3-319-26532-2_17

Learning algorithms can offer adaptable methods, often flexible too much. Their flexibility could produce visionary systems easily. In this paper, therefore, we incorporated a class of video signature tools whose standardization is in progress by ISO and IEC [1]. Thus, the similar-video retrieval method in this paper will offer a practical system. The designed system harmonizes a chosen learning algorithm and an industrially standardized method. We checked its performance through the plagiarism detection. The detection appears as ranking, not as a simple dichotomy. Its performance measured by precision-recall curves is quite high.

This paper has the following organization. In Sect. 2, we formulate the problem of the similar-video retrieval. A general concept of frame features and clustering algorithms for exemplar extraction with preliminary experiments are given. Then, we explain the M-distance for global and local alignments. Section 3 explains the Frame Signature and its related tools from the ISO/IEC Standard that can harmonize with the learning algorithms for the similar-video retrieval. Section 4 shows experimental results on the plagiarism detection measured by precision-recall curves. In Sect. 5, discussions on numerical labeling and further sophistications are given.

2   Similar-Video Retrieval

2.1   Frame Features

Each video that we can watch is a time series of still images or frames, $\{v_t\}_{t=1}^N$ that is in a raw format. Each frame v_t contains full information. Therefore, this information needs to be reduced appropriately so that similarity comparison by machines becomes possible. Let x_t be such compressed or feature-extracted data forming a time series of $\{x_t\}_{t=1}^N$. This step could be a target problem to obtain sophisticated patterns. However, the method will be ad hoc. Therefore, we set our objective of this paper to harmonize machine learning's flexibility and industrial standard's restriction. On the frame feature, we will follow the Frame Signature that is a 380-dimensional vector $x_t = [x_{t1}, \cdots, x_{tD}]^T$ with $D = 380$, where each element x_{td} assumes base 3 ternary values $\{0, 1, 2\}$. Therefore, each video makes a trajectory in a D dimensional vector space that corresponds to the feature time series $\{x_t\}_{t=1}^N$. This interpretation matches to the idea of global and local alignments between videos. On the detailed description of the Frame Signature, we will provide an independent section after the full description of the problem formulation.

2.2   Clustering Algorithms for Exemplar Extraction

Consider the similarity comparison of two videos $\{v_t^A\}_{t=1}^{N_A}$ and $\{v_t^B\}_{t=1}^{N_B}$ whose features are $\{x_t^A\}_{t=1}^{N_A}$ and $\{x_t^B\}_{t=1}^{N_B}$, respectively. Although the feature time series $\{x_t^A\}_{t=1}^{N_A}$ and $\{x_t^B\}_{t=1}^{N_B}$ have less complexity than the original raw videos $\{v_t^A\}_{t=1}^{N_A}$ and $\{v_t^B\}_{t=1}^{N_B}$, the comparison of two feature time series in the full may become a complex and wasteful task:

(a) Lengths of two videos N_A and N_B are usually large.

(b) Dragging and redundant scenes may exist in the videos.

Therefore, extraction of typical frames, or exemplars, becomes an essential task for the similarity decision by machines. In the preliminary experiments [2], we presented five clustering algorithms that are aware of frame ordering. They are (i) time-bound affinity propagation, (ii) time-partition k-means, (iii) time-split k-means, (iv) time-partition pairwise nearest-neighbor, and (v) time-bound pairwise nearest-neighbor. The first one is a variant of the affinity propagation [3]. The second and the third ones are of a class of the harmonic competition [4]. The rests are a variant of the pairwise nearest-neighbor [5]. We observed that all methods generated almost equally creditable exemplar sets. Therefore, we consider another criterion, the system complexity that is related to the computing speed. Accordingly, we set the time-partition pairwise nearest-neighbor (TP-PNN) as this paper's basic clustering algorithm for finding an exemplar set. Note that the TP-PNN chooses one end of the nearest-neighbor pair closer to the total centroid of the feature frames, unlike the basic PNN that chooses the pair's center [5]. The TP-PNN algorithm for the exemplar extraction is described as follows.

TP-PNN Algorithm

Step 1: In the start, Feature time series $\mathcal{X} = \{x_t\}_{t=1}^N$, block length b, similarity measure $d(\cdot, \cdot)$, and a stopping threshold δ are given. Before the following iteration, each data claims itself to be an exemplar; $E_i = x_i$. An order-aware partition set is $\mathcal{P} = \{p_t\}_{t=1}^M$ where $M = \lceil N/b \rceil$. For each partition p_t, we perform steps 2-1 and 2-2.

Step 2-1: Compute the centroid of p_t, say c_p.

Step 2-2: Find the nearest-neighbor pair x_i and x_j. If $d(x_i, x_j) \geq \delta$, then go to Step 3. Otherwise, eliminate x_j that is more distant from the centroid c_p than x_i, i.e., operate $E_i \leftarrow E_i \cup E_j$ and $E_j \leftarrow \emptyset$. Here, the eliminated point obtains a label E_i as a member of the remaining point. This process is repeated until the nearest pair distance exceeds δ.

Step 3: The remaining of points are exemplars. Eliminated points become a member of one exemplar.

On the block length, $b = 100$ at 30 fps is proper. The choice of the threshold δ affects the selection of the exemplar set, and consequently the performance of the total system of the similar video retrieval. It is important to understand that ranges of this value are very different depending on the choice of a distance measure, ℓ_1 or ℓ_2. Considering this property, we will make a set of preliminary experiments to decide this value in Sect. 4.2.

2.3   Global and Local Alignments

The next step is to provide methods to compare exemplar sets. Let $\{v_t^A\}_{t=1}^{N_A}$ and $\{v_t^B\}_{t=1}^{N_B}$ be two videos. For video A, we obtained a time series of triplets

$\{e_i^A, x_i^A, E_i^A\}_{i=1}^{n_A}$. Here, e_i^A is an exemplar position as a frame number, x_i^A is its frame feature, and E_i^A is the scope of the exemplar. An exemplar is a pair of (e_i^A, x_i^A), however, we will simply call e_i^A the i-th exemplar. For video B, we use the same notation. There are important observations here.

(a) An exemplar set is a numerical or a soft label. This expression will be compatible with the Frame Signature by the ISO/IEC standard that gives a class of numerical labels.
(b) The scope of E_i^A of the exemplar e_i^A is an essential existence. This structure gives a context-aware mechanism. In our later experiments, E_i^A will be an integer-valued cardinality for the sake of system lightness.
(c) There are two main methods on the matching of $\{e_i^A, x_i^A, E_i^A\}_{i=1}^{n_A}$ and $\{e_j^B, x_j^B, E_j^B\}_{j=1}^{n_B}$. One is the global alignment that finds a total matching pattern and the resulting discrepancy score. The other is the local alignment that finds the best matching sub-regions and its discrepancy score.

On item (a), we will provide Sect. 3 for detailed explanation starting from the Frame Signature. Items (b) and (c) lead to a new alignment method called M-distance. It is important to note that scopes E_i^A and E_j^B play crucial roles. Therefore, the M-distance generalizes existing alignment methods such as the Needleman-Wunsch algorithm of the global alignment [7] and the Smith-Waterman algorithm of the local alignment [8] that are essential tools in bioinformatics.

Global Alignment M-Distance
Step 1: Pick up two exemplar sets $\{e_i^A, x_i^A, E_i^A\}_{i=1}^{n_A}$ and $\{e_j^B, x_j^B, E_j^B\}_{j=1}^{n_B}$ for the global alignment. Set a gap penalty g as a design parameter.
Step 2: Make an $(n_A + 1) \times (n_B + 1)$ table.
Fill the $\{i = 0\}$-th row by $\{0, -gE_1^B, -g\sum_{j=1}^2 E_j^B, -g\sum_{j=1}^3 E_j^B, \cdots\}$.
Fill the $\{j = 0\}$-th column by $\{0, -gE_1^A, -g\sum_{i=1}^2 E_i^A, -g\sum_{i=1}^3 E_i^A, \cdots\}$.
Step 3: Starting from the position of $(i, j) = (1, 1)$, fill all elements by

$$f(i,j) = \max\{f(i-1,j)-gE_i^A, f(i-1,j-1)+r(i,j)s(i,j), f(i,j-1)-gE_j^B\}. \tag{1}$$

Here, $s(i, j)$ is the similarity measure between x_i^A and x_j^B, and $r(i, j)$ is a weight for the similarity reflecting the exemplar scope. We will use an algebraic mean $(E_i^A + E_j^B)/2$. From the cell (i, j), an arrow to the cell that gave the maximum is drawn.
Step 4: The value f_{last} in the element $(n_A + 1, n_B + 1)$ is the global alignment M-distance. Tracing back from this position following the arrows gives the global alignment.
Step 5 (Extra step for similarity ranking): For the similarity ranking on a database, a normalization of f_{last} becomes necessary.

$$u(A, B) = h(f_{\text{last}})/w(\{E_i^A\}_{i=1}^{n_A}, \{E_j^B\}_{j=1}^{n_B}) \tag{2}$$

Here, $h(x)$ is a monotone increasing function. w is an averaging function. The simplest one is an algebraic mean of the cardinalities n_a and n_B.

There is another matching method called local alignment. This method finds a strong matching between sub-regions.

Local Alignment M-Distance

Step 1: Pick up two exemplar sets $\{e_i^A, x_i^A, E_i^A\}_{i=1}^{n_A}$ and $\{e_j^B, x_j^B, E_j^B\}_{j=1}^{n_B}$ for the global alignment. Set a gap penalty g as a design parameter.

Step 2: Make an $(n_A + 1) \times (n_B + 1)$ table.
Fill the $\{i = 0\}$-th row and the $\{j = 0\}$-th column all by 0.

Step 3: Starting from the position of $(i, j) = (1, 1)$, fill all elements by

$$f(i,j) = \max\{0, f(i-1,j)-gE_i^A, f(i-1,j-1)+r(i,j)s(i,j), f(i,j-1)-gE_j^B\}. \quad (3)$$

From the cell (i, j), an arrow to the cell that gives the maximum is drawn.

Step 4: The maximum value $f_{\max}$ is identified. This value is the local alignment M-distance. The local matching region is obtained by tracing back from this position guided by the arrows.

3 Video Signature Tools

The ISO/IEC standard called multimedia content description interface is still in progress by adding and refining tool boxes. However, the standard called Frame Signature is open by [1]. This method gives a set of numerical labels for a video. A simpler method called color structure descriptor (CSD) in our previous study [2] was also a numerical label. By the preparation of Sects. 2.2 and 2.3, we can build a similar video retrieval system that harmonizes the clustering algorithm and the standardized video signature tools.

3.1 Frame Signature

The first step to obtain a video signature $\{x_t\}_{t=1}^N$ from a raw video $\{v_t\}_{t=1}^N$ is the re-sampling. By this re-sampling, each frame is divided into 32×32 sub-blocks. Each sub-block contains a plurality of pixels. Therefore, each sub-block is assigned by an average value of the luminance (Y component of YC_bC_r color space). This process gives a virtual video $\{\bar{v}_t\}_{t=1}^N$ wherein each virtual pixel is expressed by 8 bits. Then, from each $\bar{v}_t$ that is interpreted as a 1024-dimensional vector, a $D = 380$-dimensional base 3 ternary vector x_t is computed in the following way. We omit the frame index t in the following description.

(1) For the first 32 dimensions ($d = 1 \sim 32$) of x, a ternary value is decided by computing the average luminance ν_d of a sub-region whose position is pre-specified [1].

$$x_d = \begin{cases} 2, & \text{if } \nu_d - 128 > \theta_A \\ 1, & \text{if } |\nu_d - 128| \le \theta_A \\ 0, & \text{if } \nu_d - 128 < \theta_A \end{cases} \quad (4)$$

(2) For the dimensions $d = 33 \sim 380$ of $\boldsymbol{x}$, averages of the luminance of two sub-regions of whose positions are pre-specified [1]. These averages $\nu_{d,1}$ and $\nu_{d,2}$ give the following ternary values.

$$x_d = \begin{cases} 2, \text{ if } \nu_{d,1} - \nu_{d,2} > \theta_D \\ 1, \text{ if } |\nu_{d,1} - \nu_{d,2}| \leq \theta_D \\ 0, \text{ if } \nu_{d,1} - \nu_{d,2} < \theta_D \end{cases} \tag{5}$$

Thresholds θ_A and θ_D are defined depending on the virtual frame $\bar{\boldsymbol{v}}$ [1]. The above process is applied to all frames of the raw video of $\{\boldsymbol{v}_t\}_{t=1}^N$ to give the Frame Signature $\{\boldsymbol{x}_t\}_{t=1}^N$.

3.2   Word and Bag of Words

In our similar-video retrieval, the Frame Signature is the main feature to be computed. The standard by ISO and IEC [1] gives two more levels higher than the Frame Signature. They are the Word and the BagOfWords. Because we will conduct additional experiments after that of the Frame Signature, we give their definitions here.

Word
Five sets of components of a feature vector $\boldsymbol{x}_t$ is the Word [1]. In the following definition frame index t is omitted.

$$\begin{aligned} \text{Word}[0] &\Leftrightarrow \boldsymbol{w}_0 = [x_{211}, x_{218}, x_{220}, x_{275}, x_{335}] \\ \text{Word}[1] &\Leftrightarrow \boldsymbol{w}_1 = [x_{45}, x_{176}, x_{234}, x_{271}, x_{274}] \\ \text{Word}[2] &\Leftrightarrow \boldsymbol{w}_2 = [x_{58}, x_{71}, x_{104}, x_{238}, x_{270}] \\ \text{Word}[3] &\Leftrightarrow \boldsymbol{w}_3 = [x_{101}, x_{286}, x_{296}, x_{338}, x_{355}] \\ \text{Word}[4] &\Leftrightarrow \boldsymbol{w}_4 = [x_{102}, x_{103}, x_{112}, x_{276}, x_{297}] \end{aligned} \tag{6}$$

Here, Word[i] is the result of the inner product $[81, 27, 9, 3, 1] \cdot \boldsymbol{w}_i$.

BagOfWords
BoW (BagOfWords) is computed from a set of 90 consecutive frames shifted by 45 frames so that temporal information of words can be grasped. Such shifts give a histogram of Words. In its original definition by [1], this histogram is quantized to 0 or 1.

Before moving to experiments, we give a preliminary comment. Word and BoW were defined for a large-scale classification. The Frame Signature is the primary target of this paper, although characteristics of the Word and BoW will be examined in our experiments.

4 Experiments on Video Similarity Ranking

4.1 Test Video Set and Evaluation Method

We collected 2100 videos that form 100 sets of 21 videos from [6]. Collected videos are in the following class.

(a) The frame size is 640 pixels.
(b) The frame rate is 30 fps.
(c) The length is from 30 to 180 s.

In each of the 21 videos, we randomly selected one video as a query. Then, we copied and inserted more than 10 % of the query into the remaining 20 videos into random positions. Therefore, the task for the experiment is a plagiarism detection. The detection of this level was already successful by the previous system [2] based on the CSD (color structure descriptor). However, we will check to see if this paper's method can detect the plagiarism that has more venomous intentions. That is, we inserted scenes as follows after [9] with stricter constraints.

(1) Frame rate speedup by removing one frame per $\{6, 3, 2\}$ frames.
(2) Gray scale processing by

$$Y = [0.11448,\ 0.58661,\ 0.29891][R,\ G,\ B]^T \tag{7}$$

(3) Illumination change by $\{R, G, B\} \times P$ for $P = 0.6 \sim 0.9$.
(4) Size change by multiplying $0.5 \sim 2.0$.
(5) JPEG modification by CV_IMWRITE_JPEG_QUALITY$=20 \sim 80$, whose default and maximum values are 95 and 100.

For such a set of pseudo-illegal videos, we applied following methods in the order.

Step 1: Computation of the Frame Signature of Sect. 3.1.
Step 2: Exemplar selection by TP-PNN clustering of Sect. 2.2.
Step 3: Local alignment on exemplar sets by the M-distance of Sect. 2.3.
Step 4: Computation of precision-recall curves by the following equations:

$$racall = \frac{\#\ of\ correctly\ detected\ videos}{\#\ of\ videos\ in\ the\ same\ class} \tag{8}$$

$$precision = \frac{\#\ of\ correctly\ detected\ videos}{\#\ of\ top\ ranked\ videos\ to\ be\ checked} \tag{9}$$

Note that for (9), we adopt an 11-point interpolated precision.

4.2 Experimental Results

We conducted two main sets of experiments on the plagiarism detection. Inserted data were modified by the transformation of (1)–(5) of Sect. 4.1 so that the detection becomes harder than the case of the original video insertion. We compared this paper's method based on the Frame Signature with a previous MPEG-7 visual descriptor called CSD (color structure descriptor). First, we show the result by the CSD.

Experiments Using CSD: In the case of the similar-video retrieval using the CSD, we use an HSV space quantizes by $(12, 8, 8)$ levels. The similarity measure is an ℓ_2 distance in a 767-dimensional simplex $(767 = 12 \times 8 \times 8 - 1)$. As a set of preliminary experiments, we counted an average ratio of exemplars with respect to frames for test videos. We chose $\delta = 0.1$ as our default value.

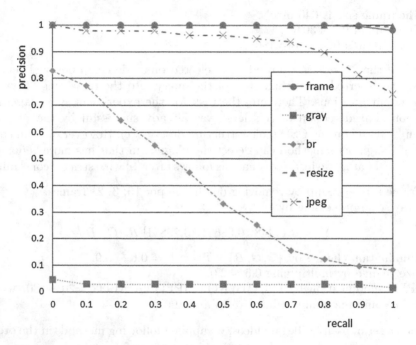

Fig. 1. Precision-recall curves for the video similarity detection by the local alignment: CSD-based method. Design parameters are $\delta = 0.1$, $g = 0.2$, and $\bar{D} = 0.05$.

Figure 1 illustrates precision-recall curves of modified video clip insertions of types (1)–(5) generated by the methods in Sect. 4.1. In this figure, "frame" means the frame rate speed up. "gray" indicates gray scale processing. "br" means illumination or brightness change. "resize" stands for the size change. "jpeg" means the compression rate change. From the result of Fig. 1, we can observe the following.

(a) Since the CSD is based on the histogram of the color concentration, this method gave high performance on the changes in the frame rate and the size. However, video clips changed to the gray scale were not detectable at all.

(b) The successful detection of the frame rate change is because of the ability of the M-distance of the local alignment.

(c) The CSD-based method can detect the insertion of size-changed videos as long as the change is within the range of viewer's perception.

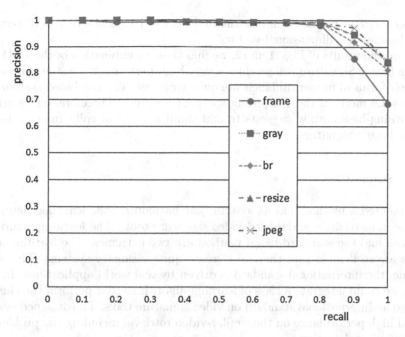

Fig. 2. Precision-recall curves for the video similarity detection by the local alignment: Frame Signature method. Design parameters are $\delta = 320$, $g = 0.2$, and $\bar{D} = 15$.

(d) The brightness decrease is also difficult to identify because changing colors darker reduced the sensitivity of the CSD.

(e) The JPEG quality degradation considerably reduced the ability of the CSD-based method because block noises affect the contents of color bins.

(f) Original video clip insertion is easier to detect than the case of item (c) (see the result in [2]).

Experiments Using Frame Signature: In the case of the similar-video retrieval using the Frame Signature, the similarity measure is an ℓ_1 distance in a 380-dimensional Euclidian space. As a set of preliminary experiments, we counted an average ratio of exemplars with respect to frames for test videos. We chose $\delta = 320$ as our default value. Figure 2 illustrates a set of resulting precision-recall curves. We could observe the following.

(a) All precision-recall curves gave creditable performances.

(b) Although the detection result of the frame rate change was satisfactory, excessive fast-forwarding makes the machine detection of plagiarism difficult like a human. In this case, we need to make the exemplar detection parameter δ smaller.

(c) Computing speed of the Frame Signature method is faster than the CSD-based method by more than two orders of magnitude. The CSD method of Fig. 1 requires the floating point ℓ_2 distance computation of 768-dimensional vectors. The Frame Signature method needs ℓ_1 distance of 1-byte tertiary values for 380-dimensional vectors.

(d) From the results of Figs. 1 and 2, we find that a combination of the CSD and the Frame Signature will produce a very high-performance system. However, their forms of numerical labels are quite different. The combined system will lose the merit of the computing speed of item (c). Therefore, our machine learning-based study suggests to add small amount of color information to the Frame Signature.

5 Discussions

We presented a method and its system that harmonizes the learning algorithm and the industrially standardized video signature tools. The learning algorithm approach and the standard-based method are two extremes. The learning algorithms are so flexible that they may create quite visionary systems. The other extreme, the international standard is driven by real world applications. In this paper, we could integrate a class of learning algorithms that perform the clustering, and an international standard on video signature tools. The designed system showed high performance on the similar-video retrieval including the problem of the plagiarism detection.

We conducted experiments on the BagOfWords. This feature value has less information than the Frame Signature so that time course property can be incorporated. However, our methods have the temporal alignment by the M-distance. Therefore, precision-recall curves by the BagOfWords was considerably inferior to Fig. 2. We omitted the illustration due to the space limitation.

The interpretation of the total method shows further possibility beyond the similar-video retrieval. What we have done in the video database processing was to find and attach numerical labels to each video. Such a class of soft labels helps to give a structure to the database. We can interpret this process as a bottom-up approach toward the language expression level. We could find many promising paths starting in the numerical labels approach of this paper. On the other hand, there will be top-down approaches from a language level. Learning algorithms will be essential there again.

References

1. Information technology – Multimedia content description interface, International Standard of ISO/IEC 15938-3, Amendment 4 (2010)
2. Matsuyama, Y., Shikano, A., Iwase, H., Horie, T.: Order-aware exemplars for structuring video sets: clustering, aligned matching and retrieval by similarity. In: Proceedings of IJCNN, TBA (2015)
3. Frey, B.J., Dueck, D.: Clustering by passing messages between data points. Science 315(5814), 972–976 (2007)
4. Matsuyama, Y.: Harmonic competition: a self-organizing multiple criteria optimization. IEEE Trans. Neural Netw. 7, 652–668 (1996)

5. Equitz, W.: A new vector quantization algorithm. IEEE Trans. ASSP **37**, 1568–1575 (1989)
6. Yahoo! Webscope dataset YFCC-100M. http://labs.yahoo.com/Academic_relations
7. Needleman, S.B., Wunsch, C.D.: A general method applicable to the search for similarities in the amino acid sequence of two proteins. J. Mol. Bio. **48**, 443–453 (1970)
8. Smith, T.F., Waterman, M.S.: Identification of common molecular subsequences. J. Mol. Biol. **147**, 195–197 (1981)
9. Pachalakis, S., Iwamoto, K., et al.: The MPEG-7 video signature tools for content identification. IEEE Trans. Circ. Syst. Video Tech. **22**, 1050–1063 (2012)

On Measuring the Complexity
of Classification Problems

Ana Carolina Lorena[1]([✉]) and Marcilio C.P. de Souto[2]

[1] Instituto de Ciência e Tecnologia, Universidade Federal de São Paulo,
Parque Tecnológico, São José dos Campos, SP, Brazil
`aclorena@unifesp.br`
[2] Univ. Orléans, INSA Centre Val de Loire, LIFO EA 4022, Orléans, France
`marcilio.desouto@univ-orleans.fr`

Abstract. There has been a growing interest in describing the difficulty
of solving a classification problem. This knowledge can be used, among
other things, to support more grounded decisions concerning data preprocessing, as well as for the development of new data-driven pattern
recognition techniques. Indeed, to estimate the intrinsic complexity of
a classification problem, there are a variety of measures that can be
extracted from a training data set. This paper presents some of them,
performing a theoretical analysis.

Keywords: Machine Learning · Complexity measures · Classification
problems

1 Introduction

The seminal work of Ho and Basu [8] proposes analyzing the difficulty of a data
classification problem using descriptors extracted from the data sets available for
learning. These descriptors can provide means to better understand the domain
of competence of different Machine Learning (ML) algorithms [6,13,14]. It is also
possible to develop new techniques for data preprocessing and pattern recognition that are more data-driven [5,7,10,15,18]. According to Ho and Basu [2,8],
the complexity of a classification problem is associated to three main factors:
(1) class ambiguity, (2) complexity of the boundary separating the classes, and
(3) sparsity of the data. There is often a combination of these three factors.

The ambiguity of the classes is present in situations where the classes cannot
be distinguished using the data provided, regardless of the classification algorithm used. This is the case of ill-defined concepts and the use of attributes
with poor discriminative power. The complexity of the classification boundary is related to the size of the smallest description needed to represent the
classes and it is due to the nature of the problem [1]. It can be characterized by

A.C. Lorena—Acknowledgements to the Brazilian Research Agencies FAPESP and
CNPq.

S. Arik et al. (Eds.): ICONIP 2015, Part I, LNCS 9489, pp. 158–167, 2015.
DOI: 10.1007/978-3-319-26532-2_18

the minimum length of an algorithm able to describe the relationships between data (Kolmogorov complexity) [11]. In practice, this property is estimated by using indicators or geometrical measurements extracted from the data sets [2,8]. Finally, an incomplete or sparse data set can also add a degree of difficulty for discriminating the data. This sparsity can make some input space regions to be arbitrarily classified.

In this paper, we introduce the main measures of complexity that can be estimated directly from the data available for learning. The contribution of this work is to bring together/present these measures and their generalizations in a uniform manner. We also present a critical analysis, discussing limitations of the measures.

2 Complexity Measures/Indices

Geometric and statistical data descriptors are often the most used to character-ize the complexity of supervised problems. Some of their main representatives were proposed in [8]. They describe the complexity of the boundary separating binary classification problems. Such measures were then extended to multiclass scenarios in [15,16]. These indices can be divided into three main groups: (1) measures of overlapping between attribute values, (2) measures of separability of classes, and (3) measures of geometry, topology and density.

To define the measures one often considers that they are estimated from a data set T (or part of it) containing n pairs of examples $(\mathbf{x}_i, y_i)$, where $\mathbf{x}_i = (x_{i1}, \ldots, x_{im})$ and $y_i \in \{1, \ldots, k\}$. That is, each example $\mathbf{x}_i$ is described by m attributes, associated with a label y_i among k classes. As most measures described in this paper are sensitive to differences in the magnitude or scales of the attributes, often, before calculating them one normalizes the attributes values so that they lie within similar ranges. Moreover, there are measures where qualitative attributes must be first mapped into quantitative values. Depending on the mapping employed, this kind of transformation only will make sense for ordinal attributes [4].

2.1 Feature/Attribute Overlapping

These measures have as goal to estimate how informative the attributes available are for separating the classes. Often, each attribute is evaluated individually. If there is at least a very discriminating attribute in the data, the problem can be considered simpler than if no such an attribute exits.

Maximum Fisher's Discriminant Ration (F1) measures the overlapping between the values of attributes in different classes:

$$F1 = \max_{i=1}^{m} r_{f_i} \tag{1}$$

where r_{f_i} is the discriminant ratio of each attribute f_i. That is, F1 takes the value of the largest discriminating ratio among all the attributes. In [16], the authors present several equations for calculating r_{f_i}. For two classes problems, one can have:

$$r_{f_i} = \frac{(\mu_{c_1}^{f_i} - \mu_{c_2}^{f_i})^2}{(\sigma_{c_1}^{f_i})^2 + (\sigma_{c_2}^{f_i})^2} \tag{2}$$

where $\mu_{c_j}^{f_i}$ and $(\sigma_{c_j}^{f_i})^2$ represent, respectively, the mean and the variance of attribute f_i in class c_j. For qualitative (ordinal) attributes, each category is first mapped into an integer number [4,16]. In this case, $\mu_{c_j}^{f_i}$ corresponds to the value of the median of f_i in c_j, and $(\sigma_{c_j}^{f_i})^2$ can be computed by:

$$\sigma_{c_j}^{f_i} = \sqrt{p_{\mu_{c_j}^{f_i}}(1 - p_{\mu_{c_j}^{f_i}}) * n_{c_j}} \tag{3}$$

where $p_{\mu_{c_j}^{f_i}}$ is the frequency of the value of $\mu_{c_j}^{f_i}$ and n_{c_j} is the number of examples in c_j. For multiclass problems ($k > 2$), r_{f_i} can be calculated as:

$$r_{f_i} = \frac{\sum_{c_j=1}^{k} \sum_{c_l=c_j+1}^{k} p_{c_j} p_{c_l} (\mu_{c_j}^{f_i} - \mu_{c_l}^{f_i})^2}{\sum_{c_j=1}^{k} p_{c_j} \sigma_{c_j}^2} \tag{4}$$

where p_{c_j} is the proportion of examples in class c_j. That is, one calculates the discriminant power of the means (or medians) for pairs of classes, weighting the result by the proportion of the number of examples in each class. Next, the results are summed up, making the measures to tend to larger values for a multiclass problem than for a binary one. This is due to the fact that the former tends to be more complex than the latter. Nonetheless, caution is required when comparing values of F1 for binary and multiclass problems. Multiclass F1 also implicitly assumes an analysis pair-by-pair of the classes. However, when solving the original multiclass problem, not necessarily the separating boundary needs to cover all classes in pairs. Large values of F1 indicate that there is at least one attribute with little overlap for the different classes. Thus, there exists an attribute for which a linear boundary perpendicular to the axis of the attribute with the largest discriminant power f_i can separate well the classes. If the problems is linearly separable, but requiring an oblique boundary for the separation of the classes, F1 will indicate an overlap between the two classes.

Volume of the Overlapping Region (F2) calculates the overlap of the distributions of the values of the attributes in each class. F2 can be determined by finding, for each attribute f_i, its minimum and maximum values in the classes. Then one calculates the length of the overlap region, normalized by the range of values in both classes. Finally, the values obtained for each attribute are multiplied.

$$F2 = \prod_i^m \frac{\min(\max(f_i, c_1), \max(f_i, c_2)) - \max(\min(f_i, c_1), \min(f_i, c_2))}{\max(\max(f_i, c_1), \max(f_i, c_2)) - \min(\min(f_i, c_1), \min(f_i, c_2))} \tag{5}$$

The maximum and minimum values of each attribute in a class are determined by $\max(f_i, c_j)$ and $\min(f_i, c_j)$ respectively. Indeed, the larger the value of F2, the larger the volume of the overlap between the class of the problem. As consequence, its complexity will be also higher. And if there is at least one non-overlapping attribute, the value of the F2 should be null. For qualitative (ordinal) attributes, each category is first mapped into an integer number [4,16]. As pointed out in [4,19], Eq. 5 can yield negative values for some situations where there is no overlap among the attributes. This occurs, for example, for the attribute illustrated in Fig. 1a. In this case, F2 takes $-\frac{1}{3}$, though it had to be null. In [16], the authors use the absolute value of the ratio in the product. However, the result would still be incorrect ($\frac{1}{3}$). [12,19] propose to perform the following modification in the numerator:

$$overlap(f_i) = \max\{0, \min(\max(f_i, c_1), \max(f_i, c_2)) - \max(\min(f_i, c_1), \min(f_i, c_2))\} \quad (6)$$

This equation provides the calculation for the overlap of the attribute values f_i in two different classes. Cummins [4] also points out a problem for the situation in Fig. 1b. There is overlap at one point, but the resulting value of F2 is null. To solve this a small ϵ can be added to the numerator when it is not null (although the choice of this value can be rather arbitrary). Two other situations where Eq. 5 yields spurious values are shown in (1) Fig. 1c, where the attribute is discriminative, but the minimum and maximum values for classes overlap; and (2) Fig. 1d, where there is noise. [4] proposes changes to deal with such cases, by counting the number of points where there is overlap, which is only suitable for discrete attributes. The presence of noise is also harmful for the F1 measure. As discussed in [10], F2 also presents problems in capturing the simplicity of an oblique linear boundary, since it assumes that the linear boundary should be perpendicular to the axes of the attributes.

Finally, the product in Eq. 5 can lead to quite small values, since it involves the product of values between the $[0, 1]$ range. This gets worse for problems with many attributes, which cannot be comparable to others with fewer attributes. In [12,19], the authors proposed to use the sum instead of the product, which partly solves the problems identified, since the value obtained for F2 will be highly dependent on the number of features the data set has. Moreover, the result would not be a volume overlapping, but the "size" of the overlapping. For multiclass problems, F2 calculation is performed for each pair of classes and produces the sum of the values returned for all pairs [16]. Again, this generalization assumes an analysis of the classes in pairs, which does not necessarily reflect the real difficulty of the problem. It also makes it difficult to compare F2 values between binary and multiclass problems.

Maximum Attribute Efficiency (F3) estimates the individual efficiency of each attribute in separating the classes, returning the maximum value found among the m attributes. For each attribute, F3 verifies whether there is overlap between examples from different classes. If this is true, one considers that the classes are ambiguous in this region. The efficiency of each attribute is given by the ratio between the number of examples that are not in the overlapping region and the total number of examples:

$$F3 = \max_{i=1}^{m} \frac{n - |overlap(f_i)|}{n} \tag{7}$$

where in $|overlap(f_i)|$ returns the number of instances where there is overlap in the values in attribute f_i in different classes. Large values of F3 indicate simpler problems, when fewer examples overlap in each dimension. The way *overlap* is computed can take into account the minimum and maximum values of f_i in different classes, as in Eq. 6. However, this causes the same problems identified for F2 with respect to: classes in which the attribute has more than one valid range (Fig. 1c), sensibility to noise (Fig. 1d), and the fact that for linearly separable problems it is assumed that the boundary is perpendicular to one of the axes, which does not always occur.

Collective Attributes Efficiency (F4) gets an overview of all the attributes together, as opposed to the indices previously defined [16]. It is based on the successive application of a procedure similar to that adopted in F3. First, the most discriminating attribute according to F3 is selected. All examples that can be separated are removed from the data set. The previous step is repeated, that is, one chooses the next most discriminative attribute, excluding the examples already separated. This whole process is repeated until all the examples are separated, or until all the attributes have been considered. The final result is the proportion of examples that were separated. Larger values of F4 indicate that it was possible to discriminate more examples and, thus, the problem is much simpler. The idea is to get the number of examples that can be correctly classified if hyperplanes perpendicular to the axes of attributes are used in the separation. It suffers, however, from the same problems of those from F1, F2 and F3 as not accepting oblique hyperplanes.

$$F4 = \frac{\sum_{i=1}^{m} |overlap(f_i)_{T_i}|}{n} \tag{8}$$

where $|overlap(f_i)_{T_i}|$ will measure the overlap in a subset of the data T_i generated by removing the examples already discriminated in T_{i-1}, while $T_1 = T$. Its calculation must be performed like that for F3, except by using increasingly reduced data sets. Thus, depending on the overlapping measure used, there could be problems in some estimates (Fig. 1c and d).

2.2 Separability of Classes

These measures try to quantify how separate classes are and estimate the shape of the boundary separating them. The intuition is that well-separated classes can be discriminated more easily.

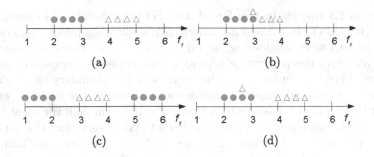

Fig. 1. Examples of situations where F2 can present problems.

Distance of Incorrect Examples to a Linear Boudary (L1) assesses how data is linearly separable, that is, whether it is possible to separate the classes by a hyperplane. L1 computes, for the examples that are incorrectly classified, the sum of the distances between the prediction of a linear classifier and the actual value of the class. If the value of L1 is zero then the problem is linearly separable. Thus, it can be considered simpler than a problem where a non-linear boundary is required. To generate the linear classifier, [16] suggest the use of a Support Vector Machine (SVM) [3] with a linear kernel.

$$L1 = \sum_{h(\mathbf{x}_i) \neq y_i} |f(\mathbf{x}_i)| \tag{9}$$

where $f(\mathbf{x})$ is a linear function $\mathbf{w} \cdot \mathbf{x} + b$ found by the SVM and $h(\mathbf{x})$ returns the class from the value of $f(\mathbf{x})$. Smaller values for L1 indicate that the problem is linearly separable (simple). However, based on the values L1, one cannot identify if a given linearly separable problem is simpler than another linearly separable problem. This occurs, for example, when the examples of a given classification problem are more distant from the linear boundary when compared to those of another problem. L1 can only be estimated for binary data sets and problems with numerical attributes. For multiple classes, [16] suggest that the problem should be first decomposed into binary sub-problems, and an average of L1 computed. Qualitative attributes must be first converted into numerical values. Another limitation of L1 is its computational cost, as it involves the induction of a classifier.

Training Error Rate of a Linear Classifier (L2) measures the error rate of a linear classifier like the one previously described. Let $h(\mathbf{x})$ be the linear classifier obtained:

$$L2 = \frac{\sum_{i=1}^{n} I(h(\mathbf{x}_i) \neq y_i)}{n} \tag{10}$$

where the $I(x)$ returns 1 if x is true and 0 otherwise. Large values of L2 indicate more errors and, thus, a greater complexity (the data cannot be separated linearly). The disadvantages of L2 are similar to those of L1.

Fraction of Examples in the Boudary (N1) first builds a Minimum Spanning Tree (MST) from the data, where each node corresponds to an example and the edges are weighted according to the distance between them. Then, one calculates the percentage of nodes that connect examples of different classes. These examples are in the regions of the boundary or overlapping between the classes. Thus, N1 provides an estimate of the size and complexity of the required boundary to separate the data through the identification of critical points (those examples of different classes very close to each other). Large values of N1 indicate a data set in which more complex boundaries could be required and/or there is more overlap between classes.

$$N1 = \frac{\sum_{i=1}^{n} I(\mathbf{x}_j \in 1NN(\mathbf{x}_i) \text{ and } y_i \neq y_j)}{n} \tag{11}$$

where the condition of the numerator is satisfied when an example has as its nearest neighbor (1-*nearest neighbor* - 1NN) an example of a different class. To work with both numerical and qualitative attributes, one should employ a heterogeneous distance, such as the Euclidean-overlap [10]. Data must also be normalized. Indeed, normalization is required for all subsequent measures based on calculating distances. N1 is sensitive to noise because the closest neighbors of these kinds of examples will normally have different classes. Nevertheless, this could be interesting in that, according to N1, a noisy data set can be considered more complex (this behavior was observed experimentally in [7,12]). An important issue discussed in [4] is that, for a given set of points, multiple valid MSTs could exist. To minimize this problem, one can generate 10 MSTs from the presentation of different orders of data and report an average. Another issue is that the value of N1 can be large even for a linearly separable problem [4]. This occurs when the distance between boundary examples are smaller than those with respect to other examples of the same class.

Ration of the Intra/Inter Class Average Distance (N2) computes the average of the ratio of the intra-class and inter-class distances of the nearest neighbor to an example:

$$N2 = \frac{\sum_{i=1}^{n} d_{intra}(\mathbf{x_i}, NN(\mathbf{x}_i))}{\sum_{i=1}^{n} d_{inter}(\mathbf{x_i}, NN(\mathbf{x}_i))} \tag{12}$$

where $d_{intra}(\mathbf{x_i}, NN(\mathbf{x}_i))$ corresponds to the distance of the example $\mathbf{x}_i$ to the nearest neighbor of its own class and $d_{inter}(\mathbf{x_i}, NN(\mathbf{x}_i))$ is the distance of $\mathbf{x}_i$ to the nearest neighbor in a different class. An heterogeneous distance function should be adopted to deal with both qualitative and numerical attributes. Small values of N2 indicate simpler problems in that the distance between examples of different classes exceeds that between examples of the same class. Thus, N2 reflects how examples in the classes are distributed and not only how the boundary between the classes is. Like N1, N2 can also be sensitive to noise in the data.

Error Rate of the Nearest Neighbor Classifier (N3) returns the error rate of a 1NN classifier by using the leaving-one-out training strategy.

$$N3 = \frac{\sum_{i=1}^{n} I(1NN(\mathbf{x}_i) \neq y_i)}{n} \tag{13}$$

Large values of N3 indicate that many examples are close to those of other classes, making the problem more complex. This measure is computationally expensive for it must access the classifier n times.

2.3 Geometry, Topology and Density

This category includes information on how dense the classes are and on their distribution. A dense class can be likely discriminated more easily than a sparse one. These characteristics are related to the topology and the geometry of the data in the input space.

Nonlinearity of the Linear Classifier (L3) is based on a method proposed by [9]. First, a new data set is created by interpolating training examples of the same class. Specifically, two examples of the same class are selected at random and a linear interpolation is done with random coefficients between them, producing a new example. Then, a linear classifier is trained with the original data and its error in the new data is computed. In [16] a linear SVM is used. This index is sensitive to how the data of a class are distributed in the boundary regions, as well as to the degree of overlapping of convex regions. Large values indicate greater complexity. Le h be the linear classifier created from the training data T, L3 can calculated as follows:

$$L3 = \frac{\sum_{i=1}^{l} I(h(\mathbf{x}_i) \neq y_i)}{l} \tag{14}$$

where l and $\mathbf{x}_i$ are, respectively, the number of points and the examples generated by the interpolation. Since L3 uses a linear classifier induced by a SVM, it can only be applied to binary problems. Another issue is that, by being generated via a linear interpolation of attribute values, L3 is only applicable to numeric attributes.

Nonlinearity of the 1NN (N4) is similar to L3 but using a 1NN, instead of the linear classifier.

$$N4 = \frac{\sum_{i=1}^{l} |1NN(\mathbf{x}_i) \neq y_i|}{l} \tag{15}$$

Large values of N4 indicate problems of greater complexity. For dealing with both quantitative and qualitative features, one should employ a heterogeneous distance. In contrast to L3, N4 can be applied directly to multiclass problems, without the need to decompose them into binary subproblems.

Fraction of Hyperspheres Covering the Data (T1) builds hyperspheres centered in each of the examples. The radius of the spheres is increased until it reaches an example of another class. The smaller spheres contained in larger spheres are eliminated. The measure then returns the ratio between the number of these hyperspheres and the total number of examples.

$$T1 = \frac{|Hyperpheres(T)|}{n} \tag{16}$$

where $|Hyperpheres(T)|$ returns the number of hyperspheres which can be built from the data set. Fewer hyperspheres are obtained for simpler data sets. It reflects the fact that the data of the same class are densely distributed and close together. Thus, this measure also captures the distribution of data within the classes and not only on its boundary.

Average Number of Examples by Dimension (T2) consists in dividing the number of samples in the data set for its dimensionality.

$$T2 = \frac{n}{m} \tag{17}$$

In some works the logarithmic function is applied to this measure for it can take arbitrarily large or small values [12]. T2 reflects the sparseness in the data. If there are many attributes and fewer examples, the examples will be probably distributed sparsely in the input space. The presence of low density regions tends to make it difficult the induction of a good classifier to the problem. Thus, large values of T2 indicate less sparseness and, therefore, a simpler problem.

3   Conclusion

This paper collected and analyzed the main measures used to characterize the complexity of classification problems. Although each measure gives an insight into the complexity of the data according to some of its characteristics, interpreting them separately is not indicated. For example, a linearly separable problem in that the hyperplane is oblique to the axes of attributes could have a small F1, indicating that it is complex. And, at the same time, it can have a small L1, denoting that it is simple. In addition, each measure has some problem associated to it. For instance, the measures of separability of attributes cannot cope with situations where an attribute can have different ranges of values for the same class (Fig. 1c). In fact, they should be considered only as an estimate of the complexity, which could have errors. This reinforces the need to examine the measures together, to provide more robustness to the conclusions reached.

There are other indices proposed in the literature to measure the complexity of the classification problems [4,7,12,17,18]. Several work extend the definitions of [8] and other propose to cover the limitations found in them. Other studies try to further define appropriate measures to specific applications. In future work, we intend to add these measures to the analysis and also conduct experiments to reinforce the observations outlined in this paper.

References

1. Antolnez, N.M.: Data complexity in supervised learning: a far-reaching implication. Ph.D. thesis, La Salle, Universitat Ramon Llull (2011)
2. Basu, M., Ho, T.K.: Data Complexity in Pattern Recognition. Springer, London (2006)
3. Cristianini, N., Shawe-Taylor, J.: An Introduction to Support Vector Machines and Other Kernel-Based Learning Methods. Cambridge University Press, Cambridge (2000)
4. Cummins, L.: Combining and choosing case base maintenance algorithms. Ph.D. thesis, National University of Ireland, Cork (2013)
5. Dong, M., Kothari, R.: Feature subset selection using a new definition of classificability. PRL **24**, 1215–1225 (2003)
6. Flores, M.J., Gámez, J.A., Martínez, A.M.: Domains of competence of the semi-naive bayesian network classifiers. Inf. Sci. **260**, 120–148 (2014)
7. Garcia, L.P.F., de Carvalho, A.C.P.L.F., Lorena, A.C.: Effect of label noise in the complexity of classification problems. Neurocomputing (accepted) (2015, in press)
8. Ho, T.K., Basu, M.: Complexity measures of supervised classification problems. IEEE Trans. Pattern Anal. Mach. Intell. **24**(3), 289–300 (2002)
9. Hoekstra, A., Duin, R.P.: On the nonlinearity of pattern classifiers. In: Proceedings of the 13th International Conference on Pattern Recognition, vol. 4, pp. 271–275. IEEE (1996)
10. Hu, Q., Pedrycz, W., Yu, D., Lang, J.: Selecting discrete and continuous features based on neighborhood decision error minimization. IEEE Trans. Syst. Man Cybern. Part B Cybern. **40**(1), 137–150 (2010)
11. Li, L., Abu-Mostafa, Y.S.: Data complexity in machine learning. Technical Report CaltechCSTR:2006.004, Caltech Computer Science (2006)
12. Lorena, A.C., Costa, I.G., Spolar, N., Souto, M.C.P.: Analysis of complexity indices for classification problems: cancer gene expression data. Neurocomputing **75**, 33–42 (2012)
13. Luengo, J., Herrera, F.: Shared domains of competence of approximate learning models using measures of separability of classes. Inf. Sci. **185**(1), 43–65 (2012)
14. Mansilla, E.B., Ho, T.K.: On classifier domains of competence. In: Proceedings of the 17th ICPR, pp. 136–139 (2004)
15. Mollineda, R.A., Sánchez, J.S., Sotoca, J.M.: Data characterization for effective prototype selection. In: Marques, J.S., Pérez de la Blanca, N., Pina, P. (eds.) IbPRIA 2005. LNCS, vol. 3523, pp. 27–34. Springer, Heidelberg (2005)
16. Orriols-Puig, A., Maci, N., Ho, T.K.: Documentation for the data complexity library in c++. Technical report, La Salle - Universitat Ramon Llull (2010)
17. Singh, S.: Multiresolution estimates of classification complexity. IEEE Trans. PAMI **25**, 1534–1539 (2003)
18. Smith, M.R., Martinez, T., Giraud-Carrier, C.: An instance level analysis of data complexity. Mach. Learn. **95**(2), 225–256 (2014)
19. Souto, M.C.P., Lorena, A.C., Spolar, N., Costa, I.G.: Complexity measures of supervised classification tasks: a case study for cancer gene expression data. In: Proceedings of IJCNN, pp. 1352–1358 (2010)

The Effect of Stemming and Stop-Word-Removal on Automatic Text Classification in Turkish Language

Mustafa Çağataylı$^{(\boxtimes)}$ and Erbuğ Çelebi

Cyprus International University, North Nicosia, North Cyprus
{mscag, ecelebi}@ciu.edu.tr

Abstract. Text classification is defined simply as the labeling of natural and unstructured language text documents using predefined categories or classes. This classification not only help organizations in improving their business communication skills and their customer satisfaction levels, but also improves the usage of unstructured data in academic and non-academic world. The aim of this study is to analyze the effect of stemming, over-sampling, and stopword-removal when doing automatic classification on Turkish content. After obtaning a Turkish Corpus, stemming, balancing, and stopword-removal is applied and the results are evaluated.

Keywords: Text classification · Turkish text classification · Stemming · Stopword removal · Over-sampling

1 Introduction

The development of information and communication technologies (ICT) allowed more people and organization to access to and benefit from more data. Increased amount of data created the need for more and better tools when dealing with it.

In 2013, the total information capacity in the digital universe was 29 PB, and only 22 % of it was considered analyzable, of which, only 5 % of that was actually analyzed. Our total digital universe is expected to increase to 290 PB by 2020. Data created by an average household in 2014 was 2080 GB. By the time 2020, this will increase to 10.176 GB [1]. Most of the mentioned data are both unclassified and unstructured. The increasing amount of data also increases the requirement for its analysis. Again in 2013, it was estimated that, 90 % of all the data in the world has been generated over the last two years. [2].

The aim of this study is to analyze the effect of Stemming, and Stop-Word-Removal when doing automatic classification on Turkish content. Text classification is defined simply as the labeling of natural and unstructured language text documents using predefined categories or classes. This classification not only help organizations in improving their business communication skills and their customer satisfaction levels, but also improves the usage of unstructured data in academic and non-academic world [3, 5].

S. Arik et al. (Eds.): ICONIP 2015, Part I, LNCS 9489, pp. 168–176, 2015.
DOI: 10.1007/978-3-319-26532-2_19

This paper is organized as follows. Background and related work is discussed in Sect. 2. In Sect. 3, the proposed work is detailed together with the corpus and used dataset. Section 4 includes the details of the methodology used throughout the study, and Sect. 5 is where the obtained results are provided. Section 6 includes the conclusion.

2 Related Work

Amasyalı and Diri [8] used author, genre, and gender when classifying documents in a corpus of 630 singly authored documents all from 3 Turkish daily newspapers written by 18 authors. The system used n-grams to achieve classification, mainly bi-grams and tri-grams. They concluded that different methods like bi-grams, tri-grams, NaiveBayes, and SVM can each be more successful on classifying on different aspects of documents like author, gender, and genre. Bi-gram model was observed as the most successful model with 83,3 % success using Naïve Bayes method when identifying author. Also SVM was observed as the most successful model with 93,6 % success when identifying genre. Again, SVM was observed as the most successful model with 96,3 % success when identifying gender. Their work did not mention the effect of stemming on these classifications.

Akkuş and Çakıcı [6] used morphological methods when analyzing 80293 news articles published in Turkish daily newspaper Milliyet. The training set consisted 5000 documents each with at least 500 words and was manually tagged. The corpus was analyzed using K-Nearest Neighbours, Naïve Bayes, and Support Vector Machine. Their work showed that different classifiers like K-Nearest Neighbour (KNN), Naïve Bayes, and SVM each have their own effects on morphological information when classifying Turkish Texts. All the three classifiers KNN, NB, and SVM showed that 5 character words with over 6000 or more features had the best scores like 92.50 %, 94.37 %, and 93.12 % respectively. Again, their work did not mention the effect of stemming on these classifications.

Torunoğlu et al. [3] analyzed the preprocessing methods on Turkish text classification. They used 3 different data sets from Turkish Daily newspapers around 1,000 documents from each, and another one with 20,000 news from newsgroups. Naïve Bayes, Naïve Bayes Multinomial, Support Vector Machines, and K-Nearest Neighbor were the classification methods used. Their work concluded that stemming and stop-word-removal has very little effect on automatic Turkish text classification because of the agglutinative property of Turkish language. Güran et al. [5] proposed a system classifying a corpus of 600 documents collected from web, using N-gram words by applying K-Nearest Neighbours, Naïve Bayes, and Decision Tree algorithms. When comparing the success rates of unigram, bi-gram, and tri-gram representations on data sets with feature selection, Multinomial Naïve Bayes (MNB) had the lead with unigram representations success rate as 95.83 %. Their work compared the results of these algorithms but did not mention the effect of stemming on any step.

Çataltepe et al. [10] analyzed the performance of classifiers when the longest and shortest roots found by a stemmer are used. They used a corpus of 200 news documents from a Turkish daily newspaper Milliyet, and another corpus with 200 documents for each of the 5 topics from the website tr.wikipedia.org. Their work concluded that stemming using words with 2 or 3 letter roots usually decreases the success rate in classification. Longer roots appeared to have around 0.2–2.0 % less successful results when compared with shorter roots. They have no results comparing a stemmed and non-stemmed corpus when classifying.

Amasyalı and Beken [12] analyzed the relation between the classification of documents and the semantic similarities of the words included in them, using a corpus of 1150 documents. They showed that the topic sports (Spor) had a success rate of 97.5 %. Their work did not involve any analysis on stemming. Some other and different classification work have also been done. Özgür et al. [9] proposed a system to classify spam e-mail messages using Artificial Neural Networks (ANN) and Bayesian Networks. Their work consisted a corpus of 750 e-mail messages of which 410 were spam, and did not involve any analysis on stemming. The best success rate were observed using the Bayesian (Binary Model) as 89 %.

Çiltik and Güngö [11] proposed another spam e-mail filter which uses N-gram and N-word heuristics. A corpus of 3680 e-mail messages were used including both English and Turkish content with success rates of 99 % and 98 % respectively, but did not involve any analysis on stemming. A broader study was done by Can et al. [4] which analyzed the effect of several stemming options and query-document matching functions on document retrieval performance using a corpus with 408,305 documents. Looking from the Information Retrieval effectiveness point of view, their work concluded that stop-word-removal has no influence that longer documents and longer queries provide better success, and that word truncation has similar effect with lemmatizer-based stemming. They concluded that the best performing matching function MF8, the stemming options F5 and LV provide, respectively, 33 % and 38 % higher performance than that of no stemming.

3 Proposed Work

The aim of this study is to analyze the effect of Stemming, and Stop-Word-Removal when doing automatic classification on Turkish content. The best corpus is assumed to be from news articles relating to different topics. For such an operation, a corpus had to be obtained or build. A research on the available corpus alternatives revealed only a few; "Bilkent News Archive" (http://bilnews.bilkent.edu.tr), "Metu Corpus" (http://ii. metu.edu.tr/metu-turkish-corpus-project), and "THY SkyLife Corpus" (http://www. skylife.com/en/magazine).

METU Turkish Corpus is a collection of post-1990 written Turkish samples. The contents were taken from 10 different genres. Each article comprises a few thousands of words. The number of words for each article and their sources makes this corpus non-suitable when doing automatic classification. Bilkent News Archive is a corpus including news only relating to Bilkent University and the events at the Bilkent University Campus. Although the number of words in each article are suitable for this study, the topics of the articles are limited only with Bilkent University and related events. Limited topics are a big disadvantage when working with text classification. The THY SkyLife Corpus is a collection of articles published in the Turkish Airlines monthly magazine. The copies of this magazine are provided to the passengers for free during the flights, and have a very limited list of topics, thus makes this corpus non-suitable too.

The only viable alternative seemed to be building the corpus from scratch with the help of a Turkish News Agency. There are many news agencies in Turkey, most being commercial. "Türk Ajansı Kıbrıs", however, is the only local and governmental agency in Northern Cyprus which not only is free but also published on web.

4   Methodology for Dataset

With the help of a web crawler coded in Java, 1000 news articles were recorded in a database. To minimize the effect of very short and very long articles, 500 of them having more than one sentence were selected to be included in the study. These 500 articles were then tagged according to 11 different topics for training purposes. The tagging was done manually by two annotators working independently from each other. The average number of words per document is 221.22, and the number of documents tagged for each topic is shown in Table 1. As seen from the number of documents of each topic, the tagged documents form an unbalanced corpus.

Table 1. Number of documents for each topic

Topic	Number of documents
Politics (Politika)	194
Sport (Spor)	8
Trade (Ticaret)	52
Health (Sağlık)	50
Entertaining (Eğlence)	9
Science and Technology (Bilim ve Teknoloji)	10
Education (Eğitim)	61
International Relations (Uluslararası İlişkiler)	95
Culture (Kültür)	62
Public and State Administration (Kamu Yönetimi)	198
Crime (Suç)	62

The articles are then saved as separate documents so that they can be used for creating a Feature Matrix. A template like "000n.txt" for naming the documents was used, in order to decrease the work done during diagnostics when creating the Feature Matrix. When analyzing the corpus a number of different methods were used. The training set was analyzed using Support Vector Machines.

Proposed by Vapnik [17], SVM includes a group of classification algorithms to solve two-class problems. They are based on finding a separation between hyperplanes defined by classes of data [15], shown in Fig. 1. The main idea is to measure the margin of separation of the data rather than measuring matches on words which are called features.

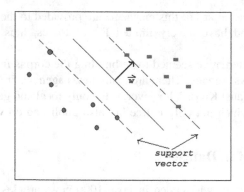

Fig. 1. Example of SVM hyper plane pattern

The SVM is trained using manually preclassified documents. Research has shown [16] that SVM scales well and has good performance even on large data sets, and also outperforms Naïve Bayes classification algorithms.

Balancing, on the other hand, is done using the Smote method. SMOTE is an algorithm used for balancing the samples by adjusting the class distribution of a data set. The main objective is to use a bias for selecting more samples from one class than from the other. Extra training data is created depending on the amount of over-sampling required [14].

5 The Experimental Results

After setting up Weka for UTF-8, the manually tagged Feature Matrix was then fed to Weka for analysis using Support Vector Machines classification technique. For each of the specified topics used during manual classification, the data was fed to Weka and analyzed according to "Correctly classified instances", "F-Measure", and "ROC Area" measures. The results are in Table 2.

Table 2. Weka analysis results

	Normal		
	Correct classification	F-measure	ROC area
Average	**88,036**	**0,859**	**0,588**

The documents are then stemmed using Zemberek-2 [7] Library with the help of another Java code. During this stemming, each and every name root is taken into consideration together with each and every verb root found. Figures 2, 3, and 4 shows the core code paragraph used for stemming, a sample input text and the resulting text, consecutively.

```
// STEMMING is done inside the paragraph below.
    List<YaziBirimi> analizDizisi =
YaziIsleyici.analizDizisiOlustur(metin);
    Zemberek zemberek = new Zemberek(new
TurkiyeTurkcesi());
        for (int i = 0; i < analizDizisi.size(); i++) {
            if (analizDizisi.get(i).tip == YaziBi-
rimiTipi.KELIME) {
                Kelime[] k = zem-
berek.kelimeCozumle(analizDizisi.get(i).icerik);
                // System.out.println(j+" : "+k.length);
                for (int t=0 ; t < k.length; t++)
                {
                // System.out.println(k.length);
                    if(!stemmedList.contains(k[t].kok().icerik()))
{
                    stemmedList.add(k[t].kok().icerik());
                    stemmed = stemmed + " "
+k[t].kok().icerik(); }
                }        }        }
```

Fig. 2. Core code

Yunanistan'da faaliyet gösteren SYRIZA partisinin Başkanı
Aleksis Çipras'ın 1 Aralık'ta Güney Kıbrıs'a gideceği
bildirildi.Filyeleftheros gazetesi Çipras'ın bu kapsamda
dün Atina'daki Rum Büyükelçi Kiriakos Kenevezos ile
görüştüğünü yazdı.

Fig. 3. Sample input text

yunanistan da faaliyet göster parti başkan başka aralık
ara ta güney kıbrıs git gazete bu kapsam dün rum ile il
görüş gör yaz

Fig. 4. Resulting text

After stemming, the Feature Matrix is rebuild again and the "Correctly classified instances", "F-Measure", and "ROC Area" are obtained once again using Support Vector Machines classification technique through Weka, as seen in Table 3.

A dataset is accepted as un-balanced if the classification categories are not equally represented. To overcome this, oversampling and undersampling techniques in data analysis are usually used to adjust the distribution classes within a corpus. When the number of documents tagged for a topic is less than the number of documents tagged for another, both the training and the analysis of the automatic classification may not be

Table 3. Comparing results of non-stemmed corpus with stemmed corpus.

	NON-stemmed (normal)			Stemmed		
	Correct classification	F-measure	ROC area	Correct classification	F-measure	ROC area
Average	88,036	0,859	0,588	88,000	0,872	0,638

biased. Document classifications with similar number of tagged documents within a corpus is accepted to provide a more trustworthy result. Increasing the number of samples or documents by counting some less occurring twice within a corpus to balance the classified documents is called oversampling. Similarly, dropping some of the more occurring documents from the corpus is called Undersampling [13]. With the help of the SMOTE filter of Weka, the stemmed Feature Matrix is analyzed once again for eliminating the difference of balanced and un-balanced sampling. The results are in Table 4.

Table 4. Comparing results of stemmed corpus with balanced corpus.

	Un-balanced			Balanced		
	Correct classification	F-measure	ROC area	Correct classification	F-measure	ROC area
Average	88,000	0,872	0,638	95,440	0,954	0,953

The last analysis for this stemmed and balanced Feature Matrix was done after applying Stop-Word-Removal. For this purpose, the Matrix was sorted according to total number of word occurrences. The columns including the words which are occurring the most were removed. The words removed are the ones above the line drawn showing the occurrences at 97.797 %. The line was drawn such that, the removed words do not include any important or relevant ones according to manual classification operation. When removing words, only "the most occurring" ones were selected. There are no "the least occurring" words removed as they included relevant ones to manual classification. Table 5 show the results obtained from this analysis.

Table 5. Comparing results of balanced corpus with stop-words-removed corpus.

	Balanced			Stop-words-removed and balanced		
	Correct classification	F-measure	ROC area	Correct classification	F-measure	ROC area
Average	95,440	0,954	0,953	95,850	0,877	0,955

To observe the combined effect of stemming, balancing, and stopword-removal, another analysis is done using all three. The comparative results showing the difference between the classification of normal corpus and the classification of corpus after stemming, balancing, and Stopwords-removing is shown in Table 6.

Table 6. Comparing results of normal corpus with stemmed, balanced and stop-words-removed corpus.

	Normal			Stemmed, stop-words-removed, and balanced		
	Correct classification	F-measure	ROC area	Correct classification	F-measure	ROC area
Average	88,036	0,859	0,588	95,850	0,877	0,955

To observe the combined effect of stemming, and stopword-removal, the analysis is done once again using both. Table 7 shows the classification results of the original normal corpus and the stemmed and Stop-Words-Removed corpus.

Table 7. Comparing results of normal corpus with stemmed and stop-words-removed corpus.

	Non-stemmed (normal)			Stemmed and stop words removed		
	Correct classification	F-measure	ROC area	Correct classification	F-measure	ROC area
Bilim-Teknoloji	98,400	0,979	0,600	98,200	0,977	0,599
Eğitim	98,200	0,973	0,500	91,400	0,907	0,732
Eğlence	90,000	0,883	0,647	98,200	0,973	0,500
Kamu Yönetim	68,800	0,669	0,642	64,600	0,643	0,623
Kültür	89,400	0,862	0,586	89,200	0,882	0,675
Politika	70,400	0,689	0,658	69,600	0,692	0,669
Sağlık	90,200	0,867	0,537	90,600	0,886	0,601
Uluslararası İlişkiler	84,600	0,815	0,627	84,000	0,834	0,708
Suç	90,600	0,888	0,656	93,800	0,936	0,833
Ticaret	89,400	0,853	0,516	90,600	0,892	0,642
Spor	98,400	0,976	0,500	98,200	0,975	0,499
Average	88,036	0,859	0,588	88,036	0,872	0,644

6   Conclusion

It has been observer that stemming has very little effect when doing text classification on Turkish corpus. When compared with the efforts required, stemming operation can easily be omitted when classifying Turkish texts. One very important benefit of stemming, however, is its effect on the number of words each document includes at the end of stemming. The lower the number of words in each document has, easier and faster the following operations are. A balanced corpus, improves the results very much. Similarly, the effect of stopword-removal, has very little effect, even when combined with stemming. The overall results when all three methods are applied, are very promising.

In order to improve the confidence of the results, future work may include the usage of a Turkish Corpus richer in terms of number of documents and number of words per article. Another improvement may be using different classification algorithms like K-Nearest Neighbours, Naïve Bayes.

References

1. Digital Universe Invaded By Sensors, Press Release, EMC 2 (2014). http://www.emc.com/about/news/press/2014/20140409-01.htm
2. Big Data, for better or worse: 90 % of world,s data generated over last two years, ScienceDaily, 2013. http://www.sciencedaily.com/releases/2013/05/130522085217.htm
3. Torunoğlu, D., Çakırman, E., Ganiz, M.C., Akyokuş, S., Gürbüz, M.Z.: Analysis of preprocessing methods on classification of Turkish texts. In: International Symposium on Innovations in Intelligent Systems and Applications (INISTA), pp. 112–117, İstanbul (2011)
4. Can, F., Kocberber, S., Balcik, E., Kaynak, C., Ocalan, H.C., Vursavas, O.M.: Information retrieval on Turkish texts. J. Am. Soc. Inform. Sci. Technol. 59(3), 407–421 (2008)
5. Güran, A., Akyokuş, S., Bayazıt, N.G., Gürbüz, M.Z.: Turkish text categorization using N-Gram words. In: International Symposium on Innovations in Intelligent Systems and Applications, Trabzon (2009)
6. Akkuş, B.K., Çakıcı, R.: Categorization of Turkish news documents with morphological analysis. In: Proceedings of the ACL Student Research Workshop, pp. 1–8, Sofia (2013)
7. Akın, A.A., Akın, M.D.: Zemberek an open source NLP framework for Turkic languages (2007)
8. Amasyalı, M.F., Diri, B.: Automatic Turkish text categorization in terms of author, genre and gender. In: Kop, C., Fliedl, G., Mayr, H.C., Métais, E. (eds.) NLDB 2006. LNCS, vol. 3999, pp. 221–226. Springer, Heidelberg (2006)
9. Özgür, L., Güngör, T., Gürgen, F.: Adaptive anti-spam filtering for agglutinative languages: a special case for Turkish. Pattern Recogn. Lett. 25(16), 1819–1831 (2004)
10. Çataltepe, Z., Turan, Y., Kesgin, F.: Turkish document classification using shorter roots. In: IEEE 15th Signal Processing and Communications Applications, Eskişehir (2007)
11. Çıltık, A., Güngör, T.: Time efficient spam e-mail filtering using n-gram models. Pattern Recogn. Lett. 29(1), 19–33 (2008)
12. Amasyalı, M.F., Beken, A.: Measurement of Turkish word semantic similarity and text categorization application. In: IEEE 17th Signal Processing and Communications Applications Conference, Antalya (2009)
13. Chawla, N.V., Bowyer, K.W., Hall, L.O., Kegelmeyer, W.P.: SMOTE: synthetic minority over-sampling technique. J. Artif. Intell. Res. Arch. 16(1), 321–357 (2002)
14. Basu, A., Walters, C., Shepherd, M.: Support vector machines for text categorization. In: Proceedings of the 36th Annual Hawaii International Conference on System Sciences (HICSS 2003), Track 4, vol. 4, pp. 103.3, Washington (2003)
15. Burges, C.J.C.: Simplified support vector decision rules. In: 13th International Conference on Machine Learning, p. 71 (1996)
16. Kwok, J.T.: Automated text categorization using support vector machine. In: Proceedings of the International Conference on Neural Information Processing (ICONIP), pp. 347–351, Kitakyushu (1998)
17. Vapnik, V.N.: The Nature of Statistical Learning Theory. Springer, Berlin (1995)

Example-Specific Density Based Matching Kernel for Classification of Varying Length Patterns of Speech Using Support Vector Machines

Abhijeet Sachdev[1], A.D. Dileep[1(✉)], and Veena Thenkanidiyoor[2]

[1] School of Computing and Electrical Engineering,
Indian Institute of Technology Mandi, Mandi 175001, Himachal Pradesh, India
abhijeet_sachdev@students.iitmandi.ac.in, addileep@iitmandi.ac.in
[2] Department of Computer Science and Engineering,
National Institute of Technology Goa, Ponda 401403, Goa, India
veenat@nitgoa.ac.in

Abstract. In this paper, we propose example-specific density based matching kernel (ESDMK) for the classification of varying length patterns of long duration speech represented as sets of feature vectors. The proposed kernel is computed between the pair of examples, represented as sets of feature vectors, by matching the estimates of the example-specific densities computed at every feature vector in those two examples. In this work, the number of feature vectors of an example among the K nearest neighbors of a feature vector is considered as an estimate of the example-specific density. The minimum of the estimates of two example-specific densities, one for each example, at a feature vector is considered as the matching score. The ESDMK is then computed as the sum of the matching score computed at every feature vector in a pair of examples. We study the performance of the support vector machine (SVM) based classifiers using the proposed ESDMK for speech emotion recognition and speaker identification tasks and compare the same with that of the SVM-based classifiers using the state-of-the-art kernels for varying length patterns.

1 Introduction

Short-time analysis of speech signal involves performing spectral analysis on each frame of about 20 milliseconds duration and representing each frame by a real valued feature vector. The speech signal of an utterance with T frames is represented as a sequential pattern $\mathbf{X} = (\mathbf{x}_1, \mathbf{x}_2, \ldots, \mathbf{x}_t, \ldots, \mathbf{x}_T)$, where $\mathbf{x}_t$ is a feature vector for frame t. The duration of the utterances varies from one utterance to another. Hence, the number of frames also differs from one utterance to another. In the tasks such as acoustic modeling of subword units of speech such as phonemes, triphones and syllables, duration of the data is short and there is a need to model the temporal dynamics and correlations among the features

© Springer International Publishing Switzerland 2015
S. Arik et al. (Eds.): ICONIP 2015, Part I, LNCS 9489, pp. 177–184, 2015.
DOI: 10.1007/978-3-319-26532-2_20

in the sequence of feature vectors. The hidden Markov models (HMMs) [1] are commonly used for sequential pattern classification. On the other hand, in the tasks such as speaker identification, spoken language identification, and speech emotion recognition, the duration of the data is long and preserving sequence information is not critical. In such cases, a speech signal is represented as a set of feature vectors. The focus of this paper is on classification of varying length patterns that are represented as sets of continuous valued feature vectors.

Conventionally, Gaussian mixture models (GMMs) [2] are used for classification of varying length patterns represented as sets of feature vectors. The maximum likelihood (ML) based method is commonly used for estimation of parameters of the GMM for each class. When the amount of the training data available per class is limited, robust estimates of model parameters can be obtained through maximum a posteriori adaptation of the class-independent GMM (CIGMM), which is also called as universal background model (UBM), to the training data of each class [3]. The CIGMM or UBM is a large GMM trained using the training data of all the classes. Classification of varying length sets of feature vectors using SVM-based classifiers requires the design of a suitable kernel as a measure of similarity between a pair of sets of feature vectors. The kernels designed for varying length patterns are referred to as dynamic kernels [4]. Fisher kernel using GMM-based likelihood score vectors [5], probabilistic sequence kernel [6], GMM supervector kernel [7], GMM-UBM mean interval kernel [8], GMM-based intermediate matching kernel [4] and GMM-based pyramid match kernel [9] are some of the state-of-the-art dynamic kernels for sets of feature vectors.

In this work, we propose the example-specific density based matching kernel (ESDMK) as a dynamic kernel to be used in building an SVM-based classifier for varying length pattern classification. We propose to construct ESDMK for a pair of examples represented as sets of feature vectors by matching the estimates of example-specific densities computed at every feature vector in the two examples. For every feature vector in the pair of examples, K nearest neighbors are computed. The ratio of the number of feature vectors of an example among the K nearest neighbors of a feature vector to K is an estimate of density specific for that example. A matching score at a feature vector is computed as the minimum of the two estimates of example-specific densities, one for each example, at that feature vector. The ESDMK for a pair of examples is computed as a combination of the matching scores at every feature vector in the pair of examples. Our studies on the speech emotion recognition and speaker identification tasks demonstrate the potential of the use of the ESDMK in building the SVM-based classifiers for the classification of varying length patterns represented as sets of feature vectors.

The rest of the paper is organized as follows. In Sect. 2, a brief review of dynamic kernels for sets of feature vectors is presented. The proposed ESDMK for sets of feature vectors is described in Sect. 3. In Sect. 4, the studies on speech emotion recognition and speaker identification tasks are presented. The conclusions are presented in Sect. 5.

2   Dynamic Kernels for Sets of Feature Vectors

In this section, we review the approaches to design dynamic kernels for varying length patterns represented as sets of feature vectors. Different approaches to design dynamic kernels are as follows: (1) Explicit mapping based approaches [6,10] that involve mapping a set of feature vectors onto a fixed-dimensional representation and then defining a kernel function in the space of that representation; (2) Probabilistic distance metric based approaches [7,8] that involve kernelizing a suitable distance measure between the probability distributions corresponding to the two sets of feature vectors; and (3) Matching based approaches [4] that involve computing a kernel function by matching the feature vectors in the pair of sets of feature vectors.

The Fisher kernel (FK) using GMM-based likelihood score vectors [5] and the probabilistic sequence kernel (PSK) [6], are the dynamic kernels for sets of feature vectors constructed using the explicit mapping based approaches. The FK uses a GMM for mapping a set of feature vectors onto a Fisher score-space. In the FK, the Fisher score-space for a class is obtained using the first order derivatives of the log likelihood of GMM for that class with respect to the GMM parameters. The probabilistic sequence kernel (PSK) maps a set of feature vectors onto a high dimensional probabilistic score space. The probabilistic score space for a class is obtained using the posterior probability of components of the GMM built for that class.

The GMM supervector kernel (GMMSVK) [7] and the GMM-UBM mean interval kernel (GUMIK) [8], are designed using the probabilistic distance metric based approaches. The GMMSVK uses example-specific adapted GMM built for each example by adapting the means of the UBM using the data of that example. The GMMSVK is then computed between the pair of examples by computing the KL-divergence between the pair of example-specific adapted GMMs. The GUMIK uses example-specific adapted GMM built for each example by adapting the mean vectors and covariance matrices of the UBM using the data of that example. The GUMIK is then computed between the pair of examples by computing the Bhattacharyya distance between the pair of example-specific adapted GMMs.

The CIGMM-based intermediate matching kernel (CIGMMIMK) [4] and the GMM-based pyramid match kernel (GMMPMK) [9] are the dynamic kernels designed using the matching based approach. An intermediate matching kernel (IMK) [4] is constructed by matching the sets of feature vectors using a set of virtual feature vectors. For every virtual feature vector, a feature vector is selected from each set of feature vectors and a base kernel (such as Gaussian kernel) for the two selected feature vectors is computed. The IMK for a pair of sets of feature vectors is computed as a combination of these base kernels. In [4], the set of virtual feature vectors considered are in the form of the components of the CIGMM. For every component of the CIGMM, a feature vector each from the two sets of feature vectors, that has the highest probability of belonging to that component (i.e., value of responsibility term) is selected and a base kernel is computed between the selected feature vectors. In the pyramid match kernel (PMK), a set of feature vectors is mapped onto a multi-resolution histogram pyramid. The kernel is computed between a pair of examples by matching the pyramids using a weighted histogram intersection match function at each level of pyramid. In [9], the CIGMMs built with increasingly larger number of components are used to construct the histograms at the different levels in the pyramid. In our studies, we compare the performance of the SVM-based classifiers using the proposed ESDMK with that of the SVM-based classifiers using kernels reviewed in this section.

3   Example-Specific Density Based Matching Kernel for Sets of Feature Vectors

In this section, we present the ESDMK designed using a matching based approach for sets of feature vectors. Let $\mathbf{X}_m = \{\mathbf{x}_{m1}, \mathbf{x}_{m2}, \ldots, \mathbf{x}_{mT_m}\}$ and $\mathbf{X}_n = \{\mathbf{x}_{n1}, \mathbf{x}_{n2}, \ldots, \mathbf{x}_{nT_n}\}$ be the sets of feature vectors for two examples. The estimates of the example-specific densities are computed in the space of two sets of feature vectors and that space is given by $\mathbf{X} = \{\mathbf{x}_1, \mathbf{x}_2, \ldots, \mathbf{x}_i, \ldots, \mathbf{x}_{T_m+T_n}\} = \{\mathbf{X}_m, \mathbf{X}_n\}$. The K nearest neighbors for $\mathbf{x}_i \in \mathbf{X}$ are obtained using Euclidean distance. An estimate of densities of feature vectors of $\mathbf{X}_m$ and $\mathbf{X}_n$ among the K nearest neighbors of $\mathbf{x}_i$ are given by $\hat{p}_m(\mathbf{x}_i) = K_m/K$ and $\hat{p}_n(\mathbf{x}_i) = K_n/K$ respectively. Here, K_m and K_n are the number of feature vectors from $\mathbf{X}_m$ and $\mathbf{X}_n$ among the K nearest neighbors of $\mathbf{x}_i$ respectively. The matching score at $\mathbf{x}_i$ is given by

$$s_i = \min\left(\hat{p}_m(\mathbf{x}_i), \hat{p}_n(\mathbf{x}_i)\right) \tag{1}$$

The matching score s_i is computed for $i = 1, 2, \ldots, (T_m + T_n)$. The ESDMK for $\mathbf{X}_m$ and $\mathbf{X}_n$ is computed as the sum of the matching score values for all $\mathbf{x}_i \in \mathbf{X}$ as follows:

$$K_{\text{ESDMK}}(\mathbf{X}_m, \mathbf{X}_n) = \frac{1}{T_m + T_n} \sum_{i=1}^{T_m+T_n} s_i \tag{2}$$

It is expected that the similarity between the two examples from the same class to be higher than that between the examples from two different classes.

The proposed ESDMK is similar to one of the state-of-the-art dynamic kernels, GMMPMK as both the kernels are computed by matching the estimates of densities. However, the procedure for computing an estimate of density is quite different in both the kernels. In the GMMPMK, an estimate of density of feature vectors of an example assigned to a component of CIGMM is computed, where CIGMM is computed using the feature vectors of all the training examples. In other words, the estimate of density is computed in the global space of all the feature vectors of all the training examples. In the ESDMK, estimate of density of feature vectors of an example is computed at every feature vector of the pair of examples among the K nearest neighbors of that feature vector. Since these estimate of densities are computed in the local space of feature vectors of pair of examples, we expect that the proposed ESDMK captures the local variability better than the GMMPMK.

The ESDMK is a valid positive semidefinite kernel. The proof for the ESDMK being a positive definite kernel is excluded due to the limitation of pages. In the next section we present our studies on speech emotion recognition and speaker identification tasks using SVM-based classifiers that use ESDMK. We also compare their performance with that of the GMM-based classifiers and SVM-based classifiers using state-of-the-art dynamic kernels.

4 Experimental Studies on Speech Emotion Recognition and Speaker Identification

In this section, effectiveness of the proposed kernel is studied for speech emotion recognition and speaker identification tasks using SVM-based classifiers. A speech utterance is represented by a set of feature vectors by extracting 39-dimensional feature vectors from every frame by performing spectral analysis. Among the 39 features, the first 12 features are the Mel frequency cepstral coefficients and the 13th feature is the log energy. The remaining 26 features are the delta and acceleration coefficients. A frame size of 20 ms and a shift of 10 ms are used for feature extraction from the speech signal of an utterance.

The Berlin emotional speech database (Emo-DB) [11] and the German FAU Aibo emotion corpus (FAU-AEC) [12] are used for studies on speech emotion recognition task. Emo-DB contains 494 utterances belonging to the following seven emotional categories with the number of utterances for the category given in parentheses: fear (55), disgust (38), happiness (64), boredom (79), neutral (78), sadness (53), and anger (127). The multi-speaker speech emotion recognition accuracy presented in this work for the Emo-DB is the average classification accuracy along with 95 % confidence interval obtained for 5-fold stratified cross-validation. We have considered an almost balanced subset of the corpus defined for these four classes by CEICES of the Network of Excellence HUMAINE funded by the European Union [12]. We perform the classification at the chunk (speech utterance) level in the Aibo chunk set. The speaker-independent speech emotion recognition accuracy presented in this study for the FAU-AEC is the average classification accuracy along with 95 % confidence interval obtained for 3-fold stratified cross validation. The 3-fold cross validation is based on the three splits defined in Appendix A.2.10 of [12].

The studies on the speaker identification task are performed on the 2002 and 2003 NIST speaker recognition (SRE) corpora [13,14]. We considered the 122 male speakers that are common to the 2002 and 2003 NIST SRE corpora. Training data for a speaker includes a total of about 3 min of speech from the single conversation in the training set of 2002 and 2003 NIST SRE corpora. The test data from the 2003 NIST SRE corpus is used for testing the speaker recognition systems. The speaker identification accuracy presented is the classification accuracy obtained for the test examples. The training and test datasets as defined in the NIST SRE corpora are used in studies.

In our studies, the SVM-based classifiers using the ESDMK are built using different values for K in K-nearest neighbor method used to obtain an estimate

Table 1. Classification accuracy (CA) (in %) of the SVM-based classifiers with ESDMK for speech emotion recognition (SER) and speaker identification (Spk-ID) tasks. Here, CA95 %CI indicates average classification accuracy along with 95 % confidence interval. K indicates the number of neighbors considered.

K	SER		Spk-ID (CA)
	Emo-DB (CA95 %CI)	FAU-AEC (CA95 %CI)	
2	90.00±0.27	57.71±0.16	88.52
3	90.60±0.24	61.93±0.14	90.54
4	**92.00±0.27**	**65.33±0.09**	**91.38**
5	91.80±0.25	64.40±0.09	89.42

of the density around each feature vector from the pair of sets of feature vectors. We consider LIBSVM [15] tool to build the SVM-based classifiers. In this study, one-against-the-rest approach is considered for 7-class and 4-class speech emotion recognition tasks and 122-class speaker identification task. The value of trade-off parameter, C in SVM is chosen empirically as 10^{-3}. The classification accuracies for the SVM-based classifier using ESDMK for different values of K are given in Table 1 for speech emotion recognition and speaker identification tasks. It is seen that the SVM-based classifiers using ESDMK with $K = 4$ give the best performance for speech emotion recognition and speaker identification tasks.

Table 2 compares the accuracies for speech emotion recognition and speaker identification tasks obtained using the GMM-based classifiers and SVM-based classifiers using the state-of-the-art dynamic kernels mentioned in Sect. 2 and the proposed ESDMK. In this study, the GMMs whose parameters are estimated using the maximum likelihood (ML) method (MLGMM) and by adapting the parameters of the UBM or CIGMM to the data of a class (adapted GMM) [3] are considered to build GMM-based classifiers. The GMMs are built using the diagonal covariance matrices. The accuracies presented in Table 2 are the best accuracies observed among the GMM-based classifiers and SVM-based classifiers with dynamic kernels using different values for Q, K, J and b. The details of the experiments can be found in [4, 9].

It is seen that the adapted GMM-based classifier gives a better performance than the MLGMM-based classifier for speech emotion recognition and for speaker identification. The better performance of the adapted GMM-based system is mainly due to robust estimation of parameters using the limited amount of training data available for an emotion class, or a speaker as explained in [3]. It is also seen that performance of the SVM-based classifiers using the state-of-the-art dynamic kernels is better than that of the GMM-based classifiers. This is mainly because a GMM-based classifier is trained using the non-discriminative learning based technique, where as an SVM-based classifier using the dynamic kernels is built using a discriminative learning based technique. It is also seen that the performance of the SVM-based classifiers using the proposed ESDMK is better than that of the SVM-based classifiers using the state-of-the-art dynamic

Table 2. Comparison of classification accuracy (CA) (in %) of the GMM-based classifiers and SVM-based classifiers using FK, PSK, GMMSVK, GUMIK, CIGMMIMK, GMMPMK and proposed ESDMK for speech emotion recognition (SER) task and speaker identification (Spk-ID) task. Here, CA95 %CI indicates average classification accuracy along with 95 % confidence interval. Q indicates the number of components considered in building GMM for each class or the number of components considered in building CIGMM or the number of virtual feature vectors considered. The pair (J, b) indicates values of J and b considered in constructing the pyramid. K indicates the number of neighbors considered in ESDMK.

Classification model		SER				Spk-ID	
		Emo-DB		FAU-AEC			
		$Q/(J,b)/K$	CA95 %CI	$Q/(J,b)/K$	CA95 %CI	$Q/(J,b)/K$	CA
MLGMM		32	66.81±0.44	128	60.00±0.13	64	76.50
Adapted GMM		512	79.48±0.31	1024	61.09±0.12	1024	83.08
SVM using	FK	256	87.05±0.24	512	61.54±0.11	512	88.54
	PSK	1024	87.46±0.23	512	62.54±0.13	1024	86.18
	GMMSVK	256	87.18±0.29	1024	59.78±0.19	512	87.93
	GUMIK	256	88.17±0.34	1024	60.66±0.10	512	90.31
	CIGMMIMK	512	85.62±0.29	1024	62.48±0.07	1024	88.54
	GMMPMK	(11,2)	88.65±0.23	(5,4)	64.73±0.16	(6,4)	90.26
	ESDMK	4	**92.00±0.27**	4	**65.33±0.09**	4	**91.38**

kernels for speech emotion recognition and speaker identification tasks. The better performance of the SVM-based classifier using the proposed kernel is mainly due to the capabilities of the ESDMK in capturing the local information better than the other dynamic kernels.

5 Conclusions

In this paper, we proposed the example-specific density based matching kernel (ESDMK) for the classification of varying length patterns represented as sets of feature vectors using the SVM-based classifiers. The ESDMK is computed between two varying length patterns based on the estimate of densities of feature vectors of the two examples among the K nearest neighbors of every feature vector from the pair of examples. If the two examples are similar, the estimates of example-specific densities used in the computation of the ESDMK must be higher than those for a pair of not so similar examples. The effectiveness of the proposed ESDMKs in building the SVM-based classifiers for classification of varying length patterns of long duration speech is demonstrated using studies on speech emotion recognition and speaker identification tasks. The performance of the SVM-based classifiers using the proposed ESDMK is better than the GMM-based classifiers and that of the SVM-based classifiers using using FK, PSK, GMMSVK, GUMIK, CIGMMIMK and GMMPMK. Though the computation complexity of the proposed ESDMK is slightly higher than that for the other dynamic kernels, it captures the local information in the data better than the other kernels. The proposed ESDMK can be used for classification of varying length patterns extracted from image, video, audio, music, and so on, represented as sets of continuous valued feature vectors using SVM-based classifiers.

References

1. Rabiner, L., Juang, B.-H.: Fundamentals of Speech Recognition. Pearson Education, New Jersey (2003)
2. Reynolds, D.A.: Speaker identification and verification using Gaussian mixture speaker models. Speech Commun. **17**, 91–108 (1995)
3. Reynolds, D.A., Quatieri, T.F., Dunn, R.B.: Speaker verification using adapted Gaussian mixture models. Digit. Signal Proc. **10**(1–3), 19–41 (2000)
4. Dileep, A.D., Chandra Sekhar, C.: GMM-based intermediate matching kernel for classification of varying length patterns of long duration speech using support vector machines. IEEE Trans. Neural Netw. Learn. Syst. **25**(8), 1421–1432 (2014)
5. Smith, N., Gales, M., Niranjan, M.: Data-dependent kernels in SVM classification of speech patterns. Technical Report CUED/F-INFENG/TR.387, Engineering Department, Cambridge University, Cambridge, April 2001
6. Lee, K.-A., You, C.H., Li, H., Kinnunen, T.: A GMM-based probabilistic sequence kernel for speaker verification. In: Proceedings of INTERSPEECH, Antwerp, Belgium, pp. 294–297, August 2007
7. Campbell, W.M., Sturim, D.E., Reynolds, D.A.: Support vector machines using GMM supervectors for speaker verification. IEEE Signal Process. Lett. **13**(5), 308–311 (2006)
8. You, C.H., Lee, K.A., Li, H.: An SVM kernel with GMM-supervector based on the Bhattacharyya distance for speaker recognition. IEEE Signal Process. Lett. **16**(1), 49–52 (2009)
9. Dileep, A.D., Sekhar Chandra, C.: Speaker recognition using pyramid match kernel based support vector machines. Int. J. Speech Technol. **15**(3), 365–379 (2012)
10. Jaakkola, T., Diekhans, M., Haussler, D.: A discriminative framework for detecting remote protein homologies. J. Comput. Biol. **7**(1–2), 95–114 (2000)
11. Burkhardt, F., Paeschke, A., Rolfes, M., Weiss, W.S.B.: A database of German emotional speech. In: Proceedings of INTERSPEECH, Lisbon, Portugal, pp. 1517–1520, September 2005
12. Steidl, S.: Automatic classification of emotion-related user states in spontaneous childern's speech. Ph.D. Thesis, Der Technischen Fakultät der Universität Erlangen-Nürnberg, Germany (2009)
13. The NIST year 2002 speaker recognition evaluation plan (2002). http://www.itl.nist.gov/iad/mig/tests/spk/2002/
14. The NIST year 2003 speaker recognition evaluation plan (2003). http://www.itl.nist.gov/iad/mig/tests/sre/2003/
15. Chang, C.-C., Lin, C.-J.: LIBSVM: a library for support vector machines. ACM Trans. Intell. Syst. Technol. **2**(3), 27:1–27:27 (2011). http://www.csie.ntu.edu.tw/cjlin/libsvm

Possibilistic Information Retrieval Model Based on Relevant Annotations and Expanded Classification

Fatiha Naouar[(✉)], Lobna Hlaoua, and Mohamed Nazih Omri

Department of Computer Sciences Faculty of Sciences of Monastir,
MARS Research Unit, University of Monastir, 5000 Monastir, Tunisia
{fatihanaouar,lobnal5ll}@yahoo.fr,
MohamedNazih.omri@fsm.rnu.tn

Abstract. The heterogeneity and the great mass of information found on the web today require an information treatment before being used. The annotations, like all other information, must be filtered to determine those that are relevant. The new concept of "relevant annotation" can be then, considered as a new source of evidence. In addition to the vast amount of annotations, we notice that annotations express generally brief ideas using some words that they cannot be comprehensible independently of his context. This is why, we thought to classify it in clusters annotations sharing the same context and semantically related. In this paper, we propose a new model based on clustering for the classification and probabilistic model for the filtering. In the experiments, we tried to consider the relevant annotation classes as a new source of information able to improve the collaborative information retrieval.

Keywords: Relevant annotation · Expanded classification · Filtering · Collaborative information retrieval

1 Introduction

Faced with the vast amount of information found on the web today, a collaborative work is seen necessary to help the user find his need. According to Dinet [8] specialist in Psychology and Ergonomics, this has demonstrated improved performance of information retrieval, including the number of relevant information found and the time taken to do the work. Indeed, work together allows to share the historical retrieval and to formulate collaborative queries. This can be done through several tools, including the annotations as a means of communication in a group.

There is today a very large amount of information on the web; some are a source and other comments and opinions on these sources. This is called annotations, initially used and have shown their interest on paper media.

These annotations are becoming increasingly attractive in the systems of collaborative retrieval. Where there has been a sharing not only information but also comments and feedback on these comments [1, 10]. These annotations show their interest in information retrieval, particularly in multimedia retrieval, where the difficulties related

© Springer International Publishing Switzerland 2015
S. Arik et al. (Eds.): ICONIP 2015, Part I, LNCS 9489, pp. 185–198, 2015.
DOI: 10.1007/978-3-319-26532-2_21

to a semantic retrieval persists [14, 17]. So it is difficult to match the user query expressed in human language to the physical characteristics of multimedia objects [15, 16, 18, 19]. The user often use annotations expressed by users to have more details on the information sought. So, the documents being annotated provide a context for the group discussions. This context enables people to find relevant discussions more easily.

One limitation of the literature works is that the annotations are considered without taking account on their relevance.

However, annotators can be expressed by experts or by novices. This explains false information present in some annotations that can disrupt the information retrieval. Indeed users can end up ruined by sometimes conflicting information. Another aspect that can be found in this context is that the returns of comments can succeed until the original topic can be diverted to other topics.

In addition to the relevance problem, the heterogeneity and the vast amount of information, an annotation may contain little information so that it cannot be treated independently of context. This is why we need to group in clusters annotation sharing the same context. Then the data processing will consider a group of semantically related annotations.

In this context, we propose a new approach able to filter and classify the annotation that will be a new resource of information. The main questions are how can we define a relative annotation, how can we detect it and which criteria will be considered to express the semantic relationship?

In the following we present in second section a brief literature. We will present our filtering and classification approaches respectively in third and fourth sections. We detail then the experimentation environment and results in the fifth section and we will end up with a conclusion.

2 Related Works

As already mentioned despite the importance of annotations under collaborative systems, it has not been analyzed before their integration into systems as a new source of information. Indeed, the annotations are expressed by different users with very diverse profiles and therefore we cannot treat them all with the same degree of importance. They were found in the literature the work of Cabanac, who tried to evaluate the relevance of an annotation based on opinions expressed in a debate that is based on the concept of "social validation" [4]. To validate his work, he tried to compare its results with the perceptions of 173 people, but the validation always depends on the opinions of the individuals.

To deal with the large volume of documents and facilitate access to information, research works have often used the classification techniques.

Mokhtari and Dieng-Kuntz in [11] suggest that the classification is divided in two types supervised and unsupervised: "Classification is the action to group objects having some commonalities without having knowledge of the form or the nature of the prior classes, it is called unsupervised learning problem or automatic classification, or the action of assigning objects to predefined classes, called supervised learning". A supervised classification is a manual classification in which a context or set of classes is specified in advance. This type of classification is not coherent since it needs to review the initial classification when the number of elements to classify increases and this by creating new classes or by changing the classification selection criteria. Many incremental learning algorithms adapted to the problem of supervised classification. Most cited in the literature are the Separators Vast Machine (SVM) [17], methods of k nearest neighbors [2]. These methods always have problems to accelerate research and learning [22, 23].

While an unsupervised clustering is an automatic classification which does not require to have previously identified classes corresponding to different clusters. The idea is to form homogeneous clusters where the elements of each cluster should be similar as possible. A calculation of intra-cluster and inter-cluster distance is performed to determine the quality of a clustering. Several existing clustering techniques in the literature and as they may be divided into two main categories: methods of partitioning and hierarchical methods. The first type of method is based on the principle to fix a priori the k clusters and using an iterative process of assigning each element to the nearest class. There are different methods based on classification by partitioning. There are methods proceeding by aggregation around mobile centers [3], in which a class is represented by a center of gravity, such as k-means and their numerous variants. This type of classification allows the classification of a large set but its disadvantage is that it requires from the number of classes.

The second type of approach is hierarchical methods. The hierarchical classification is to perform grouping classes by aggregation at each stage of the closest elements [5]. This approach uses the concept of distance, which can reflect the homogeneity or heterogeneity of classes. Thus, we considered that an element belongs to a class if it is nearer to this class than to others. There are two types of hierarchical classification: descending and ascending. The descending hierarchical classification includes all the elements in one class then divides in an iterative manner.

While the hierarchical cluster represents each element in a class, and then performs an aggregation successively two by two from the nearest clusters such that a cluster will be "absorbed" by the nearest cluster. This method allows visualizing the progressive grouping of data which can give an idea about the right number of groups in which the data can be grouped. The principal difficulty presented by the hierarchical method is the definition of two classes of grouping criterion, that is to say, the determination of a distance between the classes but the advantage of reading the tree determines the optimal number of classes. Since hierarchical ascending classifications are one of the most successful and most performed forms of classification procedures. In our work, we will choose a hierarchical cluster analysis to classify the filtered annotations considering the similarity between the annotations as grouping criteria.

In our contribution, we proposed a filtering and classification annotations based respectively on a probabilistic model and on extended hierarchical classification.

3 Filtering Annotation Approach

In our view, an annotation can have a degree of relevance if it carries information to annotated object (which can be multimedia). In addition we can say that this annotation is semantically related to the original object or other annotations with a non-zero degree of relevance.

To evaluate the weight of each annotation, we considered a probability model based on conditional probability where annotation is considered as a probabilistic event.

We considered an original document D which may be a media object in general text, image or video and a set of annotation $A = \{A_i\}$. As we have already mentioned, an annotation can be, not only a commentary on the original document, but also a reaction to previous comments. We then chose a probabilistic network to model the process of annotation.

To evaluate the weight of each annotation, we considered a probabilistic model that is based on the conditional probability in which a note is considered as a probabilistic event. According to the theorem of Bays, the conditional probability $P(A_i|D)$ of A_i given D is defined as the quotient of the probability of joint unconditional A_i and D, and the unconditional probability of D by the following Eq. 1:

$$P(A_i|D) = \frac{P(A_i \cap D)}{P(D)} \tag{1}$$

Where $P(D)$ is the probability of the original document, which is valued at 1 (since it is considered relevant).

The calculation $P(A_i \cap D)$ is given the following Eq. 2:

$$P(A_i \cap D) = P(D|A_i) * P(A_i) \tag{2}$$

With $P(D|A_i)$ is the similarity compared to D which given by the Eq. 3:

$$P(D|A_i) = Sim(D, A_i) \tag{3}$$

$P(A_i)$ is the probability of the annotation A_i linked to other existing annotations. This means, if an annotation is not connected to another annotation, it is then less susceptible to be relevant. In the other case when the annotation is only one then $P(A_i)$ $=1$ see the following Eq. 4.

$$P(A_i) = \begin{cases} 1 & if \quad N = 1 \\ \dfrac{\sum_{j=1, j \neq 1}^{N} \left[\left[Sim(A_j, A_i) + Sim(D,]\right]A_i\right)}{N}, else \end{cases} \tag{4}$$

In fact, several works have proposed measures of similarity [15, 16]. The similarity between A_j and A_i $Sim(A_j, A_i)$ is calculated as the sum of frequent terms A_j in A_i and it is given by the following Eq. 5.

$$Sim(A_j, A_i) = \sum_{t_k \in (A_j)} tf(t_k, A_i) \tag{5}$$

In addition to the evaluation of the similarity between an annotation and the previous annotations and between an annotation and the annotated object to determine the relevance of an annotation, we thought to reduce noisy annotations. We have introduced that an annotation may be considered "noisy" if it doesn't provide the information to the annotated document. We then could show that short annotations cannot be carriers of information. In this context, an annotation which its size is less than a minimum size will be considered non relevant, and then its relevance is equal to zero.

A first observation of the results allowed us to deduce the following table (Table 1).

Table 1. Filtered annotations vs. unfiltered annotations

Category of annotation	Results
invalid annotations and removed by the filtering	87.54%
valid annotations but not recognized by filtering	13.18%
valid and filtered annotations	86.82 %
invalid filtered annotations	12.46 %

By statistics, we were able to eliminate by the filtering 87.54 % of invalidated annotations by experts. We also note that some valid annotations are not known by the filtering and do not exceed the 14 % while 86.82 % of validated annotations have been filtered by our system.

After the filtering phase, the relevant annotations are always heterogeneous and their number is still high (can reach thousands). The coherent annotations can be grouped in the same class and therefore annotation class can be treated as one unit of information.

4 Classification of Annotations

In addition to the relevance problem, the heterogeneity and the vast amount of information, an annotation may contain little information so that it cannot be treated independently of context.

This is why we need to group in clusters annotation sharing the same context. Then the data processing will consider a group of semantically related annotations.

We then, propose a new method of classification, based on the method of Clustering [21], designed to determine the annotation classes according to two additional characteristics

- The highest internal homogeneity that is to say inside each annotation class,
- The highest external heterogeneity that is to say between the different classes of annotation.

In statistical language, internal homogeneity corresponds to the internal variance i.e. within cluster variance, while external heterogeneity corresponds to the external variance i.e. between cluster variance. Our classification principle is based on the algorithm of hierarchical clustering by introducing the semantic relationship between the terms.

In addition, we propose to classify in progressive manner. In fact, we consider in first time all annotations to be classified: "Initial classification" that will be explained in Sect. 4.1. Then the resulted clusters can be extended by adding new annotations: "Clusters extension" that will be detailed in Sect. 4.2.

4.1 Initial Classification

The object of our work is to classify the relevant annotations. We consider that an annotation class can contain the annotations that have strong semantic relationships: more the annotations are coherent more they are considered semantically linked. To classify the relevant annotations, several steps are performed. Initially, the annotations are considered independent. Secondly, a comparison in pairs is performed for these annotations. And in recent times, we group the semantically related annotations in the same class. The classes obtained in the previous step will be aggregated according to their semantic relationships two by two.

We consider that an annotation class consists of a set of terms $T = \{t_1, t_2, ..., t_n\}$. To give an idea of the representativeness of a term t_i of an annotation class, we calculate the combination of the factors $tf*icf$. The term frequencies tf of an annotation class allows determining how much a term is comprehensive while the inverse frequency icf determines how much a term is specific to this class. For each term t_i of an annotation class, we calculated its number of occurrences in his annotation class and the size of the annotation class.

$$tf(t_i, Ca_j) = \frac{\sum_{j=1-n} Occ(t_i, Ca_j)}{taille(Ca_j)} \tag{6}$$

With t_i represents a term of an annotation class and Ca_j represents an annotation class. The value of icf for a term t_i in all annotation classes CA, is given by the next Eq. 7:

$$icf(t_i, CA) = \log \frac{|CA|}{|\{ca_j \in CA : t_i \in Ca_i\}|} \tag{7}$$

With $|CA|$ is the cardinality of all classes in the annotation collection and $|\{ca_j \in CA : t_i \in Ca_i\}|$ is the cardinality of the set of annotation classes where the term t_i appears (that is to say $tf(t_i, Ca_j) \neq 0$).

The calculation of $tf * icf$ is given by the following Eq. 8:

$$tf * icf = tf\left(t_i, Ca_j\right) * icf(t_i, CA) \tag{8}$$

After calculating the factor of $tf*icf$ for each term t_i of an annotation class Ca_j, we move to the grouping of these classes. The grouping of annotation classes is based on the semantic relation between them. In our work, we consider that two annotations are semantically linked if they have a number of common terms higher than a threshold. The annotation classification algorithm is described below.

```
Algorithm Classification of Annotations
Input: N Annotation
Output: Annotation Classes
Begin
 for i in 1 to N-1 do
  for j in i+1 to N do
     for k who look over the terms do
        if t_k ∈ a_i = t_k ∈ a_j then
           if tf*icf_i (t_k) > s and tf*icf_j (t_k) > s  then
                      Ca_l = a_i ∪ a_j
        endif
      endif
    endfor
 endfor
endfor
```

Algorithm 1. Classification Algorithm of annotations

Take the example of a document D_k containing a set of relevant annotation A_N that is composed by a set of terms T_j. Two annotations will be grouped if they have common terms and their $tf*icf$ is above the threshold. We consider the example of annotations A_1, A_2, A_{10}, A_{11}, A_{12} and A_N composed by the following terms (Table 2).

Table 2. Terms constituting annotations

Annotations	Terms
A1	$\{T_1, T_2, T_i, T_{i+1}\}$
A2	$\{T_1, T_2, T_{i+1}, T_{i+3}\}$
A10	$\{T_1, T_i, T_{i+3}, T_j\}$
A11	$\{T_{i+2}, T_{j+2}\}$
A12	$\{T_{i+3}, T_j, T_{j+1}\}$
AN	$\{T_i, T_j, T_{j+1}, T_{j+2}\}$

The factor values $tf*icf$ of the annotation terms A_1: $tf*icf\,(T_1, A_1) > s$, $tf*icf\,(T_2, A_1) > s$, $tf*icf\,(T_i, A_1) < s$ and $tf*icf\,(T_{i+1}, A_1) > s$. With s is a threshold.

The factor values $tf*icf$ of the annotation terms A_2: $tf*icf\ (T_1,\ A_2) > s$, $tf*icf\ (T_2,\ A_2)$ $> s$, $tf*icf\ (T_{i+1},\ A_2) > s$ and $tf*icf\ (T_{i+3},\ A_2) < s$.

Since, the factor values $tf*icf$ of the terms T_1, T_2 and T_{i+1} are above the threshold for both annotations A_1 and A_2, then A_1 and A_2 are grouped in one class Ca_1: $Ca_1 = \{A_1,\ A_2\}$.

For the annotation A_{10}, the factor values $tf*icf$ of the terms are equal to: $tf*icf$ $(T_1,\ A_{10}) < s$, $tf*icf\ (T_i,\ A_{10}) < s$ and $tf*icf\ (T_{i+3},\ A_{10}) > s$.

The annotation A_{10} does not belong to the class Ca_1, but it will belong to a new independent class Ca_2: $Ca_2 = \{A_{10}\}$.

For annotations A_{12} et A_N, their factor values $tf*icf$ are respectively equal to: $tf*icf$ $(T_{i+3},\ A_{12}) > s$, $tf*icf\ (T_j,\ A_{12}) > s$, $tf*icf\ (T_{j+1},\ A_{12}) > s$, $tf*icf\ (T_i,\ A_N) < s$, $tf*icf\ (T_j,\ A_N)$ $> s$, $tf*icf\ (T_{j+1},\ A_N) > s$ and $tf*icf\ (T_{j+2},\ A_N) < s$.

By applying our classification algorithm, the annotations A_{12} and A_N will belong to the same class of the annotation Ca_2 that the annotation A_{10}: $Ca_2 = \{A_{10},\ A_{12},\ A_N\}$.

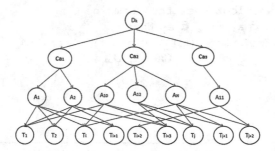

Fig. 1. Construction of annotation classes

We finish the comparison of annotations two by two to construct other classes of annotations. The annotations after classification are shown in the below figure (Fig. 1).

The emergence of new annotations should be considered and therefore an extension of the classes will be required. The principle of cluster extension will be detailed in the next section.

4.2 Clusters Extension

Our goal is to extend clusters by adding annotations without applying the new classification algorithm which remains expensive (since we must treat all annotations).

The problem of classification is to find a way to classify data into one of the predefined classes with minimum error classification. So, we propose to use a "pilot annotation" notion (see the following algorithm).

The pilot annotation concept reflects the degree of representativeness of an annotation in the annotation class. It is measured by its semantic relationship with an annotation set of annotations of his cluster. For each new annotation, comparison with pilots' annotations will be made. The new annotation will be added with the most similar pilot annotation.

```
Algorithm Extended classification
Input: M set of annotation classes Ca₁
       New arrival annotation a_new
Output: a_new added in the most appropriate class
Begin
       Determine for each class Ca₁ a pilot annotation a_pilot
       for l in 1 to M do
               Calculate Sim(a_new, a_pilot_l)
               Save the Max (Sim(a_new, a_pilot_l))
       endfor
       if (Max (Sim(a_new, a_pilot_l))) then
               Added a_new in the class annotation Ca₁
       endif
       Update the pilot annotation for the annotation class Ca₁
End
```

Algorithm 2. Extended Classification Algorithm of annotations

In our work, the semantic relationships can beings translated into similarity. For each annotation in an annotation class, we calculate his weight which characterizes it. The weight is the sum of similarities with all annotations constituting an annotation class, determined by the following Eq. 9.

$$Weight(a_i) = \sum_{j \neq 1}^{N} sim(a_i, a_j) \qquad (9)$$

The similarity of an annotation with other annotations determines the weight of this annotation in the class. The annotation that has the highest weight is considered the pilot annotation for this class and it is determined by the next Eq. 10.

$$Weight(a_{pilot}) = Max(Weight(a_i)) \qquad (10)$$

A new annotation (anew) will be then added to cluster which his pilot annotation has the highest $Sim(a_{new}, a_{pilot})$. When adding a new annotation in a class, an update must be performed to determine the new pilot annotation in this class and therefore the class index annotation pilot must be updated.

5   Experimental Evaluation and Analysis of Results

To evaluate our model, we have applied it on our relevance feedback approach in collaborative information retrieval. We proposed several approaches in this context [12, 13]. We will use the possibilistic technique detailed in [12] and used in several works [7]. Our technique consists of extraction relevant terms from annotation and added to initial query expressed by the user. To evaluate the relevance of term t, we used the possibility theory expressed by the aggregation measure ($Agg\ (t)$) given by the following Eq. 11.

$$Agg(t) = a * N(t) + \beta * P(t) \tag{11}$$

With α and $\beta \in [0, 1]$ and $\sum \alpha + \beta = 1$.

The necessity degree $N\ (t)$ is determined according to appearance the term t of an annotation class in the title of document, $e_i = \{title\}$. The necessity degree is calculated by the following formula:

$$N(t) = \frac{\sum_{i=1-n} tf(t, e_i) * ief(t, e_i, E_i)}{\sum_{i=1-n} tf(e_i)} \tag{12}$$

The possibility degree $P\ (t)$ is determined according to appearance the term t of an annotation class in the title of document, $e_i = \{body\}$. The possibility degree is calculated by the following formula 13:

$$P(t) = \frac{\sum_{i=1-n} tf(t, e_i) * ief(t, e_i, E_i)}{\sum_{i=1-n} tf(e_i)} \tag{13}$$

The factor $tf*ief$ is performed using the next Eq. 14:

$$tf * ief(t, e_i, E_i)(t, e_i) = tf(t, e_i) * ief(t, E_i) \tag{14}$$

With $e_i = \{title, body\}$. The value of $tf\ (t,\ e_i)$ is given by the following Eq. 15:

$$tf(t, e_i) = \frac{\sum_{i=1-n} Occ(t, e_i)}{size(e_i)} \tag{15}$$

With $Occ\ (t,\ e_i)$ is the occurrence of a term t of an annotation class in the element e_i, e_i represents an element $e_i = \{title\}$ or $e_i = \{body\}$, and $size\ (e_i)$ represent its size. The inverse element frequency ief of the term t for all elements E_i is defined by the next Eq. 16:

$$ief(t_1, E_i) = \log \frac{|E_i|}{1 + |\{e_i \in E_i : t_1 \in e_i\}|} \tag{16}$$

5.1   Used Collection of Data

The experiments were conducted on semi-structured document collections. This is a collection of "YouTube". It will be built by the documents returned by the system itself and that will be indexed for the extraction of the information resources already described in the previous sections. We used 10 queries in various fields and the judgment was performed by students and researchers and to evaluate our approach, we used the precision rate for the top 5, 10 and 20 documents using the residual relevance [6]. The Residual method can measure the real impact of relevance feedback since it removes the documents used for the judgment of the relevance of the collection. We calculated the improvement rates TA, for different query, by the following Eq. 17:

$$TA = \frac{new\,precision - old\,precision}{old\,precision} \qquad (17)$$

with TA (5) the rate of improvement, to show the importance of the relevance feedback of the top five documents and TA (10) and TA (20) respectively to show the performance of the top 10 and top 20 documents found. The new precision and the old precision are respectively the results of the search after the relevance feedback and the basic results, in the same residual collection.

5.2 Effects of the Classified and Filtered Annotation

We have compared the average precision obtained applying the relevance feedback approach based on annotations in the rough with obtained results after filtering and classification (see Table 3).

Table 3. Average precision with and without classified and filtered annotations

Base precision	5 Docs	10 Docs	20 Docs
	0.28	0.24	0.22
Unclassified and unfiltered annotations	0.27	0.24	0.24
(% improvement)	(-)	(0%)	(9.09%)
Classified and unfiltered annotations)	0.29	0.25	0.25
(% improvement)	(3.57 %)	(4.16%)	(13.63%)
Unclassified and filtered annotations	0.36	0.30	0.31
(% improvement)	(28.57 %)	(25%)	(40.9%)
Classified and filtered annotations	0.42	0.36	0.34
(% improvement)	(53.57%)	(50%)	(54.54%)

According to Table 3, we see that the classification and the filtering are essential for using annotations. We note that the obtained precision using the classified and filtered annotation for relevance feedback based on the possibilistic model has improved compared to the initial query especially in the top 20 documents returned. It was able to achieve an average equal 0.34 with an improvement rate equal to 54.54 %, compared to 0.24 for the unfiltered and unclassified annotations with an improvement rate equal to 9 %.

We also note that the filtered annotations showed its interest by improving the precision rate of the initial query of the user. Even the filtering of unclassified annotations improves initial accuracy which became 0.31 compared to 0.22 with an improvement rate equal to 40.9 % for the top 20 documents returned by the system.

The relevance feedback based on possibilistic model considering as a source of information the classified and unfiltered annotation gave a slight improvement compared to the original query with an improvement rate equal to 3.57 % for the top 5 documents returned by the system but using the classified and filtered annotations gave an improvement equal to 53.57 %. We obtain the corresponding recall rate and the F-measure presented in Table 4.

Table 4. Recall rate and F-measure of classified and filtered annotations

Nb Documents	Initial recall	Initial F-measure	Based on classified and filtered annotations	
			Recall after relevance feedback	F-measure after relevance feedback
5 Docs	0.07	0.11	0.13 (86 %)	0.19 (73 %)
10 Docs	0.11	0.15	0.2 (82 %)	0.25 (65 %)
20 Docs	0.16	0.19	0.32 (100 %)	0.32 (68 %)

According to Table 4, we see that the recall rate has improved from the initial query compared to the reformulated query. The reformulation performed with the extracted terms from the classified and filtered annotations based on the possibilistic model has an improvement rate about 100 % for top 20 documents. These rates are high compared to precision rates that can be explained by the fact that the initial results are too low (0.07). These notes are confirmed in the F-measure case.

6   Conclusion and Future Works

Annotation systems have shown their interest in particular in collaborative systems. These annotations can be carriers of information but also contradictions. A new challenge is then to filter and to classifier these annotations. In this paper, we define a new approach based on probabilistic model; enable to detect the relevant annotation given one source using the similarity function that can reflect the semantic relationship.

To address the heterogeneity and the great mass of sometimes incomprehensible annotations, we proposed a hierarchical classification approach to group the coherent annotations in the same cluster. The classification is based on hierarchical method where cluster can evaluate by adding new annotations using the "pilot annotation" concept.

To examine our filtering and classification model, we considered our annotations as a source of information for the relevance feedback in collaborative information retrieval based on the possibilistic model. They gave a rate of improvement of the initial request which could reach a rate equal to 55 % for the top 20 documents returned by the system.

We propose as a perspective of our work a comparison of the results provided by our filtering and classification annotations approaches with other collaborative test database.

References

1. Agosti, M., Ferro, N.: Annotations as context for searching documents. In: Crestani, F., Ruthven, I. (eds.) CoLIS 2005. LNCS, vol. 3507, pp. 155–170. Springer, Heidelberg (2005)
2. Aha, D.W. (ed.): Lazy Learning. Kluwer Academic Publishers, Norwell (1997)
3. Benzécri, J.P.: L'analyse des données. Dunod, Paris (1973)
4. Cabanac, G.: Annotation collective dans le contexte RI : définition d'une plate-forme pour expérimenter la validation sociale. In Conférence en Recherche d'Information et Applications, CORIA. pp. 385–392 (2008)
5. Celeux, G., Diday, E., Govaert, G.: Classification automatique de données environnement statistique et informatique. Dunod, Informatique (1989)
6. Chang, Y.K., Cirillo, C., Razon, J.: Evaluation of feedback retrieval using modified freezing, residual collection and test and control groups. In: the SMART Retrieval System-Experiments in Automatic Document Processing, pp. 355–370 (1971)
7. Chebil, W., Soualmia, L.F., Omri, M.N., Darmoni, S.J.: Indexing biomedical documents with a possibilistic network. J. Assoc. Inf. Sci. Technol. (2015). doi:10.1002/asi.23435,66(2)
8. Dinet, J.: Deux têtes cherchent mieux qu'une? In: Medialog, 63 (2007)
9. Frommholz, I., Fuhr, N.: Probabilistic, object-oriented logics for annotation-based retrieval in digital libraries. In: JCDL 2006: Proceedings of the 6th ACM/IEEE-CS joint conference on Digital libraries, ACM Press, New York, NY, USA, pp. 55–64 (2006)
10. Kahan, J., Koivunen, M.R., Prud'Hommeaux, E., Swick, R.R.: Annotea: an open RDF infrastructure for shared Web annotations. Comp. Netw. **32**(5), 589–608 (2002)
11. Mokhtari, N., Dieng-Kuntz, R.: Extraction et exploitation des annotations contextuelles. In: Proceedings Extraction et gestion des connaissances EGC (2008)
12. Naouar, F., Hlaoua, L., Omri, M.N.: Possibilistic model for relevance feedback in collaborative information retrieval. Int. J. Web Appl. IJWA **4**(2), 78–86 (2012)
13. Naouar, F., Hlaoua, L., Omri, M.N., Relevance feedback for collaborative retrieval based on semantic annotations. In: The International Conference on Information and Knowledge Engineering (IKE 2013), pp. 54–60 (2013)
14. Omri, M.N., Chouigui, N.: Measure of similarity between fuzzy concepts for identification of fuzzy user's requests in fuzzy semantic networks. Int. J. Uncertainty Fuzziness Knowl. Based Syst. (IJUFKS) **9**(6), 743–748 (2001)
15. Omri M.N., Chouigui, N., Linguistic variables definition by membership function and measure of similarity. In: Proceedings of the 14th International Conference on Systems Science, vol. 2, pp. 264–273 (2001)
16. Omri, M.N.: Pertinent knowledge extraction from a semantic network: application of fuzzy sets theory. Int. J. Artif. Intell. IJAIT **13**(3), 705–719 (2004)
17. Omri, M.N.: Effects of terms recognition mistakes on requests processing for interactive information retrieval. Int. J. Inf. Retrieval Res. IJIRR **2**(3), 19–35 (2012)
18. Omri, M.N.: Système interactif flou d'aide à l'utilisation de dispositifs techniques: SIFADE. Thèse de l'université Paris VI, Paris (1994)
19. Omri, M.N., Tijus, C.A.: Uncertain and approximative knowledge representation in fuzzy semantic networks. In: The Twelfth International Conference on Industrial and Engineering Applications of Artificial Intelligence and Expert Systems (IEA/AIE-99) (1999)
20. Omri, M.N., Chenaina, T.: Uncertain and approximate knowledge representation to reasoning on classification with a fuzzy networks based system. In: Fuzzy Systems Conference Proceedings, FUZZ-IEEE 1999, vol. 3, pp. 1632–1637 (1999)
21. Syed, N.A., Liu, H., Sung, K.K.: Incremental learning with support vector machines. In: Proceedings of International Joint Conference on Artificial intelligence (IJCAI) (1999)

22. Tryon, R.C.: Cluster Analysis. Edwards Brothers, Ann Arbor (1939)
23. Usunier, N., Bottou, L.: Guarantees for approximate incremental SVMs. In: International Conference on Artificial Intelligence and Statistics, pp. 884–891 (2010)
24. Vassef, H., Li, C.S, Castelli, V.: Combining fast search and learning for fast similarity search. In: Proceedings of SPIE. The International Society for Optical Engineering, vol. 3972, pp. 32–42 (2000)
25. Weinberger, K., Saul, L.: Distance metric learning for large margin nearest neighbor classification. J. Mach. Learn. Res. (JMLR) 10, 207–244 (2009)

A Transfer Learning Method with Deep Convolutional Neural Network for Diffuse Lung Disease Classification

Hayaru Shouno[1]([✉]), Satoshi Suzuki[1], and Shoji Kido[2]

[1] University of Electro-Communications, Chofugaoka 1-5-1, Chofu, Japan
shouno@uec.ac.jp
[2] Yamaguchi University, Tokiwadai 2-16-1, Ube, Japan

Abstract. We introduce a deep convolutional neural network (DCNN) as feature extraction method in a computer aided diagnosis (CAD) system in order to support diagnosis of diffuse lung diseases (DLD) on high-resolution computed tomography (HRCT) images. DCNN is a kind of multi layer neural network which can automatically extract features expression from the input data, however, it requires large amount of training data. In the field of medical image analysis, the number of acquired data is sometimes insufficient to train the learning system. Overcoming the problem, we apply a kind of transfer learning method into the training of the DCNN. At first, we apply massive natural images, which we can easily collect, for the pre-training. After that, small number of the DLD HRCT image as the labeled data is applied for fine-tuning. We compare DCNNs with training of (i) DLD HRCT images only, (ii) natural images only, and (iii) DLD HRCT images + natural images, and show the result of the case (iii) would be better DCNN feature rather than those of others.

1 Introduction

In the medical diagnosis, the performance of classification task is important for the diagnosis quality. For classifying and detecting the diffuse lung disease (DLD) pattern, which is appeared with idiopathic interstitial pneumonias (IIPs) [1,3,5,9,11], high-resolution computed tomography (HRCT) images are regarded to be effective for diagnosis since diffused DLD patterns can be observed in the lung with any cross section. Unfortunately, diagnosing the IIPs site is difficult work, because the DLDs on HRCT images show a lot of varieties in the meaning of texture patterns. The quality of diagnosis is influenced by diagnosing skill of physician, so that, objective diagnosing and improving its quality are desired for proper treatment of IIPs. In order to decrease the burden of physicians, development of the computer aimed diagnosis (CAD) system is desired for objective diagnosis in these decades. The CAD systems are designed to provide a classification function for second opinion using machine learning techniques.

In the field of machine learning and computer vision, the deep-learning style classification system arouse a notice for its classification performance [4,6,7].

© Springer International Publishing Switzerland 2015
S. Arik et al. (Eds.): ICONIP 2015, Part I, LNCS 9489, pp. 199–207, 2015.
DOI: 10.1007/978-3-319-26532-2_22

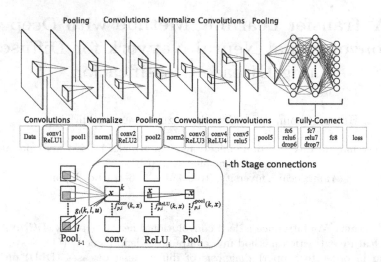

Fig. 1. Schematic diagram of DCNN architecture, which has alternate structure of "convolution", "ReLU", "pooling", and "normalize" layers. Input image is provided to the data layer and the input signal transferred from data layer to the "fc8" layer.

In their works, massive labeled example is assumed to train the system, however, the obtaining cost of such labeled data is expensive in the medical imaging since it requires physicians decision for proper disease labels. The main idea in this work is to transfer a feature extraction part in a DCNN for general object recognition, into the DLD feature extractor. The DCNN for general object recognition is trained with massive natural images that can obtain with lower cost rather than that of medical images.

This type of learning is called a kind of transfer learning [10]. In this work, we adopt a DCNN [6], which is origined from the Neocognitron [2,8], for DLD pattern analysis with transfer learning.

2   Methods

2.1   Deep Convolutional Neural Network (DCNN)

We adopt a DCNN for the feature extraction of DLD patterns. Figure 1 shows the schematic diagram of the DCNN in this experiments, which is same as Krizhevsky's DCNN [6]. The DCNN has a alternate structure of pattern transformation called "stage" [2,8], which consists of "convolution", "rectified linear(ReLU)", "pooling", and "normalize" layers. Considering the ith stage for the pth input pattern, we formulate each layer response as followings. The "convolution" layer extracts feature maps as

$$f_{p,i}^{\mathrm{conv}}(k, \boldsymbol{x}) = \sum_{l, \boldsymbol{u}} g_i(k, l, \boldsymbol{u})\, f_{p,i-1}^{\mathrm{pool}}(l, \boldsymbol{x} - \boldsymbol{u}), \qquad (1)$$

where k means the index of feature map, and $\boldsymbol{x}$ means the location of the map. The convolution kernel $g_i(k, l, \boldsymbol{u})$ means the feature including in the lth feature map of the pooling (with normalization) layer in the previous stage $f_{p,i-1}^{\text{pool}}(l, \boldsymbol{x})$. In each feature map $f_{p,i}^{\text{conv}}(k, \boldsymbol{x})$, extracted features are modulated with the rectified linear unit (ReLU):

$$f_{p,i}^{\text{ReLU}}(k, \boldsymbol{x}) = \max[0, f_{p,i}^{\text{conv}}(k, \boldsymbol{x})]. \tag{2}$$

Then, in each modulated feature map, spatially neighbor responses are gathered for calculation of representative value, which is called spatial pooling:

$$f_{p,i}^{\text{pool}}(k, \boldsymbol{x}) = \max_{\boldsymbol{\xi} \in \boldsymbol{N}(\boldsymbol{x})}[f_{p,i}^{\text{ReLU}}(k, \boldsymbol{u})], \tag{3}$$

where $N(\boldsymbol{x})$ means the spatial neighbor area around the location $\boldsymbol{x}$ in the map. Then, these response map are normalized in the same manner with Krizhevsky [6]. Hereafter, we adopt the ith stage representation as the vector which arranges whole components of the layer: $\boldsymbol{f}_{p,i}^{\text{pool}} = \{f_{p,i}^{\text{pool}}(k, \boldsymbol{x})\}_{k, \boldsymbol{x}}$.

In the Fig. 1, "fc6", which abbreviate 6th fully-connected layer, is feature extraction layer. Following layers, which are "fc7", "fc8" layers, have the multilayer perceptron (MLP) structure, which plays a role as the classier.

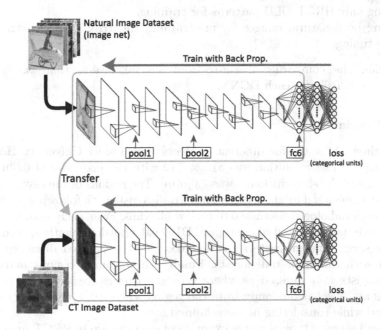

Fig. 2. Training the DCNN as general object recognition classifier with massive natural images at first. Then, tuning the pre-trained DCNN with HRCT image including DLD patterns for fine-tuning.

2.2   Transfer Learning for DCNN

We propose a transfer learning style learning for the DCNN for analyzing DLD patterns in the following. Figure 2 shows the schematic diagram of transfer learning for the HRCT DLD patterns. At First, we make a reference object classification system with DCNN. In this study, we apply it in the manner of Krizhevsky's method [6]. In this process, we expect that the connections in the lower stages are trained enough with massive natural images. After the massive natural images training, we substitute the higher MLP structure, that is, "fc7", "fc8" layers and its corresponding connections, and train the DCNN again with the HRCT DLD patterns by the conventional back propagation (BP) method with stochastic gradient decent. The problem of the conventional BP is the diffusion of the modifying information for connections along the error signal propagation from the higher layer to the lower. As the result, the lower level connections might not be enough trained in the small number of the dataset. In our approach, we trained the lower connections with massive natural image data and transfer them to the specific classification.

For evaluation of our approach, we compare the DCNN features from the followings:

(i) using only natural images for training,
(ii) using only HRCT DLD patterns for training,
(iii) using both natural images for pre-training and HRCT DLD patterns for fine-tuning.

We obtain the comparing features from several layers, that is, "pool1", "pool2", and "fc6", for each DCNN.

2.3   Materials

We acquired 117 scans for different subjects from Osaka University Hospital, Osaka, Japan. The resolution was 512×512 with the pixel size of 0.6[mm] on each slice, and the slice-thickness was 1.0[mm]. The regions of the seven types of patterns were marked by three experienced radiologists with following procedure. At first, each radiologist was asked to review all scans. From each scan, maximum of three slices were selected where typical DLD patterns dominantly spread. After that, the seven types of patterns were marked on the selected slices separately together with the other radiologists. Finally, the common regions marked by all radiologists were guessed as where typical patterns are located. The ROIs locations are selected randomly from the region where 80 % area is included in the region while considering non-overlapped area.

Figure 3 shows a typical image example of each disease in HRCT image. The left shows an overview of the axial HRCT images of lungs including lesion, and the right shows segmented images of typical examples of lesion from the left image collections. The consolidation (CON) and ground-grass opacity (GGO) patterns are often appeared with the cryptogenic organizing pneumonia diseases (COPD). The GGO pattern is also often appeared in the non-specific interstitial pneumonia (NSIP). The reticular (RET) pattern, which also imply the NSIP, sometimes appears together with partial GGO patterns. The honeycomb (HCM) pattern has more rough mesh structure rather than that of the reticular pattern, and it appeared in idiopathic pulmonary fibrosis (IPF) or usual interstitial pneumonia (UIP). Both of the nodular (NOD) and emphysema (EMP) are not DLDs, however, these patterns sometimes confuse physician, so that, we include these classes into the experiment. The normal (NOR) pattern appears in the normal tissue.

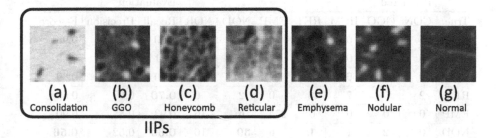

Fig. 3. Typical CT images of diffuse lung diseases: The top row shows each overview, and bottom shows magnified part (ROI) of each lesion. From (a) to (g) represents "Consolidation (CON)", "GGO", "Honeycomb (HCM)", "Reticular (RET)", "Nodular (NOD)" "Emphysema (EMP)", and "Normal (NOR)" image respectively.

The size of ROI is 32×32 pixel, which is just too small for the DCNN input, so that, we magnify the ROI image to adjust the DCNN input size. As the result, we collect 169 ROIs for CON, 655 for GGO, 355 for HCM, 276 for RET, 4702 for EMP, 827 for NOD, and 5,726 for NOR. We divide the dataset for obtaining the DCNN feature and evaluation task. For the DCNN training, we use 143 CONs, 609 GGOs, 282 HCMs, 210 RETs, 4,406 EMPs, 762 NODs, and 5371 NORs. The other 26 CONs, 46 GGOs, 73 HCMs, 66 RETs, 296 EMPs, 65 NODs, and 355 NORs are used for evaluation task.

3 Results

In the analysis, we first compare the classification performance. We compare the features of "fc6" layer, which has 4,096 units, from the DCNN with only natural images for training, with only HRCT DLD for training, and proposed method. In order to evaluate the classification performance, we introduce both the leave one out cross validation (LOOCV) method and leave one person out cross validation (LOPOCV). LOOCV method is a standard evaluation method, however, ROIs from same patients might have several correlation which conduct the classification bias in the evaluation. Thus, we introduce LOPOCV method for reducing the bias from the same patient's ROI.

Table 1. Comparison of SVM accuracy performance for the DCNN feature obtained only from natural images, obtained only from HRCT images, and the proposed method.

	Nat. Imgs	HRCT Imgs	Proposed
LOOCV accuracy	87.49 %	89.00 %	91.91 %
LOPOCV accuracy	75.78 %	74.13 %	80.04 %

Table 2. Confusion matrix: Trained with only HRCT images

	Predicted							Evaluation		
True	CON	GGO	HCM	RET	EMP	NOD	NOR	Recall	Precision	F-score
CON	26	0	0	0	0	0	0	1.00	0.93	0.96
GGO	0	19	0	15	0	10	2	0.41	0.63	0.50
HCM	0	0	66	5	2	0	0	0.90	0.88	0.89
RET	2	9	7	**46**	0	2	0	**0.70**	0.69	0.69
EMP	0	0	0	0	274	6	16	0.93	0.88	0.90
NOD	0	2	1	1	6	**39**	16	**0.60**	0.52	**0.56**
NOR	0	0	1	0	28	18	308	0.87	**0.90**	0.88

Table 3. Confusion matrix: Our method

	Predicted							Evaluation		
True	CON	GGO	HCM	RET	EMP	NOD	NOR	Recall	Precision	F-score
CON	26	0	0	0	0	0	0	1.00	0.93	0.96
GGO	0	**32**	0	9	0	2	3	**0.70**	0.63	**0.66**
HCM	0	0	**69**	4	0	0	0	**0.95**	**0.93**	**0.94**
RET	2	16	5	43	0	0	0	0.65	**0.75**	**0.70**
EMP	0	0	0	0	**283**	1	12	**0.96**	**0.94**	**0.95**
NOD	0	3	0	1	0	32	29	0.49	**0.59**	0.54
NOR	0	0	0	0	18	19	**318**	**0.90**	0.88	**0.89**

Using only DLD patterns for training (case (ii)), the DCNN shows overfitting for the training set. The classification performance using "fc8" layer for the training set becomes 93.5 %, meanwhile for the test set is 30.0 %. Thus, we use the intermediate layer expressions for the evaluation. In order to classify the expressions, we adopt linear support vector machine (SVM) with one-versus-one method for multi class classification.

The Table 1 shows the accuracy performance for the test set of our experiments. Both of the evaluation methods, that is, LOOCV and LOPOCV methods, our proposed method shows better result rather than those of others. Especially, with the LOPOCV method, the DCNN feature obtained from only HRCT images does not work as well as our method. Hereafter, we focus on the DCNN feature obtained from only HRCT images and our method. The detail confusion matrix and evaluation for LOPOCV method of these two are shown as Tables 2 and 3. In the Table 3, improved score shows as the bold, and degraded score shows as the bold in the Table 2. In these tables, we can see the improvement of our method except of NOD in the meaning of F-score.

In the DCNN, the input image is transformed local feature maps in the lower stage, and the represented features are integrated gradually. It is important to investigate the intermediate representation for construction DCNN, so that, we investigate the representation in the meaning of the projection to the orthogonal vector of the SVM decision plane. In this experiment, we generate SVMs for each layer representation. Figure 4 shows a result for the GGO versus RET classifier. In each graph, the horizontal axis shows the distance from the decision plane, where the origin indicates the decision boundary, and the vertical one shows the density of the test examples. In the figure, the top row graphs shows the result for our proposed method, and the bottom one shows the result for the DCNN with only HRCT images. The left column shows the result for the "pool1" layer representation, the middle shows for the "pool2", and the right one shows for the "fc6" layer. In our proposed method, we can see the variance

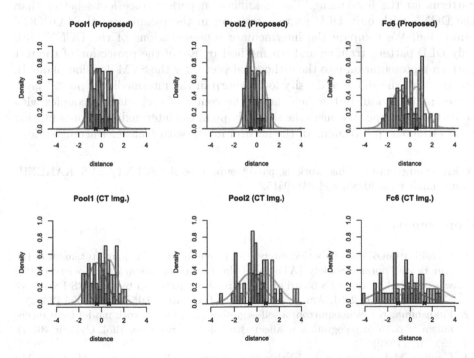

Fig. 4. Projection of each test representation into the normal vector to the SVM decision plane for the GGO vs RET classifier. The horizontal axis shows the distance from the plane. The vertical one shows the density of the test patterns. The cross marks on the horizontal axis show the mean of each class, and the curve shows the normal distribution for projection data of each class.

for the projection of each class data shrinks in the "pool2" layer, and the spread again with uncoupling of each class center in the "fc6" layer. The shapes of density shapes in the "pool1" and the "pool2" layers are established in the natural image pre-training. On the contrary, in the case of only using the HRCT image on the training, the variance of density shape does not shrink in the "pool2" layer representation. In the "fc6" layer representation, the center of each class looks uncoupled, however, the variance of density shape looks large in the representation. As the result, we guess the performance of the classification becomes worse.

4   Summary and Discussion

In this study, we propose a DLD pattern analysis method with DCNN. In general, DCNN requires massive training dataset to bring out its performance, however, it is hard to acquire in the field of medical imaging. Overcoming this problem, we introduce a transfer learning style training, that is, we train the DCNN with natural image set, which can obtain in low cost, for pre-training for the DCNN as the initial state. After that, we train again the DCNN with DLD patterns for the fine-tuning. The classification performance looks better than the DCNN with only DLD pattern training in the meaning of the LOPOCV evaluation. We compare the intermediate representations of the DCNN with only DLD pattern training and our method by use of the projection of the test pattern representation into the orthogonal vector for the SVM decision plane. In our method, the shape of density looks sharp in the intermediate "pool2" layer representation, and, in the "fc6" layer, the center of each class uncouples with spreading its shape. We guess the sharp shape in the intermediate representation comes from the pre-training of the natural image with transfer learning style.

Acknowledgments. This work is partly supported by MEXT/JSPS KAKENHI Grant number 25330285, and 26120515.

References

1. Classification of the Idiopathic Interstitial Pneumonias. This joint statement of the American Thoracic Society (ATS), and the European Respiratory Society (ERS) was adopted by the ATS board of directors, June 2001 and by the ERS Executive Committee, June 2001. Am. J. Respir. Crit. Care Med. **165**(2), 277–304 (2002)
2. Fukushima, K.: Neocognitron: a self-organizing neural network model for a mechanism of pattern recognition unaffected by shift in position. Biol. Cybern. **36**(4), 193–202 (1980)
3. Gangeh, M.J., Sørensen, L., Shaker, S.B., Kamel, M.S., de Bruijne, M., Loog, M.: A texton-based approach for the classification of Lung Parenchyma in CT images. In: Jiang, T., Navab, N., Pluim, J.P.W., Viergever, M.A. (eds.) MICCAI 2010, Part III. LNCS, vol. 6363, pp. 595–602. Springer, Heidelberg (2010)
4. Hinton, G., Salakhutdinov, R.: Reducing the dimensionality of data with neural networks. Science **313**(5786), 504–507 (2006)

5. Kauczor, H.U., Heitmann, K., Heussel, C.P., Marwede, D., Uthmann, T., Thelen, M.: Automatic detection and quantification of ground-glass opacities on high-resolution CT using multiple neural networks: comparison with a density mask. AJR Am. J. Roentgenol. **175**(5), 1329–1334 (2000)
6. Krizhevsky, A., Sutskever, I., Hinton, G.E.: Imagenet classification with deep convolutional neural networks. In: Pereira, F., Burges, C., Bottou, L., Weinberger, K. (eds.) Advances in Neural Information Processing Systems, vol. 25, pp. 1097–1105. Curran Associates, Inc., Red Hook (2012)
7. Le, Q.: Building high-level features using large scale unsupervised learning. In: 2013 IEEE International Conference on Acoustics, Speech and Signal Processing (ICASSP), pp. 8595–8598, May 2013
8. Shouno, H.: Recent studies around the neocognitron. In: Ishikawa, M., Doya, K., Miyamoto, H., Yamakawa, T. (eds.) ICONIP 2007, Part I. LNCS, vol. 4984, pp. 1061–1070. Springer, Heidelberg (2008)
9. Webb, W., Müller, N.L., Naidich, D.: High Resolution CT of the Lung, 4th edn. Lippincott Williams & Wilkins, Baltimore (2008)
10. Xiaojin, Z.: Semi-Supervised learning literature survey. Technical report, Computer Sciences, University of Wisconsin-Madison (2005)
11. Xu, R., Hirano, Y., Tachibana, R., Kido, S.: Classification of diffuse lung disease patterns on high-resolution computed tomography by a bag of words approach. In: Fichtinger, G., Martel, A., Peters, T. (eds.) MICCAI 2011, Part III. LNCS, vol. 6893, pp. 183–190. Springer, Heidelberg (2011)

Evaluation of Machine Learning Algorithms for Automatic Modulation Recognition

Muhammed Abdurrahman Hazar[1], Niyazi Odabaşioğlu[1], Tolga Ensari[2(✉)], and Yusuf Kavurucu[3]

[1] Electrical and Electronics Engineering, Istanbul University, Istanbul, Turkey
1316100006@ogr.iu.edu.tr, niyazio@istanbul.edu.tr
[2] Computer Engineering, Istanbul University, Istanbul, Turkey
ensari@istanbul.edu.tr
[3] Computer Engineering, Turkish Naval Academy, Istanbul, Turkey
ykavurucu@dho.edu.tr

Abstract. Automatic modulation recognition (AMR) becomes more important because of usable in advanced general-purpose communication such as cognitive radio as well as specific applications. Therefore, developments should be made for widely used modulation types; machine learning techniques should be tried for this problem. In this study, we evaluate performance of different machine learning algorithms for AMR. Specifically, we propose nonnegative matrix factorization (NMF) technique and additionally we evaluate performance of artificial neural networks (ANN), support vector machines (SVM), random forest tree, k-nearest neighbor (k-NN), Hoeffding tree, logistic regression and Naive Bayes methods to obtain comparative results. These are most preferred feature extraction methods in the literature and they are used for a set of modulation types for general-purpose communication. We compare their recognition performance in accuracy metric. Additionally, we prepare and donate the first data set to University of California-Machine Learning Repository related with AMR.

Keywords: Automatic modulation recognition · Nonnegative matrix factorization · Artificial neural networks · Support vector machines · Random forest tree · K-nearest neighbor · Hoeffding tree · Naive Bayes · Logistic regression

1 Introduction

Communication signals are modulated with different modulation types and are radiated at different frequencies. In some applications, signal identification and information of the signal is required. These applications are generally used for military purposes such as electronic warfare, surveillance, threat analysis, counter channel jamming or for civilian purposes such as spectrum management, interference identification and signal confirmation. In these applications, recognition of the signal's modulation type is required for demodulation and information of the signal. Modulation recognition had been manually performed in its early days and then was semi-automatic done. According

© Springer International Publishing Switzerland 2015
S. Arik et al. (Eds.): ICONIP 2015, Part I, LNCS 9489, pp. 208–215, 2015.
DOI: 10.1007/978-3-319-26532-2_23

to development of recognition technology, it has automatically performed and has become an important issue of communication intelligence (COMINT). It is investigated for analogue and digital modulation types in its early time. But depending on the expanse and development of digital electronics, digital modulation types are focused on. With steadily increasing use of intelligent modem, expanse of software defined radio, cognitive radio and systems using adaptive modulation, automatic modulation has become more important issue to reduce transmitted data as removing information of modulation type. AMR is designed based on two principles: The first one based on decision theoretic and the other one based on statistical pattern recognition. Decision theoretic is concerned many objections and difficulties. Adjusting appropriate threshold require detailed analysis. Nevertheless, increasing accuracy is really difficult. Whereas pattern recognition techniques do not require like threshold determination and are more tolerant of faults, robust against noise [1, 2].

Pattern recognition systems consist two main parts: Feature extraction and recognition subsystems. Therefore, accuracy is increased using more effective feature or machine learning techniques in a pattern recognition system except extra manipulation. In the literature, different feature set such as based signal spectral [1–4], high order cumulants [2–5] based continuous wavelet transform [3, 5, 9], constellation shape [6], based spectral analysis [7] has been used. Features based on spectral and high order cumulants are extracted from the digitally modulated signal.

There are lots of machine learning techniques for pattern recognition, support vector machines (SVM), particle swarm optimization (PSO) [3], multi-layer perceptron (MLP) [2, 7–9], probabilistic neural network (PNN) [4], genetic algorithms (GA) [2] are used for this purpose. The other important issue for AMR is modulation set. In the literature, some AMR systems are designed for more limited modulation set [1, 4]. Some of the others are applied for specific modulation set such as M-PSK and M-QAM [5], M-PSK, M-QAM and B-FSK [8] or pulse modulations [10]. The rest of them are developed for wide modulation set [2, 3, 9]. Another significant subject for AMR is channel type. In the past, almost all AMR systems are designed for AWGN channel type. In this study, some of most preferred features in the literature was extracted for 2-ASK, 4-ASK, 8-ASK, 2-PSK, 4-PSK, 8-PSK, 16-QAM, 64-QAM modulation types, and then extracted features are applied the statistical pattern recognition system used NMF, ANN, SVM, Random Forest, k-NN, Hoeffding tree, Logistic regression and Naive Bayes as benchmark. The work is performed in consideration additive white Gaussian noise (AWGN) channel type.

This paper organized as follows. Sections 2 and 3 define problem, explain signal and channel model, describe chosen features to be used in AMR. Information about NMF can be found in Sect. 4. Sections 5 and 6 consist of experimental results and conclusion, respectively.

2   System Model, Signal and Channel Representation

In AMR system, preprocessing includes data sampling, pulse shaping, estimation of carrier frequency, f_c and recovery of complex envelope through Hilbert transform. Thus, signal is prepared for feature extraction. Then feature extraction is performed and finally, the normalized features are applied to the automatic recognition subsystem.

A linearly modulated and pulse-shaped communication signal can be expressed as

$$s(t) = Re\left[v(t)\, e^{j2\pi f_c t}\right] \tag{1}$$

where $v(t)$ is pulse-shaped low-pass equivalent of the signal and can be given by:

$$v(t) = \sqrt{E_s} \sum\nolimits_{k=1}^{N} c_k g(t - kT) \tag{2}$$

where $\sqrt{E_s}$ is signal power, g is the pulse shaping function, T is symbol rate and $\{c_k\}$ is a wide sense stationary symbol sequence. A symbol can be represented as $c = x + jy$ where and y are real-valued in-phase and quadrature components of. After preprocessing of the received signal, the received complex signal can be expressed as

$$r(t) = v(t)\, e^{j2\pi f_c t} + n(t) \tag{3}$$

where $n(t)$ is the independent sample of zero mean complex Gaussian random variables with variance $N_0/2$ per dimension caused by AWGN channel. N_0 depends on signal to noise ratio (SNR) [3]. AWGN channel is a convenient model for satellite and deep space communication signals.

3   Feature Extraction

The recognition system is based on pattern recognition, so it requires suitable features for the process. A feature set consists of some spectral and statistical features. These features have been preferred much more in previous works is used.

3.1   Spectral Features

The motivation of the features used in this category is instantaneous amplitude or instantaneous phase of the signal to contain information about modulation type. The four features described in [2] and they are shown below.

1. Maximum value of the power spectral density of the normalized-centered instantaneous amplitude:

$$\gamma_{max} = \max\left|DFT\left(A_{cn}(i)\right)\right|^2 / N \tag{4}$$

where N is number of samples, $A_{cn}(i) = A_n(i) - 1$ and $A_n(i) = A(i)/m_a$, $A(i)$ is the i-th instantaneous amplitude and m_a is the sample mean value. Here, $i = t * f_s$ and f_s is sampling frequency.

2. Standard deviation of the absolute value of the centered non-linear components of the instantaneous phase:

$$\sigma_{ap} = \sqrt{\frac{1}{C}\sum_{A_n(i)>a_t}\phi_{NL}^2(i) - \left(\frac{1}{C}\sum_{A_n(i)>a_t}|\phi_{NL}(i)|\right)^2} \tag{5}$$

where $\phi_{NL}(i)$ is the non-linear component of the instantaneous phase, C is the number of samples in $\{\phi(i)\}$ for which $A_n(i) > a_t$ and a_t is the threshold for $A(i)$ below which the estimation of the instantaneous phase is very noise sensitive.

3. Standard deviation of the centered NL component of the instantaneous phase

$$\sigma_{dp} = \sqrt{\frac{1}{C}\sum_{A_n(i)>a_t}\phi_{NL}^2(i) - \left(\frac{1}{C}\sum_{A_n(i)>a_t}\phi_{NL}(i)\right)^2} \tag{6}$$

4. Standard deviation of the absolute value of the normalized-centered instantaneous amplitude

$$\sigma_{aa} = \sqrt{\frac{1}{N}\sum_{i=1}^{N}A_{cn}^2(i) - \left(\frac{1}{C}\sum_{i=1}^{N}|A_{cn}(i)|\right)^2} \tag{7}$$

3.2 Statistical Features

The signals modulated with QAM contain information at instantaneous phase and amplitude. Therefore, signals like these are regarded as complex signals. PSK and ASK signals are also regarded as complex signals. Taking advantage of high order cumulants to distinguish complex modulation types is an efficient way. High order cumulants are not affected by AWGN because the expected value of noise is zero. The features that used high order cumulants are described in [2, 11] as below.

Let X_i be a signal vector, $\{x_i^1, x_i^2, \ldots, x_i^n\}$ and $\langle\rangle$ denote the statistical expectation. The second, third and fourth-order cumulants at zero lag are then

$$C_{x_1,x_2} = \langle X_1, X_2\rangle = \frac{1}{N}\sum_{n=1}^{N}x_1^n x_2^n \tag{8}$$

$$C_{x_1,x_2,x_3} = \langle X_1, X_2, X_3\rangle = \frac{1}{N}\sum_{n=1}^{N}x_1^n x_2^n x_3^n \tag{9}$$

$$C_{x_1,x_2,x_3,x_4} = X_1, X_2, X_3, X_4 - X_1, X_2 X_3, X_4 - \langle X_1, X_3\rangle\langle X_2, X_4\rangle - \langle X_1, X_4\rangle\langle X_2, X_3\rangle$$
$$= \frac{1}{N}\sum_{n=1}^{N}x_1^n x_2^n x_3^n x_4^n - X_1, X_2 X_3, X_4 - \langle X_1, X_3\rangle\langle X_2, X_4\rangle - \langle X_1, X_4\rangle\langle X_2, X_3\rangle \tag{10}$$

It is required to obtain complex envelope of the sampled signal.

$$H_r = \left[r(t) + j\hat{r}(t)\right]e^{-j2\pi f_c t} \tag{11}$$

where H_r is the complex envelope of the sampled signal $r(t)$, $\hat{r}(t)$ is Hilbert transform of $r(t)$ and f_c is carrier frequency. When R and I is the real and imaginary part of H_r respectively. $C_{R,R}$, $C_{R,I}$, $C_{I,I}$, $C_{R,R,R}$, $C_{R,R,I}$, $C_{R,I,I}$, $C_{I,I,I}$, $C_{R,R,R,R}$, $C_{R,R,R,I}$, $C_{R,R,I,I}$, $C_{R,I,I,I}$ and $C_{I,I,I,I}$ are cumulants and cross-cumulants are used as statistical features.

4   Nonnegative Matrix Factorization (NMF)

Nonnegative matrix factorization (NMF) is one of the most used techniques in machine learning. NMF is a matrix factorization algorithm that focuses on the analysis of data matrices whose elements are nonnegative [12, 13]. NMF has a very broad range of potential applications [12] and used for different applications [12, 14–16]. Given a matrix $V \in \mathbb{R}^{MxN}$ is a set of multivariate m-dimensional data vector where n is the number of samples in the data set. This matrix is approximately factorized into a mxr matrix W and a rxn matrix H where r is the rank of V that $r \leq \min(m, n)$ [12–14].

$$V \approx W \times H \qquad (12)$$

There are some cost functions [13, 15, 16] evaluating the quality of the approximation and two of them are commonly used. The first is the square of the Euclidean distance between two matrices [13, 17].

$$\|V - W \times H\|^2 = \sum_{ij} \left(A_{ij} - (W \times H)_{ij}\right)^2 \qquad (13)$$

The second is referenced the "divergence" between two matrices.

$$D(V\|(W \times H)) = \sum_{ij} \left(V_{ij} \log \frac{V_{ij}}{(W \times H)_{ij}} - V_{ij} + (W \times H)_{ij}\right) \qquad (14)$$

Iterative update algorithms are used WxH to approximate data matrix. Multiplicative update algorithm is the most widely used due to its easy implementation and speed [13]. Other update algorithms also existed [18]. In this study, Euclidean distance (cost function) and multiplicative update rule are used for analysis:

$$H_{kj} \leftarrow H_{kj} \frac{\left(W^T \times V\right)_{kj}}{\left(W^T \times W \times H\right)_{kj}} W_{ik} \leftarrow W_{ik} \frac{\left(V \times H^T\right)_{ik}}{\left(W \times H \times H^T\right)_{ik}} \qquad (15)$$

5   Experimental Results

The data set is prepared the features that are extracted from instances which consist of 9,600 digitally modulated symbols. The system is trained with 4,000 instances that are uniformly distributed for each class and is tested with different 4,000 instances similar to the previous. Training and testing data sets are generated according to SNR level that is estimated in cognitive radio systems. We also donate (upload) this data set to University of California, Irvine - Machine Learning Repository (UCI-MLR), namely *"Digital Modulation Features Data Set for Automatic Modulation Recognition[1]"*. This data set is the first one in UCI-MLR archive related with modulation recognition. It can be downloaded and used freely in Matlab and Excel formats for all modulation recognition experiments.

We implement NMF (with multiplicative update rule [12]) on data set and give the results below. Negative data has been changed with its absolute value to fit it to the algorithm for NMF. Illustrated results are given in Fig. 1 and Table 1. We use accuracy metric to test the recognition performance. We also do the experiments for the same data set with ANN (multi-layer perceptron with one hidden layer that has 10 neurons as benchmark and gradient descent with momentum weight and bias learning function are used), SVM (with radial basis function kernel and J. Platt's minimum optimization algorithm), random forest tree (with 100 trees), k-NN (Euclidean distance k-NN = 1), Hoeffding tree (with 0.05 tie threshold and 200 grace period) and Naïve Bayes.

Table 1. The recognition rates table of NMF, ANN, SVM, Random Forest Tree, k-Nearest Neighbor, Hoeffding Tree and Naïve Bayes algorithms.

SNR values	Accuracy						
	ANN	SVM	Random Forest	k-NN	Hoeffding Tree	Naive Bayes	NMF
−10	0.87	0.75	0.77	0.61	0.83	0.83	0.62
−5	0.94	0.83	0.94	0.81	0.94	0.93	0.73
0	1.00	0.94	1.00	0.91	1.00	1.00	0.87
5	1.00	1.00	1.00	1.00	1.00	1.00	0.78
10	1.00	1.00	1.00	1.00	1.00	1.00	0.84
15	1.00	1.00	1.00	1.00	1.00	1.00	0.82
20	1.00	1.00	1.00	1.00	1.00	1.00	0.89

The recognition performance of NMF lies down between 60–90 % ranges. Especially, removing negative values of data set affect the performance. These conditions have been achieved by replacing them with their absolute values. On the other hand, ANN, SVM, random forest tree, k-NN, Hoeffding tree and Naïve Bayes give better recognition results then NMF. The accuracy reach to 1 beginning from SNR = 5. Even so, different types of NMF, cost functions, r (lower rank value) can have better performance of NMF in the future studies. The exact rates on graphic also can be seen Table 1 shown below.

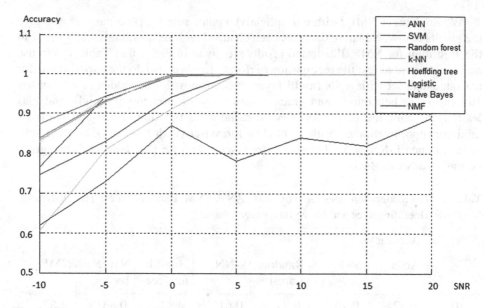

Fig. 1. The recognition performance of NMF, ANN, SVM, Random Forest Tree, k-Nearest Neighbor, Hoeffding Tree and Naïve Bayes algorithms.

6 Conclusion

In this article, we propose NMF method for AMR and make comparisons with other machine learning algorithms. Automatic recognition has never been studied with NMF in the literature yet for modulation. The system is designed for AWGN channel type consideration. Some statistical and spectral features are used for recognition. ANN, SVM, random forest tree, k-NN, Hoeffding tree and Naïve Bayes methods examined except for NMF. We show the recognition performance of the system with accuracy metric in graphic. Other algorithms has better accuracy then NMF, especially after SNR = 5. Although recognition range lies down between 60–90 % for NMF, but this situation can be improved by offering several different parametric and structural approaches in the future studies. We also prepare and donate the first "Automatic Modulation Recognition Data Set" to University of California-Machine Learning Repository for free usage.

References

1. Azzouz, E.E., Nandi, A.K.: Automatic identification of digital modulation types. Sig. Process. **47**, 55–69 (1995)
2. Wong, M.L.D., Nandi, A.K.: Automatic digital modulation recognition using artificial neural network and genetic algorithm. Sig. Process. **84**, 351–365 (2004)

3. Valipour, M.H., Homayounpour, M.M., Mehralian, M.A.: Automatic digital modulation recognition in presence of noise using SVM and PSO. In: IEEE International Symposium on Telecommunications, pp. 378–382 (2012)
4. Roganovic, M.M., Neskovic, A.M., Neskovic, N.J.: Application of artificial neural networks in classification of digital modulations for software defined radio. In: IEEE, EUROCON, pp. 1700–1706 (2009)
5. Ghasemi, S., Gangal, A.: An effective algorithm for automatic modulation recognition. In: IEEE, Signal Processing and Communications Applications Conference (SIU), pp. 903–906 (2014)
6. Mobasseri, B.G.: Digital modulation classification using constellation shape. Sig. Process. **80**, 251–277 (2000)
7. Al-Makhlasawy, R.M., Elnaby, M.M.A., El-Khobby, H.A., El-Samie, F.E.A.: Automatic modulation recognition in OFDM systems using cepstral analysis and a fuzzy logic interface. In: International Conference on Informatics and Systems (INFOS), pp. CC56–CC62 (2012)
8. El Rube, I.A.H., El-Madany, N.E.-d.: Cognitive digital modulation classifier using artificial neural networks for NGNs. In: International Conference on Wireless and Optical Communications, pp. 1–5 (2010)
9. Hassan, K., Dayoub, I., Hamouda, W., Berbineau, M.: Automatic modulation recognition using wavelet transform and neural network. In: 9th International Conference on Intelligent Transport Systems Telecommunications, pp. 234–238 (2009)
10. Sobolewski, S., Adams, W.L., Sankar, R.: Automatic modulation recognition techniques based on cyclostationary and multifractal features for distinguishing LFM, PWM and PPM waveforms used in radar systems as example of artificial intelligence implementation in test. In: IEEE AUTOTESTCON, pp. 335–340 (2012)
11. Swami, A., Sadler, B.M.: Hierachical digital modulation classification using cumulants. IEEE Trans. Commun. **48**(3), 416–429 (2000)
12. Lee, D.D., Seung, H.S.: Learning the parts of objects by nonnegative matrix factorization. Nature **401**, 788–791 (1999)
13. Lee, D.D., Seung, H.S.: Algorithms for non-negative matrix factorization. Adv. Neural Inf. Process. Syst. **13**, 556–562 (2001)
14. Zurada, J.M., Ensari, T., Asl, E.H., Chorowski, J.: Nonnegative matrix factorization and its application to pattern analysis and text mining. In: IEEE, 2013 Federated Conference on Computer Science and Information Systems (FedCSIS), pp. 11–16 (2013)
15. Ensari, T., Chorowski, J., Zurada, J.M.: Occluded face recognition using correntropy-based nonnegative matrix factorization. In: IEEE, International Conference on Machine Learning and Applications (ICMLA), pp. 606–609 (2012)
16. Ensari, T., Chorowski, J., Zurada, J.M.: Correntropy-based document clustering via nonnegative matrix factorization. In: International Conference on Artificial Neural Networks (ICANN), pp. 347–354 (2012)
17. Cai, D., He, X., Han, J., Huang, T.S.: Graph regularized non-negative matrix factorization for data representation. IEEE Trans. Pattern Anal. Mach. Intell. **33**, 1548–1560 (2010)
18. Kim, J., Park, H.: Toward faster nonnegative matrix factorization: a new algorithm and comparisons. In: IEEE International Conference on Data Mining, pp. 353–362 (2008)

Probabilistic Prediction in Multiclass Classification Derived for Flexible Text-Prompted Speaker Verification

Shuichi Kurogi[✉], Shota Sakashita, Satoshi Takeguchi,
Takuya Ueki, and Kazuya Matsuo

Kyushu Institute of Technology, Tobata, Kitakyushu, Fukuoka 804-8550, Japan
{kuro,matsuo}@cntl.kyutech.ac.jp,
{sakashita,takeguchi}@kurolab.cntl.kyutech.ac.jp
http://kurolab.cntl.kyutech.ac.jp/

Abstract. So far, we have presented a method for text-prompted multi-step speaker verification using GEBI (Gibbs-distribution based extended Bayesian inference) for reducing single-step verification error, where we use thresholds for acceptance and rejection but the tuning is not so easy and affects the performance of verification. To solve the problem of thresholds, this paper presents a method of probabilistic prediction in multiclass classification for solving verification problem. We also present loss functions for evaluating the performance of probabilistic prediction. By means of numerical experiments using recorded real speech data, we examine the properties of the present method using GEBI and BI (Bayesian inference) and show the effectiveness and the risk of probability loss in the present method.

Keywords: Probabilistic prediction · Text-prompted speaker verification · Gibbs-distribution-based extended Bayesian inference · Loss functions in multiclass classification

1 Introduction

So far, we have presented a method for text-prompted multistep speaker verification [1, 2]. Here, from [3], text-prompted speaker verification has been developed to combat spoofing from impostors and digit strings are often used to lower the complexity of processing. From another perspective, the method focuses on reducing verification error by means of multistep verification using Gibbs-distribution-based Bayesian inference (GEBI) for rejecting unregistered speakers [2], where from the analysis of the properties, it is suggested that the tuning of the thresholds for acceptance and rejection is not so easy and affects the performance. Namely, we have tuned the thresholds by the method of EER (equal error rate) for FAR (false acceptance rate) and FRR (false rejection rate) to be almost the same. Furthermore, the obtained values of the thresholds are not so easy to

© Springer International Publishing Switzerland 2015
S. Arik et al. (Eds.): ICONIP 2015, Part I, LNCS 9489, pp. 216–225, 2015.
DOI: 10.1007/978-3-319-26532-2_24

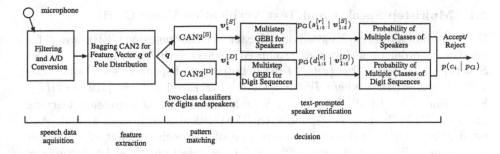

Fig. 1. Diagram of text-prompted speaker verification system using CAN2s

be modified for different security or risk level of verification. To solve this problem, this paper presents probabilistic prediction. Here, note that from [4] and our experience, we can see that the probabilistic prediction in weather and climate forecasting allows the users to decide on the level of risk they are prepared and to take appropriate action within a proper understanding of the uncertainties. For introducing probabilistic prediction into verification, we first formulate multiclass classification problem, and then apply Bayesian inference (BI) to obtain the probability. We also present loss functions to evaluate the performance of the probabilistic prediction in multiclass classification derived for verification problem, and then examine the properties and effectiveness of the present method by means of using real speech signal.

Here, note that our speech processing system employs competitive associative nets (CAN2s). The CAN2 is an artificial neural net for learning efficient piecewise linear approximation of nonlinear function [5], and we have shown that feature vectors of pole distribution extracted from piecewise linear predictive coefficients obtained by the bagging (bootstrap aggregating) version of the CAN2 reflect nonlinear and time-varying vocal tract of the speaker [6]. Although the most common way to characterize speech signal in the literature is short-time spectral analysis, such as Linear Prediction Coding (LPC) and Mel-Frequency Cepstrum Coefficients (MFCC) [7], the bagging CAN2 learns more precise information than LPC and MFCC (see [6] for details).

We show the method of probabilistic prediction in Sect. 2, experimental results and analysis in Sect. 3, and the conclusion in Sect. 4.

2 Probabilistic Prediction for Text-Prompted Speaker Verification

Figure 1 shows an overview of the present text-prompted speaker verification system using CAN2s. In the same way as general speaker recognition systems [7], it consists of four steps: speech data acquisition, feature extraction, pattern matching, and making a decision. In this research study, we use a feature vector of pole distribution obtained from a speech signal (see [6] for details).

2.1   Multistep Speaker and Text Verification Using GEBI

Here, we show a brief explanation of multistep verification using GEBI (see [2] for details). In order to achieve text-prompted speaker verification using digits, let $S = \{s_i | i \in I^{[S]}\}$ and $D = \{d_i | i \in I^{[D]}\}$ denote a set of speakers $s \in S$ and digits $d \in D$, respectively, where $I^{[S]} = \{1, 2, \cdots, |S|\}$ and $I^{[D]} = \{1, 2, \cdots, |D|\}$. Furthermore, let $\mathrm{RLM}^{[M]}$ for $M = S$ and M be a set of regression learning machines $\mathrm{RLM}^{[m]}$ ($m \in I^{[M]}$), and each $\mathrm{RLM}^{[m]}$ learns to predict a single-step verification as $v^{[m]} = 1$ for the acceptance of a speech segment of a speaker $m = s_i$ or a digit $m = d_i$, and $v^{[m]} = 0$ for the rejection. Here, let us suppose that we have speech segments of spoken digits obtained by some appropriate segmentation method and this research focuses on the multistep verification of spoken digit sequences.

For multistep verification of input sequence of spoken digits, we have proposed Gibbs-distribution-based extended Bayesian inference (GEBI) as shown below for overcoming the problem of Bayesian inference (BI) in speaker verification of unregistered speakers (see [2] for details). Let $\boldsymbol{v}_{1:T}^{[M]} = \boldsymbol{v}_1^{[M]} \boldsymbol{v}_2^{[M]} \cdots \boldsymbol{v}_t^{[M]}$ be an output sequence of $\mathrm{RLM}^{[m]}$ for the reference sequence $m_{1:T}^{[r]} = m_1^{[r]} m_2^{[r]} \cdots m_T^{[r]}$, we have recursive posterior probability for $t = 1, 2, \cdots, T$ as follows,

$$p_{\mathrm{G}}\left(m_{1:t}^{[r]} \mid \boldsymbol{v}_{1:t}^{[M]}\right) = \frac{1}{Z_t} p_{\mathrm{G}}\left(m_{1:t-1}^{[r]} \mid \boldsymbol{v}_{1:t-1}^{[M]}\right)^{\beta_t / \beta_{t-1}} p\left(\boldsymbol{v}_t^{[M]} \mid m_t^{[r]}\right)^{\beta_t}, \quad (1)$$

$$p_{\mathrm{G}}\left(\overline{m_{1:t}^{[r]}} \mid \boldsymbol{v}_{1:t}^{[M]}\right) = \frac{1}{Z_t} p_{\mathrm{G}}\left(\overline{m_{1:t-1}^{[r]}} \mid \boldsymbol{v}_{1:t-1}^{[M]}\right)^{\beta_t / \beta_{t-1}} p\left(\boldsymbol{v}_t^{[M]} \mid \overline{m_t^{[r]}}\right)^{\beta_t}. \quad (2)$$

where $\beta_t = \beta/t$ ($t \geq 1$) and $\beta_0 = 1$, and Z_t is the normalization constant. Note that the conventional BI is obtained for $\beta_t = 1 (t \geq 0)$ and we denote p_{B} instead of p_{G} for the probability obtained by the above equations with $\beta_t = 1 (t \geq 0)$, while p_{G} is obtained for $\beta_t = 1/t$ ($t \geq 1$) in the experiments shown below.

The verification by our previous method shown in [2] at $t = T$ is given by

$$V_{1:T}^{[M]} = \begin{cases} 1 & \text{if } p_{\mathrm{G}}\left(m_{1:T}^{[r]} \mid \boldsymbol{v}_{1:T}^{[M]}\right) \geq p_\theta^{[M]} \\ -1 & \text{otherwise} \end{cases} \quad (3)$$

for speaker $(m, M) = (s, S)$ and text $(m, M) = (d, D)$, respectively. Here, $p_\theta^{[M]}$ for $M = S$ and D are thresholds, and $V_{1:T}^{[M]} = 1$ and -1 indicates acceptance and rejection, respectively. The verification of text-prompted speaker is executed by $V_{1:T}^{[SD]} = V_{1:T}^{[S]} \wedge V_{1:T}^{[D]} = 1$ and -1 for acceptance and rejection, respectively. The performance of verification depends on the thresholds $p_\theta^{[M]}$ for $M = S$ and D. To execute more flexible verification than using thresholds, we introduce probabilistic probability into the verification problem in the next section.

2.2   Probabilistic Prediction for Speaker and Text Verification

We introduce multiple classes to classify the verification results, and then introduce probabilistic prediction for speaker and text verification.

Multiclass Classification for Speaker and Text Verification. For speaker verification, we consider the following three classes, where we suppose all elements in each input and reference speaker sequence, respectively, consists of the same speaker;

$c_{+1}^{[S]}$ (Class of correct speakers): class of speakers satisfying $s_{1:T} = s_{1:T}^{[r]}$ ($\in S_{1:T}$) for the input $s_{1:T}$ and the reference $s_{1:T}^{[r]}$, where S is the set of registered speakers, and $S_{1:T}$ denotes the set of $s_{1:T}$ whose all elements s_t ($t = 1, 2, \cdots, T$) are registered speaker $s_t \in S$.

$c_{-1}^{[S]}$ (Class of incorrect speakers): class of speakers satisfying $s_{1:T} \neq s_{1:T}^{[r]}$ for $s_{1:T}, s_{1:T}^{[r]} \in S_{1:T}$.

$c_0^{[S]}$ (Class of unregistered speakers): class of speakers satisfying $s_{1:T} \neq s_{1:T}^{[r]}$ for $s_{1:T} \notin S^{[T]}$.

Here, note that these classes are determined for the pair of input and reference sequences.

For text (or digit sequence) verification, we consider the following $N + 1$ classes of $T(= mN)$-length digit sequence consisting of m times of N-length subsequences:

$c_i^{[D]}$ for $i = 0, 1, 2, \cdots, N$ (Class of digit sequences with correct ratio being i/N): class of input $d_{1:T}$ and reference $d_{1:T}^{[r]}$ digit sequences, which consist of m times of N-length subsequence whose i digits are the same.

In order to simplify the explanation, let $C^{[S]} = \{c_i^{[S]} \mid i \in I^{[C^{[S]}]})\}$ be the set of speaker verification classes, $C^{[D]} = \{c_i^{[D]} \mid i \in I^{[C^{[D]}]}\}$ be the set of text verification classes, C denote $C^{[S]}$ or $C^{[D]}$, and $I^{[C]}$ denote $I^{[C^{[S]}]} = \{-1, 0, 1\}$ or $I^{[C^{[D]}]} = \{0, 1, 2, \cdots, N\}$.

Note that these classes have the ordered indices which we utilize for probabilistic prediction of multiclass classification derived for the verification. Namely, we can divide two sets of classes, where one consists of the classes with the indices from $i = i_\theta^{[C]}$ to $i_{max}^{[C]}$ and the other consists of the remaining classes, where $i_\theta^{[C]}$ and $i_{max}^{[C]}$ indicate the threshold for verification and the maximum index of the classes in C, respectively. Furthermore, as shown in Sect. 3.2, we have a possibility to have a class with a large classification error but a sum of adjacent classes has smaller error. Thus, in order to achieve a reliable probabilistic prediction, we will combine some adjacent classes so that every combined class has smaller classification error.

Probabilistic Prediction in Multiclass Classification. In order to formulate the probabilistic prediction of multiclass classification, let $X^{[test]} = \{(\boldsymbol{x}_j, t_j) \mid j \in I^{[test]}\}$ be a test dataset, where $\boldsymbol{x}_j$ is the jth data of the pair $\left(m_{1:T}^{[r]}, \boldsymbol{v}_{1:T}^{[M]}(m_{1:T}) \right)$ determined by the sequences of reference $m_{1:T}^{[r]}$ and input $m_{1:T}$, $t_j \in C$ indicates target class to be classified, and $I^{[test]} = \{1, 2, \cdots, |I^{[test]}|\}$. Furthermore, let

$p_G(x_j)$ denote the GEBI probability $p_G\left(m_{1:T}^{[r]} \mid v_{1:T}^{[M]}\right)$ given by (1). Then, from BI, we have the following posterior probability

$$p\left(c_i \mid p_G(x_j)\right) = \frac{p\left(p_G(x_j) \mid c_i\right) p(c_i)}{\sum\limits_{c_l \in C} p\left(p_G(x_l) \mid c_l\right) p(c_l)}, \tag{4}$$

where $p(c_i)$ is the prior probability of $c_i \in C$, and $p\left(p_G(x_j) \mid c_i\right)$ denotes the likelihood of the value of $p_G(x_j)$ being for c_i. Here, $p\left(p_G(x_j) \mid c_i\right)$ can be estimated from a training dataset $X^{[\text{train}]} = \{(x_j, t_j) \mid j \in I^{[\text{train}]}\}$ involving x_j independent and identically distributed (i.i.d) with respect to the data in the test dataset, and we usually use $p(c_i)$ to be equal for all c_i, while we can use specific values depending on the situation, e.g., we can use $p\left(c_0^{[S]}\right) = 0$ for the situation where there is no unregistered speaker expected.

With the above probability $p(c_i \mid p_G(x_j))$ for $c_i \in C$, the user or a decision maker is expected to make flexible decision for verification as shown in Sect. 3.2.

2.3 Loss Functions for Evaluating the Performance

We use the following loss functions to evaluate the performance of the probabilistic prediction in multiclass classification extended from the loss functions for two-class classification shown in [8]. First, we divide the multiple classes into two sets of classes: one consists of a class with the maximum probability and the other of remaining classes, where the index of the class in the former set is given by

$$i_M(j) = \underset{i \in I^{[C]}}{\operatorname{argmax}} \, p(c_i \mid p_G(x_j)). \tag{5}$$

Now, the average classification error (ACE) for $i_M(j)$ is given by

$$L_{\text{ACE}} = \frac{1}{n}\left[\sum_{j \in I^{[\text{test}]}} \mathbf{1}\{t_j \neq c_{i_M(j)}\}\right] = \frac{1}{n}\left[\sum_{\{j \mid t_j \neq c_{i_M(j)}\}} 1\right] \tag{6}$$

Here, $\mathbf{1}\{z\}$ indicates an indicator function, equal to 1 if z is true, and to 0 if z is false, $\{j \mid t_j \neq c_{i_M(j)}\}$ indicates the set of indices satisfying $t_j = c_{i_M(j)}$ for $j \in I^{[\text{test}]}$.

The negative log probability loss (NLP) for $i_M(j)$ is given by

$$L_{\text{NLP}} = -\frac{1}{n}\left[\sum_{\{j \mid t_j = c_{i_M(j)}\}} \log p\left(c_{i_M(j)} \mid p_G(x_j)\right)\right.$$

$$\left. + \sum_{\{j \mid t_j \neq c_{i_M(j)}\}} \log\left(1 - p\left(c_{i_M(j)} \mid p_G(x_j)\right)\right)\right] \tag{7}$$

The first term of the right hand side becomes smaller for larger probability of correct classification and the second term becomes smaller for smaller probability of incorrect classification.

The negative log predictive density loss (NLPD) for evaluating regression performance given by

$$L_{\mathrm{NLPD}} = -\frac{1}{n} \left[\sum_{j \in I^{[\mathrm{test}]}} \log p\left(t_j \mid p_{\mathrm{G}}(\boldsymbol{x}_j)\right) \right] \tag{8}$$

is considered to be applicable for evaluating the performance of probabilistic prediction in multiclass classification.

3 Experiments

3.1 Experimental Setting

We have recorded speech data sampled with 8 kHz of sampling rate and 16 bits of resolution in a silent room of our laboratory. They are from seven speakers (2 female and 5 mail speakers): $S = \{\mathrm{fHS, fMS, mKK, mKO, mMT, mNH, mYM}\}$ for ten Japanese digits $D = \{/\mathrm{zero}/, /\mathrm{ichi}/, /\mathrm{ni}/, /\mathrm{san}/, /\mathrm{yon}/, /\mathrm{go}/, /\mathrm{roku}/, /\mathrm{nana}/, /\mathrm{hachi}/, /\mathrm{kyu}/\}$. For each speaker and each digit, ten samples are recorded on different times and dates among two months. We denote each spoken digit by $x = x_{s,d,l}$ for $s \in S$, $w \in W$ and $l \in L = \{1, 2, \cdots, 10\}$, and the given dataset by $X = (x_{s,d,l} | s \in S, d \in D, l \in L)$.

By meas of random selection from X, we have generated training dataset $X^{[\mathrm{train}]} = \{(\boldsymbol{x}_j, t_j) \mid j \in I^{[\mathrm{train}]}\}$ for making the likelihood $p\left(p_{\mathrm{G}}(\boldsymbol{x}_j) \mid c_i\right)$ given in (4) and test dataset $X^{[\mathrm{test}]} = \{(\boldsymbol{x}_j, t_j) \mid j \in I^{[\mathrm{test}]}\}$ for evaluating the performance of probabilistic prediction. A data $\boldsymbol{x}_j$ indicates the jth data of $\left(m_{1:T}^{[r]}, \boldsymbol{v}_{1:T}^{[M]}(m_{1:T})\right)$ consists of reference and input sequences of $T(= 15)$-length spoken digits for $T = m \times N = 15$ with $m(= 3)$ times of $N(= 5)$-length digit sequences indicating some ID numbers. Of course, $\boldsymbol{x}_j \in X^{[\mathrm{train}]}$ and $\boldsymbol{x}_j \in X^{[\mathrm{test}]}$ are not the same but should be independent and identically distributed (i.i.d). To have this done, for each of training and test datasets, we have generated 1,000 data for each combination of 3 classes of speaker sequences involving correct, incorrect and unregistered speakers and 6 classes of digit sequences involving i/N correct digits for $i = 0, 1, \cdots, N = 5$. Thus, we have 18,000 data for training and test datasets.

In order to evaluate the performance of learning machines $\mathrm{RLM}^{[M]}$ for predicting unknown (untrained) data and the data of unregistered speaker, we employ a combination of LOOCV (leave-one-out cross-validation) and OOB (out-of-bag) estimate (see [2] for details). For the regression learning machines, we have used CAN2s for learning piecewise linear approximation of nonlinear functions (see [6] for details).

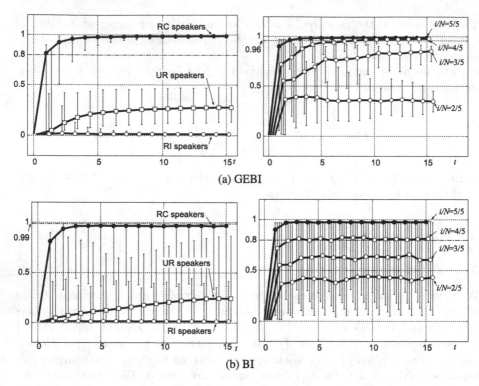

Fig. 2. Experimental result of multistep probability of (a) GEBI and (b) BI for speakers (left) and digits (right), where the curves of speakers denote RC (registered correct), UR (unregistered), RI (registered incorrect). The plus and minus error bars indicate RMS (root mean square) of positive and negative errors from the mean, respectively. The curves for different datasets are shifted slightly and horizontally to avoid crossovers.

3.2 Experimental Results and Analysis

Experimental Result of Probabilistic Prediction. First of all, we show the multistep probabilities in Fig. 2. As explained in [2], we have tuned the thresholds to be $(p_\theta^{[S]}, p_\theta^{[D]}) = (0.80, 0.96)$ for GEBI and $(0.99, 0.80)$ for BI to achieve EER (equal error rate) at $t = T = 15$ for FAR (false acceptance rate) and FRR (false rejection rate) to be almost the same. For this tuning, we also have employed thresholds $(i_\theta^{[S]}, i_\theta^{[D]}) = (1, 4)$ for deciding the security level of correct verification, i.e., we assume that the data in $c_i^{[S]}$ for $i \geq i_\theta^{[S]} = 1$ and $c_i^{[D]}$ for $i \geq i_\theta^{[D]} = 4$ should to be accepted in speaker and text verification, respectively, and the other data should be rejected. In Fig. 2, we can see that these threshold values seem reasonable but not so easy to be tuned.

By means of the probability prediction by (4), we have the probability $p\left(c_i \mid p_G\right) = p\left(c_i \mid p_G(x_j)\right)$ and $p\left(c_i \mid p_B\right) = p\left(c_i \mid p_B(x_j)\right)$ as shown in Fig. 3. From Fig. 3(a), we can estimate the probability of the classes depending on p_G. For example, from the left hand side of Fig. 3(a) for speaker verification, the

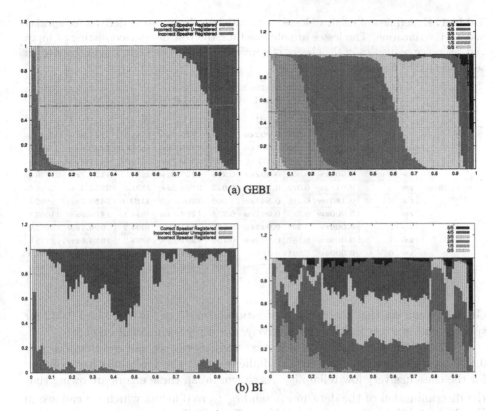

(a) GEBI

(b) BI

Fig. 3. Posterior probability $p(c_i \mid p_G)$ for (a) p_G obtained by GEBI and (b) $p_B (= p_G|_{\beta_t=1})$ by BI for speaker (left) and text (right) classification. The horizontal axis indicates p_G or p_B and the vertical length of a colored bar indicates the probability of a class c_i corresponding to the color (Color figure online).

probability of correct speaker and unregistered speaker is expected for the value of p_G larger than 0.86 and 0.04, respectively, Furthermore, from the right hand side of Fig. 3(a) for text verification, the ratio of correct digits is expected to be more than 5/5, 4/5, 3/5, 2/5, 1/5 for the value of p_G larger than 0.97, 0.93, 0.62, 0.19, 0.03, respectively, On the other hand, it is hard to obtain the property of the probability for BI as shown in Fig. 3(b). This is owing to the fluctuation of the mean value and the large variance of p_B as shown in Fig. 2 and a mathematical analysis is shown in [2].

Experimental Result of Losses and Remarks. We show experimental results of losses in Table 1, where L_{AVE_θ} indicates AVE (average verification error) obtained for the method using the thresholds given above. From the comparison of the losses between GEBI and BI, we can see that GEBI has achieved smaller losses (bold face figures) for almost all classes than BI, especially, it has achieved smaller mean values for all losses. From the columns of L_{AVE_θ} for

Table 1. Experimental result of losses for multiclass classification derived for speaker and text verification. The losses are obtained for the test dataset consisting of input and reference sequences in the classes of speakers, $c_i^{[S]}$ for $i = -1, 0, 1$, and texts (digit sequences), $c_i^{[D]}$ for $i = 0, 1, 2, \cdots, 5$.

	Class index i	L_{AVE_θ}		L_{ACE}		L_{NLP}		L_{NLPD}	
		GEBI	BI	GEBI	BI	GEBI	BI	GEBI	BI
Speaker verification	zw 1	**0.002zw**	0.046	**0.002zw**	0.029	**223.2zw**	5549.5	**224.6zw**	6538.9
	zw 0	**0.011zw**	0.050	**0.061zw**	0.748	**7730.9zw**	42052.3	**7730.8zw**	42308.0
	zw −1	**0.000zw**	0.000	0.033zw	**0.019**	4420.7zw	18796.5	**4427.7zw**	19332.4
	zw Mean	**0.004zw**	0.032	**0.032zw**	0.265	**4125.0zw**	22132.8	**4127.7zw**	22726.5
Text verification	zw 5	**0.011zw**	0.012	**0.009zw**	0.015	**2708.1zw**	5331.8	**2736.4zw**	5669.5
	zw 4	**0.180zw**	0.216	**0.844zw**	1.000	7378.3zw	**4161.6**	7455.7zw	8656.6
	zw 3	**0.000zw**	0.581	**0.061zw**	0.774	**1293.7zw**	4844.1	**1476.9zw**	11600.1
	zw 2	**0.000zw**	0.339	**0.043zw**	0.862	**911.8zw**	4694.2	**996.1zw**	12072.9
	zw 1	**0.000zw**	0.187	**0.076zw**	0.705	**1291.0zw**	5768.7	**1323.5zw**	8475.6
	zw 0	**0.000zw**	0.001	0.042zw	**0.004**	**1297.7zw**	3984.0	**1345.5zw**	4098.9
	zw Mean	**0.032zw**	0.223	**0.179zw**	0.560	**2480.1zw**	4797.4	**2555.7zw**	8428.9

GEBI, we can see that the mean verification error L_{AVE_θ} is 0.004 and 0.032 for speaker and text verification, respectively, and they seem small enough.

Next, for the class index $i = 4$ in text verification, we can see that $L_{\text{ACE}} = 0.844$ is very larger than others. This indicates that the probabilistic prediction for the class has very low reliability. As shown in [2], these errors are owing that the discrimination of the data in $c_4^{[D]}$ and $c_5^{[D]}$ are difficult which we can see in Fig. 2(a) (right) for the curves of $i/N = 5/5$ and $4/5$.

To solve this problem for more reliable classification, we combine the class $c_4^{[D]}$ and $c_5^{[D]}$ into a class $c_{4\uplus5}^{[D]}$. Then, for a test data in $c_{4\uplus5}^{[D]}$, we have $L_{\text{ACE}} = 0.021$ for the prediction using GEBI. As a result, by means of using the classes $c_i^{[D]}$ for $i = 0, 1, 2, 3, 4 \uplus 5$, we have achieved L_{ACE} less than 0.076 with the mean 0.044. These analysis and modification indicate that we have to understand and reduce the risk of probability loss in using probabilistic prediction. From this point of view, we hardly use the probabilistic prediction obtained by BI. We would like to analyse other losses in our future research.

Flexible Verification Using Probabilistic Prediction. For text-prompted speaker verification, we can use the class index thresholds $i_\theta^{[S]} = 1$ and $i_\theta^{[D]} = 4 \uplus 5$ for speaker and text verification, respectively. Here, however, when the probability $p\left(c_1^{[S]} \mid p_{\text{G}}\right)$ or $p\left(c_{4\uplus5}^{[D]} \mid p_{\text{G}}\right)$ for an input sequence is not so bigger than 0.5, a decision maker has a possibility to ask additional question to obtain much larger or much smaller probability than 0.5.

For text verification, we can tune the threshold $i_\theta^{[D]}$ for accepting the input sequence satisfying $i \geq i_\theta^{[D]}$ indicating that more than or equal to $i_\theta^{[D]}$ correct digits out of N-length sequence are expected. Here, the tuning of $i_\theta^{[D]}$ in the present method is easier and understandable than in the previous method requiring the tuning of thresholds $p_\theta^{[D]}$ in (3). Therefore, as an example of application, the tuning of $i_\theta^{[D]}$ has a possibility to be flexibly used in verifying spoken digits of a specific speaker in a recorded tape, where we do not need high security level.

4 Conclusion

We have presented a method of probabilistic prediction for flexible verification without using thresholds for acceptance and rejection. After introducing multiclass classification for classifying the verification results, the method utilizes BI to obtain the probability. The method also uses loss functions for evaluating the performance of probabilistic prediction. By means of numerical experiments using recorded real speech data, we have examined the properties of the present method using GEBI and BI, and show the effectiveness and the risk of probability loss in the present method.

References

1. Kurogi, S., Ueki, T., Mizobe, Y., Nishida, T.: Text-prompted multistep speaker verification using Gibbs-distribution-based extended Bayesian inference for reducing verification errors. In: Lee, M., Hirose, A., Hou, Z.-G., Kil, R.M. (eds.) ICONIP 2013, Part III. LNCS, vol. 8228, pp. 184–192. Springer, Heidelberg (2013)
2. Kurogi, S., Ueki, T., Takeguchi, S., Mizobe, Y.: Properties of text-prompted multistep speaker verification using Gibbs-distribution-based extended Bayesian inference for rejecting unregistered speakers. In: Loo, C.K., Yap, K.S., Wong, K.W., Teoh, A., Huang, K. (eds.) ICONIP 2014, Part II. LNCS, vol. 8835, pp. 35–43. Springer, Heidelberg (2014)
3. Beigi, H.: Fundamentals of Speaker Recognition. Springer-Verlag New York Inc., New York (2011)
4. Slingo, J., Palmer, T.: Uncertainty in weather and climate prediction. Phil. Trans. R. Soc. A **369**, 4751–4767 (2011)
5. Kurogi, S., Ueno, T. and Sawa, M.: A batch learning method for competitive associative net and its application to function approximation. In: Proceedings of SCI2004, vol. V, pp. 24–28 (2004)
6. Kurogi, S., Mineishi, S., Sato, S.: An analysis of speaker recognition using bagging CAN2 and pole distribution of speech signals. In: Wong, K.W., Mendis, B.S.U., Bouzerdoum, A. (eds.) ICONIP 2010, Part I. LNCS, vol. 6443, pp. 363–370. Springer, Heidelberg (2010)
7. Campbell, J.P.: Speaker recognition: a tutorial. Proc. IEEE **85**(9), 1437–1462 (1997)
8. Quiñonero-Candela, J., Rasmussen, C.E., Sinz, F.H., Bousquet, O., Schölkopf, B.: Evaluating predictive uncertainty challenge. In: Quiñonero-Candela, J., Dagan, I., Magnini, B., d'Alché-Buc, F. (eds.) MLCW 2005. LNCS (LNAI), vol. 3944, pp. 1–27. Springer, Heidelberg (2006)

Simple Feature Quantities for Learning of Dynamic Binary Neural Networks

Ryuji Sato and Toshimichi Saito[⊠]

Hosei University, Koganei, Tokyo 184-8584, Japan
tsaito@hosei.ac.jp

Abstract. This paper presents simple feature quantities for learning of dynamic binary neural networks. The teacher signal is a binary periodic orbit corresponding to control signal of switching circuits. The feature quantities characterize generation of spurious memories and stability of the teacher signal. We present a simple greedy search based algorithm where the two feature quantities are used as cost functions. Performing basic numerical experiments, the algorithm efficiency is confirmed.

Keywords: Dynamic neural networks · Greedy search · Switching circuits

1 Introduction

Applying a delayed feedback to a simple binary neural network, the dynamic binary neural networks (DBNN) is constructed [1–7]. The DBNN is characterized by signum activation functions, ternary connection parameters, and integer threshold parameters. Depending on parameters, the DBNN can generate various binary periodic orbits (BPOs). The dynamics is integrated into the Digital return map (Dmap) from a set of lattice points to itself. The DBNN is a digital dynamical systems represented by the cellular automata [8]. Such systems are applicable to various engineering systems: information compressors, image processors, communication systems, and switching circuits [9–13].

This paper presents two simple feature quantities for learning of DBNN. The teacher signal is one desired BPO. The first feature quantity α characterizes generation of spurious memories. The second feature quantity β characterizes basin of attraction (stability) of the stored BPO. For storage of the BPO, we have applied a simple algorithm based on the correlation learning (CL-based learning, [1,14]). For stabilization of the stored BPO, we present a simple algorithm based on the greedy search that sparsifies the connection parameters. In the algorithm, the two feature quantities α and β are used as cost functions. In order to investigate the algorithm efficiency, the algorithm is applied to simple teacher signals corresponding to control signals of switching power converters [12,13]. Performing basic numerical experiments, we have confirmed storage of teacher signal BPO, suppression of spurious memories, and reinforcement of stability of the stored BPO.

© Springer International Publishing Switzerland 2015
S. Arik et al. (Eds.): ICONIP 2015, Part I, LNCS 9489, pp. 226–233, 2015.
DOI: 10.1007/978-3-319-26532-2_25

Note that the feature quantities for the learning of DBNN and the greedy search based algorithm have not been discussed in existing literatures including our previous papers [1–5].

2   Dynamic Binary Neural Networks

In this section, we introduce DBNN and Dmaps presented in [1–7]. The dynamics of the DBNN is described by

$$x_i^{t+1} = \text{sgn}\left(\sum_{j=1}^{N} w_{ij}x_j^t - T_i\right), \quad \text{sgn}(x) = \begin{cases} +1 \text{ for } x \geq 0 \\ -1 \text{ for } x < 0 \end{cases} \quad i = 1 \sim N \quad (1)$$

where $x_i^t \in \{-1, +1\} \equiv B$ is a binary state at discrete time t. The connection parameters are ternary $w_{ij} \in \{-1, 0, +1\}$ and the threshold parameters are integer $T_i \in Z$. The domain of the DBNN is a set of binary vectors B^N that is equivalent to a set of lattice points $L_D = (C_1, \cdots, C_{2^N})$, $C_i \equiv i/2^N$. Hence the dynamics of the DBNN can be visualized by the digital return map (Dmap) from L_D to itself:

$$x^{t+1} = F_D(x^t), \quad x^t \equiv (x_1^t, \cdots, x_N^t) \in B^N \quad (2)$$

Figure 1(a) and (b) illustrate the DBNN and its Dmap for $N = 4$ where binary code is used to express $L_4 = \{C_1, \cdots, C_{16}\}$: $C_1 \equiv (-1, -1, -1, -1) \cdots C_{16} \equiv (+1, +1, +1, +1)$. Since the number of lattice points is 2^N, direct memory of all the inputs/outputs becomes hard/impossible as N increases. However, in the DBNN, the number of parameters is polynomial $N^2 + N$.

Since the number of the lattice points is finite, the steady states are BPOs defined as the following. A point $\theta_p \in L_D$ is said to be a periodic point (PEP) with period p if $F^p(\theta_p) = \theta_p$ and $F^k(\theta_p) \neq \theta_p$ for $1 \leq k < p$ where F^p is the p-fold composition of F. Especially, a PEP with period 1 is said to be a fixed point. A sequence of the PEPs, $\{F(\theta_p), \cdots, F^p(\theta_p)\}$, is said to be a binary periodic orbit (BPO).

Next, we introduce the CL-based learning for the teacher signal pairs:

$$(\xi^l, \eta^l), \quad \xi^l \in B^N, \quad \eta^l \in B^N, \quad l = 1 \sim p \quad (3)$$

The purpose of learning is to determine the parameters satisfying $\eta^l = F_D(\xi^l)$ for all l. The CL-based learning determines the parameters as the following:

$$w_{ij} = \begin{cases} +1 \text{ for } c_{ij} > 0 \\ 0 \quad \text{for } c_{ij} = 0 \\ -1 \text{ for } c_{ij} < 0 \end{cases}, \quad c_{ij} = \sum_{l=1}^{p} \eta_i^l \xi_j^l, \quad T_i = \frac{R_i + L_i}{2}$$

$$R_i = \min_l \sum_{j=1}^{N} w_{ij}\xi_j^l \text{ for } \eta_i^l = +1, \quad L_i = \max_l \sum_{j=1}^{N} w_{ij}\xi_j^l \text{ for } \eta_i^l = -1 \quad (4)$$

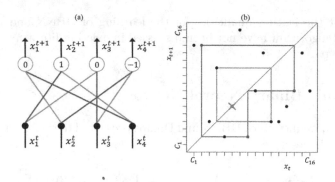

Fig. 1. (a) DBNN. Red and blue segments represent $w_{ij} = +1$ and $w_{ij} = -1$, respectively. $w_{ij} = 0$ means no connection. The threshold parameters T_i are shown in the circles. (b) Dmap. Red and blue orbits are BPOs. Blue cross is fixed point (Color figure online).

That is, the connection parameters w_{ij} are given by ternarising the correlation matrix elements c_{ij}. After w_{ij} are given, the threshold parameters T_i are determined by the quantities R_i and L_i. Note that R_i (L_i) exists if $\eta_i^l = +1$ ($\eta_i^l = -1$) for some l. If $\eta_i^l = -1$ ($\eta_i^l = +1$) for all l then R_i (L_i) does not exist and let $T_i = N + 1$ ($T_i = -N - 1$). Storage of the BPO is guaranteed if $R_i > L_i$ is satisfied for i such that both R_i and L_i exist [1].

3   Teacher Signal and Feature Quantities

This paper considers a simple teacher signal: one BPO with period p. The teacher signal BPO (TBPO) consists of p pieces of the teacher signal PEPs (TPEPs). The TBPO/TPEPs can be translated into teacher signal pairs in Eq. (3):

$$z^1, z^2, \cdots, z^p, \ z^i \neq z^j \text{ for } i \neq j, \ z^i = (z_1^i, \cdots, z_N^i) \in B^N$$
$$(\xi^l, \eta^l) \equiv (z^l, z^{l+1}), \ l = 1 \sim p, \ z^{p+1} \equiv z^1 \tag{5}$$

Although a variety of teacher signals can be considered, this paper considers two simple examples: 6 dimensional TBPOs with period 6 in Tables 1 and 2. The first and second TBPOs correspond to control signal of AC/DC and DC/AC converters in the power electonics, respectively [1,2].

Applying the CL-based learning to the two TBPOs, we obtain the DBNNs as shown in Fig. 2. The corresponding Dmaps are shown in Fig. 3. In the figures, we can see that the TBPOs can be stored. However, several spurious BPOs (different from the TBPO) exist.

In order to consider spurious BPOs and stabilization of TBPO, we present two simple feature quantities α and β. In the definition of α and β, we assume that the TBPO is stored into the DBNN. The first feature quantity is the rate of PEPs.

$$\alpha = N_p/2^N, \ p/2^N \leq \alpha \leq 1 \tag{6}$$

Table 1. Teacher signal example 1

z^1	(+1,	−1,	−1,	−1,	−1,	+1)
z^2	(+1,	+1,	−1,	−1,	−1,	−1)
z^3	(−1,	+1,	+1,	−1,	−1,	−1)
z^4	(−1,	−1,	+1,	+1,	−1,	−1)
z^5	(−1,	−1,	−1,	+1,	+1,	−1)
z^6	(−1,	−1,	−1,	−1,	+1,	+1)

Table 2. Teacher signal example 2

z^1	(+1,	−1,	−1,	−1,	+1,	+1)
z^2	(+1,	+1,	−1,	−1,	−1,	+1)
z^3	(+1,	+1,	+1,	−1,	−1,	−1)
z^4	(−1,	+1,	+1,	+1,	−1,	−1)
z^5	(−1,	−1,	+1,	+1,	+1,	−1)
z^6	(−1,	−1,	−1,	+1,	+1,	+1)

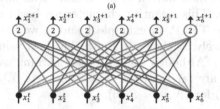 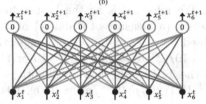

Fig. 2. DBNN after the CL-based learning. (a) example 1. (b) example 2.

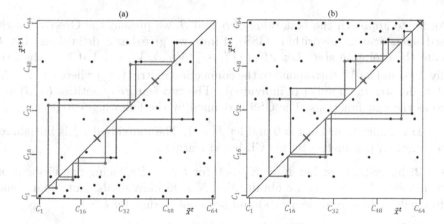

Fig. 3. Dmap after the CL-based learning. Red orbit denotes TBPO. Blue orbit denotes spurious BPO. (a) example 1. $\alpha = 0.23, \beta = 0.19$ (b) example 2. $\alpha = 0.23, \beta = 0.65$ (Color figure online)

where N_p is the number of PEPs of Dmap. It can evaluate generation of spurious memories. If no spurious memory exist, we obtain $\alpha = p/2^N b \equiv \alpha_b$. We refer to α_b as the best value. In example 1 and 2, we have $\alpha = 0.23$ as shown in Fig. 3. The best value is $\alpha = 6/2^6 \approx 0.09$.

The second feature quantity is based on the number of initial points falling into each TPEP:

$$\beta = \min_i \frac{M_i}{2^N/p}, \ i = 1 \sim p, \ 2/2^N \le \beta \le 1 \tag{7}$$

where M_i is the number of initial points falling into the i-th TPEP. This quantity is normalized by $2^N/p$. If p is a divisor of 2^N and the distribution of M_i is uniform for i then we obtain the maximum value $\beta = 1$. If p is not a divisor of 2^N then the maximum value is given by $\beta = \frac{\mathrm{INT}(2^N/p)}{2^N/p} \equiv \beta_b$, where $\mathrm{INT}(X)$ is the integer part of X. In this case, the distribution of M_i is almost uniform for i. We refer to β_b as the best value. This quantity characterizes basin of attraction (a measure of stability) of the TPEP. In example 1 and 2, we have obtained $\beta = 0.19$ and $\beta = 0.65$, respectively, as shown in Fig. 3. The best value is $\beta_b = 10/(2^6/6) \approx 0.94$ ($M_i = 10$ or 11). Although the CL-based learning can store the TBPO, the feature quantities are far from the best value: $(\alpha_b, \beta_b) = (p/2^N, \mathrm{INT}(2^N/p)/(2^N/p)) \approx (0.09, 0.94)$ for $(N, p) = (6, 6)$.

4   Greedy Search Based Sparsification Algorithm

In order to approach the best value of α and β, we present the Greedy search based sparsification algorithm (GSS). First, we give basic definitions. Let t denote the evolution step. Let $V^l \equiv (V_{11}^l, \cdots, V_{NN}^l)$, $V_{ij}^l \in -1, 0, 1$ be the l-th individual and let V^l correspond to the connection matrix (w_{ij}) where $l = 1 \sim K$ and K denotes the number of individuals. The two feature quantities (α, β) are used as the cost functions. The GSS is defined by the following 5 steps.

Step 1: Initialization. Let $t = 0$ and let $K = 1$. The individual V^1 is initialized by connection parameters by the CL-based learning.

Step 2: Sparcification. Let $K = K_1 > 0$ for $t \ge 1$. Replacing one element of each individual with zero, we obtain $N \times N \times K$ individuals. Note that some individuals include zero element(s) and does not change for $t \ge 1$.

Step 3: Evaluation by β. The individuals are evaluated by the feature quantity β. Let K_2 be the number of individuals that gives better or equal value of β in the past history. If $K_2 \ge K$ then the K individuals are selected in the elite preserving and go to Step 5. If $K_2 < K$ then K_2 individuals are preserved and go to Step 4.

Step 4: Evaluation by α. Except for the K_2 individuals selected in Step 3, the individuals are evaluated by the feature quantity α. The $(K - K_2)$ individuals are selected in the elite preserving. We obtain K individuals.

Step 5: Termination. Let $t \leftarrow t+1$, go to Step 2, and repeat until the maximum time limit t_{max}.

We apply the GSS to example 1 and 2 from control signals of switching power converters where $N = 6$. Note that, as N increases, calculation of β in Step 3

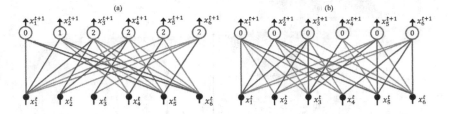

Fig. 4. DBNN after the GSS (sparsification). (a) example 1. (b) example 2.

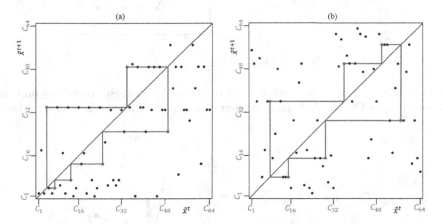

Fig. 5. Dmap of after GSS (sparcification). Red orbit denotes TBPO. (a) example 1.
$\alpha = 0.09, \beta = 0.75$ (b) example 2. $\alpha = 0.09, \beta = 0.84$

becomes hard and some approximation/prediction of β is required. However, in
practical applications such as control signals where N is not so large, evaluation
by β is effective.

After trial-and-errors, the algorithm, parameters are selected as $K_1 = 5$ and
$t_{max} = 36$. Figures 4 and 5 shows DBNNs and Dmaps after the sparsification
by the GSS. We obtain $\alpha = 0.09$ and $\beta = 0.75$ for example 1 and $\alpha = 0.09$ and
$\beta = 0.84$ for example 2. The GSS can realize the best value of α and can improve
the value of β.

Figure 6 shows evolution process of GSS for example 1. The quantity α is
improved in early stage and is reached the best value at $t = 5$. The quantity β
is improved intermittently until $t = 17$. The improvement of feature quantities
is monotone.

Figure 7 shows evolution process of GSS for example 2. For $t \leq 4$, the quantity
β is improved but α is stagnated. At $t = 5$, the GSS cannot give K individuals
and Step 4 (the evaluation by α) is applied. In Step 4, α is improved and the
improvement causes of deterioration of β. After stagnation of β and α for $6 \leq
t \leq 9$, some better individuals can be found. At $t = 10$, α reaches the best value
and β is close to the best value. This result suggest effectiveness of plural cost
functions α and β. The deterioration of β seems to be a trigger to escape from

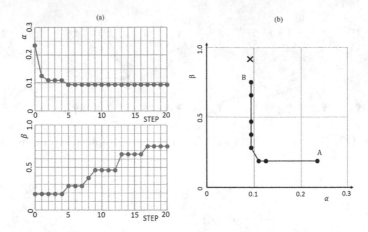

Fig. 6. Evolution process of example 1. (a) α vs t and β vs t. (b) Trajectory on α vs β plane. A: after CL-based learning. B: after GSS (sparsification). X: the best value.

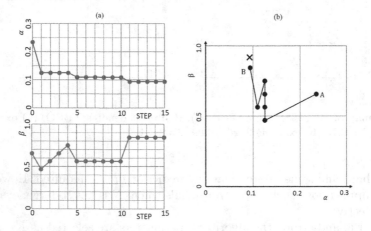

Fig. 7. Evolution process of example 2. (a) α vs t and β vs t. (b) Trajectory on α vs β plane. A: after CL-based learning. B: after GSS. X: the best value.

the stagnation of β and α. If the GGS used only one cost function β, the escape would be hard.

Note again that these two simple examples correspond to control signal of AC/DC and DC/AC converters. The larger β is suitable for robust operation against noise. The sparse connection is suitable from viewpoint of power consumption. Since one DBNN can be used for plural circuits (AC/DC and DC/AC converters), these results will be developed into an application to reconfigurable control systems of switching circuits.

5   Conclusions

Feature quantities and learning of the DBNN have been studied in this paper. Applying the CL-based learning, the TBPO can be stored. In order to characterize generation of spurious memories and stability of the stored BPO, two feature quantities α and β are presented. In order to improve the feature quantities, the GSS sparsificaton algorithm is presented. Applying the algorithm to two examples from power electronics, the algorithm efficiency is confirmed.

Future problems include analysis of sparsification effects on the stability, analysis of the evolution process of GSS, optimization of the algorithm parameters, and application to various switching circuits.

References

1. Kouzuki, R., Saito, T.: Learning of simple dynamic binary neural networks. IEICE Trans. Fundam. **E96–A**(8), 1775–1782 (2013)
2. Moriyasu, J., Saito, T.: Sparsification and stability of simple dynamic binary neural networks. IEICE Trans. Fundam. **E97–A**(4), 985–988 (2014)
3. Nakayama, Y., Kouzuki, R., Saito, T.: Application of the dynamic binary neural network to switching circuits. In: Lee, M., Hirose, A., Hou, Z.-G., Kil, R.M. (eds.) ICONIP 2013, Part II. LNCS, vol. 8227, pp. 697–704. Springer, Heidelberg (2013)
4. Moriyasu, J., Saito, T.: A deep dynamic binary neural network and its application to matrix converters. In: Wermter, S., Weber, C., Duch, W., Honkela, T., Koprinkova-Hristova, P., Magg, S., Palm, G., Villa, A.E.P. (eds.) ICANN 2014. LNCS, vol. 8681, pp. 611–618. Springer, Heidelberg (2014)
5. Moriyasu, J., Saito, T.: A cascade system of simple dynamic binary neural networks and its sparsification. In: Loo, C.K., Yap, K.S., Wong, K.W., Teoh, A., Huang, K. (eds.) ICONIP 2014, Part I. LNCS, vol. 8834, pp. 231–238. Springer, Heidelberg (2014)
6. Gray, D.L., Michel, A.N.: A training algorithm for binary feed forward neural networks. IEEE Trans. Neural Netw. **3**(2), 176–194 (1992)
7. Chen, F., Chen, G., He, Q., He, G., Xu, X.: Universal perceptron and DNA-like learning algorithm for binary neural networks: non-LSBF implementation. IEEE Trans. Neural Netw. **20**(8), 1293–1301 (2009)
8. Chua, L.O.: A Nonlinear Dynamics Perspective of Wolfram's New Kind of Science, I and II. World Scientific, Singapore (2005)
9. Wada, W., Kuroiwa, J., Nara, S.: Completely reproducible description of digital sound data with cellular automata. Phys. Lett. A **306**, 110–115 (2002)
10. Rosin, P.L.: Training cellular automata for image processing. IEEE Trans. Image Process. **15**(7), 2076–2087 (2006)
11. Iguchi, T., Hirata, A., Torikai, H.: Theoretical and heuristic synthesis of digital spiking neurons for spike-pattern-division multiplexing. IEICE Trans. Fundam. **E93–A**(8), 1486–1496 (2010)
12. Bose, B.K.: Neural network applications in power electronics and motor drives - an introduction and perspective. IEEE Trans. Ind. Electron. **54**(1), 14–33 (2007)
13. Rodriguez, J., Rivera, M., Kolar, J.W., Wheeler, P.W.: A review of control and modulation methods for matrix converters. ieee tie **59**(1), 58–70 (2012)
14. Araki, K., Saito, T.: An associative memory including time-variant self-feedback. Neural Netw. **7**(8), 1267–1271 (1994)

Transfer Metric Learning for Kinship Verification with Locality-Constrained Sparse Features

Yanli Zhang, Bo Ma$^{(\boxtimes)}$, Lianghua Huang, and Hongwei Hu

Beijing Laboratory of Intelligent Information Technology,
School of Computer Science and Technology, Beijing Institute of Technology,
Beijing, China
bma000@bit.edu.cn

Abstract. Kinship verification between aged parents and their children based on facial images is a challenging problem, due to aging factor which makes their facial similarities less distinct. In this paper, we propose to perform kinship verification in a transfer learning manner, which introduces photos of parents in their earlier ages as intermediate references to facilitate the verification. Child-young parent pairs are regarded as source domain and child-old parent ones are considered as target domain. The transfer learning scheme contains two phases. In the transfer metric learning phase, the extracted locality-constrained sparse features of images are projected into an optimized subspace where the intra-class distances are minimized and the inter-class ones are maximized. In the transfer classifier learning phase, a cross domain classifier is learned by a transfer SVM algorithm. Experimental results on UB KinFace dataset indicate that our method outperforms state-of-the-art methods.

Keywords: Kinship verification · Transfer metric learning · Cross domain · Sparse representation

1 Introduction

Kinship verification, as an emerging issue in facial image analysis, has attracted much attention in recent years [2–6,9,10,13–18]. Despite years of studies, kinship verification is still a challenging problem. Apart from the difficulties encountered in traditional facial image analysis, such as pose and illumination variations and different facial expressions, the significant age gap as well as the gender difference between parents and children make kinship verification more difficult.

Among all the factors, aging is one of the most critical factors that restrain the verification performance. It is observed that compared with aged parents, children usually look more like their parents when they were young. If the facial images of young parents are used as intermediate references, the verification between children and their old ones would be easier. Another important factor is feature representation of images. The existing methods normally use vision

© Springer International Publishing Switzerland 2015
S. Arik et al. (Eds.): ICONIP 2015, Part I, LNCS 9489, pp. 234–243, 2015.
DOI: 10.1007/978-3-319-26532-2_26

Fig. 1. Illustration of four facial image triple examples. Each triple contains the facial images of a parent at his/her young and old ages and the corresponding child which are shown from left to right. All of them are selected from the UB kinFace dataset.

features such as LBP, Gabor, SIFT or specific facial parts directly in model building, which might not be optimal for classification and could contain a lot of noisy and redundant information. It has been proved in recent studies that by coding on a dictionary under both sparsity and locality constraints, features can be projected into a more linearly separable space. This motivates us to use locality-constrained linear coding (LLC) scheme [12] to obtain the final sparse features.

Motivated by the above ideas, in this paper, we propose a transfer learning method for kinship verification based on LLC features. As shown in Fig. 1, we use facial image triples to train the model. Each triple contains the images of young parent x, old parent y and child z. We regard child-young parent pairs $\{(x_i, z_p)|i, p = 1, \cdots, N\}$ as source domain and child-old parent pairs $\{(y_j, z_p)|j, p = 1, \cdots, N\}$ as target domain, where N is the number of triples. In our transfer learning scheme, the young parent x served as an intermediate reference is only used in the training phrase. The model trained by these triples are expected to have better performance on the test child-old parent sample pairs than the traditional ones trained without young parents' facial images.

In our method, the transfer learning scheme contains two phases. In the first phase, we introduce a neighborhood repulsed transfer metric learning (NRTML) method to find a metric subspace, in which the intra-class differences of sample pairs are decreased and the inter-class differences of neighboring sample pairs are increased for each domain involved in transfer learning. At the same time, the differences of young and old parents are decreased to narrow distribution difference of domains in some degree. In the second phase, sample pairs of source and target domains are constructed and trained in a transfer SVM manner. Besides, the features of facial images are extracted by Gabor filters and encoded via LLC algorithm on a meticulously learned dictionary. At the last, the projected features of images under the learned metric subspace are used for classifier learning and predicting. Experimental results on the UB KinFace dataset demonstrate the effectiveness of our proposed method comparing with the state-of-the-art methods.

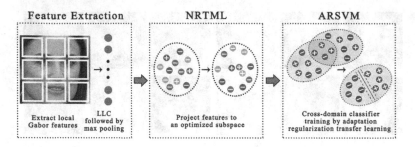

Fig. 2. The overall architecture of our proposed kinship verification method based on facial images.

2   Proposed Approach

Figure 2 shows the overview of our proposed kinship verification approach. Our method can be divided into three stages, namely, locality-constrained linear coding features extraction, transfer metric learning and cross-domain classifier learning. The process of feature extraction and the NRTML method showed in transfer metric learning stage will be described in details in Sects. 2.1 and 2.2.

The cross-domain classifier utilized in our work is Adaptation Regularization based Support Vector Machine(ARSVM) [7]. Comparing with the traditional SVM, it can efficiently solve the cross-domain classification problem by taking marginal and conditional distribution adaptation and manifold regularization into consideration. While training the ARSVM classifier, the difference between the joint probability distributions of source and target domains is measured by the projected Maximum Mean Discrepancy (MMD)[11].

The manifold regularization is devoted to further exploit the knowledge of the marginal distributions of two domains by taking their intrinsic geometry into consideration to obtain a better classification hyperplane.

2.1   Feature Extraction

The overview of our feature extraction process is illustrated in the first part of Fig. 2. Firstly, we use overlapped sliding windows on each resized facial image to obtain patches, and extract Gabor feature for each patch. Since the dimension of their Gabor features could be very high, we apply PCA to the features to avoid overfitting and reduce the computational complexity. Next, we encode these Gabor features on a dictionary using LLC scheme, and use max pooling operator to obtain the final feature of each image. The dictionary in LLC is learned by k-means algorithm. To make the solution of k-means more stable and the coefficients more discriminative, the initial cluster centers are obtained in the following way: we sort the patches in descending order by their information entropies, and select m (m is the number of dictionary items) patches in the sorted sequence at regular intervals as the initial cluster centers in k-means.

2.2   NRTML

Let $m \times N$ matrices $\mathbf{X}$, $\mathbf{Y}$ and $\mathbf{Z}$ denote young parent, old parent and child sets respectively, x_i, y_i and z_i denote the i-th young parent, old parent and child facial images respectively, where N is the number of images in each set. Then we use all the samples in the three sets to construct the source domain (child-young parent pairs) $Ss = \{(x_i, z_p)|i, p = 1, \cdots, N\}$ and target domain (child-old parent pairs) $St = \{(y_j, z_p)|j, p = 1, \cdots, N\}$ involved in our method. For the samples in the three sets with same indices, such as x_i, y_i, they represent the images of the i-th person at different ages and have kin relations with z_i.

We attempt to find a metric subspace, in which, the distances between x_i and z_i as well as y_i and z_i are minimized, and the distances between x_i and z_p $(i \neq p)$, y_j and z_p $(j \neq p)$ are maximized. Given x_i and y_i representing the same person at different ages, it is also required to minimize the distance between them. The metric matrix is denoted as W, so the distance between two samples u and v in the subspace can be formulated as

$$
\begin{aligned}
d(x,y) &= \sqrt{((u-v)^T W W^T (u-v))} \\
&= \sqrt{[W^T(u-v)]^T [W^T(u-v)]} \\
&= \sqrt{(u'-v')^T (u'-v')},
\end{aligned}
\tag{1}
$$

where $u' = W^T u$, $v' = W^T v$.

The inter-class distance between two matrices $\mathbf{U}$ and $\mathbf{V}$ is defined as follows:

$$
D_b(\mathbf{U}, \mathbf{V}) = \frac{1}{kN} \sum_{i=1}^{N} \sum_{t \in N_{v_i}} d^2(u_i, v_t) + \frac{1}{kN} \sum_{i=1}^{N} \sum_{t \in N_{u_i}} d^2(u_t, v_i),
\tag{2}
$$

where k is the number of neighbors, N_{v_i} is the set of indices corresponding to the k-nearest neighbors of v_i and N_{u_i} is the set of indices corresponding to the k-nearest neighbors of u_i. Similarly, the intra-class distance of $\mathbf{U}$ and $\mathbf{V}$ is denoted as

$$
D_w(\mathbf{U}, \mathbf{V}) = \frac{1}{N} \sum_{i=1}^{N} d^2(u_i, v_i).
\tag{3}
$$

Based on Eqs. (1), (2) and (3), the NRTML method to solve the cross-domain kinship verification problem is formulated as the following optimization problem:

$$
\max_W J(W) = D_b(\mathbf{X}, \mathbf{Z}) - D_w(\mathbf{X}, \mathbf{Z}) + D_b(\mathbf{Y}, \mathbf{Z}) - D_w(\mathbf{Y}, \mathbf{Z}) - D_w(\mathbf{X}, \mathbf{Y}).
\tag{4}
$$

By minimizing $H_w(\mathbf{X}, \mathbf{Z})$ and $H_w(\mathbf{Y}, \mathbf{Z})$, the distances between parents and their children in the metric subspace are reduced. By maximizing $H_b(\mathbf{X}, \mathbf{Z})$ and $H_b(\mathbf{Y}, \mathbf{Z})$, the distances between young and old parents and unmatched neighboring children in the metric subspace are enlarged. The term $H_w(\mathbf{X}, \mathbf{Y})$ is used for decreasing the influence of aging factor by minimizing the distances between aged and young parents. In this way, features can be projected to an optimized metric subspace through the learned projection matrix W.

For $D_b(\mathbf{U}, \mathbf{V})$, we can simplify it into the following form:

$$D_b(\mathbf{U}, \mathbf{V}) = \frac{1}{kN} \sum_{i=1}^{N} \sum_{t \in N_{v_i}} d^2(u_i, v_t) + \frac{1}{kN} \sum_{i=1}^{N} \sum_{t \in N_{u_i}} d^2(u_t, v_i)$$

$$= tr \left\{ W^T \left(\frac{1}{kN} \sum_{i=1}^{N} \sum_{t \in N_{v_i}} (u_i - v_t)(u_i - v_t)^T + \right. \right.$$

$$\left. \left. \frac{1}{kN} \sum_{i=1}^{N} \sum_{t \in N_{u_i}} (u_t - v_i)(u_t - v_i)^T \right) W \right\}$$

$$= tr(W^T H_b(\mathbf{U}, \mathbf{V})W), \tag{5}$$

where $H_b(\mathbf{U}, \mathbf{V}) = \frac{1}{kN} \sum_{i=1}^{N} \sum_{t \in N_{v_i}} (u_i - v_t)(u_i - v_t)^T + \frac{1}{kN} \sum_{i=1}^{N} \sum_{t \in N_{u_i}} (u_t - v_i)(u_t - v_i)^T$. Similarly, $D_w(\mathbf{U}, \mathbf{V})$ can be simplified as

$$D_w(\mathbf{U}, \mathbf{V}) = tr(W^T H_w(\mathbf{U}, \mathbf{V})W), \tag{6}$$

where $H_w(\mathbf{U}, \mathbf{V}) = \frac{1}{N} \sum_{i=1}^{N} (u_i - v_i)(u_i - v_i)^T$. According to Eqs. (5) and (6), the objective function of our optimization problem can be reformulated as

$$\max_{W} J(W) = tr(W^T \mathbf{H} W)$$

$$\text{subject to} \quad W^T W = I, \tag{7}$$

where $\mathbf{H} = H_b(\mathbf{X}, \mathbf{Z}) - H_w(\mathbf{X}, \mathbf{Z}) + H_b(\mathbf{Y}, \mathbf{Z}) - H_w(\mathbf{Y}, \mathbf{Z}) - H_w(\mathbf{X}, \mathbf{Y})$. The constraint on W is used to reduce the effect of its scale to the solution of our objective function.

In order to obtain the metric matrix W, we solve the following eigenvalue problem:

$$\mathbf{H}\omega = \lambda\omega. \tag{8}$$

After obtaining the eigenvalues of Eq. (8), they are sorted in descending order as $\lambda_1 \geq \lambda_2 \geq \cdots \geq \lambda_m$. the first l eigenvectors $\omega_1, \omega_2, \cdots, \omega_l$ corresponding to the sorted eigenvalues are used to form the matrix $W = [\omega_1, \omega_2, \cdots, \omega_l] \in R^{m \times l}$. Having obtained W, the original matrices $\mathbf{X}$, $\mathbf{Y}$, and $\mathbf{Z}$ are projected into the metric subspace to obtain $\mathbf{X}'$, $\mathbf{Y}'$ and $\mathbf{Z}'$:

$$\mathbf{X}' = W^T \mathbf{X}, \quad \mathbf{Y}' = W^T \mathbf{Y}, \quad \mathbf{Z}' = W^T \mathbf{Z}. \tag{9}$$

The proposed NRTML algorithm is summarized in Algorithm 1.

3   Experiments

3.1   Experimental Settings

In the UB KinFace dataset, there are 600 facial images of 400 different people, some of which are grayscale images and others are color ones. It contains three

Algorithm 1. NRTML

Input: The training samples of the children, young parents and old parents x_i, y_i, z_i $(i = 1, \cdots, N)$; neighborhood size k; iterative number T.

Output: Metric matrix W, projected sample matrices $\mathbf{X}'$, $\mathbf{Y}'$ and $\mathbf{Z}'$.

1: **initialize:** $\mathbf{X}' = \mathbf{X}$, $\mathbf{Y}' = \mathbf{Y}$, $\mathbf{Z}' = \mathbf{Z}$.
2: **for** $t = 1, 2, \cdots, T$ **do**
3:    Find the k-nearest neighbors for every x_i', y_i' and z_i' and record their indices in sets: $N_{x_i}, N_{y_i}, N_{z_i}$ respectively;
4:    Calculate $\mathbf{H}$ according to Eqs. (5), (6) and (7);
5:    Solve the eigenvalue problem in Eq. (8) to obtain ω_i $(i = 1, \cdots, l)$;
6:    Construct W with $\omega_1, \omega_2 \cdots \omega_l$;
7:    Obtain projected sample matrices $\mathbf{X}', \mathbf{Y}'$ and $\mathbf{Z}'$ according to Eq. (9);
8:    Update: $x_i = x_i'$, $y_i = y_i'$, $z_i = z_i'$.
9: **end for**

Table 1. Verification accuracy (%) of different kinship verification methods.

Method	Set1	Set2
Method in [13]	N.A	60.0
NRML	65.8	65.4
MNRML	66.8	67.3
MPDFL	67.5	67.0
Ours	N.A	73.95

Table 2. Performance comparison (%) between our method and human ability on kinship verification.

Method	Accuracy
Human A	60.75
Human B	66.00
Ours	73.95

subsets, namely, young parent set, old parent set and child set. In our experiment, we build two kinds of kinship sample pairs: child-young parent pairs and child-old parent pairs which are regarded as the samples in source and target domains respectively.

In our experiments, all the images in the UB KinFace dataset are converted into grayscale ones, then the facial parts are cropped and aligned into 64×64. We use five fold cross validation to evaluate the performance of our method, where each subset in the dataset is divided equally into five folds, four for training and one for testing. As to the training data, positive sample pairs are constructed with all the matched kinship pairs, while negative ones are obtained by carefully selecting hard negative sample pairs. When constructing negative sample pairs, for each parent sample, the child sample with the most similar feature to the it is selected from unmatched children. In this way which is different from traditional methods which randomly choose those unmatched pairs, our method can train a more stable and accurate classification hyperplane. RBF kernel is chosen for the ARSVM classifier in our experiments.

The size of patches is set to $l = 12$ and the sampling gap is $h = 5$. The Gabor filters used in our experiments have 5 scales and 8 directions. In the LLC

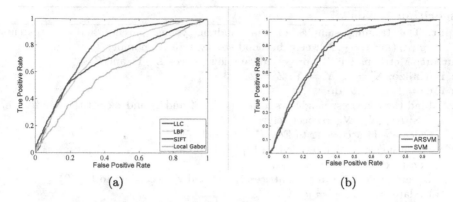

Fig. 3. The ROC curves of our method using (a) different feature descriptors and (b) different classifiers, respectively.

scheme, the number of the overcomplete dictionary items is set to $m = 630$ and the number of considered nearest neighbors is set to $k_l = 9$. After LLC coding, we reduce the dimension of LLC features to 300 using PCA algorithm. For the NRTML scheme, the number of k-nearest neighbors is set to $k_n = 9$.

3.2   Experimental Results

Comparison with the Existing State-of-the-Art Kinship Verification Methods. Table 1 compares our proposed method with the recent state-of-the-art kinship verification methods. Most of the state-of-the-art methods just use part of the UB KinFace dataset to conduct the kinship verification. To make sure the fairness of comparison, we construct two subsets of the UB KinFace dataset: (1) set 1(200 pairs of child and young parent facial images) and (2) set 2 (200 pairs of child and old parent facial images). The performance of each method is measured by its mean verification accuracy. As can be seen from Table 1, our proposed method outperforms the existing methods.

The superior performance of our method could be attributed as follows. Firstly, the proposed NRTML method projects the features of facial images to an optimized subspace, where distances between unmatched pairs are enlarged and those between matched ones are reduced. Secondly, the LLC scheme, in which not only sparsity constraint but also locality constraint are adopted, makes the features more separable. Lastly, the ARSVM classifier could further narrow the difference between distributions of source and target domains, such that it can take advantage of the intermediary set (young parent images set) to further improve the verification performance.

Comparison Between Different Features. To demonstrate the effectiveness of our locality-constrained sparse features, we choose several different feature descriptors including LBP [1], SIFT [8] and local gabor features for comparison. All the above mentioned features are all extracted locally from images, where the patch size and sampling step are the same as our features. For LBP and SIFT, the features are concatenated to obtain the final image-wise ones. For the local gabor features, they are finally processed with a max-pooling operator. Figure 3(a) illustrates the ROC curves of kinship verification performance using different feature descriptors. We can see from this figure that our locality-constrained sparse feature outperforms the other three descriptors and yields the best performance.

The LLC scheme can project the original features to an optimized space, where features are more separable. In this way, compared with directly using local LBP, SIFT and Gabor features, using LLC to generate final feature representation makes our method achieve better performance.

Comparison Between Different Classifiers. We compare the performance of our method versus different classifiers: ARSVM and SVM. Figure 3(b) illustrates the ROC curves of kinship verification performance using the two classifiers. We can see that the performance of ARSVM is better than SVM while facing the cross-domain classification task. It is mainly because that the SVM classifier regard the samples from the source and target domains as the same in the training stage while the ARSVM method narrows the difference between the two domains by taking marginal and conditional distribution adaptation and manifold regularization into consideration before classification. Hence ARSVM is more suitable for our kinship verification problem and achieves better verification performance.

Parameter Analysis. As shown in Fig. 4, the performance of our proposed algorithm under different parameter settings is compared. The impacts of neighbourhood sizes k_l in LLC and that k_n in NRTML are evaluated in the experiment. From Fig. 4(a), we observe that the impact of k_l on the verification performance is not apparent and the best performance is achieved when $k_l = 9$. The Fig. 4(b) shows that k_n has significant impact on the verification performance. When k_n is close to 1 which means the neighbors have limited effects, the verification accuracy is very poor. As k_n grows, the performance becomes better. This indicates the effectiveness of the neighborhood repulsed scheme. However, when k_n is greater than 9, the accuracy starts to degrade. Our algorithm achieves the best performance when $k_n = 9$.

Comparison with Human Observers. We also test human ability in kinship verification based on facial images. Ten candidates (five males and five females) are invited to participate in our experiment. There are two parts in this experiment. For the first one (Human A), we randomly select 80 sample pairs (40 true child-old parent pairs and 40 false child-old parent pairs) and present them to human observers. And for the second one (Human B), we add 80 corresponding pairs of child-young parent facial images (40 true pairs and 40 false pairs) into the above sample pair set. All other sample pairs are identical with the former one. Then the expanded sample pair set are shown to human observers. The verification rates of human and our method are shown in Table 2. From the results we can see, Human B gets better performance than Human A which indicates the effectiveness of young parent set in assisting distinguish the kinship of

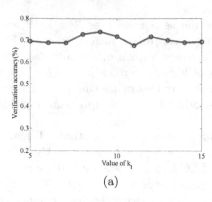 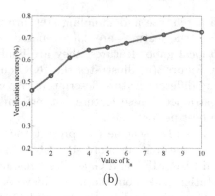

(a)   (b)

Fig. 4. Mean verification accuracy of our method versus different values of (a) parameter k_l and (b) parameter k_n respectively.

old parents and children. Besides, our method achieves better performance than both Human A and Human B.

4   Conclusion

In this paper, we propose a method for solving the problem of kinship verification between aged parents and children in a transfer learning manner. Young parent facial image set is introduced as an intermediate reference in our method to construct the source domain with the children set, which is supposed to improve the verification performance in the target domain, i.e., the child-old parent pairs. The transfer learning in our method contains two phases. The transfer metric learning phase uses the proposed NRTML scheme to project features to an optimized metric subspace, where the intra-class distances are minimized while the inter-class ones are maximized for both domains. The classifier learning phase trains a cross domain classifier using ARSVM algorithm. We employ locality-constrained coding (LLC) algorithm to construct our feature representation. Experimental results demonstrate that our algorithm performs favourably against the existing state-of-the-art methods.

Acknowledgments. This work was supported in part by the National Natural Science Foundation of China (No. 61472036).

References

1. Ahonen, T., Hadid, A., Pietikainen, M.: Face description with local binary patterns: application to face recognition. IEEE Trans. Pattern Anal. Mach. Intell. **28**(12), 2037–2041 (2006)
2. Dibeklioglu, H., Salah, A.A., Gevers, T.: Like father, like son: facial expression dynamics for kinship verification. In: 2013 IEEE International Conference on Computer Vision (ICCV), pp. 1497–1504. IEEE (2013)

3. Fang, R., Gallagher, A., Chen, T., Loui, A.: Kinship classification by modeling facial feature heredity. In: 20th IEEE International Conference on Image Processing (ICIP), pp. 2983–2987. IEEE (2013)
4. Fang, R., Tang, K.D., Snavely, N., Chen, T.: Towards computational models of kinship verification. In: 17th IEEE International Conference on Image Processing (ICIP), pp. 1577–1580. IEEE (2010)
5. Guo, G., Wang, X.: Kinship measurement on salient facial features. IEEE Trans. Instrum. Meas. **61**(8), 2322–2325 (2012)
6. Guo, Y., Dibeklioglu, H., van der Maaten, L.: Graph-based kinship recognition. In: 22nd International Conference on Pattern Recognition (ICPR) 2014, pp. 4287–4292. IEEE (2014)
7. Long, M., Wang, J., Ding, G., Pan, S.J., Yu, P.S.: Adaptation regularization: a general framework for transfer learning. IEEE Trans. Knowl. Data Eng. **26**(5), 1076–1089 (2014)
8. Lowe, D.G.: Distinctive image features from scale-invariant keypoints. Int. J. Comput. Vis. **60**(2), 91–110 (2004)
9. Lu, J., Zhou, X., Tan, Y.P., Shang, Y., Zhou, J.: Neighborhood repulsed metric learning for kinship verification. IEEE Trans. Pattern Anal. Mach. Intell. **36**(2), 331–345 (2014)
10. Qin, X., Tan, X., Chen, S.: Tri-subject kinship verification: understanding the core of a family. IEEE Trans. Multimedia **17**(10), 1855–1867 (2015)
11. Quanz, B., Huan, J.: Large margin transductive transfer learning. In: Proceedings of the 18th ACM Conference on Information and Knowledge Management, pp. 1327–1336. ACM (2009)
12. Wang, J., Yang, J., Yu, K., Lv, F., Huang, T., Gong, Y.: Locality-constrained linear coding for image classification. In: IEEE Conference on Computer Vision and Pattern Recognition (CVPR), pp. 3360–3367. IEEE (2010)
13. Xia, S., Shao, M., Fu, Y.: Kinship verification through transfer learning. In: IJCAI Proceedings-International Joint Conference on Artificial Intelligence, vol. 22, p. 2539 (2011)
14. Xia, S., Shao, M., Luo, J., Fu, Y.: Understanding kin relationships in a photo. IEEE Trans. Multimedia **14**(4), 1046–1056 (2012)
15. Yan, H., Lu, J., Deng, W., Zhou, X.: Discriminative multimetric learning for kinship verification. IEEE Trans. Inf. Forensics Secur. **9**(7), 1169–1178 (2014)
16. Yan, H.C., Lu, J., Zhou, X.: Prototype-based discriminative feature learning for kinship verification. IEEE Trans. Cybern. **45**, 2535–2545 (2014)
17. Zhou, X., Hu, J., Lu, J., Shang, Y., Guan, Y.: Kinship verification from facial images under uncontrolled conditions. In: Proceedings of the 19th ACM International Conference on Multimedia, pp. 953–956. ACM (2011)
18. Zhou, X., Lu, J., Hu, J., Shang, Y.: Gabor-based gradient orientation pyramid for kinship verification under uncontrolled environments. In: Proceedings of the 20th ACM International Conference on Multimedia, pp. 725–728. ACM (2012)

Unsupervised Land Classification by Self-organizing Map Utilizing the Ensemble Variance Information in Satellite-Borne Polarimetric Synthetic Aperture Radar

Yuto Takizawa, Fang Shang, and Akira Hirose[✉]

Department of Electrical Engineering and Information Systems,
The University of Tokyo, 7-3-1 Hongo, Bunkyo-ku, Tokyo 113-8656, Japan
takizawa@eis.t.u-tokyo.ac.jp, shangfang@secure.ee.uec.ac.jp,
ahirose@ee.t.u-tokyo.ac.jp
http://www.eis.t.u-tokyo.ac.jp/

Abstract. Polarimetric satellite-borne synthetic aperture radar is expected to provide land usage information globally and precisely. In this paper, we propose a two-stage unsupervised-learning land state classification system using a self-organizing map (SOM) based on the ensemble variance. We find that the Poincare sphere parameters representing the polarization state of scattered wave have specific features of the land state, in particular, in their dispersion (or ensemble variance). We present two-stage clustering procedure to utilize the dispersion features of the clusters as well as the mean values. Experiments demonstrate its high capability of self-organizing and discovering classification based on the polarimetric scattering features representing the land states.

Keywords: Polarimetric synthetic aperture radar · Stokes vector · Unsupervised classification

1 Introduction

Satellite-borne synthetic aperture radar (SAR) systems observe the earth continuously, globally and precisely for various purposes such as disaster monitoring and mitigation, forest biomass estimation for CO_2 reduction, glacier area and movement watching to water source protection and agricultural crop estimation in the near future [1–4]. Possible another aim is to observe the land states in artificial and natural environment. For this purpose, polarimetric SAR is expected to play an important role in the newly launched satellite systems. Decomposition of scattering matrix is the method most closest to practical use presently in the automatic land use classification [6]. Its basis lies in the linear algebra. However, it sometimes fails in meaningful classification because of its non-uniqueness in the decomposition.

© Springer International Publishing Switzerland 2015
S. Arik et al. (Eds.): ICONIP 2015, Part I, LNCS 9489, pp. 244–252, 2015.
DOI: 10.1007/978-3-319-26532-2_27

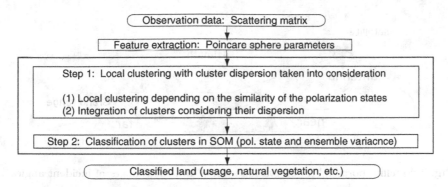

Fig. 1. Total processing flow of the proposed adaptive land state classification system using a SOM in the two-stage clustering.

Previously we proposed the use of Sotokes parameters (or Poincare sphere parameters) as the primary variables in the land classification [7]. We also constructed an adaptive classification system based on supervised learning in quaternion domain [8]. The classification performance is so high and the learning cost is so light that its practical application is strongly expected. Though the supervised learning system shows a high accuracy in its adaptive classification, it does not have the ability to discover new categories of land use. Instead, it indicates uncategorized areas as an undetermined class. Unsupervised learning system may have the ability to discover new classes adaptively.

In this paper, we propose an unsupervised adaptive method to classify land states using a self-organizing map (SOM) that deals with ensemble variance of land scattering features. We employ two-stage clustering to utilize the dispersion features, or ensemble variance features, of pre-grouped polarization clusters. In this method, first we extract the scattering feature values as Poincare sphere parameters. Then, we group the pixel parameters locally and finely into clusters. Secondly, we classify the clusters adaptively by using a SOM by taking the ensemble variance of respective clusters into consideration. The preliminary grouping realizes high robustness against the slant-angle changes in the radar observation to yield useful ensemble variance of the Poincare sphere parameters. Experiments demonstrate the effectiveness in adaptive classification corresponding to vegetation and town details.

2   Stokes Vector and Poincare Sphere Parameter

Full PolSAR system observes 2×2 complex scattering matrix $\mathbf{S}$ at each resolution area. The calculation Poincare sphere parameters requires to suppose a certain incident wave. The incident wave is expressed by a unit Jones vector $[E_H^i \; E_V^i]^T$. The scattered wave Jones vector $[E_H^r \; E_V^r]^T$ is obtained as

$$\begin{bmatrix} E_H^r \\ E_V^r \end{bmatrix} = \begin{bmatrix} S_{HH} & S_{HV} \\ S_{VH} & S_{VV} \end{bmatrix} \begin{bmatrix} E_H^i \\ E_V^i \end{bmatrix}, \tag{1}$$

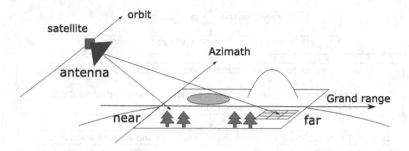

Fig. 2. Satellite-borne SAR observation system and the changes of incident angles.

where E_H and E_V stand for horizontal and vertical component, respectively. The averaged Stokes vector $[\langle g_0 \rangle \; \langle g_1 \rangle \; \langle g_2 \rangle \; \langle g_3 \rangle]^T$ is then obtained as

$$
\begin{bmatrix} \langle g_0 \rangle \\ \langle g_1 \rangle \\ \langle g_2 \rangle \\ \langle g_3 \rangle \end{bmatrix} = \begin{bmatrix} \langle E_H^r E_H^{r*} \rangle + \langle E_V^r E_V^{r*} \rangle \\ \langle E_H^r E_H^{r*} \rangle - \langle E_V^r E_V^{r*} \rangle \\ \langle E_H^r E_V^{r*} \rangle + \langle E_V^r E_H^{r*} \rangle \\ j(\langle E_H^r E_V^{r*} \rangle - \langle E_V^r E_H^{r*} \rangle) \end{bmatrix} \tag{2}
$$

where $\langle \cdot \rangle$ denotes temporal or spatial averaging process. The Poincare sphere parameter P representing polarization states in three dimension is given as

$$
P = \left(\frac{\langle g_1 \rangle}{\langle g_0 \rangle}, \; \frac{\langle g_2 \rangle}{\langle g_0 \rangle}, \; \frac{\langle g_3 \rangle}{\langle g_0 \rangle} \right) \equiv (x, y, z). \tag{3}
$$

The Poincare sphere parameter P should be on or in a unit sphere. The norm of P shows the degree of polarization, i.e., DoP $= \sqrt{\langle g_1 \rangle^2 + \langle g_2 \rangle^2 + \langle g_3 \rangle^2}/\langle g_0 \rangle$.

The Poincare sphere parameter P is a three dimensional vector relevant to the polarization states of the incident wave. If we consider P for all the possible incident polarization states, the computational cost will be huge. Instead, we use only important four polarization states, P_H, P_V, $P_{45°}$ and P_{lc}, namely, horizontal, vertical, 45° and left-handed circular polarization, as the incident polarization. The experimental data used in the following experiment is L band PALSAR 1.1 level data of ALOS (Advanced Land Observation Satellite, JAXA) observing the Mt. Fuji area having 2932×1048 pixels. The spatial averaging process in (2) is calculated for 5×5 local window.

3   Unsupervised Learning Classification of Land States by Using SOM Based on Ensemble Variance

3.1   Two-Stage Clustering to Utilize the Dispersion Feature

Figure 1 shows the processing flow in total. A rough description is given here first and, then, followed by detailed explanation in the next subsections. In this proposal, Step 1 is local clustering to realize the utilization of the variance of

Poincare sphere parameters in respective land state clusters in the following process. Regarding $Q = [P_H \; P_{lc} \; P_{45°} \; P_V]^T$ as the input vector representing features of each pixel, we conduct the clustering in Step 1 by considering the dispersion of the clusters under construction. This local clustering also mitigates the distortion in the radar observation. Figure 2 illustrates the side-looking SAR observation. A satellite-borne SAR transmits electromagnetic-wave to the earth surface with a slant angle. Hence, the scattered-wave polarization depends on the incident angle [2]. This problem mainly happens in the range direction. That is, similar land states located near or far range generates a polarization state different from each other. This concern happens also depending on the terrain shape such as slopes and mountains. Therefore, in Step 1 local clustering, we first scan in the azimuth direction, and then proceed line by line in the range direction. We calculate the mean value $Q(i)$ and the standard deviation $s(i)$ for i-th cluster data.

In Step 2, we classify the clusters adaptively by using SOM. Each cluster has a 24-dimensional feature vector. We found in the following experiments that, even when two areas are located away from each other, common features exist in their parameter distributions corresponding to land states especially in the ensemble variance. Hence, in this process, we multiply the variance by an appropriate weight in order to emphasize its contribution. We describe the details in the next subsections.

3.2 Step 1: Local Clustering by Paying Attention to the Cluster Variance

The Step 1 includes two processes, i.e. local clustering dependent on the similarity of the polarization states and integration of clusters considering their variance.

[(1): Local clustering depending on the similarity of the polarization states.]

We examine the distances between the scanning center pixel's input vector Q^c and its surrounding eight pixel vectors Q^s. We consider the difference of polarization states of scattered waves as sum of the euclidean distances in the four incidence Poincare sphere parameters to calculate distance $D(Q^c, Q^s)$ as

$$D(Q^c, Q^s) = \|P_H^c - P_H^s\| + \|P_V^c - P_V^s\| + \|P_{45°}^c - P_{45°}^s\| + \|P_{lc}^c - P_{lc}^s\|. \quad (4)$$

If $D(Q^c, Q^s)$ is less than the variable threshold D_{th} properly set based on the clustering fitness, we regard pixels c and a as the same land state cluster.

[(2): Integration of clusters considering their variance.]

It sometimes happens that a cluster j whose land state class is the same as that of an existing cluster k generates a new class faultily. We integrate them in the every azimuth direction line scanning. We use $n(j)$ to represent the number of pixels in cluster j. When $n(j)$ is less than $n(k)$ and the following condition F is fulfilled, we integrate cluster j and cluster k. The condition F is based on the standard deviation range $us(k)$. We change u decreasingly from u_s to u_e

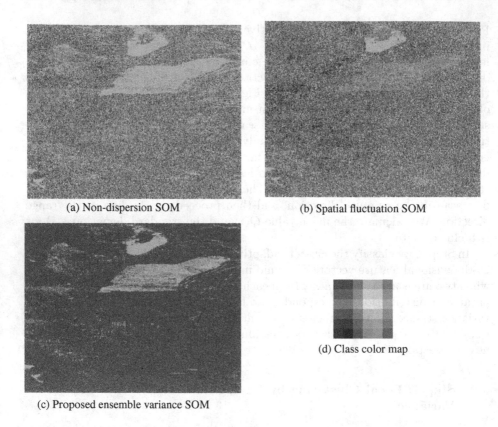

(a) Non-dispersion SOM   (b) Spatial fluctuation SOM

(d) Class color map

(c) Proposed ensemble variance SOM

Fig. 3. Land classification results for (a) no dispersion SOM, (b) spatial fluctuation SOM and (c) proposed dispersion (ensemble variation) SOM, and (d)class color map.

according to the increase of $n(k)$, that is,

$$F(j, k, u) : |\boldsymbol{Q}_{Element}(j) - \boldsymbol{Q}_{Element}(k)| < u s_{Element}(k) \; (^\forall Element) \quad (5)$$

$$u = u_e + \frac{u_s - u_e}{1 + n(k)/N} \quad (6)$$

With these processes, we can conduct the clustering based on the dispersion of Poincare sphere parameters, which is robust against the angle changes of the side-looking satellite observation.

3.3   Step 2: Classification of Clusters in SOM

We define the feature vector of i-th clusters as

$$\boldsymbol{T} = \begin{bmatrix} \boldsymbol{Q}(i) \\ \boldsymbol{s}(i) \end{bmatrix} \quad (7)$$

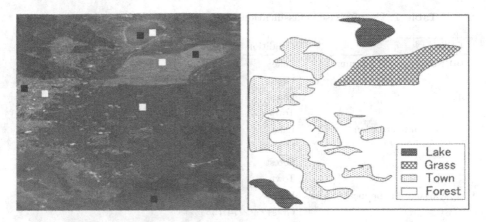

Fig. 4. (a) Optical photo example to show the numerical evaluation areas and (b) human determined four classes referenced in the evaluation of our previous supervised learning system [8].

We use a SOM to classify the clusters suitably based on the feature vectors just like we do in our land penetrating radar (GPR) systems [9–12]. In the present SAR classification system, we choose 6×6 torus as neuron topology.

The input signal is a 24-dimensional feature vector for a cluster $\boldsymbol{T}^{in}(i)$. We associate neuron g with a weight vector $\boldsymbol{T}^{w}(g)$ having the same dimension as that of the input signal. We classify an i-th input into a class represented by a neuron g when the the following distance becomes minimum for g among neuron classes:

$$H(\boldsymbol{T}^{in}(i), \boldsymbol{T}^{w}(g)) = D(\boldsymbol{Q}(i), \boldsymbol{Q}^{w}(g)) + KD(\boldsymbol{s}^{in}(i), \boldsymbol{s}^{w}(g)) \qquad (8)$$

In this distance calculation, we put an emphasis on the dispersion information in such a way that we first normalize $\boldsymbol{Q}(i)$ and $\boldsymbol{s}(i)$ and then multiply $\boldsymbol{s}(i)$ with an appropriate weight K_0 (i.e., $K = K_0\sigma(\boldsymbol{Q}(i))/\sigma(\boldsymbol{s}(i))$). Self-organization occurs as the updates of the winner $\boldsymbol{T}^{w}_{\text{win}}(g)$ and the surrounding neurons $\boldsymbol{T}^{w}_{\text{neighbor}}(g)$ with the sequential signal input. We input the features of $i = $1st cluster to $i = I$-th cluster. Then we repeat this process for a sufficient C times. We determine empirically the self-organization coefficients α_0 and β_0 properly to use them with the iteration number c, that is,

$$\boldsymbol{T}^{w}_{\text{win}}(t+1) = \boldsymbol{T}^{w}_{\text{win}}(t) + \alpha(\boldsymbol{T}^{in} - \boldsymbol{T}^{w}_{\text{win}}), \quad \alpha = \alpha_0(1 - c/C),$$
$$\boldsymbol{T}^{w}_{\text{neighbor}}(t+1) = \boldsymbol{T}^{w}_{\text{neighbor}}(t) + \beta(\boldsymbol{T}^{in} - \boldsymbol{T}^{w}_{\text{neighbor}}), \quad \beta = \beta_0(1 - c/C). \quad (9)$$

4   Experimental Results

Figure 3 shows the result of land state classification. We evaluate the result of (c) the proposed method by comparing those with other two methods, namely,

Table 1. Unsupervised classification accuracy for the Fujisusono area.

Utilizing	Condition	Accuracy	Overall Accuracy
(a) No dispersion information	Lake	99.32 %	37.33 %
	Grass	0 % (96.92 % as Lake)	
	Forest	5.06 %	
	Town	41.86 %	
(b) Spatial fluctuation	Lake	97.88 %	51.50 %
	Grass	64.36 %	
	Forest	8.98 %	
	Town	34.78 %	
(c) Variance in the local clusters	Lake	93.60 %	83.24 %
	Grass	74.72 %	
	Forest	86.60 %	
	Town	78.02 %	

(a) SOM processing using Poincare sphere parameters and (b) SOM processing Poincare sphere parameters and information about local fluctuation among
around 9×9 pixels that represents spatial variance. Figure 3(d) shows the color
map assigned to represent the classes in such a manner that the color similarity
corresponds to the class feature similarity. Table 1 shows the accuracy comparison calculated for 5000 pixels in lake, grass, forest and city areas, respectively.
Figure 4 shows (a) an optical photo example to show the numerical evaluation
areas and (b)human determined four classes referenced in the evaluation of our
previous supervised learning system [8].

It is found that the proposed method shown in Fig. 3(c) realizes the highest
precision classification. In the non-dispersion SOM result in (a) fails to distinguish lakes and grass (aqua blue for both). The spatial fluctuation SOM result
in (b) is successful in this distinction to some extent (aqua blue and pale purple). But the towns and forests are mixed in most areas. This spatial fluctuation method worked with a high performance in a supervised learning system
[8]. However, the self-organizing system having a larger number of classes may
require a different type of parameter adjustment. The spatial resolution is also
a little degraded because of the use of 9×9 spatial window to evaluate the
fluctuation.

In the proposed method, the Step 1 local clustering generates about 400 clusters, in each of which clusters the number of pixels are from one to 200,000. Then,
we conduct the SOM processing. It succeeds in basic classification clearly among
water surface (aqua blue), grass (green), forest (blue and purple), and city (red
and brown). City area is distinguished into about 10 classes, and grass is classified
into mainly four classes according to its vegetation (i.e., grass (green), grass and
tree (yellowish green), farm (jade green), and rice field (ocher)). Other classes
become exceptional classes with no more than 100 pixels. We are successful in
eliminating the effect of the incident angle changes due to the range-directional

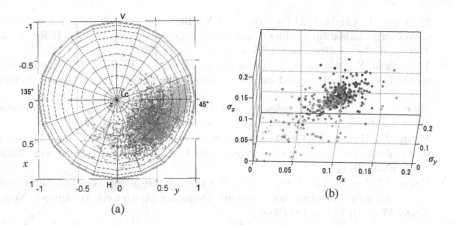

Fig. 5. (a) Distribution of Poincare sphere parameters and (b) variance of the clusters for 45° linearly polarization incidence. The colors correspond to those in Fig. 3(c).

location difference, resulting in excellent classification over a wide range of forest into a single class. In addition, we are able to detect small grass area in forest (right-hand side, lower) as well as to sharpen the lake boundary.

Figure 5 presents example information of Poincare sphere parameters after classification, that is, (a) distribution of Poincare sphere parameters and (b) variance of the clusters in 45° linearly polarization. It is found that each cluster distribution has overlaps with anisotropic dispersion. The variance works well in the self-organizing classification.

5   Conclusion

We proposed an unsupervised adaptive method to classify land states using a SOM that deals with dispersion or ensemble variance of land-scattering polarization features. The system employs two-stage clustering to utilize the dispersion features. Experimental demonstrated high accuracy and discovering land classification.

References

1. Touzi, R., Goze, S., Le Toan, T., Lopes, A., Mougin, E.: Polarimetric discriminators for SAR images. IEEE Trans. Geosci. Remote Sens. **30**(5), 973–980 (1992)
2. Ulaby, F.T., Held, D., Dobson, M.C., McDonald, K.C., Senior, T.B.A.: Relating polarization phase differencpolarSAR signals to scene properties. IEEE Trans. Geosci. Remote Sens. GE **25**(1), 83–92 (1987)
3. Evans, D.L., Farr, T.G., van Zyl, J.J., Zebker, H.A.: Radar polarimetry: analysis tools and applications. IEEE Trans. Geosci. Remote Sens. **26**(6), 774–789 (1988)
4. Touzi, R., Raney, R.K., Charbonneau, F.: On the use of permanent symmetric scatterers for ship characterization. IEEE Trans. Geosci. Remote Sens. **42**(10), 2039–2045 (2004)

5. Shirvany, R., Chabert, M., Tourneret, J.Y.: Ship and oil-spill detection using the degree of polarization in linear and hybrid/compact dual-PolSAR. IEEE J. Sel. Top. Appl. Earth Obs. Remote Sens. **5**, 885–892 (2012)

6. Yamaguchi, Y., Moriyama, T., Ishido, M., Yamada, H.: Four component scattering model for polarimetric SAR image decomposition. IEEE Trans. Geosci. Remote Sens. **43**(8), 1699–1706 (2005)

7. Shang, F., Hirose, A.: Averaged stokes vector based polarimetric sar data interpretation. IEEE Trans. Geosci. Remote Sens. **53**, 4536–4547 (2015)

8. Shang, F., Hirose, A.: Quaternion neural-network-based PolSAR land classification in poincare-sphere-parameter space. IEEE Trans. Geosci. Remote Sens. **52**(9), 5693–5703 (2014)

9. Hara, T., Hirose, A.: Plastic mine detecting system using complex-valued self-organizing map that deals with multiple-frequency interferometric images. Neural Netw. **17**(8–9), 1201–1210 (2004)

10. Masuyama, S., Hirose, A.: Walled LTSA array for rapid, high spatial resolution, and phase sensitive imaging to visualize plastic landmines. IEEE Trans. Geosci. Remote Sens. **45**(8), 2536–2543 (2007)

11. Masuyama, S., Yasuda, K., Hirose, A.: Multiple mode selection of walled-ltsa array elements for high resolution imaging to visualize antipersonnel plastic landmines. IEEE Geosci. Remote Sens. Lett. **5**(4), 745–749 (2008)

12. Nakano, Y., Hirose, A.: Adaptive identification of landmine class by evaluating the total degree of conformity of ring-SOM. Aust. J. Intell. Inf. Process. Syst. **12**, 23–28 (2010)

Algorithmic Robustness for Semi-Supervised (ϵ, γ, τ)-Good Metric Learning

Maria-Irina Nicolae[1,2]($\boxtimes$), Marc Sebban[1], Amaury Habrard[1], Eric Gaussier[2], and Massih-Reza Amini[2]

[1] Université Jean Monnet, Laboratoire Hubert Curien, Saint-Étienne, France
Irina.Nicolae@imag.fr
[2] Université Grenoble Alpes, CNRS-LIG/AMA, Saint-Martin-d'Hères, France

Abstract. Using the appropriate metric is crucial for the performance of most of machine learning algorithms. For this reason, a lot of effort has been put into distance and similarity learning. However, it is worth noting that this research field lacks theoretical guarantees that can be expected on the generalization capacity of the classifier associated to a learned metric. The theoretical framework of (ϵ, γ, τ)-good similarity functions [1] provides means to relate the properties of a similarity function and those of a linear classifier making use of it. In this paper, we extend this theory to a method where the metric and the separator are jointly learned in a semi-supervised way, setting that has not been explored before. We furthermore prove the robustness of our algorithm, which allows us to provide a generalization bound for this approach. The behavior of our method is illustrated via some experimental results.

1 Introduction

The importance of the underlying geometry of the data for improving the performance of learning algorithms has determined the expansion of a new research area termed *metric learning* [5]. From the point of view of the metric, most of these approaches focus on distance learning [3,6,7,14,16], but similarity learning has also attracted a growing interest [2,8,11,13], as the cosine similarity is more appropriate for certain problems than the euclidean distance. More recently, [1] have proposed the first framework that formalizes the relation between the quality of a metric and that of a classification algorithm making use of them. This broad framework, that can be used with a large range of similarity functions, provides generalization guarantees on a linear classifier learned from the similarity. However, to enjoy these guarantees, the similarity function is assumed to be known beforehand and to satisfy (ϵ, γ, τ)-goodness properties. The main limitation is that [1] does not provide any algorithm for learning such similarities.

In order to complete this framework, [4] have developed a method that independently learns an (ϵ, γ, τ)-good similarity. It is then plugged into the initial algorithm [1] to learn the linear separator using the metric. However, the similarity learning step is done in a completely supervised way while the (ϵ, γ, τ)-good framework opens the door to the use of unlabeled data.

© Springer International Publishing Switzerland 2015
S. Arik et al. (Eds.): ICONIP 2015, Part I, LNCS 9489, pp. 253–263, 2015.
DOI: 10.1007/978-3-319-26532-2_28

In this paper, our objective is to jointly learn the metric and the classifier in the theoretical framework of (ϵ, γ, τ)-good similarities. Furthermore, and unlike [4], the whole process is done in a semi-supervised way. To our knowledge, joint learning has not been explored before for semi-supervised metric learning. Enforcing (ϵ, γ, τ)-goodness allows us to preserve the theoretical guarantees from [1]. Lastly, proving the algorithmic robustness [17] of our method leads to consistency bounds for different types of similarity functions.

The remainder of this paper is organized as follows: Sect. 2 reviews some previous results in metric and similarity learning and presents the theory of (ϵ, γ, τ)-good similarities. Section 3 introduces our method that jointly learns the metric and the linear classifier, followed by generalization guarantees for our formulation. We show how to integrate different similarity functions in our setting. Finally, Sect. 4 features an experimental study on various standard datasets.

2   Notations and Related Work

In our developments, vectors are denoted by lower-case bold symbols ($\mathbf{x}$) and matrices by upper-case bold symbols ($\mathbf{A}$). A pairwise similarity function over $\mathcal{X} \subseteq \mathcal{R}^d$ is defined as $K : \mathcal{X} \times \mathcal{X} \to [-1, 1]$, and the hinge loss as $[c]_+ = \max(0, 1 - c)$. We note the L_1 norm by $|| \cdot ||_1$ and the L_2 norm by $|| \cdot ||_2$. The purpose of metric learning is to learn the parameters of a distance or similarity function that best fits the underlying geometry of the data. The learning is usually done using side information, expressed as pair-based ($\mathbf{x}$ and $\mathbf{x}'$ should be (dis)similar) or triplet-based constraints ($\mathbf{x}$ should be more similar to $\mathbf{x}'$ than to $\mathbf{x}''$). The metric is commonly represented by a matrix of values resulting from solving an optimization problem.

Most of state-of-the-art approaches focus on learning a Mahalanobis distance, defined as $d_{\mathbf{A}}(\mathbf{x}, \mathbf{x}') = \sqrt{(\mathbf{x} - \mathbf{x}')^T \mathbf{A} (\mathbf{x} - \mathbf{x}')}$. The distance is parameterized by the symmetric and positive semi-definite (PSD) matrix $\mathbf{A} \in \mathbb{R}^{d \times d}$. This metric implicitly corresponds to computing the Euclidean distance after linearly projecting the data to a different feature space. The PSD constraint on $\mathbf{A}$ ensures $d_{\mathbf{A}}$ is a proper metric. Setting $\mathbf{A}$ to the identity matrix gives the Euclidean distance. In this context, LMNN [16] is one of the most widely-used Mahalanobis distance learning methods. The constraints are pair- and triplet-based, derived from each instance's nearest neighbors. The optimization problem they solve is convex and has a special-purpose solver. The algorithm works well in practice, but is sometimes prone to overfitting due to the absence of regularization, especially when dealing with high dimensional data. Another limitation is that enforcing the PSD constraint on $\mathbf{A}$ is computationally expensive. Workarounds include using a specific solver or opting for information-theoretic approaches. ITML [6] was the first method to use LogDet divergence for regularization, providing an easy way for ensuring that $\mathbf{A}$ is a PSD matrix. However, the learned metric $\mathbf{A}$ is strongly influenced by the initial value $\mathbf{A}_0$, which is an important shortcoming, as $\mathbf{A}_0$ is handpicked. LRML [10] learns Mahalanobis distances with manifold regularization using a Laplacian matrix in a semi-supervised setting.

It performs particularly well compared to fully supervised methods when side information is scarce.

More generally, Mahalanobis distance learning faces two main limitations: firstly, enforcing the PSD and symmetry constraints on $\mathbf{A}$ is costly and often rules out natural similarity functions; secondly, although state-of-the-art Mahalanobis distance learning methods yield better accuracy than using the Euclidean distance, no theoretical guarantees are provided to establish a link between the quality of the metric and that of the classifier that makes use of it. [1] defined the (ϵ, γ, τ)-good similarity functions based on non PSD matrices, which uses similarities between labeled data and unlabeled reasonable points (roughly speaking, the reasonable points play the same role as that of support vectors in SVMs). Their theory was the first stone to establish generalization guarantees for a linear classifier that would be learned by making use of such similarities. Their results are derived based on the definition of a good similarity function for a given problem: considering a set of "reasonable points", a $(1 - \epsilon)$ proportion of examples $\mathbf{x}$ are on average 2γ more similar to random reasonable examples $\mathbf{x}'$ of their own label than to random reasonable examples $\mathbf{x}'$ of the other label. For this, the proportion of reasonable points from the sample must be greater than τ. In their definition, the margin violation is averaged over all reasonable points which leads to a more flexible setting than pair- or triplet-based constraints. If K is (ϵ, γ, τ)-good and enough reasonable points are available, there exists a linear separator $\boldsymbol{\alpha}$ with error arbitrarily close to ϵ in the space ϕ^S. Finding the separator is done by solving the following optimization problem:

$$\min_{\boldsymbol{\alpha}} \Big\{ \sum_{i=1}^{d_l} \Big[1 - \sum_{j=1}^{d_u} \alpha_j l(\mathbf{x}_i) K(\mathbf{x}_i, \mathbf{x}_j) \Big]_+ : \sum_{j=1}^{d_u} |\alpha_j| \leq 1/\gamma \Big\}.$$

The previous problem can be solved efficiently by linear programming. Also, tuning the value of γ (L_1 constraint) will produce a sparse solution. The main limitation of this approach is that the similarity function K is considered known.

This limitation has been partly overcome by SLLC [4] by optimizing the (ϵ, γ, τ)-goodness of a bilinear similarity function under Frobenius norm regularization. The learned metric is then used to build a global linear classifier with guarantees. Moreover, a bound on the generalization error of the associated classifier through uniform stability can be obtained. More recently, [9] derived generalization bounds for similarity learning formulations that are regularized with more general matrix-norms, based on the Rademacher complexity and Khinchin-type inequalities.

There are three main distinctions between these approaches and our work. Firstly, we propose a method that jointly learns the metric and the linear separator at the same time. This allows us to make use of the semi-supervised setting presented by [1] to learn well with only a small amount of labeled data. Secondly, our setting uses the algorithmic robustness to establish bounds, which enables us to characterize our algorithm by exploiting the geometry of the data; that is not the case with the Rademacher complexity. Lastly, regularization is integrated through constraints in our setting, as explained in the following sections, which leads to a formulation with less hyperparameters.

3 Learning Consistent Good Similarity Functions

In this section, we present our semi-supervised framework for jointly learning a similarity function and a linear separator from data. We also provide a generalization bound for our approach based on the recent algorithmic robustness framework [17]. We end this section by presenting some particular similarity functions that can be used in our setting.

3.1 Optimization Problem

Let S be a sample set of d_l labeled examples $(\mathbf{x}, l(\mathbf{x})) \in \mathcal{Z} = \mathcal{X} \times \mathcal{Y}$ ($\mathcal{X} \subseteq \mathcal{R}^d$) and d_u unlabeled examples. We assume that $\mathcal{X}$ is bounded, which can be expressed, after normalization, by $||\mathbf{x}||_2 \leq 1$. Let $K_{\mathbf{A}}(\mathbf{x}, \mathbf{x}')$ be a generic (ϵ, γ, τ)-good similarity function, parameterized by the matrix $\mathbf{A} \in \mathbb{R}^{d \times d}$. We want to optimize the goodness of $K_{\mathbf{A}}$ w.r.t. the empirical loss of a finite sample. To this end, we must find the matrix $\mathbf{A}$ and the global separator $\boldsymbol{\alpha} \in \mathbb{R}^{d_u}$ that minimize the loss function (in <our case, the hinge loss) over the training set S. Our learning algorithm takes the form of the following constrained optimization problem.

$$\min_{\boldsymbol{\alpha}, \mathbf{A}} \frac{1}{d_l} \sum_{i=1}^{d_l} \left[1 - \sum_{j=1}^{d_u} \alpha_j l(\mathbf{x}_i) K_{\mathbf{A}}(\mathbf{x}_i, \mathbf{x}_j) \right]_+ \tag{1}$$

$$\text{s.t.} \sum_{j=1}^{d_u} |\alpha_j| \leq 1/\gamma \tag{2}$$

$$\mathbf{A} \text{ diagonal}, \quad |A_{kk}| \leq 1, \quad 1 \leq k \leq d, \tag{3}$$

The novelty of this algorithm is the *joint optimization* over $\mathbf{A}$ and $\boldsymbol{\alpha}$: by solving problem (1), we are learning the metric and the separator at the same time. A significant advantage of this formulation is that it extends the semi-supervised setting from the separator learning step to the metric learning, and the two problems are solved using the same data. This method can naturally be used in situations where one has access to few labeled examples and many unlabeled ones: the labeled examples are used in this case to select the unlabeled examples that will serve to classify new points. Another important advantage of our technique is that the constraints on the pair of points do not need to be satisfied entirely, as the loss is averaged on all the reasonable points. In other words, this formulation is less restrictive than pair or triplet-based settings.

Constraint (2) takes into account the desired margin γ and is the same as in [1]. The new Constraint (3) serves two purposes: first, it restricts the similarity $K_{\mathbf{A}}$, thus preserving its (ϵ, γ, τ)-goodness; second, as it bounds the values in the matrix $\mathbf{A}$, it limits the risk of overfitting, and thus plays the role of regularization without imposing sparsity. Regularizing metrics through standard L_1 or $L_{(1,2)}$ norms would slowly push the values in the matrix towards zero, which is not necessarily desirable. Indeed, let $f(\mathbf{x}) = \sum_{j=1}^{d_u} \alpha_j K_\mathbf{A}(\mathbf{x}, \mathbf{x}_j)$ be the output of the linear separator w.r.t. $\mathbf{x}$. For some linear similarities $K_\mathbf{A}(x, x')$, such as the bilinear form $K_\mathbf{A}(\mathbf{x}, \mathbf{x}') = \mathbf{x}^T \mathbf{A} \mathbf{x}'$, computing $f(\mathbf{x})$ boils down to calculating the similarity between $\mathbf{x}$ and the barycenter of the (weighted) unlabeled points, making sparsity superfluous.

3.2 Consistency Guarantees

We now present a theoretical analysis of our approach. For the purpose of discussing the algorithmic robustness of the method, let us rewrite the minimization problem (1) with a more generalized notation of the loss function:

$$\min \frac{1}{d_l} \sum_{i=1}^{d_l} \ell(\mathbf{A}, \boldsymbol{\alpha}, \mathbf{z}_i = (\mathbf{x}_i, l(\mathbf{x}_i))),$$

where $\ell(\mathbf{A}, \boldsymbol{\alpha}, \mathbf{z}_i = (\mathbf{x}_i, l(\mathbf{x}_i))) = \left[1 - \sum_{j=1}^{d_u} \alpha_j l(\mathbf{x}_i) K_\mathbf{A}(\mathbf{x}_i, \mathbf{x}_j)\right]_+$ is the instantaneous loss estimated at point $(\mathbf{x}_i, l(\mathbf{x}_i))$. Therefore, the optimization problem (1) under constraints (2) and (3) reduces to minimizing the empirical loss $\hat{\mathcal{R}}^\ell = \frac{1}{d_l} \sum_{i=1}^{d_l} \ell(\mathbf{A}, \boldsymbol{\alpha}, \mathbf{z}_i)$ over the training set $\mathcal{S}$. To begin with, let us recall the notion of robustness of an algorithm $\mathcal{A}$.

Definition 1 (Algorithmic Robustness [17]). *Algorithm $\mathcal{A}$ is $(M, \epsilon(\cdot))$-robust, for $M \in \mathbb{N}$ and $\epsilon(\cdot) : \mathcal{Z}^{d_l} \rightarrow \mathbb{R}$, if $\mathcal{Z}$ can be partitioned into M disjoint sets, denoted by $\{C_i\}_{i=1}^M$, such that the following holds for all $\mathcal{S} \in \mathcal{Z}^{d_l}$:*

$$\forall \mathbf{z} = (\mathbf{x}, l(\mathbf{x})) \in \mathcal{S}, \forall \mathbf{z}' = (\mathbf{x}', l(\mathbf{x}')) \in \mathcal{Z}, \forall i \in [M] :$$
$$if\ \mathbf{z}, \mathbf{z}' \in C_i, then |\ell(\mathbf{A}, \boldsymbol{\alpha}, \mathbf{z}) - \ell(\mathbf{A}, \boldsymbol{\alpha}, \mathbf{z}')| \leq \epsilon(\mathcal{S}).$$

Roughly speaking, an algorithm is robust if for any example $\mathbf{z}'$ falling in the same subset as a training example $\mathbf{z}$, the gap between the losses associated with $\mathbf{z}$ and $\mathbf{z}'$ is bounded. Subsets are constructed using a partitioning of $\mathcal{Z}$ based on covering numbers [12]. Two examples are close if they belong to the same region, implying that the norm between them is lesser than a fixed quantity ρ (see [17] for details about building the covering). Now we can state the first theoretical contribution of this paper.

Theorem 1. *Given a partition of $\mathcal{Z}$ into M subsets $\{C_i\}$ such that $\mathbf{z} = (\mathbf{x}, l(\mathbf{x}))$ and $\mathbf{z}' = (\mathbf{x}', l(\mathbf{x}')) \in C_i$ and $l(\mathbf{x}) = l(\mathbf{x}')$, and provided that $K_\mathbf{A}(\mathbf{x}, \mathbf{x}')$ is l-lipschitz w.r.t. its first argument, the optimization problem (1) with constraints (2) and (3) is $(M, \epsilon(\mathcal{S}))$-robust with $\epsilon(\mathcal{S}) = \frac{1}{\gamma} l\rho$, where $\rho = \sup_{\mathbf{x}, \mathbf{x}' \in C_i} ||\mathbf{x} - \mathbf{x}'||$.*

Proof

$$|\ell(\mathbf{A}, \boldsymbol{\alpha}, \mathbf{z}) - \ell(\mathbf{A}, \boldsymbol{\alpha}, \mathbf{z}')| \le \left| \sum_{j=1}^{d_u} \alpha_j l(\mathbf{x}') K_{\mathbf{A}}(\mathbf{x}', \mathbf{x}_j) - \sum_{j=1}^{d_u} \alpha_j l(\mathbf{x}) K_{\mathbf{A}}(\mathbf{x}, \mathbf{x}_j) \right|$$
(4)

$$= \left| \sum_{j=1}^{d_u} \alpha_j (K_{\mathbf{A}}(\mathbf{x}', \mathbf{x}_j) - K_{\mathbf{A}}(\mathbf{x}, \mathbf{x}_j)) \right| \le \sum_{j=1}^{d_u} |\alpha_j| \cdot |K_{\mathbf{A}}(\mathbf{x}', \mathbf{x}_j) - K_{\mathbf{A}}(\mathbf{x}, \mathbf{x}_j)| \quad (5)$$

$$\le \sum_{j=1}^{d_u} |\alpha_j| \cdot l\|\mathbf{x} - \mathbf{x}'\| \le \frac{1}{\gamma} l\rho$$
(6)

Setting $\rho = \sup_{\mathbf{x}, \mathbf{x}' \in C_i} \|\mathbf{x} - \mathbf{x}'\|_1$, we get the Theorem. We get Inequality (4) from the 1-lipschitzness of the hinge loss; Inequality (5) comes from triangle inequality; the first inequality on line (6) is due to the l-lipschitzness of $K_{\mathbf{A}}(\mathbf{x}, \mathbf{x}_j)$ w.r.t. its first argument, and the result follows from Condition (2).

We now give a PAC generalization bound on the true loss making use of the previous robustness result. Let $\mathcal{R}^\ell = \mathbb{E}_{\mathbf{z} \sim \mathcal{Z}} \ell(\mathbf{A}, \boldsymbol{\alpha}, \mathbf{z})$ be the true loss w.r.t. the unknown distribution $\mathcal{Z}$ and $\hat{\mathcal{R}}^\ell = \frac{1}{d_l} \sum_{i=1}^{d_l} \ell(\mathbf{A}; \boldsymbol{\alpha}, \mathbf{z}_i)$ be the empirical loss over the training set $\mathcal{S}$. We have the following concentration inequality that allows one to capture statistical information coming from the different regions of the partition of $\mathcal{Z}$.

Proposition 1 *[15] Let* $(|N_1|, \dots, |N_M|)$ *be an i.i.d. multinomial random variable with parameters* $d_l = \sum_{i=1}^{M} |N_i|$ *and* $(p(C_1), \dots, p(C_M))$. *By the Bretagnolle-Huber-Carol inequality we have:* $\Pr \left\{ \sum_{i=1}^{M} \left| \frac{|N_i|}{d_l} - p(C_i) \right| \ge \lambda \right\} \le 2^M \exp\left(\frac{-d_l \lambda^2}{2}\right)$, *hence with probability at least* $1 - \delta$, $\sum_{i=1}^{M} \left| \frac{N_i}{d_l} - p(C_i) \right| \le \sqrt{\frac{2M \ln 2 + 2\ln(1/\delta)}{d_l}}$.

We are now able to present our generalization bound in the following theorem.

Theorem 2 *Considering that problem (1) is* $(M, \epsilon(\mathcal{S}))$-*robust, and that* $K_{\mathbf{A}}$ *is* l-*lipschitz w.r.t. to its first argument, for any* $\delta > 0$ *with probability at least* $1 - \delta$, *we have:*

$$|\mathcal{R}^\ell - \hat{\mathcal{R}}^\ell| \le \tfrac{1}{\gamma} l\rho + B\sqrt{\frac{2M \ln 2 + 2\ln(1/\delta)}{d_l}},$$

where $B = 1 + \frac{1}{\gamma}$ *is an upper bound of the loss* ℓ.

The proof of Theorem 2 follows the one described in [17]. Note that the cover radius ρ can be arbitrarily small at the expense of larger values of M. As M appears in the second term, decreasing to 0 when d_l tends to infinity, this bound provides a standard $O(1/\sqrt{d_l})$ asymptotic convergence.

As one can note, our main theorems strongly depend on the l-lipschitzness of the similarity function. We provide below three standard similarity functions

Table 1. Properties of the datasets used in the experimental study.

	Balance	Ionosphere	Iris	Liver	Pima	Sonar	Wine
# Instances	625	351	150	345	768	208	178
# Dimensions	4	34	4	6	8	60	13
# Classes	3	2	3	2	2	2	3

together with their lipschitz property. K_A^1 and K_A^2 are linear w.r.t. their arguments, and have the advantage of keeping Problem (1) convex. K_A^3 is gaussian-like kernel based on the Mahalanobis distance, and is non linear.

Ex. 1. Let K_A^1 be the bilinear form $K_A^1(\mathbf{x}, \mathbf{x}') = \mathbf{x}^T \mathbf{A} \mathbf{x}'$. $K_A^1(\mathbf{x}, \mathbf{x}')$ is 1-lipschitz w.r.t. its first argument.

Ex. 2. We define $K_A^2(\mathbf{x}, \mathbf{x}') = 1 - (\mathbf{x} - \mathbf{x}')^T \mathbf{A}(\mathbf{x} - \mathbf{x}')$, a similarity derived from the Mahalanobis distance. $K_A^2(\mathbf{x}, \mathbf{x}')$ is 4-lipschitz w.r.t. its first argument.

Ex. 3. Let $K_A^3(\mathbf{x}, \mathbf{x}') = \exp\left(-\frac{(\mathbf{x}-\mathbf{x}')^T \mathbf{A}(\mathbf{x}-\mathbf{x}')}{2\sigma^2}\right)$. $K_A^3(\mathbf{x}, \mathbf{x}')$ is l-lipschitz w.r.t. its first argument with $l = \frac{2}{\sigma^2}\left(\exp\left(\frac{1}{2\sigma^2}\right) - \exp\left(\frac{-1}{2\sigma^2}\right)\right)$.

Plugging $l = 1$ (resp. $l = 4$ and $l = \frac{2}{\sigma^2}\left(\exp\left(\frac{1}{2\sigma^2}\right) - \exp\left(\frac{-1}{2\sigma^2}\right)\right)$) in Theorem 2, we obtain consistency results for Problem (1) using $K_A^1(\mathbf{x}, \mathbf{x}')$ (resp. $K_A^2(\mathbf{x}, \mathbf{x}')$ and $K_A^3(\mathbf{x}, \mathbf{x}')$). As the gap between empirical and true loss presented in Theorem 2 is proportional with the l-lipschitzness of each similarity function, we would like to keep this parameter as small as possible. We notice that the generalization bound is tighter for K_A^1 than for K_A^2. The bound for K_A^3 depends on the additional parameter σ, that adjusts the influence of the similarty value w.r.t. the distance to the landmarks.

4 Experiments

Metric learning state-of-the-art algorithms are mostly designed for a supervised setting, and usually optimize a metric for kNN classification. It is thus difficult to propose a totally fair comparative study. We compare our method (JSL – Joint Similarity Learning) with algorithms from different categories (supervised, kNN-oriented). The experimental study is conducted on 7 classic datasets taken from the UCI Machine Learning Repository (Table 1).

4.1 Experimental Setup

For a complete comparison, we analyse two main families of approaches: first, linear classifiers, for which we consider BBS [1], SLLC [4], linear SVM with L_2 regularization and our method, JSL; second, nearest neighbor approaches: ITML [6], LMNN [16] and LRML [10], for which we report accuracies for 3NN classification. All attributes are centered around zero and scaled to ensure $||\mathbf{x}||_2 \leq 1$. We randomly choose 15 % of the data for validation purposes,

Table 2. Average accuracy (%) with conf. interval at 95 %, 5 labeled points/class.

Lmks	Sim	Balance	Ionosphere	Iris	Liver	Pima	Sonar	Wine
All pts.	K_A^1	85.7±3.5	88.5±2.6	**74.5±4.4**	63.9±5.3	71.1±3.8	**72.3±4.1**	**87.7±5.0**
	K_A^2	**87.1±2.5**	**91.0±2.0**	71.4±5.9	**69.2±3.2**	**72.9±3.9**	71.9±4.2	84.2±6.9
	K_A^3	81.1±8.5	86.2±2.8	68.2±8.5	58.6±6.3	71.1±4.3	63.9±10.0	83.5±6.2
15 pts.	K_A^1	84.9±2.6	**86.7±1.6**	**75.5±2.3**	63.1±5.9	71.1±4.1	72.9±4.6	87.3±5.5
	K_A^2	**87.5±2.7**	85.0±3.8	74.1±6.3	**67.3±4.3**	**74.3±4.1**	**77.4±6.3**	76.9±10.5
	K_A^3	79.6±10.0	76.3±7.4	72.7±6.3	59.6±6.0	69.0±8.6	68.7±10.0	**88.5±5.0**

and another 15 % as a test set. The training set and the unlabeled data are chosen from the remaining 70 % of examples not employed in the previous sets. We illustrate the classification using a restricted quantity of labeled data by limiting the number of labeled points to 5, 10 or 20 examples per class, as this is usually a reasonable minimum amount of annotation to rely on. The number of landmarks is either equal to the size of the training set, either set to 15 points (corresponding to k-means++ cluster centroids). When all the available data is used as landmarks, the L_1 constraint on α forces the algorithm to choose the most valuable of them by adapting their respective weights. All of the experimental results are averaged over 10 runs, for which we compute a 95 % confidence interval. We tune the following parameters by cross-validation: $\gamma \in \{10^{-4}, \ldots, 10^{-1}\}$ for BBS and JSL, $\lambda_{ITML} \in \{10^{-4}, \ldots, 10^4\}$, $\gamma_{SLLC}, \beta_{SLLC} \in \{10^{-7}, \ldots, 10^{-2}\}$, $\lambda_{SLLC} \in \{10^{-3}, \ldots, 10^2\}$, while for LRML we consider $\gamma_s, \gamma_d, \gamma_i \in \{10^{-2}, \ldots, 10^2\}$. For LMNN, we set $\mu = 0.5$, as done in [16]. We solve BBS and JSL using projected gradient descent. In JSL, we alternate the optimization between α and $\mathbf{A}$.

4.2 Results

Choice of Similarity. We first study the influence of the similarity function on the proposed framework. We plug into JSL the three similarities studied previously (see Sect. 3.2) and present the results for classification in Table 2. For both unlabeled configurations, 15 points or the whole training set, K_A^2 yields the best results on 4 out of 7 datasets, while K_A^3 performs best in only one case. We explain this by the topology of the involved datasets, which make the Mahalanobis distance a better dicriminant for classification than the other similarities. In the case of K_A^3, there is a trade-off between the tightness of the bound in Theorem 2 and the stability of the results. Large values of σ will lead to tighter bounds (as l is smaller), but the resulting similarity function becomes linear and less discriminative. As a consequence, the results vary more for this similarity function, leading to larger confidence intervals, as can be seen on almost all the collections. When comparing the two unlabeled settings, we notice that there are only a few cases when the best accuracy is attained with less unlabeled points for the same similarity, but that when this happens the improvement is significant. This is due to the fact that the 15 unlabeled points are not chosen randomly, but contain relevant information w.r.t. data topology.

Table 3. Average accuracy (%) over all datasets with confidence interval at 95%.

Data	LMNN-dg	LMNN	ITML	SVM	BBS	SLLC	LRML	JSL
5 pts./cl	65.1±5.5	69.4±5.9	75.8±4.2	76.4±4.9	77.2±7.3	70.5±7.2	74.7±6.2	**78.4±2.3**
10 pts./cl	68.2±5.6	70.9±5.3	76.5±4.5	76.2±7.0	77.0±6.2	75.9±4.5	75.3±5.9	**78.7±1.9**
20 pts./cl	71.5±5.2	73.2±5.2	76.3±4.8	77.7±6.4	77.3±6.3	75.8±4.8	75.8±5.2	**78.3±1.6**

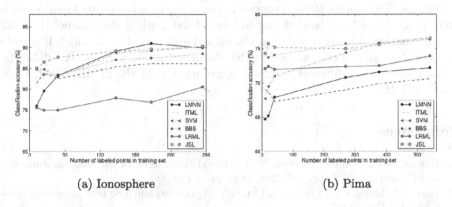

      (a) Ionosphere   (b) Pima

Fig. 1. Average accuracy w.r.t. the number of labeled points with 15 landmarks.

Comparison of Different Methods. Following the previous analysis, we now propose to study the classification performance of our method. For this purpose, we focus on JSL with K_A^2 using 15 unlabeled landmarks. We compare our approach to state-of-the-art methods when a limited amount of labeled data is used and present the results in Table 3. In order to ensure fairness, we fix the similarity function in BBS to the Euclidean distance. On average over all datasets, JSL obtains the best performance in all the settings. The only other methods with comparable results are BBS and SVM. We mention that JSL using also K_A^2 and all the training set as unlabeled landmarks performs similarly to the setting presented in Table 3. This result proves that we can learn well with a small amount of both labeled and unlabeled data, when the unlabeled points are informative (*e.g.*, correspond to cluster centroids, as it is the case here).

Quantity of Labeled Data. We now study the method's behavior when the level of supervision varies. For this we keep on using JSL with K_A^2 and set the number of unlabeled points to 15. Figure 1 presents the accuracies on two representative datasets, Ionosphere and Pima, with an increasing number of labeled examples. JSL obtains the best performance in both cases when less than 50% of the labeled data is used, which is coherent with the results presented in Table 3. For greater amounts of data, JSL performs similarly to the best state-of-the-art methods: SVM and LMNN for Ionosphere, and SVM and BBS for Pima; these results also correspond to those presented in the previous subsection.

5   Conclusion

In this paper, we extend the (ϵ, γ, τ)-good similarity theory to a method where the metric and the separator are jointly learned in a semi-supervised way, setting that has not been explored before. We show that our joint approach is theoretically founded using results from [1] and new results based on algorithmic robustness. The approach we propose is particularly adapted to learning with small amounts of both labeled and unlabeled data, when the unlabeled points are informative. This is revealed in the experiments conducted which illustrate the good behavior of our method in the above setting on various UCI datasets, in comparison with different standard approaches (LMNN, ITML, SVM, BBS, SLLC, LRML).

References

1. Balcan, M.F., Blum, A., Srebro, N.: Improved guarantees for learning via similarity functions. In: COLT, pp. 287–298. Omnipress (2008)
2. Bao, J.P., Shen, J.Y., Liu, X.D., Liu, H.Y.: Quick asymmetric text similarity measures. ICMLC (2003)
3. Baoli, L., Qin, L., Shiwen, Y.: An adaptive k-nearest neighbor text categorization strategy. ACM TALIP **3**, 215–226 (2004)
4. Bellet, A., Habrard, A., Sebban, M.: Similarity learning for provably accurate sparse linear classification. ICML **2012**, 1871–1878 (2012)
5. Bellet, A., Habrard, A., Sebban, M.: A survey on metric learning for feature vectors and structured data. arXiv preprint arXiv:1306.6709 (2013)
6. Davis, J.V., Kulis, B., Jain, P., Sra, S., Dhillon, I.S.: Information-theoretic metric learning. In: ICML, pp. 209–216. ACM, New York (2007)
7. Diligenti, M., Maggini, M., Rigutini, L.: Learning similarities for text documents using neural networks. In: ANNPR (2003)
8. Grabowski, M., Szałas, A.: A technique for learning similarities on complex structures with applications to extracting ontologies. In: Szczepaniak, P.S., Kacprzyk, J., Niewiadomski, A. (eds.) AWIC 2005. LNCS (LNAI), vol. 3528, pp. 183–189. Springer, Heidelberg (2005)
9. Guo, Z.C., Ying, Y.: Guaranteed classification via regularized similarity learning. CoRR abs/1306.3108 (2013)
10. Hoi, S.C.H., Liu, W., Chang, S.F.: Semi-supervised distance metric learning for collaborative image retrieval and clustering. TOMCCAP **6**(3), 18 (2010)
11. Hust, A.: Learning similarities for collaborative information retrieval. In: Machine Learning and Interaction for Text-Based Information Retrieval Workshop, TIR-04, pp. 43–54 (2004)
12. Kolmogorov, A., Tikhomirov, V.: ϵ-entropy and ϵ-capacity of sets in functional spaces. Am. Math. Soc. Transl. **2**(17), 277–364 (1961)
13. Qamar, A.M., Gaussier, É.: Online and batch learning of generalized cosine similarities. In: ICDM, pp. 926–931 (2009)
14. Shalev-Shwartz, S., Singer, Y., Ng, A.Y.: Online and batch learning of pseudo-metrics. In: ICML. ACM, New York (2004)
15. van der Vaart, A., Wellner, J.: Weak Convergence and Empirical Processes. Springer Series in Statistics. Springer, New York (1996)

16. Weinberger, K., Saul, L.: Distance metric learning for large margin nearest neighbor classification. JMLR **10**, 207–244 (2009)
17. Xu, H., Mannor, S.: Robustness and generalization. In: COLT, pp. 503–515 (2010)

Patchwise Tracking via Spatio-Temporal Constraint-Based Sparse Representation and Multiple-Instance Learning-Based SVM

Yuxia Wang[✉] and Qingjie Zhao

Beijing Lab of Intelligent Information Technology, School of Computer Science,
Beijing Institute of Technology, Beijing 100081, People's Republic of China
yuxiawang2006@163.com, zhaoqj@bit.edu.cn

Abstract. This paper proposes a patch-based tracking algorithm via a hybrid generative-discriminative appearance model. For establishing the generative appearance model, we present a spatio-temporal constraint-based sparse representation (STSR), which not only exploits the intrinsic relationship among the target candidates and the spatial layout of the patches inside each candidate, but also preserves the temporal similarity in consecutive frames. To construct the discriminative appearance model, we utilize the multiple-instance learning-based support vector machine (MIL&SVM), which is robust to occlusion and alleviates the drifting problem. According to the classification result, the occlusion state can be predicted, and it is further used in the templates updating, making the templates more efficient both for the generative and discriminative model. Finally, we incorporate the hybrid appearance model into a particle filter framework. Experimental results on six challenging sequences demonstrate that our tracker is robust in dealing with occlusion.

Keywords: Patchwise tracking · Hybrid generative-discriminative appearance model · MIL&SVM · Spatio-temporal constraint · Sparse representation

1 Introduction

Visual tracking is an active field of research in computer vision. While numerous tracking methods have been proposed with demonstrated success in recent years, designing a robust tracking method is still an open problem, due to factors such as scale and pose change, illumination variation, occlusion, etc. Especially, occlusion is a core issue. One of the main reasons is the lack of the effective object appearance models, which play a significant role in visual tracking.

For designing a robust tracker, most tracking algorithms employ generative learning or discriminative learning based appearance models. Generative learning based appearance models mainly concentrate on how to fit the data accurately from the object class using generative methods. Among them, sparse representation is a widely used generative method. Jia et al. [3] developed a

© Springer International Publishing Switzerland 2015
S. Arik et al. (Eds.): ICONIP 2015, Part I, LNCS 9489, pp. 264–271, 2015.
DOI: 10.1007/978-3-319-26532-2_29

local appearance model by utilizing the sparse representation of the overlapped patches. Zhang et al. [10] proposed a structural sparse tracking algorithm to exploit the relationship among the target candidates and spatial layout of the patches inside each candidate. Zarezade et al. [9] presented a joint sparse tracker by assuming that the target and the previous best candidates have a common sparsity pattern. Although these methods achieve convincing performance, they either lack of the description of the target spatial layout or ignore the temporal consistency constraint of successive frames. In this paper, we propose a spatio-temporal constraint-based sparse representation (STSR), which not only exploits the spatial layout of the local patches inside each candidate and the intrinsic relationship among the candidates and their local patches, but also preserves the temporal consistency of the sparsity pattern in consecutive frames.

In comparison, discriminative appearance models pose visual tracking as a binary classification issue, aiming to maximize the inter-class separability between the object and non-object regions via discriminative learning techniques. Babenko et al. [2] introduced the multiple-instance learning technique into online object tracking where training samples can be labeled more precisely. In [4], Kalal et al. proposed to train a binary classifier using the P-N learning algorithm with both labeled and unlabeled samples. Despite the convincing performance, most of these methods use holistic representation to represent the object and hence do not handle occlusion well. In this paper, we utilize the patch-based discriminative appearance model proposed by [6] to locate the target from the background, in which the multiple-instance learning-based support vector machine (MIL&SVM) is used as the classifier and it can predict the occlusion state and alleviate the drifting problem. According to the occlusion state, we update the template set as mentioned in [6], making the templates more effective both for the generative and discriminative appearance model.

2 Patchwise Tracking via a Hybrid Generative-Discriminative Appearance Model

In our tracker, we utilize s_t to denote the object state at time t, and construct our tracker in the particle filter framework (PF). For the dynamic model of PF, $p(s_t|s_{t-1})$, we assume a Gaussian distributed model. For the appearance model in PF, $p(y_t|s_t)$, we use our patch-based hybrid generative-discriminative appearance model, which will be introduced below.

2.1 Generative Appearance Model Based on STSR

Given the image set of the target templates $\mathbf{T} = [\mathbf{T}_1, \mathbf{T}_2, ..., \mathbf{T}_m]$, where m is the number of target templates, we sample K overlapped local patches inside each target region. The sampled patches are used to form a dictionary $\mathbf{D} = [\mathbf{d}_1^{(1)}, ..., \mathbf{d}_m^{(1)}, ..., \mathbf{d}_1^{(K)}, ..., \mathbf{d}_m^{(K)}]$, each column in $\mathbf{D}$ is obtained by ℓ_2 normalization on the vectorized gray scale image observations extracted from $\mathbf{T}$.

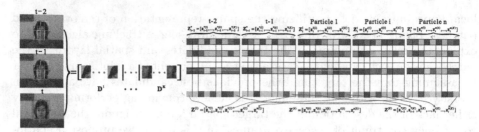

Fig. 1. Spatio-temporal constraint-based sparse representation

Let $\{\mathbf{x}_{t-i}^*\}_{i=1}^N$ and $\{\mathbf{x}_t^i\}_{i=1}^n$ represent the best candidates obtained in the previous tracking and particles from the current frame respectively. For $\{\mathbf{x}_{t-i}^*\}_{i=1}^N$ and $\{\mathbf{x}_t^i\}_{i=1}^n$, we also sample K overlapped local patches as done in the template set and denote $\mathbf{x}_{t-i}^* = [\mathbf{x}_{t-i}^{*(1)}, ..., \mathbf{x}_{t-i}^{*(K)}]$ and $\mathbf{x}_t^i = [\mathbf{x}_t^{i(1)}, ..., \mathbf{x}_t^{i(K)}]$. Let $\mathbf{X}_t^{(k)} = [\mathbf{x}_t^{1(k)}, ..., \mathbf{x}_t^{n(k)}]$ denote the k-th local patches of n particles at time t. In order to represent this observations matrix $\mathbf{X}_t^{(k)}$, we not only consider the spatial constraint of the particles and local patches, but also utilize the temporal constraint in consecutive frames.

Spatio-Temporal Constraint. Based on the fact that n particles at current frame are densely sampled at and around the target of the previous frame and the target's appearance changes smoothly, it is reasonable to assume that these particles are likely to be similar and they have the similar sparse pattern with previous tracking results over a period of time. Thus the k-th image patches of n particles and previous tracking results are expected to be similar. In addition, for patches extracted from a candidate particle or a previous tracking result, their spatial layout should be preserved.

Spatio-Temporal Constraint-Based Sparse Representation (STSR). Based on the above observations, we use $\mathbf{X}^{(k)} = [\mathbf{x}_{t-i}^{*(k)}, ..., \mathbf{x}_{t-1}^{*(k)}, \mathbf{x}_t^{1(k)}, ..., \mathbf{x}_t^{n(k)}]$ to represent the k-th local patches of previous tracking results and n particles in current frame, $\mathbf{D}^{(k)} = [\mathbf{d}_1^{(k)}, \mathbf{d}_2^{(k)}, ..., \mathbf{d}_m^{(k)}]$ to express the k-th patches of m templates, and $\mathbf{Z}^{(k)} = [\mathbf{z}_{t-i}^{*(k)}, ..., \mathbf{z}_{t-1}^{*(k)}, \mathbf{z}_t^{1(k)}, ..., \mathbf{z}_t^{n(k)}]$ to denote the representations of the k-th local patch observations of $\mathbf{X}^{(k)}$ with respect to $\mathbf{D}^{(k)}$. Then the joint sparse appearance model for the object tracking under the spatio-temporal constraint can be obtained by using the $\ell_{2,1}$ mixed norm as

$$\min_{\mathbf{Z}} \frac{1}{2} \sum_{k=1}^K ||\mathbf{X}^{(k)} - \mathbf{D}^{(k)}\mathbf{Z}^{(k)}||_F^2 + \lambda||\mathbf{Z}||_{2,1} \tag{1}$$

where, $\mathbf{Z} = [\mathbf{Z}^{(1)}, \mathbf{Z}^{(2)}, ..., \mathbf{Z}^{(K)}]$, $|| \cdot ||_F$ denotes the Frobenius norm, λ is a regularization parameter which balances reconstruction error with model complexity, $||\mathbf{Z}||_{2,1} = \sum_i (\sum_j |[\mathbf{Z}]_{ij}|^2)^{\frac{1}{2}}$ and $[\mathbf{Z}]_{ij}$ denotes the entry at the i-th row and j-th column of $\mathbf{Z}$. The $\ell_{2,1}$ mixed norm regularizer is optimized using an

Accelerated Proximal Gradient (APG) method. The illustration of the spatio-temporal constraint-based sparse representation is shown in Fig. 1.

Generative Appearance Model Based on STSR. After learning the $\mathbf{Z}$, the observation likelihood of the tracking candidate i is defined as

$$p_g(\mathbf{y}_t|\mathbf{s}_t) = \frac{1}{\beta} \exp(-\alpha \sum_{k=1}^{K} ||\mathbf{x}_t^{i(k)} - \mathbf{D}^k \mathbf{z}_t^{i(k)}||_F^2) \tag{2}$$

where, $\mathbf{z}_t^{i(k)}$ is the coefficient of the k-th image patch of the i-th particle corresponding to the target templates, and α and β are normalization parameters.

2.2 Discriminative Appearance Model Based on MIL&SVM

Despite the robust performance of the generative appearance model achieved, it is not effective in dealing with the background distractions. Therefore, we introduce a discriminative appearance model based on MIL&SVM to improve the performance of our tracker.

We denote the overlapped image patches extracted from the target templates as the positive pathes p^+, and the overlapped patches extracted from the background (which is an annular region and the distance from the center-point of the target object to the edge of the negative patch sampling area is set to R) are denoted as negative patches p^-. As we all known, some positive patches obtained above may contain some noisy pixels from background because the bounding box is rectangular whereas the shape of the target may not be a standard rectangle. In order to deal with this problem, we adopt the patch-based MIL&SVM to train a robust classifier. In the training procedure, a row of patches are defined as a positive bag b^+ if they extracted from the target templates, or negative bag b^- if they come from background. The training procedure is illustrated in Fig. 2.

With this classifier, we can classify each patch of a candidate object at time t. For a candidate, we use r^+ to denote the local patches which are classified as positive and use r^- to denote patches classified as negative. Then the probability of a candidate being the tracking result can be defined as

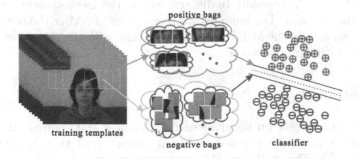

Fig. 2. Illustration for the patch-based MIL&SVM

$$p_d(\mathbf{y}_t|\mathbf{s}_t) = \frac{|r^+|}{|r^-| + |r^+|} \tag{3}$$

where $|r^+|$ and $|r^-|$ are the number of positive patches and negative patches.

Furthermore, according to the classification result, the occlusion state of a candidate can be obtained as

$$O = \frac{|r^-|}{|r^-| + |r^+|} \tag{4}$$

2.3   Adaptive Hybrid Generative-Discriminative Appearance Model

Based on the likelihood obtained from the spatio-temporal constraint-based sparse representation and the probability got via multiple-instance learning-based SVM, we construct our final observation model as:

$$p(\mathbf{y}_t|\mathbf{s}_t) = \eta p_g(\mathbf{y}_t|\mathbf{s}_t) + (1 - \eta)p_d(\mathbf{y}_t|\mathbf{s}_t) \tag{5}$$

where $\eta \in [0,1]$ is a control parameter, which can adjust weights of the two methods according to the occlusion state and can be defined as $\eta = \frac{1}{2}(1 + O)$.

In order to deal with appearance variation with time, we need to update our templates. We divide the templates $\mathbf{T}$ into two groups according to the occlusion state. The group without occlusion is denoted as $\mathbf{T}_{unocc} = [\mathbf{T}_1, ..., \mathbf{T}_{m_1}]$, and the occluded template set is denoted as $\mathbf{T}_{occ} = [\mathbf{T}_{m_1+1}, ..., \mathbf{T}_m]$, where m_1 is the number of unoccluded patches. The templates in $\mathbf{T}_{unocc}$ are ordered by time and the templates in $\mathbf{T}_{occ}$ are ordered reversely by time. We use two increasing interval sequences and a random number $r \in [0,2]$ to determine the sequence number of the template needed to be deleted as Eq. 6.

$$f(r) = \begin{cases} i, & r \in [\dfrac{(i-1)^2 + (i-1)}{m_1^2 + m_1}, \dfrac{i^2 + i}{m_1^2 + m_1}], & 0 < r \le 1 \\[3mm] j, & r \in [1 + \dfrac{(j-1)^2 + (j-1)}{m_2^2 + m_2}, 1 + \dfrac{j^2 + j}{m_2^2 + m_2}], & 1 < r \le 2 \end{cases} \tag{6}$$

where $m_2 = m - m_1$.

After selecting the template to discard, we use the method mentioned in [3] to update the template. For more detail, please refer [3]. After the templates $\mathbf{T}$ is updated, we retrain the MIL&SVM classifier only with the templates without occlusion or with light occlusion.

3   Experiments

We validate our tracker on six challenging sequences and compare it with six state-of-the-art methods proposed in recent years. All of these sequences are publicly available. The challenges of these sequences include severe occlusion and drastic shape deformation. In order to test the effectiveness and robustness of our

Fig. 3. Comparative experimental tracking results of 7 methods on six sequences, from top to bottom are *basketball*, *DavidOutdoor*, *girl_move*, *woman_sequence*, *face_sequence* and *girl_head*

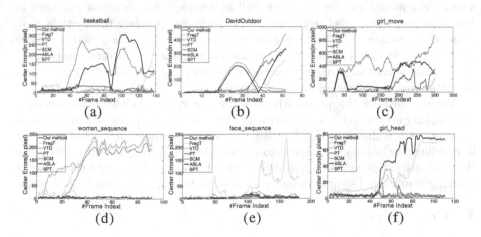

Fig. 4. Center error plots for 7 methods on six video sequences

tracker, we compare it with FragT [1], VTD [5], PT [8], SCM [11], ASLA [3] and SPT [7]. For our tracker, we set the number of templates $m = 10$, the number of local patches $K = 9$, the number of particles $n = 400$, and we use 2 previous tracking results in STSR. We resize all the targets or candidates as (32, 32).The size of the sampling patch is (16,16) and the sampling step is 8 pixels.

Table 1. Location errors (in pixel, the bold font indicates the best performance)

Sequences	FragT	VTD	PT	SCM	ASLA	SPT	Ours
basketball	16.3	9.0	19.1	126.4	112.6	17.4	**8.2**
DavidOutdoor	63.5	70.0	88.0	101.7	105.2	50.0	**8.5**
girl_move	8.9	45.4	110.2	414.8	214.5	30.0	**5.7**
woman_sequence	138.1	163.1	**3.8**	122.9	4.4	9.0	4.7
face_sequence	**4.4**	8.5	5.4	4.5	5.4	48.1	5.5
girl_head	3.6	7.2	3.1	3.3	38.0	30.0	**3.1**
Overall	39.1	50.5	38.3	128.9	80.0	30.8	**6.0**

Comparative tracking results of selected frames are shown in Fig. 3, from which we can find that our proposed tracker performs very well on all these challenging sequences. FragT is designed for dealing with occlusion and performs well in *face_sequence* and *girl_head* when the target is large enough, but it cannot get good results in other sequences when there exists sever occlusion in a small target. VTD adopts multi-trackers to track the target and it achieves satisfactory results in *face_sequence* and *basketball* but also shows less effective in dealing with the situation when there exists both rigid shape deformation and occlusion. PT is a part-based tracker and it performs well in dealing with partial occlusion, but it fails when the target is full occluded. Both SCM and ASLS adopt sparsity-based appearance model and they perform well in dealing with occlusion as shown in *face_sequence*, but cannot get satisfactory performance when there exists rigid shape deformation. SPT achieves good results on *DavidOutdoor* and *girl_move* as shown in Fig. 3, but cannot obtain stable performance in clutter scene or when there exists severe and frequent occlusion as shown in screenshots of sequences *basketball*, *woman_sequence* and *face_sequence*.

We also measure the quantitative tracking error, the Euclidean distance from the tracking center to the ground-truth. The center error plots of 7 methods on 6 sequences are shown in Fig. 4, which demonstrates that our tracker is robust in handling occlusion and shape deformation even in a complex scene. We show the location errors in Table 1, which shows that our tracker achieves the best tracking results on 4 sequences and gives the the best tracking result on average.

4   Conclusion

In this paper, we have proposed a novel patch-based tracking method based on the combination of spatio-temporal constraint-based sparse representation

(STSR) and multiple-instance learning-based SVM (MIL&SVM). By utilizing the STSR, our tracker effectively captures the structure cues of the target and the temporal similarity in consecutive frames. Furthermore, we utilize MIL&SVM as our discriminative appearance model, which is robust in cluttered background and can predict the occlusion state. Based on the occlusion state, we update the template set separately, making the generative method obtain more precise templates and the discriminative method maintain correctness. Qualitative and quantitative experimental results on different challenging sequences demonstrate that our tracker is very robust to the occlusion.

Acknowledgments. This work is supported by the National Natural Science Foundation of China (No. 61175096 and 61273273), Specialized Fund for Joint Building Program of Beijing municipal Education Commission.

References

1. Adam, A., Rivlin, E., Shimshoni, I.: Robust fragments-based tracking using the integral histogram. In: 2006 IEEE Computer Society Conference on Computer Vision and Pattern Recognition, vol. 1, pp. 798–805, IEEE (2006)
2. Babenko, B., Yang, M.H., Belongie, S.: Visual tracking with online multiple instance learning. In: IEEE Conference on Computer Vision and Pattern Recognition, CVPR 2009, pp. 983–990, IEEE (2009)
3. Jia, X., Lu, H., Yang, M.H.: Visual tracking via adaptive structural local sparse appearance model. In: 2012 IEEE Conference on Computer Vision and Pattern Recognition (CVPR), pp. 1822–1829, IEEE (2012)
4. Kalal, Z., Matas, J., Mikolajczyk, K.: PN learning: bootstrapping binary classifiers by structural constraints. In: 2010 IEEE Conference on Computer Vision and Pattern Recognition (CVPR), pp. 49–56, IEEE (2010)
5. Kwon, J., Lee, K.M.: Visual tracking decomposition. In: 2010 IEEE Conference on Computer Vision and Pattern Recognition (CVPR), pp. 1269–1276, IEEE (2010)
6. Li, X., He, Z., You, X., Chen, C.P.: A novel joint tracker based on occlusion detection. Knowl. Based Syst. **71**, 409–418 (2014)
7. Yang, F., Lu, H., Yang, M.H.: Robust superpixel tracking. IEEE Trans. Image Process. **23**(4), 1639–1651 (2014)
8. Yao, R., Shi, Q., Shen, C., Zhang, Y., van den Hengel, A.: Part-based visual tracking with online latent structural learning. In: 2013 IEEE Conference on Computer Vision and Pattern Recognition (CVPR), pp. 2363–2370, IEEE (2013)
9. Zarezade, A., Rabiee, H., Soltani-Farani, A., et al.: Patchwise joint sparse tracking with occlusion detection. IEEE Trans. Image Process. **23**(10), 4496–4510 (2014)
10. Zhang, T., Liu, S., Xu, C., Yan, S., Ghanem, B., Ahuja, N., Yang, M.H.: Structural sparse tracking. In: Proceedings of the IEEE Conference on Computer Vision and Pattern Recognition, pp. 150–158 (2015)
11. Zhong, W., Yang, M., et al.: Robust object tracking via sparse collaborative appearance model. IEEE Trans. Image Process. **23**(5), 2356–2368 (2014)

An Autonomous Mobile Robot with Functions of Action Learning, Memorizing, Recall and Identifying the Environment Using Gaussian Mixture Model

Masanao Obayashi[1(✉)], Taiki Yamane[1], Takashi Kuremoto[1],
Shingo Mabu[1], and Kunikazu Kobayashi[2]

[1] Yamaguchi University, 2-16-1 Tokiwadai, Ube, Yamaguchi, 755-8611, Japan
`{m.obayas,s050vk,wu,mabu}@yamaguchi-u.ac.jp`
[2] Aichi Prefectural University, Nagakute, Aichi, 480-1198, Japan
`kobayashi@ist.aichi-pu.ac.jp`

Abstract. In this paper, behavior scheme of autonomous mobile robots to achieve the objectives of them in environments are proposed, having function of identifying the current environment in which they are placed and making use of learning, memorizing and recalling behaviors of corresponding to each of plural different environments. Specifically, each robot has the function of identifying the environment using some behavioral statistical data for each environment, and if the robot has already experienced the environment, it behaves by making use of own experienced data stored in the database, otherwise it performs a new behavior learning and adds the learning results into the database.

Keywords: Intelligent robot · Chaotic neural network · Reinforcement learning · Identification of the environment · Gaussian mixture model

1 Introduction

In recent years, researches on intelligent robots and their practical application have been attracting attention. For example, meal transport robot, workplace patrol robot, document delivery robot, etc. and their applications have expanded dramatically.

However, such robots are not called so intelligent, because that almost all of them are given the path to the place of destination and only move along the same path repeatedly. Furthermore, in the field of discrimination of the environment for agent action learning, we can hardly see papers about such a subject. As previous researches on identification method of the environment, there are a few ones that are our papers published in the past, for example, the method [1] that check in order the environments in the database whether the series of actions in the environment information in the database are appropriate for the current environment or not. The other is the method [2] identifying the environment using the feedback SOM with high efficiency and high precision. In this paper, behavior scheme of autonomous mobile robots to achieve the objectives of them in environments are proposed, having functions of identifying the current environment in which they are placed and making use of learning and

© Springer International Publishing Switzerland 2015
S. Arik et al. (Eds.): ICONIP 2015, Part I, LNCS 9489, pp. 272–282, 2015.
DOI: 10.1007/978-3-319-26532-2_30

memorizing behaviors of corresponding to each of plural different environments. Specifically, each robot has the function of identifying the environment using some behavioral statistical data for each environment, and if the robot has already experienced the environment, it behaves by making use of own experienced data memorized in the database constructed by chaotic neural networks (CNNs) and Gaussian mixture models, otherwise it performs a new action learning and adds them into the database. In this paper, as the database to perform the memorization and recall of pairs of action and perceptual information corresponding to the action, we also adopt chaotic neural networks (CNNs) [3] which is a coupled system of chaotic neuron that models the refractory and analog input and output characteristics, as an action learning method, we adopt reinforcement learning, a kind of representative machine learning.

2 Whole Structure of the System

The whole structure of the system including the proposed agent and environment are shown in Fig. 1. The structure of the agent is inside surrounded by a broken line, it consists of three functions: identification of the environment, action learning and data-base with pairs of state of the environment and action corresponding to it. At first, the agent receives state information from the environment in which the agent is placed, it identifies the environment in the database as same as the facing environment, in other word, the agent investigates the database whether there is the environment that matches the facing environment in it, or not. If the information about the facing environment has already memorized in the environment, it behaves by making use of own experienced data memorizing in the database, otherwise it performs a new action learning and adds the results of action learning into the database.

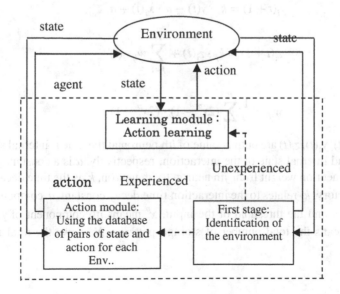

Fig. 1. Whole structure of the proposed autonomous agent

3 Action Learning

In this paper, as an action learning approach, Q learning method is used, which is one of typical reinforcement learning method. For more information, see reference [4]. After end of action learning, pairs of action and perceptual information corresponding to the action and Gaussian mixture model (GMM) about the environment are added as labeled information into the database which consisting of CNNs and GMMs.

4 State-Action Storage System

After end of action learning, pairs of useful state and action sequences by trial and error learning are stored in the state-action storage unit, that is, in the database. There are two kinds of memory; short term memory and long term memory. In this paper, the short-term memory corresponds to the transient memory when an action learning, while long-term memory does to the memory gotten at the end of action learning, which is transferred to the state-action storage unit called database. CNNs are adopted as the long-term memory. We describe CNN below shortly.

4.1 Chaotic Neural Network (CNN)

CNN is configured by interconnecting chaotic neuron models, represented by the following formula [3].

$$x_i(t+1) = f(y_i(t+1) + z_i(t+1)) \tag{1}$$

$$y_i(t+1) = k_r \cdot y(t) - \alpha \cdot x_i(t) + a_i \tag{2}$$

$$z_i(t+1) = k_f \cdot z_i(t) + \sum_{j=1}^{N} w_{ij} x_j(t) \tag{3}$$

$$w_{ij} = \frac{1}{P} \sum_{p=1}^{P} (2x_i^p - 1) \cdot (2x_j^p - 1) \tag{4}$$

Here, $x_i(t), y_i(t), z_i(t)$ are output value of ith neuron at the time t, internal state on the refractory and internal state of the interaction, respectively. α is a constant parameter, w_{ij} is the connection weight from jth neuron to ith neuron, k_r is the time decay constant for the refractory, k_f relates to the interaction time decay constant, a_i compound term of external input and the threshold of the input, x_i^p means ith component of pth storage pattern, P means the total number of storage patterns, N shows the total number of neurons.

4.2   Memorizing and Recall of an Environmental Data Using Mutual Associative Type CNN

While our mutual associative CNN (MACNN) has same structure of auto-associative CNN (AACNN) and MACNN operates same as AACNNs, it looks working as mutual associative memory model. Figure 2 shows the structure of the MACNN, consisting of three parts; input, hidden and output part. In the hidden part of neurons, random data are stored in order to make correlations between input data and output data weaken. When pairs of both environment and learned action corresponding to the environment gotten by the action learning are memorized into MACNN, the perceptual patterns are memorized into the input part of neurons and action patterns are memorized into the output part of the neurons using Eq. (4). When making use of the MACNN, the current environment perceptual pattern is set as the output (x in Sect. 4.1) of the input part of neurons, fixing their values, let MACNN operate, after the MACNN converges, a corresponding action pattern is recalled on the output part of neurons. Please refer to chapter 11 in the reference [4] for detail explanation.

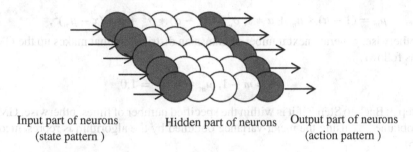

Input part of neurons   Hidden part of neurons   Output part of neurons
   (state pattern)   (action pattern)

Fig. 2. Structure of mutual associative type CNN (MACNN)

5   Identification of the Environment

When being stored the features of plural different environments into the database, environmental state patterns of first several random steps from the start position are gotten many times for each environment, and statistical information for these patterns is memorized by Gaussian mixture model (GMM). On the other side, when identifying the facing environment, its GMM is gotten in the same manner. Comparing the GMM in the database and the GMM for the facing environment, we judge whether the agent has already experienced the facing environment or not. Details are described below.

5.1   Identification Method of the Environment

In this section, we describe the method to identify the environment. We propose the identification method using GMM. GMM is represented by mean μ and variance σ^2. GMM is used as a distribution function for identification of the environment. GMM is built up using the amount of observation data s ($\in R^1$) resulting from (T × X) times actions where X means the number of trials, one trial is T step random actions from start position. In 5.1.1, building up algorithm of GMM is explained.

5.1.1 Mean and Variance Configuration Algorithm

The algorithm for determining the mean and variance of constructing a GMM is shown below.

Step 1: (Initialize) Using the first observation state data s ($\in R^1$), mean and variance of the first Gaussian distribution $f_1(s)$ are set as follows,

$$\mu_m = s, \sigma_m^2 = 1.0 \ (m = 1)$$

Here, m is number of Gaussian distribution $f_m(s)$ that makes up the MMG.

Step 2: Modify the mean and variance using the next data s as follows,

(1) when $-3\sigma_m < \min | \mu_m - s | < 3\sigma_m$,

$$\mu_m = (1 - \alpha) * \mu_m + \alpha * s, \ \sigma_m^2 = (1 - \alpha) * \sigma_m^2 + \alpha * (x - \mu_m)^2, \tag{5}$$

(2) otherwise, generate next number of Gaussian distribution that makes up the GMM as follows,

$$m = m + 1, \ \mu_m = s, \ \sigma_m = 1.0$$

Step 3: Back to Step 2 if it is within the specified number of times, otherwise, GMM, $f(s)$, obtained by using the mean-variance obtained by the algorithm is represented by Eq. (6).

$$f(s) = \sum_{m=1}^{M} \frac{w_m}{K} * f_m(s) \tag{6}$$

$$f_m(s) = \frac{1}{\sqrt{2\pi\sigma_m^2}} \exp(-\frac{(s - \mu_m)^2}{2\sigma_m^2}) \tag{7}$$

Here, K: the number of data, w_m: the number of data used during mth Gaussian distribution creation, M is the number of Gaussian distributions.

5.1.2 Identification Algorithm of the Environment

At first, the feature vector for each environment in the database to identify the environment are defined as U-dimensional vector consisting of U-dimensional probability values calculated using GMM (Eq. (6)) by observation resulting from T times random actions from start position in the current environment. Here U is the number of different kinds of the states within the T observation states. We identify the environment by the degree of similarity between each of all the feature vectors in the database and the current probability vector. The algorithm is as follows,

Step 1: Using the observation state vector, calculate all the feature vectors corresponding to each environment presented by GMM in the database and the current probability vector for the observation state vector.

Step 2: Calculate the degrees of similarity $\cos \theta$ between the feature vector $\vec{a}$ and the current probability vector $\vec{b}$, here θ is angle between the feature vector and the current probability vector.

$$\cos \theta = \frac{\vec{a} \cdot \vec{b}}{|\vec{a}| \cdot |\vec{b}|} = \frac{\sum\limits_{i=1}^{U} (a_i \cdot b_i)}{\sqrt{\sum\limits_{i=1}^{U} (a_i)^2} \sqrt{\sum\limits_{i=1}^{U} (b_i)^2}} \tag{8}$$

Here, $\vec{a} = (a_1, \cdots, a_U)$, $\vec{b} = (b_1, \cdots, b_U)$, $l = \arg \max\limits_{j} \theta_j$, $j \in L$. Here, L is the number of the storage environments in the database, that is, they have already learned and stored.

(i) When $\theta_l \leq \theta_{th}$, the agent judges that the storage environment l is most similar to the current environment, and pairs of stored state and action corresponding to the environment l is determined to be available, moves to action module. (θ_{th} is the threshold of the similarity of the environment.)

(ii) When $\theta_l > \theta_{th}$, the agent judges that there is no storage environments to be available and moves to action learning module.

6   Computer Simulation

In the maze problem, a performance comparison of the proposed method and conventional method [2] through the following two simulations is carried out. The proposed method identifies the mazes using Gaussian mixture models, and also the conventional method identifies using feedback SOMs. As data for construction of identification systems and identification of unknown mazes for each maze, when first 5 actions (T = 5 in Sect. 5.1) from the start position is called one trial, we get the 150 perceptional patterns through 30 trials (X = 30).

6.1   Preparation

Put the following assumptions at GMM construction of the environment.

(i)   GMM is constructed by the data during several actions from the start position in the maze.
(ii)   The agent acts T steps randomly without use of the storage information.
(iii) The agent stays in the same position when colliding with the obstacle.
(iv) It is possible for the agent to take only one of next three actions: forward, left and right without backward per a step.

In this simulation, it is assumed that the agent is able to perceive an aisle or an obstacle toward the five directions up to one square away around it as shown in Fig. 3.

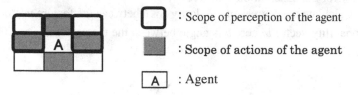

Fig. 3. Scope of perception and action of the agent

6.2   Observation State Description

The feature extraction algorithm of the environment is as follows,

Step 1: Put the agent at the start position.

Step 2: Repeat Step 3 to Step 4 T (number of actions) times.

Step 3: Move one cell to random direction and get the observation information according to the following equation,

$$evi \ (\text{direction}) = \begin{cases} 1: & \text{adjacent is obstacle,} \\ 0: & \text{adjacent is aisle,} \end{cases} \tag{9}$$

evi (direction): observation information toward the observation direction.

Step 4: Get a feature value, a perceptual pattern information named "state", for each step according to the following equation,

$$\begin{aligned} \text{state} = evi \ (\text{right}) + 2 \cdot evi \ (\text{front right}) + 4 \cdot evi \ (\text{front right}) \\ + 8 \cdot evi \ (\text{left}) + 16 \cdot evi \ (\text{front left}). \end{aligned} \tag{10}$$

Here, state ($0 \le state \le 31$) represents one of the situations ($=2^5$) around the agent (See Fig. 3).

6.3   Simulation 1

We consider multiple maze problems. The robot has perception-action capabilities as shown in Fig. 3, assuming that the robot always faces the upward direction and is able to move only one cell in an action. After end of action learning for each maze, pairs of perceptual pattern and action along optimal path from the start to the goal are stored into the database created by the CNNs group. In this simulation, after one of the stored mazes is given to the robot, we investigate whether the maze is correctly identified or not. The predefined number of mazes are generated by the following procedure.

(i)   The size of a maze is randomly determined

(ii)   A start and goal positions in the maze are randomly decided.

(iii) Walls in the maze are randomly placed

Here, white cell is aisle, black cell is wall, S is a start position, G indicates a goal. Among the mazes created by above procedure, mazes that can't reach the goal, or have shorter path less than T steps are excluded. As a result, the discrimination rates in the conventional method and the proposed method obtained in the simulation are shown in Table 1. It shows the average discrimination rate of 10 trials (the number of available mazes per trial is 25). Table 1 says our method is better than the conventional one (Fig. 4).

Table 1. Results of simulation 1

	The conventional method	The proposed method
Discrimination rate	65.20 [percent]	71.40 [percent]

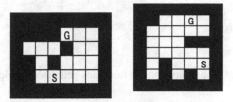

Fig. 4 Failure examples of identification using the conventional method

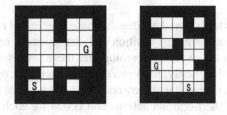

Fig. 5 Left: success example of identification, right: failure example of identification using the proposed method

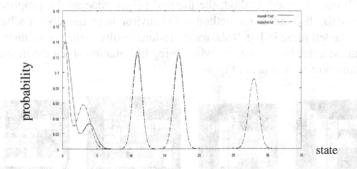

Fig. 6. Gaussian mixture probability density functions generated by the left maze in Fig. 5 (two times construction)

Figures 6 and 7 show Gaussian mixture probability density functions generated by left maze and right maze in Fig. 5, respectively. These are constructed by the proposed method two times in the same manner. The former is success example and the latter is failure example. The former case looks less difference between two probability density functions than the latter one.

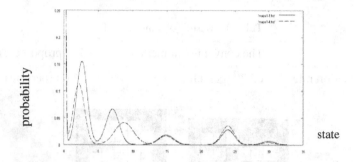

Fig. 7. Gaussian mixture probability density functions generated by right maze in Fig. 5 (two times construction)

6.4 Simulation 2

Figure 8 shows 4 mazes to be stored into the database after end of action learning. White rhombus and black one mean a start position and goal position, respectively. Triangles means the optimal path at the end of learning and these results are memorized in the database. In this simulation, firstly, for 4 mazes shown in Fig. 8, optimal paths from the start to the goal through Q learning are learned. Next, after end of learning both pairs of perceptual patterns and their optimal actions and GMM for each maze constructed by the proposed method are memorized into the database. Lastly, we try to solve a little large-size maze shown in Fig. 9, while making use of these information of 4 mazes in the database of the agent with much less of additional action learning. As comparing the method with our proposed method, the method [2] as same as the proposed system without use of the discrimination method of the environment, namely "feedback SOM" is adopted. The left maze in Fig. 9 shows the failure results of the conventional method because that the agent behaved along while using information about the incorrect maze (b) instead of correct maze (a) in Fig. 8.

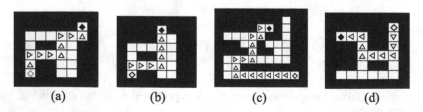

(a)	(b)	(c)	(d)

Fig. 8. The 4 optimal paths to be stored into the database after action learning

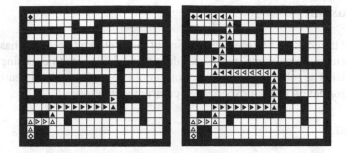

Fig. 9. Simulation result (left: for the conventional method, right: for the proposed method), making use of storage information about mazes in Fig. 8

Figure 9 shows Simulation result (left: for the conventional method, right: for the proposed method) for a little large-scale maze, making use of storage information about mazes in Fig. 8. In Fig. 9, the white 5 (=T) triangles mean the action series of the agent when identifying the environment and black ones mean the action series by making use of the stored information. At the start position, the proposed method correctly identified the facing environment as maze (a) and it succeeds to get the goal. However, the conventional method failed to identify the facing the maze as maze (b) in shown Fig. 8.

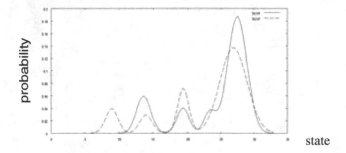

Fig. 10. Gaussian mixture probability density function obtained using maze (a) and (b) in Fig. 8 (maze (a): solid line, maze (b): broken line)

Figure 10 shows two Gaussian mixture probability density functions for maze (a) and (b) in Fig. 8. It says that they are very similar. Table 2 shows similarities between the maze (a) and (b) in Fig. 8 and the maze of Fig. 9 for the first T steps from the start position, respectively. The table shows the maze (a) is more similar than maze (b) for the maze (9) at the start point.

Table 2. Similarities between maze (a), (b) in Fig. 8 and the first T step action from the start position of maze of Fig. 9.

maze (a)	maze (b)
0.988	0.856

7   Conclusion

We proposed the construction method of a kind of the intelligent agent that has functions of action learning, memorizing, recall and identifying the environment using Gaussian mixture model. The robot with these functions could go to the goal efficiently, making use of the information of their experienced environment before.

References

1. Obayashi, M., Narita, K., Kuremoto, T., Kobayashi, K.: A reinforcement learning system with chaotic neural networks-based adaptive hierarchical memory structure for autonomous robots. In: Proceedings of International Conference on Control, Automation and Systems 2008 (ICCAS 2008), pp. 69–74 (2008)
2. Obayashi, M., Narita, K., Okamoto, Y., Kuremoto, T., Kobayashi, K., Feng, L.: A reinforcement learning system embedded agent with neural network-based adaptive hierarchical memory structure. In: Advances in Reinforcement Learning, chapter 11, pp.189–208, IN-TECH (2011)
3. Adachi, M., Aihara, K.: Associative dynamics in a chaotic neural network. Neural Netw. **10**(1), 83–98 (1997)
4. Sutton, R.S., Barto, A.G.: Reinforcement Learning: An Introduction. MIT PRESS, Cambridge (1988)

SEIR Immune Strategy for Instance Weighted Naive Bayes Classification

Shan Xue[1,2](✉), Jie Lu[1], Guangquan Zhang[1], and Li Xiong[2]

[1] Lab of Decision Systems & e-Service Intelligence (DeSI),
Centre for Quantum Computation & Intelligent Systems (QCIS),
Faculty of Engineering and Information Technology,
University of Technology Sydney, Ultimo, NSW 2007, Australia
shan.xue@student.uts.edu.au, {jie.lu,guangquan.zhang}@uts.edu.au
[2] School of Management, Shanghai University, Shanghai 200436, China
xiongli8@shu.edu.cn

Abstract. Naive Bayes (NB) has been popularly applied in many classification tasks. However, in real-world applications, the pronounced advantage of NB is often challenged by insufficient training samples. Specifically, the high variance may occur with respect to the limited number of training samples. The estimated class distribution of a NB classier is inaccurate if the number of training instances is small. To handle this issue, in this paper, we proposed a SEIR (Susceptible, Exposed, Infectious and Recovered) immune-strategy-based instance weighting algorithm for naive Bayes classification, namely SWNB. The immune instance weighting allows the SWNB algorithm adjust itself to the data without explicit specification of functional or distributional forms of the underlying model. Experiments and comparisons on 20 benchmark datasets demonstrated that the proposed SWNB algorithm outperformed existing state-of-the-art instance weighted NB algorithm and other related computational intelligence methods.

Keywords: Naive bayes · Classification · Immune strategy · SEIR

1   Introduction

As a special case of a Bayesian network, Naive Bayes (NB) [3] has been popularly applied in many real-world learning tasks, such as text classification [7,17], web mining [19] and other computational approach [10]. Specifically, the high variance may occur with respect to the limited number of training samples [6,14], where the estimated class distribution of a NB classifier is inaccurate if the number of training instances is small.

To address this research problem, instance weighted naive Bayes (IWNB), as an effective solution, assign different weight values to instances for probability value estimation can improved the performance of NB. For example, an instance-cloned naive Bayes, which produces an expanded training set by cloning some training instances based on their similarities to the test instance

S. Arik et al. (Eds.): ICONIP 2015, Part I, LNCS 9489, pp. 283–292, 2015.
DOI: 10.1007/978-3-319-26532-2_31

is proposed in [5]. Moreover, Jiang [4] proposed to use instance weighting to improve the performance of Averaged One-Dependence Estimators [12], which is another Bayesian model. For this type of weighting method, each training instance is eagerly weighted according to the similarity with the "model" of training dataset. These instance weight setting methods have achieved good performance to solve domain specific problems [15]. However, for all these methods, the instance weights are determined without taking the NB objective function into consideration and the underlying sample distributions should be known in advance for the former approaches.

In this paper we propose a SEIR Immune based algorithm, which automatically calculates the optimal instance weight values for IWNB, by directly working on IWNB's objective function based on SEIR immune strategy. Specifically, our method uses SEIR procedures to design an automated search strategy to find optimal instance weight for each dataset. The SEIR immune strategy, including initialization, clone, mutation and selection, ensures that our method can adjust itself to the data without any explicit specification of functional or distributional form for the underlying model.

In contrast to the conventional statistical probabilistic evaluation in NB, the SWNB algorithm is a self-learning algorithm by utilizing the immunological properties, such as memory property and clonal selection. The advantages of SWNB can be understood from the following three aspects: SWNB is a data-driven self-adaptive method because it does not requires explicit specification of functional or distributional form for the underlying model. The SWNB algorithm is a nonlinear model and is flexible in modeling complex real world relationships. It inherits the memory property of human immune mechanism and can recognize the same or similar antigen quickly at different times.

The rest of the paper is organized as follows. In Sect. 2, we present a new SEIR immune strategy and our SWNB algorithm. In Sect. 3, we describe the experimental conditions, process, and results in details. Section 4 concludes the paper and outlines several directions for future study.

2   SEIR Immune Strategy Based Instance Weighted Bayes

2.1   Instance Weighted Naive Bayes

Given a training set $\mathcal{D} = \{\mathbf{x_i}\}$ with N instances, each instance contains n attribute values and corresponds a class label. We use $\mathbf{x}_i = \{x_{i1}, \cdots, x_{ij}, \cdots, x_{in}, y_i\}$ to stand for the ith instance, with x_{ij} denoting the jth attribute value and y_i denoting the class label of the instance. Meanwhile, the class space $\mathcal{Y} = \{c_1, \cdots, c_k, \cdots, c_L\}$ denotes the set of labels that each instance belongs to and c_k denotes the kth label of the class space. We use $\mathbf{a} = \{a_j\}$ to include all the attributes of all instances, with a_j representing the jth attribute.

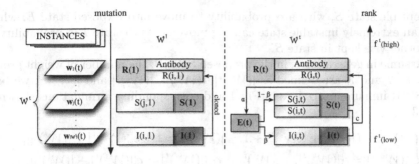

Fig. 1. The SEIR immune procedure for IWNB

For a training instance $\mathbf{x}_i \in \mathcal{D}$ with its class label satisfies $y_i \in \mathcal{Y}$, the training method is based on the IWNB model, which is formally defined as

$$c(\mathbf{x}_i) = \arg\max_{c_k \in \mathcal{Y}} P_{\mathbf{w}}(c_k) \prod_{j=1}^{n} P_{\mathbf{w}}(x_{ij}|c_k). \tag{1}$$

In Eq. (1), $P_{\mathbf{w}}(c_k)$ denotes the probability of class c_k with a certain weights set $\mathbf{w} = \{w_i, i \in [1, \cdots, N]\}$. $P_{\mathbf{w}}(x_{tj}|c_k)$ denotes the joint distribution of $\mathbf{x}_i$ conditioned by the given class c_k based on $\mathbf{w}$.

In this paper, we focus on the calculation of the priori probability $P_{\mathbf{w}}(c_k)$ and the conditional probability $P_{\mathbf{w}}(x_{ij}|c_k)$ by using optimal instance weight value $\mathbf{w}$. The Laplace-estimate instance weighted strategy is introduced and shown in Eq. (2).

$$\begin{cases} P_{\mathbf{w}}(c_k) = \dfrac{n_k + 1}{N_k + L}, \quad P_{\mathbf{w}}(x_{ij}|c_k) = \dfrac{n_k^{ij} + 1}{n_k + |a_j|}, \\[2mm] n_k = \displaystyle\sum_{\mathbf{x}_i \in \mathcal{D}, y_i = c_k} w_i, N_k = \displaystyle\sum_{\mathbf{x}_i \in \mathcal{D}} w_i, \quad n_k^{ij} = \displaystyle\sum_{\mathbf{x}_i \in \mathcal{D}, y_j = c_k, x_{pq} \in \mathbf{x}_p \in \mathcal{D}, x_{pq} = x_{ij}} w_p \end{cases} \tag{2}$$

where $|a_j|$ is the number of distinct values of attribute a_j, and L is the number of classes.

2.2   SEIR Immune Strategy for Instance Weighting

The classic model for microparasite dynamics is the flow of hosts between Susceptible (S), Exposed (E), Infectious (I) and Recovered (R) compartments [20]. However, the SEIR immune strategy could not directly applied for the IWNB classification problem. In this paper, we introduce SEIR model to describe the immune strategy proposed for instance weight (*i.e.*, antibody) optimization. By rating the performance of instance weights, the low scorekeepers are defined in Infectious state I, which are planned to be cloned with a set clone factor c and then move to Recovered state R. Through mutation, weights are driven into

Susceptible state S, with a α probability to move into Exposed state E, where is in an extremely unstable state causing movements to state I in probability β, and otherwise kept in state S.

Assume in generation t, in the proposed SWNB, the number of weight groups in S, E, I and R are defined as $|(\mathcal{W}^s)^t|$, $|(\mathcal{W}^e)^t|$, $|(\mathcal{W}^i)^t|$ and $|(\mathcal{W}^r)^t|$, we have the SEIR immune procedure in Fig. 1 and corresponding formulations shown in Eq. (3).

$$\begin{cases} |\mathcal{W}^t| = |(\mathcal{W}^s)^t| + |(\mathcal{W}^i)^t| + 1, \quad |(\mathcal{W}^r)^t| = 1 + c[|(\mathcal{W}^s)^t| + |(\mathcal{W}^i)^t|], \\ |(\mathcal{W}^e)^t| = \alpha|(\mathcal{W}^s)^t|, \quad |(\mathcal{W}^i)^t| = \alpha\beta|(\mathcal{W}^s)^t| = c[|(\mathcal{W}^s)^t| + |(\mathcal{W}^i)^t|], \quad (3) \\ |(\mathcal{W}^{s(j)})^t| = \alpha(1-\beta)|(\mathcal{W}^s)^t|, \quad |\mathcal{W}^t| = |(\mathcal{W}^{s(j)})^t| + |(\mathcal{W}^i)^t| + 1 \end{cases}$$

where $|\mathcal{W}^t|$ is the total amount of weight groups in the tth immune generation. From Eq. (3), we have $\alpha = \frac{1}{1-c}$, $\beta = c$.

2.3   SWNB Classifier

In this section, we first introduce some important notations and definitions, then propose our solutions.

Definition 1 (Calculation of Affinity Function). The affinity of the jth individual in the tth generation $\mathbf{w}_j^t$ is the classification accuracy that is obtained by SWNB using the $\mathbf{w}_j^t$ to carry out the probability estimation. The calculation of affinity function is defined as

$$f(\mathbf{w}_j^t) = \frac{1}{N^S} \sum_{i=1}^{N^S} s(c^t(\mathbf{x}_i), y_i), \qquad (4)$$

where $c^t(\mathbf{x}_i)$ is the classification result of each ith instance in the tth training generation using the SWNB classifier based on each individual $\mathbf{w}_j^t$, $i \in [1, \cdots, N]$. $s(c^t(\mathbf{x}_i), y_i)$ defines the similarity between $c^t(\mathbf{x}_i)$ and y_i, where $s(c^t(\mathbf{x}_i), y_i)$ is 1 if $c^t(\mathbf{x}_i) = y_i$ and 0 otherwise.

Definition 2 (Antibody Clone). We select $\mathbf{w}_r^t$ as the antibody of the tth generation with the best affinity performance of $f(\mathbf{w}_j^t)$ sorting from $\mathcal{W}^t$. After that, we use the antibody to replace the weight groups $(\mathcal{W}^i)^t$ with low affinity according to the same rate of a set clone factor c. As a result, $\mathbf{w}_i^t \in (\mathcal{W}^i)^t$ are cloned by $\mathbf{w}_r^t$ and move to $(\mathcal{W}^r)^t$.

Definition 3 (Antibody Mutation). The mutate operation is used to treat all individuals in the tth population $\mathcal{W}^t$ and for training preparation on $\mathcal{W}^{t+1}$ in the $t+1$th generation. For any individual $\mathbf{w}_j^t$, the new variation $\mathbf{v}_j^{t+1}$ is generated as

$$\mathbf{v}_j^{t+1} = \mathbf{w}_j^t + [1 - f(\mathbf{w}_j^t)] * N(0,1) * (\mathbf{w}_r^t - \mathbf{w}_j^t), \qquad (5)$$

where $N(0,1)$ is a normally distributed random variable within the range of $[0,1]$, and

$$\mathbf{w}_j^{t+1} = \mathbf{v}_j^{t+1}, \ f(\mathbf{v}_j^{t+1}) > f(\mathbf{w}_j^t); \quad or \quad \mathbf{w}_j^t, \ f(\mathbf{v}_j^{t+1}) \leq f(\mathbf{w}_j^t). \qquad (6)$$

Algorithm 1. SWNB Classifier

Input:

    Training dataset $\mathcal{D}$; Clone Factor c; Threshold ε; Maximum Evolution Generation $MaxGen$.

Output:

    The target class label $c(\mathbf{x}_t)$ of a test instance $\mathbf{x}_t$.

1: $\mathcal{W}^0 \leftarrow$ Initial population, which is a set of random number distributed between $(0, 1]$.

2: **while** $t \leq MaxGen$ and $f(\mathbf{w}_c^{t+1}) - f(\mathbf{w}_c^t) \leq \varepsilon$ **do**

3:     $f(\mathbf{w}_j^t) \leftarrow$ Apply Eq. (4) to calculate affinity of $\mathbf{w}_j^t$ for the tth generation.

4:     $\mathbf{w}_r^t \leftarrow$ The selected antibody with highest $f(\mathbf{w}_j^t)$.

5:     $\mathcal{W}^t \leftarrow$ The population of the tth generation.

6:     $(\mathcal{W}^i)^t \leftarrow$ The individual sets with low $f(\mathbf{w}_j^t)$ selected as $I(t)$ on clone factor c within $\mathcal{W}^t$ but excluding $\mathbf{w}_r^t$.

7:     $(\mathcal{W}^r)^t \leftarrow$ The $R(t)$ individuals consist of the antibody and the cloned $(\mathcal{W}^i)^t$ by $\mathbf{w}_r^t$.

8:     **for all** each $\mathbf{w}_j^t$ in $\mathcal{W}^t$ **do**

9:         $\mathbf{v}_j^{t+1} \leftarrow$ Apply Eq. (5) to do mutation on each $\mathbf{w}_j^t$.

10:        $\mathbf{w}_j^{t+1} \leftarrow$ Apply Eq. (6) to obtain the new weights in the $t + 1$th generation.

11:     **end for**

12:     $(\mathcal{W}^s)^{t+1} \leftarrow$ The $S(t+1)$ individuals.

13:     $(\mathcal{W}^e)^{t+1} \leftarrow$ The $E(t+1)$ individuals contribute to $(\mathcal{W}^i)^{t+1}$ on β and $(\mathcal{W}^{s(j)})^{t+1}$ on $1 - \beta$.

14:     $\mathcal{W}^{t+1} \leftarrow (\mathcal{W}^s)^{t+1} \cup (\mathcal{W}^e)^{t+1} \cup (\mathcal{W}^i)^{t+1} \cup (\mathcal{W}^r)^{t+1}$.

15: **end while**

16: $c(\mathbf{x}_t) \leftarrow$ Apply $\mathbf{w}_t$ to instance $\mathbf{x}_t$ to predict the underlying class label via Eqs. (1) and (2).

---

3   Experiments

3.1   Experimental Conditions and Baselines

We validate the performance of the proposed method on 20 benchmark datasets from UCI data repository [2]. Because naive Bayes based classifiers are designed for categorical attributes, in our experiments, we first replace all missing attribute values using unsupervised attribute filter *ReplaceMissingValues* in WEKA [13]. Then, we apply unsupervised filter *Discretize* in WEKA to discretize numeric attributes into nominal attributes. In our experiments, the algorithms are evaluated in terms of classification accuracy via 10 runs of 10-fold cross validation. Besides, the three parameters maximum iteration *MaxGen*, threshold ε, and the clone factor c in Algorithm 1 are set to 50, 0.001 and 0.1 respectively. Moreover, all experiments are conducted on a Linux cluster node with an Intel(R) Xeon(R) @3.33 GHZ CPU and 3 GB fixed memory size.

    For comparison purposes, we use the following baseline algorithms in our experiments.

(1) NB: The standard naive Bayes classifier with conditional attribute independence assumption [3].

Table 1. Experimental results for SWNB vs. baselines: classification accuracy %

Dataset	SWNB	IWNB [4]	NB [3]	SBC [9]	C4.4 [11]	KNN [1]
Anneal	**97.04**	95.93	93.70	91.48	80.37	92.96
Anneal.ORIG	**91.11**	88.52	90.37	84.07	88.89	84.07
Balance-scale	**92.91**	89.84	89.84	89.84	64.17	83.42
Breast-cancer	**81.76**	71.76	70.59	70.59	69.41	75.29
Colic.ORIG	**84.09**	74.55	79.09	71.82	73.64	70.00
Credit-a	**88.54**	84.54	84.54	85.02	85.02	84.54
Credit-g	**78.67**	74.67	74.33	72.00	70.00	72.00
Diabetes	**85.65**	76.09	77.39	76.96	71.30	70.87
Heart-c	**92.42**	86.81	86.81	85.71	80.22	82.42
Heart-h	**93.76**	91.01	86.52	85.39	78.65	85.39
Heart-statlog	**83.95**	83.95	83.95	82.72	71.60	81.48
Hepatitis	**91.11**	85.11	89.36	78.72	80.85	82.98
Ionosphere	**88.67**	86.67	88.57	84.76	83.81	88.57
kr-vs-kp	**98.67**	81.67	81.67	94.00	93.00	88.00
Labor	**94.12**	88.24	87.34	88.24	88.24	82.35
Letter	**69.56**	61.00	57.67	60.67	50.00	45.67
Lymph	**88.64**	84.09	84.09	81.82	79.55	79.55
Segment	**89.74**	87.42	84.11	85.43	83.44	80.46
Soybean	**93.20**	92.68	91.71	88.78	89.76	87.80
Waveform	**85.67**	79.67	80.33	80.33	65.33	72.00

(2) $IWNB$: Instance weighted naive Bayes with the weighting method based on the instance similarity [4].
(3) SBC: A bagged decision-tree based attribute selection filter for naive Bayes [9].
(4) $C4.4$: A specially designed tree to improve C4.5 performance [11] on classification ranking.
(5) KNN: The k-Nearest Neighbors algorithm [1] with k value been set to 10.

3.2   UCI Standard Classification Task

The initial important task is to analyze the performance between the IWNB with the related instance weighting strategy in literature and NB, in terms of classification accuracy, which is calculated by the percentage of successful predictions on domain specific problems [16,18]. Besides, some other types of algorithms that have been well used in real-world applications have also been used for comparison. Specifically, we compare the effect of IWNB [4] with the standard NB [3], SBC [9], C4.4 [11], and KNN [1]. The purpose of the second experiment is

Table 2. Winning or losing statistical analysis on 20 UCI datasets

	KNN [1]	C4.4 [11]	SBC [9]	NB [3]	IWNB [4]
C4.4 [11]	9/1/10				
SBC [9]	13/3/4	14/2/4			
NB [3]	16/2/2	17/0/3	12/3/5		
IWNB [4]	16/1/3	16/1/3	14/2/4	8/6/6	
SWNB	**20/0/0**	**20/0/0**	**20/0/0**	**19/1/0**	**19/1/0**

All analyses are under two-tailed t-test with a 95 % confidence
level.

to compare the proposed self-adaptive instance weighted Naive Bayes, namely
SWNB, with each other types of baseline approaches in literature.

Instance Weighted NB Vs. Standard NB. Table 1 reports the detailed
results of SWNB and other baseline algorithms, respectively. Besides, Table 2
illustrates the compared results about the winning-or-losing statistical analysis
(*i.e.,* two-tailed t-test with a 95 % confidence level) on those benchmark datasets.
Based on the statistical theory, the difference is statistically significant only if
the probability of significant difference is at least 95 percent, *i.e.,* the p-value for
a t-test between two algorithms is less than 0.05. In Table 2, each entry $w/t/l$
means that the algorithm in the corresponding row wins in w datasets, ties
in t datasets and loses in l datasets on the 20 UCI datasets, compared to the
algorithm in the corresponding column. Overall, the results can be summarized
as follows:

(1) Instance weighting IWNB outperforms NB (8 wins and 6 losses). In partic-
    ular, for the dataset "letter" with 20000 instances and 26 classes, the classi-
    fication accuracy for IWNB (61.00 %) is higher than NB (57.67 %). Because
    the 26 classes make the classification task particularly difficult, the 3.33 %
    superiority on 20000 samples become significant.
(2) Instance weighting IWNB greatly outperforms SBC with (14 wins and 4
    losses) on the 20 UCI benchmark datasets.
(3) Instance weighting IWNB significantly outperforms decision tree C4.4 and
    lazy learning approach KNN both with (16 wins and 3 losses).

SWNB Vs. Baselines. Our experimental results from Tables 1 and 2 indicate that SWNB has very significant gain compared to the state-of-art instance
weighting strategy and other types of methods. In summary, our experimental
results can be listed as:

(1) SWNB significantly outperforms existing instance weighted IWNB and
    unweighted NB both with (19 wins and 0 losses). For the dataset "let-
    ter" with 26 classes, the SWNB can achieve a high classification accuracy

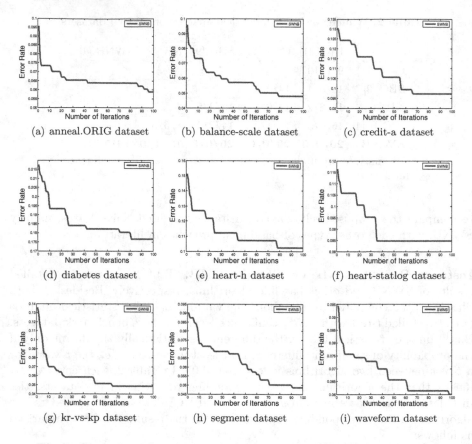

Fig. 2. Convergence learning curves by error rate of SWNB for 9 datasets

(69.56 %), which is 8.56 % and 11.89 % higher than IWNB and NB, respectively.

(2) SWNB greatly outperforms selected naive Bayes SBC with (20 wins and 0 losses) on the 20 UCI benchmark datasets.

(3) SWNB also significantly outperforms decision tree C4.4 and lazy learning approach KNN both with (20 wins and 0 losses).

3.3   Convergence and Learning Curves

In order to investigate the convergence of the SWNB algorithm, we report the relationship between the number of iterations and error rate on the 9 datasets, and the results are shown in Fig. 2. Each point in the curves corresponds to the accuracy under the underlying iteration with the current optimal instance weight values. Figure 2 shows that SWNB converges quickly. Although the curves are not quite smooth, they converge well, which accords with the immunization strategy in SWNB.

For additional insight into our experiment, we observe the "kr-vs-kp" for example. It is a high-dimensional dataset (37 attributes) with 3196 instances. In addition, strong attribute dependencies have been found in this dataset by Kohavi [8]. Our results show that SWNB achieves 98.67 % classification accuracy, which is significantly higher accuracy than instance weighted and unweighted NB on the same dataset (81.67 %). The accuracy of the final convergence is also much better than selected naive Bayes SBC (94.00 %), decision tree C4.4 (93.00 %), and lazy learning KNN (88.00 %). Similar levels of improvement can also be observed from other datasets.

4 Conclusion and Future Work

In this paper, we proposed a novel algorithm to train weighted instances for naive Bayes classification, namely SWNB, by extending the classical SEIR immune strategy. The SWNB algorithm calculates the probability values by using the adaptively instance weighting approach. Considering real-world applicabilities, experiments and comparisons taken on the 20 benchmark UCI datasets, with respect to the classification accuracy performance, show that SWNB outperforms existing NB instance weighting models and other related algorithms, such as classification ranking tree, k-Nearest Neighbors, etc.

The proposed immune strategy based instance weighting for naive Bayes can also be extended to Bayesian networks and applied to other dynamic social networks. Our further study will focus on dynamic Bayesian network applications.

Acknowledgments. We thank the Australian Research Council (ARC) Discovery Project under Grant No. DP140101366, Shanghai Education Commission under grant No. 14ZS085 and Education Ministry of China under grant No. 12YJA630158, support this work.

References

1. Aha, D.W., Kibler, D., Albert, M.K.: Instance-based learning algorithms. Mach. Learn. **6**(1), 37–66 (1991)
2. Bache, K., Lichman, M.: UCI machine learning repository (2013). http://archive. ics.uci.edu/ml
3. Friedman, N., Geiger, D., Goldszmidt, M.: Bayesian network classifiers. Mach. Learn. **29**(2–3), 131–163 (1997)
4. Jiang, L., Cai, Z., Wang, D.: Learning averaged one-dependence estimators by instance weighting. J. Comput. Inf. Syst. **4**, 2753–2760 (2008)
5. Jiang, L., Zhang, H.: Learning instance greedily cloning naive bayes for ranking. In: Proceedings of ICDM, pp. 202–209 (2005)
6. Jiang, L., Zhang, H., Cai, Z.: A novel bayes model: hidden naive bayes. IEEE Trans. Knowl. Data Eng. **21**(10), 1361–1371 (2009)
7. Kim, S.B., Han, K.S., Rim, H.C., Myaeng, S.H.: Some effective techniques for naive bayes text classification. IEEE Trans. Knowl. Data Eng. **18**(11), 1457–1466 (2006)

8. Kohavi, R.: Scaling up the accuracy of naive-bayes classifiers:a decision-tree hybrid. In: Proceedings of KDD, pp. 202–207 (1996)
9. Langley, P., Sage, S.: Induction of selective bayesian classifiers. In: Proceedings of UAI, pp. 339–406 (1994)
10. Naderpour, M., Lu, J., Zhang, G.: A fuzzy dynamic bayesian network-based situation assessment approach. In: Proceedings of IEEE FUZZ, pp. 1–8 (2013)
11. Quinlan, J.R.: C4.5: Programs for Machine Learning. Morgan Kaufmann Publishers Inc., San Francisco (1993)
12. Webb, G.I., Boughton, J.R., Wang, Z.: Not so naive bayes: aggregating one-dependence estimators. Mach. Learn. **58**(1), 5–24 (2005)
13. Witten, I.H., Frank, E.: Data Mining: Practical Machine Learning Tools and Techniques. The Morgan Kaufmann Series in Data Management Systems, 2nd edn. Morgan Kaufmann Publishers, San Francisco (2005). http://www.cs.waikato.ac.nz/ml/weka/
14. Wu, J., Cai, Z.: A naive bayes probability estimation model based on self-adaptive differential evolution. J. Intell. Inf. Syst. **42**(3), 671–694 (2014)
15. Wu, J., Pan, S., Cai, Z., Zhu, X., Zhang, C.: Dual instance and attribute weighting for naive bayes classification. In: 2014 International Joint Conference on Neural Networks (IJCNN), pp. 1675–1679, IEEE (2014)
16. Wu, J., Pan, S., Zhu, X., Cai, Z.: Boosting for multi-graph classification. IEEE Trans. Cybern. **45**(3), 430–443 (2015)
17. Wu, J., Pan, S., Zhu, X., Cai, Z., Zhang, P., Zhang, C.: Self-adaptive attribute weighting for naive bayes classification. Expert Syst. Appl. **42**(3), 1487–1502 (2015)
18. Wu, J., Zhu, X., Zhang, C., Yu, P.S.: Bag constrained structure pattern mining for multi-graph classification. IEEE Trans. Knowl. Data Eng. **26**(10), 2382–2396 (2014)
19. Zhang, C., Xue, G.R., Yu, Y., Zha, H.: Web-scale classification with naive bayes. In: Proceedings of WWW, pp. 1083–1084 (2009)
20. Zhang, T., Liu, J., Teng, Z.: Existence of positive periodic solutions of an seir model with periodic coefficients. Appl. Math. **57**(6), 601–616 (2012)

Enhancing Competitive Island Cooperative Neuro-Evolution Through Backpropagation for Pattern Classification

Gary Wong[1,2] and Rohitash Chandra[1,2](✉)

[1] School of Computing Information and Mathematical Sciences,
University of South Pacific, Suva, Fiji
[2] Artificial Intelligence and Cybernetics Research Group,
Software Foundation, Nausori, Fiji
{gary.wong.fiji,c.rohitash}@gmail.com

Abstract. Cooperative coevolution is a promising method for training neural networks which is also known as cooperative neuro-evolution. Cooperative neuro-evolution has been used for pattern classification, time series prediction and global optimisation problems. In the past, competitive island based cooperative coevolution has been proposed that employed different instances of problem decomposition methods for competition. Neuro-evolution has limitations in terms of training time although they are known as global search methods. Backpropagation algorithm employs gradient descent which helps in faster convergence which is needed for neuro-evolution. Backpropagation suffers from premature convergence and its combination with neuro-evolution can help eliminate the weakness of both the approaches. In this paper, we propose a competitive island cooperative neuro-evolutionary method that takes advantage of the strengths of gradient descent and neuro-evolution. We use feedforward neural networks on benchmark pattern classification problems to evaluate the performance of the proposed algorithm. The results show improved performance when compared to related methods.

1 Introduction

Cooperative coevolution (CC) decomposes a problem into subcomponents that are implemented as sub-populations which cooperatively evolves while mating is restricted within sub-populations [1]. The process of breaking a problem down into subcomponents is called problem decomposition. In the case of neuro-evolution, efficient problem decomposition depends on the network architecture and nature of the application problem in terms of separability [2]. Cooperative coevolution has been mostly used for large scale optimisation [3] and evolution of feedforward and recurrent neural networks in pattern classification and time series prediction [4–7]. The use of cooperative coevolution for neuro-evolution is referred to as cooperative neuro-evolution.

In cooperative neuro-evolution, much attention has been given to problem decomposition, i.e. how to break the neural network into sub-problems through

© Springer International Publishing Switzerland 2015
S. Arik et al. (Eds.): ICONIP 2015, Part I, LNCS 9489, pp. 293–301, 2015.
DOI: 10.1007/978-3-319-26532-2_32

the interconnected weights that contain inter-dependencies [2]. The major problem decomposition methods involve those that fully or partially decompose the network, i.e. in full decomposition, the neural network is decomposed into the lowest level where a single subcomponent represents a weight connection, this is also called synapse level decomposition [8]. In partial decomposition, the network is decomposed with reference to weight connections linked to each hidden and output neurons that is also called neuron level decomposition [2,9]. The performance of a decomposition method varies on different types of problems, for instance, synapse level decomposition showed very good results in pole balancing [8] but have been unsuccessful in pattern classification [9]. Both synapse and neuron level decomposition have shown competitive performance for time series problems [5]. There has been much focus on adaptation of the problem decomposition method during the learning process in order to take advantage global - local search and inter-dependencies [10–12]. In competitive island-based cooperative neuro-evolution (CICN), two or more problem decomposition methods are implemented as islands that compete and collaborate at different phases of evolution [4]. The competitive feature gives subcomponents the ability to compete for resources. There is altruism feature in the algorithm where the winner island shares its solution with the losing islands so that they can catch up in the next phase of evolution. The competitive and collaborative features enables strong solutions to be retained and has been very promising for training neural networks for time series and pattern classification problems [13,14]

Neuro-evolution has limitations in terms of training time although they are known as global search methods. Backpropagation algorithm employs gradient descent which helps in faster convergence. Backpropagation suffers from premature convergence and its combination with neuro-evolution can help eliminate the weaknesses of both the approaches. In this paper, we propose a competitive island cooperative neuro-evolutionary method that takes advantage of the strengths of gradient descent and neuro-evolution. Integrating backpropagation in competitive island cooperative coevolution can help in achieving faster convergence to a near global optimum solution. We implement backpropogation as an island in competitive island-based cooperative neuro-evolution (CICN) and use it for training feedforward networks for selected pattern classification problems.

The remaining sections of the paper are structured as follows. Section 2 provides the details of the proposed method that features backpropagation in CICN. Section 3 presents the results with discussion and Sect. 4 concludes the paper with discussion on future work.

2   Proposed Method

2.1   Backpropagation in CICN

In Competitive Island Cooperative Neuro-evolution (CICN), two or more decomposition methods are implemented as islands that compete and provide altruism where the winning islands share solutions with the losing islands over a period

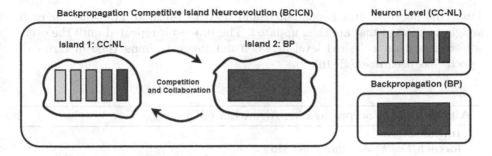

Fig. 1. The three algorithms employed in this study. These are standalone Backpropagation, CC-NL and the 2-Island BCICN algorithm.

until termination. In an environment with multiple species, the competitive feature relates to the ability of the species to outperform each other for possession of resources [15]. In the proposed method, two standalone methods are used that include backpropogation and cooperative neuro-evolution that employs neuron level problem decomposition as shown in Fig. 1. The details of each island are given below.

1. **Backpropagation Algorithm (BP):** Standard backpropagation algorithm where the entire network is used 'as-is' without decomposition.
2. **Cooperative Coevolution with Neuron Level Problem Decomposition (CC-NL):** The number of neurons in the hidden and output layer determine the number of subcomponents [2,5].

The proposed backpropagation competitive island cooperative neuro-evolution (BCICN) method is given in Algorithm 2. In Stage 1, the sub-populations are randomly initialised and cooperatively evaluated using neuron level problem decomposition. After evaluation, the current best individual in the cooperative neuro-evolution island is copied to the Backpropagation Island. This is to ensure that both islands start from the same set of initial solution(s). In Stage 2, the islands are evolved per the island evolution time and in Stage 3 the best solutions from both islands are compared and the winner is selected to be transferred to the losing island in Stage 4 of the algorithm and then the process is repeated for the next phase of evolution. As presented in previous work [4], the respective islands need to be given the same number of function evaluations for each phase in evolution and this is due to the requirement that each island be evaluated for complete cycles.

2.2   Backpropagation Island

The conventional backpropagation procedure consists of forward pass where information is propagated forward through neurons using their activation function that computes weighted sum of incoming weight-connections to the respective neurons. Once the information is propagated from input, hidden to output later, the network error is computed and used to calculate gradients for each weight connected that are then updated. The process is repeated until the overall error reaches a desired level or when maximum training time in terms of epochs has been reached [16].

Algorithm 1. Backpropagation Algorithm (BP)

Initialisation:
foreach *Epoch until Max-Epoch* **do**
    foreach *Training-Sample until Total-Training-Samples* **do**
        Forward propagation through network
        Backward propagation through network
    Increment Epoch

If the backpropagation island wins a phase of competition, it transfers the solution to the island that features cooperative neuro-evolution taking into account that the solution needs to be decomposed as defined by neuron problem decomposition in order to maintain solution validity as shown in Fig. 2. In the case where the Backpropagation Island loses the competition, the solution from the winner island will be concatenated by combining the best solutions from all its respective sub-populations. This individual is then refined using backpropogation and then the competition continues.

3   Experiments and Results

In this section, we apply the proposed BCICN to pattern classification problems. In our previous work, we applied competitive neuro-evolution to pattern classification and time series prediction [4]. We use the same classification problems from the UCI Machine Learning Repository [17]. The problems are Cleveland Heart Disease, Wisconsin Breast Cancer, Iris and the 4-Bit parity problem. They have been used in other studies to evaluate performances of new methods [2,14]. The details of problems tested are provided in Table 1.

Table 1. Data set information and neural network configuration

Problem	Input	Output	Min. train (%)	Max. time	Samples
Wisconsin breast cancer	9	1	95	15000	699
4-Bit	4	1	1E-3	30000	16
Wine	13	3	95	15000	178
Iris	4	3	95	15000	150
Cleveland heart disease	13	1	88	50000	303

Algorithm 2. BCICN for Pattern classification

Stage 1: Initialisation:
i. Generate and cooperatively evaluate NL Island
ii. Copy Best Individual from NL Island to BP Island
Stage 2: Evolution:
while *FE ≤ Global-Evolution-Time* **do**
    while *FE ≤ Island-Evolution-Time* **do**
        foreach *Sub-population at NL Island* **do**
            foreach *Depth of n Generations* **do**
                Create new individuals using genetic operators.
                Cooperative evaluation.

    while *FE ≤ Island-Evolution-Time* **do**
        Execute BP (**Algorithm 1**) .
    Stage 3: Competition: Compare NL Island fitness with BP Island fitness.
    Stage 4: Collaboration: Inject the best individual from the island with
    better fitness into the other island.
    if *NL Island fitness ≤ BP Island fitness* **then**
        Copy NL Island best individual into the BP Island.
    else
        Copy BP Island Individual to NL Island

The termination condition for an unsuccessful run is provided in Table 1 as maximum time (*Max. Time*). Each problem is set to have 50 independent runs where the evaluation time, generalisation performance and success rate is given. We evaluate the performance on different number of hidden neurons (H) in order to test robustness and scalability of BCICN. For all the 3 methods employed, the maximum time or island evolution time remained the same regardless of the number of islands used (in this case, we used 2 islands).

3.1   Results and Discussion

The results of the experiment are presented in Tables 2–3. A comparison is made between standalone cooperative coevolution with neuron level decomposition (CC-NL) and the BCICN.

The results show that the method that performed best in terms of convergence time for the Iris, Cancer, Wine and Heart problems was the standard backpropogation algorithm while the worst performance was that of CC-NL. BCICN obtained faster convergence and outperformed CC-NL as shown in Fig. 3 where results for 10 hidden neurons are compared. The success rate of BCICN improved in some problems but was the same when the standalone methods had an average success rate of 100 %. BCICN performed best in the 4-Bit problem where

---

Keys for Tables 2 and 3
x̄ev = Mean Fitness Evaluations, x̄er = Mean Generalisation Performance, (**H**) = No. Hidden Neurons, and (**sr**) = Success Rate

Table 2. Performance for the Iris, Cancer and Heart classification problems

| Method | | Iris | | | Cancer | | | Heart | | | |
	H	$\bar{x}$ev	$\bar{x}$er	sr	H	$\bar{x}$ev	$\bar{x}$er	sr	H	$\bar{x}$ev	$\bar{x}$er	sr
BP	4	687	87.50	100	4	**246**	**97.99**	**100**	6	**4034**	**81.29**	**96**
	6	684	87.50	100	6	165	98.16	100	8	3390	81.17	100
	8	**680**	**87.50**	**100**	8	145	98.38	100	**10**	**1440**	**81.28**	**50**
	10	692	87.50	100	10	129	98.39	100	12	1569	81.53	100
	12	**700**	**87.50**	**100**	12	**124**	**98.42**	**100**	14	1451	81.33	100
CC-NL	**4**	**4356**	**95.50**	**100**	**4**	**5562**	**96.98**	**94**	6	19097	79.50	90
	6	5184	94.88	100	**6**	4519	**97.70**	**100**	8	15719	79.88	100
	8	5430	96.75	100	8	5227	97.96	100	**10**	**35760**	**80.00**	**50**
	10	5860	96.00	100	10	5174	98.08	100	12	24445	80.55	100
	12	**6636**	**96.20**	**100**	12	5475	98.31	98	14	21051	79.11	100
BCICN	**4**	**2204**	**95.50**	**100**	**4**	**2034**	**96.71**	**94**	6	8990	80.94	100
	6	3618	95.00	100	6	1454	97.33	100	**8**	**12580**	**79.55**	**100**
	8	4117	94.00	100	8	1485	97.34	100	10	9675	79.95	100
	10	3632	94.75	100	**10**	**1140**	**97.45**	**100**	12	7321	81.53	100
	12	**4369**	**95.25**	100	12	1248	97.62	100	**14**	**6340**	**81.02**	**100**

once again the worst performance was that of CC-NL. This is the only problem BCICN method outperformed backpropogation. The success rate of BCICN improved over backpropogation in all cases given by number of hidden neurons (H). The focus of this study was to reduce the convergence time of CC-NL and

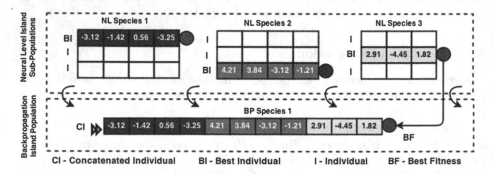

Fig. 2. Concatenation of the best individuals from the neuron level island and injection into the backpropagation island. Note the fitness of the concatenated individual is acquired from the fitness of the last best individual from the neural level island. When transferring, backpropagation's solution is decomposed as defined by neuron level problem decomposition.

Table 3. Performance for the Wine and 4-Bit classification problems

Method		Wine				4-Bit		
	H	$\bar{x}$ev	$\bar{x}$er	sr	H	$\bar{x}$ev	$\bar{x}$er	sr
BP	4	262	98.12	100	4	30010	-	0
	6	279	98.50	100	6	13995	100.00	70
	8	282	98.75	100	8	9442	100.00	95
	10	300	99.62	100	10	5228	100.00	95
	12	323	99.75	100	12	4795	100.00	95
CC-NL	4	6573	94.73	95	4	11151	100.00	100
	6	7371	92.75	100	6	6001	100.00	100
	8	7293	94.25	100	8	5772	100.00	100
	10	8268	94.00	100	10	7012	100.00	100
	12	8730	94.12	100	12	6318	100.00	100
BCICN	4	1959	95.25	100	4	9324	100.00	100
	6	1994	94.87	100	6	4944	100.00	100
	8	1728	95.50	100	8	3298	100.00	100
	10	921	95.12	100	10	3067	100.00	100
	12	1023	94.37	100	12	3967	100.00	100

this was achieved in all the problems tested. The performance measure was in terms of minimising the function evaluations and improving the success rates.

This improved performance is due to the collaborative feature employed here where the two islands shared best solutions throughout the island evolutionary phases. In the backpropagation island, gradient information is used for weight update whereas in the cooperative neuron-evolution island, genetic operators are used. Gradient information features local search and ensures faster convergence when compared to neuro-evolution that features global search which is slower in convergence. Backpropagation island does not require network decomposition and hence does not face the problems of grouping interacting variables. BCICN provides the balance between global and local search and also features network decomposition. It approaches the problem as partially separable through neuron level problem decomposition and non-separable through backpropagation.

In terms of scalability, we look at the mean evaluations at each total number of hidden neurons used. It is observed that increasing the number of hidden neurons in the Cancer, 4-Bit and Heart problems decreased the mean evaluations needed. On the other hand, mean evaluation performance improved when more hidden neurons were used in the Wine and Iris problems. The BCICN method showed good scalability in the Wine and Iris problems, but poor scalability in Cancer, 4-Bit and Heart problems. It can be generalised that that scalability features depend greatly on the problem nature, which is in terms of the size of the problem, noise, number of attributes and level of inter-dependencies amongst them.

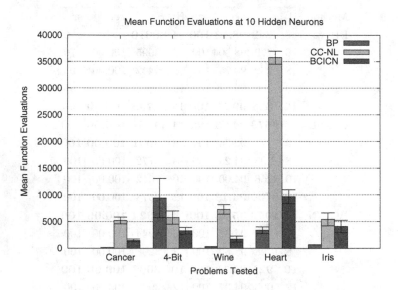

Fig. 3. Visualisation of the performance of BCICN with the standalone methods for 10 hidden neurons taken from Tables 2 and 3.

4 Conclusions and Future Work

This paper proposed an algorithm that incorporates backpropagation in competitive island cooperative neuro-evolution for pattern classification. The results show that the proposed method outperformed the standalone methods through faster convergence. The backpropagation algorithm provided neuro-evolution gradient information that led to faster convergence. This can be very beneficial in the use of neuro-evolution for big data related problems that require faster learning. Evolutionary computation methods have limitations in the field of big data due to time required for convergence. The proposed method can motivate the development of other hybrid algorithms that speed up evolutionary learning methods for big data problems.

In future work, the proposed method can be used for training recurrent neural networks for time series prediction problems such as renewable energy load forecasting. It can also be used in selected big data problems.

References

1. Potter, M., De Jong, K.: A cooperative coevolutionary approach to function optimization. In: Davidor, Y., Schwefel, H.-P., Mnner, R. (eds.) PPSN III. LNCS, vol. 866, pp. 249–257. Springer, Heidelberg (1994)

2. Chandra, R., Frean, M., Zhang, M.: On the issue of separability for problem decomposition in cooperative neuro-evolution. Neurocomputing **87**, 33–40 (2012)
3. Omidvar, M., Li, X., Yao, X.: Cooperative co-evolution with delta grouping for large scale non-separable function optimization. IEEE Congr. Evol. Comput. (CEC) **2010**, 1762–1779 (2010)
4. Chandra, R.: Competitive two-island cooperative coevolution for training Elman recurrent networks for time series prediction. In: International Joint Conference on Neural Networks (IJCNN), Beijing, China, pp. 565–572, July 2014
5. Chandra, R., Zhang, M.: Cooperative coevolution of Elman recurrent neural networks for chaotic time series prediction. Neurocomputing **186**, 116–123 (2012)
6. Garcia-Pedrajas, N., Hervas-Martinez, C., Munoz-Perez, J.: COVNET: a cooperative coevolutionary model for evolving artificial neural networks. IEEE Trans. Neural Netw. **14**(3), 575–596 (2003)
7. Potter, M.A., De Jong, K.A.: Cooperative coevolution: an architecture for evolving coadapted subcomponents. Evol. Comput. **8**, 1–29 (2000)
8. Gomez, F., Schmidhuber, J., Miikkulainen, R.: Accelerated neural evolution through cooperatively coevolved synapses. J. Mach. Learn. Res. **9**, 937–965 (2008)
9. Chandra, R., Frean, M., Zhang, M.: An encoding scheme for cooperative coevolutionary feedforward neural networks. In: Li, J. (ed.) AI 2010. LNCS, vol. 6464, pp. 253–262. Springer, Heidelberg (2010)
10. Chandra, R.: Adaptive problem decomposition in cooperative coevolution of recurrent networks for time series prediction, In: International Joint Conference on Neural Networks (IJCNN), Dallas, TX, USA, pp. 1–8, August 2013
11. Chandra, R., Frean, M., Zhang, M.: Adapting modularity during learning in cooperative co-evolutionary recurrent neural networks. Soft Comput. A Fusion Found. Methodol. Appl. **16**(6), 1009–1020 (2012)
12. Chandra, R., Frean, M.R., Zhang, M.: Crossover-based local search in cooperative co-evolutionary feedforward neural networks. Appl. Soft Comput. **12**(9), 2924–2932 (2012)
13. Chandra, R.: Competition and collaboration in cooperative coevolution of Elman recurrent neural networks for time-series prediction. IEEE Trans. Neural Netw. Learn. Syst. p. (2015, in press)
14. Chandra, R., Wong, G.: Competitive two-island cooperative coevolution for pattern classification problems. In: International Joint Conference on Neural Networks (IJCNN), Killarney, Ireland, pp. 393–400, July 2015
15. Bella, G.: A bug's life: competition among species towards the environment. ser. Fondazione Eni Enrico Mattei Working Papers. Fondazione Eni Enrico Mattei (2007)
16. Rumelhart, D.E., Hinton, G.E., Williams, R.J.: Learning internal representations by error propagation. In: Rumelhart, D.E., McClelland, J.L., CORPORATE PDP Research Group (eds.) Parallel Distributed Processing: Explorations in the Microstructure of Cognition, vol. 1, pp. 318–362. MIT Press, Cambridge (1986). http://dl.acm.org/citation.cfm?id=104293
17. Asuncion, A., Newman, D.: UCI machine learning repository (2007). http://archive.ics.uci.edu/ml/datasets.html

Email Personalization and User Profiling Using RANSAC Multi Model Response Regression Based Optimized Pruning Extreme Learning Machines and Gradient Boosting Trees

Lavneet Singh[(✉)] and Girija Chetty

Faculty of ESTEM, University of Canberra, Canberra, Australia
{Lavneet.singh,Girija.chetty}@canberra.edu.au

Abstract. Email personalization is the process of customizing the content and structure of email according to member's specific and individual needs taking advantage of member's navigational behavior. Personalization is a refined version of customization, where marketing is done automated on behalf of customer's user's profiles, rather than customer requests on his own behalf. There is very thin line between customization and personalization which is achieved by leveraging customer level information using analytical tools. E-commerce is growing fast, and with this growth companies are willing to spend more on improving the online experience.

Thus, in this study, we propose a new architectural design of email personalization and user profiling using gradient boost trees and optimized pruned extreme learning machines as base estimators. We also conducted an in-depth data analysis to find each member's behavior and important attributes which plays a significant role in increasing click rates in personalized emails. From the experimental validation, we concluded that our prosed method works much better in predicting customer's behavior on deals send in personalized emails compared to other methods in past literature.

Keywords: Email personalization · Gradient boosting · Optimized pruning extreme learning machines

1 Introduction

Personalization is the idea of interactive marketing with respect to the customization of some or all elements of the marketing strategy to an individual level. Personalization is a refined version of customization, where marketing is done automated on behalf of customer's user's profiles, rather than customer requests on his own behalf. There is very thin line between customization and personalization which is achieved by leveraging customer level information using analytical tools [1, 2].

Personalization is a key component for adapting a standardized product or service to an individual customer's needs. Application of personalization fits nicely into notions of Internet to provide a rich environment for well suited interaction and segmentation.

© Springer International Publishing Switzerland 2015
S. Arik et al. (Eds.): ICONIP 2015, Part I, LNCS 9489, pp. 302–309, 2015.
DOI: 10.1007/978-3-319-26532-2_33

The exponential growth of World Wide Web enables new methods of development and design of online information services. Most web development is large and complicated and users receive ambiguous results during web navigation. But on the other hand, rapidly evolving and need of e-commerce marketplaces in World Wide Web anticipate the need of customers to distinct level. Therefore, the requirement for predicting user needs in order to improve the usability and user retention of a Web site can be addressed by personalizing it [6]. Web personalization is defined as any action that adapts the information or services provided by a Web site to the needs of a particular user or a set of users, taking advantage of the knowledge gained from the users' navigational behavior and individual interests, in combination with the content and the structure of the Web site. The objective of a Web personalization system is to "provide users with the information they want or need, without expecting from them to ask for it explicitly" [7].

2 Email Personalization

Email personalization strongly relies on of statistical and data mining methods to the Web log data, resulting in a set of useful patterns that indicate users' navigational behavior. In past, the data mining methods that are employed are: association rule mining, sequential pattern discovery, clustering, and classification [7]. This knowledge is then used from the system in order to personalize the site according to each user's behavior and profile.

2.1 Gradient Boost Trees

Gradient Tree Boosting or Gradient Boosted Regression Trees (GBRT) [12] is a generalization of boosting to arbitrary differentiable loss functions. GBRT is an accurate and effective off-the-shelf procedure that can be used for both regression and classification problems. Gradient Tree Boosting models are used in a variety of areas including Web search ranking and ecology. GBM trains many models turn by turn and each new model gradually minimizes the loss function of the whole system using Gradient Descent method. Assuming each individual model i is a function $h(X;p_i)$ (which we call "base function" or "base learner") where X is the input and p_i is the model parameter. Now let's choose a loss function $L(y, y')$ where y is the training output, y' is the output from the model. In GBM, $y' = \sum_{i=1}^{M} \beta_i h(X_i p_i)$, where M is the number of base learners. This can be more defined as

$$\beta, P = \arg \min_{\{\beta_i p_i\}_1^M} L(y, \sum_{i=1}^{M} \beta_i h(X;p_i)) \tag{1}$$

In further section, we discuss about proposed optimized extreme learning machines based on RANSAC regularization as base estimator in gradient boosting trees.

The whole classification model is implemented on customer database provided by OURDEAL database to design user profiles and email personalization for their marketing strategy. The model spit up the customer plan each morning at 8.00 am morning with personalized customer plan for each customers and the personalized deals can be send to each customers with respect to their web usage behavior on website and clicks and open rate of emails. Due to sparse database available for new and non-engaging users, model works perfectly in terms of defining the personalized emails. The whole strategy is briefly defined in research methodology section.

2.2   Optimized Extreme Learning Machines as Base Estimators for Gradient Boosting Trees

To resolve these limitations of ELM, constructive and heuristic approaches have proposed in the literature. In most recent years, regularization or penalty approach seems to be significant in resolving the ELM limitations. As in extreme machine learning, there is linear behavior between hidden layer and output layer, thus as a problem of linear regression, regularization helps to reduce the number of predictors in hidden layer by using sparse model.

2.2.1   RANSAC Multi Model Response Regularization

RANSAC Multi Model Response Regularization for Regression Problems. To implement the RANSAC on regression problems, we proposed a RANSAC multi model response regularization which implements the sequential RANSAC on multiple models [16–19]. To take out the outliners from data, which in our case, are the irrelevant hidden nodes as predictor variables, and H is the hidden matrix as input from equation. In our case, the output weights follow a linear regression between hidden and output layer defined as

$$Y = mx + \epsilon \tag{2}$$

Where Y is the output of instances of data, m is the predictor's weights or slope and x is the input data and c is the constant. $Y = $ Output weights $* H + \epsilon$

$$H = \begin{bmatrix} g(w_1 \cdot x_1 + b_1) & \cdots & g(w_{\tilde{N}} \cdot x_1 + b_{\tilde{N}}) \\ \vdots & \cdots & \vdots \\ g(w_1 \cdot x_N + b_1) & \cdots & g(w_{\tilde{N}} \cdot x_1 + b_{\tilde{N}}) \end{bmatrix}_{N \times \tilde{N}} \tag{3}$$

Where

$$OutputWeights = RANSAC - Multi \begin{bmatrix} g(w_1 \cdot x_1 + b_1) & \cdots & g(w_{\tilde{N}} \cdot x_1 + b_{\tilde{N}}) \\ \vdots & \cdots & \vdots \\ g(w_1 \cdot x_N + b_1) & \cdots & g(w_{\tilde{N}} \cdot x_1 + b_{\tilde{N}}) \end{bmatrix}_{N \times \tilde{N}} * T \tag{4}$$

$$CS = RANSAC \left(\sum_{W=1}^{m} \left(\sum_{j=1}^{n} D_{W_j} \right) \right) \tag{5}$$

$Dx_{wj} = \{x_{11}, x_{12} \ldots x_{wj}\}$ be the sets of data H with wth observations in rows and j^{th} as hidden nodes predictors in columns of D matrix. For regression problems, sequential

RANSAC implements on the set of all inliners, D1 that are generated by W different models where $W_m = rand(D_M)$. The numbers of models are randomly generated using 20 % of the input data.

To estimate the parameters of W models, each one is represented by k dimensional parameter vector θ_w at each iteration iter. CS is estimated using MSS of each W model. The iteration run M times which is calculated before after removing the inliners from data D. the total number of inliners at iteration iter is less than total number of inliners at iteration iter-1. The whole formulation of multiple RANSAC response is defined as

The set of all inliners D is generated by W different models has cardinality CS as

$$N_I = (N_{I,1} + N_{I,2} + \ldots N_{I,W}) \tag{6}$$

Let $M_w(\theta_w)$ defines the manifold of dimension k_w of all points with respect to parameter $\theta_w \in R^{kw}$ for the specified model for $1 \leq w \leq W$ with a subset S_w from D of k_w elements at iteration i called minimal sample set (MSS). To estimate the parameters of W models, each one represented parameter vector θ_w. At each i iterations, MSS for each W model is defined and CS is estimated removing all outliners.

The proposed RANSAC multi model response regularization for binary and multi-class problems for ELM is implemented using one against all (OAA) method. As in OAA method, j binary classifiers will be constructed in which all the training examples will be used a teach time of training. The training data having the original class label $j_n = (1 \ldots n)$ have each j_n elements of positive one class and the remaining training data will be of zero class, creating j_n models implementing proposed RANSAC multi model response regularization on (j_n) binary classes. Finally, CS defined as $S^i(\theta)$ of j(n) classes is computed as

$$S^i(\theta)_j = \sum_{j=1}^{n} S^{(i)}\theta_j \cup S^{(i-1)}\theta_j \tag{7}$$

$$S^i(\theta) = \sum_{j=1}^{n} S^{(i)}\theta_j \cup S^{(i)}\theta_n \tag{8}$$

For this, consider the ELM for multi-class classification problem, formulated as k binary ELM classification problem with the following form:

$$Hw_1 = y_1 \ldots Hw_j = y_{j;}$$

Where for each j, w_j is the output weights from the hidden layer to the output layer with output vector $y_j = (y_{1j}, \ldots y_{mj})^t \in R_m$. Thus the output of the hidden layer as H hidden matrix defines with respect to multiclass binary classifiers as

$$H_j = \sum_{j=1}^{n} H * Y \left(\begin{array}{c} Y_j = 1 \\ o \end{array} \right) \tag{9}$$

where H is the hidden layer output matrix and Y is the j binary classes with m^{th} observations of training data and n binary classes as columns vectors. Thus, we get H_j hidden matrix where each H_j belong to each binary class and RANSAC multi response regularization is implemented to acquire CS for each binary class as $S^i(\theta)_j$. It can be concluded that RANSAC multi response regularization for binary and multiclass problems work in similar fashion as OAA-ELM with j binary classes with a difference of jth label with positive class and rest other classes with -1 class.

3   Experimental Results

To implement the email personalization, firstly data is extracted from data warehouse and fed into pandas Dataframe in python. Sql queries are integrated in python for accessing the data from redshift data warehouse and uploading the final plan for email personalization. Figure 1 represent the distribution of member's data with respect to subscribers and non-subscribers in 1 year tenure scale. Subscribers and non-subscribers in the figure represent members who are subscribers for receiving emails from their starting tenure until this experiment. Non subscriber are the members which unsubscribes during this experimental study from their starting tenure. Global parameters are initialized along with exception handling foe extracting data from data warehouse. As we can see in Fig. 1, distribution of data seems to be normal distribution with skewed towards right. Figure 2 depicts the member's data density distribution with respect to subscriber and non-subscribers in different states of Australia as Sydney, Melbourne ranked in scale of 1-10. As we can see in all figures, there is a very thin line to differentiate members behavior in terms of they will keep subscribe the emails or unsubscribe it (Fig. 1).

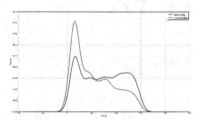

Fig. 1. Density Distribution of member's data with respect to subscribers and non-subscribers in 1 year tenure scale;

Figure 3 depicts the comparison of training data and testing data variance for 300 iterations using regularization and proposed optimized extreme learning machines as base estimators for gradient boosting trees. A we can see in the Fig. 4, there is less variance in training and testing data using optimized extreme learning machines as base estimator for grad boost classifiers. Compared to Fig. 4, where we do not use regularization and base estimators, grad boost classifiers by itself doesn't performs well in term of low deviance and the model is over fitted creating huge difference of deviance between

Fig. 2. Density Distribution of member's data with respect to subscribers and non-subscribers in 1-10 location scale

High Regularized, Less variance, High Bias
Not good Probability Estimates

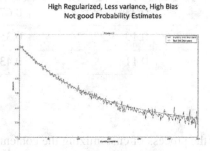

Fig. 3. Distribution of deviance for training and testing data for 0-300 iterations. With regularization and using optimized extreme learning machine as base estimators for gradient boosting trees;

Less Regularized, High variance, less Bias
Good Probability Estimates

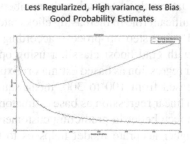

Fig. 4. Distribution of deviance for training and testing data for 0-300 iterations with no regularization and no base estimators for gradient boosting trees

training and testing data. In such case, regularization with optimized RANSAC extreme learning machines performs significantly well with grad boost trees in terms of less deviance among training and testing data.

Table 1 depicts the comparative analysis of grad boost classifier on Customer database provided by OURDEAL database using base estimators and non-base estimators. As we can see from table as iterations increases from 100 to 300, Grad boost classifier with optimized RANSAC pruned extreme learning machines (OPRELM) perform better than grad boost with linear regression (LR) as base estimators. The ROC curve using base estimator as OPRELM is estimated as 0.96 and confusion matrix shows less false positive and false negative rate. Later, deals recommended by the classifiers are ranked with respect to click rate and send to each member with personalized deals chosen by classifier.

Table 1. Comparative analysis using grad boost classifier with proposed base estimator and non-base estimator

Iterations	Mean Square Error (MSE)		
	Grad-Boost	Grad-boost with ORELM	Grad-boost with LR
100	0.75	0.23	0.67
200	0.56	0.19	0.35
300	0.45	0.11	0.43

4   Conclusions

Email personalization is the process of customizing the content and structure of email according to members specific and individual needs taking advantage of members navigational; behavior. In this paper, we proposed the architectural design for email personalization using Customer database provided by OURDEAL database based on grad boost with optimized pruned extreme learning machines as base estimators. We also conducted a depth dive in data analysis to find each members behavior and important attributes which plays a significant role in increasing clicks rates in personalized emails.

After data is extracted using step up process according to defined architectural design, we treat the data with grad boost classifier using optimized pruned extreme learning machine and linear regression as based estimator. Experimental results showed that as the iterations increased from 100 to 300, mean square error is much lower compare to grad boost with linear regression as base estimators. We can concluded that our prosed method works much better in predicting customers behavior on deals send in personalized emails. Higher accurate model helps us to find out better customer behavior and design the customer plan according to accurate customer's needs. Currently this model is implemented in Customer database provided by OURDEAL, and every morning set of deals are selected using proposed method for personalized emails and defining user profiles.

Later, further work can be explore by adding more web traffic data and more optimization to increase the accuracy in predicting the probability of clicking emails in personalized emails.

References

1. Montgomery, A.L., Smith, M.D.: Prospects for personalization on the internet. J. Interact. Mark. **23**(2), 130–137 (2009)

2. Maxwell, J.C.: A Treatise on Electricity and Magnetism, vol. 2, 3rd edn, pp. 68–73. Clarendon, Oxford (1892)
3. Ansari, S., Kohavi, R., Mason, L., Zheng, Z.: Integrating e-commerce and data mining: architecture and challenges. In: ICDM 2001 Proceedings IEEE International Conference on Data Mining, 2001, pp. 27–34. IEEE (2001)
4. Schmitt, E., Manning, H., Paul, Y., Roshan, S.: Commerce software takes off. Forrester report, March 2000
5. Schmitt, E., Manning, H., Paul, Y., Tong, J.: Measuring web success. Forrester report, November 1999
6. Miceli, G., Ricotta, F., Costabile, M.: Customizing customization: a conceptual framework for interactive personalization. J. Interact. Mark. 21(2), 6–25 (2007)
7. Venasen, J.: What is personalization? A conceptual framework. Eur. J. Mark. 41(5–6), 409–418 (2007)
8. Mobasher, B., Cooley, R., Srivastava, J.: Automatic personalization based on web usage mining. Commun. ACM 43(8), 142–151 (2000)
9. Srivastava, J., Cooley, R., Deshpande, M., Tan, P.N.: Web usage mining: discovery and applications of usage patterns from web data. SIGKDD Explor. 1(2), 12–23 (2000)
10. Eirinaki, M., Vazirgiannis, M.: Web mining for web personalization. ACM Trans. Internet Technol. (TOIT) 3(1), 1–27 (2003)
11. Montgomery, A.L., Li, S., Srinivasan, K., Liechty, J.: Modeling online browsing and path analysis using clickstream data. Mark. Sci. 23(4), 579–595 (2004)
12. Friedman, J.H.: Greedy function approximation: a gradient boosting machine, February 1999
13. Hastie, T., Tibshirani, R., Friedman, J.H.: Boosting and additive trees (Chap. 10). In: The Elements of Statistical Learning, 2nd edn. pp. 337–384. Springer, New York. ISBN 0-387-84857-6
14. Huang, G.-B., Zhu, Q.-Y., Siew, C.-K.: Extreme learning machine: theory and applications. Neurocomputing 70(1–3), 489–501 (2006). doi:10.1016/j.neucom.2005.12.126
15. Abid, S., Fnaiech, F., Najim, M.: A fast feedforward training algorithm using a modified form of the standard backpropagation algorithm. IEEE Trans. Neural Networks 12(2), 424–430 (2001). doi:10.1109/72.914537
16. Singh, L., Chetty, G.: Pruned annular extreme learning machine optimization based on RANSAC multi model response regularization. In: Mao, K., Cambria, E., Cao, J., Man, Z., Toh, K.-A. (eds.) Proceedings of ELM-2014 Volume 1. PALO, vol. 3, pp. 163–182. Springer, Heidelberg (2015)
17. Singh, L., Chetty, G.: An optimal approach for pruning annular regularized extreme learning machines. In: 2014 IEEE International Conference on Data Mining Workshop (ICDMW), pp. 80–87, 14 December 2014
18. Singh, L., Chetty, G.: RANSAC multi model response regression based pruned extreme learning machines for multiclass problems. Australian Journal of Intelligent Information Processing Systems 14(1) (2014)
19. Singh, L., Chetty, G.: Understanding the brain via fMRI classification. In: Kasabov, N. (ed.) Springer Handbook of Bio-/Neuroinformatics, pp. 703–711. Springer, Berlin Heidelberg (2014)

An Auto-Encoder for Learning Conversation Representation Using LSTM

Xiaoqiang Zhou[✉], Baotian Hu, Qingcai Chen, and Xiaolong Wang

Intelligent Computing Research Center,
Harbin Institute of Technology Shenzhen Graduate School, Shenzhen, China
{xiaoqiang.jeseph,baotianchina,qingcai.chen}@gmail.com,
wangxl@insun.hit.edu.cn

Abstract. In this paper, an auto-encoder is proposed to learn conversation representation. First, the long short term memory (LSTM) neural network is used to encode the sequence of sentences in a conversation. The interactive context is encoded into a fixed-length vector. Then, through the LSTM-decoder, the learnt representation is used to reconstruct the sentence vectors of a conversation. To train our model, we construct one corpus with 32,881 conversations from the online shopping platform. Finally, experiments on topic recognition task demonstrate the effectiveness of the proposed auto-encoder on learning conversation representation, especially when training data of topic recognition is relatively small.

Keywords: Auto-encoder · LSTM · Conversation representation

1 Introduction

Many artificial intelligence tasks are essentially the problem of understanding sequences, such as speech recognition [13], machine translation [8] and community question answering (cQA) [15]. Likewise, human conversation is a temporal process [11] consisting of a sequence of sentences, and involves in many challenging problems such as discourse segmentation [1], topic recognition [2], act classification [3], and answer or summary generation [4, 5]. These tasks largely depend on how to find and represent useful features and structures in conversations, but the expression in human conversation is too flexible to learn good conversation representations by task-driven supervised learning. Traditionally, it always relies on human intensive work in collecting labeled data [22]. On the other hand, it is more feasible to learn useful features from a large scale of unlabeled data by building unsupervised learning model [19].

Recently, deep learning has been used to learn representations of natural language such as sentence representation [17, 18], document representation [20]. And the encoder-decoder framework based on recurrent neural networks has demonstrated powerful ability on various of sequence learning tasks such as short text generation [9], summarization [21]. To learn representation of paragraph or document, Li et al., [16] introduced a hierarchical neural auto-encoder based on LSTM. However, the human interactions in conversation highlight the need for capturing the interactive context to learn

S. Arik et al. (Eds.): ICONIP 2015, Part I, LNCS 9489, pp. 310–317, 2015.
DOI: 10.1007/978-3-319-26532-2_34

conversation representation, except for the temporal context. Take the efficient of hierarchical model into consider, our approach represents sentence by mean pooling word vectors, instead of learning sentence embedding from word embeddings.

In this paper, we explore unsupervised learning model to learn conversation representation, and propose an auto-encoder, which consists of LSTM-encoder and LSTM-decoder, to capture the interactive context in a conversation. Unlike the sequence to sequence learning framework described in [7], the LSTM-encoder in our model uses the interactive scheme to run through the sequence of sentences, and learns one hidden representation of the conversation. Based on the learnt representation, the LSTM-decoder generates the reconstructed sentence vectors of a conversation. To achieve the quantitative evaluation of the auto-encoder, we initialize the LSTM-classifier with the learned weights of our model, and fine tune it for the supervised task of topic recognition. Experimental results demonstrate the auto-encoder is able to learn good conversation representation for the supervised task.

2   The Dataset for Our Work

In this paper, the model is trained on a collected corpus, which consists of 32,881 human-human conversations on the online shopping platform[1].

The human-human conversation is a temporal and interactive sequence of sentences, but quite free-form. Firstly, most of sentences in conversation are short-message. Secondly, the user tends to use misspelling words and emoticons (e.g., "/: ^_^" and "/:>_<"). In addition, this data in human-human conversation is less structured than human-computer conversation, in which clients' input and agents' responses alternate consecutively. Due to the asynchronism of the interactive process, the conversation is segmented into uncertain turns, e.g., the seller's response may take multiple turns as the sentence 2 and 3 shown in Table 1.

Table 1. Human-human conversation on online shopping

Client	Merchant
1: 你好,你家的小吃能够包邮吗? Hi, will you deliver the snack free of charge?	2: 你好,你先拍下,我看下 Hi, i will check it as soon as you take the order.
\\\\	3: 宁波,是吧?改好价格了 Ninbo City, right? OK, i have charged the fee for you.
4: 再便宜 2 元吧 How about two yuan cheaper?	5: 包邮后没有什么利润了 It is less profit margin for us after free shipping.
6: /: ^_^ ~ 凑个整数呗 /: ^_^ ~ Make the deal an even figure.	7: 亲,不能了,包邮已经很吃力了 Honey, i can't. Free shipping is the best we can offer.
8: 就是赠几颗糖也是可以的啊 It's ok to have a few candies for free gifts.	9: 不好意思,没有的哦 Sorry, no gifts to deliver.

---

[1] http://www.taoboo.com/.

To construct this dataset, we first collect thousands of logs of real web-based conversations from 3 online shopping merchants, and then filter out redundant black spaces in conversations, finally, we replace the privacy information with pre-defined semantic labels (e.g., "<Human-name>", "<Cell-phone>" and "<Email>").

In our work, 5,000 conversations are used to train unsupervised models, and 1,000 conversations are labeled for supervised learning task – topic recognition. 500 conversations are used to fine tune unsupervised models and train topic classifiers, 500 conversations are used as the test dataset. Table 2 shows the statistics of the dataset.

Table 2. The statistics of the dataset

Dataset	#Conversation	#Sentence
Collected data	32,881	683,157
Unsupervised learning	5,000	102,783
Fine tune/Training	500	10,381
Test	500	9,862

3   The Auto-Encoder Model

To learn the conversation representation, we propose interactive auto-encoder to unsupervised learning by using LSTM. The basic idea is to use the sequence learning framework [9] to reconstruct sentence vectors of a conversation. First, our model runs through an input sequence to learn one hidden representation, which is the interactive context of a conversation, and then generate the corresponding vectors of the target sequence based on the learnt representations. The target sequence is the reverse input sequence, which makes the optimization easier for our model by looking at low range correlation. Figure 1 summarizes the process of learning conversation representation.

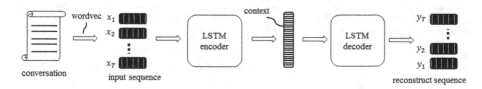

Fig. 1. The framework of learning conversation representation

3.1   LSTM

As the basic building block of our model, The LSTM unit, which has been successfully to perform sequence learning [9, 10], is used to learn the context and structure in conversation. Unlike to traditional recurrent unit, LSTM unit modulates the memory at each step, instead of overwriting the states. This makes it better to exploit long range dependencies [12] and discover long-range features in the sequence of sentences. The key component of LSTM unit is the cell which has a state c_t over time, and the LSTM unit decides to modify and add the memory in the cell via the sigmoidal gates – input gate i_t, forget gate f_t and output gate o_t. These updates for LSTM unit are chose the one discussed in [6], and summarized as Eq. 1.

$$i_t = \sigma\left(W_{xi}x_t + W_{hi}h_{t-1} + W_{ci}c_{t-1} + b_i\right)$$
$$f_t = \sigma\left(W_{xf}x_t + W_{hf}h_{t-1} + W_{cf}c_{t-1} + b_f\right)$$
$$c_t = f_t c_{t-1} + i_t tanh\left(W_{xc}x_t + W_{hc}h_{t-1} + b_c\right) \qquad (1)$$
$$o_t = \sigma\left(W_{xo}x_t + W_{ho}h_{t-1} + W_{co}c_t + b_o\right)$$
$$h_t = o_t tanh\left(c_t\right)$$

(W_*, b_*) is the parameters of LSTM, W_{cf}, W_{ci}, and W_{co} are diagonal matrices.

3.2 The Interactive Scheme in LSTM-Encoder

Using LSTM units, general sequence learning framework [7] is able to capture temporal context as global encoding scheme, shown in Fig. 2a, but fail to learn the interactive context of a conversation. Inspired by this point, our model uses the interactive scheme, shown in Fig. 2b, to encode the human interactions of a conversation.

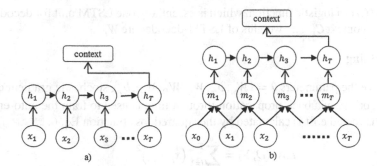

a)   b)

Fig. 2. Encoding schemes in LSTM-encoder. Figure (a) shows the global scheme to learn the global context; Figure (b) shows the interactive scheme to learn the interactive context.

Given an input sequence of sentences $I = [x_1, x_2, \ldots, x_T]$ of a conversation, we extend general LSTM-encoder with one hidden layer H_m to achieve interactive encoding. The joint representation m_t learned by LSTM-encoder is computed as Eq. 2:

$$m_t = Fm\left(\left(x_{t-1}, x_t\right); W_m, b_m\right) = \sigma\left(W_m i_t + b_m\right) \qquad (2)$$

Where i_t is the input vector of semantic matching by concatenating x_{t-1} and x_t, and x_0 is one all-zero padding vector. $\left(W_m, b_m\right)$ are the parameters of the matching layer.

Based on the joint representation m_t, the LSTM-encoder of our model captures the matching patterns over time. The final output h_T is the global representation after encoding multi-turns interactions, which is essentially the learnt interactive context C_T of a conversation and calculated as Eq. 3:

$$C_T = e\left(m_T, h_{T-1}, c_{T-1}\right) \tag{3}$$

Here $e(.)$ is a logistic function, which is one LSTM unit for encoding the conversation. The weights of LSTM-encoder are W_E.

3.3 The Computation in LSTM-Decoder

As shown in Fig. 1, the LSTM-decoder decodes the learned contexts $C_T = h_T$ to generate one sequence of sentence vectors $Y = [y_1, y_2, \dots, y_T]$, which is the reconstructed sequence of sentence representations at the output units, but in reverse order. Each sentence representation y_t is given by Eq. 4.

$$y_t = \sigma\left(W_{hy}h_t + b_y\right) \tag{4}$$

where σ is the activation function, (W_{hy}, b_y) is the parameters of generating the sentence vector. The hidden state h_t of decoder at time t is calculated as Eq. 5:

$$h_t = d\left(y_{t-1}, h_{t-1}, c_t\right) \tag{5}$$

Here $d(.)$ is a logistic function, which is essentially one LSTM unit for decoding the interactive context C_t. The weights of LSTM-decoder are W_D.

3.4 Training

To optimize the parameters $\theta = \left(W_m, b_m, W_E, W_D, W_{hy}, b_y\right)$, stochastic gradient descent (SGD) algorithm via back-propagation through time is used to train the auto-encoder. The reconstructed cost is calculated by the squared loss function Eq. 6.

$$Loss\left(O, Y\right) = \sum_{i=1}^{T} \left(x_i - y_i\right)^2 \tag{6}$$

where $O = [x_T, x_{T-1}, \dots, x_1]$ is the target sequence of sentence vectors, each vector is computed by mean pooling the corresponding word vectors. The word embedding is pre-trained with the unsupervised neural language model [14] on Collected Data, and the length of each word vector is 300.

4   Experiments

We design experiments to accomplish two objectives: First, evaluate the benefit of initializing LSTM-classifiers with the weights trained by unsupervised learning, especially with very few training samples. Second, evaluate our auto-encoder in learning conversation representation, which is given by interactive encoding scheme.

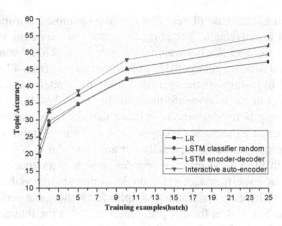

Fig. 3. Comparison of different models on topic recognition

4.1 Experimental Setting

Table 2 in Sec. 2 shows the experimental dataset. To evaluate our model, we conduct incremental experiments on labeled data, in which each batch is 20 conversations.

Baselines and Model Setting: In this work, the baseline of unsupervised learning model is LSTM encoder-decoder [7]. For the topic recognition, LTSM-classifier initialized with the weights of auto-encoder is compared to the one initialized with the weights of baseline. In addition, the logistic regression (LR) classifier and LTSM-classifier with randomly initialized weights are two supervised models for this task.

For our model, the size of *input gate* in LSTM units is set to 300. The sizes of *forget gate*, *output gate*, and *memory cells* in LSTM units are all set to 500.

Evaluation Metric: We use the accuracy of topic recognition to evaluate the models on supervised learning task.

4.2 Experimental Results and Analysis

Figure 3 shows the experimental results of competitor models on incremental experiments, and the summarized results are shown in Table 3.

Table 3. Summary of results on supervised task (%).

Model	One-batch-train (#20)	Full-train (#500)
LR	19.42	47.32
LSTM-classifier random	21.77	49.53
LSTM encoder-decoder	24.63	52.17
Interactive auto-encoder	**25.99**	**54.96**

We can see that for the case of very few training examples, unsupervised learning gives a substantial improvement. The interactive auto-encoder achieves the improvement from 19.42 % to 25.99 % when training on only one batch (20 conversations). For the full-train labeled data, we get a considerable improvement from 47.32 % to 54.96 %. The improvement by using our unsupervised learning model was not as big as we expected, however, our model outperforms than the strong baseline - LSTM encoder-decoder model. These promising results indicate that the conversation representation learned from sequence to sequence unsupervised learning is helpful to improve the performance of topic recognition, especially our auto-encoder.

Comparing with the LSTM encoder-decoder, which is another one unsupervised model, the classification performance of our model is almost same as the LSTM encoder-decoder model when there are few training samples (2 batches), but performs better with more training data. The reason for this phenomenon is that the interactive scheme for encoding conversation is difficult to find interactive context in the case of very few training data, the model still can capture the temporal context in conversations. But as the size of the labeled data grows, our model performs the advantage of interactive scheme in capturing and representing the interactive context by modeling the interactions in conversations.

5   Conclusions and Future Work

In this paper, the auto-encoder is proposed to learn interactive conversation representation based on the sequence to sequence learning framework. We evaluate our approach by fine tuning the model for topic recognition task. Experimental results demonstrate that the unsupervised models using LSTM units can learn useful conversation representation to improve the classification performance. Moreover, the improvements performed by auto-encoder shows that the interactive encoding scheme is able to capture interactive context and structure in a conversation. In the future, we plan to extend the auto-encoder with multi RNN layers to exploit the semantic relationships of the sentences in a conversation. Additionally, we will integrate the interactive scheme into hierarchical auto-encoder [16] to explore answer generation for multi-rounds conversation.

Acknowledgements.   This paper is supported in part by grants: National 863 Program of China (2015AA015405), National Natural Science Foundation of China (61473101 and 61272383).

References

1. Hsueh, P.-Y., Moore, J.D., Renals, S.: Automatic segmentation of multi-party dialogue. In: Proceedings of EACL, pp. 273–280 (2006)
2. Purver, M., Körding, K., Griffiths, T.L., Tenenbaum, J.: Unsupervised topic modelling for multi-party spoken discourse. In: Proceedings of COLING-ACL, pp. 17–24 (2006)

3. Stolcke, A., Coccaro, N., Bates, R., Taylor, P., Van Ess-Dykema, C., Ries, K., Shriberg, E., Jurafsky, D., Martin, R., Meteer, M.: Dialogue act modeling for automatic tagging and recognition of conversational speech. Comput. Linguist. **26**(3), 339–374 (2000)
4. Rieser, V., Lemon, O.: Natural language generation as planning under uncertainty for spoken dialogue systems. In: Proceedings of EACL, pp. 683–691 (2009)
5. Liu, J., Seneff, S., Zue, V.: Dialogue-oriented review summary generation for spoken dialogue recommendation systems. In: Proceedings of NAACL, pp. 64–72 (2010)
6. Graves, A.: Generating sequences with recurrent neural networks. CoRR, abs/1308.0850 (2013)
7. Sutskever, I., Vinyals, O., Le, Q.V.V.: Sequence to sequence learning with neural networks. In: Advances in NIPS, pp. 3104–3112, (2014)
8. Cho, K., van Merrienboer, B.., Gulcehre, C., Bahdanau, D., Bougares, F., Schwenk, H., Bengio, Y.: Learning phrase representations using RNN encoder-decoder for statistical machine translation. In: Proceedings of EMNLP, pp. 1724–1734 (2014)
9. Shang, L., Lu, Z., Li, H.: Neural responding machine for short-text conversation. CoRR, abs/1503.02364 (2015)
10. Srivastava, N., Mansimov, E., Salakhutdinov, R.: Unsupervised learning of video representations using LSTMs. CoRR, abs/1502.04681 (2015)
11. Zhai, K., Williams, J.: Discovering latent structure in task-oriented dialogues. In: Proceedings of ACL. pp. 36–46 (2014)
12. Hochreiter, S., Bengio, Y., Frasconi, P., Schmidhuber, J.: Gradient flow in recurrent nets: the difficulty of learning long-term dependencies. In: Kolen, J.F., Kremer, S. (eds.) A Field Guide to Dynamical Recurrent Neural Networks, vol. 28, pp. 297–318. IEEE Press, New York (2001)
13. Graves, A., Jaitly, N.: Towards end-to-end speech recognition with recurrent neural networks. In Proceedings of ICML, pp. 1764–1772, (2014)
14. Mikolov, T.., Chen, K., Corrado, G., Dean, J.: Efficient estimation of word representations in vector space. CoRR, abs/1301.3781 (2013)
15. Zhou, X., Hu, B., Chen, Q., Tang, B., Wang, X.: Answer sequence learning with neural networks for answer selection in community question answering. In: Proceedings of ACL-IJCNLP, pp. 713–718 (2015)
16. Li, J., Luong, M.-T., Jurafsky, D.: A hierarchical neural autoencoder for paragraphs and documents. In: Proceedings of ACL-IJCNLP, pp. 1106–1115 (2015)
17. Hu, B., Lu, Z., Li, H., Chen, Q.: Convolutional neural network architectures for matching natural language sentences. In: Advances in NIPS, pp. 2042–2050 (2014)
18. Socher, R., Pennington, J., Huang, E.H., Ng, A.Y., Manning, C.D.: Semi-supervised recursive autoencoders for predicting sentiment distributions. In: Proceedings of EMNLP, pp. 151–161 (2011)
19. Hinton, G.E., Salakhutdinov, R.R.: Reducing the dimensionality of data with neural networks. Science **313**(5786), 504–507 (2006)
20. Le, Q., Mikolov, T.: Distributed representations of sentences and documents. In: Proceedings of ICML, pp. 1189–1196 (2014)
21. Hu, B., Chen, Q., Zhu, F.: LCSTS: a large scale chinese short text summarization dataset. CoRR, abs/1506.05865 (2015)
22. Jurafsky, D., Shriberg, E., Biasca, D.: Switchboard SWBD-DAMSL shallowdiscourse-function annotation coders manual. Institute of Cognitive Science Technical report, pp. 97–02 (1997)

On the Use of Score Ratio with Distance-Based Classifiers in Biometric Signature Recognition

Carlos Vivaracho-Pascual[✉], Arancha Simon-Hurtado,
and Esperanza Manso-Martinez

Department Informática, Universidad de Valladolid, Valladolid, Spain
{cevp,arancha,manso}@infor.uva.es

Abstract. Biometric user verification or authentication is a pattern recognition problem that can be stated as a basic hypothesis test: X is from client C (H_0) vs. X is not from client C (H_1), where X is the biometric input sample (face, fingerprint, etc.). When probabilistic classifiers are used (e.g., Hidden Markov Models), the decision is typically performed by means of the likelihood ratio: $P(X/H_0)/P(X/H_1)$. However, as far as we know, this ratio is not usually performed when distance-based classifiers (e.g., Dynamic Time Warping) are used. Following that idea, we propose, here, to perform the decision based not only on the score ("score" being the classifier output) supposing X is from the client (H_0), but also using the score supposing X is not from the client (H_1), by means of the ratio between both scores: the score ratio. A first approach to this proposal can be seen in this work, showing that to use the score ratio can be an interesting technique to improve distance-based biometric systems. This research has focused on the biometric signature, where several state of the art systems based on distance can be found. Here, the score ratio proposal is tested in three of them, achieving great improvements in the majority of the tests performed. The best verification results have been achieved with the use of the score ratio, improving the best ones without the score ratio by, on average, 24 %.

Keywords: Score ratio · Signature verification · Distance-based classifier

1 Introduction

This work focuses on the use of score ratio in biometric person verification systems which use distance-based classifiers.

The goal in biometric person verification is to authenticate the user (client or Target Class, TC) by means of unique human characteristics (biometrics, e.g., iris, fingerprint, etc.). Given a test sample X, the problem of biometric verification can be stated as a basic hypothesis test between two hypotheses:

$$H_0 : X \text{ is from client } C \quad H_1 : X \text{ is not from client } C$$

© Springer International Publishing Switzerland 2015
S. Arik et al. (Eds.): ICONIP 2015, Part I, LNCS 9489, pp. 318–327, 2015.
DOI: 10.1007/978-3-319-26532-2_35

The decision between the two hypotheses can be performed as shown in Eq. (1), using client information only, or it can be performed by means of the likelihood ratio test given by Eq. (2), also using impostor information. $p(X/H_0)$ and $p(X/H_1)$ are, respectively, the probability density functions for the hypotheses H_0 and H_1 evaluated for the observed biometric sample X, and θ is the decision threshold.

$$p(X/H_0) \begin{cases} \geq \theta \; Accept \; H_0 \\ < \theta \; Reject \; H_0 \end{cases} \tag{1}$$

$$\frac{p(X/H_0)}{p(X/H_1)} \begin{cases} \geq \theta \; Accept \; H_0 \\ < \theta \; Reject \; H_0 \end{cases} \tag{2}$$

Biometric verification in general, and signature verification in particular, is a pattern recognition problem, where each client C is represented by means of a model λ_C, e.g., Hidden Markov Model (HMM), Gaussian Mixture Model (GMM), Support Vector Machine, etc. Then, $p(X/H_0)$, whose calculation is not a straightforward task, is estimated (approximated) by means of the classifier output (score) $s(X/\lambda_C)$. $p(X/H_1)$ is estimated by means of the score $s(X/\lambda_{\overline{C}})$, where $\lambda_{\overline{C}}$ is the impostor (Non Target Class) model, "impostor" being anybody different from the client. Since an accurate representation of the impostor class is impossible, different $\lambda_{\overline{C}}$ model estimation approaches can be found in the literature; several of them will be shown in Sect. 2.

Following this pattern recognition approximation, Eq. (1) becomes Eq. (3), and Eq. (2) becomes the **score ratio** shown in Eq. (4).

$$s(X/\lambda_C) \begin{cases} \geq \theta \; Accept \; H_0 \\ < \theta \; Reject \; H_0 \end{cases} \tag{3}$$

$$\frac{s(X/\lambda_C)}{s(X/\lambda_{\overline{C}})} \begin{cases} \geq \theta \; Accept \; H_0 \\ < \theta \; Reject \; H_0 \end{cases} \tag{4}$$

When probability-based classifiers (e.g., HMM or GMM) are used, the classifier output can be interpreted as a probability, $p(X/\lambda_C)$, and the decision has been typically performed by means of the likelihood ratio $\frac{p(X/\lambda_C)}{p(X/\lambda_{\overline{C}})}$ test [1,2,4], since better performance is achieved. However, when distance-based classifiers are used, as far as we know, this score ratio (Eq. 4) is not usually performed.

Here, a first approximation to this proposal is successfully approached, showing that to use the score ratio in biometric systems based on distance classifiers can improve the system, that is, it can be an interesting alternative.

The proposal has been tested in biometric signature recognition. Of the several biometrics, signature is the second most important of the behavioral biometrics. Here, on-line signature (the signature is written in a digitizing device) is used. Depending on the test conditions, two types of forgeries can be established: (i) skilled forgery, where the impostor imitates the client signature, and (ii) random forgery, where the impostor uses his/her own signature as a forgery.

The score ratio proposal has been tested in three different approaches of systems based on distance classifiers [3,6,8]. For the impostors representation, the "cohort" approach (this will be seen in greater depth in Sect. 2) was used: a set of M signatories (called the cohort set) is used as impostor representatives, choosing a subset of N of them to get $s(X/\lambda_{\overline{C}})$. Different values of M and N have been tested.

The rest of the paper is organized as follows. Section 2 gives a brief theoretical background of the score ratio problem, showing the approach chosen in this work. After describing the experimental setup in Sect. 3, the results achieved with and without score ratio in all of the tested scenarios can be seen in Sect. 4. The conclusions and future works are shown in Sect. 5.

2   Score Ratio

The score ratio proposal (Eq. 4) can generally be found associated with systems based on probabilistic classifiers in the biometric literature. In these works, as it has been pointed out in the introduction, the classifier output (score) can be interpreted as a probability, performing the likelihood ratio test: $\frac{p(X/\lambda_C)}{p(X/\lambda_{\overline{C}})}$.

Two ways to get $p(X/\lambda_{\overline{C}})$ can be found: using a cohort set (representatives set) of the Non Target Class (NTC) [2,4], or using a single model to explain the behavior of the NTC [1].

When the cohort set alternative is used, $p(X/\lambda_{\overline{C}})$ is calculated as the product of the scores over the cohort set models, $\lambda_{\overline{C}}^i$: $p(X/\lambda_{\overline{C}}) = \prod_i p(X/\lambda_{\overline{C}}^i)$, where i is each element of the cohort set. For an adequate representation of the NTC (impostors), the cohort set size must be large, then, a big amount of calculations over the cohort set is necessary. Several methods are applied to simplify this calculation. The most often used is to select the N highest likelihoods from the cohort set [2] ($p(X/\lambda_{\overline{C}}) = \prod_i^N p(X/\lambda_{\overline{C}}^i)$), that means, using only the N cohort set elements closest to the client (λ_C).

When a single model is used to estimate $p(X/\lambda_{\overline{C}})$, the model is trained using samples provided by many different users. For example, in [1], a User Adapted Universal Background Model (UA-UBM) based on a discrete HMM is proposed. First, they obtain a UBM trained using signatures from many signatories. Then, the client model is achieved by adapting the UBM with the enrollment (training) client signatures. The score ratio is performed by means of the log likelihood: $\log \frac{p(X/\lambda_C)}{p(X/UBM)}$.

With distance-based classifiers, it is impossible to get a single impostors model, so, here, the cohort set approximation is adopted:

$$\frac{s(X/\lambda_C)}{\sum_{i=1}^N s(X/\lambda_{\overline{C}}^i)} \tag{5}$$

Following the approach of selecting the cohort set elements closest to the client, here, the N cohort set signatories with the minimal distances with regard to the client are selected to perform the score ratio (Eq. 5). We have performed the score ratio test by crossing two parameters: the cohort set size (50, 100 and 150) and different values of N (1, 3, 5 and 10).

3 Experimental Setup

3.1 Score Normalization

Given a learning paradigm, its Match (client scores) and Non-Match (impostors scores) distributions vary from the classifier trained to distinguish one user from another (see Fig. 1(a)). For this reason, score normalization is essential to transform the scores of the client matchers into a common domain (Fig. 1(b)).

The main score normalization techniques for signature recognition [7] have been tested here:

Impostor-Centric Techniques:

IC1: $s_{norm} = s - \hat{\mu}_C^N$ **IC2:** $s_{norm} = s - (\hat{\mu}_C^N + \hat{\sigma}_C^N)$ **IC3:** $s_{norm} = (s - \hat{\mu}_C^N)/\hat{\sigma}_C^N$

where $\hat{\mu}_C^N$ and $\hat{\sigma}_C^N$ are the mean and the standard deviation of the Non-Target Class (impostors) scores distribution for the client C classifier, estimated by means of a *Cohort Gallery* (see Sect. 3.3).

Target-Centric Techniques:

TC1: $s_{norm} = s - \hat{\mu}_C^M$ **TC2:** $s_{norm} = s - (\hat{\mu}_C^M + \hat{\sigma}_C^M)$ **TC3:** $s_{norm} = (s - \hat{\mu}_C^M)/\hat{\sigma}_C^M$

where $\hat{\mu}_C^M$ and $\hat{\sigma}_C^M$ are the mean and the standard deviation of the Target Class (client) scores distribution for the client C classifier, estimated by means of the client model training signatures using the leave-one-out technique.

Target-Impostor Technique:

TI1: $s_{norm} = s - s_{EER_C}$

where s_{EER_C} is the a priori decision threshold of the client C at the Equal Error Rate point (see Sect. 3.5), obtained from the Non-Target and Target class scores distributions estimation.

3.2 Corpus

The MCYT database [5] has been used. This database is one of the most popular and largest in signature verification, and can be considered a benchmark. The signatures were acquired with a graphic tablet WACOM, obtaining from each sampling instant: position in x-axis and y-axis, pressure, azimuth and altitude

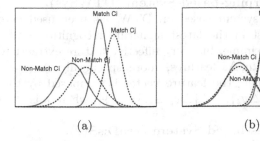

(a)   (b)

Fig. 1. Example of Match and Non-Match distributions of two users, (a) without score normalization and (b) with score normalization.

angles. Samples of 333 different people were acquired. Each target user produced 25 genuine signatures, and 25 skilled forgeries were also captured for each user.

3.3 Experimental Sets

The corpus was split into the following three different and separate subsets:

- **Cohort Set (ChS).** This was applied to get $s(X/\lambda_{\overline{C}})$ in order to perform the score ratio. Three different sizes were tested: (i) 50 signatories (**ChS-50**), (ii) 100 signatories (**ChS-100**) and (iii) 150 signatories (**ChS-150**). From this set, the N nearest to each client and test sample X were used to perform the score ratio (Sect. 2).
- **Cohort Gallery (ChG).** This was used to get $\hat{\mu}_C^N$ and $\hat{\sigma}_C^N$ to perform the score normalization (Sect. 3.1). The signatories from the ChS not used to perform the score ratio were included in this set.
- **Test Set (TS).** This was used to test. This consists of 183 signatories not used in the previous sets. The same TS was used in all of the tests performed for an objective comparison. From each user in this set: (i) his/her five first signatures were used to build the signatory model, the remaining twenty being used for *genuine tests* (3660 in total), (ii) his/her 25 skilled forgeries captured were used for *skilled tests* (4575 in total) and (iii) one signature out of 100 randomly selected users in the TS (different from the client) were used for *random forgery tests* (18300 in total).

3.4 Signature Verification Systems

The following three systems based on distances were tested:

Vector Quantization-based System (VQSys).

The system shown in [6] is used. This system achieves very good performance with a reduced computational requirement, which is around 47 times lower than DTW. In addition, the system improves the database storage requirements due to vector compression. The codebook size is 256. The features vector is composed by the X and Y coordinates, the pressure, the azimuth, the altitude and each point timestamp.

Dynamic Time Warping-based System (DTWSys).

Our state of the art system based on DTW has been used here [3]. This has been among the best in the latest signature recognition evaluation performed (ESRA'2011). A simple, but very effective feature extraction is accomplished by means of $delta, \Delta$, features, more specifically $\Delta x_i = x_{i+1} - x_i$ and $\Delta y_i = y_{i+1} - y_i$. Then, the feature vector is composed by the sequence: $F = \{\{\Delta x_1, \Delta y_1\}, \{\Delta x_2, \Delta y_2\}, \ldots, \{\Delta x_{p-1}, \Delta y_{p-1}\}\}$, where p is the number of signature points.

Fractional Distances-based System (FraDisSys).

Our low-cost proposal shown in [8] is used. This system has less computational and storage requirements than the previous ones. The signature points number is normalized to a fixed value (15 here), then, signatures can be matched by means of a simple distance calculation. Due to its better performance, fractional distances are used. An improved feature extraction has been accomplished here, adding the following new ones to the features in [8] (X, Y, Pressure, Azimuth and Altitude): point number, signature section length, direction changes per section and signature duration.

3.5 Performance Measure

Verification systems can be evaluated using the False Match Rate (FMR, those situations where an impostor is accepted) and the False Non Match Rate (FNMR, those situations where a user is incorrectly rejected). The performance can be plotted on an curve (e.g. ROC), but if the number of comparisons is high, the use of a single number measure is more useful and easier to understand. The most used one in the literature is the Equal Error Rate (EER), that is, the system error when FMR equals the FNMR. This is the measure used here.

4   Results

The score normalization (Sect. 2) performance is system dependent. So, firstly, we will choose the best score normalization technique with and without score ratio for each system. A representative cohort set size of 100 users has been chosen for these tests, using $N = 5$ for the score ratio. A criterion based on the lowest average between random forgery EER and skilled forgery EER has been used to choose the best score normalization technique in each case.

4.1   Vector Quantization-Based System (VQSys)

When VQSys is used, the best system configuration is, exceptionally, without using the score normalization for both with and without score ratio (Table 1). Once the best score normalization technique has been fixed (here, none), the comparative study for this system with and without score ratio can be seen in Table 2. The smallest error is achieved using the score ratio for both random and skilled forgeries, with a cohort set of 100 signatories and $N = 1$ and a cohort set of 150 signatories and $N = 3$, respectively.

4.2   Dynamic Time Warping-Based System (DTWSys)

When DTWSys is used, the best score normalization techniques are TI1 without the score ratio and TC1 when the score ratio is applied (Table 3). Once the best score normalization technique has been fixed, the comparative study for this system with and without score ratio is shown in Table 4. The smallest error for random forgery is achieved using the score ratio with a cohort set of 150 signatories and $N = 10$. However, for skilled forgery, the best error value has been achieved without the score ratio and with a cohort set of 50 signatories.

Table 1. Choosing the best score normalization technique for VQSys. Best results are bold face emphasized.The system error is measured by means of the EER (%).

| | NoScoreRatio | | ScoreRatio | |
| | ChS-100 | | ChS-100, N=5 | |
Technique	Random	Skilled	Random	Skilled
No norm.	**1.15**	**6.56**	**0.71**	**4.56**
IC1	0.85	9.97	1.20	13.01
IC2	0.87	11.24	1.61	13.92
IC3	0.87	14.18	1.15	13.53
TC1	3.37	12.16	2.35	9.45
TC2	3.36	12.24	2.43	9.68
TC3	18.36	26.56	19.78	26.45
TI1	1.04	7.08	0.63	5.33

Table 2. With and without ($N = 0$ row) score ratio performance comparison for VQSys. The results for all of the cohort set sizes and values of N are shown. The system error is measured by means of the EER (%). Best results are bold face emphasized.

| | ChS-50 | | ChS-100 | | ChS-150 | |
N	Random	Skilled	Random	Skilled	Random	Skilled
0	1.15	6.56	1.15	6.56	1.15	6.56
1	0.71	5.16	**0.63**	4.95	0.68	4.87
3	0.72	4.85	0.67	4.57	0.66	**4.45**
5	0.74	4.86	0.71	4.56	0.68	4.51
10	0.74	4.87	0.72	4.61	0.71	4.54

Table 3. Choosing the best score normalization technique for DTWSys. Best results are bold face emphasized. The system error is measured by means of the EER (%).

| | NoScoreRatio | | ScoreRatio | |
| | ChS-100 | | ChS-100, N=5 | |
Technique	Random	Skilled	Random	Skilled
No norm	14.45	20.55	0.79	9.01
IC1	1.80	7.37	20.46	38.95
IC2	2.49	9.51	34.58	44.63
IC3	1.88	8.02	1.22	15.54
TC1	2.54	5.87	**1.01**	**6.48**
TC2	2.80	6.04	1.53	8.17
TC3	7.95	9.70	8.83	16.01
TI1	**1.83**	**4.67**	1.31	10.45

Table 4. With and without ($N = 0$ row) score ratio performance comparison for DTWSys. The results for all of the cohort set sizes and values of N are shown. The system error is measured by means of the EER (%). Best results are bold face emphasized.

N	ChS-50		ChS-100		ChS-150	
	Random	Skilled	Random	Skilled	Random	Skilled
0	1.72	**4.66**	1.83	4.67	1.94	4.67
1	1.23	7.13	1.18	7.21	1.20	7.49
3	1.05	6.64	1.07	6.54	1.07	6.83
5	0.96	6.54	1.01	6.48	0.96	6.62
10	0.85	6.37	0.85	6.39	**0.83**	6.51

Table 5. Choosing the best score normalization technique for FraDisSys. Best results are bold face emphasized. The system error is measured by means of the EER (%).

Technique	NoScoreRatio		ScoreRatio	
	ChS-100		ChS-100, N=5	
	Random	Skilled	Random	Skilled
No norm	2.73	8.63	1.39	6.83
IC1	2.70	10.05	6.21	20.27
IC2	3.58	11.89	11.10	26.34
IC3	2.47	10.01	2.38	13.06
TC1	**2.16**	**6.08**	**1.31**	**6.15**
TC2	2.73	7.34	1.89	7.43
TC3	10.08	14.29	8.47	14.45
TI1	2.19	7.02	1.97	8.13

4.3   Fractional Distance-Based System (FraDisSys)

When FraDisSys is used, the best score normalization technique is TC1 for both with and without score ratio (Table 5). Once the best score normalization technique has been fixed, the comparative study for this system with and without score ratio can be seen in Table 6. The smallest error is achieved using the score ratio for both random and skilled forgeries, with a cohort set of 100 signatories and $N = 10$ and a cohort set of 50 signatories and $N = 10$, respectively.

4.4   Results Analysis

From the previous results, the first important consideration is that the use of the score ratio has improved all of the cases studied except one, skilled forgeries with DTW. For random forgeries, great improvements have been achieved with

Table 6. With and without ($N = 0$ row) score ratio performance comparison for FraDisSys. The results for all of the cohort set sizes and values of N are shown. The system error is measured by means of the EER (%). Best results are bold face emphasized.

N	ChS-50		ChS-100		ChS-150	
	Random	Skilled	Random	Skilled	Random	Skilled
0	2.16	6.08	2.16	6.08	2.16	6.08
1	1.94	7.02	1.64	6.94	1.72	7.16
3	1.53	6.31	1.45	6.39	1.49	6.23
5	1.37	6.17	1.31	6.15	1.42	6.10
10	1.28	**5.85**	**1.23**	5.93	1.34	5.93

all of the classifiers: 45 % for VQSys (ChS-100, $N = 1$), 52 % for DTWSys (ChS-150, $N = 10$) and 43 % for FraDisSys (ChS-100, $N = 10$), with regard to the best results without score ratio, respectively. For skilled forgeries, the following improvements have been achieved: 32 % for VQSys (ChS-150, $N = 3$) and 4 % for FraDisSys (ChS-50, $N = 10$) with regard to the best results without score ratio, respectively. It is interesting to note that the skilled forgery tests are used only in signature recognition, while, for the rest of the biometrics, the impostor tests are performed by means of random samples (i.e., samples of other users), which is where the biggest improvements have been achieved with the use of the score ratio.

From the systems without score ratio, that based on DTW has achieved the best results for skilled forgeries (this can also be seen in the international competitions), while for random ones, the best results have been achieved with VQ. This is typical in signature recognition: improvements in a forgery type usually worsen the other. However, here, the use of the score ratio has allowed a system to be achieved with the best results for both forgeries on average: 0.66 % in random forgery and 4.45 % in skilled forgery for VQSys, with ChS-150 and $N = 3$.

The size of the cohort set does not seem to have any great influence on the results with score ratio. In general, better results have been achieved with sizes greater than 50, but this is classifier dependent and the differences in the results are very small. With regard to the value of N, the worst results have been achieved, in general, with 1, but for the rest of the values tested, the differences in the results are also very small. These results show that the score ratio proposal can be applied with small cohort sets and with small values of N, which is very important for real applications.

5   Conclusions and Future Works

The score ratio, generally applied in biometric probabilistic-based systems, has not been used, as far as we know, with distance-based ones. Here, we have shown that it can also be an interesting proposal for these systems. In this first approximation, the use of the score ratio has been successfully tested with three different biometric signature systems based on distance classifiers. Except for one case, improvements have been achieved in all of the tested scenarios. The best results have been achieved with the use of the score ratio for both random and skilled forgeries, improving the results achieved with the reference state of the art systems tested.

Several cohort set sizes and values of N (number of users used to get the impostor score) have been tried, showing that even with small values of both, good results can be achieved with the score ratio application. This is important for real applications, where limited data are usually available.

These results encourage us to continue with the proposal tried in this work, testing new proposals to get the score ratio denominator. It can also be interesting to test other biometrics, where, besides, the impostor tests are accomplished only with random forgeries, the case where greater improvements have been achieved with the score ratio application.

Acknowledgments. Thanks to A. F. Hynds B.A. Dip. TEFL for revising the English grammar.

References

1. Rua, E.A., Castro, J.L.A.: Online signature verification based on generative models. IEEE Trans. Syst. Man Cybern. B Cybern. **42**(4), 1231–1242 (2012)
2. Furui, S.: An overview of the speaker recognition technology. In: Proceedings of the Workshop on Automatic Speaker Recognition Identification and Verification, Martigny, Switzerland, pp. 1–9, 5–7 Apr 1994
3. Houmani, N., et al.: Biosecure signature evaluation campaign (BSEC'2009): evaluating online signature algorithms depending on the quality of signatures. Pattern Recogn. **45**(3), 993–1003 (2012)
4. Nanni, L., Lumini, A.: A supervised method to discriminate between impostors and genuine in biometry. Expert Syst. Appl. **36**(7), 10401–10407 (2009)
5. Ortega-Garcia, J., Fierrez, J., Simon, D., Gonzalez, J., Faundez-Zanuy, M., Hernaez, I., Espinosa, V., Satue, A., Igarza, J.J., Vivaracho, C., Escudero, D., Moro, Q.I.: MCYT baseline corpus: a bimodal biometric database. IEE Proc. Vis. Image Signal Process. **150**(6), 395–401 (2003)
6. Pascual-Gaspar, J.M., Faundez-Zanuy, M., Vivaracho, C.: Fast on-line signature recognition based on VQ with time modeling. Eng. Appl. Artif. Intell. **24**(2), 368–377 (2011)
7. Vivaracho-Pascual, C., Simon-Hurtado, A., Manso-Martinez, E., Pascual-Gaspar, J.M.: A new proposal for score normalization in biometric signature recognition based on client threshold prediction. In: 2012 IEEE 12th International Conference on Data Mining (ICDM), pp. 1128–1133, December 2012
8. Vivaracho-Pascual, C., Faundez-Zanuy, M., Pascual, J.M.: An efficient low cost approach for on-line signature recognition based on length normalization and fractional distances. Pattern Recogn. **42**(1), 183–193 (2009)

A Multifactor Dimensionality Reduction Based Associative Classification for Detecting SNP Interactions

Suneetha Uppu$^{(\boxtimes)}$, Aneesh Krishna, and Raj P. Gopalan

Department of Computing, Curtin University, Perth, Australia
Suneetha.uppu@postgrad.curtin.edu.au,
{A.Krishna,R.Gopalan}@curtin.edu.au

Abstract. Identification and characterization of interactions between genes have been increasingly explored in current Genome-wide association studies (GWAS). Several machine learning and data mining approaches have been proposed to identify the multi-locus interactions in higher order genomic data. However, detecting these interactions is challenging due to bio-molecular complexities and computational limitations. In this paper, a multifactor dimensionality reduction based associative classifier is proposed for detecting SNP interactions in genetic epidemiological studies. The approach is evaluated for one to six loci models by varying heritability, minor allele frequency, case-control ratios and sample size. The experimental results demonstrated significant improvements in accuracy for detecting interacting single nucleotide polymorphisms (SNPs) responsible for complex diseases when compared to the previous approaches. Further, the approach was successfully evaluated by using sporadic breast cancer data. The results show interactions among five polymorphisms in three different estrogen-metabolism genes.

Keywords: Epistasis · Genome wide association studies · Associative classification · SNP interactions · Multifactor dimensionality reduction

1 Introduction

Genome-wide association studies (GWAS) are increasingly being used to identify SNPs that underlay the genetic architecture of complex diseases. A SNP is a variation of a single nucleotide (A, C, G, and T) that occurs in coding and non-coding regions of a DNA sequence [1]. On average, SNPs occur once in every 300 nucleotides of the DNA. It has been estimated that about 10 million SNPs occur along the 3-billion-base human genome [1]. A number of GWAS have focused on the role of SNPs and their associations in revealing the genetic epidemiology of disease susceptibility. However, complex diseases occurring in biological systems are unknown due to multiple genetic factors, environmental factors, and their interaction effects [2]. The magnitude and prevalence of interactions and/or the joint actions of SNPs are being increasingly recognized in recent studies. A number of recent reviews have covered the current methods and the related software packages to detect these interacting SNPs that contribute to a disease [2–4]. The data mining and machine learning methods used to

© Springer International Publishing Switzerland 2015
S. Arik et al. (Eds.): ICONIP 2015, Part I, LNCS 9489, pp. 328–336, 2015.
DOI: 10.1007/978-3-319-26532-2_36

detect SNP interactions include random forests (RFs) [5], regression models [6], Bayesian models [7], multifactor dimensionality reduction (MDR) [8], neural networks (NNs) [9], support vector machines (SVMs) [4], cellular automata (CAs) [10], and pathway approaches [11].

However, detecting SNP interactions using these approaches are still challenging. The challenges include the computational burden imposed by the high dimensional search space, the complexity of genetic architecture, and the choice of evaluation measures to determine the contribution of the interactions to a disease [2]. Flexible approaches and the use of appropriate software play a vital role in revealing the complexity of diseases [12]. Among the recently emerging methods, MDR is the most prominent as it reduces the dimensionality of multi-locus information to a single dimension. It can also be used as a preprocessing or interleaving step (combining attribute selection with the construction method into an iterative process) [13]. Interleaving allows hierarchical interaction models to be constructed and the newly constructed attributes can be placed back into the dataset. Interleaving is a valid approach for GWA studies with more than 300,000 SNPs due to the computational limitations of exhaustive searches [13]. Many researchers have shown that Associative classification (AC) improves the classification performance by generating strong association rules with high confidence and low support [14]. The classifier uses up to k rules (ranging from one to ten) to determine the class of the test data. Hence, the accuracy rates of selected Class association rules (CARs) corresponding to the classes are higher than traditional classifiers [15]. The interactions between SNPs associated with the disease were explored by integrating association rules and classification in the previous work [16]. The approach was evaluated for two-locus interactions using balanced and imbalanced simulated data. The experimental results were encouraging in imbalanced data compared to balanced data in terms of accuracy [16].

In this paper, the research has progressed by proposing an MDR based associative classifier (MDRAC) for revealing the unexplained features of a complex disease due to interaction effects. The approach constructs the new attribute with high or low risk factors using MDR (constructive induction method) [8] and classifies using an associative classifier (CPAR) [18]. The approach was evaluated for one to six loci models on the simulated data by varying heritability, minor allele frequencies, case-control ratios, and sample size. The experimental results demonstrate the substantial improvements in accuracy over existing MDR based methods by reducing classification errors. The proposed approach identifies interesting multi-locus SNPs that were not identified by previous MDR based techniques. Further, the method was successfully evaluated over sporadic breast cancer data. The approach identified the five-locus interaction model and its association with the disease.

2 Methods

2.1 MDR Based Associative Classifier (MDRAC)

Multifactor dimensionality reduction (MDR) is a widely used data mining approach for identifying interactions between genes. It is a non-parametric model which reduces high dimensional genetic data to a single dimensional data using the constructive induction approach [8]. It exhaustively searches all the possible n-locus SNP interactions associated with a complex genetic disease. The proposed approach implements a series of six steps to find the best model for a genetic trait.

Step 1 performs ten-fold cross validation on dataset of sample size N (with n_0 controls and n_1 cases). The dataset is divided into ten equal subsets in which, nine subsets are used as training data and one subset is used as testing data.

Step 2 enumerates all possible combinations of n loci. For each combination of loci, the number of cases and controls is counted for 3^n genotype combinations. For example, each SNP has two possible alleles (A and a). The possible genotypes are AA (common homozygous subjects), Aa/aA (heterozygous subjects) and aa (variant homozygous subjects) due to duplication of DNA in each cell of the individual. Statistically, AA is represented as zero, Aa/aA is represented as one and aa is represented as two. Hence, the interaction between two SNPs with three genotypes will have $3^2 = 9$ two locus genotypes. That is, a three by three contingency table is created and subjects are placed in their corresponding cells (each cell is referred to as a multifactor cell).

In step 3, the ratio of cases to controls is calculated for each factor and their corresponding value is compared with the threshold to classify the factor as either high risk or low risk. The genotype combinations that are not represented in the dataset are left blank. In balanced data, where the number of cases and controls are equal, the threshold is equal to one. A multifactor class is classified as high risk when their corresponding case to control ratio is higher than one. The multifactor cell is classified as low risk if the corresponding case to control ratio is lower than one. In imbalanced data (the number of cases and controls are not equal), the threshold is adjusted and its value is used to classify each multifactor cell as either high risk or low risk. The high risk factors are pooled together and are labeled as G1. The low risk factors are pooled together and are labeled as G0. Hence, the data is reduced from n dimensional space to one dimensional space by forming a new attribute with G0 and G1 levels.

In step 4, the new attribute is constructed by the MDR method [8] and it uses the weighted FOIL gain associative classifier for the classification [18]. Laplace training accuracy for each rule of the n-locus model is calculated. The rules with the highest expected training accuracy are selected. The best rules generated for each class in a rule set are used for the prediction. Overall classification accuracy is calculated from the observed data.

For step 5, steps from 2 to 4 are repeated for each possible combination and its overall classification accuracy is calculated. Finally, the best model based on the highest classification accuracy is selected. The goal of this procedure is to minimize the misclassification rates.

In step 6, to avoid over fitting of the data, steps 1 to 5 are repeated for all ten cross validation intervals. The Laplace expected accuracy is averaged for each n-locus model in all ten cross validations. Finally, the overall best n locus model is selected with high cross validation consistency (CVC) and high prediction accuracy, where CVC is defined as the number of times the n-locus model is selected as the best model during the cross validation. The balanced accuracy is calculated when cases and controls are not equal and is defined as the mean of sensitivity and specificity [19].

$$Balanced\ Accuracy = \frac{Sensitivity + Specificity}{2}$$

$$where, \quad Sensitivity = \frac{TP}{TP + FN} \quad and \quad Specificity = \frac{TN}{TN + TP}$$

where TP are true positives, FN are false negatives, and TN are true negatives. The statistical significance of the predicted model is evaluated using permutation testing. The more parsimonious model is selected if the models with highest prediction accuracy and highest CVC are different [20].

2.2 Data Simulation

In this section, a simulation based study is performed before applying to the real data. The n locus interaction models are generated from publicly available tool GAMETES [21]. The datasets are generated for single-locus to six locus models with 20 SNPs. Each genetic model is distributed across seven heritability 0.01, 0.025, 0.05, 0.1, 0.2, 0.3 and 0.4 along with two different minor allele frequencies 0.2 and 0.4 respectively [19]. Each model is created for 14 heritability-allele frequency combination in accordance with Hardy-Weinberg proportions. The case-control ratios of 1:1, 1:2, and 1:4 are generated for each sample size of 400, 800 and 1600. 100 datasets are generated for each model. Hence, 12,600 datasets are generated for each locus model. In total, 75,600 datasets has to be generated for one to six loci models. However, only 54,900 datasets were generated due to limited ability of GAMETES to generate models with higher heritability [21]. These datasets are analyzed using the proposed method to identify the interactions responsible for diseases.

2.3 Sporadic Breast Cancer Data

The data comprise of 410 samples obtained according to the requirements of the Institutional Review Board of Vanderbilt University Medical School [8]. The study is based on 207 white women with sporadic primary invasive breast cancer and 204 controls that were treated at Vanderbilt University Medical Centre. The genetic variants in five genes (COMT, CYP1A1, CYP1B1, GSTM1 and GSTT1) affected the metabolism of estrogens, which could increase the risk of breast cancer [8]. Hence, the analysis focused on the genes COMT (Catechol-O-methyl transferase) on chromosome 22q11.2, CYP1A1 (Cytochrome P450 1 A1 enzyme) on chromosome 15q22-q24, CYP1B1 (Cytochrome P450 1 B1 enzyme) on chromosome 2p21-22, GSTM1 (Glutathione S-transferase Mu 1) on chromosome 1p13.3 and GSTT1 (Glutathione S-transferase theta 1) on chromosome 22q11.2. The polymorphisms in these genes are summarised and reported in the research [8]. The dataset considered 10 SNPs (Cyp1A1m1, Cyp1A1m2, Cyp1A1m4, Cyp1B1-48, Cyp1B1-119, Cyp1B1-432, Cyp1B1-453, COMT, GSTM1, and GSTT1) in five genes for the analysis. Statistically, the possible genotypes are numerically represented as zero for AA, one for Aa/aA, and two for aa. There are 19 missing values and these are represented numerically by three.

3   Results

Several experiments are performed over simulated datasets and real breast cancer data, to evaluate the accuracy of MDRAC over other approaches. The goal of this study is to determine whether MDRAC is a better approach for identifying the higher order SNP interactions in the absence of main effect. The approach considers the ratio of cases and controls for each SNP combination at different loci. It generates statistically significant genotype combinatorial associations in terms of rules based on cases and controls. Predicting class labels of test objects from these rules retains higher accuracy in genetic combinations that contribute to a disease.

3.1   Analysis of Simulated Data

MDRAC identifies the interactions between SNPs that contributes to a phenotype. Further, the accuracy of MDRAC is evaluated and validated over the original MDR method and associative classification. The methods are evaluated with and without adjusting threshold levels. Table 1 summarises the evaluation results from single locus to six loci interactions. Figure 1 graphically represents the multi-locus analysis of SNPs. In the single-locus models, the results show that MDRAC performed better than MDR and AC both in balanced and imbalanced datasets. In a 1:1 ratio, AC and MDR performed almost equally across all models. In 1:2 ratio, the accuracy of AC is slightly better than MDR by about 5 %. In ratio 1:4, the accuracy of MDRAC is about 23 % to 31 % higher than MDR for both balanced and imbalanced datasets. In two-locus models, MDR performed better than AC in all 14 models of 1:1 case-control ratio. MDR and AC performed almost equally in 1:2 ratio. In 1:4 ratio, AC performed well (11 % to 16 %) compared to MDR. However, accuracy of MDRAC is higher than MDR and AC by up to 27 %. In three-locus models, the results demonstrated maxi-mum accuracy for MDRAC in all 12 models. MDR performed better than AC in 1:1 case-control ratios and performed equally well with AC in 1:2 case-control ratios. However, MDRAC predominantly performed well compared to MDR and AC with approximately 30 % increase in prediction accuracy. In four-locus interaction models, as expected AC performed poorly in balanced datasets and performed well in

Table 1. Average balanced accuracy of one to six locus models

Sample Size	Ratio	Single Locus MDR	AC	MDRAC	Two Locus MDR	AC	MDRAC	Three Locus MDR	AC	MDRAC	Four Locus MDR	AC	MDRAC	Five Locus MDR	AC	MDRAC	Six Locus MDR	AC	MDRAC
	1:1	65.8571	64.4643	94.3571429	64.9779	61.4821	94.5714286	62.2292	55.8958	94.875	57.225	52.15	95.65	53	54.6875	97.1875	50.7143	51.9286	96.0714286
	1:2	66.85	71.66	93.9464286	64.9586	68.8036	95.25	61.7083	64.1458	94.6458333	56.567	62.025	95.7	51.4225	62.8125	97.3125	52.8057	57.2143	96.2142857
400	1:4	65.22	80.79	96.5714286	63.2036	79.3929	95.7142857	59.205	79.2917	96	55.141	78.25	96.55	54.96	78.0625	98.5625	52.3014	76	97.4642857
	1:1	65.62	61.65	90.4285714	66.8114	64.0107	93.2767857	62.8342	53.4688	92.6770833	58.75	50.425	93.5125	53.81	49.375	96.78125	53.0543	51.1071	93.875
800	1:2	65.84	66.53	91.7946429	66.9293	66.0179	93.0446429	62.6233	60.9896	93.2708333	57.821	59.5625	94.3125	54.945	57	95.875	53.8186	58.1786	93.7142857
	1:4	64.85	78.63	93.9285714	65.47	76.5368	94.8482143	62.6875	75.2708	95.3333333	58.001	74.5	96.05	54.6475	73.8438	97.15625	52.7014	73.9286	97.1428571
	1:1	65.83	61.64	87.9464286	67.4386	62.0045	88.2991071	62.6567	55.4479	87.4635417	59.781	51.9438	89.025	53.53	50.3594	91.5625	54	49.5982	91.3214286
	1:2	65.43	67.93	89.34375	66.0164	67.1161	90.7901786	64.0192	63.1979	90.8489583	58.775	59.6188	91.2	54.385	58.8438	93.703125	53.8786	58.5089	90.625
1600	1:4	65.75	79.50	92.5401786	65.2357	77.9375	92.9017857	62.605	76.3073	92.7135417	58.434	75.0875	93.04375	53.8175	74.0625	95.84375	54.7214	74.535	94.6875

imbalanced datasets compared to MDR. Analysis of results from MDRAC showed maximum prediction accuracies ranging from 89 % to 96 %. This indicates that MDRAC can effectively eliminate prediction errors by increasing prediction accuracies. Similarly, MDRAC performed well for both five-locus and six-locus analysis. The results are also compared with and without adjusting the threshold value. It is concluded that MDRAC consistently had higher prediction accuracy for both balanced and imbalanced datasets. Hence, the rules generated from the model had better ability to identify the correct interaction model.

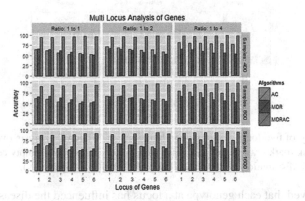

Fig. 1. Summary of multi-locus analysis for ratios 1:1, 1:2, and 1:4, and sample size 400, 800, and 1600.

3.2 Analysis of Breast Cancer Data

Table 2 summarises the prediction accuracy obtained from MDR and MDRAC analysis of the sporadic breast cancer case-control dataset. Both methods were evaluated for each number of loci from one to nine. The five-locus model was identified as the best model with the highest prediction accuracy and highest cross validation consistency (CVC). The best prediction model identified by MDR analysis includes the polymorphisms of COMT, GSTM1, CYP1A1m1, CYP1B1-codon 48, and CYP1B1-codon 432. The model had a maximum prediction accuracy of 53.41 and had a maximum CVC of ten. Statistical significance is determined by permutation testing under the null hypothesis of no association with disease. The $\chi2$ value of the model is 95.4553, whose p value is less than 0.05 (p < 0.05). Hence, the identified five locus model is statistically significant. It is suggested that the interactions between five SNPs that occurs in four genes contributes to the association of the disease.

The best prediction model identified by MDRAC analysis includes the polymorphisms of COMT, CYP1A1m1, CYP1B1-codon 119, CYP1B1-codon 432 and CYP1B1-codon 453. Figure 2, illustrates the five-locus genotype combinations associated with sporadic breast cancer. Each cell in the multi-locus genotype combinations are distributes with cases (left bars in the cell) and controls (right bars in the cell). The high risk patterns of cells are represented by the dark colour and low risk cells by the light colour. Finally, empty cells represent no genotype combinations of case-control

Table 2. Summary of results for breast cancer

Algorithms	Best model	No. of loci	Prediction accuracy	CVC
MDR	Cyp1A1m1, Cyp1B1-48, Cyp1B1-432, COMT, GSTM1	5	53.41	10/10
MDRAC	Cyp1B1-119, Cyp1B1-432, COMT, Cyp1A1m1, Cyp1B1-453	5	75.61	10/10

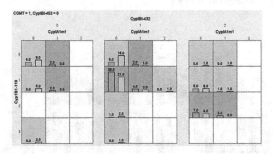

Fig. 2. Summary of five locus genotype combinations of MDRAC analysis over breast cancer data with high risk (dark gray shaded), low risk (light gray shaded), and empty cells (white) with the corresponding distribution of cases (left bars), and controls (right bars).

data. It is observed that each genotype at a locus has influenced the disease risk with the interactions of other two genes at different locus. The visual illustration of five interacting polymorphisms in three genes is graphically represented in Fig. 3. Tan lines indicate independence between SNPs. Red lines indicate synergistic relationship between polymorphisms in Cyp1B1-453, and COMT genes. The blue line indicates strong interactions of Cyp1B1-119 with COMT, Cyp1B1-453, and Cyp1A1m1. The identified model had highest prediction accuracy of 75.61 with highest CVC of ten. The accuracy of the predicted model by MDRAC is about 22 % higher than the model identified by MDR analysis. The statistical significance of the model is evaluated by chi square test with a p value less than 0.05. Hence, it is evident that the interactions between the five SNPs that occurred in three genes contributed to the disease and had better prediction ability than the MDR approach.

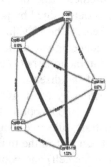

Fig. 3. Entropy graph for five-locus interaction model of MDRAC analysis over breast cancer data.

4 Conclusions and Future Work

In this paper, the MDR based associative classifier (MDRAC) was described for the detection and interpretation of epistasis in genetic and epidemiologic studies of complex diseases. The approach was evaluated for one to six loci models of both balanced and imbalanced simulated case-control data. The experimental results demonstrated substantial improvements in the accuracy for detecting interaction effects associated with the phenotype. Further, the approach was successfully evaluated over sporadic breast cancer data. The results showed that the five-locus interaction model was responsible for the disease. As these results are reported in terms of rules, their interpretation was straightforward. Further studies will investigate strategies such as parallel computational algorithms and incorporating better optimization algorithms to reduce the computational time.

References

1. Sheet, S.F., Human genome project. US Department of Energy Genome Program's Biological and Environmental Research Information System (BERIS). http://www.ornl.gov/sci/techresources/Human_Genome/. Accessed 28 July 2010
2. Padyukov, L.: Between the Lines of Genetic Code: Genetic Interactions in Understanding Disease and Complex Phenotypes. Academic Press, Waltham, MA (2013)
3. Cordell, H.J.: Detecting gene–gene interactions that underlie human diseases. Nat. Rev. Genet. **10**(6), 392–404 (2009)
4. Koo, C.L., et al.: A review for detecting gene-gene interactions using machine learning methods in genetic epidemiology. In: BioMed Research International (2013)
5. Qi, Y.: Random Forest for Bioinformatics. In: Zhang, C., Ma, Y. (eds.) Ensemble Machine Learning, pp. 307–323. Springer, New York (2012)
6. Chen, C.C., et al.: Methods for identifying SNP interactions: a review on variations of logic regression, random forest and Bayesian logistic regression. IEEE/ACM Trans. Comput. Biol. Bioinform. **8**(6), 1580–1591 (2011)
7. Zhang, Y., Liu, J.S.: Bayesian inference of epistatic interactions in case-control studies. Nat. Genet. **39**(9), 1167–1173 (2007)
8. Ritchie, M.D., et al.: Multifactor-dimensionality reduction reveals high-order interactions among estrogen-metabolism genes in sporadic breast cancer. Am. J. Hum. Genet. **69**(1), 138–147 (2001)
9. Motsinger-Reif, A.A., et al.: Comparison of approaches for machine-learning optimization of neural networks for detecting gene-gene interactions in genetic epidemiology. Genet. Epidemiol. **32**(4), 325–340 (2008)
10. McKinney, B.A., et al.: Machine learning for detecting gene-gene interactions. Appl. Bioinform. **5**(2), 77–88 (2006)
11. Ramanan, V.K., et al.: Pathway analysis of genomic data: concepts, methods, and prospects for future development. Trends Genet. **28**(7), 323–332 (2012)
12. Upstill-Goddard, R., et al.: Machine learning approaches for the discovery of gene–gene interactions in disease data. Briefings Bioinform. **14**(2), 251–260 (2013)

13. Moore, J.H., et al.: A flexible computational framework for detecting, characterizing, and interpreting statistical patterns of epistasis in genetic studies of human disease susceptibility. J. Theor. Biol. **241**(2), 252–261 (2006)
14. Thabtah, F.: A review of associative classification mining. Knowl. Eng. Rev. **22**(01), 37–65 (2007)
15. Yu, P., Wild, D.J.: Fast rule-based bioactivity prediction using associative classification mining. J. Cheminformatics **4**(1), 1–10 (2012)
16. Uppu, S., Krishna, A., Gopalan, R.P.: Detecting SNP Interactions in balanced and imbalanced datasets using associative classification. Aust. J. Intell. Inf. Process. Syst. **14**(1), 7–18 (2014)
17. Uppu, S., Krishna, A., Gopalan, R.P.: An associative classification based approach for detecting SNP-SNP interactions in high dimensional genome. In: IEEE International Conference on Bioinformatics and Bioengineering (BIBE). IEEE (2014)
18. Han, J.: CPAR: Classification based on predictive association rules. In: Proceedings of the Third SIAM International Conference on Data Mining (2003)
19. Velez, D.R., et al.: A balanced accuracy function for epistasis modeling in imbalanced datasets using multifactor dimensionality reduction. Genet. Epidemiol. **31**(4), 306–315 (2007)
20. Hahn, L.W., Ritchie, M.D., Moore, J.H.: Multifactor dimensionality reduction software for detecting gene–gene and gene–environment interactions. Bioinformatics **19**(3), 376–382 (2003)
21. Urbanowicz, R.J., et al.: GAMETES: a fast, direct algorithm for generating pure, strict, epistatic models with random architectures. BioData Min. **5**(1), 1–14 (2012)

Distributed *Q*-learning Controller
for a Multi-Intersection Traffic Network

Sahar Araghi[✉], Abbas Khosravi, and Douglas Creighton

Center for Intelligent Systems Research (CISR), Deakin University,
Waurn Ponds, VIC 3216, Australia
saraghi@deakin.edu.au

Abstract. This paper proposes a *Q*-learning based controller for a network of multi intersections. According to the increasing amount of traffic
congestion in modern cities, using an efficient control system is demanding. The proposed controller designed to adjust the green time for traffic signals by the aim of reducing the vehicles' travel delay time in a
multi-intersection network. The designed system is a distributed traffic
timing control model, applies individual controller for each intersection.
Each controller adjusts its own intersection's congestion while attempt
to reduce the travel delay time in whole traffic network. The results of
experiments indicate the satisfied efficiency of the developed distributed
Q-learning controller.

1 Introduction

Huge amount of traffic congestion in cities and especially modern areas needs
professional managing system. Different studies has been done in this regards.
Applying artificial intelligence (AI) techniques has significant influence in different application and traffic control system as well.

The ability of learning from experience is one of the characteristics of AI
methods that makes these methods useful to address real world problems. In
the case of transportation systems, concepts of intelligent agents is applied in
different areas such as traffic signals [8,9,16], vehicles [3], pedestrians [19], and
also to model the behavior of traffic system and detecting critical cases and
violations [10].

Among different approaches to handle increasing amount of traffic, controlling traffic signal's timing recognized as one of the beneficial methods to decrease
daily travel times and its side effects.

During last two decades several related publications has been released. Parts
of these methods handle control of traffic signals by predefined rule-based system
[11], fuzzy rules [12], and centralized techniques [7]. In addition, there are studies
that applied AI to control the signal timing just for an isolated intersection
[5,13,14,21]. However, techniques using for controlling single intersection may
not efficient enough for a multi-intersection network.

Current traffic control systems usually need predefined model of traffic condition in order to provide a prediction for incoming traffic flow. Among AI techniques, *Q*-learning is a learning method that does not require pre-specified model

© Springer International Publishing Switzerland 2015
S. Arik et al. (Eds.): ICONIP 2015, Part I, LNCS 9489, pp. 337–344, 2015.
DOI: 10.1007/978-3-319-26532-2_37

of the environment. A Q-learning controller is able to learn relationship between actions, states, and environment by interaction with the environment.

For the first time Thorpe [20] used reinforcement learning for traffic signal control [22]. SARSA [17] was the reinforcement technique applied by Thorpe.

Abdulhai has several studies in this field. For example, Abdulhai et al. [2], applied Q-learning as a traffic controller and performed the experiment for an isolated intersection.

Prashanth and Bhatnagar, in [15] proposed the feature based reinforcement learning for traffic signal controlling. It is claimed that using feature based state-action algorithms made the technique useful to be applied in high-dimensional setting of a multi-intersection network. In this work, it is mentioned that the proposed method is against the prior work like Abdulhai et al. [2], that required full state representation and was not practically possible to be implemented. We have presented a review on applying reinforcement learning in traffic signal controlling in [4].

In one of our previous work [6], Q-learning is applied for controlling traffic signal timing for an isolated intersection. During [6], a Q-learning controller developed based on Abdoos et al. [1] focusing on improving its deficiency. In current paper, we extend our previous work for a multi-intersection network while considering new options for improving its efficiency. Flexible cycle time, broader range of possible green phase time, possibility to extend for different size of urban traffic networks are the characteristics of the developed controller.

The rest of this paper is organized as: Section 2 for related background and proposed traffic signal timing controller, explanation about experiments environment and results discussion provided in Sect. 3, In Sect. 4 we conclude the paper.

2 Q-learning for Traffic Signal Timing

2.1   The Q-learning Algorithm

Q-learning is an incremental reinforcement learning method. This method does not need a model of the environment during learning, and it can be performed online [18].

In Q-learning the agent choses action a, while considering relative value of all possible actions in current state s. This value represents the Q-value or $Q(s,a)$ of action a in state s and leads to transition to state s'. The Q-value is obtained gradually during the learning. To reach the value the Q-learning agent needs to explore randomly various possible actions in each state. The agent receives reward $r(s,a)$ by performing action a in state s. The reward completely depends on the effect of the action on the environment. During the learning process, the agent aims to find the optimal policy that maximize the accumulative reward. Considering reward or punishment depends on the problem. In the case of punishment-based problem the agent aim to minimizes the accumulative punishment over time.

```
1: Initialize Q(s, a) arbitrarily
2: for all episode do
3:      Initialize s
4:      for all step of episode do
5:          Choose a from s using policy derived from Q (e.g., ϵ-greedy)
6:          Take action a, observe r, s'
7:              Q(s, a) ← Q(s, a)+
8: α [r + γ maxₐ' Q(s', a') − Q(s, a)]
9:              s ← s';
10:     end for
11: end for
```

Fig. 1. The standard Q-learning algorithm based on [18].

The other factor that is generally considered in Q-learning is the discount factor γ $(0 \leq \gamma \leq 1)$. γ is applied for bounding the reward. This factor is useful to consider higher value for short term reward compared to the long term ones in problem domain with continuous episodes. The updating formula of Q-value in learning process is presented in Eq. 1, and Fig. 1 describes the standard Q-learning algorithm.

$$Q(s_t, a_t) \leftarrow Q(s_t, a_t) + \alpha \left(r_t + \gamma \max_{a \in \mathcal{A}} Q(s_{t+1}, a) - Q(s_t, a_t) \right) \tag{1}$$

where α $(0 \leq \alpha \leq 1)$ is the learning rate and γ $(0 \leq \gamma \leq 1)$ is the discount factor.

2.2 Proposed Developed Q-learning Controller

During our previous work [6], we developed a Q-learning controller for an isolated intersection based on Abdoos et al.'s work [1] while it is considered to improve the performance of their controller. In our current work, we designed the enhanced version of our previous controller and extend it to be adopted in a multi-intersection network.

The proposed systems in [1,6] were cycle-based, it means at the beginning of each cycle the controller provide the green time for all phases. Noting this fact that traffic situation is always changing during the time, the proposed green time at the beginning of a cycle may not be suitable for next phases of the cycle. In this regard, the current Q-learning controller is designed phase-based.

Here, a tabular Q-learning controller is designed. Each intersection has its own controller. In this situation, each controller can has its own Q-table as well. By this distributed design smaller Q-tables are required that leads to increase of the convergence speed in Q-learning.

States are defined based on the number of vehicles. This number includes the vehicles make queue at approaching links and to have collaboration with neighbor intersections, each individual controller also considers the number of vehicles are coming from the neighbor intersections. Four groups are considered

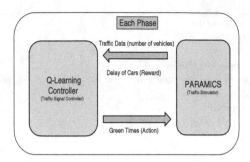

Fig. 2. This figure shows the learning process during each traffic phase in a Q-learning controller.

to categorized the vehicles' length in different states: *low, medium, high,* and *very high*. All states are made of combinations of these four groups for all approaching links. As an example, for a junction with k approaching links we will have 4^k members in the state list. In our current case, we have 256 states for each isolated junction with four approaching links.

The action list of the developed tabular Q-learning controller is composed of these numbers: $\{10, 20, 30, 40, 50, 60, 70, 80\}$. Actions are the proposed green time for the next phase. The Q-table for each Q-learning controller contains 256×80 cells.

The Q-learning controller uses inversely proportional to the average delay time per vehicle in the traffic network at the end of each phase for all the approaching links as the reward of the Q-learning. It means there is a higher value for cases with a lower average delay.

During the process of training the obtained traffic information from the traffic detectors (here this information is obtained from PARAMICS as a simulator) are sent to Q-learning controller and the proposed green time for each phase is provided in Fig. 2. The value of green times are selected from the possible green time in action list previously mentioned.

The Q-learning controller is implemented for a multi-intersection traffic network. The designed control system does not need a central controller for cooperation. Each controller receives the traffic data from its neighbors and set the suitable green time for the next phase based on its current traffic condition and the incoming vehicles from the neighbors.

3   Experiments Environment and Results Discussion

Evaluation is done in a network of nine intersections designed in PARAMICS Fig. 3.b. All intersections are 4-way with four phases, as shown in Fig. 3.a. The simulation model is set up in PARAMICS version 6.8. and all controllers are implemented in Matlab R2011b.

Three different traffic scenarios with different number of vehicles in the network are considered for evaluating the performance of the controllers.

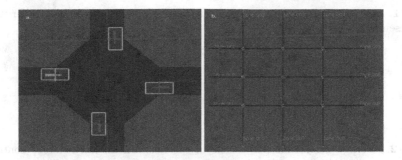

Fig. 3. a. An intersection with four phases.; b. A network of nine intersections. Zones are the area vehicles released to the intersection.

Scenario 1, 2, and 3 has about 700, 1000, and 1400 vehicles for 15 min of simulation respectively.

In addition, three different seed numbers are involved in each testing scenario. Accordingly, there are nine different test simulations performed for each controller.

There are in total nine controllers in the network each designed and dedicated to one intersection. As it is mentioned previously, to balance the traffic in the network, each controller consider its own traffic congestion and the number of arrival vehicles from neighbor intersections in its estimation of next phase green time calculation.

As it is mentioned in previous section, the developed Q-learning control system has 256 states for each controller with four approaching links. In addition, action list is composed of eight values 10, 20, ..., 80 which shows possible green times for Q-learning controller. For the Q-learning controller with ϵ-greedy method, a decreasing ϵ between 0.9 and 0.1 is considered, α (learning rate) is set to 0.1, and γ (discount factor) is set to 0.9. The reward function is defined as:

$$Reward = \frac{1}{mean(\sum_{i=1}^{4} d_i) + 1} \qquad (2)$$

where $i = 1, \ldots, 4$ is the number of approaching links, d is the calculated delay time for each link, and $+1$ is to refuse zero in denominator.

Fixed-time controller or pre-timed controller is a common benchmark for evaluating the performance of controllers. Therefore, three fixed-time controllers are also developed as benchmarks. Their phase green times are set to 20, 50, and 80 seconds respectively. Each fixed-time controller, benefits constant amount of time for all of its green phase. This property of fixed-time controllers, decreases its flexibility. During this work, by considering three different fixed-time controllers it is attempted to have a more comprehensive comparison.

In Tables 1, 2, 3, 4, the average delay per vehicle for each of the test scenario is presented. Obtained results show the superiority of Q-learning controller over the three fixed-time controllers. Furthermore, it can be concluded that fixed-time controllers with different green phase time present different performances.

Table 1. Average delay time per vehicle in the network for Q-learning controller (times are presented in second)

Q-learning	Scenario1	Scenario2	Scenario3
*Seed*1	34.7	35.1	38.1
*Seed*2	52.1	46.3	54.9
*Seed*3	77.5	83.6	86.1

Table 2. Average delay time per vehicle in the network for fixed-time20 controller (times are presented in second)

Fixed-time20	Scenario1	Scenario2	Scenario3
*Seed*1	55.5	53.8	58.3
*Seed*2	69.5	65.5	76.8
*Seed*3	99.8	98.2	104.5

Table 3. Average delay time per vehicle in the network for fixed-time50 controller (times are presented in second)

Fixed-time50	Scenario1	Scenario2	Scenario3
*Seed*1	69.7	68.0	72.9
*Seed*2	78.6	77.6	83.2
*Seed*3	103.5	102.2	106.6

Table 4. Average delay time per vehicle in the network for fixed-time80 controller (times are presented in second)

Fixed-time80	Scenario1	Scenario2	Scenario3
*Seed*1	84.7	86.3	91.1
*Seed*2	90.6	92.1	92.4
*Seed*3	110.91	105.6	108.8

Accordingly, results prove the lack of flexibility for fixed-time controller to adapt to traffic demand.

Calculating the performance of the Q-learning against the fixed-time controllers in three different scenarios shows its superiority by 49.48 %, 36.65 %, and 21.12 % against scenario 1,2, and 3 in respect. For obtaining these results we compare the average performance of Q-learning controller for different seeds against the average performance of different fixed-time controllers in different seed numbers. The bars show the performance of Q-learning controller in three different scenarios is higher than average performance of different fixed-time controllers.

In overall, it can be calculated that the performance of the developed Q-learning controller in nine different test traffic simulation is about 35.75 % higher than the designed fixed-time controllers.

4   Conclusions

One of the important issues regarding to the problems use Q-learning is about managing the huge amount of state-action space. Here, by categorizing possible states in groups we reduced the number of states. In addition, we developed a distributed controller and each intersection controls its own traffic congestion while considering arrival traffic from neighbor intersections. In future, we will propose an efficient method for optimal categorizing in state definition instead of the manual one used here. In addition, finding the optimal reward definition is another future plan.

References

1. Abdoos, M., Mozayani, N., Bazzan, A.: Traffic light control in non-stationary environments based on multi agent q-learning. In: International IEEE Conference on Intelligent Transportation Systems (ITSC), pp. 1580–1585, Washington, DC, USA (2011)
2. Abdulhai, B., Pringle, R., Karakoulas, G.: Reinforcement learning for true adaptive traffic signal control. J. Transp. Eng. **129**(3), 278–285 (2003)
3. Adler, J.L., Satapathy, G., Manikonda, V., Bowles, B., Blue, V.J.: A multi-agent approach to cooperative traffic management and route guidance. Transp. Res. B Methodol. **39**(4), 297–318 (2005)
4. Araghi, S., Khosravi, A., Creighton, D.: A review on computational intelligence methods for controlling traffic signal timing. Expert Syst. Appl. **42**(3), 1538–1550 (2015)
5. Araghi, S., Khosravi, A., Johnstone, M., Creighton, D.: Intelligent traffic light control of isolated intersections using machine learning methods, pp. 3621–3626 (2013)
6. Araghi, S., Khosravi, A., Johnstone, M., Creighton, D.: Q-learning method for controlling traffic signal phase time in a single intersection. In: IEEE Conference on Intelligent Transportation Systems, Proceedings, pp. 1261–1265 (2013)
7. Balaji, P., Sachdeva, G., Srinivasan, D., Tham, C.K.: Multi-agent system based urban traffic management. In: IEEE Congress on Evolutionary Computation, CEC 2007, pp. 1740–1747, IEEE (2007)
8. Cai, C., Wong, C.K., Heydecker, B.G.: Adaptive traffic signal control using approximate dynamic programming. Transp. Res. C Emerg. Technol. **17**(5), 456–474 (2009)
9. Chin, Y., Bolong, N., Kiring, A., Yang, S., Teo, K.: Q-learning based traffic optimization in management of signal timing plan. Int. J. Simul. Syst. Sci. Technol. **12**(3), 29–35 (2011)
10. Doniec, A., Mandiau, R., Piechowiak, S., Espié, S.: A behavioral multi-agent model for road traffic simulation. Eng. Appl. Artif. Intell. **21**(8), 1443–1454 (2008)
11. Hirankitti, V., Krohkaew, J., Hogger, C.J.: A multi-agent approach for intelligent traffic-light control. In: World Congress on Engineering, vol. 1, Citeseer (2007)

12. Kosonen, I.: Multi-agent fuzzy signal control based on real-time simulation. Transp. Res. C Emerg. Technol. **11**(5), 389–403 (2003)
13. Kronborg, P., Davidsson, F.: MOVA and IHOVRA: traffic signal control for isolated intersections. Traffic Eng. Control **34**(4), 195–200 (1993)
14. Nair, B., Cai, J.: A fuzzy logic controller for isolated signalized intersection with traffic abnormality considered. In: 2007 IEEE Intelligent Vehicles Symposium, pp. 1229–1233 (2007)
15. Prashanth, L., Bhatnagar, S.: Reinforcement learning with function approximation for traffic signal control. IEEE Trans. Intell. Transp. Syst. **12**(2), 412–421 (2011)
16. Spall, J.C., Chin, D.C.: Traffic-responsive signal timing for system-wide traffic control. Transp. Res. C Emerg. Technol. **5**(3–4), 153–163 (1997)
17. Sutton, R.S.: Generalization in reinforcement learning: successful examples using sparse coarse coding. In: Touretzky, D.S., Mozer, M.C., Hasselmo, M.E. (eds.) Advances in Neural Information Processing Systems, vol. 8, pp. 1038–1044. MIT Press, Cambridge (1996)
18. Sutton, R.S., Barto, A.G.: Reinforcement Learning an Introduction. MIT Press, Cambridge (1998)
19. Teknomo, K.: Application of microscopic pedestrian simulation model. Transp. Res. F Traffic Psychol. Behav. **9**(1), 15–27 (2006)
20. Thorpe, T.L., Anderson, C.W.: Traffic light control using SARSA with three state representations. Technical report, IBM Corporation (1996)
21. Vincent, R., Peirce, J.: MOVA: traffic responsive, self-optimising signal control for isolated intersections. Technical report RR 170, Crowthorne: TRRL, Transport and Road Research Laboratory (1988)
22. Wiering, M., Vreeken, J., van Veenen, J., Koopman, A.: Simulation and optimization of traffic in a city. In: IEEE Intelligent Vehicles Symposium, pp. 453–458 (2004)

Learning Rule for Linear Multilayer Feedforward ANN by Boosted Decision Stumps

Mirza Mubasher Baig[1]([⊠]), El-Sayed M. El-Alfy[2], and Mian M. Awais[1]

[1] School of Science and Engineering (SSE), Lahore University of Management
Sciences (LUMS), Lahore 54792, Pakistan
mubasher.baig@umt.edu.pk, awais@lums.edu.pk
[2] College of Computer Sciences and Engineering, King Fahd University of Petroleum
and Minerals, Dhahran 31261, Saudi Arabia
alfy@kfupm.edu.sa

Abstract. A novel method for learning a linear multilayer feedforward
artificial neural network (ANN) by using ensembles of boosted decision
stumps is presented. Network parameters are adapted through a layer-
wise iterative traversal of neurons with weights of each neuron learned
by using a boosting based ensemble and an appropriate reduction. Performances of several neural network models using the proposed method
are compared for a variety of datasets with networks learned using three
other algorithms, namely Perceptron learning rule, gradient decent back
propagation algorithm, and Boostron learning.

1 Introduction

The single-layer Perceptron of Rosenblatt [6], as shown in Fig. 1(a), is a simple mathematical model for binary classification of patterns. It takes an input
feature vector pattern $\bar{x} = [x_0, x_1, x_2, \ldots, x_m]$ and computes the class of the
input pattern by taking its dot product with an internally stored weight vector,
$\bar{W} = [w_0, w_1, w_2, \ldots, w_m]$. The input component x_0 is permanently set to -1
and represents the external bias with weight w_0 representing the magnitude of
the bias. The output, y, of a Perceptron is mostly computed using a non-linear
activation function such as *sign* and can be written as:

$$y = sign\left(\bar{W}.\bar{x}\right) = sign\left(\sum_{i=0}^{m} w_i.x_i\right) \tag{1}$$

A supervised-learning algorithm uses n labeled training examples of the form
$(\bar{x}_i, y_i)\ i = 1 \ldots n$ to select an optimal set of weights $\bar{W}$ of a given Perceptron. For
example, the well-studied Perceptron learning algorithm initializes the weight
vector to zeros or small random values and iteratively modifies these weights for
each misclassified training example $(\bar{x}_i, y_i)$ using the Perceptron learning law:

$$\bar{W}_{new} = \bar{W}_{old} + \eta.(y_i - y_i').\bar{x}_i \tag{2}$$

© Springer International Publishing Switzerland 2015
S. Arik et al. (Eds.): ICONIP 2015, Part I, LNCS 9489, pp. 345–353, 2015.
DOI: 10.1007/978-3-319-26532-2_38

where η is a pre-specified constant known as learning rate. The algorithm continues until either the mean-squared error over the training data becomes less than a given threshold, γ, or a maximum number of iterations is reached [7]. To handle a multi-class problem, a Perceptron structure similar to that shown in Fig. 1(b) is used. Each output neuron represents a single class and its weights are adapted independently using the binary encoding of the classes.

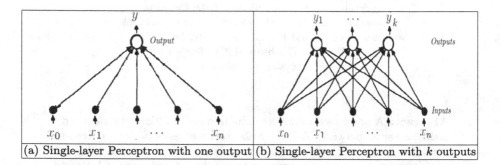

(a) Single-layer Perceptron with one output (b) Single-layer Perceptron with k outputs

Fig. 1. Typical structure of a single-layer Perceptron

Recently an effective learning algorithm, known as Boostron [1], has been proposed for learning weights of a single-layer Perceptron using a real-valued variant of AdaBoost algorithm [8]. AdaBoost algorithm constructs an accurate classifier ensemble using a moderately accurate classification algorithm. AdaBoost selects a number of base classifier instances h_k by modifying a weight distribution maintained on the training examples. It also computes a weight, α_k, of each selected classifier, h_k, and constructs a classifier ensemble using a linear combination of the selected classifiers as given by:

$$H(\bar{x}) = sign\left(\sum_{k=1}^{T} \alpha_k.h_k(\bar{x})\right) \tag{3}$$

Boostron uses homogenous coordinates to represent a decision stump [4] as a dot product of a weight vector $\bar{w}$ and the homogenous representation of the instance $\bar{x}$ as $\bar{X}$ using $s^w(x) = \bar{w}^T.\bar{X}$. It uses this form of a decision stump as base learner in AdaBoost [1] to learn a Perceptron given by

$$H(\bar{x}) = sign\left(\bar{W}.\bar{x}\right) \tag{4}$$

where $\bar{W} = \sum_{k=1}^{T}\left(\alpha_k.\bar{w_k^T}\right) = [w_1, w_2, \ldots, w_{m+1}]$, is the weighted sum of weight vectors $\bar{w_k^T}$ learned using decision-stump learning algorithm.

A generalization of Boostron to learn parameters of a feedforward **multilayer** neural network is not obvious. This problem is addressed in this paper and a more general approach for learning a two-layer linear ANN is presented. The proposed extension of the Boostron uses a transformed set of examples and an iterative approach to optimize weights of neurons in the network.

The proposed method has been used to learn ANN for several binary and multiclass classification tasks taken from the UCI machine learning repository [5] and the results are compared to ANN trained using Boostron, Perceptron learning and back-propagation learning algorithms.

Section 2 presents a detailed description of the proposed method. Experimental settings and the corresponding results are presented in Sect. 3. Finally, Sect. 4 concludes the paper and highlights some future directions.

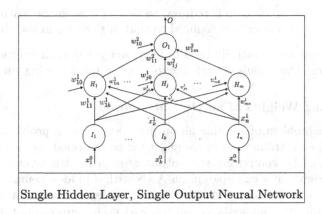

Single Hidden Layer, Single Output Neural Network

Fig. 2. Multilayer feedforward neural network

2   Generalized Boostron

This section presents the proposed extension to learn a feedforward ANN having a single hidden layer of neurons, a linear activation function, and a single neuron at the output layer. A more general form is presented later to handle multiclass problems.

To present the proposed method, it is assumed that inputs of a neuron at layer l are denoted by $x_0^l, x_1^l, ..., x_k^l, ..., x_n^l$ respectively where x_0^l is permanently set to -1 and represents the bias term. In this notation, the superscript denotes the layer number and the subscript denotes the input feature number where n is the total number of neurons in the *previous* layer (i.e. layer $l-1$). The corresponding weights of the jth neuron at layer l are denoted by $w_{j0}^l, w_{j1}^l, ...,$ $w_{jk}^l, ..., w_{jn}^l$ where the term w_{jk}^l for $k \in \{1, ..., n\}$ is the weight of the connection from the kth input from the previous layer to j^{th} neuron, and w_{j0}^l is the bias term of jth neuron at layer l.

A two-layer feedforward neural network with a set of n input neurons $\{I_1, ..., I_n\}$ at layer 0, m hidden neurons $\{H_1, ..., H_m\}$ at layer 1 and a single output neuron O_1 at layer 2 is shown in Fig. 2. Assume the function f_k^l denotes the activation function of the kth neuron in the lth layer. The output, O, of this neural network is computed as follows:

$$O = f_1^2 \left(\sum_{k=0}^{m} w_{1k}^2 . x_k^2 \right) \tag{5}$$

where f_1^2 denotes the activation function of neuron 1 at layer 2. Each neuron in the network is either an output neuron or a hidden neuron, therefore the proposed algorithm uses two reductions:

– Learning an output neuron is reduced to that of Perceptron learning.
– Learning a hidden neuron is reduced to that of Perceptron learning.

These reductions are iteratively used to learn weights of each neuron in a given neural network. The details of each is explained in the following subsections.

2.1   Learning Weights of an Output Neuron

To reduce the problem of learning an output neuron into a problem of learning a Perceptron, each training example $(\bar{x}_i, y_i)$ is transformed into a new training example (x_i^2, y_i) by computing the output of each hidden layer neuron using its present weights. For example, in an ANN with d hidden neurons in a single hidden layer, for each training instance $\bar{x}_i \in R^m$ a new training instance $x_i^2 \in R^d$ is computed using the hidden-layer neurons with each component of the vector x_i^2 corresponding to the output of exactly one of the hidden neuron. The Boostron algorithm is then used to learn the weights of the output neuron using the transformed training examples $(x_i^2, y_i), i = 1...N$.

2.2   Learning Weights of a Hidden Neuron

To learn the weights, $\{w_{j0}^1, w_{j1}^1, \ldots, w_{jn}^1\}$, of the jth hidden neuron H_j while keeping the rest of network fixed, Eq. 5 is written as:

$$O = f_1^2 \left(w_{1j}^2 . x_j^2 + \sum_{k=0, k \neq j}^{m} w_{1k}^2 . x_k^2 \right) \tag{6}$$

Here, the term x_j^2 is the output of the hidden neuron H_j and can be written as a combination of the inputs from layer 0 and the weights of the neuron H_j as:

$$x_j^2 = f_j^1 \left(\sum_{i=0}^{n} w_{ji}^1 . x_i^1 \right) \tag{7}$$

Substituting this value of x_j^2 in Eq. 6 gives:

$$O = f_1^2 \left(w_{1j}^2 . f_j^1 \left(\sum_{i=0}^{n} w_{ji}^1 . x_i^1 \right) + \sum_{k=0, k \neq j}^{m} w_{1k}^2 . x_k^2 \right) \tag{8}$$

When both activation functions are linear, the above equation can be written as:

$$O = w_{1j}^2 \cdot \sum_{i=0}^{n} w_{ji}^1 \cdot f_1^2 \left(f_j^1 \left(x_i^1 \right) \right) + f_1^2 \left(\sum_{k=0, k \neq j}^{m} w_{1k}^2 \cdot x_k^2 \right) \tag{9}$$

If $C = f_1^2 \left(\sum_{k=0, k \neq j}^{m} w_{1k}^2 \cdot x_k^2 \right)$ denotes the output contribution of all hidden neurons other than the neuron H_j and $X_i^j = f_1^2 \left(f_j^1 \left(x_i^1 \right) \right)$ denotes the inputs transformed using the activation functions, Eq. 9 can be written as:

$$O = w_{1j}^2 \cdot \sum_{i=0}^{n} w_{ji}^1 \cdot X_i^j + C \tag{10}$$

A method of learning the weights of the hidden neuron H_j can be obtained by ignoring the effect of fixed constant term C and the scale term w_{1j}^2 on the overall output by rewriting Eq. 10 as:

$$O = \sum_{i=0}^{n} w_{1i}^1 \cdot X_i^j \tag{11}$$

As the form of this equation is exactly equivalent to the equation of a Perceptron, the Boostron learning algorithm [1] can be used to learn the required weights, $w_{ji}^1, i = 0, 1, \ldots, n$, of a hidden neuron.

Algorithm 1 uses the above reductions and outlines a method of iterating over the neurons of a linear feedforward ANN. The algorithm randomly initializes all weights in the interval $(0, 1)$ and assigns a randomly selected subset of features to each hidden-layer neuron so that the hidden neuron uses only these features to compute its output. Such a random assignment is the key difference between a simple Boostron and the proposed method. It causes each hidden neuron to use a different segment of the feature space to learn its weights. After this initialization, the algorithm iterates between the hidden layer and the output layer neurons in order to learn the ANN. At the hidden layer, the algorithm iterates over the hidden neurons and compute their weights in step 5 and step 6 by using the transformed training examples computed in step 4. The transformed training examples are computed using the reduction given in Eq. 11. The weights of each hidden neuron are computed using the Boostron algorithm while keeping the weights of all remaining neurons fixed. These hidden neuron weights are then used to transform the training examples $(\bar{x}_i, y_i), i = 1 \ldots N$ into new training examples $(x_i^2, y_i), i = 1 \ldots N$ which are subsequently used to learn the output neuron using Boostron. This whole process is repeated a number of times specified by the input parameter T.

2.3   Learning a Multiclass Classifier

The algorithm for learning a feedforward ANN, as presented above, can only be used with networks having a single output neuron and working as binary

---

Algorithm 1. ANN learning using AdaBoost and fixed targets

Require: l Training examples $(\bar{x}_1, y_1) \ldots (\bar{x}_l, y_l)$ where
　　$\bar{x}_k$ is a training instance and $y_i k \in \{-1, +1\}$
　　T is the number of iterations over ANN layers

1: Randomly initialize all weights in the range (0 1)

2: Randomly assign features to each hidden neuron.

3: **for** j $= 1$ **to** T **do**

4:　　Compute Transformed training examples $(\bar{X}_k, y_k), k = 1, 2, \ldots, l$ where $\bar{X}_k = [X_0^j, X_1^j, \ldots, X_n^j]$ and the component $X_i^j = f_1^2 \left(f_j^1 \left(x_i^1 \right) \right)$

5:　　**for** Each Neurons H_j in the Hidden Layer **do**

6:　　　Use the Boostron algorithm and the training examples $(\bar{X}_k, y_k)$, to learn the weights w_{1i}^1 for $i = 0, 1, \ldots, n$

7:　　**end for**

8:　　Compute Transformed training examples $(\bar{X}_k^2, y_k), k = 1, 2, \ldots, l$ where $\bar{X}_k^2 = [x_0^2, x_1^2, \ldots, x_n^2]$ and the component $x_j^2 = f_j^1 \left(\sum_{i=0}^n w_{1i}^1 . x_i^1 \right)$

9:　　Use the Boostron algorithm and the training examples $(\bar{X}_k^2, y_k) \, k = 1, 2, \ldots, l$, to learn the weights $w_0^2, w_1^2, \ldots, w_m^2$, of the output neuron O_1

10: **end for**

11: Output the learned ANN weights.

---

classifiers. Several simple methods for reducing a multiclass learning problem into a set of problems involving binary classification are in common use. Such methods include the binary encoding of classes using error correcting codes [2], the all-pairs approach of Hastie and Tibshirani [3] and a simple approach of one-versus-remaining coding of classes. For each bit in the binary code of classes, a binary classifier is trained and the outputs of all binary classifiers are combined, e.g. using hamming distance, to produce a final multiclass classifier. This paper uses one-versus-remaining coding of classes (+1 for the class and −1 for the remaining classes) to reduce a k-class classification into k binary classification problems.

3　Experiments and Results

A description of datasets, experimental settings and results obtained is presented in this section. Performance of the proposed method is compared to networks trained using three learning algorithms including back propagation learning, Boostron algorithm, and Perceptron learning algorithm.

3.1　Datasets Description and Experimental Settings

A summary of six binary and five multiclass classification datasets, from the UCI machine learning repository [5], used to evaluate the proposed method, is

Table 1. Summary of datasets used in our experiments

Data set Name	Total Feature	Training Set	Test Set	Total classes
Balance scale	4	625	c.v	3
Breast cancer	30	569	c.v	2
Spambase	57	4601	c.v	2
Two norm	20	2000	2000	2
Three norm	20	1000	2000	2
Ring norm	20	2479	2442	2
Iris	3	150	c.v	3
Lung cancer	56	27	c.v	3
Forest fire	5	500	c.v	3
Glass	10	214	c.v	7
Vowel	10	528	468	11

presented in Table 1. To estimate the test error rate, 10-fold cross validation (c.v.) has been used for datasets without explicit division into training and test sets.

For each learning problem, an ANN having 15 neurons in a single hidden layer and linear activation function (i.e., $f(x) = x$) for each neuron has been trained.

3.2   Results

The first set of results, shown in Table 2, compares the performance of the proposed method for the six binary classification problems given in Table 1. Comparisons are based on the overall error rate of trained ANN using the proposed method, back propagation learning algorithm, Boostron learning algorithm and Perceptron learning algorithm.

Table 2. Overall error rate for linear ANN with 15 hidden neurons

Data set	Proposed	Back-Propagation	Boostron	Perceptron
Balance scale	4.82	**4.79**	7.32	9.73
Spambase	**10.58**	11.45	24.67	39.43
Ionosphere	**13.31**	13.68	24.09	17.10
Two norm	2.15	**2.05**	2.25	5.0
Three norm	18.25	**17.4**	29.05	35.7
Ring norm	**22.29**	24.53	46.71	31.58

Both the proposed method and the well-studied back propagation perform much better than the Perceptron and Boostron learning algorithms. Moreover,

the proposed method has very similar results to the well-studied back propagation algorithm with slightly better performance in case of Spambase, Ionosphere
and Ringnorm dataset.

For multiclass classification, a separate neural network was trained for each
class using one-versus-remaining encoding of classes, therefore k different neural
networks are created for a k-class learning problem. For an instance x, the class
corresponding to the neural network producing the highest positive output is
predicted as the class of x. To test and compare the performance in this case, a
second set of results compares the error rate for the proposed method versus the
back propagation algorithm for five multiclass learning problems. The attained
results are shown in Table 3.

Table 3. Comparison of overall average error rate for ANN with 15 hidden neurons

Data set	Proposed	Back-Propagation
Iris	**5.33**	7.33
Forest fire	**13.6**	18.4
Glass	18.39	**10.32**
Vowel	6.25	**1.73**
Lung cancer	**26.67**	41.67

Figure 3 presents a further comparison of the proposed method with the back
propagation algorithm for datasets having no explicit division into training and

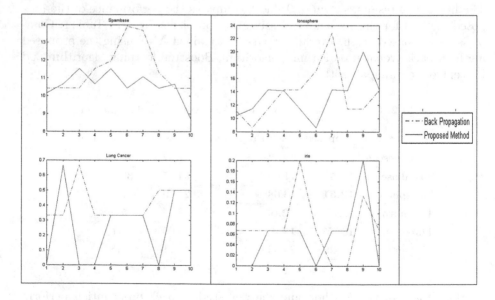

Fig. 3. Test error rate per fold

test sets. The figure plots a fold-wise comparison of the two methods for each of the 10 folds of training data. It reveals that the proposed method mostly gives improved accuracy. For binary classification problems, the improvement is significant for the spambase dataset. For multiclass problems, the proposed method has better accuracy in the case of iris and lung cancer datasets.

4 Conclusions

This paper presented a method for learning a two-layer feedforward ANN with linear activation functions. The proposed method reduces the problem of learning weights of a hidden layer neuron into that of learning a single-layer Perceptron, then it generalizes the Boostron algorithm to determine the weights. The proposed method has been modified to work with the multiclass problems as well. Using eleven datasets of various characteristics, the performance of the proposed method is empirically evaluated and compared to that of back-propagation, Boostron, and Perceptron learning algorithms. The results demonstrate the effectiveness of the proposed method. As future work some modifications need to be considered for learning an ANN with arbitrary non-linear activation function and multiple hidden layers.

References

1. Baig, M.M., Awais, M.M., El-Alfy, E.-S.M.: BOOSTRON: boosting based perceptron learning. In: Loo, C.K., Yap, K.S., Wong, K.W., Teoh, A., Huang, K. (eds.) ICONIP 2014, Part I. LNCS, vol. 8834, pp. 199–206. Springer, Heidelberg (2014)
2. Dietterich, T.G., Bakiri, G.: Solving multiclass learning problems via error-correcting output codes. J. Artif. Intell. Res. **2**(1), 263–286 (1995)
3. Hastie, T., Tibshirani, R.: Classification by pairwise coupling. Ann. Stat. **26**(2), 451–471 (1998)
4. Iba, W., Langley, P.: Induction of one-level decision trees. In: Proceedings of the 9th International Conference on Machine Learning, pp. 233–240 (1992)
5. Lichman, M.: UCI machine learning repository (2013). http://archive.ics.uci.edu/ml
6. Rosenblatt, F.: The perceptron: a probabilistic model for information storage and organization in the brain. Psychol. Rev. **65**(6), 386 (1958)
7. Rumelhart, D.E., McClelland, J.L., PDP Research Group.: Parallel Distributed Processing: Explorations in the Microstructures of Cognition, vol. 1. MIT Press, Cambridge, MA, USA (1986)
8. Schapire, R.E., Singer, Y.: Improved boosting algorithms using confidence-rated predictions. Mach. Learn. **37**(3), 297–336 (1999)

Class-Semantic Color-Texture Textons
for Vegetation Classification

Ligang Zhang[1(✉)], Brijesh Verma[1], and David Stockwell[1,2]

[1] Central Queensland University, Rockhampton, QLD, Australia
{l.zhang,b.verma,d.stockwell}@cqu.edu.au
[2] Rockhampton, QLD, Australia

Abstract. This paper proposes a new color-texture texton based approach for roadside vegetation classification in natural images. Two individual sets of class-semantic textons are first generated from color and filter bank texture features for each class. The color and texture features of testing pixels are then mapped into one of the generated textons using the nearest distance, resulting in two texton occurrence matrices – one for color and one for texture. The classification is achieved by aggregating color-texture texton occurrences over all pixels in each over-segmented superpixel using a majority voting strategy. Our approach outperforms previous benchmarking approaches and achieves 81% and 74.5% accuracies of classifying seven objects on a cropped region dataset and six objects on an image dataset collected by the Department of Transport and Main Roads, Queensland, Australia.

Keywords: Object classification · Roadside vegetation · Image segmentation · Texture feature

1 Introduction

Roadside vegetation analysis is of great significance for many applications, such as vegetation growth monitoring and fire-prone area identification. Although vegetation has received extensive attention in the fields of remote sensing and agriculture using satellite and aerial data, vegetation classification on roadside data is a relatively less investigated field. Previous work can be broadly grouped into visible approaches which analyze visual characteristics of vegetation and invisible approaches which focus on the use of the spectral properties of chlorophyll-rich vegetation such as Vegetation Index (VI). It is still a challenging task to select suitable visible features or design reliable VIs for real-world conditions, due to substantial variations in the environment.

Recently, textons have demonstrated a compact and effective representation of visual characteristics for object categorization. The textons are often generated from filter bank features and further aggregated to form a histogram profile for each image. Texton based approaches have been successfully applied into natural scene understanding [1, 2], but have seldom used for roadside vegetation classification [3].

A new color-texture texton based approach is proposed in this paper for roadside vegetation classification. It generates class-semantic textons of color and filter bank

© Springer International Publishing Switzerland 2015
S. Arik et al. (Eds.): ICONIP 2015, Part I, LNCS 9489, pp. 354–362, 2015.
DOI: 10.1007/978-3-319-26532-2_39

responses from the training data, which represent the intrinsic features learnt for each class. Features in a testing image are quantized into one of the learnt textons. The classification is then performed by aggregating texton occurrences over each superpixel using a simple yet effective majority voting strategy. We demonstrate promising performance of the approach in classifying objects on two natural datasets.

The rest of the paper is organized as follows. Section 2 discusses related work. Section 3 introduces the proposed approach. The experiments are presented in Sect. 4 and finally Sect. 5 draws the conclusions.

2 Related Work

The visible approaches utilize visual characteristics of vegetation in the visible spectrum to distinguish them from others. An early study was described in [4], which extracted color, texture, shape, size, centroid and contextual features of segmented regions for object classification including vegetation. Except for color, a popular feature is the intensity difference between pixels, which was combined with a 3D Gaussian model of YUV channels for the detection of grass regions [5], and with L, a, and b components for the segmentation of roadside objects such as grass and tree [3]. Other features include neighborhood statistic [6], entropy [7], superpixel-based texture [8], and 2D-Discrete Fourier Transform [9]. Most visible approaches are restricted to vegetation versus non-vegetation classification, and there still lacks a common feature set that can work well for natural conditions.

The invisible approaches use the spectral properties of chlorophyll-rich vegetation in the invisible spectrum, particularly VIs such as red and near infrared ray (NIR) reflectance [10]. Extensions have been made to the NIR, including the normalized difference vegetation index (NDVI) [10], the modification of NDVI (MNDVI) [11], and combination of NDVI and MNDVI [12] to achieve robustness against illumination variations. Invisible and visible features were also fused for vegetation detection [13, 14]. Invisible approaches often require specialized data capturing equipment, which limits its direct adoption in applications, and also face the challenge of designing reliable VIs for natural conditions.

Texton based approaches employ textons of filter bank responses or other feature descriptors for object representation. A set of universal texture textons was generated using 17-D filter banks in [15], showing high accuracy of classifying real-world objects. The textons have been extended to textonboost [16], hierarchical bag-of-textons from multibands [2], and histograms of textons of color and intensity differences [3], etc. Most texton based approaches build generic textons for all classes and then form a texton histogram profile for each image. However, generic textons may not be effective to capture class-specific characteristics and histogram representations may fail in small images due to too sparse histogram bins. This paper investigates the use of class-specific textons for vegetation classification. The most similar work to ours is [1], which created a set of class-semantic SIFT words for scene understanding. We extend this work by integrating color textons.

3   Proposed Approach

The proposed approach consists of two stages (1) Training Stage and (2) Testing Stage. An overview of the framework for the proposed approach is shown in Fig. 1. During the training stage, an equal set of local regions is manually cropped from the training images for each class. Color and filter bank responses are then extracted from these regions, which are further fed into K-means clustering to generate two individual sets of class-semantic color and texture textons for each class. During the testing stage, the input image is first segmented into a group of superpixels. The color and filter bank features are then extracted and mapped separately into one of the learnt textons using the Euclidean distance, forming color-texture texton occurrence matrices – one for color and one for texture. A superpixel based majority voting is employed to assign each superpixel to a class which has the maximum occurrence of color-texture textons.

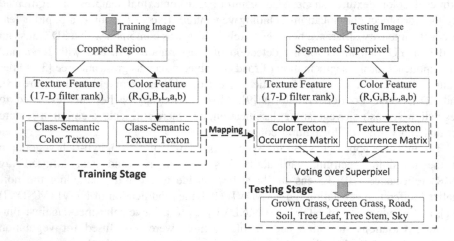

Fig. 1. Framework of the proposed approach.

3.1   Color and Texture Feature Extraction

The selection of a suitable color space for effective vegetation representation remains a challenging task. This paper adopts the CIELab, which has high perceptually consistency with human vision and good performance on scene understanding [2]. We also include RGB as it may contain complementary information for specific objects. For a pixel at the coordinate (x, y) in an image, its color feature is composed of:

$$V_{x,y}^c = \langle R, G, B, L, a, b \rangle \tag{1}$$

For a more discriminative feature representation, we also include texture features. This paper uses 17-D filter banks [15], due to their high accuracy for texture-rich object classification. The filter banks include Gaussians with 3 different scales (1, 2, 4) applied to L, a, and b channels, Lapacians of Gaussians with 4 different scales (1, 2, 4, 8) and the derivatives of Gaussians with two different scales (2, 4) for each axis (x and y) on the L channel. For a pixel at (x, y), its texture feature is composed of:

$$V_{x,y}^t = \left\langle G_{1,2,4}^L, G_{1,2,4}^a, G_{1,2,4}^b, LOG_{1,2,4,8}^L, DOG_{2,4,x}^L, DOG_{2,4,y}^L \right\rangle \tag{2}$$

3.2   Class-Semantic Color-Texture Texton Generation

Unlike existing approaches that generate a universal vocabulary for all classes, this paper extracts a set of class-semantic textons to reduce confusions between classes.

Assume there are C classes and n training pixels in the ith class ($i = 1, 2, ..., C$). Let V_i^c and V_i^t be the color and texture features respectively for the ith class, the K-means algorithm is used to learn a set of textons for each of V_i^c and V_i^t by minimizing:

$$J_c = \sum_{j=1}^{n} \min_k \left| V_{i,j}^c - T_{i,k}^c \right|^2 \tag{3}$$

where, $V_{i,j}^c$ is color features of the jth pixel in V_i^c, $T_{i,k}^c$ is the kth color textons learnt ($k = 1, 2, ..., K$) for the ith class, and J_c is the error function. The ith class-semantic color and texture texton vectors are composed of:

$$T_i^c = \left\langle T_{i,1}^c, T_{i,2}^c, ..., T_{i,k}^c \right\rangle \quad \text{and} \quad T_i^t = \left\langle T_{i,1}^t, T_{i,2}^t, ..., T_{i,k}^t \right\rangle \tag{4}$$

For all C classes, a color and a texture texton matrix can be formed respectively:

$$T^c = \left\{ \begin{matrix} T_{1,1}^c, T_{1,2}^c, ..., T_{1,K}^c \\ T_{2,1}^c, T_{2,2}^c, ..., T_{2,K}^c \\ \vdots \\ T_{C,1}^c, T_{C,2}^c, ..., T_{C,K}^c \end{matrix} \right\} \quad \text{and} \quad T^t = \left\{ \begin{matrix} T_{1,1}^t, T_{1,2}^t, ..., T_{1,K}^t \\ T_{2,1}^t, T_{2,2}^t, ..., T_{2,K}^t \\ \vdots \\ T_{C,1}^t, T_{C,2}^t, ..., T_{C,K}^t \end{matrix} \right\} \tag{5}$$

The two matrices contain color and texture textons for all C classes learnt from training data, and are used for distinguishing between objects in testing data.

3.3   Texton Occurrence and Superpixel-Based Voting

For all pixels in a testing image, they are first mapped into one of the learnt textons and a superpixel based majority voting is then adopted to label each superpixel with the class which has the maximum occurrence of textons over all classes.

For an image I, it is first segmented into a set of superpixels using [17]: $S = \langle S_1, S_2, ..., S_L \rangle$, where, L is the number of segmented superpixels, and S_l stands for the lth superpixel. For S_l, its color and texture features can be extracted using (1) and (2) respectively, i.e. $V_{S_l}^c = \bigcup_{x,y \in S_l} V_{x,y}^c$ and $V_{S_l}^t = \bigcup_{x,y \in S_l} V_{x,y}^t$. The $V_{S_l}^c$ and $V_{S_l}^t$ are mapped to the learnt class-semantic color and texture textons respectively by finding the closest texton using the Euclidean distance:

$$f\left(V_{x,y}^c, T_{i,k}^c\right) = \begin{cases} 1, & if\|V_{x,y}^c - T_{i,k}^c\| = \min_{q=1,2,...C;p=1,2,...,K} \|V_{x,y}^c - T_{q,p}^c\| \\ 0, & otherwise \end{cases} \tag{6}$$

A color texton occurrence matrix which accounts for the number of the mapped textons $T_{i,k}^c$ for all pixels in S_l can be obtained: $A_{i,k}^c(S_l) = \sum_{x,y \in S_l} f(V_{x,y}^c, T_{i,k}^c)$. The values in the matrix are accumulated for the ith class, yielding the color confidence of S_l belonging to this class: $A_i^c(S_l) = \sum_{k=1}^K A_{i,k}^c(S_l)$. The color confidence is further combined with the texture confidence using a simple summation strategy: $A_i = a * A_i^c(S_l) + b * A_i^t(S_l)$, where, a and b are weights and fixed to 0.5, and $A_i^t(S_l)$ is the texture confidence. The S_l is finally assigned to the cth class which has the maximum confidence over all classes: $S_l \in$ cth class if $A_c = \max_{i=1,2,...C} A_i$.

4   Experiments and Comparative Analysis

The proposed approach is evaluated on two datasets using four metrics – global accuracy over all classes, class accuracy over each class, as well as pixel- and region-level accuracies which are the percentages of pixels and regions correctly classified.

4.1   Evaluation Datasets

Two datasets were created from the video collected by the Department of Transport and Main Roads, Queensland, Australia. The video was captured using a left-view camera mounted on a vehicle, which runs across main roads in Queensland. The first dataset is composed of 650 regions manually cropped from video frames for seven objects, including brown grass, green grass, road, soil, tree leaf, tree stem, and sky. As shown in Fig. 2, it is ensured that each region belongs to only one object and there are substantial appearance variations between regions. The second dataset includes 50 frames that contain different types of vegetation and other objects. Pixelwise ground truths regarding the presence of seven objects, including brown grass, green grass, road, soil, tree, sky, and unknown objects, were manually annotated. The two datasets are publicly accessible at https://sites.google.com/site/cqucins/projects.

4.2   Global Accuracy

Figure 3 shows the global accuracy on two datasets. For both datasets, using color-texture textons leads to slightly higher accuracy than using color or texture textons alone. The highest global accuracy of 81.0% is obtained for the cropped dataset using 80 color-texture textons, and 74.5% for the image dataset using 90 color-texture textons. When color textons are used alone, the highest accuracy is obtained using 80 and 70 textons respectively on the two datasets. The number of textons for the highest accuracy is the same (i.e. 90) for the two datasets when texture textons are used alone. When tested on the cropped data, region-level classification has much higher accuracy

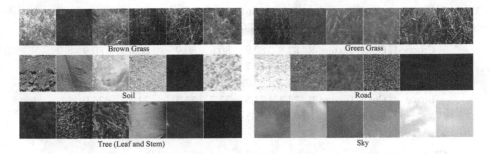

Fig. 2. Samples of cropped regions for seven objects. Note substantial variations in the appearance within the same object, as well as between objects (e.g. green grass and tree leaf).

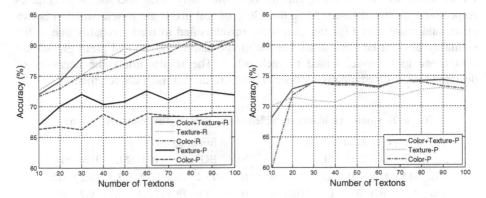

Fig. 3. Pixel- and Region-level accuracies of using color textons, texture textons, and their fusion on the cropped region (left), and Pixel-level accuracy on the natural image data (right). The same number of color and texture textons is used, and Gaussian filters are 7 * 7 pixels.

(about 7%) than pixel-level classification for both the cases of using color or texture textons. This is within our expectation as region-level classification assigns a class label to a region by utilizing collective decision information over all pixels in the region, which is more robust than pixel-level classification.

4.3 Global Accuracy Versus Size of Gaussian Filters

Table 1 reveals the impact of the size of Gaussian filters on the global accuracy. Four methods are included: pixel- and region-level classification using texture textons (Texture-P and Texture-R) and using color-texture textons (Color-Texture-P and Color-Texture-R). For all methods and both datasets, there are only small differences in the accuracy using different sizes of Gaussian filters, but small sizes appear to slightly outperform higher sizes. The highest accuracy is achieved using the sizes of 5 * 5 and 7 * 7 respectively for the two datasets. This implies that a small size of Gaussian filters is capable of capturing the most discriminative features.

Table 1. Accuracy (%) versus size of Gaussian filters (color no. = 70 or 80, texture no. = 90).

Dataset	Approach	5 * 5	7 * 7	9 * 9	11 * 11	13 * 13	15 * 15
Cropped	Color-Texture-R	82.8	80.9	80.6	80.9	80.8	80.9
	Texture-R	82.5	79.7	80.6	81.2	81.1	79.6
	Texture-P	72.7	72.7	71.7	72.2	72.6	71.7
Image	Color-Texture-P	74.2	74.5	73.7	73.7	74.4	74.0
	Texture-P	73.3	73.3	72.2	72.4	72.7	72.5

4.4 Class Accuracy

Table 2 displays the confusion matrix for different objects. For both datasets, sky is the easiest one for classification with around 96% accuracies, which is followed by road. By contrast, soil is the most difficult one with only 60% and 44.7% accuracies respectively on two datasets. More than 30% of soil pixels are misclassified to brown grass, and more than 17% tree pixels are misclassified to road, probably due to the similarity of color between them. For the image dataset, brown and green grasses are also prone to be misclassified to each other. The results imply the necessity of designing more discriminative texture features to distinguish between them.

Table 2. Confusion matrix for different classes using color-texture textons.

	Cropped dataset							Image dataset					
	BG	GG	RD	SL	TL	TS	Sky	BG	GG	RD	SL	Tree	Sky
BG	**94.0**	0	2.0	4.0	0	0	0	**73.2**	15.5	2.8	3.3	5.2	0.0
GG	0	**90.0**	0	0	10.0	0	0	7.9	**79.9**	2.3	0.7	9.2	0.0
RD	0	0	**88.0**	0	0	10.0	2.0	6.8	0.6	**85.0**	6.7	0.2	0.7
SL	30.0	0	2.0	**60.0**	2.0	4.0	2.0	37.8	5.7	10.3	**44.7**	1.5	0.0
TL	0	10.0	0	0	**84.0**	6.0	0	5.4	5.4	17.9	0.3	**68.1**	2.9
TS	2.0	0	20.0	2.0	2.0	**74.0**	0						
Sky	0	0	0	4.0	0	0	**96.0**	0.2	0.0	2.6	0.2	0.4	**96.6**

Note: BG – brown grass; GG – green grass; RD – road; SL- soil; TL – tree leaf; TS – tree stem.

4.5 Performance Comparisons

Table 3 compares the proposed approach with two other approaches on the cropped dataset. The generic texton approach generates a universal set of textons for all classes using K-means clustering, and classifies an image by comparing its texton histogram with that of each class. For fair comparison, the same color and texture features as in this paper are used. The second approach is based on color intensity and moment features [18]. It can be seen that using class-semantic textons leads to more than 21% and 2% higher accuracies than using generic textons and color features respectively, confirming the benefit of considering class-semantic textons.

Table 3. Performance comparisons (color no. = 70, texture no. = 90, 7 * 7 Gaussian filters).

Approach	Class no.	Accuracy (%)
Class-semantic texton (proposed)	7	81.0
Generic texton	7	59.7
Color intensity and moment [18]	6	79.0

5 Conclusions

This paper proposes a texton based approach for vegetation classification on natural roadside images. It learns class-semantic color-texture textons for more effective representation of class specific features from training data, and then map features of all pixels into the learnt textons. A superpixel based majority voting is used to label each superpixel by aggregating the occurrence of color-texture textons. We achieve the highest accuracies of 81% and 74.5% on two natural datasets. For robust classification in natural conditions, our results indicate that it is desirable to consider more discriminative features for the classification of road and tree, as well as soil and brown grass. Our future work will further improve the results by using summary statistical features over superpixels to generate more robust descriptors of objects.

Acknowledgments. This research was supported under Australian Research Council's Linkage Projects funding scheme (project number LP140100939).

References

1. Haibing, Z., Shirong, L., Chaoliang, Z.: Outdoor scene understanding using SEVI-BOVW model. In: 2014 International Joint Conference on Neural Networks, pp. 2986–2990 (2014)
2. Kang, Y., Yamaguchi, K., Naito, T., et al.: Multiband image segmentation and object recognition for understanding road scenes. IEEE Trans. Intell. Transp. Syst. **12**, 1423–1433 (2011)
3. Blas, M.R., Agrawal, M., Sundaresan, A., et al.: Fast color/texture segmentation for outdoor robots. In: 2008 IEEE/RSJ International Conference on Intelligent Robots and Systems (IROS), pp. 4078–4085 (2008)
4. Campbell, N.W., Thomas, B.T., Troscianko, T.: Automatic segmentation and classification of outdoor images using neural networks. Int. J. Neural Syst. **08**, 137–144 (1997)
5. Zafarifar, B., de With, P.H.N.: Grass field detection for TV picture quality enhancement. In: 2008 International Conference on Consumer Electronics (ICCE), pp. 1–2 (2008)
6. Schepelmann, A., Hudson, R.E., Merat, F.L., et al.: Visual segmentation of lawn grass for a mobile robotic lawnmower. In: 2010 IEEE/RSJ International Conference on Intelligent Robots and Systems (IROS), pp. 734–739 (2010)
7. Harbas, I., Subasic, M.: Detection of roadside vegetation using features from the visible spectrum. In: 2014 37th International Convention on Information and Communication Technology, Electronics and Microelectronics (MIPRO), pp. 1204–1209 (2014)
8. Balali, V., Golparvar-Fard, M.: Segmentation and recognition of roadway assets from car-mounted camera video streams using a scalable non-parametric image parsing method. Autom. Constr. **49**(Part A), 27–39 (2015)

9. Yu, T., Muthukkumarasamy, V., Verma, B., et al.: A texture extraction technique using 2D-DFT and hamming distance. In: 2003 Fifth International Conference on Computational Intelligence and Multimedia Applications (ICCIMA), pp. 120–125 (2003)

10. Bradley, D.M., Unnikrishnan, R., Bagnell, J.: Vegetation detection for driving in complex environments. In: 2007 IEEE International Conference on Robotics and Automation, pp. 503–508 (2007)

11. Nguyen, D.V., Kuhnert, L., Kuhnert, K.D.: Structure overview of vegetation detection. A novel approach for efficient vegetation detection using an active lighting system. Robot. Auton. Syst. **60**, 498–508 (2012)

12. Nguyen, D.V., Kuhnert, L., Thamke, S., et al.: A novel approach for a double-check of passable vegetation detection in autonomous ground vehicles. In: 2012 15th International IEEE Conference on Intelligent Transportation Systems (ITSC), pp. 230–236 (2015)

13. Nguyen, D.V., Kuhnert, L., Jiang, T., et al.: Vegetation detection for outdoor automobile guidance. In: 2011 IEEE International Conference on Industrial Technology, pp. 358–364 (2011)

14. Nguyen, D.V., Kuhnert, L., Kuhnert, K.D.: Spreading algorithm for efficient vegetation detection in cluttered outdoor environments. Robot. Auton. Syst. **60**, 1498–1507 (2012)

15. Winn, J., Criminisi, A., Minka, T.: Object categorization by learned universal visual dictionary. In: 2005 Tenth IEEE International Conference on Computer Vision (ICCV), vol. 1802, pp. 1800–1807 (2005)

16. Shotton, J., Winn, J., Rother, C., et al.: Textonboost for image understanding: multi-class object recognition and segmentation by jointly modeling texture, layout, and context. Int. J. Comput. Vis. **81**, 2–23 (2009)

17. Felzenszwalb, P., Huttenlocher, D.: Efficient graph-based image segmentation. Int. J. Comput. Vis. **59**, 167–181 (2004)

18. Zhang, L., Verma, B., Stockwell, D.: Roadside vegetation classification using color intensity and moments. In: 2015 11th International Conference on Natural Computation, pp. 1250–1255 (2015)

Towards Unsupervised Learning for Arabic Handwritten Recognition Using Deep Architectures

Mohamed Elleuch[1,2(✉)], Najiba Tagougui[3], and Monji Kherallah[2]

[1] National School of Computer Science (ENSI), University of Manouba, Manouba, Tunisia
mohamed.elleuch.2015@ieee.org
[2] Advanced Technologies for Medicine and Signals (ATMS), University of Sfax, Sfax, Tunisia
monji.kherallah2014@ieee.org
[3] The Higher Institute of Management of Gabes, University of Gabes, Gabès, Tunisia
najiba.tagougui@isggb.rnu.tn

Abstract. In the pattern recognition field and especially in the Handwriting recognition one, the Deep learning is becoming the new trend in Artificial Intelligence with the sheer size of raw data available nowadays. In this paper, we highlights how Deep Learning techniques can be effectively applied for recognizing Arabic handwritten script, our field of interest, and this by investigating two deep architectures: Deep Belief Network (DBN) and Convolutional Neural Networks (CNN). The two proposed architectures take the raw data as input and proceed with a greedy layer-wise unsupervised learning algorithm. The experimental study has proved promising results which are comparable or even superior to the standard classifiers with an efficiency of DBN over CNN architecture.

Keywords: Recognition · Arabic handwritten script · DBN · CNN · Unsupervised learning

1 Introduction

In recent years the classification systems based on deep networks and their derivatives such as Deep Boltzmann Machine (DBM), Stacking Auto-Encoder (SAE), Deep Neural Networks (DNN), Deep Belief Network (DBN) and Convolutional Neural Networks (CNN) have proven their performance and accuracy in a broad area of applications namely in speech recognition [1] and image recognition [2–5]. Recently, it has gained success in optical character recognition in Latin and Asian languages [6, 7].

We noted that such approaches have not been applied yet to the handwritten Arabic field. Therefore, we study here two of the above methods: the DBN and CNN, to study the potential benefits of this Deep learning on working with raw data without feature extraction. And thus, by analyzing the error classification rates on the Arabic handwritten characters classification task.

We start our study by overviewing existing research works on handwritten Arabic script recognition. Indeed, a fast review of the literature shows that relatively limited applications are based on Deep Neural Networks [8, 9]. The major works were using only fuzzy approaches or statistical approaches [10]. During the last three decades,

© Springer International Publishing Switzerland 2015
S. Arik et al. (Eds.): ICONIP 2015, Part I, LNCS 9489, pp. 363–372, 2015.
DOI: 10.1007/978-3-319-26532-2_40

HMMs were widely used. Indeed, El Abed and Margner [11] presented a system based on the HMM classifier. They applied sliding approach for extracting the pixel features. They used the skeleton direction based feature extraction technique where each word image was splitting into uniform vertical frames and each word image was split into five horizontal zones with equal height. Al-Hajj et al. [12] used a similar sliding window for features extraction. Their system relies on combining three homogeneous HMM classifiers having the same topology in order to make the system more efficient. Benouareth et al. [13, 14] presented a system based on semi-continuous hidden Markov models using statistical and structural features.

Another classifier which is used extensively is Support Vector Machines (SVM). Chen et al. [15] presented a recognition system using SVM. They demonstrate the efficiency of Gabor features over the previous used features techniques. Later, Hamdi et al. [16] proposed a new system for handwritten isolated Arabic characters using Principal component analysis and SVM classifier. The features used in this work consist of moment and Fourier descriptor of profile projection and centroid distance.

The Long Short-Term Memory and the Connectionist Temporal Classification neural networks were used by Chherawala et al. [17]. They argue their use by their ability to automatically learn features from the input image. The system has been used for recognition Arabic isolated handwritten word and validated using the IFN/ENIT database. And recently, Porwal et al. [18] proposed a system using DBN for feature enhancement which incrementally learns complex structure of the data by representing it in a more compact and abstract manner. They applied this model for Arabic character recognition on raw data and on extracted features using GSC and Gabor features.

Thereby, works using Deep architectures are relatively scarce on handwriting recognition. That is why we investigate on using such approaches for recognizing Arabic characters.

The rest of the paper is organized as follows: In Sects. 2 and 3, we describe the basic concepts behind Deep Belief Network and Convolutional Neural Networks respectively. Our experimental study using DBN and CNN is next presented in the Sect. 4. Discussion of the results is made in Sect. 5. And finally, some concluding remarks are presented in Sect. 6.

2   Deep Belief Networks (DBN)

The DBN introduced by Hinton et al. [19], is a particular type of deep architecture that can gradually learn complex structures of the data by learning its probability distribution. It takes raw data of handwritten text as input with the expectation that subsequent layers would learn good feature representation. Learning in the network takes two steps, an unsupervised feature learning followed by a supervised learning of discriminating function. The unsupervised learning is performed by using contrastive divergence algorithm which is an approximation of maximum likelihood estimation [20], while in the later stage the network parameters are fine-tuned with the gradient based back propagation algorithm [21].

Indeed, a DBN is a generative graphical model composed of multiple layers of stochastic hidden variables and one layer of visible units [19] (see Fig. 1). It can be trained with an efficient algorithm, by greedily training each layer as a Restricted Boltzmann Machines (RBM) [22].

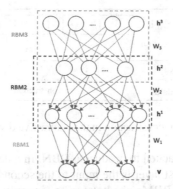

Fig. 1. Illustration of a DBN with three hidden layers

The energy function and the probabilistic semantics for an RBM are defined as:

$$E(v,h;\theta) = -\sum_{i=1}^{V} \sum_{j=1}^{H} w_{ij} v_i h_j - \sum_{i=1}^{V} b_i v_i - \sum_{j=1}^{H} a_j h_j \tag{1}$$

$$p(v;\theta) = \frac{1}{Z} \sum_{h} e^{-E(v,h;\theta)} \tag{2}$$

Where $\theta = (W, b, a)$ and w_{ij} represents the weights between visible units v_i and hidden units h_j and b_i and a_j are their biases. Z is the partition function. V and H are the number of visible and hidden units.

For binary (or real-valued) visible and hidden units, the probability that hidden unit h_j is activated given visible vector **v** and the probability that visible unit v_i is activated given hidden vector **h** are given by:

$$P(h_j|v;\theta) = \sigma\left(a_j + \sum_i w_{ij} v_i\right) \tag{3}$$

$$P(v_i|h;\theta) = \sigma\left(b_i + \sum_j w_{ij} h_j\right) \tag{4}$$

Where σ denotes the logistic sigmoid. This logistic function $\sigma(x) = 1/1 + e^{-x}$, is a common choice for the activation function.

A RBM is trained to learn probability distribution of the data with the help of hidden units. Using Markov Chain Monte Carlo (MCMC) method to maximize the likelihood is presumably more accurate and stable. In contrast, running an MCMC algorithm to convergence at each iteration of gradient descent taking a long time, therefore an approximation called Contrastive Divergence (CD) is generally employed [23]. The training process of a single layer RBM to minimize the reconstruction error using CD is shown in Fig. 2.

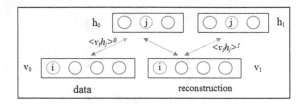

Fig. 2. Data and reconstruction in the contrastive divergence training

Several RBMs can be stacked to produce a DBN. In a deep network, the activation of the hidden units in the first layer is the input to the second layer.

After training a stack of RBMs, the bottom-up recognition weights of the resulting DBN can be used to initialize the weights of a Multi-Layer Feed-Forward Neural Network, which can then be discriminatively fine-tuned by back propagating error derivatives. The Feed-Forward Network is given a final "soft-max" layer that calculates a probability distribution over class labels and the derivative of the log-probability of the correct class is back propagated to train the incoming weights of the final layer and to discriminatively fine-tune the weights in all lower layers [24].

3   Convolutional Neural Networks (CNN)

The second architecture, the CNN was first developed by LeCun et al. [25]. It is a specialized type of neural network which learns the good features at each layer of the visual hierarchy via back propagation (BP). Ranzato et al. [5] achieves improvements in performance when they applied an unsupervised pre-training to a CNN.

In fact, CNN are hierarchical, multi-layer neural networks trained with the back propagation algorithm [25]. CNN are used to learn complex, high-dimensional data. A variation on convolutional subsampling layers is investigated. The difference is inside their architecture. Several architectures of CNN are proposed for different problems such as object recognition [26] and handwritten digit/character recognition [5, 25] and achieve the best performance on pattern recognition task.

4   Experimental Results and Discussion

We conducted our experimental studies using DBN and CNN for recognizing offline Arabic character. These both classifiers were tested on HACDB database [27]. Results are detailed and discussed in the next subsections.

4.1   HACDB Database

The HACDB database [27] contains 6.600 shapes of handwritten characters written by 50 persons (Fig. 3). Each writer has generated two forms for 66 shapes: 58 shapes of characters and 8 shapes of overlapping characters (representing 24 basic characters/ overlapping characters without dots). The images are normalized 28 by 28 pixels and are in the gray scale. The dataset is divided into a training set of 5.280 images and a test set of 1.320 images.

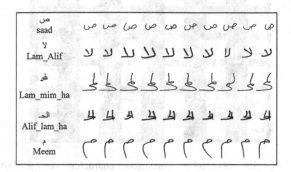

Fig. 3. Samples from the HACDB database written by 10 different writers

4.2 Experiments Using Deep Belief Network

In order to evaluate the effectiveness of unsupervised feature learning approach using DBN, we investigated its performance for training and recognizing characters of HACDB database. To best train the model on further data in order to better take account the variability of handwriting, we expand the size of the training set ten times by the technique of elastic deformation proposed by Simard et al. [28]. Technical implementation details of the adopted system are given below.

- **Model selection and training process**

An experimental study was established to evaluate the number of hidden layers (N_{hl}) and the number of neurons (N_n) by each hidden layer. Our choice of these parameters is based on the criterion of the error rate on the test set. To calculate this criterion, a softmax layer with 66 output units is first added to the top of each unsupervised trained DBN and later learned with BP. The output of each unit is just the probability of assigning the corresponding label. When the number of hidden layers is fixed, the optimal number of hidden units is selected depending on the number of epoch giving the minimum value of error rate.

The experiments are based on the HACDB Database with elastic distortion were performed, the input images consist of 28 × 28 pixels giving an input dimensionality of 784, and the inputs were scaled between 0 and 1. The adopted architecture of DBN was 784-1000-1000-66, i.e., it consisted of two RBMs each one with 1000 hidden neurons and the number of epoch is fixed at 200 for DBN train. Each RBM of the DBN was trained using greedy layer-wise unsupervised learning algorithm [9] with contrastive divergence [29]. The obtained error rate was 3.64 % (see Fig. 4). Again the same architecture is used with the 24 class problem; error rate is two times smaller (1.67 %).

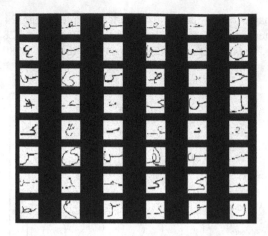

Fig. 4. Samples of 48 incorrectly classified characters using DBN architecture; 784-1000-1000-66

4.3 Experiments Using Convolutional Neural Networks

In this section, we investigated the performance of the CNN for training and recognizing Arabic characters. For the setting architecture, a convolutional layer is parameterized by the size and the number of the maps, kernel sizes, skipping factors, and the connection table.

Figure 5 shows a typical CNN architecture for handwritten character recognition. It consists of a set of several layers. First, the input is convoluted with a set of filters (C hidden layers) to get the values of the feature map. After, each convolution layer is followed by a sub-sampling layer to reduce the dimensionality (S hidden layers) of the spatial resolution of the feature map. The lowest level of the CNN architecture is the input layer. It receives the gray-level image containing the character to recognize.

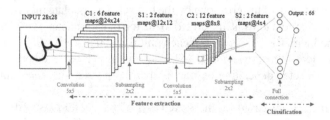

Fig. 5. A Typical Convolutional Neural Network architecture composed of layers for feature maps

The description of the CNN architecture used in experiments applied to HACDB database with elastic distortion is given in the following way: $1 \times 28 \times 28$-6C2S-12C2S-66 represents a net with input images of size 28×28 with four Convolutional-Subsampling layers which can be viewed as a trainable feature extractor. The highest level of the CNN architecture composed of universal classifier characterized with a full connection layer, which consists of 66 output neurons corresponding to the 66 class labels of character patterns.

The first convolutional layer "C1" has 6 feature maps each having 25 weights, constituting a 5×5 trainable kernel, and a bias. The feature maps' size is 24×24. The second hidden layer "S1" named sub-sampling consists of 2 features maps with size 12×12. The third layer "C2" has 12 convolutional maps of size 8×8 and the fourth layer "S2" has 2 sub-sampling maps of size 4×4. When training this architecture, the feature maps of the 4th layer are merged into a feature vector which feeds into the final layer.

The CNN architecture already described was trained using gradient descent for 200 epochs; our obtained result is 14.71 % error classification rate on the test set with 66 classes. Using only the 24 basic characters from HACDB database, error rate decrease approximately three times giving 5 %.

5   Discussion

In the last experimental section, two deep architectures were compared to recognize Arabic handwritten characters. The DBN architecture outperforms the CNN one tested on the same database as illustrated in Table 1. It showed that there is a significant gain in error classification rate (ECR) compared to CNN system where the absolute recognition rate was improved by 11.07 % with the proposed DBN system for the 66 class problem.

Table 1. ECR for our proposed systems applied on HACDB database

Approach	ECR	
	24 classes (only basic characters)	66 classes (all shapes)
Deep belief network	**1.67 %**	**3.64 %**
Convolutional neural network	5 %	**14.71 %**

The ECR obtained with the DBN system (with 66 outputs) on the HACDB database equals to 3.64 %. This rate is statistically significantly important in comparison with character recognition accuracies obtained from state-of-the-art offline Arabic systems [18] (see Table 2). The tested CNN architecture presented an ECR equivalent to 14.71 %. It is to say that CNN are prone to get trapped in local optima of a non-convex objective function, however DBN resolve this problem by learning feature representations with two stages: unsupervised pre-training and supervised fine-tuning strategies to construct the models.

Table 2. Performance comparisons using DBN approach

Authors	Feature extraction	Databases (class)	ECR
Present work	Automatic	HACDB (**24**)	**1.67 %**
		HACDB (**66**)	**3.64 %**
Hinton et al. [6]	Automatic	MNIST (**10**)	1.25 %
Sazal et al. [30]	Automatic	BANGLA CHAR-ACTERS (**60**)	9.73 %
Porwal et al. [18]	Automatic	AMA Arabic PAW (**34**)	19.48 %

The obtained results are more sufficiently significant in comparison with the research studies using other classification techniques especially that they were obtained with raw data without any feature extraction step (see Tables 2 and 3). This represents a real revolution in the pattern recognition domain since it will be a real motivation to encourage the use of deep learning strategies with big data analytics.

Table 3. Performance comparisons using CNN approach

Authors	Feature extraction	Database (class)	ECR
Present work	Automatic	HACDB (**24**)	**5 %**
		HACDB (**66**)	**14.71 %**
Ciresan et al. [31]	Automatic	NIST (**62:** 52 letters and 10 digits)	11.88 %
		NIST (**52:** 26 upper and 26 lower case letters)	21.41 %

6 Conclusion

In this work, we presented two deep learning approaches for handwritten Arabic script recognition namely DBN and CNN. The purpose was to take advantages of the power of these deep networks that are able to manage large dimensions input, which allows the use of raw data inputs rather than to extract a feature vector and learn complex decision border between classes. In an experimental section we showed that the results were promising with an ECR of 3.64 % using DBN and 14.71 % with the CNN classifier when applied to the HACDB database with 66 class labels of character patterns. As perspective, we have to reconfigure our proposed DBN network architecture to be able to deal with high-dimension data like words and integrate hand-craft features in order to increase the recognition rate. It is a serious challenge to generalize the use of the deep learning to several recognition applications.

References

1. Mohamed, A., Dahl, G., Hinton, G.: Acoustic modeling using deep belief networks. IEEE Trans. Audio Speech Lang. Process. **20**(1), 14–22 (2011)
2. LeCun, Y., Bottou, L., Bengio, Y., Haffner, P.: Gradient-based learning applied to document recognition. Proc. IEEE **86**(11), 2278–2324 (1998)
3. Zhong, S., Liu, Y., Liu, Y.: Bilinear deep learning for image classification. In: Proceedings of the 19th ACM International Conference on Multimedia, pp. 343–352 (2011)
4. Karpathy, A., Joulin, A., Fei-Fei, L.: Deep fragment embeddings for bidirectional image sentence mapping. In: Computer Vision and Pattern Recognition, pp. 1889–1897 (2014)
5. Ranzato, M., Huang, F., Boureau, Y., LeCun, Y.: Unsupervised learning of invariant feature hierarchies with applications to object recognition. In: Proceedings of Computer Vision and Pattern Recognition Conference (CVPR), pp. 1–8. IEEE Press (2007)
6. Ciresan, D.C., Meier, U., Schmidhuber, J.: Transfer learning for latin and Chinese characters with deep neural networks. In: Proceedings of International Joint Conference on Neural Networks (IJCNN), pp. 1–6. IEEE Press (2012)
7. Ciresan, D.C., Schmidhuber, J.: Multi-Column Deep Neural Networks for Offline Handwritten Chinese Character Classification. IDSIA Technical Report No. IDSIA-05-13 (2013)
8. Salakhutdinov, R., Tenenbaum, J.B., Torralba, A.: Learning with hierarchical-deep models. Pattern Anal. Mach. Intell. **35**(8), 1958–1971 (2013)
9. Bengio, Y., Lamblin, P., Popovici, D., Larochelle, H.: Greedy layer-wise training of deep networks. In: Advances in Neural Information Processing Systems (NIPS), pp. 153–160 (2007)
10. Tagougui, N., Kherallah, M., Alimi, A.M.: Online Arabic handwriting recognition: a survey. Int. J. Doc. Anal. Recogn. (IJDAR) **16**(3), 209–226 (2013)
11. El Abed, H., Margner, V.: Comparison of different preprocessing and feature extraction methods for offline recognition of handwritten Arabic words. In: Ninth International Conference on Document Analysis and Recognition (ICDAR), pp. 974–978 (2007)
12. Al-Hajj Mohamad, R., Likforman-Sulem, L., Mokbel, C.: Combining slanted-frame classifiers for improved HMM-based Arabic handwriting recognition. IEEE Trans. Pattern Anal. Mach. Intell. **31**, 1165–1177 (2009)
13. Benouareth, A., Ennaji, A., Sellami, M.: HMMs with explicit state duration applied to handwritten Arabic word recognition. In: 18th International Conference on Pattern Recognition (ICPR), pp. 897–900 (2006)
14. Benouareth, A., Ennaji, A., Sellami, M.: Semi-continuous HMMs with explicit state duration for unconstrained Arabic word modeling and recognition. Pattern Recogn. Lett. **29**, 1742–1752 (2008)
15. Chen, J., Cao, H., Prasad, R., Bhardwaj, A., Natarajan, P.: Gabor features for offline arabic handwriting recognition. In: Proceedings of the 9th IAPR International Workshop on Document Analysis Systems (DAS), pp. 53–58 (2010)
16. Hamdi, R., Bouchareb, F., Bedda, M.: Handwritten Arabic character recognition based on SVM Classifier. In: 3rd International Conference on Information and Communication Technologies: From Theory to Applications (ICTTA), pp. 1–4. IEEE Press (2008)

17. Chherawala, Y., Roy, P.P., Cheriet, M.: Feature design for offline Arabic handwriting recognition: handcrafted vs automated?. In: 12th International Conference on Document Analysis and Recognition (ICDAR), pp. 290–294 (2013)
18. Porwal, U., Zhou, Y., Govindaraju, V.: Handwritten Arabic text recognition using deep belief networks. In: 21st International Conference on Pattern Recognition (ICPR), pp. 302–305. IEEE Press (2012)
19. Hinton, G., Osindero, S., Teh, Y.W.: A fast learning algorithm for deep belief nets. Neural Comput. **18**(7), 1527–1554 (2006)
20. Tieleman, T., Hinton, G.: Using fast weights to improve persistent contrastive divergence. In: Proceedings of the 26th Annual International Conference on Machine Learning (ICML), pp. 1033–1040 (2009)
21. Rumelhart, D.E., Hinton, G.E., Williams, R.J.: Learning representations by back-propagating errors nature. Nature **323**(9), 533–536 (1986)
22. Mohamed, A., Sainath, T.N., Dahl, G., Ramabhadran, B., Hinton, G.E., Picheny, M.A.: Deep belief networks using discriminative features for phone recognition. In: Proceedings of the IEEE International Conference on Acoustics, Speech and Signal Processing, pp. 5060–5063 (2011)
23. Hinton, G.E.: Training products of experts by minimizing contrastive divergence. Neural Comput. **14**(8), 1771–1800 (2002)
24. Hinton, G.E., Salakhutdinov, R.R.: Reducing the dimensionality of data with neural networks. Science **313**, 504–507 (2006)
25. LeCun, Y., Bottou, L., Bengio, Y., Haffner, P.: Gradient-based learning applied to document recognition. Proc. IEEE **86**(11), 2278–2324 (1998)
26. LeCun, Y., Huang, F.J., Bottou, L.: Learning methods for generic object recognition with invariance to pose and lighting. In: Proceedings of Computer Vision and Pattern Recognition Conference (CVPR). IEEE Press (2004)
27. Lawgali, A., Angelova, M., Bouridane, A.: HACDB: handwritten Arabic characters database for automatic character recognition. In: European Workshop on Visual Information Processing (EUVIP), pp. 255–259 (2013)
28. Simard, P., Steinkraus, D., Platt, J.C.: Best practices for convolutional neural networks applied to visual document analysis. In: International Conference on Document Analysis and Recognition (ICDAR), pp. 958–962 (2003)
29. Carreira-Perpinan, M.A., Hinton, G.E.: On contrastive divergence learning. In: Proceedings of the tenth International Workshop on Artificial Intelligence and Statistics, pp. 33–40 (2005)
30. Sazal, M.M.R., Biswas, S.K., Amin, M.F., Murase, K.: Bangla handwritten character recognition using deep belief network. In: Proceedings of the 2013 International Conference on Electrical Communication and Information Technology (ECIT), pp. 1–5 (2013)
31. Ciresan, D.C., Meier, U., Gambardella, L.M., Schmidhuber, J.: Convolutional neural network committees for handwritten character classification. In: 11th International Conference on Document Analysis and Recognition (ICDAR), pp. 1135–1139. IEEE Press (2011)

Optimum Colour Space Selection for Ulcerated Regions Using Statistical Analysis and Classification of Ulcerated Frames from WCE Video Footage

Shipra Suman[1(✉)], Nicolas Walter[1], Fawnizu Azmadi Hussin[1], Aamir Saeed Malik[1],
Shiaw Hooi Ho[2], Khean Lee Goh[2], and Ida Hilmi[2]

[1] Electrical and Electronic Engineering, Universiti Teknologi PETRONAS, 31750 Tronoh,
Perak, Malaysia
suman.shipra@ieee.org, walter.nicolas.pro@gmail.com,
{fawnizu,aamir_saeed}@petronas.com.my
[2] Department of Medicine, University of Malaya, 50603 Kuala Lumpur, Malaysia
shooiho@yahoo.com, klgoh56@gmail.com, i_hilmi@um.edu.my

Abstract. The Wireless Capsule Endoscopy (WCE) is a painless and non-invasive procedure that allows clinicians to visualize the entire Gastrointestinal Tract (GIT) and detect various abnormalities. During the inspection of GIT, numerous images are acquired at a rate of approximately 2 frames per second (fps) and recorded into a video footage (containing about 55,000 images). Inspecting the WCE video is very tedious and time consuming for the doctors, resulting in limited application of WCE. Therefore, it is crucial to develop a computer aided intelligent algorithm to process the huge number of WCE frames. This paper proposes an ulcerated frame detection method based on RGB and CIE Lab colour spaces. In order to select and provide the classifier with the bands containing most ulcer information, a statistical analysis of ulcerated images pixel based is proposed. The resulting band selection will enhance the classification results and increase the sensitivity and specificity with regards to ulcerated frame identification.

Keywords: WCE image processing · Ulcer frame detection · Statistical analysis · Colour spaces selection · Classification

1 Introduction

Gastrointestinal tract (GIT) includes the oesophagus, stomach, small bowel, and large bowel. In order to examine the whole tract, traditional endoscopes have played an important role. For example, upper endoscopy, push enteroscopy and colonoscopy have been used to visualise the upper and lower parts of the digestive system. However, they are not capable of examining the entire small bowel due to the high risk encountered of penetrating the small bowel membrane. Resulting for patients in more harmful and life threatening complications such as bleeding and scars. To overcome these drawbacks, Wireless Capsule Endoscopy (WCE) has been proposed by Given Imaging and was

© Springer International Publishing Switzerland 2015
S. Arik et al. (Eds.): ICONIP 2015, Part I, LNCS 9489, pp. 373–381, 2015.
DOI: 10.1007/978-3-319-26532-2_41

approved by Food and Drug Administration (FDA) in 2002. Since then, WCE has become a crucial tool for small bowel examination.

1.1 Wireless Capsule Endoscopy

WCE is a technological breakthrough that allows visualisation of the whole GIT in a comfortable and efficient way for the patients [1]. Before a patient swallows the capsule, 8 skin antennas are taped to the anterior abdominal wall (see Fig. 1). While moving inside GIT, the capsule acquire images at a rate of approximately 2 frames per second (fps) and sends them through radio frequency (RF) transmitter to the data logger (DL) through the 8 sensor arrays that have been fixed on the anterior abdominal wall (see Fig. 1). The DL is attached to the patient and stored the complete footage of images in specific format. After the completion of examination (i.e. WCE exits patient's body after 8 h), the images are downloaded to a computer from DL and inspected by doctors through dedicated software. The use of the real time viewer may shorten the global procedure, as the patient can be disconnected once the cecum is visualised [2]. WCE produces approximately 55,000 images per examination and clinicians need about 2–3 h to examine the complete footage carefully, usually frame-by-frame in some cases to detect the abnormalities. This time consuming and tedious process is the major disadvantage of WCE imaging that prevents and limits its use as the clinicians spend more time in finding the abnormalities than diagnosing or grading the diseases. Hence, image processing with automatic frame detection step is a priority key to save lots of analysis time and enhance the diagnostic results.

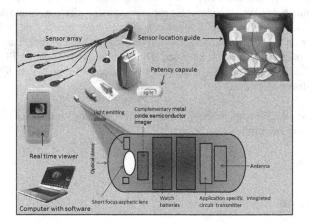

Fig. 1. Capsule endoscopy components including schematic parts representation and sensor location on patient body [5]

Many efforts and computational approaches towards the direction of automatic inspection and analysis of WCE images have been reported in the literature. In particular, detection of abnormal patterns is achieved by employing texture spectrum for feature extraction and neural network techniques [3, 4].

In the same direction, image enhancement [6] image registration [7], along with segmentation techniques [8] were also used. Co-occurrence matrices [9] and local binary patterns (LBPs) [10] contributed to polyps and tumours detection. Moreover, blood detection was performed by taking advantage of chromaticity moments [11] and colour spectrum transformation [12]. However, there is a lack of research towards ulcer recognition, despite the high importance and occurrence of the disease. The techniques proposed include MPEG7 descriptors [13], RGB pixel values evaluation [14], curvelet transform with uniform LBP [15], and chromaticity moments [16]. The different classification accuracy achieved with these methodologies varies from 36 % to 88 %.

1.2 Problem Definition: Ulcer

Peptic ulcer is described as an area where tissues have been destroyed by gastric juices. In particular, gastric juices are produced by the stomach and the intestine to digest the starch, fat, and protein in food. Although most peptic ulcers appear in the stomach (gastric ulcers) and the duodenum (duodenal ulcers), they may also appear in the small bowel. Some image samples of precise ulcerated regions found in different parts of GIT are shown in Fig. 2.

Fig. 2. Examples of Ulcerated regions that can be found in GIT

Basically ulcers have three types of complications such as bleeding ulcer, perforated ulcer and narrowing ulcer, usually known as stomach ulcer. Bleedings occur in the wall of stomach or duodenal wall when blood vessels are damaged [17]. When abdominal or duodenal wall have holes, the sterile abdominal cavity is invaded by gastric juices causing severe inflammations, also known as perforated ulcer [18]. Usually, a narrowing ulcer is diagnosed and located at the end of stomach at duodenal attachment area, causing scars and swelling which in turns narrow or sometimes close the opening of intestine [19].

2 Related Works

Usually, classification results depend on the precision of the features given with respect to the object to be detected. Hence, the feature extraction is crucial for exhibiting best classification results. Related works have extracted different features from different colour spaces to detect ulcers in the GIT image sequences.

2.1 Colour Space Transformation and Information Content

RGB colour space is one of the most well-known colour spaces. It is commonly used in most of the devices for image representation. Although RGB colour space resembles or mimics the human visual system in a similar way, it suffers from a major disadvantage:

for applications on natural images, a high correlation between its components can be observed [20]. Hence, the highly correlated channels contain redundant information [11]. In order to overcome this issue and find colour bands that will provide less correlated features, we selected Lab colour space.

Unlike RGB colour space, Lab is described to approximate human vision system. It pursues to perceptual uniformity and its L component closely matches human perception of lightness. This can lead to a more realistic visualisation in specific applications. The 'a' and 'b' parts are mixed colour representations of magenta-green and yellow-blue colours, respectively. The representation obtained mimics the non-linear response of human eyes achieving a more informative description of colours and objects.

2.2 Machine Learning

Support Vector Machines (SVMs) is a supervised learning method, broadly used to classify different kind of data. The fundamental idea is that an SVM maps the input informative data of a n-dimensional space, where it tries to locate the ideal hyper planes to differentiate between the different classes associated to datasets [21]. The popularity of SVMs lies on their application capacity for an extensive variety of pattern recognition issues.

The next section will describe the methodology employed and the specific use of different colour spaces and machine learning to achieve ulcerated frame extraction.

3   Methodology

Contrary to other researches that are focusing on processing all images in order to detect the different abnormalities, the proposed work is focused on a divide and conquers strategy. Indeed, the processing of 55,000 images is very long and our goal is to reduce the analysis time.

3.1 Image Enhancement

Image enhancement is one of the essential steps in image processing. It is a technique that is referred to as highlighting key data while reducing or removing non-essential or auxiliary information in an image. Moreover, the inherent noise of the WCE images needs to be processed. In this work, wavelet de-noising using 3 level of decomposition has been used. Soft fixed thresholding method using db2 wavelet has been applied to reduce noise and enhance ulcer information in WCE images.

3.2 Colour Space Selection for Enhanced Representation of Ulcer Information

RGB and CIE Lab colour spaces have been investigated in terms of ulcer information content. The goal is to provide the classifier with most informative bands. Indeed, the different bands obtained (as shown in Fig. 3) highlights the various and different information contained in the RGB and Lab bands. In particular, we are interested to provide best bands for the classification step.

	Colour Image	R band	G band	B band	L band	a band	b band
Image 1							
Image 2							

Fig. 3. Sample colour images of ulcerated regions visualized in separated bands. The best band set for classification cannot be inferred directly.

Hence, the choice of best band, in terms of ulcer and foreground separation, will be based on statistical analysis of ulcer pixel values against foreground pixel values in each different band. A normal distribution is deduced from each set of pixel values in order to highlight the separation produced in each separated band. The overlapping area obtained for each band serves as an index of separation. Indeed, a minimised overlapping area will express a better separation capability of a specific band (Table 1).

Table 1. Overlapping area results for separated band capability of ulcer and non-ulcer separation.

Colour band	R	G	B	L	a	b
Overlapping values	0.47	0.30	0.67	0.39	0.24	0.75

4   Experimental Results and Discussions

Our work has been done in collaboration with gastroenterologist from Endoscopy unit at University of Malay Medical Center (UMMC), Kuala-Lampur, Malaysia. They provide us with a huge set of WCE footages. Moreover, they provide us ground truth images from these videos with highlighted Ulcerated regions. As per our explained strategy of divide and conquer, our goal is to use the image set provided as references to process the video and extract ulcerated frames. Samples of highlighted ulcerated regions by the doctors can be seen in Fig. 2. It has been very difficult for the doctors to gather a large number of real ulcer cases in WCE videos as most of them contained a few frames of significant ulcer findings. The WCE pill used to acquire the different image sets and videos is of OLYMPUS type. The resolution provided is 576 × 576.

4.1   Data Set

As per availability of data at UMMC, we currently have 6 videos of diagnosed ulcer patients. From the different images of ulcer extracted by our collaborators (with highlights of ulcer regions), training and testing set have been created for classification purpose. Our training set consists in 50 ulcerated and 60 non-ulcerated images forming a 110 image set. For the testing set 60 ulcerated and 80 non-ulcerated images have been used, summing a total number of frames equal to 140 frames. Labelled images provided by our collaborators have all been used in the training set for ulcer candidate. Among

60 non-ulcer images, 19 of these images are contaminated with food residues. 22 images are contaminated with bubbles and high illumination due to the rolling of the camera in the GIT. The rest of the images consist in non-ulcerated images from the different video footages.

4.2   Results of Statistical Analysis

As shown in Fig. 4, each band analysed present a different separation of ulcer and non-ulcer pixel values. The overlapping areas extracted from the normal distribution of ulcer and non-ulcer pixels in each band, indicates that the band provide the best separation between ulcer and non-ulcer pixels. The second band is the Green channel from RGB colour space. Indeed, this result could be expected as the non-ulcer pixels present mostly red information and ulcer pixels whitish values. Finally, the Luminance information from L band of Lab colour space present as well a good separation of ulcer and non-ulcer pixels. These three bands are believed to provide best separation of ulcer and non-ulcer pixels compared to the normal colour space representation, i.e. RGB or Lab. These three selected bands have been fed to the SVM classifier in order to enhance classification results.

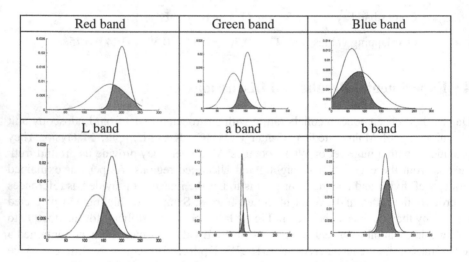

Fig. 4. Normal distribution curve for RGB and Lab channels.

The SVM classifier has been used with a Radial Basis Function (RBF) kernel. The classification results in terms of sensitivity and specificity are presented in Table 2 in which we tested for various combinations to select the best channels to achieve high sensitivity and specificity. The band has been selected as high to low overlapping values. For first phase, 'b' band of Lab color space has been chose for high has been coosen as high overlapping value and 'a' from Lab as least overlapping values to compute sensitivity and specificity for 'a, G, L, R, B, b' combination as input vector to SVM. Similarly in next phases we choose vectors such as 'a, G, L, R, B', 'a, G, L, R', 'a, G, L' and 'a,

G,'. So, we can conclude that channel 'a', 'L' of Lab color space and channel 'G' of RGB color space are best channels to achieve high impact result for disease finding.

Table 2. Sensitivity and specificity For various bands

Colour Bands used	Sensitivity	Specificity
All 6 Bands (a, G, L, R, B and b)	87.08	89.11
a, G, L, R and B	87.60	89.32
a, G, L and R	87.63	90.00
a, G and L	**88.09**	**90.88**
a and G	87.03	90.21

Also classification result with other author has been compared in Table 3. As compared to the results of the other methods, our proposed method shows a trade-off between sensitivity and specificity. Indeed, compared to [22], our method surpasses both the sensitivity and specificity. Compared to [15, 23], our method shows similar sensitivity but exhibits an improved specificity. It is believed that these results could be enhanced by the use of other bands from different colour spaces. Moreover, a high sensitivity is of crucial importance for the doctors as they do not want to miss any disease.

Table 3. sensitivity and specificity results comparison

Authors	Colour Bands	Classifier	Sensitivity	Specificity
[22]	Various colour space	Joint boost	82.3	89.10
[23]	HSV	SVM	88	84
[15]	RGB	MLP	88.8	84.1
Proposed method	a, G, L	SVM	88.09	90.88

To compare with other research works, the same dataset could not be used due to the inexistence of public capsule endoscopy datasets. Our methodology has focused on perforated ulcers since these are characteristic ulcers [22], whereas the authors of [15] do not clarify the type of ulcers they detect. Moreover, there is a big difference in patterns of various types of ulcers.

5 Conclusion

This paper proposed a novel approach to ulcer detection in GIT. Specifically, our proposed methodologies are based on divide and conquer strategy to extract ulcer frames from the huge video footage obtained after WCE examination. Statistical analysis of ulcer and non-ulcer pixel values has been successfully investigated to provide the classifier with the best separating bands from RGB and Lab colour spaces. The classification results suggest that the approach is efficient and can be enhanced in the future by adding more bands of other colour spaces.

Acknowledgements. This research work is supported by Graduate Assistantship (GA) scheme, Universiti Teknologi PETRONAS, Perak, Malaysia. We would like to thank our collaborators in UMMC for their endless help and support in the realisation of this project.

References

1. Avgerinos, A., Kalantzis, N.: Endoscopy with wireless capsule (in Greek: Ενδοσκόπηση με ασύρματη κάψουλα), Athens: Vita (2006)
2. Lai, L.H., Wong, G.L., Lau, J.Y., Sung, J.J., Leung, W.K.: Initial experience of real-time capsule endoscopy in monitoring progress of the videocapsule through the upper GI tract. Gastrointest Endosc. **66**, 1211–1214 (2007)
3. Kodogiannis, V.S., Boulougoura, M., Lygouras, J.N., Petrounias, I.: A neuro-fuzzy-based system for detecting abnormal patterns in wireless-capsule endoscopic images. Neurocomputing **70**, 704–717 (2007)
4. Iakovidis, D.K.: Unsupervised summarization of capsule endoscopy video. In: Proceedings of the 4th International Conference IEEE on Intelligent Systems, vol. 1, pp. 3-15–3-20, September 2008
5. Ghoshal, U.C.: Capsule Endoscopy: A New Era of Gastrointestinal Endoscopy. INTECH Open Access Publisher (2013)
6. Suman, S., Hussin, F.A., Malik, A.S., Walter, N., Goh, K.L., Hilmi, I., Ho, S.H.: Image enhancement using geometric mean filter and gamma correction for WCE Images. In: Loo, C.K., Yap, K.S., Wong, K.W., Beng Jin, A.T., Huang, K. (eds.) ICONIP 2014, Part III. LNCS, vol. 8836, pp. 276–283. Springer, Heidelberg (2014)
7. Bourbakis, N.: Detecting abnormal patterns in WCE images. In: 5th IEEE Symposium on BIBE, pp. 232–238, October 2005
8. Dhandra, B.V., Hegadi, R., Hangarge, M., Malelath, V.S.: Analysis of abnormality in endoscopic images using combined HSI color space and watershed segmentation. In: 18th International Conference on IEEE ICPR, vol. 4, pp. 695–698, August 2006
9. Ameling, S., Wirth, S., Paulus, D., Lacey, G., Vilarino, F.: Texture-based polyp detection in colonoscopy. In: Meinzer, H.-P., Deserno, T.M., Handels, H., Tolxdorff, T. (eds.) Bildverarbeitung fur die Medizin, pp. 346–350. Springer, Berlin (2009)
10. Iakovidis, D.K., Maroulis, D., Karkanis, S.A.: An intelligent system for automatic detection of gastrointestinal adenomas in video endoscopy. Comp. Biol. Med. **36**, 1084–1103 (2006)
11. Meng, M.Q.-H., Li, B.: Computer aided detection of bleeding in capsule endoscopy images. In: Canadian Conference on Electrical and Computer Engineering, pp. 1963–1966, May 2008
12. Jung, Y.S., Kim, Y.H.., Lee, D.H., Kim, J.H.: Active blood detection in a high resolution capsule endoscopy using color spectrum transformation. In: International Conference on IEEE on Biomedical Engineering and Informatics, vol. 1, pp. 859–862, May 2008
13. Cunha, J.P.S., Coimbra, M.: MPEG-7 visual descriptors contributions for automated feature extraction in capsule endoscopy. IEEE Trans. Circuits Syst. Video Technol. **16**, 628–637 (2006)

14. Gan, T., Wu, J.-C., Rao, N.-N., Chen, T., Liu, B.: A feasibility trial of computer-aided diagnosis for enteric lesions in capsule endoscopy. World J. Gastroenterol. **14**(45), 6929–6935 (2008)
15. Meng, M.Q.-H., Li, B.: Ulcer recognition in capsule endoscopy images by texture features. In: Proceedings of 7th IEEE World Congress on Intelligent Control and Automation, pp. 234–239 (2008)
16. Meng, M.Q.-H., Li, B.: Computer-based detection of bleeding and ulcer in wireless capsule endoscopic images by chromaticity moments. Comp. Biol. Med. **39**, 141–147 (2009)
17. Peptic Ulcers, Harvard Medical School, Well-Connected reports, September 2001
18. eMedicineHealth—Practical Guide to Health. http://www.emedicinehealth.com/peptic_ulcers/article_em.htm. January 2009
19. Karnam, U.S., Rosen, C.M., Raskin, J.B.: Small bowel ulcers. Curr. Treat. Options Gastroenterol. J. **4**(1), 15–21 (2001)
20. Joblove, G.H., Greenberg, D.: Color spaces for computer graphics. In: ACM Siggraph Computer Graphics. ACM (1978)
21. Vapnik, V.N.: Statistical Learning Theory. Wiley, Hoboken (1989)
22. Liu, X., et al.: A new approach to detecting ulcer and bleeding in Wireless capsule endoscopy images. In: 2012 IEEE-EMBS International Conference on Biomedical and Health Informatics (BHI). IEEE (2012)
23. Karargyris, A., Bourbakis, N.: Detection of small bowel polyps and ulcers in wireless capsule endoscopy videos. IEEE Trans. Biomed. Eng. **58**(10), 2777–2786 (2011)

Learning the Optimal Product Design Through History

Victor Parque[✉] and Tomoyuki Miyashita

Department of Modern Mechanical Engineering, Waseda University,
3-4-1 Okubo, Shinjuku-ku, Tokyo 169-8555, Japan
parque@aoni.waseda.jp, tomo.miyashita@waseda.jp

Abstract. The search for novel and high-performing product designs is a ubiquitous problem in science and engineering: aided by advances in optimization methods the conventional approaches usually optimize a (multi) objective function using simulations followed by experiments.

However, in some scenarios such as vehicle layout design, simulations and experiments are restrictive, inaccurate and expensive. In this paper, we propose an alternative approach to search for novel and high-performing product designs by optimizing not only a proposed novelty metric, but also a performance function learned from historical data. Computational experiments using more than twenty thousand vehicle models over the last thirty years shows the usefulness and promising results for a wider set of design engineering problems.

Keywords: Design · Vehicle · Optimization · Genetic programming · Particle swarm

1 Introduction

How to build optimal product design layouts? A natural answer to this question is through *metaphoring*, which consists of mimicking the functions and features of pre-existing referents, and *tooling*, which consists of building through modularity principles to scale toward sophisticated and hierarchical entities. Both principles rely on heuristics, either artificial or natural, to find articulated rules that map the space of needs and functions to the product entities with concrete parametric representations [1–6].

However, in certain circumstances it is difficult to perform the above: either real-world experiments are expensive or dangerous, or simulations are inaccurate to consider the real-world invariants. Furthermore, neither real-world experiments nor simulation tests consider novelty metrics such as the work of Grignon et al. [7], which is meaningful for a complete and uniform sampling of the design search space.

In order to tackle the above problems, we propose a simple and alternative approach to search for optimal product designs under restrictive simulations-experimentations. We use historical data to approximate a surrogate function

© Springer International Publishing Switzerland 2015
S. Arik et al. (Eds.): ICONIP 2015, Part I, LNCS 9489, pp. 382–389, 2015.
DOI: 10.1007/978-3-319-26532-2_42

that explains the performance of observed past design variables reasonably; and use the learned surrogate function as a performance metric, along with a proposed novelty metric, to optimize the product design layouts. The main goal and advantage of our approach is to map the fast approximations of the off-line learning (based on historical data) to the global refinements of parameter optimization (both in continuous space), so that the overall optimization is focused on searching over the most promising (novel and high-performing) regions of the product layout space. Our approach aims at contributing towards the holistic design of things.

The rest of this paper is organized as follows. Section 2 describes the problem. Section 3 shows a case study using real world data for a vehicle design layout problem and Sect. 4 concludes the paper.

2 The Proposed Method

We aim at tackling the following problem:

$$\text{Maximize } N(x).f(x)$$
$$x \in \mathbf{X} \tag{1}$$

where x is the product design variable, $\mathbf{X}$ is the box-bounding restriction on the variables, $N(x)$ is the novelty function of the design variable x, and $f(x)$ is the performance of the design variable $x \in R^n$. In order to tackle the above, the following sub-sections describe both how to compute the novelty factor $N(x)$ and the performance factor $f(x)$.

2.1 Computing Novelty

Let $x_0 \in R^n$ be the design variable we are interested in measuring the novelty from a referential set S of points.

A distance metric is a simple heuristic to compute how novel x is compared to the set S: either by using Euclidean or Manhattan approaches to the mean or centroid of the set of points S. However, if one is interested in learning optimal design configurations, the distance metrics using a single referential point give inaccurate metrics of novelty. We propose a more robust approach to measure novelty: using the minimum distance to the convex hull of the given input data S. The problem can be stated as follows:

$$N(x_0) = Min \, ||\mathbf{I}x - x_0||^2$$
$$Ax \leq b \tag{2}$$
$$A_{eq}x = b_{eq}$$

where $\mathbf{I}$ is the identity matrix, $||.||$ is the norm function, and A, b, A_{eq} and b_{eq} are the linear constraints defining the convex hull $Conv(S)$ of the points in the set S so that any point $x \in Conv(S)$ satisfies the following: $Ax \leq b$ and

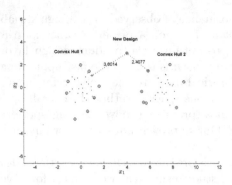

Fig. 1. Example of Measuring Novelty between a New Design point and two referential sets of points. The convex hull points are represented with circles, and any other point is represented with dots.

$A_{eq}x = b_{eq}$. The statement above is a constrained linear least squares problem and convergence to the optimal is guaranteed. Figure 1 shows an example of two sets of data and the minimum distance to two differentiated sets of data.

One would take the distance of x_0 to each point in S and take the mean (or any statistic) to compute a representative novelty metric for x_0; but this approach is: (1) *inefficient*: the time complexity is $O(|S|)$ per sampled point, which is sufficiently large for problems involving rich historical data; and (2) *inaccurate*: any point sufficiently close to each point in S brings an incorrect metric for novelty. Thus, the benefit of using the convex hull $Conv(S)$ instead of the set of points S is to improve accuracy and robustness to points close to the set S.

Figure 1 shows differentiated convex hulls. How to deal with arbitrary data points? A simple solution is to use the K-means clustering algorithm, but it requires having a-priori knowledge of the number of clusters. A more robust approach is to use the hierarchical clustering with single (minimum) linkage, in which the convex hull is computed using the leaves of the tree generated by the clustering algorithm.

Formally, we follow the two steps: First, cluster the set S, using the single-linkage hierarchical clustering, into tree T_o. Second, cut the tree T_o using the threshold TH. The result is the set of trees T, each of which defines a subset of points of S. Note that $|T|$ is unknown a-priori (it is a consequence of choosing the threshold TH). Thus, smaller values of the threshold TH increases the granularity of the observed clusters.

To show an example of the above, Fig. 2 shows an example of a tree T_o generated by the hierarchical clustering of 100 arbitrary points in R^2: In x-axis we plot the ordering of the points, in y-axis we plot the minimum distance among clusters. If one decides some threshold TH (Cutting Line) to separate the clusters, it is possible to obtain the trees T (clusters) automatically. Thus, by using the points on the leaves of the trees T, it is possible to obtain the convex hull for every cluster. Figure 2-(b) shows the Convex Hulls using the trees

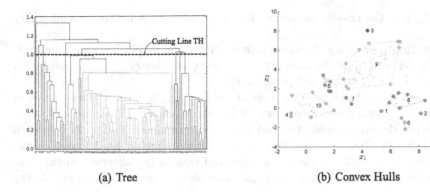

(a) Tree   (b) Convex Hulls

Fig. 2. Tree and convex hulls generated from a set of arbitrary points.

T obtained from Fig. 2-(a), in which the Convex Hulls are enumerated using the numbers from 1 to 10.

The reader may note that both computing the Convex Hull and the hierarchical clustering are time-consuming tasks. Indeed, for a set of data inputs consisting of m observations in R^n ($m = |S|$), computing the convex hulls with the *gift wrapping* algorithm takes time complexity $O(m^{n+1})$ [10]. Though, the *quick hull* algorithm provides reasonable approximations with average time complexity $O(mlog(m))$ [9]. In contrast, computing the hierarchical clustering with single linkage takes time complexity $O(m^2)$ [8].

Since the complexity is polynomial in the number of observations, and considering that the hierarchical clustering is relatively faster; an alternative method to compute the degree of novelty of a variable x with respect to a set of observed points in a set of clusters T is the following:

$$N(x) = min\ V_i(x) \tag{3}$$

$$V_i(x) = \left[1 - \frac{1}{n}\sum_{j=1}^{n} v_{i,j}(x)\right]^{1/a} \tag{4}$$

$$v_{i,j}(x) = \begin{cases} 1 & \text{if } x \in [L_{i,j}, U_{i,j}] \\ 0 & otherwise \end{cases} \tag{5}$$

$$L_{i,j} = min\ (T_{i,j})$$
$$Ui, j = max\ (T_{i,j}) \tag{6}$$
$$i \in [1, k], j \in [1, n]$$

where i and j are index suffixes for clusters and dimensions, respectively; $L_{i,j}$ is the lower bound in the j-th dimension of the i-th cluster T_i; $U_{i,j}$ is the upper bound in the j-th dimension of the i-th cluster T_i; $v_{i,j}(x)$ is an n-dimensional binary variable indicating whether x is inside or outside the boundaries of the i-th cluster T_i; $V_i(x)$ is the novelty degree of variable x with respect to the i-th cluster T_i; a is the novelty normalization constant; and n is the dimensionality of x.

Given that computing the single linkage clustering is relatively fast, computing the lower and upper bounds $[L_{i,j}, U_{i,j}]$ for each cluster $i \in [1, k]$, and each dimension $j \in [1, n]$ is fast and parallelizable.

The above equations have the role of measuring the dissimilarity degree of the variable x with respect to a set of identified clusters T_i: more dissimilar elements per dimension make the variable x more unique in comparison to each observed cluster. The reason of taking the *min* function in Eq. 3 is due to the fact of aiming to minimize the worst expected case in the observed distance to each cluster. Other functions such as the *quantile* or *mean* are also applicable to the case.

2.2   Computing Performance

Generally speaking, there exists two widely studied approaches to measure the performance of a design variable x: by real-world experimental tests or by simulation experiments. However, in a number of circumstances not only real-world experiments are expensive and dangerous, but also simulations are inaccurate to consider the real-world noises. An alternative approach is to use historical data to approximate a surrogate function that explains the variances on the performances of the past design variables.

Concretely speaking,

$$Find \; f : S \to Y \tag{7}$$

where S is the set of historical data (observed points) of the design variables, Y is the metric with historical performance associated with the set of observed points in S, and f is the function that approximates the mapping between the design variable $x \in S$ to its performance metric Y.

Basically, without knowledge on the convexity of Y, the above is the regression:

$$Minimize \; \sqrt{\frac{1}{|S|} \sum_{x \in S} [f(x) - Y]^2} \tag{8}$$

$$f \in \mathbf{F}$$

where $\mathbf{F}$ is the space of function encodings. There exists various suitable heuristics able to get reasonable approximations of the function f, depending on which space $\mathbf{F}$ we use. For simplicity and without loss of generality, we use Genetic Programming due to its feature to offer understandability of the modeled functions. Concretely speaking, to model the function f we used Genetic Programming with multi-trees as follows:

$$f(x) = w_0 + \sum_{i=1}^{nGP} w_i . t_i(x) \tag{9}$$

where w_0 is the bias term, t_i represents a tree in Genetic Programming, w_i is the weight of the i-th tree used in the linear combination, and nGP is the maximum number of trees set by the user.

Table 1. List of variables

Variable	Variable name	Units	Min	Max	Std
x_1	Torque	Nm	4	92.4	8.474
x_2	Maximum output	Ps	28	550	59.259
x_3	Engine displacement	cc	0.533	4.996	0.751
x_4	Total length	m	2.735	5.460	0.456
x_5	Full width	m	1.390	1.970	0.096
x_6	Height	m	1.040	2.285	0.184
x_7	Interior length	m	0.630	4.250	0.424
x_8	Interior width	m	1.150	1.730	0.089
x_9	Interior height	m	0.935	1.615	0.079
Y	Fuel efficiency	km/l	0	61	4.181

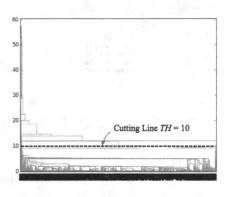

Fig. 3. Hierarchical Clustering

3   Computational Experiments

This section describes studies on finding high-performing and novel vehicle layouts. We collected 23599 observations of car models from 1982 to 2013 considering the vehicle design variables as shown by Table 1. Thus, we aim at searching for the vehicle parametric layouts x being both novel and high performing, in which the novelty function $N(x)$ is computed by Eq. 3 and the performance function $f(x)$ is computed by the ratio of fuel efficiency.

The hierarchical clustering with single linkage rendered the tree as shown in Fig. 3, and using the threshold $TH = 10$ we rendered the set T consisting of 17 subtrees. To generate the performance function $f(x)$, as shown in Eqs. 7, 8 and 9, we used Genetic Programming with $nGP = 3$ trees with the following parameters: *population size* 500, 500 *generations*, *tournament selection* of size 15, trees with *maxdepth* 3 and the following *operators*: $+, -, *, /, ()^2, tanh, sin, cos, exp, iflte$. The best function f minimizing Eq. 8 after 20 independent runs was the following:

$$f(x) = 0.03862x_1 - 5.54x3 - 16.5x_6 + 32.88tanh(x_9^2) + 32.88cos(x_5 + 2.05)$$
$$-5.501cos(x_9) + 32.88tanh(x_6)$$
$$-0.01931(cos(x_6 + sin(x_3)))iflte(x_5, x_1 - 5.912, x_2 + 7.258, -x_2 - 8.84)$$
$$-0.01931iflte(5.637x_1, x_2 + x_7, exp(x_4), -x1 - 9.488) + 20.8$$
$$(10)$$

To solve Eq. 1, we used Particle Swarm Optimization (PSO) with Niching properties [11] due to its feature to build internal clusters while searching for the optimal solution. The parameters used include *particle size* 20, *acceleration constant* 2.05 (both local and global) and *inertia weight* 0.7298. Fine tuning of

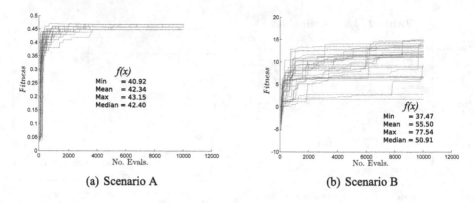

(a) Scenario A   (b) Scenario B

Fig. 4. Convergence behaviour in the studied scenarios.

Table 2. Best results in the studied scenarios

Case	x_1	x_2	x_3	x_4	x_5	x_6	x_7	x_8	x_9	Y
A	92.33	508.81	0.63	3.6	1.97	1.59	3.13	1.22	1.57	43.15
	90.4	506.61	0.62	4.31	1.96	1.6	2.46	1.48	1.6	43.04
	90.98	505.79	0.65	3.51	1.97	1.7	3.5	1.7	1.61	43.03
B	133.17	636	0.64	1.72	2.93	1.81	0.58	0.74	1.81	77.54
	138.6	599.86	0.82	7.46	2.91	2.03	4.01	0.58	2.03	75.76
	128.56	644.97	0.47	3.34	2.93	2.18	0.66	1.07	1.97	75.54

these parameters is out of the scope of this paper. For experiments, we considered the following scenarios:

- Scenario A, a tight scenario, in which the search space is restricted to the boundaries of the observed data points, that is the box-bounding constraint **X** is set to $[min(S),\ max(S)]$.
- Scenario B, a relaxed scenario, in which the search space is allowed to expand by some small factor, that is the box-bounding constraint **X** is set to $[\frac{min(S)}{2},\ \frac{3max(S)}{2}]$.

The convergence of PSO over 30 independent runs with 10000 evaluations in both scenarios is shown by Fig. 4. Table 2 shows the 3 best solutions in both scenarios. We can observe that in Scenario A (B), fast (slow) convergence to optimal fitness values (Eq. 1) is achieved because of having a narrow (wider) search space. Also, we can confirm that it is possible to obtain improved fitness values of fuel efficiency $f(x)$ in the *relaxed* scenario B as a result of (1) widening the search space, and (2) obtaining different values of x_1 and x_5 (interestingly, the learned function f has a direct relationship with the variables x_1 and x_5 as shown by Eq. 10).

4   Conclusion

Inspired by the domain-general process of designing, we have proposed a new approach to search for novel and high-performing design layouts given observations of historical data. The basic idea is to first learn a surrogate performance function that approximates the real-world invariants from past historical data, and then maximize the learned function along with the proposed novelty metric, which is defined as the dissimilarity degree with respect to historical clusters.

The application using real world data of more than twenty thousand vehicles from the last 30 years has shown the usefulness of our proposed approach. Results show that (1) under a narrow search space, it is possible to obtain novel car layouts with fairly good fuel efficiency ratios, and (2) under a wider search space, it is possible to obtain novel car layouts with increased fuel-efficiency compared to the historical upper bounds.

Our future work aims at exploring the generalizable abilities of our proposed approach to tackle nonlinear design problems, as well as investigating the trade-off between novelty and performance in the vehicle design problem. We believe that advances on learning and pattern recognition algorithms for real world product design problems will bring further insights to build unique and high-performing products.

References

1. Tay, F.E.H., Gu, J.: A methodology for evolutionary product design. Eng. Comput. **19**(2–3), 160–173 (2003)
2. Miyashita, T., Satoh, D.: An improvement method of conceptual design ideas using Data Envelopment Analysis, ICED, pp. 6.13–6.22 (2009)
3. Karniel, A., Reich, Y.: Multi-level modelling and simulation of new product development processes. J. Eng. Des. **24**(3), 185–210 (2013)
4. Ostrosi, E., Bi, S.T.: Generalised design for optimal product configuration. Int. J. Adv. Manuf. Technol. **43**(1–4), 13–25 (2010)
5. Li, Z., Cheng, Z., Feng, Y., Yang, J.: An integrated method for flexible platform modular architecture design. J. Eng. Des. **24**(1), 25–44 (2013)
6. Parque, V., Kobayashi, M., Higashi, M.: Reinforced explorit on optimizing vehicle powertrains. In: Lee, M., Hirose, A., Hou, Z.-G., Kil, R.M. (eds.) ICONIP 2013, Part II. LNCS, vol. 8227, pp. 579–586. Springer, Heidelberg (2013)
7. Grignon, P.M., Fadel, G.M.: Configuration design optimization method. In: Proceedings of ASME Design Engineering Technical Conferences DETC99, 1999, Las Vegas, Nevada, September 12–15 (1999)
8. Sibson, R.: SLINK: an optimally efficient algorithm for the single-link cluster method. Comput. J. (British Computer Society) **16**(1), 30–34 (1973)
9. Barber, C.B., Dobkin, D.P., Huhdanpaa, H.: The quickhull algorithm for convex hulls. ACM Trans. Math. Softw. **22**(4), 469–483 (1996)
10. Skiena, S.S.: "Convex Hull". 8.6.2 in The Algorithm Design Manual, pp. 351–354. Springer, New York (1997)
11. Qu, B.Y., Liang, J.J., Suganthan, P.N.: Niching particle swarm optimization with local search for multi-modal optimization. Inf. Sci. **197**, 131–143 (2012)

Learning Shape-Driven Segmentation Based on Neural Network and Sparse Reconstruction Toward Automated Cell Analysis of Cervical Smears

Afaf Tareef[1]([⊠]), Yang Song[1], Weidong Cai[1], Heng Huang[2],
Yue Wang[3], Dagan Feng[1,4], and Mei Chen[5,6]

[1] School of Information Technologies, University of Sydney, Sydney, Australia
atar8654@uni.sydney.edu.au
[2] Department of Computer Science and Engineering,
University of Texas at Arlington, Arlington, USA
[3] Department of Electrical and Computer Engineering,
Virginia Tech Research Center - Arlington, Arlington, USA
[4] Med-X Research Institute,
Shanghai Jiaotong University, Shanghai, China
[5] Department of Informatics,
University of Albany State University of New York, Albany, USA
[6] Robotics Institute, Carnegie Mellon University, Pittsburgh, USA

Abstract. The development of an automatic and accurate segmentation approach for both nuclei and cytoplasm remains an open problem due to the complexities of cell structures resulting from inconsistent staining, poor contrast, and the presence of mucus, blood, inflammatory cells, and highly overlapping cells. This paper introduces a computer vision slide analysis technique of two stages: the 3-class cellular component classification, and individual cytoplasm segmentation. Feed forward neural network along with discriminative shape and texture features is applied to classify the cervical cell images in the cellular components. Then, a learned shape prior incorporated with variational framework is applied for accurate localization and delineation of overlapping cells. The shape prior is dynamically modelled during the segmentation process as a weighted linear combination of shape templates from an over-complete shape repository. The proposed approach is evaluated and compared to the state-of-the-art methods on a dataset of synthetically generated overlapping cervical cell images, with competitive results in both nuclear and cytoplasmic segmentation accuracy.

Keywords: Cervical cell segmentation · Overlapping cells · Neural network · Sparse reconstruction · Level set evolution

1 Introduction

Cervical cancer is the second most commonly diagnosed gynaecological cancer, with around 530 thousand new cases diagnosed and more than 270 thousand

© Springer International Publishing Switzerland 2015
S. Arik et al. (Eds.): ICONIP 2015, Part I, LNCS 9489, pp. 390–400, 2015.
DOI: 10.1007/978-3-319-26532-2_43

deaths every year in the world [19]. The signs and symptoms of cervical cancer often do not begin until a pre-cancer becomes a true invasive cancer and grows into nearby tissues during the course of several years, reducing the chance of survival. Fortunately, cervical cancer can be detected by Papanicolau (Pap) smear test, where the abnormalities in the morphology of the cellular nuclei and cytoplasm can be identified. Currently, Pap test is a manual screening procedure, for which a sample of cells is gathered from the vagina and the neck of the uterus, and deposited onto a microscope slide for visual examination by a pathologist. This procedure is a time-consuming and error-prone job. Therefore, the development of computer-aided cervical cancer diagnosis system has become important to ensure the quality of the Pap test. Accurate segmentation of cervical cells is the most critical precursor step to effective diagnosis system.

The cervical cell in the cytology images consists of two subcellular components: nucleus and cytoplasm. Based on the segmented components, recent research can be classified into two main categories: (1) localizing only the nuclei boundaries in isolated or overlapping smear cells [3,10–12,22]; (2) localizing the boundaries of both nucleus and cytoplasm in either isolated or overlapping smear cells [2,4,7,13,15,16]. For nuclei localization, several methods have been proposed based on active contours [10], level set [3], adaptive active shape model [21], watershed transform [11,22], and unsupervised classification [12].

To segment single or touching cytoplasm from cervical smear images, earlier researchers used thresholding techniques (e.g., [27]), which lead to unsatisfactory results due to the complex structure of cervix cells resulted from the poor contrast and variable staining. Marker-based and multi-scale watersheds have also been used to segment the cytoplasm [7]. However, it could be difficult to find a representative marker for each cell and would result in over-segmentation. Unsupervised classification is another option that has been applied to single cell segmentation [4]. Other widely used segmentation methods include active contour models (ACM) with edge- and region-based models [5,15] due to their ability to recover closed object boundaries with pixel accuracy. These techniques extract the whole cellular clusters consisting a number of cells, which however is insufficient for shape analysis.

It is more challenging to segment nuclei and cytoplasm from overlapping cells, and this has attracted increasing research interests. Algorithms based on geodesic active contour [9] and watershed transform [2,13] have been developed to segment cells with small overlapping areas. Nevertheless, these methods may not be effective for highly overlapping clusters. Ushizima et al. [26] have proposed a segmentation approach based on nuclear narrow-band seeding, graph-based region growing and Voronoi diagrams to segment the overlapping cells by straight lines, which however do not represent the true cell boundaries. Tareef et al. [25] have designed a segmentation framework based on gradient thresholding and morphological operations, but their results are limited to overlapping cells with noticeable differences in intensities. Some recent methods have tended to incorporate shape priors with parametric segmentation procedures for more accurate cell segmentation. Lu et al. [16] and Nostrati et al. [17] incorporated

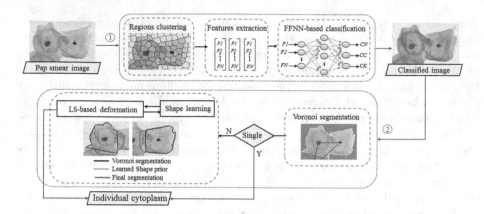

Fig. 1. The workflow of the proposed learning methodology. (1) shows the 3-class cellular components classification phase, including clustering, region-based feature extraction, and FFNN-based classification explained in Fig. 2. (2) is the individual cytoplasm segmentation phase, including Voronoi segmentation and LS-based deformation.

elliptical shape priors, and Nosrati et al. [18] incorporated a star shape prior with the level set method. These shape priors however are too simplified to approximate the real shape of the cervical cells. We have proved in this study that using a learned shape prior is more effective to segment cells with large shape variations.

In this paper, a novel segmentation approach for the nuclei and cytoplasm of overlapping cervical cells is proposed based on the neural network and learning shape-based variational method. The proposed approach performs in two phases: (1) cellular component classification by applying feed forward neural network to classify image clusters into three groups: background, nuclear, and cytoplasmic clusters, and (2) individual cytoplasm segmentation inside each cellular mass by incorporating a learned shape prior that is iteratively updated into a variational segmentation framework. The remainder of this paper is organized as follows. In Sect. 2, the proposed learning segmentation methodology is presented. Section 3 describes the material and experimental setting. The experimental results and discussion are given in Sect. 4, and the conclusion is provided in Sect. 5.

2   Methodology

The workflow of the proposed segmentation framework is illustrated in Fig. 1.

2.1   Cellular Components Classification Based on Feed Forward Neural Network

This phase aims to divide the Pap image into background, nuclei, or cellular cytoplasmic mass without separating the cytoplasm of different cells by three steps:

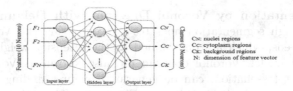

Fig. 2. Building the neural network classifier.

Clustering. In order to get easier, faster, and more accurate classification, we proposed to segment the image into small coherent regions taking into account the intensity similarities and spatial proximity. There are many clustering methods that can be used to divide the image into local regions or clusters. In our approach, we chose to utilize the simple linear iterative clustering algorithm [1] to get regular clusters (See region clustering step in Fig. 1).

Feature Extraction. For each region, a number of features with good discriminative ability are extracted. In particular, Gaussian filtering is first applied on the original Pap image to reduce the noise and smooth the inconsistent regions, and contrast-limited adaptive histogram equalization is performed to improve the local contrast in the image. Then, ten shape and texture features (i.e., compactness, circularity, minor axis length, eccentricity, average gray level, standard derivation, entropy, median value, number of edge pixels, and the average intensity difference between the region of interest and the surrounding regions) are extracted from each region and arranged into a vector.

Classification. A 3-class feed forward neural network (FFNN) classifier was proposed to classify image clusters into three cellular components: background, nuclei, and cytoplasmic mass. A back-propagation neural network was created and trained with three layers; the input layer, the hidden layer, and the output layer as shown in Fig. 2. The inputs of the network are the features extracted from each cluster in the Pap image, and the target outputs are the three cellular component classes. After classification, the contour of the nuclei and cytoplasmic mass identified by FFNN is refined by level set evolution to retrieve the missed pixels around their actual contour.

Neural networks have been recently used in many medical diagnosis systems, including cervical cancer diagnosis systems to classify the normal and abnormal cervical cells [4,24]. FFNN classifier is chosen in this research due to its reliability in classifying data with complex relationships between input and output.

2.2   Individual Cytoplasm Segmentation

In this stage, each individual cytoplasm inside the cellular mass was separated with two steps: rough segmentation by Voronoi diagram with Delauney Triangulation, followed by final segmentation based on level set with learned shape prior. Figure 3 illustrates the processes of the individual cytoplasm segmentation, where (I) shows the shape repository generating process, and (II) represents the segmentation process with its two phases: rough and final segmentation.

Rough Segmentation by Voronoi Diagram with Delauney Triangulation. The rough segmentation operates by computing the Voronoi regions of the image seeds, which are determined as the intensity weighted centroids of the detected nuclei. Voronoi diagram [6], also known as Voronoi tessellation or Dirichlet tessellation, can be generated by first finding Delauney triangles between the seeds (i.e., red lines in Fig. 3 (II)). Then, the convex regions, whose boundary points have the shortest Euclidean distances to the corresponding nucleus centroid, are computed (i.e., black dash lines in Fig. 3 (II)). Let C denote a set of n nuclei centroids $\{C_1, C_2, ..., C_n\}$. The Delauney triangulation can be generated by connecting C together, and the Voronoi region V for each $C_i, i = 1, ..., n$ can be written as:

$$V(C_i) = \cap_{1 \leq j \leq n, j \neq i} \{p | \mathcal{D}(p, C_i) < \mathcal{D}(p, C_j)\} \tag{1}$$

where $\mathcal{D}(p, C_i)$ is the Euclidean distance between the pixel p and corresponding centroid C_i. The segmentation process for images with a single cell is completed by the end of this step, whereas, the overlapping cells are passed to the final segmentation step by dilating the corresponding Voronoi cells to increase the search areas.

Final Segmentation Using Learned Shape Prior with Regularized Level Set Evolution. To get the final segmentation of the overlapping cells, two shape-driven deformation steps are designed: shape re-initialization, and shape prior-based level set evolution. The first step re-initializes the cell shape based on the shape repository, whereas the second step combines local characteristics of the cell boundary and a prior knowledge about the expected cell shape.

Specifically, an over-complete shape repository is generated based on a training set of annotated cell images. The annotated cells are first Procrustes transformed [8] to the coordinate system of the mean cell shape to remove the geometrical translation, scale, and rotation effects. Each shape is then represented by the coordinates of its boundary points. The shape repository is thus represented by a matrix $\Phi \in \mathbb{R}^{K \times M}$, where each column refers to a single shape represented by $K/2$ boundary points with each point denoted by its xy coordinates, and M is the total number of cell shapes in the repository Φ.

The segmentation process starts by shape re-initialization. This is conducted via the reconstruction of the obtained Voronoi cell shape as a weighted combination of the shape repository Φ, and can be formulated as a sparse reconstruction problem:

$$\alpha_i = \arg \min_{\alpha_i} \lambda \|\alpha_i\|_1 + \|v_i - \Phi\alpha_i\|_2^2 \tag{2}$$

where v_i of dimension K is the boundary coordinates of the ith Voronoi cell, $\lambda > 0$ is the regularization parameter automatically selected, and α_i is a weighting vector with few significant entries corresponding to the most representative shape templates and their weights in approximating v_i computed by [28]. The reconstructed boundary coordinates are then obtained by $\widehat{v}_i = \Phi\alpha_i$.

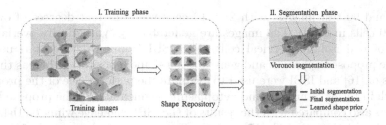

Fig. 3. Overlapping cytoplasm segmentation processes: (I) the shape repository generation process, (II) the rough and final segmentation process.

Then, the level set function (LSF) $\phi : \Omega \to \mathbb{R}$ on a domain Ω is separately built for each initial cell shape. The energy functional $\mathcal{E}$ to be minimized consists of several terms including the shape prior, data-driven, regularization, and area terms. $\mathcal{E}(\phi)$ can be written as:

$$\mathcal{E}(\phi) = \sigma \mathcal{E}_{\mathcal{S}}(\phi) + \beta \mathcal{E}_{\mathcal{D}}(\phi) + \mu \mathcal{E}_{\mathcal{R}}(\phi) + \gamma \mathcal{E}_{\mathcal{A}}(\phi) \tag{3}$$

where $\mathcal{E}_{\mathcal{S}}$ is the shape prior term that constrains possible cell shapes, and defined as $\mathcal{E}_{\mathcal{S}}(\phi) = \int_{\Omega} g(x) H(-S(\phi)) dx$ [23], where $g(x)$ is the stopping function defined as $g(x) = 1/1 + (|\nabla G_{\sigma} I|^2)$, G_{σ} is the Gaussian kernel with standard deviation σ, and I is the image on a domain Ω. $H(.)$ is the Heaviside function, and $S(\phi)$ is the LSF of the approximated shape prior for ϕ. The shape prior is dynamically constructed using the shape repository Φ and the weighting vector α obtained by Eq. (2) under sparsity constraint, where v_i refers to the deformed cell shape in the previous iteration, and then Procrustes transformed to the cell coordinate system. $\mathcal{E}_{\mathcal{D}}(\phi) = \int_{\Omega} g(x) \delta(\phi(x)) |\nabla \phi(x)| dx$ is the data-driven term driving the segmenting curve to the object boundaries by having a lower energy when the zero level contour of ϕ is located at the cell boundaries. $\delta(.)$ is the Dirac delta function. $\mathcal{E}_{\mathcal{R}}(\phi) = \int_{\Omega} p(|\nabla \phi(x)|) dx$ is the regularization term defined by [14], where $p : [0, \infty) \to \mathbb{R}$ is a potential or energy density function, which guarantees the smoothness of the segmentation boundaries by maintaining the signed distance property $|\nabla \phi| = 1$. $\mathcal{E}_{\mathcal{A}}(\phi) = \int_{\Omega} g(x) H(-\phi(x)) dx$ is the area term computing the segmentation area of $\phi < 0$. $\sigma, \beta, \mu,$ and γ are constant weights determining the contribution of each energy term. Finally, the energy functional $\mathcal{E}$ is minimized by solving the gradient descent flow for each LSF as $\frac{\partial \phi_i}{\partial t} = -\frac{\partial \mathcal{E}(\phi_i)}{\partial \phi_i}$, where $\partial \mathcal{E}/\partial \phi_i$ is the Gâteaux derivative of the functional $\mathcal{E}$ with respect to ϕ_i.

3   Material and Experimental Setting

The performance of the proposed methodology was evaluated using two datasets: the dataset provided by [16] consisting 18 gray-scale cervical cytology images with 2 to 5 cells with different degrees of overlap (i.e., 60 cells in total), and the dataset of the ISBI 2014 challenge [20] consisting of 135 images (45 training images and 90 test images) with 2 to 10 cells with different degrees of overlap (i.e., 810 cells in total). The images are generated by synthetically overlapping images of real isolated cervical cells. The ISBI training dataset was used to train the proposed approach and generate the shape repository, whereas the test datasets of [16] and ISBI were used to evaluate the performance of the proposed approach in terms of nuclear and cytoplasmic segmentation. The proposed app-roach was also compared with the results of the ISBI challenge winners: Ushizima et al. [26], and Nosrati et al. [17] and their newly proposed approach [18] on the same ISBI test dataset. Furthermore, the proposed approach was compared with the baseline approach proposed by the ISBI challenge organizers [16] on their provided dataset.

There were two sets of evaluation measures used to assess the segmentation results: object-level and pixel-level measures. To evaluate the performance of the nuclear segmentation, we used the criteria developed by Gentav et al. [7] which used the ground truth objects (O_{gt}) to categorize all segmented objects (O_d) into true detection (TP_o) , false negative detection (FN_o), or false alarm (F_a) with respect to a threshold $\Lambda = 0.6$ on the proportion of overlap between O_d and O_{gt}. For each true detection instance, the numbers of true-positive pixels (TP_p), false-positive(FP_p), and false-negative (FN_p) pixels were counted. Based on this information, the object-level precision and recall (i.e., P_o and R_o, respectively), and pixel-level precision and recall (i.e., P_p and R_p, respectively) were computed as described in [7].

The Zijdenbos similarity index (ZSI), also known as the Dice similarity coefficient, is also employed in our evaluation. The values of FN_o, TP_p, FP_p, and ZSI, computed by the evaluation code provided by ISBI challenge, are used to evaluate the performance of the cytoplasm segmentation.

4   Experimental Results

Nuclear Segmentation Evaluation. The nuclear segmentation of our pro-posed approach is assessed and compared with the approaches proposed in [7,16]. As shown in Table 1, the performance of our nuclear segmentation is the best in terms of both object-based and pixel-based segmentation. We yielded a very high improvement in object-level precision value of 0.97 (i.e., 31 %, and 41 % improvement), compared with 0.74, and 0.69 obtained by [7,16], respectively. We also achieved object-level recall of 0.96 with on average improvement of 8 %. Our pixel-level results were also better than those obtained by other approaches. There were only two missed nuclei (i.e., out of 60 nuclei) by our approach. These results proved the effectiveness of our proposed method in differentiating the nuclear regions from the other regions. Dividing the image into small clusters, the selected shape and texture features, and using neural network succeeded in achieving the highest nuclear segmentation results over the recent approaches, with promising improvement.

Cytoplasmic Segmentation Evaluation. Table 2 shows a comparison of the performance of our approach with [17,18,26] for cytoplasm segmentation. Our

Table 1. Quantitative results of the nuclear segmentation. Bold numbers indicate superior results.

	Po	Ro	Pp	Rp	ZSI
[7]	0.74	0.93	0.91($\pm$0.08)	0.88($\pm$00.07)	0.89($\pm$0.04)
[16]	0.69	0.90	0.97($\pm$0.04)	0.88($\pm$0.08)	0.92($\pm$0.04)
Our approach	**0.97**	**0.96**	**0.98($\pm$0.03)**	**0.90($\pm$0.08)**	**0.93($\pm$0.04)**

Table 2. Quantitative results of the cytoplasm segmentation. Bold numbers indicate superior results.

ISBI test dataset			
FNo	**TPp**	**FPp**	**ZSI**
[26]	0.17	0.83	**0.001**	0.87
[17]	0.14	0.90	0.005	0.87
[18]	**0.11**	0.93	0.005	0.88
Our approach	0.16	**0.94**	0.005	**0.89**
Test dataset of [16]			
[16]	0.21	0.92	**0.002**	0.88
Our approach	**0.00**	**0.93**	0.005	**0.91**

proposed approach achieved the highest ZSI value of 0.89 and TP_p of 0.94 among the compared approaches. Our approach also outperformed [26] in FN_o. These high ZSI and TP_p values demonstrated the capability of our approach to accurately segment the cytoplasm from highly overlapping cells. We also compared our approach with the elliptical shape prior-based approach proposed by Lu et al. [16] using their test dataset. The results showed a substantial improvement in performance consisting of zero false negative, TP_p of 0.93, and ZSI of 0.91; compared with FN_o of 0.21, TP_P of 0.92, and ZSI of 0.88 obtained by [16]. Our optimal object-level true positive detection (i.e., 1.00) led to an increase of the FP_p value (i.e., 0.005) over that obtained by [16] (i.e., 0.002). However, this FP_p is still small and has minimal impact on the reliability of our approach. Overall, we suggest that the proposed learned shape prior succeeded in improving the object-level and pixel-level segmentation performance over the elliptical shape prior [16,17] and the star shape prior [18], as it was dynamically generated based on the most representative shape templates from similar training cells. Qualitative segmentation results for cells with different shapes are shown in Fig. 4.

Finally, the average computational time of our proposed approach was ~ 40 seconds per image using non-optimized MATLAB code on a PC with Intel Core i5 3.2 GHz and 8 GB RAM, which was 25 times faster than [16] whose average computational time was ~ 1000 seconds per image.

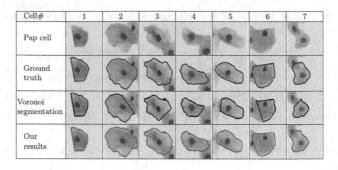

Fig. 4. Qualitative segmentation results of single (column 1) and overlapping (columns 2 to 7) cervical cells.

5   Conclusions

Cell segmentation is the critical step toward the development of automated analysis of Pap smears. This study addresses this issue by designing a new slide analysis technique based on a neural network with cluster-based shape and texture features, and a variational framework incorporating a learned shape prior dynamically generated based on Voronoi cells and shape templates from an overcomplete shape repository. The proposed learning method has been tested and compared to the state-of-the-art methods on two cervical cell databases for a total of 870 cells. The overall segmentation accuracy and efficiency of the proposed approach have been shown to be better than the compared techniques in both nuclear and cytoplasmic segmentation performance.

References

1. Achanta, R., Shaji, A., Smith, K., Lucchi, A., Fua, P., Susstrunk, S.: SLIC superpixels compared to state-of-the-art superpixel methods. IEEE Trans. Pattern Anal. Mach. Intell. **34**(11), 2274–2282 (2012)
2. Béliz-Osorio, N., Crespo, J., García-Rojo, M., Muñoz, A., Azpiazu, J.: Cytology imaging segmentation using the locally constrained watershed transform. In: Soille, P., Pesaresi, M., Ouzounis, G.K. (eds.) ISMM 2011. LNCS, vol. 6671, pp. 429–438. Springer, Heidelberg (2011)
3. Bergmeir, C., Silvente, G.M., Benítez, J.M.: Segmentation of cervical cell nuclei in high-resolution microscopic images: a new algorithm and a web-based software framework. Comput. Methods Programs Biomed. **107**(3), 497–512 (2012)
4. Chankong, T., Theera-Umpon, N., Auephanwiriyakul, S.: Automatic cervical cell segmentation and classification in pap smears. Comput. Methods Programs Biomed. **113**(2), 539–556 (2014)
5. Fan, J., Wang, R., Li, S., Zhang, C.: Automated cervical cell image segmentation using level set based active contour model. In: 12th International Conference on Control Automation Robotics & Vision (ICARCV), 2012, pp. 877–882. IEEE (2012)

6. Fu, T., Yin, X., Zhang, Y.: Voronoi algorithm model and the realization of its program. Comput. Simulation **23**, 89–91 (2006)
7. Genctav, A., Aksoy, S., Onder, S.: Unsupervised segmentation and classification of cervical cell images. Pattern Recogn. **45**(12), 4151–4168 (2012)
8. Goodall, C.: Procrustes methods in the statistical analysis of shape. J. Roy. Stat. Soc. B (Methodological) **53**, 285–339 (1991)
9. Harandi, N.M., Sadri, S., Moghaddam, N.A., Amirfattahi, R.: An automated method for segmentation of epithelial cervical cells in images of ThinPrep. J. Med. Syst. **34**(6), 1043–1058 (2010)
10. Hu, M., Ping, X., Ding, Y.: Automated cell nucleus segmentation using improved snake. In: International Conference on Image Processing 2004, ICIP 2004. vol. 4, pp. 2737–2740. IEEE (2004)
11. Jung, C., Kim, C.: Segmenting clustered nuclei using H-minima transform-based marker extraction and contour parameterization. IEEE Trans. Biomed. Eng. **57**(10), 2600–2604 (2010)
12. Jung, C., Kim, C., Chae, S.W., Oh, S.: Unsupervised segmentation of overlapped nuclei using bayesian classification. IEEE Trans. Biomed. Eng. **57**(12), 2825–2832 (2010)
13. Kale, A., Aksoy, S.: Segmentation of cervical cell images. In: 20th International Conference on Pattern Recognition (ICPR), 2010, pp. 2399–2402. IEEE (2010)
14. Li, C., Xu, C., Gui, C., Fox, M.D.: Distance regularized level set evolution and its application to image segmentation. IEEE Trans. Image Process. **19**(12), 3243–3254 (2010)
15. Li, K., Lu, Z., Liu, W., Yin, J.: Cytoplasm and nucleus segmentation in cervical smear images using radiating GVF snake. Pattern Recogn. **45**(4), 1255–1264 (2012)
16. Lu, Z., Carneiro, G., Bradley, A.P.: Automated nucleus and cytoplasm segmentation of overlapping cervical cells. In: Mori, K., Sakuma, I., Sato, Y., Barillot, C., Navab, N. (eds.) MICCAI 2013, Part I. LNCS, vol. 8149, pp. 452–460. Springer, Heidelberg (2013)
17. Nosrati, M., Hamarneh, G.: A variational approach for overlapping cell segmentation. In: ISBI Overlapping Cervical Cytology Image Segmentation Challenge, pp. 1–2. IEEE (2014)
18. Nosrati, M., Hamarneh, G.: Segmentation of overlapping cervical cells: a variational method with star-shape prior. In: IEEE International Symposium on Biomedical Imaging (ISBI), IEEE (2015)
19. World Health Organization.: Who Guidance Note: Comprehensive Cervical Cancer Prevention and Control: A Healthier Future for Girls and Women, WHO Press, Geneva (2013)
20. Overlapping Cervical Cytology Image Segmentation Challenge ISBI 2014: http:// cs.adelaide.edu.au/carneiro/isbi14_challenge/
21. Plissiti, M.E., Nikou, C.: Overlapping cell nuclei segmentation using a spatially adaptive active physical model. IEEE Trans. Image Process. **21**(11), 4568–4580 (2012)
22. Plissiti, M.E., Nikou, C., Charchanti, A.: Combining shape, texture and intensity features for cell nuclei extraction in Pap smear images. Pattern Recogn. Lett. **32**(6), 838–853 (2011)
23. Rousson, M., Paragios, N.: Shape priors for level set representations. In: Heyden, A., Sparr, G., Nielsen, M., Johansen, P. (eds.) ECCV 2002, Part II. LNCS, vol. 2351, pp. 78–92. Springer, Heidelberg (2002)

24. Sokouti, B., Haghipour, S., Tabrizi, A.D.: A framework for diagnosing cervical cancer disease based on feedforward MLP neural network and ThinPrep histopathological cell image features. Neural Comput. Appl. **24**(1), 221–232 (2014)
25. Tareef, A., Song, Y., Cai, W., Feng, D., Chen, M.: Automated three-stage nucleus and cytoplasm segmentation of overlapping cells. In: 13th International Conference on Control Automation Robotics & Vision (ICARCV), 2014, pp. 865–870. IEEE (2014)
26. Ushizima, D., Bianch, A., Carneiro, C.: Segmentation of subcellular compartiments combining superpixel representation with voronoi diagrams. In: ISBI Overlapping Cervical Cytology Image Segmentation Challenge, pp. 1–2. IEEE (2014)
27. Wu, H.S., Gil, J., Barba, J.: Optimal segmentation of cell images. In: IEE Proceedings: Vision, Image and Signal Processing, vol. 145, pp. 50–56. IET (1998)
28. Zhang, Z., Rao, B.D.: Sparse signal recovery with temporally correlated source vectors using sparse bayesian learning. IEEE J. Sel. Top. Signal Process. **5**(5), 912–926 (2011)

Adaptive Differential Evolution Based Feature Selection and Parameter Optimization for Advised SVM Classifier

Ammara Masood[✉] and Adel Al-Jumaily

School of Electrical, Mechanical and Mechatronic Engineering, University of Technology,
Sydney, Australia
ammara.masood@student.uts.edu.au, Adel.Ali-Jumaily@uts.edu.au

Abstract. This paper proposes a pattern recognition model for classification. Adaptive differential evolution based feature selection is used for dimensionality reduction and a new advised version of support vector machine is used for evaluation of selected features and for the classification. The tuning of the control parameters for differential evolution algorithm, parameter value optimization for support vector machine and selection of most relevant features form the datasets all are done together. This helps in dealing with their interdependent effect on the overall performance of the learning model. The proposed model is tested on some latest machine learning medical datasets and compared with some well-developed methods in literature. The proposed model provided quite convincing results on all the test datasets.

Keywords: Feature selection · Optimization · Classification · Support vector machine · Differential evolution · Dimensionality reduction

1 Introduction

Machine learning methods play a great role these days in analyzing and extracting useful information from datasets in various fields like medical diagnosis, image recognition, and many other applications. However, affectivity of these models heavily depends on efficiency of underlying feature selection and classification algorithms.

Feature selection is important for dimensionality reduction and improving accuracy of the predictive model. This is achieved by identifying features that offer complementary information to differentiate the target classes. Finding best feature subset is usually difficult and has led to the development of variety of techniques for selecting optimal subset of features from larger sets of possible features [1, 2]. Feature selection methods can be categorized based on the search strategy and evaluation measure used.

Searching strategy is important aspect of feature selection methods. It can be sorted in three major types, i.e. the exhaustive search, sequential search [3] and stochastic/evolutionary [4] search. The exhaustive search, can guarantee the optimal solution, but it is impractical to run, even with moderate size feature sets. Sequential search methods can be simpler but are prone to nesting effect. Stochastic search methods especially the evolutionary algorithms like Genetic methods (GA), Ant Colony Optimization (ACO), and Particle Swarm Optimization (PSO) and differential evolution (DE) have got a lot

© Springer International Publishing Switzerland 2015
S. Arik et al. (Eds.): ICONIP 2015, Part I, LNCS 9489, pp. 401–410, 2015.
DOI: 10.1007/978-3-319-26532-2_44

of attention as a search strategy for feature selection in the past decade [5–7]. They can include some randomness in the search process and makes it less sensitive to dataset and avoid local minima but they should be able to explore and exploit the search space properly to get optimum solutions. Among these methods, DE outperformed other optimization algorithms in terms of robustness over common benchmark problems and real world applications. It has less tuning parameters and show better potential of increasing its explorative and exploitation capabilities [8, 9]. Thus, in this paper differential evolution based search strategy is used.

Typical evaluation methods can be categorized as filters or wrappers. Filters approaches evaluate quality of selected features independently, without using classification algorithm. While, wrapper approach use a classifier trained for given feature set to evaluate the quality. Filter based methods are faster in general than wrapper based method, however, wrapper based methods are found to be more accurate. This work will be using the wrapper based approach for evaluation of selected feature subsets. The learning model proposed here is based on an improved version of support vector machine [10] which has the ability to generate advised weights to deal with misclassified data and outliers present in data to lessen their effect on classifier performance.

It must be noted here that the control parameter of search strategy, tuning parameter for the classifier and the selected feature subsets all contribute at the same time towards the overall performance of the model. For example, control parameters and learning strategies involved in DE are highly dependent on the problem under consideration and must be set adaptively. Similarly, optimal feature subset selection and setting best kernel parameter for SVM are crucial for high predictive accuracy and the choices are interdependent and must occur simultaneously. In addition to this, the need to deal with the outlier in the training phase is also important, in order to prevent their effect in the development of predictive model. Although, there are a number of ways proposed in literature for setting the control parameters for differential evolution [11], choosing appropriate parameter for SVM [5] and selecting optimal feature subsets independently. However, our research objective is to adaptively adjust the control parameter for the problem under consideration and optimize classifier parameters and select the feature subset simultaneously, without degrading classification accuracy.

The paper is organized as follows; Sect. 2 provides the overview of the proposed methodology, while Sect. 3 provides the details of the adaptive differential evolution algorithm and the advised support vector machine algorithm. Section 4 presents the experimental results and finally conclusion is given in Sect. 5.

2 Proposed Model

The proposed model is presented in Fig. 1. Firstly, data set under consideration is linearly scaled to the range [−1, +1] or [0, 1] to avoid domination of features in greater numeric ranges on the ones with smaller numeric ranges. Optimization of feature subsets and control parameters is done based on adaptive DE algorithm explained in next section. After evolutionary operations are done trail vector provides the selected feature subset and optimized SVM C and Kernel parameters. Using selected feature sets, the data divided into training and testing sets are fed into classification stage where advised SVM classifier tuned on the basis of optimized parameters is used. The advised SVM algorithm is also explained in next section. Each trail vector is evaluated by fitness function which is the classification accuracy of the classifier. If termination criterion (maximum number of generations) is satisfied, the process ends, otherwise it proceeds to the next generation and the evolution process is repeated.

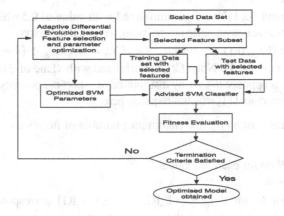

Fig. 1. Proposed learning model

3 Details of Proposed Algorithms

3.1 Adaptive Differential Evolution

Differential evolution (DE) is a population based optimization method, which has attracted an increased attention in the past few years [12, 13]. It is capable of handling nonlinear objective functions with parallel and direct search approach and has good convergence. Many extended versions of Differential Evolution are presented to use it as a feature selector in pattern recognition process [7, 8]. Although it showed quite promising results in various applications but some recent publications indicated that DE may face challenges in complex applications and search performance get highly depended on the mutation strategy, crossover operation and control factors including scale factor (F), Cross over rate (Cr) and population size (NP) [11, 14]. The paper proposes a feature selection method that an extension of DE- based feature selection technique proposed in [8] with adaptive approach to make the feature selection process more dynamic to be applied for different pattern recognition applications. It will use advised support vector machine explained in following section for evaluation of selected feature subset. The steps of the feature selection procedure are as follows.

1. Initially the generation number is set as $G = 0$ and a population of NP individuals is randomly initialized say $PopG = \{\vec{X}\_1_G, ..., \vec{X}\_NP_G\}$ where $\vec{X}_i_G = $ [x1_iG, x2_iG, x3_iG, ..., xD_iG], with $i = [1, 2, ..., NP]$ and D is the number of parameter

to be optimized including features and setting parameter for the classifier. Each individual parameter is uniformly distributed in the range [$\vec{X}$ min, $\vec{X}$ max], where $\vec{X}$ min = {x1 min, x2 min, ..., xDmin} and $\vec{X}$ max = {x1max, x2max, ..., xDmax}

2. At every generation, the mutation and cross over control parameter are generated independently for each target vector using following equations

$F_i = $ Cauchy $(F_m, 0.1)$ with $F_m = w_F.F_m + (1 - w_F)$ and $w_F = 0.8 + 0.2 \times$ rand $(0, 1)$ Note F_m is initialized with value of 0.5 while Cauchy distribution prevent premature convergence due to its far wider tail property.

$Cr_i = $ Gaussian $(Cr_m, 0.1)$ with $Cr_m = w_{Cr}.Cr_m + (1 - w_{Cr})$ and $w_{Cr} = 0.9 + 0.1 \times$ rand $(0, 1)$ Note Cr_m is initialized with vlaue of 0.6 and Gaussian distribution is used as opposite to Cauchy distribution, its short tail property help in keeping the value of Cr within unity [11] which is required here.

3. WHILE the termination criterion (maximum number of iterations) is not satisfied

DO
for i = 1 to NP//do for each individual
Step 1 – Mutation
Create a mutant vector $\vec{V}\_i_G = \{v1\_iG, ..., vD\_iG\}$ corresponding to the ith target vector $\vec{X}\_i_G$ by merging three different randomly selected vectors i.e. using the DE/rand/1 Mutation strategy.

$$\vec{V}\_i_G = \overrightarrow{X_{r1}}\_i_G + \text{Fi} \cdot \left(\overrightarrow{X_{r2}}\_i_G - \overrightarrow{X_{r3}}\_i_G\right) \tag{1}$$

Step 2 – Cross over
Employ binomial crossover on each of the D variable as follows for building trial vector.

$$uj\_i_G = \begin{cases} vj\_i_G & if(randi, j\,[0, 1] \leq Cr_i or j = j_{rand} \\ xj\_i_G & otherwise \end{cases} \tag{2}$$

Here jrand ∈ [1,2,..D] is a randomly selected index to ensure that $\vec{U}\_i_G$ gets at least some component from $\vec{V}\_i_G$.

Step 3 – Selection
Evaluate the trial vector $\vec{U}\_i_G$ with the fitness function f = accuracy of classifier if f $(\vec{U}\_i_G) \geq$ f $(\vec{X}\_i_G)$, then $\vec{X}\_i_{G+1} = \vec{U}\_i_G$ else $\vec{X}\_i_{G+1} = \vec{X}\_i_G$.
end if
end for

It must be noted here that as DE is a real number optimizer, two dimensions can settle at the same feature coordinates after rounding off. So before going to the next generation, the roulette wheel weighing scheme [15] is utilized in order to overcome the problem of duplicate features.

For this a cost weighting is implemented where the probabilities of individual features are calculated from the distribution factor that is associated with each feature. The distribution factor of feature f_i within the current generation G is calculated as follows:

$$FD_{j,g} = a_1 \times \left(\frac{PD_j}{PD_j + ND_j} \right) + \frac{NF - DNF}{NF} \times \left(1 - \frac{(PD_j + ND_j)}{\max (PD_j + ND_j)} \right) \tag{3}$$

where NF is the total number of features and DNF number of desired features. PD_j and ND_j is the number of times feature f_i has been used in the good subsets and less competitive subsets respectively. Whereas, a1 is the constant that reflects the importance of features in PD. Here $\frac{PD_j}{PD_j + ND_j}$ factor shows the degree to which feature f_i contributes in forming good subsets and the second term help in favouring exploration as this term will get close to 1 when the overall use of a particular feature is too low. Thus, based on this ranking the duplicated features of the trial vector are replaced by the next available top ranked features.

Secondly, for supressing the domination of certain features on the distribution factor, the relative difference in distribution factor is calculated using (4) [16].

$$T = \left(FD_{g+1} - FD_g \right) \times FD_{g+1} + FD_g) \tag{4}$$

This gives higher weigh to features that are making noticeable improvement in the current generation as compared to previous one. It also helps in intentionally keeping features that are found to be highly relevant for a particular application, even if they do not show noticeable improvement.

3.2 Advised Support Vector Machine Based Classification

In this paper, a non-iterative self-advising approach for SVM is adapted that extracts subsequent knowledge from the misclassified data in training phase that can be a result of outliers or the data that have not been separated correctly. This is done by generating advice weights based on the distance of misclassified training data from the correctly classified training data, and use of these weights together with decision values of SVM in the test phase. These weights also help the algorithm to eliminate the outlier data. The details of Advised SVM algorithm (Fig. 2) is as follows:

1. The classifying hyperplane is found by using decision function $f(x) = sign(\sum_{\alpha_{l>0}} y_l \alpha_l k(x, x_l) + b)$, here xl is the input vector corresponding to the lth sample and labelled by yl depending on its class and αl is the nonnegative Lagrange multiplier that is inconsistence with standard SVM training.
   Note that in order to use SVM to produce non-linear decision functions as the data is comprised of nonlinearly separable cases, radial basis function kernel $K\left(x_l x_m\right) = e^{-\gamma |x_l - x_m|^2}$ is used to make all necessary operations in the input space.

2. The data samples that are misclassified in the initial training phase are identified. The misclassified data sets (MD) in the training phase is determined as

$$MD = \bigcup_{l=1}^{N} x_l | y_l \neq sign \left(\sum_{\alpha_m > 0} y_l \alpha_m k \left(x_l, x_m \right) + b \right) \quad (5)$$

The MD set can be null, but experimental results revealed that the occurrence of misclassified data in training phase is a common occurrence.

3. If the MD is null, go to the testing phase, else compute neighborhood length (NL) for each member of MD. NL is given as

$$NL \left(x_l \right) = minimum_{x_m} \left(\| x_l - x_m \| \, | y_l \neq y_m \quad (6) \right)$$

Where xm, m = 1, ..., N are the training data that do not belong to the MD set.

4. For each sample xn from the test set advised weight AW(xn) is computed. Where AW is computed as (7), These AWs represent how close the test data is to the misclassified data

$$\begin{cases} 0 & \forall x_l \in MD, x_n - x_l > NL \left(x_l \right) \, or MD = NUL, \\ \sum 1 - \frac{\sum_{x_l} x_n - x_l}{\sum_{x_i} NL(x_i)} & x_l \in MD, x_m - x_l \leq NL \left(x_l \right) \end{cases} \quad (7)$$

5. The absolute value of the SVM decision values for each xn from the test set are calculated and scaled to [0, 1].

6. For each xk from the test set, If (AW (xk) < decision value (xk) then $y_k = sign \left(\sum_{\alpha_m > 0} y_m \alpha_m k \left(x_k, x_m \right) + b \right)$ which is in consistence with normal SVM labelling, otherwise $y_k = y_l | \left(x_k - x_l \leq NL \left(x_l \right) \, and x_l \in MD \right)$.

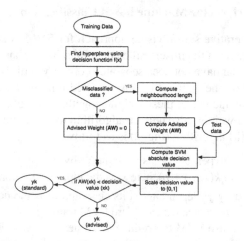

Fig. 2. Advised support vector machine

4   Experimental Results

In order to analyze the effectiveness of the proposed model, 4 medical datasets with varying dimensionalities are utilized and the classification accuracies are calculated. The first three datasets used are available online from https://archive.ics.uci.edu/ml/machine-learning-databases, while the fourth one is own data set based on histopathological images of skin cancer collected from Sydney Melanoma Diagnostic Centre, Royal Prince Alfred Hospital. The details of extracted features for this data set can be found in our previous publication [17]. The classification accuracy of the proposed model is also compared with KNN, linear SVM, and LDA used for these data sets. The datasets information and the average classification accuracies (based on 2 fold cross validation) for respective datasets are presented in Table 1. For testing the effect of feature selection method and the number of selected features on the overall performance of the model, the performance of the proposed model is also compared with the ones based on well-established binary Genetic Algorithm BGA [17], Binary PSO (BPSO) [18], improved BPSO [19] and hybrid GA [20], see Fig. 3.

Table 1. Classifier performance evaluation based on accuracy

Data sets	No of attributes	No of Classes	No of instances	Classification Accuracy (%)			
				KNN	LDA	SVM	ADSVM
Lung Cancer	56	3	32	77	82.3	81	84.5
Breast cancer	30	2	569	87.2	89.8	90	93
Dermatology	33	6	366	86	87.9	88	90
Skin Cancer	50	2	42	88.6	89.1	89	92

For GA probability of mutation = 0.02 and probability of crossover was chosen as 0.5 after running several tests. This is used to make sure to have the number of '1's in the strings matching a predefined number of desired features. For BPSO the inertia weight was made to decrease linearly from 0.9 to 0.4 while the maximum velocity was set to be clipped within 20 % of the corresponding variable; and acceleration constants were set to 2.0. Both of BGA and BPSO utilize binary strings representing a feature subset with ones and zeros to indicate the selection and neglecting of features respectively. Improved binary particles warm (IBPSO) was implemented according to the algorithm described in [19]. Hybrid genetic search algorithm (HGA) was implemented as proposed in [20] to search for subsets of fixed sizes. It should be noted HGA is computationally very expensive for larger datasets, as the number of subsets to be formed and evaluated increases with the number of features in the dataset.

All methods were made to start from the same initial population with the population size set to 50 and terminated at the same number of iterations set to 100. The chosen fitness function was set to the classification accuracy. It can be seen that the proposed method attained comparable or better classification accuracies as compared to other methods for comparatively lesser number of selected features. This shows that if

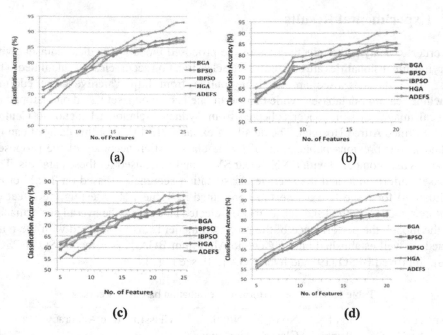

Fig. 3. Average classification accuracies vs. feature subset sizes for Dataset (a) Skin cancer (b) Dermatology (c) Lung Cancer (d) Breast Cancer

parameter tuning and feature selection is done simultaneously and effect of misclassified data/outliers is minimized, it can improve the classification accuracies of the learning models. In addition to that, it can also help in minimizing the use of redundant/irrelevant features in the final optimized model, which will reduce the computational complexity and chances of having over fitted models.

5   Conclusion and Future Work

This paper presents a novel learning model based on adaptive differential evolution based feature selection and advised support vector machine based classification, where the parameter tuning for feature selector and classifier and feature selection process for the dataset are done simultaneously. Experimental analysis shows that the proposed model works well and provides an optimal feature set with higher classification rate when compared with some other popular methods used in literature. The model is more adaptive and can work for various different types of datasets and help in choosing more relevant features that can help in classification and reducing the number of feature used in the final learning model. It also limits the effect of outliers and misclassified data values of the training set over the final optimized model. In future, this method can be made more effective by choosing more advanced mutation and selection strategies developed for differential evolution based feature selection.

References

1. Chandrashekar, G.S., Ferat, S.: A survey on feature selection methods. Comput. Electr. Eng. **40**(1), 16–28 (2014)
2. Khalid, S., Khalil, T., Nasreen, S.: A survey of feature selection and feature extraction techniques in machine learning. In: Science and Information Conference (SAI) (2014)
3. Nakariyakul, S., Casasent, D.P.: Improved forward floating selection algorithm for feature subset selection. In: International Conference on Wavelet Analysis and Pattern Recognition (2008)
4. Spall, J.C.: Introduction to Stochastic Search and Optimization: Estimation, Simulation, and Control. Wiley, New York (2003)
5. Huang, C.-L.W., Chieh-Jen, A.: GA-based feature selection and parameters optimization for support vector machines. Expert Syst. Appl. **31**(2), 231–240 (2006)
6. Khushaba, R.N., Ahmed A., Al-Jumaily, A.: Swarm intelligence based dimensionality reduction for myoelectric control. In: 3rd International Conference on in Intelligent Sensors, Sensor Networks and Information (2007)
7. Al-Ani, A.A., Akram Khushaba, R.N.: Feature subset selection using differential evolution and a wheel based search strategy. Swarm and Evol. Comput. **9**, 15–26 (2013)
8. Khushaba, R.N., Ahmed, A., Al-Jumaily, A.: Feature subset selection using differential evolution and a statistical repair mechanism. Expert Syst. Appl. **38**(9), 11515–11526 (2011)
9. Bhadra, T., Bandyopadhyay, S., Maulik, U.: Differential evolution based optimization of SVM parameters for meta classifier design. Procedia Technol. **4**, 50–57 (2012)
10. Maali, Y., Al-Jumaily, A.: Self-advising support vector machine. Knowl. Based Syst. **52**, 214–222 (2013)
11. Islam, S.M.D., Swagatam, G., Saurav, R., Subhrajit, S., Ponnuthurai, N.: An adaptive differential evolution algorithm with novel mutation and crossover strategies for global numerical optimization. IEEE Trans. Syst. Man Cybern. Part B Cybern. **42**(2), 482–500 (2012)
12. Price, K.S., Rainer, M., Lampinen, J.A.: Differential Evolution: a Practical Approach to Global Optimization. Springer Science & Business Media, Berlin (2006)
13. Das, S.S., Ponnuthurai, N.: Differential evolution: a survey of the state-of-the-art. IEEE Trans. Evol. Comput. **15**(1), 4–31 (2011)
14. Kumar, P., Pant, M.: A self adaptive differential evolution algorithm for global optimization. In: Panigrahi, B.K., Das, S., Suganthan, P.N., Dash, S.S. (eds.) SEMCCO 2010. LNCS, vol. 6466, pp. 103–110. Springer, Heidelberg (2010)
15. Khushaba, R.N., Al-Ani, A., Al-Jumaily, A.: Feature subset selection using differential evolution. In: Köppen, M., Kasabov, N., Coghill, G. (eds.) ICONIP 2008, Part I. LNCS, vol. 5506, pp. 103–110. Springer, Heidelberg (2009)
16. Bharathi, P.T., Subashini, D.P.: Optimal feature subset selection using differential evolution and extreme learning machine. Int. J. Sci. Res. **3**(7), 1898–1905 (2012)
17. Masood, A., Al-Jumaily, A., Anam, K.: Texture analysis based automated decision support system for classification of skin cancer using SA-SVM. In: Loo, C.K., Yap, K.S., Wong, K.W., Teoh, A., Huang, K. (eds.) ICONIP 2014, Part II. LNCS, vol. 8835, pp. 101–109. Springer, Heidelberg (2014)
18. Haupt, R.L.H., Sue, E.: Practical Genetic Algorithms. Wiley, New York (2004)
19. Firpi, H.A.G., Erik, D.: Swarmed feature selection. In: Proceedings of the 33rd Applied Imagery Pattern Recognition Workshop (2004)

20. Khushaba, R.N., AlSukker, A., Al-Ani, A., Al-Jumaily, A., Zomaya, A.Y.: A novel swarm based feature selection algorithm in multifunction myoelectric control. J. Intell. Fuzzy Syst. **20**(4–5), 175–185 (2009)

21. Oh, I.S., Lee, J.S., Moon, B.R.: Hybrid genetic algorithms for feature selection. IEEE Trans. Pattern Anal. Mach. Intell. **26**(11), 1424–1437 (2004)

TNorm: An Unsupervised Batch Effects Correction Method for Gene Expression Data Classification

Praisan Padungweang[✉], Worrawat Engchuan, and Jonathan H. Chan

School of Information Technology, King Mongkut's University of Technology Thonburi,
Bangkok, Thailand
{praisan.pad,worrawat.eng,jonathan}@sit.kmutt.ac.th

Abstract. In the field of biomedical research, gene expression analysis helps to identify the disease-related genes as genetic markers for diagnosis. As there is a huge number of publicly available gene expression datasets, the ongoing challenge is to utilize those available data effectively. Merging microarray datasets from different batches to improve the statistical power of a study is one of the active research topics. However, various works have addressed the issue of batch effects variation, which describes variation in gene expression levels induced by different experimental environments. Ignoring this variation may result in erroneous findings in a study. This work proposes a method for batch effect correction by mapping underlying topology of different batches. The mapping process for cross-batch normalization is examined using basic linear transformation. The comparative study of three cancers is conducted to compare the proposed method with a proven batch effects correction method. The results show that our method outperforms the existing method in most cases.

Keywords: Gene expression · Batch effects · Clustering · Linear regression · Classification · Cancer · Diagnosis

1 Introduction

Gene expression analysis helps to improve the understanding of cell behavior or how cell response to stimulus. It is also commonly applied to identify genetic markers of disease [1, 2]. With advances in molecular biology technology, a huge number of gene expression datasets has been generated and publicly provided on the online databases. This is a great opportunity for the research community in gene expression studies. Many researches have been conducted to improve the efficiency in analyzing those available data [3–5]. Some complex diseases have high complexity in development mechanisms. So, analyzing a particular gene expression dataset is inadequate to understand those complex mechanisms, as the number of samples may be limited [6–8]. Therefore, incorporating many datasets from multiple experiments is required. To incorporate multiple datasets, two main approaches have been proposed. Meta-analysis is one of those approaches, which analyze each dataset independently. The result of each analysis is then combined to produce the final results [9, 10]. Unlike meta-analysis, one of the hottest issues is to merge microarray datasets from different batches to enlarge the sample size. As the number of samples is increased, studying

© Springer International Publishing Switzerland 2015
S. Arik et al. (Eds.): ICONIP 2015, Part I, LNCS 9489, pp. 411–420, 2015.
DOI: 10.1007/978-3-319-26532-2_45

on merged datasets will improve the statistical power and reliability of a study [11]. However, studying combined gene expression data from different batches can be detrimental due to the occurrence of batch effects variation [12]. Batch effects variation is the altering of gene expression levels driven by experimental environment. Ignoring the batch effects variations can result in an improper analysis, as this confounding factor is not controlled. Our previous work also found that this batch effects variation directly affects to the performance of disease classification [13]. Many batch effects correction methods have been proposed to remove the variation in gene expression levels caused by different batches [12, 14]. Combatting Batch Effects (COMBAT) is a well-known batch effects correction method, which implements an empirical Bayes framework to normalize the dataset across genes [14]. However, this method tries to correct the batch effects by treating all samples in each batch the same way. With the heterogeneity of gene expression profile in each dataset, it is better to adjust expression levels of different groups differently. This work tries to improve on the current batch effects correction method by developing a novel method namely "Topology-based Normalization with Linear transformation (TNorm)". The topology of a given gene expression data is viewed as a structure of a set of representative samples which are present in the data distribution. Then, by matching the representative samples for each batch, basic linear transformation can be effectively used as a data normalizer. The comparative study is conducted in this work to compare the performance of proposed method with COMBAT. Six gene expression datasets of three cancer types are used as case study in this work. The gene-set activity transformation and cross dataset classification are applied here to evaluate the performance of each batch effects correction method.

2   The Proposed Method

Topology preserving map is one of widely used technique in Machine learning. Normally the high dimensional data set is mapped to a lower one [15] for dimension reduction purpose. However, this work, the mapping is used for normalization purpose. In order to analyze data from different source with different experimental environments, the data set should be preprocessed to be in the same appropriate space. That is, the mapped data should remain the same dimension as the original data set but the data are transformed using the same appropriate space and appropriate structure matching. This paper proposes a batch effect correction called Topology-based Normalization with Linear transformation (TNorm). The topology structure can be viewed as a graph of representatives. Node of the graph is a representative sample, which is present in the distribution of the original dataset. Let $\mathbf{X}^{(b)} = \left\{ \mathbf{x}_i^{(b)} | 1 \le i \le n \right\}, b = 1, 2, \dots$ be a matrix of gene expression with n samples which can be viewed as a matrix of column vectors. $\mathbf{x}_i^{(b)} = [x_{i,1}^{(b)}, x_{i,2}^{(b)}, \dots, x_{i,m}^{(b)}]^t \in \mathbb{R}^{m \times 1}$ is an ith sample vector with m genes from experimental environments b. $\mathbf{X}^{(b1)}$ and $\mathbf{X}^{(b2)}$ are gene expression from experimental environments b1 and b2 respectively. $\mathbf{C}^{(b)} = \left\{ \mathbf{c}_i^{(b)} | 1 \le i \le k \right\}$ denotes the k representative samples of $\mathbf{X}^{(b)}$ here $\mathbf{c}_i^{(b)} = [c_{i,1}^{(b)}, c_{i,2}^{(b)}, \dots, c_{i,m}^{(b)}]^t \in \mathbb{R}^{m \times 1}$. Suppose $\mathbf{X}^{(b1)}$ is used for model training and used as a reference data space. $\mathbf{X}^{(b2)}$ is a data set from different experimental environments.

The batch-adjusted $\mathbf{X}^{(b2)}$ can be viewed as a linear transformation from $\mathbf{X}^{(b2)}$ space to the reference data space. In order to match the topology of different batch data set, the same number of representative samples of each batch are considered. The distance between pair of representative samples is used to determine which representative samples from different batch data set should be mapped with which one of the reference data. Then, the mapping process can be considered as a linear transformation between $\mathbf{C}^{(b1)}$ and $\mathbf{C}^{(b2)}$ that is a map $\mathbf{T}:\mathbf{C}^{(b2)} \rightarrow \mathbf{C}^{(b1)}$ and can be computed as

$$\mathbf{C}^{(b1)} = \mathbf{T} \times \mathbf{C}^{(b2)}$$

$$\mathbf{T} = \mathbf{C}^{(b1)} \times (\mathbf{C}^{(b2)})^{-1} \tag{1}$$

then T is a linear transformation from $\mathbb{R}^m$ to $\mathbb{R}^m$. The batch-adjusted data are given by

$$\hat{\mathbf{X}}^{(b2)} = \mathbf{T} \times \mathbf{X}^{(b2)} \tag{2}$$

Figure 1 is an illustrative example of adjusting batch data set from different sources, batch 1 and batch 2 data set, by using the batch 1 data set as a reference. We randomly generated four normally distributed clusters as shown in Fig. 1(a) and (b). The representative samples in each dataset, which are used for topology mapping are also shown. The Fig. 1(c) and (d) shows the merged data without and with batch effect correction by our algorithm respectively. It can be seen that using topology mapping, the data set from different source is also mapped into the same space with the reference one (Fig. 1(d)).

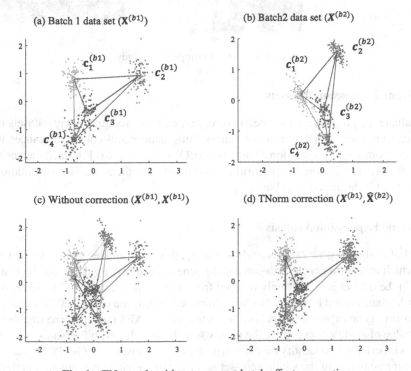

Fig. 1. TNorm algorithm to correct batch effects correction

3   Experimental Design

The comparative study conducted in this study is divided into several processes (Fig. 2). First, the gene expression data is pre-processed by z-transformation in order to standardize the data. Secondly, gene-set activity transformation method is applied to transform gene expression to gene-set activity data. Then, the gene-set activity data is adjusted to remove batch effects variation. Finally, the adjusted data is used to build a classification model and the performance of the model is evaluated by cross dataset validation scheme.

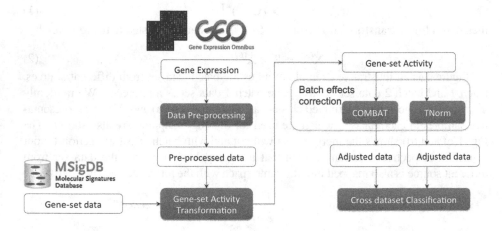

Fig. 2. Overview of comparative study.

3.1   Gene Expression Datasets

To evaluate the proposed batch effects correction method, six microarray datasets from three cancer types, which are breast cancer, lung cancer and colorectal cancer were obtained from Gene Expression Omnibus (GEO) datasets [16]. For each cancer type two datasets were obtained to perform cross dataset classification. The information of each dataset is described in Table 1.

3.2   Gene Expression Analysis

Instead of analyzing gene expression data directly, this study applied gene-set based analysis, which transform gene expression data to gene-set activity. By doing so, the features space in the dataset is significantly reduced from gene expression data [23]. Here, Analysis of Variance based Feature Sets (AFS) proposed by Engchuan et al. 2015 was applied to transform gene expression data [24]. In each gene-set, AFS rank the gene members by the F-value of ANOVA test. Then, the greedy search algorithm with Pearson's correlation analysis is performed to identify the phenotype-correlated genes (PCOGs). PCOGs are used as the representative of the whole gene-set and their expression level are summarized as gene-set activity. AFS is available as web-based application and java library on Gene-set

Activity Toolbox (GAT) at http://gat.sit.kmutt.ac.th. The gene-set data used in this study was obtained from Molecular Signature Database (MSigDB) [25]. The curated data of canonical pathways containing 1,320 gene-sets and 8,428 genes was chosen.

Table 1. Gene expression datasets used in this study

Accession	Name	Publication	Samples
GSE5764	Breast1	Turashvili *et al.* [17]	normal: 20 samples tumor: 10 samples
GSE7904	Breast2	Richardson *et al.* [18]	normal: 19 samples tumor: 43 samples
GSE4107	Colorectal1	Hong *et al.* [19]	normal: 10 samples tumor: 12 samples
GSE8671	Colorectal2	Sabates-Bellver [20]	normal: 32 samples tumor: 32 samples
GSE4115	Lung1	Spira *et al.* [21]	normal: 90 samples tumor: 97 samples
GSE10072	Lung2	Landi *et al.* [22]	normal: 49 samples tumor: 58 samples

3.3 Batch Effects Correction

The gene-set activity datasets transformed from each two datasets of the same cancer types are then adjusted to remove batch effects variation. In this study, we compare two batch effects correction methods, which are COMBAT and our proposed TNorm. The detail of each method is described as follows.

COMBAT or Combatting Batch Effects was proposed by Johnson and Li in 2007 [14]. The empirical Bayes framework is implemented with COMBAT. So, it makes this approach robust to high-dimensional data like microarray data. This method pools the information and estimate batch effects across genes so it is also robust to small sample datasets compare to other existing batch effects correction methods.

TNorm or Topology-based Normalization with Linear transformation is developed in this study. Here, we implemented the simply and powerful clustering algorithm, K-Means clustering. The number of clusters is set to three, where first two clusters are for case and control samples, and the last cluster is for the outliers. After clustering, the representative samples from two datasets are then mapped according to their topological aspect. The Euclidean distance between each pair of representative sample is used to determine which representative sample of testing data should be mapped with which one of the training data. Then, the linear transformation is applied to adjust the testing data to make two datasets have same data distribution topology.

3.4 Cross Dataset Classification

Cross dataset classification was applied in this study to evaluate the performance of our proposed batch effects correction method and COMBAT. For each cancer type, one dataset is used to transform gene expression levels to gene-set activity as mentioned in Sect. 2.2 and built the classification model by using the transformed gene-set activity. Then, the model is tested by classifying test instances in another dataset of the same cancer. To transform gene-set activity in test data, the PCOGs identified from training data are used. To build the classification model, Support Vector Machine (SVM) was used as classifier. SVM has been commonly applied as classifier and feature ranker for gene expression analysis [26]. In addition to classifier, Correlation-based Feature Subset selection (CFSSubset) was used to select subset of features to improve the classification performance [27]. In this study, 1 %, 10 % and all features were selected to build three different models. The Area Under Receiver Operating Characteristic (AUC) was reported as classification performance. This performance measure has been proved to be robust against class imbalance issue [28].

4 Results and Discussion

We conducted a case study on six microarray datasets of three cancer types. The case study was used to assess the performance of batch effects correction methods and compare against the dataset without batch effect correction. The performance was reported as AUC of cross dataset validation as mentioned in previous section. The example of cross dataset validation result can be notated as Breast1-2, which representing the result of model built by Breast1 dataset and tested by Breast2 dataset. The transformation of the Breast1-2 is a map $\mathbf{T}:\mathbf{C}^{(Breast2)} \rightarrow \mathbf{C}^{(Breast1)}$. Figures 3, 4, and 5 present the cross dataset validation results of this study. In breast cancer classification (Fig. 3), using TNorm as batch effects correction method show significantly better performance than COMBAT (p-value = 0.0175) and without correction (p-value = 0.0434).

For colorectal cancer classification, TNorm is significantly better in Colorectal1-2 validation than COMBAT (p-value = 0.0132) and without correction (p-value = 0.006). However, in Colorectal2-1 validation, using COMBAT achieves higher classification performance (p-value = 0.0282) but none of Colorectal2-1 validation can achieves AUC over 0.5.

For lung cancer classification, TNorm is worse than COMBAT (p-value = 0.0411) and without correction (p-value = 0.0498) in Lung1-2 validation. In Lung2-1 validation, however, TNorm achieves significantly higher AUC than COMBAT (p-value = 0.0039) and without correction (p-value = 0.0077).

From results in Figs. 3, 4, and 5, while the improvement from COMBAT is unclear, it can be seen that TNorm can be used to correct the batch effects variation. As a result, the classification performance is also improved in most cases. In the case that TNorm is worse in performance, we found that the other methods also performed poorly (AUC ~ 0.5). It reflects that those models cannot be applied for classifying testing instances, which can be caused by the heterogeneity in cancer itself or the bad choice

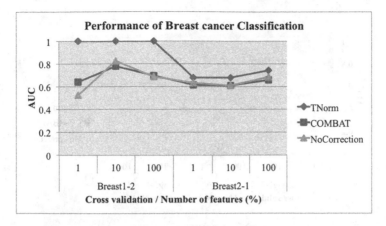

Fig. 3. Comparative result of classification performance in breast cancer between TNorm, COMBAT and without batch effect correction.

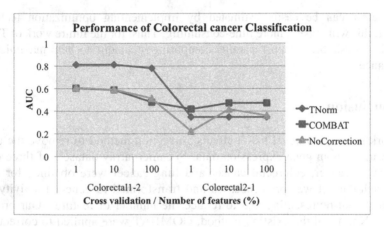

Fig. 4. Comparative result of classification performance in colorectal cancer between TNorm, COMBAT and without batch effect correction.

in data collection. So, the batch effects correction would be unable to help improving classification performance in this situation.

Besides that hypothesis, the implementation of clustering technique can be considered for improvement. Currently, all 1,320 features were used to build clusters. With K-Means clustering algorithm, all features are treated the same way. So, if there were a lot of noisy features, the data points would be poorly clustered and result in a failure in further analysis. Thus, in the future, the unsupervised feature selection or Weighted K-means clustering should be applied to address this issue.

In linear transformation, it is still improvable. This work calculates the transformation matrix without estimating of random error (ε) to make it fast and simply to implement. Calculating transformation matrix with random error is hypothesized to be more precise as the topology of testing data will be more fitted to training data topology. This

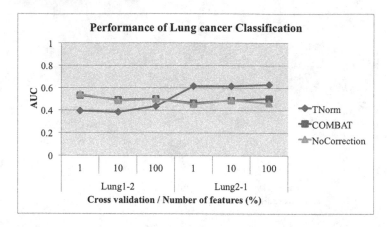

Fig. 5. Comparative result of classification performance in lung cancer between TNorm, COMBAT and without batch effect correction.

random error can be easily estimated by implementing optimization technique. However, this will make it more time-consuming. Thus, for the future work of TNorm, it would be desirable to apply parallel computing technique to help maximizing its performance.

5   Conclusion

This work proposes a novel batch effects correction method to remove the uncontrolled variations in gene expression data. Six microarray datasets of three cancer types: breast cancer, colorectal cancer and lung cancer were obtained for a case study. Each dataset was pre-processed and transformed to gene-set activity to be more functional-representing and to reduce the number of features. Our proposed method, TNorm and the existing method, COMBAT were applied to correct batch effects variation in the dataset. Cross dataset validation was used to evaluate the performance of each method and the results were reported as AUC. The results show that TNorm outperforms COMBAT and significantly improves the classification performance from original gene-set activity data in most cases. Incidentally, TNorm still needs several improvements such as implementing feature selection, choice of clustering techniques, parallel computing and optimization for estimating random error. With these improvements, TNorm should stand out and can be a good choice for batch effects correction in future gene expression analysis.

References

1. Su, A.I., Welsh, J.B., Sapinoso, L.M., Kern, S.G., Dimitrov, P., Lapp, H., Schultz, P.G., Powell, S.M., Moskaluk, C.A., Frierson Jr., H.F., Hampton, G.M.: Molecular classification of human carcinomas by use of gene expression signatures. Cancer Res. **61**, 7388–7393 (2001)

2. Lu, Y., Han, J.: Cancer classification using gene expression data. Inf. Syst. **28**, 243–268 (2008)
3. Wang, Y., Klijn, J.G., Zhang, Y., Sieuwerts, A.M., Look, M.P., Yang, F., Talantov, D., Timmermans, M., Meijer-van Gelder, M.E., Yu, J., Jatkoe, T., Berns, E.M.J.J., Atkins, D., Foekens, J.A.: Gene-expression profiles to predict distant metastasis of lymph-node-negative primary breast cancer. Lancet **365**, 671–679 (2005)
4. Dupuy, A., Simon, R.M.: Critical review of published miroarray studies for cancer outcome and guidelines on statistical analysis and reporting. J. Natl Cancer Inst. **99**, 147–157 (2007)
5. Michiels, S., Koscielny, S., Hill, C.: Prediction of cancer outcome with microarrays a multiple random validation strategy. Lancet **365**, 488–492 (2005)
6. Ein-Dor, L., Suk, O., Domany, E.: Thousands of samples are needed to generate a robust gene list for predicting outcome in cancer. Proc. Natl. Acad. Sci. U.S.A. **103**, 5923–5928 (2006)
7. Xu, L., Tan, A.C., Winslow, R.L., Geman, D.: Merging microarray data from separate breast cancer studies provides a robust prognostic test. BMC Bioinf. **9**, 125 (2008)
8. Shabalin, A.A., Tjelmeland, H., Fan, C., Perou, C.M.: Merging two gene-expression studies via cross-platform normalization. Bioinformatics **24**, 1154 (2008)
9. Wang, Y., Joshi, T., Zhang, X.S., Xu, D., Chen, L.: Inferring gene regulatory networks from multiple microarray datasets. Bioinformatics **22**, 2413 (2006)
10. Choi, H., Shen, R., Chinnaiyan, A.M., Ghosh, D.: A latent variable approach for meta-analysis of gene expression data from multiple microarray experiments. BMC Bioinf. **8**, 364 (2007)
11. Warnat, P., Eils, R., Brors, B.: Cross-platform analysis of cancer microarray data improves gene expression based classification of phenotypes. BMC Bioinf. **6**, 265 (2005)
12. Larsen, M.J., Thomassen, M., Tan, Q., Sorensen, K.P., Kruse, T.A.: Microarray-based RNA profiling of Breast cancer: batch effect removal improves cross-platform consistency. BioMed Res. Int. **2014**, 11 (2014)
13. Engchuan, W., Meechai, A., Tongsima, S., Chang, J.H.: Handling batch effect on cross-platform classification of microarray data. Int. J. Adv. Intell. Paradigms (in press)
14. Johnson, W.E., Li, C.: Adjusting batch effects in microarray expression data using empirical Bayes methods. Biostatistics **8**, 118 (2007)
15. Marian, P., Wesam, B., Colin, F.: Topology-preserving mappings for data visualization, pp. 131–150. Principal Manifolds for Data Visualization and Dimension Reduction. Springer, Berlin Heidelberg (2008)
16. Edgar, R., Domrachev, M., Lash, A.E.: Gene expression omnibus: NCBI gene expression and hybridization array data repository. Nucleic Acids Res. **30**, 207–229 (2002)
17. Turashvili, G., Bouchal, J., Baumforth, K., Wei, W., Dziechciarkova, M., Ehrmann, J., Klein, J., Fridman, E., Skarda, J., Srovnal, J., Hajduch, M., Murray, P., Kolar, Z.: Novel markers for differentiation of lobular and ductal invasive breast carcinomas by laser microdissection and microarray analysis. BMC Cancer **7**, 55 (2007)
18. Richardson, A.L., Wang, Z.C., De Nicolo, A., Lu, X., Brown, M., Miron, A., Liao, X., Iglehart, J.D., Livingston, D.M., Ganesan, S.: X chromosomal abnormalities in basal-like human breast cancer. Cancer Cell **9**, 121–132 (2006)
19. Hong, Y., Ho, K.S., Eu, K.W., Cheah, P.Y.: A susceptibility gene set for early onset colorectal cancer that integrates diverse signaling pathways: implication for tumorigenesis. Clin. Cancer Res. **13**, 1107–1114 (2007)
20. Sabates-Bellver, J., Van der Flier, L.G., de Palo, M., Cattaneo, E., Maake, C., Rehrauer, H., Laczko, E., Kurowski, M.A., Bujnicki, J.M., Menigatti, M., Luz, J., Ranalli, T.V., Gomes, V., Pastorelli, A., Faggiani, R., Anti, M., Jiricny, J., Clevers, H., Marra, G.: Transcriptome profile of human colorectal adenomas. Mol. Cancer Res. **5**, 1263–1275 (2007)

21. Spira, A., Beane, J.E., Shah, V., Steiling, K., Liu, G., Schembri, F., Gliman, S., Dumas, Y.M., Calner, P., Sebastiani, P., Sridhar, S., Beamis, J., Lamb, C., Anderson, T., Gerry, N., Keane, J., Lenburg, M.E., Brody, J.S.: Airway epithelial gene expression in the diagnostic evaluation of smokers with suspect lung cancer. Nat. Med. **13**, 361–366 (2007)

22. Landi, M.T., Dracheva, T., Rotunno, M., Figueroa, J.D., Liu, H., Dasgupta, A., Mann, R.E., Fukuoka, J., Hames, M., Bergen, A.W., Murphy, S.E., Yang, P., Pesatori, A.C., Consonni, D., Bertazzi, P.A., Wacholder, S., Shih, J.H., Caporaso, N.E., Jen, J.: Gene expression signature of cigarette smoking and its role in lung adenocarcinoma development and survival. PLoS ONE **3**, e1651 (2008)

23. Sootanan, P., Prom-on, S., Meechai, A., Chan, J.H.: Pathway-based microarray analysis for robust disease classification. Neural Comput. Appl. **21**, 649–660 (2011)

24. Engchuan, W., Chan, J.H.: Pathway activity transformation for multi-class classification of Lung cancer datasets. Neurocomputing **165**, 81–89 (2014)

25. Subramanian, A., Tamayo, P., Mootha, V.K., Mukherjee, S., Ebert, B.L., Gillette, M.A., Paulovich, A., Pomeroy, S.L., Golub, T.R., Lander, E.S., Mesirov, J.P.: Gene set enrichment analysis, a knowledge-based approached for interpreting genome-wide expression profiles. PNAS **102**, 15545–15550 (2005)

26. Li, T., Zhang, C., Ogihara, M.: A comparative study of feature selection and multiclass classification methods for tissue classification based on gene expression. Bioinformatics **20**, 2429–2437 (2004)

27. Hall, M.A.: Correlation-Based Feature Subset Selection for Machine Learning. Hamilton, New Zealand (1998)

28. Kotsiantis, S., Kanellopoulos, D., Pintelas, P.: Handling imbalanced dataset: A review. GESTS Int. Trans. ComSci. Eng. **30**, 25–36 (2006)

Finger-Vein Quality Assessment
by Representation Learning
from Binary Images

Huafeng Qin[1,2(✉)] and Mounîm A. El-Yacoubi[2]

[1] Chongqing Engineering Laboratory of Detection Control
and Integrated System, Chongqing Technology and Business University,
Chongqing 400067, China
huafeng.qin@telecom-sudparis.eu
[2] Department of EPH, Telecom-SudParis, 91011 Paris, Evry Cedex, France

Abstract. Finger-vein quality assessment is an important issue in finger-vein verification systems as spurious and missing features in poor quality images may increase the verification error. Despite recent advances, current solutions depend on domain knowledge and are typically driven by visual inspection. In this work, we propose a deep Neural Network (DNN) for representation learning from binary images to predict vein quality. First, driven by the primary target of biometric quality assessment, i.e. verification error minimization, we assume that low quality images are false rejected finger-vein images in a verification system. Based on this assumption, the low and high quality images are labeled automatically. Second, as image processing approaches such as enhancement and segmentation may produce false features and ignore actual ones thus degrading verification accuracy, we train a DNN on binary images and derive deep features from its last hidden layer for quality assessment. Our experiments on two large public finger-vein databases show that the proposed scheme accurately identifies high and low quality images and significantly outperform existing approaches in terms of the impact on equal error rate (EER) improvement.

Keywords: Biometrics · Finger-vein quality assessment · Deep learning · Deep neural network · Representation learning · Feature representation

1 Introduction

Biometrics authentication consists of verifying a person based on his/her physiological or behavioral traits. Compared to traditional identification means such as cards or passwords, biometrics are more secure and convenient to users [1]. Biometrics, however, are prone to performance degradation due to poor image acquisition. For this reason, various quality assessment approaches have been developed [2–5]. Contrary to extrinsic biometric traits (e.g., fingerprint, face, iris), finger-vein is intrinsic and is thus difficult to copy and forge. However, like other biometric traits, finger-vein image quality is inherently affected by a number of factors that can be split into two categories: (1) Extrinsic factors associated with environmental illumination, ambient

© Springer International Publishing Switzerland 2015
S. Arik et al. (Eds.): ICONIP 2015, Part I, LNCS 9489, pp. 421–431, 2015.
DOI: 10.1007/978-3-319-26532-2_46

temperature, physiological changes, light scattering, and user behavior; (2) Intrinsic factors as associated with inaccurate parameter estimation during the image preprocessing stage. For example, the finger-vein image enhancement and segmentation schemes have been developed to extract vein pattern in most finger-vein verification systems. Incorrect estimation of orientation, scale and rotation angle of vein pattern may produce false finger-vein features and fail to detect some genuine vein features.

Various finger-vein quality schemes have been recently developed to solve the problem. Qin *et al.* [6] combine the Radon transform and curvature to describe the contrast and quality of finger-vein features. Their experiments on a private database show their method can detect low quality finger-vein images and reduce the verification error rate (EER). In [7], quality assessment based on the number of detected vein points is shown to reduce the EER when rejecting finger-vein images with low quality score. In [8], Yang *et al.* combine gradient, image contrast, and information capacity based on SVM to assess finger-vein quality. Instead of SVM, Peng *et al.* [9] use Triangular norm to fuse attributes described in [8] for performance improvement.

Current solutions, nonetheless, have serious issues. First, quality vein attributes are therein determined by human intuition and a *priori* knowledge and are detected by hand-crafted descriptors. Second, all finger-vein quality assessment methods have only considered the condition of input images because the models to describe vein attributes are built based on finger-vein grayscale images. However, in most finger-vein verification systems, the image processing approaches such as normalization, enhancement and segmentation are employed to extract the vein pattern for verification. Therefore, the quality of finger-vein grayscale image can be further affected by inaccurate parameter estimation during the preprocessing stage. Furthermore, the vein network which is directly related to EER is extracted from binary and not grayscale images. Thus the vein network may not match the temp late despite a grayscale-based high quality score, and vice versa. To sum it up, the limited performance of the quality assessment methods in [6–9] is due to the fact that the quality attributes are not objectively linked to the EER but are based on visual inspection.

In this paper, we propose a novel quality assessment method based on Deep Neural Networks (DNN) to predict low-quality finger-vein image on unseen samples. DNNs have been successfully applied to several vision tasks [10, 11], as they are capable of learning robust features from raw pixel images. To the best of our knowledge, this is the first DNN-based approach on quality assessment for vein images, and actually for any other biometrics. Our contributions are fourfold: (1) unlike traditional methods, the low (high) quality finger-vein images in our scheme are associated with the false non-match/rejected (true match/accepted) images by a finger-vein verification system. Our assumption can guide the DNN to automatically extract robust quality features that directly affect the EER rather than being designed by human intuition or knowledge. (2) We automatically label the high and low quality images, which avoids heavy manual labeling and human errors. Our label scheme is directly related to biometrics verification performance rather than to human image quality judgment. (3) Unlike existing approaches based on handcrafted features, an effective scheme based on DNN is employed to automatically learn features from raw pixel images. (4) Unlike current approaches considering only extrinsic factors, we propose an effective quality assessment scheme that considers both extrinsic and intrinsic factors.

First, vein networks are extracted from grayscale images using a state of the art system and are input to a DNN for training. The DNN's last hidden layer is taken as input deep features to a probabilistic SVM (P-SVM) for finger-vein quality assessment. As vein networks are associated with extrinsic and intrinsic factors and directly relate to EER, it makes more sense to evaluate finger-vein quality based on binary rather than grayscale image. We carry out experiments on two public databases to verify the efficiency of proposed method.

2 DNN for Finger-Vein Image Quality Assessment

The proposed DNN (Convolutional Neural Network (CNN)) (Fig. 1) is trained to extract a deep feature representation from each binary finger-vein image. Then, the deep representation is taken from the last hidden layer of DNN and input to P-SVM for finger-vein image assessment.

2.1 Deep Neural Networks

Our DNN contains three convolutional layers and two max-pooling layers to extract features hierarchically, followed by two fully-connected layers and the softmax output layer indicating identity classes. Figure 1 shows the detailed architecture of the DNN which takes an image of size 20×96 as input and outputs two classes (high quality and low quality). The dimension of the last hidden layer is fixed to 1000. The width and height of map in each layer vary according to input size. After forwarding the input image though the DNN, compact and high level discriminating features are inferred.

For convolutional layers, we use as hidden neuron activations Rectified Linear Units (RELUs) ($y = max(0, x)$) that have shown good fitting abilities in many DNN architectures [10, 11]. We also employ max-pooling to extract location information that ensures robustness to translation while reducing the filter responses to a lower dimension. In addition, the dropout technique [10] is applied in the two fully-connected layers to prevent overfitting. The output of the last fully-connected layer is taken as finger-vein deep feature representation. Finally, the posterior probability distribution over n (here $n = 2$) different classes is estimated using an output softmax layer. Stochastic gradient descent is performed by back-propagation.

2.2 Feature Extraction

In most finger-vein verification systems, vein networks are extracted in a binary form from grayscale images and are input for verification matching. As finger-vein grayscale image quality is further degraded during the binarization stage, we train a DNN for finger-vein feature representation from binarized images. For a given binary image F, its label is denoted as $q \in \{0, 1\}$ based on the baseline method [12] (the detail of labeling is described in Sect. 3.2), where 0 and 1 denote low and high quality finger-vein images, respectively. Each image and its label are then combined to form a dataset $\{(F\ 1, q_1), (F\ 2, q\ 2), \ldots, (F\ N, q\ N)\}$ on which we can train DNN to learn

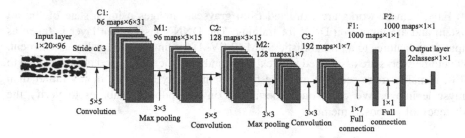

Fig. 1. Our DNN architecture.

discriminating information, with the N number of training finger-vein images. The output of the last hidden layer generates a 1000-dimensional deep feature vector for each image. In this way, new deep feature based training and test sets are generated and used to train a classifier to predict the quality of a finger-vein image.

2.3   Generating Quality Score

Let v be the deep feature representation extracted from a training finger-vein image F with a label $q \in \{ 0, 1 \}$. We train a P-SVM [13] to obtain a probabilistic value p (0 to 1: from low to high image quality).

$$p(q = 1|\xi(v)) = \frac{1}{1 + \exp(\omega \cdot \xi(v) + \gamma)} \tag{1}$$

where $\xi(v)$ denotes the output of a two-class SVM classifier [14] which takes v as the input feature vector; ω and γ are probability fitting parameters estimated by P-SVM.

3   Experiments

3.1   Database

Database A: Hong Kong Polytechnic University provides a public database [12] including 3132 contactless finger-vein images from 156 subjects. The first 105 subjects provide 2520 images (105 subjects × 2 fingers × 6 images × 2 sessions) in 2 sessions, separated by 66.8 days in average. The remaining 51 subjects only provide 612 images (51 subjects × 2 fingers × 6 images) captured in one session. To test our approach, we use the subset consisting of the first 105 subjects as it is more realistic. All images are normalized to 50 × 240 using the preprocessing method in [12].

Database B: The Universiti Sains Malaysia finger-vein database [15] contains 5904 images from 123 subjects at two sessions with interval of more than two weeks' time. In each session, each volunteer provided four fingers such as left index, left middle, right index and right middle fingers, and each finger was captured six times, resulting in a total of 2952 (123 subjects × 4 fingers × 6 images) images. So, for two sessions, there are a total of 5904 (2952 × 2 sessions) images. ROI images are extracted by the algorithm described in [15] and proportionally resized to 80 × 240 in our work.

To test our approach, we employ the state of the art method, named "Even Gabor with Morphological" in [12] to segment the images into two finger-vein databases. Figure 2 shows the vein network (binary image) and its corresponding grayscale image. The performance of our approach is evaluated based on binary images.

3.2   Experiment Settings on Database A

Image quality assumption: As the EER is the primary performance indicator in biometric authentication systems, image quality algorithms should be assessed by their effect on EER rather than human perception judgment. Therefore, we assume low quality finger-vein images are those falsely rejected by the verification system.

Template selection: For each finger image, its average distance score with respect to the other 11 images from the same finger is computed based on the matching approach in [12]. We select the image with lowest average score as the finger template. As different fingers from the same subject are treated as different classes, we get 210 (2 fingers × 105 persons) templates and 2310 (210 × 11) query images.

Labeling high and low quality images: For each query image, we determine its quality label based on the matching distance against its template. 210 × 11 = 2310 genuine matches are obtained by matching each template against the remaining 11 finger-vein images, based on which the False Rejection Rate (FRR) is computed. We compute the False Acceptance Rate (FAR) by matching each template against the other 209 templates. Overall, there are 210 × 209/2 = 21945 impostor matches, as the symmetric imposter matches are not executed. The FAR indicates the system security level in a biometric verification system. At the predefined system security level (typically, FAR = 0.1 %), we set label q to 0 for a query finger-vein image that is falsely rejected by the system, and set q to 1 for accepted genuine finger-vein images. Note that 0 and 1 denote low and high quality finger-vein images respectively.

Generating the training and test sets: After labeling each query finger-vein image at FAR = 0.1 %, we select half the fingers associated with 1155 images (105 × 11) for training and the remaining fingers for test. We obtain, in this way, 101 low-quality and 1054 high-quality images in the training set, and 110 low-quality and 1045 high-quality images in the test set. To increase the gap between high-quality and low-quality training images, we keep only 406 high-quality images from the 1054 above by setting FAR = 0 %. In general, low-quality images are much fewer than high-quality ones in biometric systems. To overcome the class-imbalanced problem, we generate additional low-quality images based on our image quality assumption: we pick out 305 samples from the remaining 648 (1054-406) high-quality training images, and change their illumination, scale and rotation angle so that the corresponding binary images get falsely rejected by the verification system. We thus obtain an overall low-quality image set of 406 (305 + 101) images for training. Figure 3 illustrates some synthesized low-quality finger-vein binary images using this mechanism.

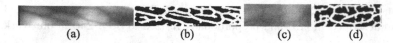

      (a)   (b)   (c)   (d)

Fig. 2. Sample results based on Even Gabor with Morphological: (a) Finger-vein grayscale image from database A, (b) Vein network from (a), (c) Finger-vein grayscale image from database B, and (d) Vein network from (c).

3.3 Experiment Settings on Database B

The training and test sets on database B are built in the same way as above. $492 \times 491/2 = 120786$ impostor scores and $492 \times 11 = 5412$ genuine scores are produced to compute the FAR and FRR, respectively. Subsequently, the quality of all images is labeled at FAR = 0.1 %. We select 2706 images from half the fingers for training and remaining images for testing. As a result, we obtain 182 low-quality and 2524 high-quality training images, and 140 low-quality and 2566 high-quality test images. We keep only 635 high-quality images from the 2524 above by setting FAR = 0 % (highest system security level). Likewise, we generate 495 additional low-quality images from the remaining 1889 (2524-635) high-quality ones. Therefore, the size of overall low-quality image training set is also 635 (140 + 495).

3.4 Selection of CNN Architecture

One of the major issues with DNNs is that the size of hidden layers should grow significantly with input size for robust representation learning, which may lead to overfitting and to expensive computing and memory costs. In this work, we seek a DNN with a minimum number of parameters with a small finger-vein image size for efficient quality assessment. We use a similar CNN architecture to that of Krizhevsky et al. [10] successfully used on the large-scale ImageNet classification challenge. Our DNN is trained on binary finger-vein images for subsequent quality assessment. The DNN parameters such as input size, depth, and dimensionality of the output representation are optimized in a greedy way, step by step. Table 1 shows identifying accuracies on Database A of various DNNs for different input image sizes. We observe that the best performance is achieved based on DNN (96-128-192-1000-1000-2), consisting of an input layer of size of 20×96, 3 convolutional layers, 2 max-pooling layers, 2 fully-connected layers and the Softmax output layer with 2 classes, as shown Fig. 1. For database B, we proportionally resized all images to 32×96 before DNN training.

3.5 Evaluation of Quality Image Assessment

To identify high and low quality images, we train a DNN with softmax output layer and a DNN + P-SVM. In the DNN + P-SVM classification setting, the DNN last feature representation is input to P-SVM for image quality prediction. The parameters of DNN such as architecture and size of filters are selected heuristically and the two-class SVM

Fig. 3. Low quality images generated by artificially degrading high quality images.

classifier [14] is optimized through cross-validation on the training dataset. To further evaluate the effectiveness of the proposed method, we train a DNN based on grayscale images on both databases and input the inferred deep feature representation from the DNN last hidden layer to Softmax and SVM classifiers. We also evaluate two state of the art finger-vein quality assessment methods, namely Radon transform [6] and hand-crafted features +SVM [8]. The hand-crafted features we considered for this purpose are gradient, contrast, and information capacity as implemented in [8]. As these two methods are based on grayscale images, we only use them to classify the quality of grayscale finger-vein images. Tables 2 and 3 show the quality assessment performance of these various methods on databases A and B, respectively. Table 2 shows that deep representation is robust for identifying low and high quality finger-vein images on database A. Based on *grayscale* finger-vein image, DNN and DNN + P-SVM identify more than 80 % low and high quality images respectively using deep representation, which significantly outperforms current state of the art approaches based on Hand-craft features. The accuracy of DNN and DNN + P-SVM dramatically increases and reaches an identification rate above 88 % when the deep representation is inferred from *binary* finger-vein images.

The results in Table 3 are consistent with the trends observed in Table 2. When using the binary images to train DNN, the performance for identifying high and low quality image is boosted significantly for database B. The highest quality identification accuracies are 75.02 % and 70.71 % for DNN-Softmax, and 74.98 % and 70.07 % for DNN + P-SVM, which is again significantly higher those of other methods. There experimental results in Tables 2 and 3 consistently show that the proposed deep learning representation is effective in identifying finger-vein image quality and systematically outperform methods that take grayscale finger-vein image as input.

3.6   Effects on the Finger-Vein Verification System

In this section, we evaluate finger-vein verification performance [12] and show how much it can be improved by adopting our quality assessment scheme. Without filtering, $105 \times 11 = 1155$ genuine matches and $105 \times 104/2 = 5460$ impostor matches on database A, and $246 \times 11 = 2706$ genuine and $246 \times 245/2 = 30135$ impostor matches on database B, are available for computing the FRR and FAR (the same protocol as described in Sects. 3.2 and 3.3), respectively. When the templates are selected based on the method described in Sect. 3.2, the EER of the finger-vein verification system on databases A and B are 4.97 % and 1.70 %. Harnessing our DNN-based image quality assessment, we use a filtering mechanism by automatically rejecting genuine matches associated with low quality query images. Figure 4(a) and (b) show the finger-vein EER on the two databases after filtering low quality query images using various approaches at different levels. For example, 1 %, 2 %,...,30 % of test samples

Table 1. the accuracy for identifying low and high quality finger-vein image under different architectures and size of input image

Size of input image	50 × 240		40 × 192		30 × 144		20 × 96		10 × 48	
Architecture	96 × 16 × 79- 128 × 8×39 -192 × 4×19 -192 × 2×9 -4000-4000-2		96 × 12 × 63 -128 × 6×31 -192 × 3×15 -192 × 1×7 -4000-2000-2		96 × 9×47 -128 × 4×23 -192 × 2×11 -1000-1000-2		96 × 6×31 -128 × 3×15 -192 × 1×7 -1000-1000-2		96 × 6×44 -128 × 3×22 -192 × 1×11 -500-500-2	
Different quality image	High	Low	High	Low	High	Low	High	Low	High	Low
Accuracy (%)	85.26	86.36	86.60	85.45	86.12	86.36	88.13	88.18	86.22	82.73

(1155 images in Database A and 2706 images in Database B) are automatically removed, based on quality prediction. It can be seen from Fig. 4 that the EER on databases A and B decreases after filtering. For database A, after rejection of just 10 % of the samples using the proposed approach, the EER significantly decreases to less than 1.5 %, which is better than the best performance achieved by the two state of the art approaches using hand-crafted features. The EER is further reduced after rejecting more than 10 % images. For database B, the approaches based on hand-crafted features decrease the EER to about 1 % while ours reduce the EER to about 0.7 %. From Fig. 4, we can also see that the proposed method significantly reduces the ERR and performs consistently better than the other approaches at all levels.

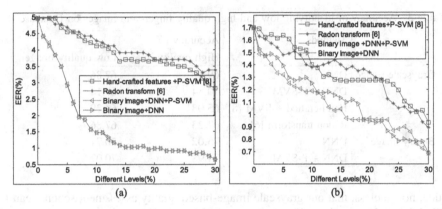

Fig. 4. Finger-vein verification EER on (a) database A and (b) database B after filtering the low quality query finger-vein images at different levels using different approaches

The hand-crafted feature based approaches [6, 8] have shown promising results on their private finger-vein databases, but they do not perform well on the public datasets A and B. This may be explained by the existence of additional factors affecting image quality in the contactless finger-vein image capturing system [12], that are not considered in [6, 8]. This is not surprising as the approaches above are based on a priori human knowledge of quality degrading factors in a finger-vein verification system. Different capturing systems, however, may be associated with different quality degrading factors. Deep feature representations, by contrast, achieve better performance, as DNNs directly learn robust abstract features for finger-vein image representation.

From Tables 2 and 3, we see that using binary instead of grayscale finger-vein images to train DNN and DNN + P-SVM achieve higher identifying accuracy on both databases. This is due to following facts: (1) Compared to grayscale images, binary images are prone to additional quality degradation factors such as intrinsic factors. (2) Finger-vein binary image are more related to EER degradation, since the EER in verification systems is usually computed by matching two binary images. (3) Compared to grayscale images, the patterns in binary images are easier to learn by DNN to assess finger-vein image quality as noise and background information are removed.

Table 2. Accuracy of identifying low and high quality finger-vein images on Database A

	Methods	Accuracy (%)	
		High quality image	Low quality image
Grayscale image	DNN	85.17	85.45
	DNN + P-SVM	83.16	83.64
	Hand-craft + SVM [8]	68.80	66.36
	Radon transform [6]	71.96	67.27
Binary image	DNN	88.13	88.18
	DNN + P-SVM	88.99	88.18

Table 3. Accuracy of identifying low and high quality finger-vein images on Database B

	Methods	Accuracy (%)	
		High quality image	Low quality image
Grayscale image	DNN	69.80	67.86
	DNN + P-SVM	70.54	67.14
	Hand-crafted + SVM [8]	65.08	64.29
	Radon transform [6]	67.23	67.86
Binary image	DNN	75.02	70.71
	DNN + P-SVM	74.98	70.07

Notice, nonetheless, that our grayscale image-based quality assessment scheme can be applied for all finger-vein verification systems, while our binary image-based approach can be used only for verification systems that match binary image pairs.

Overall, all approaches achieve better performance on database A than on database B. This may be explained by the fact that the former is faced with additional quality degradation factors. For example, unlike the work in [15] (Database B) requiring the user to put the finger inside an envelop box and touch the device wall during the whole imaging process, the images in Database A are collected by a contactless and open imaging device. In addition, the average interval time of collecting images in database A is more than two months which is longer than that in database B.

4   Conclusions and Future Work

This paper proposed a novel approach to predict finger-vein image quality using deep representation learning from finger-vein binary inputs. Our image quality definition targets the reduction of EER in biometric authentication systems rather than human perception judgment. Based on this definition, high and low quality images are auto-matically labeled. Our DNN directly learns abstract feature representations from raw pixel images. Experimental results show that learning deep features from *binary* finger-vein images significantly outperforms current state of the art methods in terms of predicting high and low finger-vein images and of reducing the EER accordingly.

Acknowledgments. This work is supported by the Direction générale des Entreprises (DGE) of Ministère de l'économie, de l'industrie et du numérique(Project IDEA4SWIFT 12028), the National Natural Science Foundation of China(Grant No. 61402063), the Natural Science Foundation Project of Chongqing (Grant No. cstc2013kjrc-qnrc40013), and the Scientific Research Foundation of Chongqing Technology and Business University(Grant No. 1352019; Grant No. 2013-56-04).

References

1. Jain, A.K., Ross, A., Pankanti, S.: Biometrics: a tool for information security. IEEE Trans. Inf. Forensics Secur. **1**(2), 125–143 (2006)
2. Grother, P., Tabassi, E.: Performance of biometric quality measures. IEEE Trans. Pattern Anal. Mach. Intell. **29**(4), 531–543 (2007)
3. Alonso-Fernandez, F., Fierrez, J., Ortega-Garcia, J., Gonzalez-Rodriguez, J., Fronthaler, H., Kollreider, K., Bigun, J.: A comparative study of fingerprint image-quality estimation methods. IEEE Trans. Inf. Forensics Secur. **2**(4), 734–743 (2007)
4. Chen, J.S., Deng, Y., Bai, G.C., Su, G.D.: Face image quality assessment based on learning to rank. IEEE Signal Process. Lett. **22**(1), 90–94 (2015)
5. Proenc, H.: Quality assessment of degraded iris images acquired in the visible wavelength. IEEE Trans. Inf. Forensics Secur. **6**(1), 82–95 (2011)
6. Qin, H.F., Li, S., Kot, A.C., Qin, L.: Quality assessment of finger-vein image. In: APSIPA ASC, pp. 1–4 (2012)
7. Nguyen, D.T., Park, Y.H., Shin, K.Y., Park, K.R.: New finger-vein recognition method based on image quality assessment. TIIS **7**(2), 347–365 (2013)
8. Yang, L., Yang, G., Yin, Y., Xiao, R.Y.: Finger vein image quality evaluation using support vector machines. Opt. Eng. **52**(2), 027003 (2013)
9. Peng, J., Li, Q., Niu, X.: A novel finger vein image quality evaluation method based on triangular norm. In: IIH-MSP, pp. 239–242 (2014)
10. Krizhevsky, A., Sutskever, I., Hinton, G.E.: ImageNet classification with deep convolutional neural networks. In: NIPS (2012)
11. Sun, Y., Wang, X., Tang, X.: Deep learning face representation from predicting 10,000 classes. In: CVPR, pp. 1891–1898 (2014)
12. Kumar, A., Zhou, Y.: Human identification using finger images. IEEE Trans. Image Process. **21**(4), 2228–2244 (2012)
13. Platt, J.C.: Probabilistic outputs for support vector machines and comparisons to regularized likelihood methods. In: Advances in Large Margin Classifiers. MIT Press, Cambridge (1999)
14. Chang, C., Lin, C.: LIBSVM. http://www.csie.ntu.edu.tw/cjlin/libsvm
15. Asaari, M.S.M., Suandi, S.A., Rosd, B.A.: Fusion of band limited phase only correlation and width centroid contour distance for finger based biometrics. Expert Syst. Appl. **41**(7), 3367–3382 (2014)

Learning to Predict Where People Look
with Tensor-Based Multi-view Learning

Kitsuchart Pasupa[1]([⊠]) and Sandor Szedmak[2]

[1] Faculty of Information Technology,
King Mongkut's Institute of Technology Ladkrabang,
Bangkok 10520, Thailand
kitsuchart@it.kmitl.ac.th
[2] Institute of Computer Science, University of Innsbruck,
6020 Innsbruck, Austria
sandor.szedmak@uibk.ac.at

Abstract. Eye movements data collection is very expensive and laborious. Moreover, there are usually missing values. Assuming that we are collecting eye movements data on a set of images from different users (views). There is a possibility that we are not able to collect eye movements of all users on all images. One or more views are not represented in the image. We assume that the relationships among the views can be learnt from the complete items. The task is then to reproduce the missing part of the incomplete items from the relationships derived from the complete items and the known part of these items. Using the properties of tensor algebra we show that this problem can be formulated consistently as a regression type learning task. Furthermore, there is a maximum margin based optimisation framework where this problem can be solved in a tractable way. This problem is similar to learning to predict where human look. The proposed algorithm is proved to be more effective than well-known saliency detection techniques.

Keywords: Multi-view learning · Missing data · Tensor algebra · One rank tensor approximation · Maximum margin learning · Eye movements

1 Introduction

It is important to learn where human looks at scenes or images as this can facilitate designers to evaluate their visual design quality. Therefore, many works on saliency modelling have been proposed [1,2]. These methods are investigated on bottom-up visual saliency (i.e. low level image feature) but human gaze does not usually match the map [3]. The reason is because task can influence human gazes. If users are requested to view images without given a particular task, the gaze will automatically direct by low-level image feature. In the case that users are given a clear and specific task, the eye movements will be controlled by the content of images. Consequently, the top-down visual features should be considered [4].

© Springer International Publishing Switzerland 2015
S. Arik et al. (Eds.): ICONIP 2015, Part I, LNCS 9489, pp. 432–441, 2015.
DOI: 10.1007/978-3-319-26532-2_47

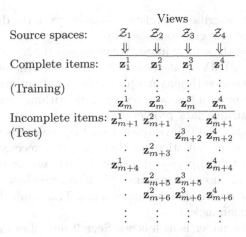

Fig. 1. Graphical representation of the multi-view learning framework.

In real-world scenarios, eye movement data collection is tedious, laborious, and expensive. Moreover, data loss is inevitable as (i) the eye tracker temporarily loses track of the subjects because subject is moving during the experiment, and (ii) participants can fail to respond all the tasks. Consequently, we aim to estimate the missing eye movement data with available data on the same task. It is similar to learning to predict where human look based on their previous eye movement data on other images and other users eye movement data on a considered image. This leads to the learning scenario focused in this paper. It is on a general setting in which multiple views of a problem exists. As previously mentioned, it is not always possible to observe all views in realistic. Therefore, this can be cast as a multi-view learning problem with missing data as shown in Fig. 1.

The goal of the learning task is to estimate the values of the missing views from each sample. This scenario usually occurs. It can be seen that the problem generalises classical supervised learning problems such as regression. Face recognition is one of applications which can be considered in this framework (when some parts–the views–of the faces are unknown, e.g. through occlusion). In developing the learning framework, we make two mild assumptions: (i) There is a reasonable large number of observations (samples) where all views are known. Thus, learning procedure can be made. (ii) In the incomplete observations, at least one view is available but there are no assumptions in which views are missing. However, any prior knowledge about the distribution of the missing data can be exploited to improve the estimation of their values.

In this paper, we introduce a formulation which can be considered as a generalised regression problem whereby the missing values are estimated from the relationship amongst the views as well as the known views. Assuming that the missing views of a sample item can be handled as output $\mathbf{y}$ and the known part as input $\mathbf{x}$, then we have $\mathbf{y} \Leftarrow \mathbf{W}\mathbf{x}$, where $\mathbf{W}$ is a linear operator learned from the complete data which describes the relationships between the different views. The difficulty of this kind of regression arises from the fact that the output and the input can vary among the sample items. We proposed the so-called "Tensor-based Multi-view Learning (TMVL)"algorithm. It is based on the properties of tensor algebra in a conjunction with maximum margin based optimization framework. Thus, it is able to provide a tractable learning algorithm. Tensor decomposition has been used in missing data problem e.g. [5,6], but they are different from our settings. Liu et al. (2013) investigated on low rank tensor technique based on tensor trace norm minimisation problem in image reconstruction [5], while Chen and Grauman (2014) proposed a probabilistic tensor model for inferring human appearance in unseen viewpoints [6]. Here, we show that the proposed method can estimate the missing eye movements. This implied that we can learn to predict where human look.

The outline of the paper is as follows: Sect. 2 describes an algebraic framework, followed by the corresponding optimisation problem in Sect. 3. In Sect. 4, we evaluate our proposed algorithm on real-world dataset. Finally, we conclude our study in Sect. 5.

2   Algebraic Framework

Let us denote $\mathcal{R} = \{1, \ldots, n_R\}$ as the set of indices of the views considered. In our model each of these views has a corresponding linear vector space $\mathcal{Z}_r$, $r \in \mathcal{R}$ over the real numbers, and the dimensions of these spaces are denoted by $\mathrm{Dim}(\mathcal{Z}_r) = d_r$, $r \in \mathcal{R}$. The set $\mathcal{J}_R = \{j_1, \ldots, j_{n_R}\}$ comprises of the indices of the sample examples within each of the space corresponding to the views, enumerating the components of the vectors chosen from the space corresponding to the views. The range of these indices is equal to the dimension of the corresponding spaces.

The sample is chosen out of the direct product of these spaces and each sample item consists of as many vectors as the number of views,

$$
\begin{array}{c}
\text{Views:} \\
\text{Linear vector spaces: } \mathcal{Z}_1 \ldots \mathcal{Z}_{n_R} \\
\Downarrow \ldots \Downarrow \\
\text{Sample:} \qquad \mathbf{z}_i^1 \ldots \mathbf{z}_i^{n_R} \;\; i = 1, \ldots, m.
\end{array}
$$

The product space of the views is given by the tensor product of the spaces, $\mathcal{Z} = \bigotimes_{r \in \mathcal{R}} \mathcal{Z}_r$. This construction forms the algebraic framework of our solution, see [7,8] and the references therein for more details.

If we are given two tensor products of vectors then the following contraction operator $[.,.]$ can be defined over them as

$$
[\textstyle\bigotimes_{q \in \mathcal{Q}} \mathbf{u}^q, \bigotimes_{r \in \mathcal{R}} \mathbf{v}^r] = \prod_{q \in \mathcal{Q} \cap \mathcal{R}} \langle \mathbf{u}^q, \mathbf{v}^q \rangle \bigotimes_{q \in \mathcal{Q} \setminus \mathcal{R}} \mathbf{u}^q \bigotimes_{r \in \mathcal{R} \setminus \mathcal{Q}} \mathbf{v}^r,
$$

where the inner product is computed for all common indices. When the two index sets are coincident then the following well known identity can be used to unfold the inner products of the tensor products as

$$
\langle \textstyle\bigotimes_{q \in \mathcal{Q}} \mathbf{z}_i^r, \bigotimes_{q \in \mathcal{Q}} \mathbf{z}_j^q \rangle = \prod_{q \in \mathcal{Q}} \langle \mathbf{z}_i^q, \mathbf{z}_j^q \rangle.
$$

This identity states that the inner product of tensor products of vectors is equal to the product of the inner product of these vectors.

This interpretation of the indices is compatible with the notations used in tensor algebra, namely, with the so-called "Einstein summation convention". The symbol of the summation $\sum$ is omitted and the summation has to be carried out over all indices which are denoted with the same symbol. Because we use this to denote views and algorithmic iterations, which are not tensor indices, therefore, we choose to handle the summations with a special care by making them explicit via the contraction operator $[.,.]$. Furthermore we assume an orthogonal representation of the indices, and in turn, there is no need to make distinction between covariant and contravariant indices.

In the learning problem, we look for a linear operator, a tensor, which is an element of the dual space of $\mathcal{Z}$, the space of the linear functionals defined on $\mathcal{Z}$, namely

$$\mathbf{W} \in \mathcal{Z}^*, \ \mathbf{W} = [W_{\mathcal{J}_R}] = [W_{j_1,\ldots,j_{n_R}}],$$

where $\mathcal{Z}^*$ denotes the dual space of all possible linear functionals defined on $\mathcal{Z}$.

We can write up Frobenius type inner products between the linear operator $\mathbf{W}$ and the tensor product of a vectors of the views by

$$\langle \mathbf{W}, \otimes_{r \in \mathcal{R}} \mathbf{z}_i^r \rangle_F = \sum_{j_1,\ldots,j_{n_R}} W_{j_1,\ldots,j_{n_R}} \prod_{r \in \mathcal{R}} z_{ij_r}^r.$$

In similar fashion we can compute the Frobenius norm of $\mathbf{W}$ by

$$\|\mathbf{W}\|_F = \left(\sum_{j_1,\ldots,j_{n_R}} W_{j_1,\ldots,j_{n_R}}^2 \right)^{\frac{1}{2}}.$$

In the next step the set of views is partitioned into two arbitrary parts

$$\mathcal{R}_X \subset \mathcal{R}, \ \mathcal{R}_Y = \mathcal{R} \setminus \mathcal{R}_X,$$

where we term the views occurring in $\mathcal{R}_X$ as inputs, and the views in $\mathcal{R}_Y$ can be handled as outputs. Corresponding to this partition, the set of indices belonging to each view has to be split,

$$\mathcal{J}_X \subset \mathcal{J}_R, \ \mathcal{J}_X = \{j_r, \ r \in \mathcal{R}_X\}, \ \mathcal{J}_Y = \mathcal{J}_R \setminus \mathcal{J}_X.$$

Fixing a partition, a contraction of $\mathbf{W}$ can be defined by

$$W_{\mathcal{J}_Y} = W_{\mathcal{J}_R \setminus \mathcal{J}_X} = \mathbf{W} \otimes_{r \in \mathcal{R}_X} \mathbf{z}_i^r \overset{\text{def}}{=} \sum_{j_r \in \mathcal{J}_X} W_{\mathcal{J}_R} \prod_{r \in \mathcal{R}_X} z_{ij_r}^r,$$

where the components of $\mathbf{W}$ are summed over the input views only.

Consequently, the relationship between the inputs and the outputs can be described by the following inner product

$$\langle \otimes_{s \in \mathcal{R}_Y} \mathbf{z}_i^s, \mathbf{W} \otimes_{r \in \mathcal{R}_X} \mathbf{z}_i^r \rangle_F \overset{\text{def}}{=} \sum_{\mathcal{J}_Y} \prod_{s \in \mathcal{R}_Y} z_{ij_s}^s \sum_{\mathcal{J}_X} W_{\mathcal{J}_R} \prod_{r \in \mathcal{R}_X} z_{ij_r}^r,$$

which provides a similarity measure between the outputs and the projection of the inputs by the linear operator $\mathbf{W}$. If the norm of $\mathbf{W}$ is fixed then this inner product takes a greater value if the angle between the direction of outputs and the projection of the inputs is smaller, thus the correlation between them is greater. If both the inputs and the outputs are normalised to the same length then this similarity measure implies small distance as well.

Based on these definitions we can derive a simple but fundamental Lemma:

Lemma 1. *For all partitions* $\mathcal{R}_X, \mathcal{R}_Y$ *of* $\mathcal{R}$ *the inner products*

$$\langle \bigotimes_{s\in\mathcal{R}_Y} \mathbf{z}_i^s, \mathbf{W} \bigotimes_{r\in\mathcal{R}_X} \mathbf{z}_i^r \rangle_F$$

have the same value, namely

$$\langle \mathbf{W}, \bigotimes_{r\in\mathcal{R}} \mathbf{z}_i^r \rangle_F.$$

Proof. We need to unfold only the corresponding definitions of the inner products which give the next chain of equalities

$$\begin{aligned}
\langle \bigotimes_{s\in\mathcal{R}_Y} \mathbf{z}_i^s, \mathbf{W} \bigotimes_{r\in\mathcal{R}_X} \mathbf{z}_i^r \rangle_F &= \sum_{J_Y} \prod_{s\in\mathcal{R}_Y} z_{ij_s}^s \sum_{J_X} W_{J_R} \prod_{r\in\mathcal{R}_X} z_{ij_r}^r \\
&= \sum_{j_1,\ldots,j_{n_R}} W_{j_1,\ldots,j_{n_R}} \prod_{r\in\mathcal{R}} z_{ij_r}^r \\
&= \langle \mathbf{W}, \bigotimes_{r\in\mathcal{R}} \mathbf{z}_i^r \rangle_F.
\end{aligned}$$

This Lemma shows that the value of the inner product of the tensor products is invariant on the partition of the views into inputs and outputs.

3   The Optimisation Problem

To force the high similarity between the projected inputs and the outputs taken out of a fixed partition of the views, a "Support Vector Machine"-style, maximum margin based optimisation problem is formulated for the regression task, see earlier application of the framework [9,10]:

$$\begin{aligned}
\min \quad & \tfrac{1}{2}\|\mathbf{W}\|_F^2 + C\sum_{i=1}^m \xi_i \\
\text{w.r.t.} \quad & \mathbf{W} \text{ tensor} \in \mathcal{Z}^*, \ \boldsymbol{\xi} \in \mathbb{R}^m, \\
\text{s.t.} \quad & \langle \underbrace{\bigotimes_{s\in\mathcal{R}_Y} \mathbf{z}_i^s}_{\text{Outputs}}, \mathbf{W} \underbrace{\bigotimes_{r\in\mathcal{R}_X} \mathbf{z}_i^r}_{\text{Inputs}} \rangle_F \geq 1 - \xi_i, \\
& \xi_i \geq 0, \ i = 1,\ldots,m,
\end{aligned} \qquad (1)$$

where $C > 0$ is penalty constant.

The form is similar to the Support Vector Machine case with two notable exceptions: (i) the outputs are no longer binary labels, $\{-1, +1\}$, but vectors of an arbitrary linear vector space, and (ii) the normal vector of the separating hyperplane is reinterpreted as a linear operator projecting the inputs into the space of the outputs.

The regularisation term in the objective function forces the projections of the inputs and the outputs to be similar with respect to their inner products. When the inputs and the outputs are normalised they live on a sphere in both corresponding spaces then we solve a problem between spaces with structure of a Spherical rather then Euclidean geometry.

Based on Lemma 1 we state the next theorem:

Theorem 1. *For all partitions $\mathcal{R}_X, \mathcal{R}_Y$ of $\mathcal{R}$ the optimisation problem (1) is equivalent to the following one:*

$$
\begin{aligned}
\min \quad & \tfrac{1}{2}\|\mathbf{W}\|_F^2 + C\sum_{i=1}^{m}\xi_i \\
\text{w.r.t.} \quad & \mathbf{W} \text{ tensor} \in \mathcal{Z}^*, \ \boldsymbol{\xi} \in \mathbb{R}^m, \\
\text{s.t.} \quad & \langle \mathbf{W}, \bigotimes_{r\in\mathcal{R}} \mathbf{z}_i^r \rangle_F \geq 1 - \xi_i, \ i = 1, \ldots, m, \\
& \xi_i \geq 0, \ i = 1, \ldots, m.
\end{aligned}
\tag{2}
$$

This equivalence holds true if the inputs and the outputs are partitioned independently in every sample item.

Proof. We can reformulate the constraints following Lemma 1 which proves the statement.

This fact guarantees that the linear operator $\mathbf{W}$ has an universal property, namely *it is independent on how the views are grouped into inputs and outputs*, thus, it consistently characterises the underlying multi-view learning problem.

The seemingly complex problem (2) leads to a simple Lagrangian dual:

$$
\begin{aligned}
\min \quad & \tfrac{1}{2}\boldsymbol{\alpha}'(\mathbf{K}_1 \bullet \cdots \bullet \mathbf{K}_{n_R})\boldsymbol{\alpha} - \mathbf{1}'\boldsymbol{\alpha} \\
\text{w.r.t.} \quad & \boldsymbol{\alpha} \in \mathbb{R}^m \\
\text{s.t.} \quad & \mathbf{0} \leq \boldsymbol{\alpha} \leq C\mathbf{1},
\end{aligned}
\tag{3}
$$

where

$$
(\mathbf{K_r})_{ij} = \langle \mathbf{z}_i^r, \mathbf{z}_j^r \rangle, \ r \in \mathcal{R}, \ i, j \in \{1, \ldots, m\}
\tag{4}
$$

are kernels corresponding to each of the views. The $\bullet$ expresses the element wise product of matrices. This dual can be solved in a straightforward way for very large scale applications[1]. After computing the dual variables the optimum solution for the universal linear operator is given by

$$
\mathbf{W} = \sum_{i=1}^{m} \alpha_i \bigotimes_{r\in\mathcal{R}} \mathbf{z}_i^r.
$$

In test phase, known and unknown views are considered as inputs and outputs, respectively. The output can be estimated in the following way:

$$
\begin{aligned}
\left(\bigotimes_{s\in\mathcal{R}_y} \mathbf{z}^s\right) \sim \mathbf{W} \bigotimes_{r\in\mathcal{R}_X} \mathbf{z}^r &= \sum_{i=1}^{m} \alpha_i [\bigotimes_{r\in\mathcal{R}} \mathbf{z}_i^r, \bigotimes_{r\in\mathcal{R}_X} \mathbf{z}^r] \\
&= \sum_{i=1}^{m} \alpha_i \prod_{r\in\mathcal{R}_X} \langle \mathbf{z}_i^r, \mathbf{z}^r \rangle \bigotimes_{s\in\mathcal{R}_Y} \mathbf{z}_i^s \\
&= \sum_{i=1}^{m} \beta_i \bigotimes_{s\in\mathcal{R}_Y} \mathbf{z}_i^s,
\end{aligned}
\tag{5}
$$

where

$$
\beta_i = \alpha_i \prod_{r\in\mathcal{R}_X} \langle \mathbf{z}_i^r, \mathbf{z}^r \rangle, \ i = 1 \ldots, m.
$$

Thus the prediction is a linear combination of the corresponding outputs.

---

[1] The website of the authors provides an open source implementation to this problem.

4   Performance Evaluations

We evaluate our method, TMVL, on the public available eye tracking dataset [3]. The dataset contains eye tracking data of 15 different users on 1003 images. Each image consists of three-second free viewed trajectory of different user. In order to encourage users to pay attention on the task, users were memory tested at the end of the data collection on 100 images.

In our experiment, only eight users are randomly selected, hence, there are eight views in this setting. Each view is represented by a heatmap for each user. Heatmap quantifies the degree of importance of part of image; the higher probability of the importance of part of image is implied by the higher density of eye movements on that part of the image. A users heatmap is created by convolving a Gaussian kernel on each eye movement point. Here, linear kernel function is used. The model selection is performed by five-fold cross validation based on the area under the receiver operating characteristic. All heatmaps are resized to 50×50 and normalised to unit norm.

We examined on two scenarios on test sets: (i) randomly select $\{1\text{--}7\}$ missing views and (ii) one fixed view is missing on each run. The experiments were run 10 times with different random data splits. We compare our method with well-known saliency map model, i.e. Graph-Based Visual Saliency (GBVS) [2] and Conventional Visual Saliency (CVS) [1]. The performance matrices used in this work are as follows:

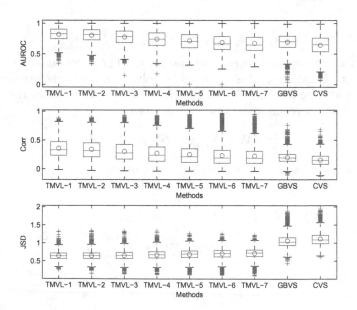

Fig. 2. A comparison of all methods when randomly select $\{1\text{--}7\}$-missing-view cases are considered.

1. Area under the receiver operating characteristic (AUROC) is one of the commonly used performance metric. It is based on the ROC curve which can be computed by varying the threshold of predicted heatmap. A pixel is predicted as a target when its heatmap value is greater than a threshold. It is classified as a background when the value is below the threshold. AUROC ranges from 0 (perfect match) to 1 (complete mismatch).
2. Correlation measures the degree of linear correlation between two maps. It ranges from -1 (perfect correlation but in opposite direction) to $+1$ (perfect correlation). Zero indicates no correlation between two maps.
3. Jensen-Shannon divergence (JSD) is used to identify the dissimilarity of two distributions. It is based on Kullback-Leibler divergence (KLD) which can capture a certain kind of non-linear, entropy type and dependency. JSD is symmetric while KLD is not [11]. Square root of JSD yields a matric properties. The more similar two objects are, the smaller the value of the JSD is, and vice versa.

4.1 Randomly Select Missing Views

According to Fig. 2, when the number of missing view increases, AUROC and correlation decrease. On the other hand, JSD increases when there is an increasing number of missing view. The performances of TMVL on correlation and JSD are better than GBVS and CVS performances in all cases. In the case of AUROC, TMVL can outperforms CVS in all cases but is only better than GBVS in {1–5}-missing-view cases. TMVL is comparable to GBVS when six views are missing but is worse than GBVS in the case of seven missing views.

Figure 3 shows an example of prediction by GBVS, CVS, and our proposed algorithm when two views are missing. It can be seen that GBVS and CVS fail to predict where users look at. Both algorithm distributed their attention on woman's arm and windows while users focus on her face. Clearly, TMVL is more effective than GBVS and CVS on all three performance matrices as shown in Table 1.

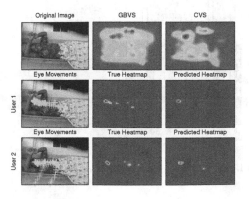

Fig. 3. An example of 8-view problem with 2-view missing.

Table 1. Performance matrices of all methods on figure 3.

User	Method	AUROC	Corr	JSD
1	TMVL	**0.8065**	**0.7160**	**0.3726**
	GBVS	0.6543	0.0783	1.2755
	ITTI	0.7358	0.0792	1.2454
2	TMVL	**0.7812**	**0.6900**	**0.4064**
	GBVS	0.6707	0.0985	1.1863
	ITTI	0.6983	0.0981	1.1627

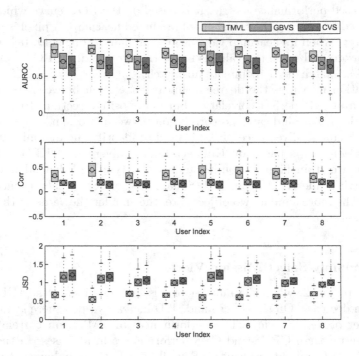

Fig. 4. A comparison of all methods when only one fixed view/user are missing.

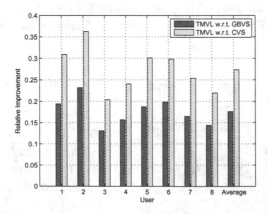

Fig. 5. Relative improvement of TMVL on AUROC w.r.t. GBVS and CVS.

4.2 Fixing a Missing View

In this scenario, we want to predict where a user looks at a (test) set of images based on the other seven's eye movements on test set. The overall picture is much the same as previous scenario according to Fig. 4. TMVL is still the best contender in all users, followed by GBVS and CVS, respectively.

TMVL displays improvement on AUROC with respect to GBV and CVS in all users by 17.6 % and 27.3 % on average, respectively, as shown in Fig. 5. Our proposed algorithm achieves the highest improvement with respect to both baselines in user 2. This means that other users' eye movements/behaviours is very useful to user 2. However, the lowest improvements are found in user 3. This indicates that user adaptation might be useful as the performance is improved but have not yet attained high performance as on user 3.

5 Conclusion

In this paper we introduced a missing value prediction schema built upon maximum margin based learning and invariances of the tensor algebra. The proposed algorithm was examined on eye movement dataset to identify where human look. We have demonstrated that it can perform better than well-known saliency detection techniques. Some initial results show that user adaptation might be useful, thus user information should be investigated in the future.

References

1. Itti, L., Koch, C., Niebur, E.: A model of saliency-based visual attention for rapid scene analysis. IEEE Trans. Pattern Anal. Mach. Intell. **20**, 1254–1259 (1998)
2. Harel, J., Koch, C., Perona, P.: Graph-based visual saliency. In: Advances in Neural Information Processing Systems, pp. 545–552 (2006)
3. Judd, T., Ehinger, K., Durand, F., Torralba, A.: Learning to predict where humans look. In: IEEE 12th International Conference on Computer Vision, pp. 2106–2113 (2009)
4. Henderson, J.M., Brockmole, J.R., Castelhano, M.S., Mack, M.: Visual saliency does not account for eye movements during visual search in real-world scenes. In: Eye Movements: A Window on Mind and Brain, pp. 537–562 (2007)
5. Liu, J., Musialski, P., Wonka, P., Ye, J.: Tensor completion for estimating missing values in visual data. IEEE Trans. Pattern Anal. Mach. Intell. **35**, 208–220 (2013)
6. Chen, C.Y., Grauman, K.: Inferring unseen views of people. In: IEEE Conference on Computer Vision and Pattern Recognition, pp. 2011–2018 (2014)
7. Itskov, M.: Tensor Algebra and Tensor Analysis for Engineers With Applications to Continuum Mechanics. 2nd edn. Springer, Heidelberg (2009)
8. Synge, J., Schild, A.: Tensor Calculus. Dover, New York (1978)
9. Astikainen, K., Holm, L., Pitkänen, E., Szedmak, S., Rousu, J.: Towards structured output prediction of enzyme function. In: BMC Proceedings, vol. 2(Suppl 4:S2) (2008)
10. Szedmak, S., De Bie, T., Hardoon, D.R.: A metamorphosis of canonical correlation analysis into multivariate maximum margin learning. In: The 15th European Symposium on Artificial Neural Networks, pp. 211–216 (2007)
11. Briët, J., Harremoës, P.: Properties of classical and quantum Jensen-Shannon divergence. Phys. Rev. A **79**, 052311 (2009)

Classification of the Scripts in Medieval Documents from Balkan Region by Run-Length Texture Analysis

Darko Brodić[1]([✉]), Alessia Amelio[2], and Zoran N. Milivojević[3]

[1] Technical Faculty in Bor, University of Belgrade, V.J. 12, 19210 Bor, Serbia
dbrodic@tf.bor.ac.rs
[2] Institute for High Performance Computing and Networking,
National Research Council of Italy, CNR-ICAR,
Via P. Bucci 41C, 87036 Rende, CS, Italy
amelio@icar.cnr.it
[3] College of Applied Technical Sciences,
Aleksandra Medvedeva 20, 18000 Niš, Serbia
zoran.milivojevic@vtsnis.edu.rs

Abstract. The paper presents a script classification method of the medieval documents originated from the Balkan region. It consists in a multi-step procedure which includes the text mapping according to typographical features, creation of equivalent image patterns, run-length pattern analysis in order to establish a feature vector and state-of-the art classification method Genetic Algorithms Image Clustering for Document Analysis (GA-ICDA) which successfully disseminates the documents written in different scripts. The proposed method is evaluated on custom oriented document databases, which include the handprinted or printed documents written in old Cyrillic, angular and round Glagolitic, ancient Latin and Greek scripts. The experiment demonstrates very good results.

Keywords: Classification · Historical document · Optical character recognition · Pattern recognition · Run-length statistics · Script identification

1 Introduction

Script recognition represents an important stage in document image analysis [1]. The techniques for the script recognition are usually divided into global and local methods. Global methods process statistically the large image areas. In contrast, the local methods treat smaller image areas like characters, words or lines. Then, the black pixel runs analysis is performed [2].

Medieval documents from the Balkan region are especially interesting due to the variety of scripts. Pre to the Slavs population of the Balkan region, the church books were written in Greek and ancient Latin. However, the creation of new

© Springer International Publishing Switzerland 2015
S. Arik et al. (Eds.): ICONIP 2015, Part I, LNCS 9489, pp. 442–450, 2015.
DOI: 10.1007/978-3-319-26532-2_48

alphabet(s), i.e. Glagolitic and old Cyrillic scripts accustomed to the Church Slavonic language enabled the translation of church books. It contributed to the variety of handprinted and printed books in the Church Slavonic language. Hence, the complexity of the problem to be solved raised due to the presence of five different scripts. The main purpose of the paper is to propose an algorithm for the classification of the scripts present in the medieval documents from the Balkan region. A similar approach was proposed in Refs. [3,4] in order to differentiate the South Slavic scripts. However, we extend the method by introducing the run-length statistical pattern analysis and enlarging the custom oriented document database, which includes all five scripts. Furthermore, it should be noticed that the documents include a small number of characters, i.e. up to 150, which makes the problem more complex to solve. Until now, there was no attempt to solve such a script recognition problem containing five scripts from medieval documents originated from the Balkans. Although, these old historical documents are not in good shape, the algorithm is partly prone to errors, which leads to good script classification results.

The paper is organized as follows. Section 2 describes the proposed algorithm. Section 3 illustrates the experiments, presents the obtained results and discusses them. Section 4 gives the conclusions.

2   The Algorithm

2.1   The Proposed Algorithm

The proposed algorithm can discriminate different scripts present in medieval hand-printed or printed documents of the Balkan region. At the beginning, we suppose that the skew and slant corrections were previously processed. The algorithm consists of the following stages: (i) horizontal projection profile of the text in order to segment text lines, (ii) extraction of the blob by bounding boxing, (iii) distribution of the blob heights and its center point, (iv) classification and mapping according to typographical features into image pattern, (v) feature extraction of the image pattern according to the run-length texture analysis, (vi) classification of the extracted features by specialized GA-ICDA algorithm.

2.2   Mapping According to Typographical Features

In the first stage of the algorithm, the initial text is converted into the image pattern according to its typographical characteristics. Figure 1 illustrates the samples of the medieval documents, which are written in different scripts.

The aforementioned algorithm's stage includes different steps. First, the horizontal projection profile is applied to the text in order to segment the text lines. It is used to establish a reference (central) line of each text line. Then, each blob is framed by the bounding box. Accordingly, the distribution of the blob heights and its center point can be extracted. These features are mandatory in order to make a classification according to typographical features. Figure 2 illustrates these steps of the algorithm.

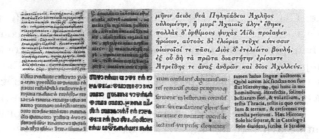

Fig. 1. The excerpts from the medieval documents written in different scripts.

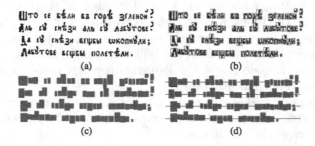

Fig. 2. Pre-mapping steps: (a) initial text, (b) bounding box extraction, (c) bounding box filling, and (d) reference line with the center point of the bounding box.

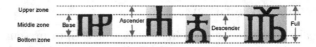

Fig. 3. Classification of the letters according to typographical features in the text line.

Classification with respect to typographical features means the division of all the blobs, i.e. letters into the following groups [5]: (i) base letter, (ii) ascender letter, (iii) descendent letter, and (iv) full letter. Figure 3 illustrates the classification according to typographical features.

It can be noticed that there exist three different letter heights. The base letters have the smallest height, while the full letters have the largest height. Furthermore, the ascender and descendent letters have similar heights in between the base and full letters. Still, the ascender and descendent letters can be discriminated by different center positions. Accordingly, ascender letter has the center point above the reference line, while the descendent letter has the center point below the reference line. The algorithm maps each letter according to the typographical features into an equivalent image pattern. In this way, the following mapping is carried out: base letter to 0, ascender letter to 1, descendent letter to 2, and full letter to 3 [3]. Figure 4 illustrates the procedure of mapping initial text into a gray-level image pattern.

300 00 0100 00 0021 2000001
300 01 00120 000 01 00010000
30 01 00120 002000 00000100
30010000 002000 00000100

Fig. 4. Mapping steps: coding text according to typographical features (top left), equivalent image pattern (top right), and full 1-D pattern of the initial text (bottom).

It should be pointed out that each text line is continued in the next line up to the end of the text. Also, each of four different numbers corresponds to a different level of gray in the image. Hence, the image is given as long 1-D image with four gray levels.

2.3   Feature Extraction with the Run-Length Statistics

Texture is a measure of the intensity variation in the image surface. Hence, it is suitable for information extraction, which can be used to quantify the properties like image smoothness, coarseness, and regularity. Accordingly, the texture can be used to calculate statistical measures of the image and make discrimination between images.

Run-length based statistical analysis can be used to extract texture features in order to quantify the image (di)similarity. A run is a set of consecutive pixels characterized with the same gray-level intensity in a specific direction of the texture. Accordingly, various types of textures characterize different runs. To define a run-length statistic element, we start with a gray-level image G. It is determined by X rows, Y columns and M levels of gray. As a first step, the extraction of run-length matrix $p(i, j)$ is carried out. The matrix $p(i, j)$ is determined by counting the occurrence of runs for each gray-level and length. The number of rows represents the number of gray levels $i = 1, \ldots, M$, while the number of columns corresponds to the number of run lengths $j = 1, \ldots, N$. In this way, a set of consecutive pixels with identical gray levels constitutes a gray-level run. The aforementioned matrix is a starting point for the extraction of the following eleven run-length features: (i) Short run emphasis (SRE), (ii) Long run emphasis (LRE), (iii) gray-level non-uniformity (GLN), (iv) Run length non-uniformity (RLN), and (v) Run percentage (RP) [6], (vi) Low gray-level run emphasis (LGRE), (vii) High gray-level run emphasis (HGRE) [7], (viii) Short run low gray-level emphasis (SRLGE), (ix) Short run high gray-level emphasis (SRHGE), (x) Long run Low gray-level emphasis (LRLGE), and (xi) Long run high gray-level emphasis (LRHGE) [8]. Table 1 shows the definition of run-length texture features.

Table 1. Eleven run-length texture feature definition.

Run length texture features	Equation
Short run emphasis (SRE)	$SRE = \dfrac{1}{n_r}\sum\limits_{i=1}^{M}\sum\limits_{j=1}^{N}\dfrac{p(i,j)}{j^2}$
Long run emphasis (LRE)	$LRE = \dfrac{1}{n_r}\sum\limits_{i=1}^{M}\sum\limits_{j=1}^{N}p(i,j)\cdot j^2$
Gray-level non-uniformity (GLN)	$GLN = \dfrac{1}{n_r}\sum\limits_{i=1}^{M}\left(\sum\limits_{j=1}^{N}p(i,j)\right)^2$
Run length non-uniformity (RLN)	$RLN = \dfrac{1}{n_r}\sum\limits_{j=1}^{N}\left(\sum\limits_{i=1}^{M}p(i,j)\right)^2$
Run percentage (RP)	$RP = \dfrac{n_r}{n_p}$
Low gray-level run emphasis (LGRE)	$LGRE = \dfrac{1}{n_r}\sum\limits_{i=1}^{M}\sum\limits_{j=1}^{N}\dfrac{p(i,j)}{i^2}$
High gray-level run emphasis (HGRE)	$HGRE = \dfrac{1}{n_r}\sum\limits_{i=1}^{M}\sum\limits_{j=1}^{N}p(i,j)\cdot i^2$
Short run low gray-level emphasis (SRLGE)	$SRLGE = \dfrac{1}{n_r}\sum\limits_{i=1}^{M}\sum\limits_{j=1}^{N}\dfrac{p(i,j)}{i^2\cdot j^2}$
Short run high gray-level emphasis (SRHGE)	$SRHGE = \dfrac{1}{n_r}\sum\limits_{i=1}^{M}\sum\limits_{j=1}^{N}p(i,j)\cdot i^2\,/\,j^2$
Long run Low gray-level emphasis (LRLGE)	$LRLGE = \dfrac{1}{n_r}\sum\limits_{i=1}^{M}\sum\limits_{j=1}^{N}p(i,j)\cdot j^2\,/\,i^2$
Long run high gray-level emphasis (LRHGE)	$LRHGE = \dfrac{1}{n_r}\sum\limits_{i=1}^{M}\sum\limits_{j=1}^{N}p(i,j)\cdot i^2\cdot j^2$

2.4   Feature Extraction by GA-ICDA

To discriminate eleven dimensional run-length feature vectors obtained from medieval documents written in different scripts, we adopt a framework called Genetic Algorithms Image Clustering for Document Analysis (GA-ICDA), previously discussed in [9]. It represents an extension of the Genetic Algorithms Image Clustering (GA-IC) method, proposed in [10] and performs clustering of documents written in different scripts. Here we recall it briefly. GA-ICDA is a bottom-up evolutionary method modelling the document database as a weighted graph $G = (V, E, W)$. The nodes V are the documents and the edges E connect the nodes to each other, with associated weights W representing the similarity degree among the nodes. Each node $v \in V$ is connected only to a subset of its h-nearest neighbor nodes $nn_v^h = \{nn_v^h(1), \ldots, nn_v^h(k)\}$, representing the documents most similar to that node v, in terms of the L_1 norm of the corresponding feature vectors. h is a parameter related to the size of the neighborhood. Now, let f be a node ordering determined by a one-to-one association from nodes to integer labels $f : V \to \{1, 2, \ldots, n\}$, where n is the number of nodes. For each node

$v \in V$ with associated label $f(v)$, the difference is computed between $f(v)$ and the labels F of the nodes in nn_v^h. Then, for each node v, only the edges between v and the nodes in nn_v^h are considered, whose label difference $|f(v) - f(nn_v^h(j))|$ is less than or equal to a given value T. This concept is derived from the Matrix Bandwidth definition [11]. In the first step, the node clustering is performed on the graph G by a genetic algorithm. After that, a merging procedure is repeated on the clusters found from the algorithm, until a fixed cluster number is determined. Specifically, for each cluster c_x, another cluster c_y is selected with the minimum distance from it. The distance between c_x and c_y is evaluated as the L_1 norm computed between the two farthest document feature vectors, one for each cluster. Then, the pair of the nearest clusters (c_x, c_y) is merged into a single cluster.

3 Experimentation, Results and Discussion

The proposed framework is evaluated on two custom-oriented databases of short documents (labels) written in old Cyrillic, angular Glagolitic, round Glagolitic, ancient Latin and Greek scripts. Each of the documents has less than 150 characters. The first database is composed of 26 text excerpts from documents, where 16 are in Cyrillic script, 5 are in ancient Latin and 5 are in ancient Greek. The second database is composed of 15 text excerpts from documents, where 5 are in Cyrillic script, 5 are in angular Glagolitic and 5 are in round Glagolitic.

The experiment consists in evaluating the ability of the proposed framework in correctly discriminating among different kinds of scripts. A first test employes the GA-ICDA classifier on the first database, for clustering of the eleven dimensional coded run-length feature vectors of labels in Cyrillic, ancient Latin and Greek scripts. Then, in order to confirm the efficacy of our method, a second more complex test is provided where the second database of feature vectors of labels in Cyrillic, angular Glagolitic and round Glagolitic scripts is required to be clustered by GA-ICDA. The task consists in distinguishing the labels written in one kind of script (i.e. Cyrillic) from the labels written in the other kinds of scripts (i.e. angular or round Glagolitic).

The classification task presents some difficulties in this context and it is a real challenge, because the analyzed labels are hand-engraved in stone and hand-printed on paper. Furthermore, in the second test, the proposed angular and round Glagolitic scripts are very similar to each other and very difficult to discriminate. This is mainly because they are both from the Glagolitic script, but belong to different historical periods and are spread in different regions.

A trial and error procedure is employed for tuning the parameters of the GA-ICDA classifier on benchmark documents. Consequently, the h value of the node neighborhood has been fixed to 10 and the T value to 5 in the first test, and the h value of the node neighborhood to 7 and the T value to 4, in the second test.

Our framework is compared with other four unsupervised classifiers, well-known in text document classification, for which the same run-length feature

Table 2. Classification results obtained from the first database of Cyrillic, ancient Latin and ancient Greek scripts. n_c is the number of detected clusters.

	F-Measure	NMI	ARI	nc
GA-ICDA	**1.0000 (0.0000)**	**1.0000 (0.0000)**	**1.0000 (0.0000)**	3
EM	0.6908 (0.1407)	0.5153 (0.2208)	0.3476 (0.3109)	3
K-Means	0.7532 (0.0202)	0.6561 (0.0722)	0.4669 (0.0779)	3
Hierarchical	0.8519 (0.0000)	0.7645 (0.0000)	0.7512 (0.0000)	3
SOM	0.5922 (0.1385)	0.3837 (0.1687)	0.1452 (0.2326)	3

vector representation of labels is used. They are Expectation Maximization (EM), K-Means, Single Linkage Hierarchical clustering and Self-Organizing Map (SOM) [12–14]. F-Measure, Normalized Mutual Information (NMI) and Adjusted Rand Index (ARI) have been adopted as the performance indexes for the classifiers evaluation [15,16]. Experiments have been executed on a Desktop computer quad-core 2.3 GHz with 4 Gbyte of RAM and Linux operative system. Algorithms have been run 100 times on each dataset and the average values together with the standard deviation (in parenthesis) of the performance indexes have been calculated. Table 2 shows the results of the first test for our framework compared with EM, K-Means, hierarchical and SOM. It is interesting to note as GA-ICDA outperforms all the other classifiers, by also reaching the maximum value of 1 for F-Measure, NMI and ARI (in bold). Furthermore, the standard deviation is always 0, demonstrating the stability of the procedure. Hierarchical clustering obtains good results but not comparable with GA-ICDA. In fact, it reaches an F-Measure value of 0.8519, an NMI value of 0.7645 and an ARI value of 0.7512. Also, it outperforms EM, K-Means and SOM on this database, for discrimination of Cyrillic, ancient Latin and Greek. Table 3 reports the results of the second test for our framework compared with EM, K-Means, hierarchical and SOM. It is worth to observe that, although the task is much more complex than before, GA-ICDA always obtains the best classification, by reaching the 1 value for all the performance indexes (in bold). Also in this case, the standard deviation is 0, confirming again the very good results of our procedure. On the contrary, the other classifiers perform poorly. Among them, the best classification result is provided by SOM, with an F-Measure value of 0.6821, an NMI value of 0.5960, an ARI value of 0.3441 and high values of standard deviation. On the other hand, hierarchical clustering is not sufficiently robust to a more difficult

Table 3. Classification results obtained from the second database of Cyrillic, angular Glagolitic and round Glagolitic scripts. n_c is the number of detected clusters.

	F-Measure	NMI	ARI	nc
GA-ICDA	**1.0000 (0.0000)**	**1.0000 (0.0000)**	**1.0000 (0.0000)**	3
EM	0.6707 (0.0939)	0.4531 (0.1601)	0.2695 (0.1748)	3
K-Means	0.6386 (0.0688)	0.3710 (0.1225)	0.1458 (0.0998)	3
Hierarchical	0.5826 (0.0000)	0.3198 (0.0000)	0.0973 (0.0000)	3
SOM	0.6821 (0.0270)	0.5960 (0.1097)	0.3441 (0.1028)	3

scenario. In fact, it performs more poorly than before, with an F-Measure value of 0.5826, an NMI value of 0.3198 and a very low ARI value of 0.0973. On the contrary, we observe as our framework maintains high performance values independently from the proposed database, also when a more difficult classification task is proposed.

4   Conclusion

The paper proposed a method for the script dissemination on the example of the old medieval documents from the Balkan region, which were written in old Cyrillic, angular and round Glagolitic, ancient Latin and Greek scripts. The algorithm aggregated the text coding procedure in order to create image patterns, run-length analysis of image patterns which established 11 elements feature vectors, and classification of the feature vectors by Genetic Algorithms Image Clustering for Document Analysis (GA-ICDA) in order to disseminate documents written in different scripts. The proposed algorithm was evaluated on two custom-oriented databases which include short documents written in old Cyrillic, angular and round Glagolitic, ancient Latin and Greek scripts. The experiments gave very good results.

Acknowledgments. This work was partially supported by the Grant of the Ministry of Science of the Republic Serbia within the project TR33037.

References

1. Ghosh, D., Dube, T., Shivaprasad, A.: Script recognition - a review. IEEE Trans. Pattern Anal. Mach. Intell. **32**(12), 2142–2161 (2010)
2. Joshi, G.D., Garg, S., Sivaswamy, J.: A generalised framework for script identification. Int. J. Doc. Anal. Recogn. **10**(2), 55–68 (2007)
3. Brodić, D., Milivojević, Z.N., Maluckov, Č.A.: An approach to the script discrimination in the Slavic documents. Soft Comput. **19**(9), 2655–2665 (2015). doi:10.1007/s00500-014-1435-1
4. Brodić, D., Maluckov, Č.A., Milivojević, Z.N., Draganov, I.R.: Differentiation of the script using adjacent local binary patterns. In: Agre, G., Hitzler, P., Krisnadhi, A.A., Kuznetsov, S.O. (eds.) AIMSA 2014. LNCS, vol. 8722, pp. 162–169. Springer, Heidelberg (2014)
5. Zramdini, A.W., Ingold, R.: Optical font recognition using typographical features. IEEE Trans. Pattern Anal. Mach. Intell. **20**(8), 877–882 (1998)
6. Galloway, M.M.: Texture analysis using gray level run lengths. Comput. Graph. Image Process. **4**(2), 172–179 (1975)
7. Chu, A., Sehgal, C.M., Greenleaf, J.F.: Use of gray value distribution of run lengths for texture analysis. Pattern Recogn. Lett. **11**(6), 415–419 (1990)
8. Dasarathy, B.R., Holder, E.B.: Image characterizations based on joint gray-level run-length distributions. Pattern Recogn. Lett. **12**(8), 497–502 (1991)
9. Brodić, D., Amelio, A., Milivojević, Z.N.: Characterization and distinction between closely related south Slavic languages on the example of Serbian and Croatian. In: Azzopardi, G., Petkov, N., Yamagiwa, S. (eds.) CAIP 2015. LNCS, vol. 9256, pp. 654–666. Springer, Heidelberg (2015)

10. Amelio, A., Pizzuti, C.: A new evolutionary-based clustering framework for image databases. In: Elmoataz, A., Lezoray, O., Nouboud, F., Mammass, D. (eds.) ICISP 2014. LNCS, vol. 8509, pp. 322–331. Springer, Heidelberg (2014)
11. Marti, R., Laguna, M., Glover, F., Campos, V.: Reducing the bandwidth of a sparse matrix with tabu search. Eur. J. Oper. Res. **135**(2), 450–280 (2001)
12. Marinai, S., Marino, E., Soda, G.: Self-organizing maps for clustering in document image analysis, machine learning in document analysis and recognition. In: Marinai, S., Fujisawa, H. (eds.) Machine Learning in Document Analysis and Recognition. LNCS (SCI), vol. 90, pp. 193–219. Springer, Heidelberg (2008)
13. Pu, Y., Shi, J., Guo, L.: A hierarchical method for clustering binary text image. In: Yuan, Y., Wu, X., Lu, Y. (eds.) ISCTCS 2012. CCIS, vol. 320, pp. 388–396. Springer, Heidelberg (2013)
14. Rigutini, L., Maggini, M.: A semi-supervised document clustering algorithm based on EM. In: Proceedings of the International Conference on 2005 IEEE/WIC/ACM on Web Intelligence, pp. 200–206 (2005)
15. Hu, X., Yoo, I.: A comprehensive comparison study of document clustering for a biomedical digital library medline. In: Proceedings of the 6th ACM/IEEE-CS Joint Conference on Digital Libraries, pp. 220–229 (2006)
16. De Vargas, R.R., Bedregal, B.R.C.: A way to obtain the quality of a partition by adjusted rand index. In: Workshop-School on Theoretical Computer Science, pp. 67–71 (2013)

Accelerating Artificial Bee Colony Algorithm for Global Optimization

Xinyu Zhou[✉], Mingwen Wang, and Jianyi Wan

School of Computer and Information Engineering, Jiangxi Normal University,
Nanchang 330022, China
xyzhou@whu.edu.cn

Abstract. As an efficient optimization technique, artificial bee colony (ABC) algorithm has attracted a lot of attention for its good performance. However, ABC is good at exploration but poor at exploitation for its solution search equation. Thus, how to enhance the exploitation becomes an active research trend. In this paper, we propose a trigonometric search equation in which a hypergeometric triangle is formed to generate offspring. Additionally, the orthogonal learning strategy is integrated into the scout bee phase for generating new food source. Experiments are conducted on 23 well-known benchmark functions, and the results show that our approach has promising performance.

Keywords: Artificial bee colony · Exploitative capability · Trigonometric search equation · Orthogonal learning

1 Introduction

Artificial bee colony (ABC) algorithm, proposed by Karaboga in 2005, is one of the most popular evolutionary algorithms (EAs) for solving optimization problems [1,2]. A recent comparative study has shown that the performance of ABC is competitive to that of some other popular EAs [3], such as particle swarm optimization (PSO) and differential evolution (DE). In solving complex optimization problems, however, ABC has an insufficiency regarding its solution search equation which is good at exploration but poor at exploitation [4]. To overcome this drawback, some improved ABC variants have been proposed in the last few years. For example, inspired by PSO, Zhu *et al.* [5] proposed a gbest-guided ABC (GABC) algorithm in which the global best individual is incorporated into the solution search equation. Based on the idea of utilizing the global best individual, Gao *et al.* [6] designed an ABC/best/1 search equation in their modified ABC (MABC) algorithm. Very recently, Wang *et al.* [16] proposed a multi-strategy ensemble ABC (MEABC) algorithm. In MEABC, three solution search strategies with different characteristics compete to produce offspring for a tradeoff between exploration and exploitation.

Obviously, from the above representative ABC variants, it's not difficult to realize that the global best individual (solution) can be utilized to enhance the

© Springer International Publishing Switzerland 2015
S. Arik et al. (Eds.): ICONIP 2015, Part I, LNCS 9489, pp. 451–458, 2015.
DOI: 10.1007/978-3-319-26532-2_49

exploitation of ABC. Following this active research trend, in this paper, we propose a trigonometric search equation to replace the original solution search equation. In the trigonometric search equation, the global best individual and other two randomly selected individuals are included to form a hypergeometric triangle in the search space. By applying this search equation, the offspring can be biased strongly in the direction where the global best individual is, which benefits the exploitation of ABC. This proposed search equation is inspired by the concept of trigonometric mutation DE (TDE) [7]. Furthermore, the orthogonal learning (OL) strategy is employed to generate a new food source for the scout bee. In ABC, both of the abandoned food source and the global best individual may contain useful information (experience), which can be used for the search of new food source. Hence, in the OL strategy, these two solutions are considered as the learning exemplars, and the orthogonal experimental design (OED) method is used to construct an efficient combination of them as a new food source. Our approach is tested on 23 well-known benchmark functions, and it is compared with other three state-of-the-art ABC variants. The comparative results show that our approach has superior performance.

2   Basic ABC Algorithm

The ABC algorithm is a swarm intelligence-based algorithm that simulates the intelligent foraging behavior of a honeybee swarm. The swarm consists of three different kinds of bees: employed bees, onlooker bees and scout bees. Accordingly, the search process of ABC can be divided into the following three phases. Similar to other EAs, at first, ABC starts with an initial population of SN randomly generated food sources. Each food source $X_i = (x_{i,1}, x_{i,2}, \cdots, x_{i,D})$ represents a candidate solution, and D denotes the dimension size. After initialization, these three phases can be described as follows [8].

(1) Employed bee phase
    In this phase, each employed bee generates a new food source $V_i = (v_{i,1}, v_{i,2}, \cdots, v_{i,D})$ in the neighborhood of its parent position $X_i = (x_{i,1}, x_{i,2}, \cdots, x_{i,D})$ by using the following solution search equation.

$$v_{i,j} = x_{i,j} + \phi_{i,j} \cdot (x_{i,j} - x_{k,j}) \tag{1}$$

    where $k \in \{1, 2, \cdots, SN\}$ and $j \in \{1, 2, \cdots, D\}$ are randomly chosen indexes, k has to be different from i. $\phi_{i,j}$ is a random number in the range $[-1, 1]$. If the new food source V_i is better than its parent X_i, then X_i is replaced with V_i.
(2) Onlooker bee phase
    After the employed bees finish their search work, the onlooker bees would continue to select part of the food sources to exploit by using the same solution search equation listed in Eq. (1). The select probability p_i depends

on the nectar amounts of a food source, the following Eq. (2) is usually used to calculate the probability.

$$p_i = \frac{f(X_i)}{\sum_{j=1}^{SN} f(X_j)} \tag{2}$$

where $f(X_i)$ is the fitness value of the ith food source. As in the case of the employed bees, the greedy selection method is also employed to retain a better one from the old food source and the new food source.

(3) Scout bee phase

If a food source cannot be further improved for at least *limit* times, it is considered to be exhausted. In ABC, *limit* is the only one single specific control parameter needed to be tuned. For the scout bee, the following Eq. (3) is used to generate a new food source to replace the abandoned one.

$$x_{i,j} = a_j + rand_j \cdot (b_j - a_j) \tag{3}$$

where $[a_j, b_j]$ is the boundary constraint for the jth variable, and $rand_j \in [0,1]$ is a random number.

3   Our Approach

3.1   Trigonometric Search Equation in the Onlooker Bee Phase

In ABC, the solution search equation (see Eq. (1)) is used in both the employed bee and onlooker bee phases, thus the performance of ABC mainly depends on it. However, some pervious works have pointed out that this search equation is good at exploration but poor at exploitation [5,9,10]. An active research trend of solving this demerit is to utilize the globe best individual, such as GABC [5] and MABC [6]. Although the utilization of the globe best individual can indeed enhance the exploitation, it may also run the risk of being too greedy or getting be trapped by local optimum. So how to design a mechanism of reasonably utilizing the globe best individual plays an important role.

In TDE, Fan *et al.* [7] proposed a trigonometric mutation to speed up the convergence rate of DE. In this operator, three individuals are randomly selected as vertexes to form a hypergeometric triangle in the search space at first, and then the center point of the hypergeometric triangle is perturbed to produce a trial solution. The perturbation is made up with three weighted differentials among these three individuals. This mutation operator can be expressed by the following equation.

$$V_i = (X_{r1} + X_{r2} + X_{r3})/3 + (p_2 - p_1)(X_{r1} - X_{r2}) \\ + (p_3 - p_2)(X_{r2} - X_{r3}) + (p_1 - p_3)(X_{r3} - X_{r1}) \tag{4}$$

where $p_1 = |f(X_{r1})|/p$, $p_2 = |f(X_{r2})|/p$, $p_3 = |f(X_{r3})|/p$, $p = |f(X_{r1})| + |f(X_{r2})| + |f(X_{r3})|$, and $f(\cdot)$ is the fitness function. As seen, the center point can move along in the direction from a vertex with a higher fitness value towards another with a lower fitness value. It implies that the new trial solution can be located in the relatively good search landscape.

Motivated by these observations, we propose a trigonometric search equation in the onlooker bee phase. In contrast with the mutation operator of TDE, the proposed trigonometric search equation has a minor difference, i.e., the global best individual is included to substitute one of the three randomly selected individuals. Accordingly the trigonometric search equation is modified as follows.

$$
\begin{aligned}
V_i = (X_b + X_{r1} + X_{r2})/3 + (p_1 - p_b)(X_b - X_{r1}) \\
+ (p_2 - p_1)(X_{r1} - X_{r2}) + (p_b - p_2)(X_{r2} - X_b)
\end{aligned}
\tag{5}
$$

where X_b is the global best individual, $p_b = |f(X_b)|/p$, $p_1 = |f(X_{r1})|/p$, $p_2 = |f(X_{r2})|/p$, and $p = |f(X_b)| + |f(X_{r1})| + |f(X_{r2})|$. By utilizing this search equation, the offspring can be biased strongly in the direction where the global best individual is. It's worth to note that the proposed search equation is only used in the onlooker bee phase, while the original search equation is still used in the employed bee phase. By using both of these two search equations simultaneously, it's expected to balance the exploration and exploitation of ABC.

In addition to the trigonometric search equation, another modification is made in the onlooker bee phase. In the original ABC, a fitness-based selection mechanism is used to select the food sources with relatively good nectar amounts for further exploitation. In this case, a food source may be selected for more than once and then be modified by the same search equation. Unlike this mechanism, however, we propose an average-fitness-based approach. In this approach, the average fitness value f_{mean} of all food sources is calculated at first, if a food source has better fitness value than f_{mean}, then it will be selected and modified by the proposed search equation. For the remaining food sources whose fitness value is worse than f_{mean}, the following search equation will be used for them.

$$
V_i = X_{r1} + \phi \cdot (X_{r2} - X_{r3})
\tag{6}
$$

where r_1, r_2 and r_3 are random indexes within the range $[1, SN]$, and they are mutually exclusive. ϕ has the same role as well as in the Eq. (1). This search equation is derived from the basic DE, which aims to expand the available search space for avoiding being trapped in local optimum.

3.2   Orthogonal Learning in the Scout Bee Phase

In the original scout bee phase, if a food source is considered to be abandoned, then it would be replaced with a new one which is generated in a random manner by the scout bee. Although this mechanism is relatively simple, it may cause a problem that the search experience contained in the abandoned food source would be lost. Hence, we introduce the OL strategy to overcome this shortcoming. In the OL strategy, the abandoned food source X_a and the global best individual X_b are considered as the learning exemplars, and the OED method is used to construct an efficient combination of these two exemplars as the new

---

Algorithm 1. The procedure of OL strategy in the scout bee phase

---

1: Determine which food source would be abandoned;
2: Select an appropriate orthogonal array $L_M(Q^N)$;
3: Use the quantization technique to divide the dimensions of X_a and X_b into N factors, and define Q levels among them;
4: Construct M orthogonal combinations according to the selected orthogonal array;
5: Use the factor analysis to construct the best combination of levels;
6: Pick out the best one from M orthogonal combinations and the best combination of levels as the new food source.

---

food source. Due to the limited paper space, more details about the OED method can be referred to the literatures [4, 11–13]. The procedure of OL strategy in the scout bee phase is described in Algorithm 1, where $L_M(Q^N)$ denotes a predefined orthogonal array, it has N factors and Q levels per factor, M is the number of combinations.

3.3 The Procedure of Our Approach

To better clarify our approach, called TABC-OL, its procedure is described in Algorithm 2. *FEs* is the number of used fitness function evaluations, and *MaxFEs*, as the stopping criterion, is the maximal number of fitness function evaluations.

Algorithm 2. The procedure of our approach

1: Randomly initialize SN food sources $\{X_i \mid i = 1, 2, \cdots, SN\}$;
2: **while** $FEs \leq MaxFEs$ **do**
3:     /* Employed bee phase */
4:     Update the SN food sources using the original solution search equation Eq. (1);
5:     $FEs = FEs + SN$;
6:     /* Onlooker bee phase */
7:     Calculate the average fitness value f_{mean} of all food sources;
8:     Determine the relationship of $f(X_i)$ and f_{mean} for each food source;
9:     Update X_i with the trigonometric search equation Eq. (5) in the case of $f(X_i) \leq f_{mean}$, otherwise the search equation Eq. (6);
10:     $FEs = FEs + SN$;
11:     /* Scout bee phase */
12:     Use the OL strategy described in Algorithm 1 to generate a new food source;
13:     $FEs = FEs + M + 1$;
14: **end while**

---

4 Experiments

4.1 Benchmark Functions

To verify the performance of our approach, 23 well-known benchmark functions are used, these functions are also widely used in other works. Functions F01–F13

Table 1. The comparative results of GABC, MABC, MEABC, and TABC-OL.

Functions	GABC	MABC	MEABC	TABC-OL
F01	2.12E−36+	1.07E−25+	1.12E−38+	3.33E−56
F02	1.67E−19+	4.31E−14+	8.35E−21+	6.08E−34
F03	7.46E+03+	1.59E+04+	9.43E+03+	2.28E+03
F04	1.93E+01+	1.27E+01+	4.56E+00+	2.34E+00
F05	2.07E+01+	3.69E+00+	8.14E−01≈	1.78E+00
F06	0.00E+00≈	0.00E+00≈	0.00E+00≈	0.00E+00
F07	8.16E−02+	3.79E−02+	3.00E−02+	6.23E−03
F08	3.82E−04≈	3.82E−04≈	3.82E−04≈	3.82E−04
F09	5.92E−17≈	0.00E+00≈	0.00E+00≈	0.00E+00
F10	4.09E−14+	6.38E−13+	2.98E−14+	1.39E−14
F11	1.95E−05+	3.23E−14+	0.00E+00≈	0.00E+00
F12	1.57E−32≈	1.03E−27+	1.57E−32≈	1.57E−32
F13	1.35E−32≈	1.84E−26+	1.35E−32≈	1.35E−32
F14	1.12E−13+	1.17E−13+	1.23E−13+	5.68E−14
F15	8.11E+03+	1.37E+04+	9.98E+03+	4.07E+03
F16	1.50E+07+	1.31E+07+	1.03E+07+	2.40E+06
F17	3.73E+04+	4.65E+04+	3.55E+04+	2.02E+04
F18	7.34E+03+	9.40E+03+	8.98E+03+	3.20E+03
F19	4.16E+01≈	1.93E+00−	3.27E+01≈	4.46E+01
F20	5.47E−02+	5.83E−02+	3.32E−02+	9.78E−03
F21	2.09E+01+	2.08E+01+	2.08E+01+	2.03E+01
F22	9.28E−14+	1.23E−13+	1.17E−13+	5.68E−14
F23	1.57E+02+	1.44E+02+	1.62E+02+	5.08E+01
+/−/≈	17/0/6	19/1/3	15/0/8	− −

are classic scalable problems, and they are the first 13 functions in Table 1 of Yao's literature [14]. Functions F14–F23 are shifted and/or rotated problems, and they are the first 10 functions in the CEC 2005 competition [15]. For the sake of limited paper space, details of these functions can be found in [14,15], respectively.

4.2   Comparison with Other ABC Variants

This section presents a comparative study of TABC-OL with other three state-of-the-art ABC variants, they are GABC [5], MABC [6], and MEABC [16]. Short reviews about these three ABC variants have been given in Sect. 1. For a fair comparison, the parameter settings of the three ABC variants are kept the same as in their original literatures. For the stopping criterion, according to

Table 2. Average rankings of GABC, MABC, MEABC, and TABC-OL, and the best value is shown in **boldface**.

Algorithms	Average rankings
GABC	2.91
MABC	3.24
MEABC	2.35
TABC-OL	**1.50**

the suggestions in [15], *MaxFEs* is set to $5000 \cdot D$ and $10000 \cdot D$ for the first 13 functions and the remaining 10 functions, respectively. The dimension size D is set to 30. The orthogonal array in TABC-OL is $L_9(3^4)$. Each algorithm is run 30 times, the mean error $(f(X) - f(X^*)$, X^* is the global optimum) values are recorded. Table 1 presents the final results.

In this test, the paired Wilcoxon signed-rank test is used to compare the significance between two algorithms, and the signs "+", "−", and "≈" in Table 1 indicate our approach is, respectively, better than, worse than, and similar to its competitor according to the Wilcoxon signed-ranked test at $\alpha = 0.05$. The last row in Table 1 summarizes the comparative results. It can be seen from this table, our approach shows the best overall performance. To be specific, compared with GABC and MEABC, TABC-OL wins on 17 and 15 functions respectively. For MABC, TABC-OL achieves better results on 19 function, while on the F19 function MABC shows the best performance among the involved four algorithms. To compare the performance of multiple algorithms, the Friedman test is also conducted to obtain the average rankings. Table 2 shows the average rankings, it can been that TABC-OL achieves the best average ranking, and the rest of algorithms can be sorted into the following order: MEABC, GABC, and MABC.

5   Conclusions

To enhance the exploitive capability of ABC, in this paper, a trigonometric search equation is proposed to replace the original solution search equation. In the proposed search equation, the global best individual and other two randomly selected individuals are included to form a hypergeometric triangle in the search space. As a result, the offspring can be biased strongly in the direction where the global best individual is, which benefits the exploitation of ABC. Furthermore, the orthogonal learning strategy is employed in the scout bee phase to generate new food source. By using this strategy, it's helpful to preserve the good information of the abandoned food source. The experiments are conducted on 23 well-known benchmark functions including shifted and rotated problems. The results are compared with other three state-of-the-art ABC variants, which show that our approach has superior performance. In the future, our approach will be tested on more benchmark functions and applied to solve some real-world problems, such as data clustering.

Acknowledgments. This work was supported by the Foundation of State Key Laboratory of Software Engineering (No. SKLSE2014-10-04), the National Natural Science Foundation of China (Nos. 61272212 and 61462045), the Science and Technology Foundation of Jiangxi Province (Nos. 20132BAB201030 and 20151BAB217007), and the Application Research Project of Nantong Science and Technology Bureau (No. BK2014057).

References

1. Karaboga, D., Basturk, B.: A powerful and efficient algorithm for numerical function optimization: artificial bee colony (ABC) algorithm. J. Glob. Optim. **39**, 459–471 (2007)
2. Karaboga, D., Gorkemli, B., Ozturk, C., Karaboga, N.: A comprehensive survey: artificial bee colony (ABC) algorithm and applications. Artif. Intell. Rev. **42**, 21–57 (2014)
3. Karaboga, D., Akay, B.: A comparative study of artificial bee colony algorithm. Appl. Math. Comput. **214**, 108–132 (2009)
4. Gao, W., Liu, S., Huang, L.: A novel artificial bee colony algorithm based on modified search equation and orthogonal learning. IEEE Trans. Cybern. **43**, 1011–1024 (2013)
5. Zhu, G., Kwong, S.: Gbest-guided artificial bee colony algorithm for numerical function optimization. Appl. Math. Comput. **217**, 3166–3173 (2010)
6. Gao, W., Liu, S.: A modified artificial bee colony algorithm. Comput. Oper. Res. **39**, 687–697 (2012)
7. Fan, H.Y., Lampinen, J.: A trigonometric mutation operation to differential evolution. J. Glob. Optim. **27**, 105–129 (2003)
8. Karaboga, D., Gorkemli, B.: A quick artificial bee colony (QABC) algorithm and its performance on optimization problems. Appl. Soft Comput. **23**, 227–238 (2014)
9. Gao, W., Liu, S., Huang, L.: A novel artificial bee colony algorithm with Powell's method. Appl. Soft Comput. **13**, 3763–3775 (2013)
10. Gao, W.F., Liu, S.Y., Huang, L.I.: Enhancing artificial bee colony algorithm using more information-based search equations. Inf. Sci. **270**, 112–133 (2014)
11. Leung, Y.W., Wang, Y.: An orthogonal genetic algorithm with quantization for global numerical optimization. IEEE Trans. Evol. Comput. **5**, 41–53 (2001)
12. Zhan, Z., Zhang, J., Li, Y., Shi, Y.: Orthogonal learning particle swarm optimization. IEEE Trans. Evol. Comput. **15**, 832–847 (2011)
13. Wang, Y., Cai, Z., Zhang, Q.: Enhancing the search ability of differential evolution through orthogonal crossover. Inf. Sci. **185**, 153–177 (2012)
14. Yao, X., Liu, Y., Lin, G.: Evolutionary programming made faster. IEEE Trans. Evol. Comput. **3**, 82–102 (1999)
15. Suganthan, P.N., Hansen, N., Liang, J.J., Deb, K., Chen, Y.P., Auger, A., Tiwari, S.: Problem definitions and evaluation criteria for the CEC 2005 special session on real-parameter optimization. Technical report, Nanyang Technological University, Singapore (2005)
16. Wang, H., Wu, Z., Rahnamayan, S., Sun, H., Liu, Y., Pan, J.S.: Multi-strategy ensemble artificial bee colony algorithm. Inf. Sci. **279**, 587–603 (2014)

Classification of High and Low Intelligent Individuals Using Pupil and Eye Blink

Giyoung Lee, Amitash Ojha, and Minho Lee[✉]

School of Electronics Engineering, Kyungpook National University, 1370 Sankyuk-Dong, Puk-Gu, Taegu, 702-701, South Korea
{giyoung0606,amitashojha,mholee}@gmail.com

Abstract. A commonly used method to determine the intelligence of an individual is a group test. It checks accuracy and response time while they solve a series of problems. However, it takes long time and is often inaccurate if the difficulty level of problems is high or the number of problems is too small. Therefore, there is an urgent need to find an objective, readily available, fast and more reliable method to determine the intelligence level of individuals. In this paper, we propose an alternative method to distinguish between high and low intelligent individuals using pupillary response and eye blink pattern. Studies have shown that these measures indicate the cognitive state of an individual more accurately and objectively. Our experimental results show that the bio-signals between high and low intelligent individuals are significantly different and proposed method has good performance.

Keywords: Bio-signals · Pupil dilation · Eye blink · Intelligence · Classification

1 Introduction

For developing effective and customizable learning methods, it is important to distinguish high intelligent individuals from low intelligent individuals. Several subjective and objective assessment methods have been proposed in this regard but they have their own limitations. For example, most commonly used group intelligence tests to check behavioral data such as accuracy and response time are economical to administer but fail to identify individual differences. Moreover, it takes long time because participants need to solve a long list of problems. Similarly individual tests like 'self-assessment' and 'interview' provide subjective information but they are costly and equally time consuming. Therefore, considering problems of existing assessment methods, it is important to find reliable, cheaper, fast and objective assessment methods that can distinguish between high and low intelligent individuals.

Several studies have shown that cognitive state of humans can be recognized by analyzing bio-signals such as pupil dilation, eye blink, gaze pattern, facial expressions, heart rate, skin conductance, etc. Kaliouby and Robinson [1] conducted a study in which they tried to infer complex mental states of users by analyzing their head gestures and facial expressions in a real-time video stream. D'Mello and his colleagues [2] specifically showed that by analyzing students' gaze on a commercial eye tracker, they could

S. Arik et al. (Eds.): ICONIP 2015, Part I, LNCS 9489, pp. 459–466, 2015.
DOI: 10.1007/978-3-319-26532-2_50

detect if the student is bored, disengaged, or is zoning out. In particular, pupillary response and eye blinking pattern can be reliably associated with higher cognitive functions such as thinking, problem solving, memory retrieval, etc., and may also indicate the intelligence of an individual [3, 4]. Usually the pupil is larger under conditions of higher attentional allocations, memory use or interpretation of more difficult material. Increase in pupil diameter also correlates with sustained processing, which means that pupil dilation persists as long as the demand of information processing sustains [3]. Eye blink, which is a regular part of daily waking life, too indicates cognitive processing and does not occur randomly. Blinks potentially reflect preparation and very short blinks are associated with errors on cognitive tasks, consistent with adequate preparation [5].

Considering above-mentioned findings, we suggest an alternative method (which can replace conventional methods) to distinguish between high and low intelligent individuals by analyzing 'pupil' and 'eye blink' patterns of individuals while they solve our pre-designed problems. The rest of the paper is organized as follows: In the next section, we explain related work for determining intelligence level using bio-signals in detail. In Sect. 3, we explain the proposed method. In Sect. 4, we report our experimental results and finally in last section, we present our conclusions.

2   Bio-Signal Analysis for Determining Intelligence

Cognitive activities require consumption of resources [6]. Resources here mean the amount of activation available for information storage and processing in the underlying cortical neural system. This pool of activation is assumed to be limited. In last few years, different measures of activity as indices of resource allocation have been verified [6]. Particularly, studies have shown that assessment of variation of pupil size and eye blink pattern provides complementary indices of information processing. In general, pupil dilation indicates sustained information processing. For example, as individuals are asked to remember larger number of digits, pupil dilation increases proportionally [7]. Pupil dilation also reflects resource allocation, interpretation of complex material, deception and affective processing.

Along with pupil dilation, eye blinks also indicate cognitive processing. Studies show that eye blink bursts follow high cognitive load or information processing [8]. This suggests that eye blinks reflect the release of resources used in stimulus related cognition [9]. Blinks potentially reflect preparation and very short blinks are associated with errors on cognitive tasks, consistent with adequate preparation [5]. Other studies have pointed out that pupil dilation often follows blinks in a cognitive task [8]. Moreover, It has been suggested that blinking is avoided to maximize stimulus perception during high attention tasks and blink occurrence is reduced for increasing information content and task demands [10].

In our earlier study [4], we showed that pupil dilation and eye blink together can be used as a measure to determine the intelligence of individuals while they solve different types of tasks with different level of difficulties. Based on our experiment results we concluded that high intelligent individuals modulate their resource allocation mechanism while low intelligent individuals do not. Moreover, high intelligent individuals have higher attention just before the task begins and that period could be the key to determine the intelligence of an individual. Based on our findings, here we propose an alternative method to categorize individuals into high intelligent and low intelligent by analyzing their pupil and eye blink pattern.

3   Proposed Method

The overall process of proposed method is shown in Fig. 1. In the experimental setup, we collect and extract pupil size using eye tracker (Tobii 1750) and eye blink using captured face images from web camera. To classify high and low intelligent individuals by existing group test results, which can be used for teaching signal of classifier, we also obtain behavioral data such as accuracy and response time using computer keyboard. Finally, we classify high and low intelligent individuals using pupil size and eye blink variation with support vector machine (SVM) classifier [11].

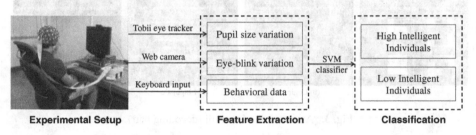

<table>
<tr><td>Tobii eye tracker</td><td>Pupil size variation</td><td></td><td>High Intelligent
Individuals</td></tr>
<tr><td>Web camera</td><td>Eye-blink variation</td><td>SVM
classifier</td><td></td></tr>
<tr><td>Keyboard input</td><td>Behavioral data</td><td></td><td>Low Intelligent
Individuals</td></tr>
</table>

Experimental Setup   **Feature Extraction**   **Classification**

Fig. 1. Overview of proposed method

3.1   Experimental Setup and Data Acquisition

Participants were called and seated in a comfortable chair. Stimuli were presented on a 21-inch screen monitor (1280 × 1024). The distance between participants and the screen was around 60–80 cm. At the beginning of the experiment, participants filled out a questionnaire that ascertained demographic information as well as other factors that are known to affect pupil dilation (e.g., psychiatric and neurological dysfunction, drug consumption, medication). Then they were calibrated to Tobii eye tracker. A calibration procedure was conducted prior to every task instruction to obtain accurate pupil data of participants.

At first, the baseline task was conducted to obtain baseline pupil size and eye blink of participants. Participants were asked to fixate on a plus (+) sign, presented 5 times for 20 s. The duration between fixations was 40 s when participants close the eyes. The individual average pupil diameter and the number of eye blink of 100 s of fixation was taken as pupil and eye blink baseline not influenced by any instructional and expectation effects. Then, participants did actual tasks as shown in Fig. 2. The stimulus material included four different types of tasks namely (1) spatial, (2) logic, (3) linguistic, and (4) natural. These tasks were divided into two parts: encoding and decoding. In encoding part, there were 10 problems in each type of task, which users had to memorize or learn. Problems of encoding part included unfamiliar or new information. For example, in case of logic tasks, we defined new operators. In case of spatial, linguistic and natural tasks, we chose unfamiliar figure patterns, idioms and flower pictures. In decoding part, there were 5 problems in each type of task, which users had to solve based on memorized/learnt information from the encoding part (see Fig. 3). The medium of instruction as well as presentation of problems was in Korean Language. Stimuli were presented in white foreground and black background.

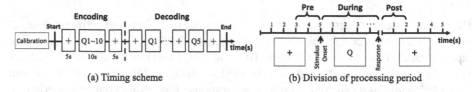

(a) Timing scheme   (b) Division of processing period

Fig. 2. Experimental procedure in each task

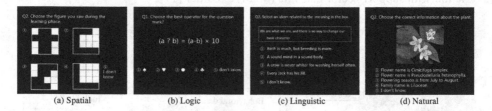

(a) Spatial   (b) Logic   (c) Linguistic   (d) Natural

Fig. 3. An example of stimuli (decoding part)

3.2   Feature Extraction

Division of Processing Period. The standard methodology in a pupillometry study is to analyze the data during the processing of stimulus (after the onset of stimulus). Pre-stimulus data is considered to be affected by expectation of task demand [3]. However, in our study for better assessment of expectation effect (as we argue, is the key to determine the intelligence of individuals), we divided the whole processing period into three stages: (1) pre-stimulus, (2) during-stimulus and (3) post-stimulus as shown in Fig. 2(b). We assumed that an analysis of pupil size during pre-stimulus stage would indicate the pattern of preparation and attention [5], post-stimulus would indicate release of cognitive resources [9] and during-stimulus would indicate time course of processing [7].

Pupil Size Variation. After obtaining left and right pupil size from eye tracker, we removed artifacts due to excessive blinking and replaced very small blinks by linear interpolation. Also, we removed the effect of luminance and noise by method in [4]. Then, average pupil size of all problems of decoding part was calculated for feature. However, bio-signals like pupil dilation are different for different individuals. To compensate individual characteristic, average pupil size of experimental task was divided by average pupil size of baseline task using Eq. (1). Change of pupil dilation is represented in percentage assuming the baseline to be 100 %.

$$(Data_{task}/Data_{baseline} - 1) \times 100 \qquad (1)$$

Eye-Blink Variation. The eye blink was detected in captured face images obtained from web camera by method in [12], which detected face and eye region using MCT-based AdaBoost then detected eye blinks using the difference between center and surround of Hough circle transform image. After detecting eye blink, the average number of eye blinks per second of all problems of decoding part was calculated. To consider subjective characteristics of bio-signals, we subtracted baseline value from calculated data of each experimental task by using Eq. (2).

$$Data_{task}/Data_{baseline} \tag{2}$$

3.3 Classification of High and Low Intelligent Individuals

Participants were divided into two groups namely (1) High Intelligent Individuals (HIs) and (2) Low Intelligent Individuals (LIs) by traditional method for teaching signal (target or ground truth) of classifier. However, intelligence, because of its multiple dimensions, is a complex phenomenon. David Wechsler insisted that intelligence is the aggregate or global capacity of the individual to act purposefully and to think rationally [13]. Howard Gardner, on the other hand, proposed theory of Multiple Intelligences (MI), according to which everyone possess some level of eight aspects of intelligences namely Linguistic, Logical, Musical, Bodily, Spatial, Interpersonal, Intrapersonal and Naturalist [14] and people have outstanding skills in more than one intelligence. Therefore, in our study, we divided HIs and LIs groups using both lines of thought following two criteria.

1. **Global group:** Participants were divided into HIs and LIs according to Wechsler's definition. We calculated average accuracy and response time of all problems in decoding part across four different types of tasks. Participants who scored more than average accuracy of all participants and had less than average response time of whole participants were categorized as HIs. Others were categorized as LIs.
2. **MI group:** Participants were also divided into HIs and LIs groups in each intelligence following Gardner's definition. Using the same method as above, participants were categorized as HIs using average accuracy and response time in each type of intelligence. So for each intelligence we had HIs and LIs.

To train the classifier, we extracted pupil and eye-blink features and made input vector with 6 dimensions: 2 features (pupil size variation and eye blink variation) × 3 processing period (pre-stimulus, during-stimulus, and post-stimulus). To choose proper classifier for distinguishing HIs and LIs, we tested several classifiers experimentally and selected SVM classifier [11] (see Sect. 4.3).

4 Experimental Results

4.1 Participants

Thirty-five university students (6 females and 29 males), with a mean age of 23.6 years (SD = 1.9) participated in the study. Their participation was voluntary and their socio-economic background was controlled. It was also confirmed that participants were not taking any medications. Data of 5 participants (2 females and 3 males) could not be analyzed because of missing and/or noisy data. Table 1 shows the divided group information of participants using two criteria, namely Global group and MI group.

Table 1. Number of participants in divided groups

	Global group	MI group			
		Spatial	Logic	Linguistic	Natural
HIs	11	16	20	16	15
LIs	19	14	10	14	15

4.2   Statistical Analysis

To check difference of the pupil size and eye blink between high and low intelligent individuals, we conducted independent t-test using IBM SPSS Statistics 21 software. The pupil size variation was found to be statistically significant in all processing period as shown in Fig. 4. The pupil size of HIs was bigger than LIs. Also, the eye-blink variation was found to be statistically significant in post-stimulus period. HIs blink more than LIs. The results of Global group and MI group were found to be similar.

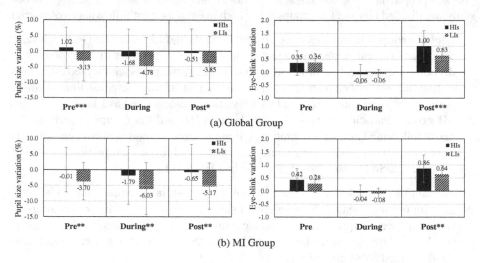

(a) Global Group

(b) MI Group

Fig. 4. Pupil and eye blink patterns of high and low intelligent individuals. (*p < .05, **p < .01, ***p < .001)

4.3   Comparison with Other Classifiers

To choose the proper classifier for determining intelligence level, a leave-one-out cross validation was performed, where the classifier was trained on data of 29 participants and tested on data of the 30th one. It was repeated 30 times that all data of participants are used once as the validation data. All experiments were conducted using MATLAB functions. The compared classification performance of various classifiers (adaptive neuro fuzzy inference system (ANFIS) [15], naïve Bayesian [16], k-nearest neighbor (kNN) [17], support vector machine (SVM) [11]) is shown in Table 2. To get the maximum performance, the parameters were tuned. In ANFIS, we uses 'Sugeno' type and fuzzy c-means clustering. The number of clusters was set automatically. In naïve Bayesian, we used normal distribution as data distribution and empirical class prior probabilities that the software uses the class relative frequencies distribution for the prior probabilities. In kNN, the number of neighbors was 15. We used Mahalanobis distance and equal distance weights. In SVM, we used Gaussian radial basis function kernel with sigma of 3. The SVM had the best test performance. Therefore, SVM was chosen as the proper classifier to determine intelligence level.

Table 2. Classification performance (%)

		ANFIS	Naïve Bayesian	kNN	SVM
Global Group	Train	90.10	79.39	72.97	81.75
	Test	67.14	75.00	67.38	76.67
MI Group	Train	90.32	66.18	66.99	68.65
	Test	56.36	63.64	61.82	65.45

5 Conclusion and Future Work

The purpose of our experiment was to find objective, fast and more reliable method to determine the intelligence level of individuals. For this, we analyzed pupil size and eye blink in four intelligence tasks. Our results indicate that there exists a significant difference in pupil size and eye blink between high and low intelligent individuals while they perform problem-solving tasks. Also, the SVM classifier was found to be best in classifying two groups using pupil size and eye blink. Based on these results, we argue that the proposed method can be used as a new measure over conventional ones to determine the Global intelligence level of an individual. However, determining the intelligence of an individual in different domains more accurately requires identification of various factors. In our future work, we investigate difference of results between encoding and decoding part and increase performance using additional features such as heart rate, skin conductance, EEG, etc.

Acknowledgements. This work was supported by the ICT R&D program of MSIP/IITP. [10041826, Development of emotional features sensing, diagnostics and distribution S/W platform for measurement of multiple intelligence from young children]

References

1. el Kaliouby, R., Robinson, P.: Generalization of a vision-based computational model of mind-reading. In: Tao, J., Tan, T., Picard, R.W. (eds.) ACII 2005. LNCS, vol. 3784, pp. 582–589. Springer, Heidelberg (2005)
2. D'Mello, S., Olney, A., Williams, C., Hays, P.: Gaze tutor: a gaze-reactive intelligent tutoring system. Int. J. Hum Comput Stud. **70**, 377–398 (2012)
3. Van Der Meer, E., Beyer, R., Horn, J., Foth, M., Bornemann, B., Ries, J., Kramer, J., Warmuth, E., Heekeren, H.R., Wartenburger, I.: Resource allocation and fluid intelligence: insights from pupillometry. Psychophysiology **47**, 158–169 (2010)
4. Lee, G., Ojha, A., Kang, J.-S., Lee, M.: Modulation of resource allocation by intelligent individuals in linguistic, mathematical and visuo-spatial tasks. Int. J. Psychophysiol. **97**, 14–22 (2015)
5. Sirevaag, E.J., Rohrbaugh, J.W., Stern, J.A., Vedeniapin, A.B., Packingham, K.D., LaJonchere, C.M.: Multi-dimensional characterizations of operator state: a validation of oculomotor metrics (1999)
6. Just, M.A., Carpenter, P.A., Miyake, A.: Neuroindices of cognitive workload: neuroimaging, pupillometric and event-related potential studies of brain work. Theor. Issues Ergon. Sci. **4**, 56–88 (2003)
7. Granholm, E., Asarnow, R.F., Sarkin, A.J., Dykes, K.L.: Pupillary responses index cognitive resource limitations. Psychophysiology **33**, 457–461 (1996)
8. Fukuda, K.: Eye blinks: new indices for the detection of deception. Int. J. Psychophysiol. **40**, 239–245 (2001)
9. Ohira, H., Winton, W.M., Oyama, M.: Effects of stimulus valence on recognition memory and endogenous eyeblinks: further evidence for positive-negative asymmetry. Pers. Soc. Psychol. Bull. **24**, 986–993 (1998)
10. Irwin, D.E., Thomas, L.E.: Eyeblinks and cognition. In: Tutorials in Visual Cognition pp. 121–141 (2010)
11. Cristianini, N., Shawe-Taylor, J.: An Introduction to Support Vector Machines and Other Kernel-Based Learning Methods. Cambridge University Press, Cambridge (2000)
12. Jo, H., Lee, M.: In-attention state monitoring for a driver based on head pose and eye blinking detection using one class support vector machine. In: Loo, C.K., Yap, K.S., Wong, K.W., Teoh, A., Huang, K. (eds.) ICONIP 2014, Part II. LNCS, vol. 8835, pp. 110–117. Springer, Heidelberg (2014)
13. Wechsler, D.: Wechsler Adult Intelligence Scale–Fourth Edition (WAIS–IV). NCS Pearson, San Antonio (2008)
14. Gardner, H.: Frames of Mind: The Theory of Multiple Intelligences. Basic Books, New York (2011)
15. Jang, J.-S.: ANFIS: adaptive-network-based fuzzy inference system. IEEE Trans. Syst. Man Cybern. **23**, 665–685 (1993)
16. John, G.H., Langley, P.: Estimating continuous distributions in Bayesian classifiers. In: Proceedings of the Eleventh Conference on Uncertainty in Artificial Intelligence, pp. 338–345. Morgan Kaufmann Publishers Inc. (1995)
17. Friedman, J.H., Bentley, J.L., Finkel, R.A.: An algorithm for finding best matches in logarithmic expected time. ACM Trans. Math. Softw. (TOMS) **3**, 209–226 (1977)

Learning Task Specific Distributed Paragraph Representations Using a 2-Tier Convolutional Neural Network

Tao Chen[1], Ruifeng Xu[1]([✉]), Yulan He[2], and Xuan Wang[1]

[1] Shenzhen Engineering Laboratory of Performance Robots at Digital Stage,
Shenzhen Graduate School, Harbin Institute of Technology, Shenzhen, China
xuruifeng@hitsz.edu.cn
[2] School of Engineering and Applied Science, Aston University, Birmingham, UK

Abstract. We introduce a type of 2-tier convolutional neural network model for learning distributed paragraph representations for a special task (e.g. paragraph or short document level sentiment analysis and text topic categorization). We decompose the paragraph semantics into 3 cascaded constitutes: word representation, sentence composition and document composition. Specifically, we learn distributed word representations by a continuous bag-of-words model from a large unstructured text corpus. Then, using these word representations as pre-trained vectors, distributed task specific sentence representations are learned from a sentence level corpus with task-specific labels by the first tier of our model. Using these sentence representations as distributed paragraph representation vectors, distributed paragraph representations are learned from a paragraph-level corpus by the second tier of our model. It is evaluated on DBpedia ontology classification dataset and Amazon review dataset. Empirical results show the effectiveness of our proposed learning model for generating distributed paragraph representations.

Keywords: Natural language processing · Distributed representation · Convolutional neural network

1 Introduction

Paragraph or short document representations are important for a variety of NLP tasks including sentiment analysis, text classification, and document retrieval. Many applications use distributional representation of text, such as Bag-Of-Words (BOW) representation, as the input of their algorithms. With the rapid development of deep neural networking and parallel computing, especial after distributed representations of word and sentence provides the basis for many state-of-the-art approaches [1,3,7,11,16], there is rising interest in learning distributed representational formats of paragraph.

Distributed representations of paragraph usually refers to dense and real-valued vectors in a low-dimensional space to represent paragraph or short document, which are assumed to convey semantic information contained in

© Springer International Publishing Switzerland 2015
S. Arik et al. (Eds.): ICONIP 2015, Part I, LNCS 9489, pp. 467–475, 2015.
DOI: 10.1007/978-3-319-26532-2_51

paragraph. Many current approaches try to learn distributed paragraph representations without the knowledge of text structures and task-specific annotation information (e.g. sentiment labels for sentiment analysis task) which is often important for supervised approaches to improve performance in a task. For example, Le and Mikolov [8] propose an unsupervised framework to learn continuous distributed vector representations for pieces of texts ranging from sentences to documents. They only use the context of words in a large unlabeled corpus to learn paragraph vectors. Zhang and LeCun [17] apply a 9 layers deep temporal ConvNets to text understanding from character-level inputs all the way up to abstract text concepts without any syntactic or semantic structures. Dos Santos and Gatti [15] propose a deep convolutional neural network that also exploits sentence-level information directly from character-level information. Both of them perform sentiment analysis of short texts regarding a short document as a long sentence.

This paper proposes a supervised framework that learns distributed paragraph representations by learning word, sentence and paragraph representations hierarchically using the structure and task-specific annotation information. Firstly, distributed word representations are learned by a Continuous Bag-Of-Words (CBOW) model [11] from a large unstructured text corpus. Secondly, task specific distributed sentence representations are learned from a sentence level corpus by a one dimensional CNN [7] using previous learned word embeddings as pre-trained input vectors. Finally, task specific distributed paragraph representations are learned from a paragraph or document level corpus by the same CNN using previous learned sentence vectors as pre-trained input vectors. The obtained paragraph representations can be used as the input of a machine learning classifier or clustering algorithm for further processing.

We evaluate our model on DBpedia ontology classification dataset [2] and Amazon review dataset from Stanford Network Analysis Project (SNAP) [10]. We achieve new state-of-the-art results, better than complex methods, yielding a relative improvement of more than 20.61 % and 5.19 % in terms of error rate, respectively.

2   Our Approach

In this study, we propose to learn task specific distributed paragraph representations by learning distributed word, sentence and paragraph representations hierarchically. The system framework with three main components is shown in Fig. 1. The first component, the *Distributed Word Representation Model*, takes a large collection of raw text to train a distributed word representation model to generate word embeddings. These word embeddings are then fed into the *Distributed Sentence Representation Tier*, in which a one-dimensional CNN takes a sentence level corpus to train sentence composition model which is used to train distributed sentence representation for sentences in target document level corpus. The third component, the *Distributed Paragraph Representation Tier* takes sentence representations as input to train distributed document representation of the target document level corpus.

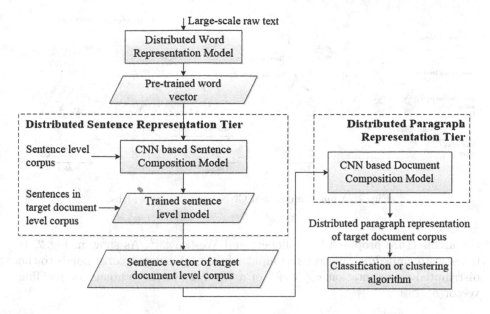

Fig. 1. Framework of our approach.

2.1 Distributed Word Representation Model

The CBOW model introduced by Mikolov et al. [11] is used to learning distributed word representations. The training objective of CBOW model is to use the surrounding words of the target word in a sentence or a document to predict word representations. It can capture a large number of syntactic and semantic word relationships from unstructured text data.

Given a sequence of training words $w_1, w_2, w_3 \ldots w_T$, the training objective is to maximize the average log probability:

$$\frac{1}{T} \sum_{t=1}^{T} \sum_{-c \leq i \leq c, i \neq 0} \log p(w_t | w_{t+i}) \tag{1}$$

where c is the size of the training context, w_t is the center word, and $\log p(w_t | w_{t+i})$ is the conditional log probability of the center word w_t given the surrounding words w_{t+i}. The prediction task is performed via softmax. The *hierarchical softmax* [13,14] process which uses a binary tree representation of the output layer with the words as leaves, is used to reduce computational complexity.

2.2 Distributed Sentence Representation Tier

The one-dimensional CNN proposed by Kim [7][1] is used to learn distributed sentence representation from a sentence level corpus. It is a slight variant of

---

[1] https://github.com/yoonkim/CNN_sentence.

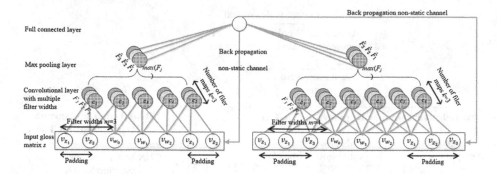

Fig. 2. A one-dimensional CNN with 2 filter widths.

the architecture proposed by Collobert and Weston [3][2]. As show in Fig. 2, It takes word embedding matrix s as input where each column corresponds to the distributed representation $v_{w_i} \in \mathbb{R}^d$ of a word w_i in the sentence or padding vector $v_{z_i} \in \mathbb{R}^d$:

$$s = [v_{z_1}, \cdots, v_{z_{m-1}}, v_{w_1}, \cdots, v_{w_n}, v_{z_1}, \cdots, v_{z_{m-1}}] \tag{2}$$

where v_{w_i} is a d dimensional pre-trained word vector. v_{z_i} is d dimensional zero vector. m is the size of filter window. n is defined as the max length of sentences in the training set.

The idea behind the one-dimensional convolution is to take the dot product of the vector w with each m-gram in the sentence s to obtain another sequence c. In the convolutional layer, one-dimensional convolution is taken between a filter vector $w \in \mathbb{R}^{md}$ and a vector $s_{i:i+m-1} \in \mathbb{R}^{md}$ of m concatenated columns in s. The i-th feature $c_i \in \mathbb{R}$ of a feature map $F_j \in \mathbb{R}^{n+m-1}$ is generated as follows:

$$c_i = f(w \cdot s_{i:i+m-1} + b) \tag{3}$$

Where $b \in \mathbb{R}$ is a bias term and f is a point-wise non-linear function such as the hyperbolic tangent. $s_{i:i+m-1}$ refers to the i-th to $(i+m-1)$-th column of s. A feature map $F_j \in \mathbb{R}^{n+m-1}$ is defined as

$$F_j = [c_1, c_2, \cdots, c_{n+m-1}] \tag{4}$$

In the pooling layer, a max-overtime pooling operation [4], which forces the network to capture the most useful local features produced by the convolutional layers, is applied over F_j. The maximum value $\hat{F}_j = max(F_j)$ taken as the feature corresponding to a particular filter w. k-$\hat{F}_j$ concatenates to a vector $\hat{F} \in \mathbb{R}^k$.

The model uses multiple filters (with varying window sizes) to obtain multiple features. These features form the penultimate layer and are passed to a fully connected softmax layer whose output is the probability distribution over

---

[2] http://ronan.collobert.com/senna/.

labels. In non-static channel, the training error propagates back to fine-tune the parameters (w, b) and the input word vectors.

The vector generated in the penultimate layer of the CNN architecture is regarded as distributed sentence representation which captures the semantic content of the input sentence, in some degree.

After training distributed sentence representation tier of 2-TCNN on a sentence level corpus, the CNN model are saved. Sentences in the target document level corpus are fed to the saved CNN model and corresponding sentence vectors are generated in the penultimate layer of the CNN.

If there are not sentence level corpus for a special task, we can choose sentences in the document level corpus as positive training samples and randomly replace part of the words (for example, half of the words) in a positive sample to construct a negative training sample.

2.3   Distributed Paragraph Representation Tier

The distributed paragraph representation tier of 2-TCNN has the same architecture with the sentence representation tier. The difference is that it takes sentence vectors of the target document level corpus as input instead of using word embeddings. The vector generated in the penultimate layer of the CNN in this tier is regarded as distributed document representation which captures the semantic content of the target document level corpus.

Both the training set and the test set of the target document level corpus can be represented as paragraph vectors. These vectors can used as features in a machine learning classification or clustering algorithm.

3   Evaluation and Discussion

3.1   Experiment Settings

In all experiments, we use the publicly available word embeddings trained by CBOW model on 100 billion words from Google News as a pre-trained word embeddings. Words not present in the set of pre-trained words are initialized randomly in the same way as in [11].

For sentence level corpus used in sentence representation tier of our model, in the DBpedia ontology classification experiment, we use the glosses in WordNet [12] as positive training samples. There are 117,791 glosses in the WordNet 3.1 version. We randomly replace half of words in a positive sample to construct a negative training sample. In Amazon reviews sentiment analysis experiment, we use Stanford sentiment tree bank [16] to train the sentence composition model.

In all experiments, we train sentence vector and paragraph vector using the same setting of CNN. We use: rectified linear units, filter windows of 3, 4, 5 with 100 feature maps each, AdaDelta decay parameter of 0.95, dropout rate of 0.5.

For paragraph vectors classification, we use LIBLINEAR tools [5] with default parameter settings in Weka [6].

For metrics, we use error rate in all experiments.

We benchmark four models on distributed paragraph representations learning: word2vec, Bag of Words, ConvNet and 2-TCNN. *word2vec* refers to a bag-of-centroids model via word2vec [11], in which k-means algorithm are applied on word vectors learned from Google News corpus with $k = 5000$, and then a bag of these centroids are used for multinomial logistic regression. *Bag of Words* refers to the bag-of-words model, in which appearance frequency of each word in the training dataset are counted, and 5,000 most frequent ones are chose as the bag of features for multinomial logistic regression. *ConvNet* model proposed by Zhang and LeCun [17] are divided into four models by scale and using or not using thesaurus. Both the large and small ConvNet are 9 layers deep with 6 convolutional layers and 3 fully-connected layers, with different number of hidden units and frame sizes. For more detail, please see the reference. *2-TCNN* is our proposed model.

3.2 DBpedia Ontology Classification

We use the same dataset provided by Zhang and LeCun [17]. The DBpedia ontology classification dataset is constructed by picking 14 non-overlapping classes from DBpedia 2014 [9]. From each class, 40,000 training samples and 5,000 testing samples are chose randomly. In Table 1, we report the error rates of different models on this dataset. The results of other models are also reported in [17]. It is observed that *Bag of Words* model goes beyond the barrier of 5 % error rate. It is a very significant improvement comparing with *word2vec*. Another significant improvement comes when distributed representation are used in ConvNet model. *Large ConvNet with thesaurus* achieves better performance. Our model outperforms the best ConvNet model for 0.33 % on error rate. This means a relative improvement of 20.61 %.

Table 1. Experimental results on the DBpedia ontology classification datasets.

Model	Error rate
word2vec	10.59 %
Bag of words	3.57 %
Small ConvNet without thesaurus	2.01 %
Small ConvNet with thesaurus	1.85 %
Large ConvNet without thesaurus	1.74 %
Large ConvNet with thesaurus	1.60 %
2-TCNN	**1.27%**

3.3 Amazon Review Sentiment Analysis

The Amazon review sentiment analysis dataset from the Stanford Network Analysis Project (SNAP) contains review texts with 5 score labels from 1 to 5.

In order to construct sentiment polarity dataset, labels 1 and 2 are converted to negative, 4 and 5 positive. 3,000,000 samples for each positive or negative label as training set and 450,000 samples for testing are selected randomly by Zhang and LeCun [17]. The error rates of different models on this dataset are shown in Table 2. The results of other models are also reported in [17]. It is observed that *Bag of Words* model perform better than *word2vec* again. ConvNet models make a very significant improvement over the two models. It is a little surprising that *Large ConvNet without thesaurus* achieves better performance than *Large ConvNet with thesaurus*. It shows that data augmentation techniques don't always work well. Our model outperforms the best ConvNet model for 0.19 % on error rate. This means a relative improvement of 5.19 %. It shows the effectiveness of our proposed learning model for distributed paragraph representations.

Table 2. Experimental results on the Amazon review sentiment analysis datasets.

Model	Error rate
word2vec	16.93 %
Bag of words	14.46 %
Small ConvNet without thesaurus	4.16 %
Small ConvNet with thesaurus	3.99 %
Large ConvNet without thesaurus	3.66 %
Large ConvNet with thesaurus	3.92 %
2-TCNN	**3.47%**

3.4 Discussion

All of comparing models regard input paragraph or short document as a long sentence without any structures and sentence-level annotation information. By make use of knowledge about words, sentences and paragraphs, our model achieves the best performance on both the two datasets.

Both the large and small ConvNet are 9 layers deep with 6 convolutional layers and 3 fully-connected layers. Our model is only 2 layers deep. [17] reports that using an NVIDIA Tesla K40, training takes about 5 hours per epoch for the large model, and 2 hours for the small model. Using GeForce GTX TITAN X, training the DBpedia ontology classification dataset takes less than 10 min per epoch in our model.

4   Conclusion and Future Directions

This paper presents a 2-tier convolutional neural network model to learn distributed paragraph representations. For each paragraph in a corpus, the model generates the high quality distributed paragraph representations by using the text structure and task-specific annotation information. It is shown helpful to improve the performance of distributed paragraph representations. The obtained paragraph representations achieve state-of-the-art results on DBpedia ontology classification dataset and Amazon review sentiment analysis dataset. It shows the effectiveness of our proposed learning framework for distributed paragraph representations. Our model is only 2 layers deep. It is much more efficient than the ConvNet models. Future works include further improving the proposed method and applying the paragraph vectors to more NLP tasks.

Acknowledgments. This work is supported by the National Natural Science Foundation of China (No. 61370165, 61203378), National 863 Program of China 2015AA015405, the Natural Science Foundation of Guangdong Province (No. S2013010014475), Shenzhen Development and Reform Commission Grant No.[2014]1507, Shenzhen Peacock Plan Research Grant KQCX20140521144507925 and Baidu Collaborate Research Funding.

References

1. Bengio, Y., Ducharme, R., Vincent, P.: A neural probabilistic language model. J. Mach. Learn. Res. **3**, 1137–1155 (2003)
2. Bizer, C., Lehmann, J., Kobilarov, G., Auer, S., Becker, C., Cyganiak, R., Hellmann, S.: Dbpedia-a crystallization point for the web of data. Web Semant. Sci. Serv Agents World Wide Web **7**(3), 154–165 (2009)
3. Collobert, R., Weston, J.: A unified architecture for natural language processing: Deep neural networks with multitask learning. In: Proceedings of the 25th International Conference on Machine Learning (ICML), pp. 160–167. ACM (2008)
4. Collobert, R., Weston, J., Bottou, L., Karlen, M., Kavukcuoglu, K., Kuksa, P.: Natural language processing (almost) from scratch. J. Mach. Learn. Res. **12**, 2493–2537 (2011)
5. Fan, R.E., Chang, K.W., Hsieh, C.J., Wang, X.R., Lin, C.J.: Liblinear: a library for large linear classification. J. Mach. Learn. Res. **9**, 1871–1874 (2008)
6. Hall, M., Frank, E., Holmes, G., Pfahringer, B., Reutemann, P., Witten, I.H.: The weka data mining software: an update. ACM SIGKDD explorations newsletter **11**(1), 10–18 (2009)
7. Kim, Y.: Convolutional neural networks for sentence classification. In: Proceedings of the 2014 Conference on Empirical Methods in Natural Language Processing (EMNLP), pp. 1746–1751 (2014)
8. Le, Q.V., Mikolov, T.: Distributed representations of sentences and documents. In: Proceedings of the 31th International Conference on Machine Learning (ICML). pp. 1188–1196 (2014)
9. Lehmann, J., Isele, R., Jakob, M., Jentzsch, A., Kontokostas, D., Mendes, P.N., Hellmann, S., Morsey, M., van Kleef, P., Auer, S., et al.: Dbpedia-a large-scale, multilingual knowledge base extracted from wikipedia. Semant. Web J. **5**, 1–29 (2014)
10. McAuley, J., Leskovec, J.: Hidden factors and hidden topics: understanding rating dimensions with review text. RecSys (2013)
11. Mikolov, T., Chen, K., Corrado, G., Dean, J.: Efficient estimation of word representations in vector space. In: Proceedings of Workshop at the International Conference on Learning Representations (ICLR) (2013)

12. Miller, G.A.: Wordnet: a lexical database for English. Commun. ACM **38**(11), 39–41 (1995)
13. Mnih, A., Hinton, G.E.: A scalable hierarchical distributed language model. In: Advances in Neural Information Processing Systems (NIPS), pp. 1081–1088 (2009)
14. Morin, F., Bengio, Y.: Hierarchical probabilistic neural network language model. In: Proceedings of the International Workshop on Artificial Intelligence and Statistics, pp. 246–252. Citeseer (2005)
15. dos Santos, C.N., Gatti, M.: Deep convolutional neural networks for sentiment analysis of short texts. In: Proceedings of the 25th International Conference on Computational Linguistics (COLING). Dublin, Ireland (2014)
16. Socher, R., Perelygin, A., Wu, J.Y., Chuang, J., Manning, C.D., Ng, A.Y., Potts, C.: Recursive deep models for semantic compositionality over a sentiment treebank. In: Proceedings of the Conference on Empirical Methods in Natural Language Processing (EMNLP). pp. 1631–1642. Citeseer (2013)
17. Zhang, X., LeCun, Y.: Text understanding from scratch. arXiv preprint arXiv:1502.01710 (2015)

A Comparison of Supervised Learning Techniques for Clustering

William Ezekiel[✉] and Umashanger Thayasivam[✉]

Mathematics Department, Rowan University, Glassboro, NJ, USA
ezekie97@students.rowan.edu, thayasivam@rowan.edu

Abstract. The significance of data mining has experienced dramatic growth over the past few years. This growth has been so drastic that many industries and academic disciplines apply data mining in some form. Data mining is a broad subject that encompasses several topics and problems; however this paper will focus on the supervised learning classification problem and discovering ways to optimize the classification process. Four classification techniques (naive Bayes, support vector machine, decision tree, and random forest) were studied and applied to data sets from the UCI Machine Learning Repository. A Classification Learning Toolbox (CLT) was developed using the R statistical programming language to analyze the date sets and report the relationships and prediction accuracy between the four classifiers.

Keywords: Classification technique · Supervised learning · Data mining · Naive Bayes · Decision tree · Random forest · Support vector machine

1 Introduction to Data Mining and Classification

Data mining is a common field of study for many practices. Businesses have generated customer profiles using data that describes when and where customers' purchases are made [1]. Mining of this data allows businesses to target their product to customers more effectively [1]. The most important application of data mining in the 21st century is Big Data, where data mining is utilized immensely. Massive amounts of genomic data are used by molecular biology researchers to compare the behavioral traits and molecular architecture of countless genes [1,2]. With this knowledge, genes that cause particular illnesses could be determined and isolated [1]. These are only some of the multiple applications of data mining, and they provide ample evidence that supports the usefulness of data mining.

Data mining is a broad topic that can be split into several subcategories, one of which is the classification problem. Classification is a form of supervised machine learning; in this type of machine learning, previously labeled data, training data, is used to generate a function or model that can classify future instances [3]. Classification is the process of taking new instances and predicting their class labels using a model [1]. This paper will focus on four common classification algorithms: naive Bayes, support vector machine, decision tree, and random forest.

© Springer International Publishing Switzerland 2015
S. Arik et al. (Eds.): ICONIP 2015, Part I, LNCS 9489, pp. 476–483, 2015.
DOI: 10.1007/978-3-319-26532-2_52

With so many classification techniques to choose from, it takes data miners some time to apply each technique to their data set, especially for the aforementioned Big Data. No technique is perfect, but some are better suited for certain data than others. The number of categorical attributes, the number of continuous attributes,and the number of training instances are all factors that affect the best classification technique to use on a data set.

In the following sections, a brief description of each classification technique, an explanation of the procedure used to perform the analysis, and the results and findings of the analysis will be given.

2   A Brief Overview of the Classification Techniques

Before examining the analysis itself, the common applicable techniques will briefly be discussed so that readers may have a better comprehension each method's different features.

2.1   Naive Bayes

Bayes Theorem provides a formula for determining the probability of a previously defined hypothesis remaining true under a new set of data or evidence [4]. The naive Bayes classifier uses Bayes Theorem and assumes that the conditional independence assumption holds true [1]. In the conditional independence assumption, two attributes are considered independent when a third attribute's observations are considered [5].

The posterior probabilities [1], or the probability of an event occurring given a new data [6], is calculated for each class 'Y' given a set of attributes X using Eq. (1).

$$P(Y|X) = \frac{P(Y) \prod_{i=1}^{d} P(X_i|Y)}{P(X)} \tag{1}$$

2.2   Support Vector Machine

Support vector machines utilize decision boundaries, specifically the maximal margin hyperplane [1,7]. Maximal margin hyperplanes create a decision boundary that accurately splits the data points into their corresponding classes (represented by different sides of the boundary) while maintaining the longest possible distance from closest possible training records for each class [1,7]. The maximum distance is chosen to reduce the classification error as much as possible [1].Training instances known as support vectors are used to represent the decision boundary [1]. Support vector machines, or SVM's for short, can be used for linear or nonlinear decision boundaries [1,7]. When outliers are present, a slack variable may be used to allow some amount of training error [1]. This prevents noise points from influencing the decision boundary and generating a less accurate model [1].

2.3 Decision Tree

Decision tree is a simple classifier. As the name implies, decision trees are tree like structures made up of nodes and edges. Edges are assigned values from the attribute in the node above them [1]. Nodes fall into three categories: decision nodes,leaf nodes, and the root node [1,7]. The root node is the very first node in the tree; it has no edges coming into it, but can have 0 or more edges coming. out of it [1]. Decision nodes represent an attribute and branches off into two or more values [7]. Leaf nodes are the ends of a decision tree; it is at these nodes a classification is made. When creating a decision tree, the goal is to break down the training data into smaller and smaller subsets as it moves down the tree [1]. Optimal decision trees are generated by determining the best split for all nodes by determining the best impurity measurement, such as gini or entropy [1].

2.4 Random Forest

Random forest is an ensemble method, or a technique that combines two or more classification methods into one [1]. In this case, two or more decision trees are combined [1]. Each decision tree makes its own classification , and after all classifications are made, the predictions are tallied to a vote [1]. The class that has the most votes becomes the classification for a new instance [1]. To keep the decision trees in the forest mostly unique, different bootstrapped samples [8] of the set with only some of the attributes are used for each tree [1]. As an ensemble method, the random forest is much better than a decision tree in terms of accuracy because of its ability to prevent one of decision tree's biggest issues, model overfitting [1]. This occurs when a decision tree becomes too big and complex [1]. Overly complex trees may have minimal error with the training data, but will perform poorly with new instances [1]. Decision trees that suffer from overfitting can even end up utilizing noise points [1]. Random forests overcome this issue by ensuring that each decision tree in the forest is unique [1,8]. By reducing the number of instances and attributes in each tree, the decision trees have far less of a chance of becoming too complicated [1].

3 The Analysis

3.1 Computational Algorithm

To analyze the performances of each classifier, twelve data sets from the UCI Machine Learning Repository [9–11] were acquired. The R statistical programming language was used to build the CLT, so tests could be run on each data set. The steps of the CLT's procedure are outlined below.

1. Load all the data sets into the R environment.
2. Run a classifier on the data three times using three different splits. The three values were 70, 80, and 90. A value of 80 for the split means that 80 % of the original data set is used as training records. The other 20 % is treated as new instances.

3. Compare the predictions for the new instances with their actual classifications using a confusion matrix.
4. Produce a '.csv' file containing the accuracies, time, number of training records, and number of new instances for each split of each data set.

Several R packages were implemented into the software in order to utilize the different techniques. The package 'e1071' contains functions for naive Bayes and SVM [12]. Two packages were used for random forest: 'randomForest' [13] and 'caret' [14]. The better accuracy between the two was chosen for random forests. The 'caret' package was also used to build the confusion matrices [14]. Decision trees were implemented using the 'rpart' package [15]. The package allows for the use of two impurity measurements: gini and entropy. One tree for each measurement was generated and the tree with the better accuracy was chosen for the results.

Table 1. UCI machine learning repository data sets

Data Set	T	N	Categorical	Continuous	Classes
adult	26049	6512	8	6	2
blood	599	149	0	4	2
credit	552	138	9	6	2
ionosphere	281	70	0	34	2
mushroom	6500	1624	21	0	2
parkinsons	156	39	0	2	2
pima	615	153	0	8	2
rvkp	2557	639	36	0	2
sonar	167	41	0	60	2
spambase	3681	920	0	57	2
telescope	15216	3804	0	10	2
tic-tac-toe	767	191	9	0	2

3.2 Data Analysis

The values of "T" and "N" are based on each data set's 80|20 split. "T" is the number of training records. "N" is the number of new instances. "Categorical" and "Continuous" are the number of categorical and continuous variables in the set respectively (Table 1).

3.3 Results

The overall classification error and rankings for each classifier are found in Table 2. The overall classification error was calculated [16] using Eq. 2, where

Table 2. Overall classification error and rankings of each classifier

Data Set	NB Error	SVM Error	DT Error	RF Error
adult	0.172 **(3)**	0.145 **(1)**	0.152 **(2)**	0.174 **(4)**
blood	0.3 **(4)**	0.28 **(2)**	0.247 **(1)**	0.287 **(3)**
credit	0.203 **(4)**	0.188 **(3)**	0.159 **(2)**	0.123 **(1)**
ionosphere	0.085 **(3)**	0.042 **(2)**	0.127 **(4)**	0.028 **(1)**
mushroom	0.052 **(4)**	0.003 **(2)**	0.006 **(3)**	0 **(1)**
parkinsons	0.41 **(4)**	0.154 **(2)**	0.205 **(3)**	0.128 **(1)**
pima	0.201 **(1)**	0.214 **(2)**	0.292 **(4)**	0.24 **(3)**
rvkp	0.139 **(4)**	0.058 **(3)**	0.031 **(2)**	0.014 **(1)**
sonar	0.405 **(4)**	0.238 **(3)**	0.214 **(2)**	0.19 **(1)**
spambase	0.284 **(4)**	0.061 **(2)**	0.098 **(3)**	0.052 **(1)**
telescope	0.278 **(4)**	0.134 **(2)**	0.18 **(3)**	0.121 **(1)**
tic-tac-toe	0.302 **(4)**	0.089 **(2)**	0.109 **(3)**	0.01 **(1)**
Mean error	0.236	0.134	0.152	0.114

TP, TN, FP, FN represent the True Positive, True Negative, False Positive and False Negative rates of the confusion matrix respectively.

$$Error = 1 - (\frac{TP + TN}{TP + TN + FP + FN}) \tag{2}$$

In Table 2, only the results of the 80|20 split were recorded, but the other two splits showed similar results.

Note: if a classifier's bar is not shown for a data set, the overall classification error was 0 (Fig. 1).

Random forests surpassed the other techniques, having the lowest overall classification error in most data sets. SVM performed second best, followed by decision tree. The worst performing classifier is naive Bayes, which had to worst error values in nearly all of the sets. Random forests low error value is most likely due to the fact that random forest is an ensemble method. Ensemble methods are mean to perform better than their singleton counterparts [1].

The types of attributes impact the overall error classifications of these classifiers. Random forest and decision trees, for example, perform best when the data sets consists entirely of categorical attributes, but only if there is a high amount (2000+) of instances (*see* mushroom, rvkp, and tic-tac-toe). SVM's overall classification error decreases as the total number of attributes (both types) increases. SVM also has better performance when more continuous attributes are present. no discernible patterns could be found for naive Bayes.

Another important factor when measuring the performance of a classifier is the false positive (FP) rate. The false positive rate is defined as the proportion of minority (negative) cases that were inaccurately classified as a majority

Overall Classification Error for UCI Data Sets

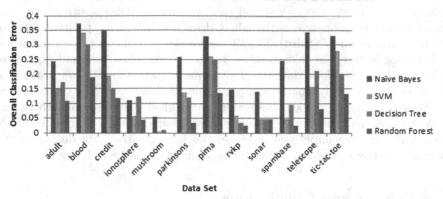

Fig. 1. Overall classification error (Color figure online).

Table 3. False positive rates for each classifier

Data Set	NB FP	SVM FP	DT FP	RF FP
adult	0.501 (3)	0.42 (1)	0.485 (2)	0.724 (4)
blood	0.814 (2)	0.837 (3)	0.698 (1)	0.837 (3)
credit	0.434 (4)	0.132 (1)	0.208 (2)	0.226 (3)
ionosphere	0.06 (3)	0 (1)	0.12 (4)	0 (1)
mushroom	0.097 (4)	0.006 (2)	0.011 (3)	0 (1)
parkinsons	0.414 (4)	0 (1)	0.103 (3)	0 (1)
pima	0.298 (1)	0.439 (4)	0.368 (2)	0.421 (3)
rvkp	0.1 (4)	0.044 (3)	0.035 (2)	0.018 (1)
sonar	0.333 (2)	0.429 (4)	0.19 (1)	0.381 (3)
spambase	0.043 (1)	0.102 (3)	0.159 (4)	0.086 (2)
telescope	0.626 (4)	0.299 (2)	0.374 (3)	0.223 (1)
tic-tac-toe	0.183 (4)	0 (1)	0.023 (3)	0 (1)
Mean value	0.325	0.226	0.231	0.243

(positive) case [17]. Equation 3 shows the calculations for the false positive rate [17].

$$FP = \frac{\#\,of\,Negative\,Instances\,Predicted\,as\,Positive}{\#\,of\,Actual\,Negative\,Instances} \qquad (3)$$

The false positive rate finds its significance in the classification of the minority class. A classifier may be able to perform well overall, but if it has a high false positive rate, the classifier has trouble classifying new instances belonging to the minority. This is never good, especially in situations of great importance.

Table 3 shows the false positive rates for each classifier. Despite random forest ranking first in most cases, SVM had the best average false positive rate. The second best is decision tree, although it does not have false positive rates of zero like random forest (third). Finally, naive Bayes is shows the overall worst performance in terms of false positive rates. All the classifiers had high false positive rates in the blood data set. This is most likely due to is low amount of attributes and relatively small training sample. Decision trees had the best false positive rates when all the attributes were categorical (*see* rvkp, mushroom and tic-tac-toe). A trend seen in the naive Bayes classifier is that as total attribute count and training sample size increase, the false positive rate decreases. No discernible patterns could be noted for random forest and SVM.

4   Summary and Conclusions

Four supervised classification techniques were tested with twelve different data sets to determine any relationships between their accuracies and the data sets. The results of this presented some ideas about which techniques work best for which data sets. Further investigations aim to apply this to more data sets, see if the notions given here continue to hold true for more data sets, including multi-class sets, and to add support for data set balancing techniques, such as the Synthetic Minority Over-sampling Technique (SMOTE), to the CLT. With these discoveries, the optimization of classification could be realized.

Acknowledgement. This work was supported via Grant provided by the Rowan University Mathematics Department.

References

1. Tan, P., Steinbach, M., Kumar, V.: Introduction to Data Mining. Pearson Addison Wesley, Boston (2006)
2. Khoury, M.J.: Public health approach to big data in the age of genomics: how can we separate signal from noise? Centers for Disease Control and Prevention (2014). http://blogs.cdc.gov/genomics/2014/10/30/public-health-approach/
3. Donalek, C.: Supervised and Unsupervised Learning (2011). http://www.astro. caltech.edu/~george/aybi199/Donalek_Classif.pdf
4. An Introduction to Bayes Theorem. Bayes Theorem: Introduction. http://www. trinity.edu/cbrown/bayesweb/
5. Leyton-Brown, K.: Reasoning Under Uncertainty: Marginal and Conditional Independence. http://www.cs.ubc.ca/~kevinlb/teaching/cs322%20-%202006-7/ Lectures/lect25.pdf
6. Investopedia US, A Division of IAC.http://www.investopedia.com/terms/p/ posterior-probability.asp
7. Sayad, S.: An Introduction to Data Mining (2015). http://www.saedsayad.com/
8. Ho, T.: The random subspace method for constructing decision forests. IEEE Trans. Pattern Anal. Mach. Intell. **20**(8), 832–844 (1998)

9. Lichman, M.: UCI Machine Learning Repository (2013). http://archive.ics.uci.edu/ml
10. Little, M.A., McSharry, P.E., Roberts, S.J., Costaello, D., Moroz, I.: Exploiting nonlinear recurrence and fractal scaling properties for voice disorder detection. BioMed. Eng. Online **6**(23) (2007). doi:10.1186/1475-925X-6-23
11. Yeh, I., Yang, K.J., Ting, T.: Knowledge discovery on RFM model using Bernoulli sequence. Expert Syst. Appl. **36**, 5866–5871 (2008)
12. Meyer, D., Dimitriadou, E., Hornik, K., Weingessel, A., Leisch, F.: Misc Functions of the Department of Statistics (e1071). Institute for Statistics and Mathematics of WU, TU Wien (2014). http://cran.r-project.org/web/packages/e1071/e1071.pdf
13. Cutler, A., Breiman, L., Liaw A., Wiener, M.: RandomForest: Breiman and Cutler's Random Forests for Classification and Regression. Institute for Statistics and Mathematics (2015). http://cran.r-project.org/web/packages/randomForest/randomForest.pdf
14. Khun, M., Wing, J., Weston, S., Williams, A., Keefer, C., Engelhardt, A., Cooper, T., Mayer, Z., Kenkel, B., Benesty, M., Lescarbeau, R., Ziem, A., Scrucca, L.: Classification and Regression Training. Institute for Statistics and Mathematics (2015). http://cran.r-project.org/web/packages/caret/caret.pdf
15. Therneau, T., Atkinson, B., Ripley, B.: Recursive partitioning for classification, regression and survival trees. An implementation of most of the functionality of the 1984 book by Breiman, Friedman, Olshen and Stone. Institute for Statistics and Mathematics (2015). http://cran.r-project.org/web/packages/rpart/rpart.pdf
16. Chawla, N.V., Bowyer, K.W., Hall, L.O., Kegelmeyer, W.P.: SMOTE: synthetic minority over-sampling technique. J. Artif. Intell. **16**, 321–357 (2002)
17. Hamilton, H.: Confusion Matrix (2012). http://www2.cs.uregina.ca/~dbd/cs831/notes/confusion_matrix/confusion_matrix.html

Radar Pattern Classification Based on Class Probability Output Networks

Lee Suk Kim[1], Rhee Man Kil[1(✉)], and Churl Hee Jo[2]

[1] College of Information and Communication Engineering, Sungkyunkwan University, 2066, Seobu-ro, Jangan-gu, Suwon, Gyeonggi-do 440-746, Korea
{lktime,rmkil}@skku.edu
[2] Electronic Warfare R&D Laboratory, LIGNex1 Co., Ltd., 333, Pangyo-ro, Bundang-gu, Seongnam City, Gyeonggi-do 463-400, Korea
churlhee.jo@lignex1.com

Abstract. Modern aircraft and ships are equipped with radars emitting specific patterns of electromagnetic signals. The radar antennas are detecting these patterns which are required to identify the types of emitters. A conventional way of emitter identification is to categorize the radar patterns according to the sequences of frequencies, time of arrivals, and pulse widths of emitting signals by human experts. In this respect, this paper presents a method of classifying the radar patterns automatically using the network of calculating the p-values of testing the hypotheses of the types of emitters referred to as the class probability output network (CPON). Through the simulation for radar pattern classification, the effectiveness of the proposed approach has been demonstrated.

Keywords: Radar pattern · Classification · One class · Class probability · Beta distribution

1 Introduction

In modern days, radars are essential devices to detect objects such as aircraft or ships. For detecting objects emitting specific patterns of electromagnetic signals, the detected signal patterns should be analyzed and categorized according to the types of emitters. This emitter identification plays an important role especially in the electronic warfare [1]. The robust performances of emitter identification becomes more important in complex environments of emitters and landscapes. In the conventional approach of emitter identification, the key features of radar patterns such as the sequences of radar frequencies (RFs), time of arrivals (TOAs), and pulse widths (PWs) are used to extract the emitter parameters and these parameters are compared with tabulated emitter parameters. However, this process usually requires high computational complexity and needs to be verified by human experts. In this respect, an approach of automatic classification of radar patterns is proposed to obtain the conditional class probability for the given radar pattern.

© Springer International Publishing Switzerland 2015
S. Arik et al. (Eds.): ICONIP 2015, Part I, LNCS 9489, pp. 484–491, 2015.
DOI: 10.1007/978-3-319-26532-2_53

There are various ways of implementing pattern classifiers. The most popular way is using the discriminant function whose value indicates the degree of confidence for the classification; that is, the decision of classification is made by selecting the class that has the greatest discriminant value. In this direction, the support vector machines (SVMs) [2] are widely used in many classification problems because they provide reliable performances by maximizing the margin between the positive and negative classes. However, more natural way of representing the degree of confidence for classification is using the conditional class probability for the given pattern. In this context, the class probability output network (CPON) in which the conditional class probability is estimated using the beta distribution parameters, was proposed [3]. This method is implemented on the top of a classifier; that is, many-to-one nonlinear function such as the linear combination of kernel functions. Then, the classifier's output is identified by beta distribution parameters and the output of CPON; that is, the conditional class probability for the given pattern is calculated from the cumulative distribution function (CDF) of Beta distribution parameters. In this computation, the output of CPON represents the p-value of testing a certain class. For the final decision of classification, the class which has the maximum conditional class probability is selected. As a result, the suggested CPON method is able to provide consistent improvement of classification performances for the classifiers using discriminant functions alone. For the detailed descriptions of CPONs and CPON applications, refer to [3,4]. In this approach, the selected features of radar patterns are used as the input to the classifier of many-to-one mapping nonlinear function and the output distribution is identified by beta distribution parameters to obtain the p-value of testing the type of emitters. As a result, the proposed method provides the p-values of testing hypotheses of the types of emitters and the better performances of classification than other classifiers using the discriminant function.

The rest of this paper is organized as follows: in Sect. 2, the problem of radar classification is described, Sect. 3 presents the method of radar pattern classification using the CPON, Sect. 4 shows simulation results for radar pattern classification, and finally, Sect. 5 presents the conclusion.

2   Key Features for Radar Pattern Classification

The proposed method is intended to identify radar patterns from various emitters. In this approach, it is assumed that the radar has the ability to monitor a region of microwave spectrum and to extract pulse patterns. The whole process of emitter identification (or radar pattern classification) is illustrated in Fig. 1. In this diagram, the feature extractor receives pulses from the microwave radar receiver and processes each pulse into feature values such as azimuth, elevation, intensity, frequency, and pulse width. These data are then stored and tagged with the time of arrival of the pulse. Then, the clustering block is grouping radar pulses into groups in which each group represents radar pulses from a single emitter. For each group of radar pulses, the pulse extraction block is analyzing the pulse repetition patterns of an emitter by using the information of time of arrivals. Finally, from the information of pulse repetition patterns, input features for the classifier are computed and the decision for the classification of emitters is made based on extracted key features.

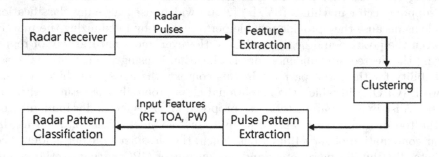

Fig. 1. Process of emitter identification

In the proposed approach, the selected key features are RFs, TOAs, and PWs. Then, for each sequence of key feature values x_i, $i = 1, \cdots, n$, the statistical measures such as the mean $\bar{x}$, variance s^2, skewness, and kurtosis are determined by

$$\bar{x} = \frac{1}{n} \sum_{i=1}^{n} x_i, \tag{1}$$

$$s^2 = \frac{1}{n-1} \sum_{i=1}^{n} (x_i - \bar{x})^2, \tag{2}$$

$$\text{skewness} = \frac{\frac{1}{n} \sum_{i=1}^{n} (x_i - \bar{x})^3}{(\frac{1}{n-1} \sum_{i=1}^{n} (x_i - \bar{x})^2)^{\frac{3}{2}}}, \quad \text{and} \tag{3}$$

$$\text{kurtosis} = \frac{\frac{1}{n} \sum_{i=1}^{n} (x_i - \bar{x})^4}{(\frac{1}{n-1} \sum_{i=1}^{n} (x_i - \bar{x})^2)^2} - 3. \tag{4}$$

These statistical feature values are calculated for every sweep of received radar signals. Then, as a result, 12 feature values are used as the input to the classifier and the decision of emitter identification is made by using the CPON.

In this approach, the distributions of these feature values are analyzed and the centroids representing the types of emitters are determined as the center points for the distributions of radar patterns. Then, these distributions are used for the decision of determining the specific emitter type in the CPON.

3   Class Probability Output Networks for Emitter Identification

In many classification problems, it is desirable that the output of a classifier represents the conditional class probability. For the conditional class probability, the distribution of classifier's output can be well approximated by the beta distribution under the assumption that the output of classifier lies within a finite range and the distribution of classifier's output is unimodal; that is, the distribution has one modal value with the greatest frequency. This assumption is quite reasonable for many cases of classification problems with the proper selection of kernel parameters of a classifier. Here, we consider the following discriminant function y as the classifier's output for the input pattern $\mathbf{x}$:

$$\hat{y}(\mathbf{x}) = \sum_{i=1}^{m} w_i \phi_i(\mathbf{x}|\theta), \tag{5}$$

where m represents the number of kernels and w_i, ϕ_i, and θ represent the ith weight, the ith kernel function, and the kernel parameter, respectively.

In the proposed CPON, the probability model represents the conjugate prior of the binomial distribution; that is, in our case, the conditional class probability in binary classification problems. In this context, we consider the following Beta probability density function (PDF) of a random variable Y as the normalized classifier's output:

$$f_Y(y|a,b) = \frac{1}{B(a,b)} y^{a-1}(1-y)^{b-1}, \quad 0 \le y \le 1, \tag{6}$$

where a and b represents the parameters of beta distribution, and $B(a,b)$ represents a Beta function defined by

$$B(a,b) = \int_0^1 y^{a-1}(1-y)^{b-1} dy. \tag{7}$$

Here, we assume that the classifier's output value; that is, $\hat{y}$ is normalized between 0 and 1. One of the advantages of the Beta distribution is that the distribution parameters can be easily guessed from the mean $E[Y]$ and variance $Var(Y)$ as follows:

$$a = E[Y]\left(\frac{E[Y](1-E[Y])}{Var(Y)} - 1\right) \tag{8}$$

and

$$b = (1 - E[Y])\left(\frac{E[Y](1-E[Y])}{Var(Y)} - 1\right). \tag{9}$$

Although this moment matching (MM) method is simple, these estimators usually don't provide accurate estimations especially for smaller number of data. In such cases, the maximum likelihood estimation (MLE) or the simplex method for searching parameters [5] can be used for more accurate estimation of Beta parameters. If the data distribution follows a Beta distribution and the optimal Beta parameters are obtained, the ideal cumulative distribution function (CDF) values of the data $u = F_Y(y)$ follow an uniform distribution; that is,

$$f_U(u) = \frac{f_Y(y)}{|dF_Y/dy|} = \frac{f_Y(y)}{|f_Y(y)|} = 1. \tag{10}$$

To check whether the data distribution fits with the proposed Beta distribution, the Kolmogorov-Smirnov (K-S) test [6] of data distribution can be considered as follows:

- First, determine the distance D_n between the empirical and ideal CDF values:

$$D_n = \sup_u |F_U^*(u) - F_U(u)|, \tag{11}$$

where $F_U^*(u)$ and $F_U(u)$ represent the empirical and theoretical CDFs of $u = F_Y(y)$; that is, the CDF values of the normalized output of a classifier. In this case, $F_U(u) = u$ since the data $u = F_Y(y)$ follow an uniform distribution if the data y follows the presumed (or ideal) Beta distribution.
- Determine the p-value of testing the hypothesis of Beta distribution:

$$p\text{-value} = P(D_n \geq t/\sqrt{n}) = 1 - H(t), \tag{12}$$

where $t = \sqrt{n}d_n$ (the value of a random variable D_n) and the CDF of the K-S statistic $H(t)$ is given by

$$H(t) = \frac{\sqrt{2\pi}}{t} \sum_{i=1}^{\infty} e^{-(2i-1)^2 \pi^2/(8t^2)}. \tag{13}$$

- Make a decision of accepting the hypothesis of beta distribution H_0 using the p-value according to the level of significance δ:
  accept H_0, if p-value $\geq \delta$; reject H_0, otherwise.

In the construction of CPON for radar pattern recognition, first, the centroids as the representative of the radar pattern data are obtained in the feature space by a clustering algorithm such as the learning vector quantization (LVQ) method [7]. Then, the kernel functions are located at the positions of centroids and linearly combined as the form of (5). The output of (5) is normalized between 0 and 1 by using the linear scale and the normalized classifier's output distribution is approximated by the Beta distribution parameters. In this training of classifiers, the Beta distribution parameters as well as the kernel parameters are adjusted in such a way that the classifier's output distributions become closer to the ideal Beta distributions. The algorithm of constructing the CPON for radar pattern classification is described as follows:

Step 1. For the features of radar patterns, centroids are determined by the clustering algorithm such as the LVQ method. In this application, one centroid is assigned to a specific emitter. For more complicated distributions in the feature space, more than one centroids can be assigned.
Step 2. Then, for each centroid, a kernel function is assigned.
Step 3. Determine the classifier's output for each kernel function and normalize the output value between 0 and 1 using the linear scale.
Step 4. The distribution of classifier's normalized output is identified by Beta distribution parameters. In this estimation of Beta parameters, the kernel parameters such as the kernel widths are adjusted in such a way of maximizing the p-value of (12). For the detailed description of estimating parameters, refer to [3].

After the CPON is trained, the classification for an unknown pattern can be determined by the beta distribution for each class. First, for the unknown pattern, the normalized output y for the classifier is computed. Here, if the normalized value is greater than 1, we set that value as 1; on the other hand, if the value is less than 0, we set that value as 0. Then, the conditional class probability is determined by the CPON output as the CDF value for the classifier's normalized output.

For multi-class classification problems, the CPON can be constructed for each classifier's output. Then, the following conditional probability for the kth class C_k; that is, the output of the kth CPON $F_k(y_k)$ for kth classifier's normalized output y_k is calculated as

$$F_k(y_k) = P(C_k|Y_k \leq y_k) = F_{Y_k}(y_k), \tag{14}$$

where Y_k represents a random variable for the kth class C_k and $F_{Y_k}(y_k)$ represents its CDF. This output implies the p-value of testing hypotheses of the kth class C_k. Then, the final decision can be made by selecting the class with the maximum p-value; that is, for K classes, the selected class C_l is determined by

$$l = \arg \max_{1 \leq k \leq K} F_k(y_k). \tag{15}$$

From the above equation, the final decision of the type of emitter is made.

4   Simulation

To demonstrate the effectiveness of the proposed method, the simulation for radar pattern classification was performed for the radar data patterns generated from the emitter simulator developed by LIGNex1. This simulator was designed to accommodate the variation of key features such as the RFs, TOAs, and PWs of real emitters. In this benchmark data, there were 50 sets of emitter types (or classes) in which each data set included 100 sequences of emitter patterns containing the features of RFs, TOAs, and PWs. For the evaluation of the proposed method, 10-fold evaluation method was used: 10 disjoint sets of 90% of data as training data and the rest 10% of data as test data were used. Then, the average performances of the following accuracy for 50 classes were determined:

$$\text{Accuracy} = \frac{1}{50} \sum_{i=1}^{50} \frac{TP_i + TN_i}{TP_i + TN_i + FP_i + FN_i}, \tag{16}$$

where TP_i, TN_i, FP_i, and FN_i represent the true positive, true negative, false positive, and false negative of the ith classifier, respectively.

However, in this evaluation of classification performances, the accuracy measure can be misleading particularly in the multi-class and/or unbalanced data. From this point of view, the following exact match ratio (EMR) for n test patterns were also determined:

$$\text{EMR} = \frac{1}{n} \sum_{i=1}^{n} I(\mathbf{x}_i), \tag{17}$$

where $\mathbf{x}_i$ represents the ith test pattern and $I(\mathbf{x}_i)$ represents the following indicator function:

$$I(\mathbf{x}_i) = \begin{cases} 1 & \text{if } L(\mathbf{x}_i) = D(\mathbf{x}_i) \\ 0 & \text{otherwise.} \end{cases}$$

Here, $L(\mathbf{x}_i)$ and $D(\mathbf{x}_i)$ represent the class label and the decision label of the classifier for the ith test pattern, respectively; that is, the EMR represents the ratio of correct decision of the classifier.

To compare the performances of the CPON-based method, the k-nearest neighbor (kNN) and SVM classifiers using the Scikit-learn package [8] were also trained for the same training data and evaluated for the same test data. In this simulation, the same features of RFs, TOAs, and PWs were also used for the training and testing the classifiers. In the case of SVM, 50 binary (one-against-rest) classifiers were trained and tested for the given data. The simulation results for emitter identification were summarized in Table 1.

Table 1. Simulation results for emitter identification

Classifier	kNN	SVM	CPON
Accuracy	**0.9944**	0.9934	0.9861
EMR	0.7952	0.6734	**0.9622**

These simulation results have shown that (1) three classifiers were good in the accuracy measure, (2) the SVM was not a good choice for these multi-class data from a view point of the EMR measure, and (3) the proposed method provided the best performance in the EMR measure compared with other classifiers. This implies that the proposed statistical features of the RFs, TOAs, and PWs are quite effective to identify the characteristics of the emitter types and the proposed CPON-based classification is also an effective approach for the problems of emitter identification. Furthermore, the proposed CPON-based method is also able to provide p-values for testing the types of emitters. In practice, this information of p-values helps us to make a decision whether the received radar pattern is a new type of emitter or one of known types of emitters. For example, if the maximum p-value is less than some threshold value (the usual value is 0.05), then there is a high probability that the received radar pattern comes from a new type of emitter. This ability of finding a new type of emitter is also an important issue in emitter identification problems.

5   Conclusion

A new method of radar pattern classification was proposed based on the class probability output network (CPON). In the proposed method, the sequences of key features such as the frequencies, time of arrivals, and pulse widths of emitting signals are analyzed and statistical measures of these features such as the mean, variance, skewness, and kurtosis are extracted and used as the input to the CPON. Then, the CPON is used to construct a hypothesis of specific emitter from the distributions of these features. As a result, the proposed CPON provides the p-values of testing the hypotheses of the types of emitters. Through the simulation for radar pattern classification, it has been demonstrated that the proposed method provides the better performance of classification than other classifiers using the discriminant function. From the classification performance of the proposed CPON-based emitter identification, it is expected to be comparable with human experts. Furthermore, the proposed CPON-based method is able to provide the information for making a decision whether the received radar pattern comes from a new type of emitter.

References

1. Schleher, D.: Introduction to Electronic Warfare. Artech House, New York (1986)
2. Vapnik, V.: Statistical Learning Theory. Wiley, New York (1998)
3. Park, W., Kil, R.: Pattern classification with class probability output network. IEEE Trans. Neural Netw. **20**(10), 1659–1673 (2009)
4. Rosas, H., Kil, R., Han, S.: Automatic media data rating based on class probability output networks. IEEE Trans. Cons. Electron. **56**(4), 2296–2302 (2010)
5. AbouRizk, S., Halpin, D., Wilson, J.: Fitting beta distributions based on sample data. J. Constr. Eng. Manag. **120**(2), 288–305 (1994)
6. Rohatgi, V., Saleh, A.: Nonparametric statistical inference. In: An Introduction to Probability and Statistics, 2nd edn. Wiley, New York (2001)
7. Kohonen, T.: Improved versions of learning vector quantization. In: IEEE International Joint Conference on Neural Networks, vol. 1, pp. 545–550 (1990)
8. Pedregosa, F., et al.: Scikit-learn: machine learning in Python. J. Mach. Learn. Res. **12**, 2825–2830 (2011)

Hierarchical Data Classification Using Deep Neural Networks

Sreenivas Sremath Tirumala[✉] and A. Narayanan

Auckland University of Technology, Auckland, New Zealand
ssremath@aut.ac.nz

Abstract. Deep Neural Networks (DNNs) are becoming an increasingly interesting, valuable and efficient machine learning paradigm with implementations in natural language processing, image recognition and hand-written character recognition. Application of deep architectures is increasing in domains that contain feature hierarchies (FHs) i.e. features from higher levels of the hierarchy formed by the composition of lower level features. This is because of a perceived relationship between on the one hand the hierarchical organisation of DNNs, with large numbers of neurons at the bottom layers and increasingly smaller numbers at upper layers, and on the other hand FHs, with comparatively large numbers of low level features resulting in a small number of high level features. However, it is not clear what the relationship between DNNs hierarchies and FHs should be, or whether there even exists one. Nor is it clear whether modelling FHs with a hierarchically organised DNN conveys any benefits over using non-hierarchical neural networks. This study is aimed at exploring these questions and is organized into two parts. Firstly, a taxonomic FH with associated data is generated and a DNN is trained to classify the organisms into various species depending on characteristic features. The second part involves testing the ability of DNNs to identify whether two given organisms are related or not, depending on the sharing of appropriate features in their FHs. The experimental results show that the accuracy of the classification results is reduced with the increase in 'depth'. Further, improved performance was achieved when every hidden layer has the same number of nodes compared with DNNs with increasingly fewer hidden nodes at higher levels. In other words, our experiments show that the relationship between DNNs and FHs is not simple and may require further extensive experimental research to identify the best DNN architectures when learning FHs.

Keywords: Deep Neural Networks · Hierarchical data classification

1 Introduction

Artificial Neural Networks (ANNs) are 'once again' popular due to the success of 'Deep Learning' involving multi-layer neural networks for solving tasks that are too complex for single-layer or dual-layer neural networks. Common ANN

© Springer International Publishing Switzerland 2015
S. Arik et al. (Eds.): ICONIP 2015, Part I, LNCS 9489, pp. 492–500, 2015.
DOI: 10.1007/978-3-319-26532-2_54

learning problems persist in Deep Neural Networks (DNNs). If training is too long, test results can be poor because the weights have become too specialized (overfitting). If training is too short, training results can be poor, leading to poor overall results on the full dataset even when improved test results are taken into account (underfitting). Introducing a recalibration training dataset (i.e. extra training data introduced after an initial training set) to help deal with overfitting or underfitting can lead to oscillation of weights and unlearning of initial samples. Recently the problem of overfitting was addressed in DNNs to some extent [1]. The idea is to randomly drop units during learning to prevent unit over-adaptation. But a number of different DNNs need to be trained and then converged through averaging at the final stage to result in a thinned DNN. Also, it is not clear what the implications of thinned DNNs are for data representation. It may not be clear what exactly is being thinned when hierarchical features are present in the data.

Historically, the concept of DNNs was proposed in 1989 as Convolutional Neural Networks (CNNs) without using the word 'Deep'. Back Propagation (BP) was used to train CNNs and was shown to be not so effective because of the limitations of BP. For instance, feedback is to only the immediately previous layer. After the introduction of a new greedy-layer-wise training, ANNs once again became popular in the form of DNNs [2]. The learning mechanism of DNNs is called Deep Learning and has so far proved to be successful over SVM based systems on the same problems [3]. DNNs have become increasingly successful with applications in natural language processing [4], image recognition [5], visual recognition [6], computer vision [7], text mining [8] and hand-written character recognition [9]. Corporate giants like Apple, Google and Microsoft are using deep learning principles for data mining and Facebook and Twitter have invested in DNN research for understanding aspects of social interactions. Further implementation of DNNs using evolutionary principles is also gaining momentum [10]. One of the major problems with all neural networks, however, remains understanding why the ANN is performing the way it does and formalising this understanding in terms of mathematical formulae or rule-based models.

Application of deep architectures is increasing in the domains that contain feature hierarchies (FHs) i.e., features from higher levels of the hierarchy formed by the composition of lower level features. While previous research has indicated that DNNs are quite successful when using flat representations, the capability of DNNs for FH learning has not been explored. The aim of this paper is to undertake some exploratory analysis and evaluation of DNNs using synthetic data known to contain FHs and to evaluate DNNs to identify exactly how DNNs handle data expressing an FH. It is important to understand whether an effective conjugation occurs between two (hidden) layers [11]. As a study, we try to understand this by experimenting with an equal number of nodes for hidden layers versus unequal number of nodes.

A synthetic dataset is generated with 6 classes (species) of organisms organized by FHs. To measure the FHs of a dataset, the cophenetic correlation coefficient was calculated from the plotted dendrogram of the data. The resulting coefficient value of 0.9934 is considered to be sufficiently precise for experimental purposes. For the first task of classifying organisms into various species based on their FHs, experiments are conducted with two strategies: varying DNN depth and changing the number of nodes for every DNN hidden layer. The results show that varying the depth is effective in both cases. Further, a DNN topology with the same number of nodes in all hidden layers is better than having different number of nodes at different layers, despite the reducing hierarchical aspects of the FHs involved 'from bottom to top'. Detailed observations are presented in experimental results section.

The paper is organized as follows. Section 2 introduces various types of Deep Architectures. Section 3 briefly explains the data representation and synthetic data used in this research. Experimental results are presented in Sect. 4 followed by Conclusion and Future Directions in Sect. 5.

2   History of Deep Neural Networks

The number of layers of an ANN constitutes its depth. If the number of layers is more than 1, the ANN architecture can be Deep if multiple layers are used to form DNNs [12]. Feed-forward ANNs with more than one layer of connections can solve more problems and be more accurate than one-layer [12]. We define a 'hidden layer' in a DNN as any layer of connections or units/nodes apart from those at the input and output stages. We let the context determine whether we refer to hidden connections or hidden units/nodes. Theoretical studies also support the statement that DNNs have the advantage of more efficient representation compared with shallow networks and with fewer hidden units [12]. Unlike ANNs, the layers of CNNs have neurons arranged in 3 dimensions for overlapping purposes. CNNs were first proposed by Fukushima as Neocognition [13]. CNNs, being the first form of a DNN, use the standard BP algorithm for training and weights are updated using $\Delta w_{ij}(t+1) = \Delta w_{ij}(t) + \eta \frac{\partial C}{\partial w_{ij}}$ where η represents the learning rate, C is the associated cost function , w_{ij} represents the weight between unit i and unit j and t represents time.

The major problem with DNNs is that BP results in slow learning with increasing dimensionality of the data, specially with large volumes of training data, where the process of training DNNs is done in two steps. Firstly, the data is divided into small batches of data followed by the batch-wise training process [4]. With this, there will be an increase in the number of parameters to be optimized. To overcome this problem, a new greedy-layer-wise training procedure was introduced. The layer wise training mechanism was first implemented successfully by Lecun on CNNs, Hinton on Deep Belief Networks (DBNs) and Bengio on stacked auto-encoders. Lecun extended Fukushima's work by training each layer of CNN with BP algorithm training [5].

DBN is a multi-layer perceptron (MLP) based DNN with connections existing only between the layers but not between the neurons within the layer [14]. DBNs have multiple interconnected hidden layers where each layer acts as an input to the next layer without lateral connection between the nodes present in that layer. DBN is constructed using supervised models like Restricted Boltzmann Machines (RBMs). An RBM is formed by applying a restriction on Boltzmann machines that the units within the layers are not connected with each other but can have connections with the units of other layers. DBN uses probabilistic logic nodes and Softmax as activation function. Bengio's stacked auto-encoders implements encoding and decoding mechanisms using ANNs. The main aim of auto-encoders is to reproduce the input [2]. Initially both encoder and decoder networks are assigned with random weights and trained by observing the discrepancy between original data and the reconstructed output. The error is back propagated through the decoder network followed by encoder network. A stacked de-noising auto-encoder algorithm was proposed in 2010 that reduced the performance gap between RBM based DBNs and auto-encoder based DNNs [15].

3   Data Representation

Binary number representation generates datasets using just two digits 0 and 1. Connectionist methods of data representation can be categorized into specific (localist) or spread out (distributed). In a localist representation each neuron or unit is associated with a single feature and each feature is represented by one and only one neuron or unit [16]. Localist representation is simple, easy to code and understand. However, localist representation cannot be used for componential structure-based data such as FHs. In distributed representation a single concept is represented by a combination of neurons or units and each neuron or unit can be a part of multiple concept [16]. In a distributed representation, an isolated neuron has no meaning by itself and the existence of a neuron has meaning only when it is present in a group.

Deep learning is based on feature learning which is also known as representation learning and we use gradient descent as the most suitable procedure to train DNNs. Since, distributed representation is considered efficient for representing taxonomic (hierarchical) data, it makes gradient based learning as ideal choice to test FHs in distributed representation. With binary encoding, n neurons can produce 2^n patterns when distributed representation patterns are used. A localist representation represents a single feature or concept on its own, such as an organism having backbone, hair etc., which determines the uniqueness of the organism. If the organism has multiple features which are represented as 8 localist bits, each bit will stand for the presence or absence of a concept or feature. For instance, an organism that has a backbone is coded as C1 with last bit as 1 and is represented as 0 0 0 0 0 0 0 1. Similarly, other features may be represented as shown in Fig. 1 (a). The remaining bits of a localist representation can be used to identify individual organisms.

FHs classification of organisms into taxa is based on the features they possess. Since each feature of an organism is represented in bits, organisms with multiple features are represented in localist form as a combination of binary bits. For instance, organism O1 has backbone and hair which are C1 - 00000001 and C2 - 00000010 as presented in Fig. 1 (a). So, the features of organism O1 are

Features	Representation
C1 (Backbone)	00000001
C2 (Hair)	00000010
C3 (Hands and Feet)	00000100
C4 (hair on hands)	00001000

Organism	Features	Representation
O1	C1 and C2	00000011
O2	C1, C3, C4	00001101

(a) Localist representation   (b) Multiple features

Fig. 1. Localist representation of features

Organism	Rank	Group	Sub-Group	Features
1	0001	1101	1101	11011101
2	0010	1010	1010	10101010
3	0011	0101	0101	01010101
4	0100	0100	0100	01000100
5	0101	1001	1001	10011001
6	0110	1011	1101	10111001

Fig. 2. Binary Representation of Organism

represented as 00000011 with combined features as shown in Fig. 1 (b). Similarly, organism O2 has backbone, hands and feet, and hair on the hands (C1, C3 and C4) which is represented as 00001101. This pattern of features of the organism determines its sub-group.

Hierarchical data can be defined as data units with hierarchical based inter relations among them. A taxonomic dataset is taxa based data with FHs to represent organisms organized by species for easy and efficient management of data as well as retrieval. Our hierarchical tree is constructed from a synthetic dataset of organisms. We represent an organism as a stream of binary data of 20 bits categorized into Rank (4 bits), Group (4 bits), Sub-Group (4 bits) and features (8 bits) as shown in Fig. 2. The taxonomic Rank is determined by the shared features, Group and Sub-Group making this a hierarchical representation. The cophenetic correlation coefficient determines the efficiency of hierarchical structure by determining similarity of the data between two values by calculating the distance between a pair of unmodelled data within the dendrogram [17]. The typical value for this is around 0.8 with values above 0.95 considered as more efficient [18]. The cophenetic correlation coefficient for our data is 0.9934. This values highlights that the synthetic data is efficiently structured with considerable accuracy.

4   Experimental Results and Discussion

There are 90 organisms in the dataset categorised into 6 different species. For all experiments, the dataset is divided randomly with the first 60 % for training the next 10 % for calibration/validation and the remaining 30 % for testing. After some initial trials to identify appropriate parameters, the initial learning rate is determined as 0.01 with a step-ratio (incremental learning step size) of 0.001 and the momentum as 0.3. To reduce the complexity and irregularity which may be caused by large weights, weight-decay, a simple penalty function is introduced to penalize large weights. Weight-decay is calculated as the half of the sum of squared weights times a coefficient termed as weight-cost which is 0.0002 for this experiment (a typical starting value for weight-cost is 0.0001). The objective function for the experiments is 'Cross Entropy'. Each experiment is performed 10 times with 100 epochs.

This experimental study is divided into two categories. In Experiment I, a DNN is trained to classify the species depending on the features. In the second experiment (Experiment II), a second set of data is used to identify whether any two given organisms are related (belongs to same species) or not. 4 types of scenarios are adopted for each experiment. For scenario A and B, a 5-layer ANN with hidden nodes 30,30,30 and 30,40,50 is used whereas for C and D, a 6-layered ANN with hidden nodes 30,30,30,30 and 30,40,50,60 is used. This scenario would help to determine the influence of symmetric and asymmetric node count.

Table 1. Results of experiment - I: confusion matrix values

No	Hidden layers	Training	Validation	Testing	All	Avg.training error
1A	3 (30,30,30)	100 %	100 %	100 %	100 %	0.023
1B	3 (30,40,50)	100 %	100 %	81.5 %	94.4 %	0.021
1C	4 (30,30,30,30)	100 %	100 %	92.6 %	97.8 %	0.0499
1D	4 (30,40,50,60)	13 %	55.6 %	14.8 %	17.8 %	0.0448

For Experiment I, 20 inputs representing the 20 bits of the organism are used with 6 separate outputs for determining the species of the organism and the results obtained are presented as Table 1. Experiment 1A, in which the hidden nodes are 30,30,30, shows 100 % results for training, validation and testing. When the number of nodes in the hidden layers is changed to 30,40,50 there was variation in the testing results which is 81.5 % constituting the overall results as 94.4 % as shown in Table 1. However, when the depth of the ANN is increased to 4, the confusion matrix showed a little variation compared with earlier ANN with depth as 3, whereas the results of the experiment with different number of hidden nodes (1D) showed a drastic fall in the accuracy rate with 17.8 % as an overall percentage. Inspection of the confusion matrix reveals that classification error has occurred for species 5 with 3 of class 5 being classified as class 4 due to similarity in most of their features.

The performance difference between experiment 1A and 1C is 2.2 % in the favor of 1A. However, the difference between 1B and 1D is 76.6 % in favor of 1C. On the other hand, if we analyze the significance of the same number of nodes and different number of nodes with depth being same, the difference between 1 A and 1B is 5.6 % in favor of 1 A and 1 C and 1D is 80 % in favor of 1C.

Table 2. Results of Experiment - II: Confusion Matrix values

No	Hidden layers	Training	Validation	Testing	All	Avg. Training Error
2A	3 (30,30,30)	100 %	100 %	88.9 %	96.7 %	0.0491
2B	3 (30,40,50)	100 %	100 %	100 %	100 %	0.0027
2C	4 (30,30,30,30)	100 %	100 %	83.3 %	95.0 %	0.0497
2D	4 (30,40,50,60)	100 %	100 %	93.4 %	98.3 %	0.0428

The second set of experiments (Experiment II) were carried out to identify whether two organisms are related or not. For example, tiger is related to cat they belongs to the same species whereas rat which belong to different species, is not related to cat as defined in our synthetic data. The input in this case is a 40 bit binary number vector fed to the network (20 each for two organisms) resulting in either '0' for not related or '1' if related. 60 data samples, (10 from each species) are used for this experiment and the results are shown in Table 2.

Summarising, in Experiment I, better results are achieved with the topology having the same number of hidden nodes whereas in Experiment II, better results are achieved with the topology having different number of hidden nodes. The difference between overall accuracy for experiments with 3 hidden layers, 2A and 2B is 3.3 % in favor of 2B. In case of experiments with 4 hidden layers, experiment 2D is 3.3 % more accurate than 2C. When the performance difference is analyzed in terms of depth, the topology with 3 hidden layers (2A and 2B) has better performance than the 4 hidden layered topology (2C and 2D) with an average difference of 5.6 % and 6.4 % respectively.

5 Conclusion and Future Work

The specific aim of this paper is to identify the ability of DNNs to classify organisms into various groups depending on characteristic features and to identify whether two given organisms are related or not. A taxonomic FH with associated data is generated with 6 classes (species) of organisms. The cophenetic correlation coefficient value of 0.9934 confirms the hierarchical nature of the dataset. Two experiments are conducted by varying the depth and changing the number of hidden nodes. Experiment I evaluates the classification into species and Experiment II evaluates the relationships between species. The experimental results show that increasing the 'depth' of the DNN has negative effects on the accuracy of the results, especially in the case of classification into species. Interestingly, experiments with same number of hidden nodes at each layer better results compared with that of different number of hidden nodes. In other words, a homogeneous node network appears, from our experiments, to be better than a hybrid node network for dealing with data that contains FHs.

With regard to limitations of our study, these results are derived from synthetic datasets. It will be interesting to observe the results with high volume and naturally acquired data as well as DNNs with more hidden layers. Also, further representational experiments are required to test the efficacy of representation in relation to architecture.

In summary, while the results reported here are based on simple experiments, the principles being explored and experimented on are far from simple. The relationship between architecture and representation is still an open issue in ANN research, as is the optimal relationship of localist with distributed representation. There has been little work so far on the relationship between depth and learning accuracy, especially when the level of depth and the number of nodes at each depth can influence learning. If deep learning with DNNs is to be better understood, further systematic experiments such as ours will be required.

References

1. Srivastava, N., Hinton, G., Krizhevsky, A., Sutskever, I., Salakhutdinov, R.: Dropout: a simple way to prevent neural networks from overfitting. J. Mach. Learn. Res. **15**, 1929–1958 (2014)
2. Bengio, Y., Lamblin, P., Popovici, D., Larochelle, H.: Greedy layer-wise training of deep networks. In: Advances in Neural Information Processing Systems, vol. 19, pp. 153–160. MIT Press (2007)
3. Bengio, Y., LeCun, Y.: Scaling learning algorithms towards AI. In: Large Scale Kernel Machines. MIT Press (2007)
4. Hinton, G.E., Deng, L., Yu, D., Dahl, G.E., Mohamed, A.R., Jaitly, N., Senior, A., Vanhoucke, V., Nguyen, P., Sainath, T.N., Kingsbury, B.: Deep neural networks for acoustic modeling in speech recognition: the shared views of four research groups. IEEE Sig. Process. Mag. **29**(6), 82–97 (2012)
5. LeCun, Y., Kavukcuoglu, K., Farabet, C.: Convolutional networks and applications in vision. In: Proceedings of 2010 IEEE International Symposium on Circuits and Systems (ISCAS), pp. 253–256 (2010)
6. He, K., Zhang, X., Ren, S., Sun, J.: Spatial pyramid pooling in deep convolutional networks for visual recognition. IEEE Trans. Pattern Anal. Mach. Intell. **99**, 1–1 (2015)
7. Xiong, C., Liu, L., Zhao, X., Yan, S., Kim, T.: Convolutional fusion network for face verification in the wild. IEEE Trans. Circuits Syst. Video Technol. **99**, 1–1 (2015)
8. Hingu, D., Shah, D., Udmale, S.S.: Automatic text summarization of wikipedia articles. In: 2015 International Conference on Communication, Information Computing Technology (ICCICT), pp. 1–4 (2015)
9. Graves, A., Schmidhuber, J.: Offline handwriting recognition with multidimensional recurrent neural networks. In: Advances in Neural Information Processing Systems, pp. 545-552 (2009)
10. Tirumala, S.S.: Implementation of evolutionary algorithms for deep architectures. In: Proceedings of the 2nd International Workshop on Artificial Intelligence and Cognition (AIC), Torino, Italy, November 2014, pp. 164–171 (2014)
11. Ban, J.-C., Chang, C.-H.: The learning problem of multi-layer neural networks. Neural Netw. **46**, 116–123 (2013)
12. Schmidhuber, J.: Deep learning in neural networks: an overview. Neural Netw. **61**, 85–117 (2015). Published online 2014; TR arXiv:1404.7828

13. Fukushima, K.: Neocognitron: a self-organizing neural network model for a mechanism of pattern recognition unaffected by shift in position. Biol. Cybern. **36**, 193–202 (1980)
14. Hinton, G.E., Salakhutdinov, R.R.: Reducing the dimensionality of data with neural networks. Science **313**, 504–507 (2006)
15. Vincent, P., Larochelle, H., Lajoie, I., Bengio, Y., Manzagol, P.-A.: Stacked denoising autoencoders: learning useful representations in a deep network with a local denoising criterion. J. Mach. Learn. Res. **11**, 3371–3408 (2010)
16. Hinton,G.E., McClelland, J.L., Rumelhart, D.E.: Parallel distributed processing: explorations in the microstructure of cognition. In: Distributed Representations, vol. 1, pp. 77–109. MIT Press (1986)
17. Sokal, R.R., Rohlf, F.J.: The comparison of dendrograms by objective methods. Taxon **11**(2), 33–40 (1962)
18. Rohlf, F.J., Fisher, D.R.: Tests for hierarchical structure in random data sets. Syst. Biol. **17**(4), 407–412 (1968)

Model Inclusive Learning for Shape from Shading with Simultaneously Estimating Illumination Directions

Yasuaki Kuroe[1]([✉]) and Hajimu Kawakami[2]

[1] Faculty of Information and Human Sciences, Kyoto Institute of Technology,
Matsugasaki, Sakyo-ku, Kyoto 606-8585, Japan
kuroe@kit.ac.jp
[2] Department of Electronics and Informatics, Ryukoku University,
1-5, Yokotani, Ohe-cho, Seta, Otsu 520-2194, Japan
kawakami@rins.ryukoku.ac.jp

Abstract. The problem of recovering shape from shading is important in computer vision and robotics and several studies have been done. We already proposed a versatile method of solving the problem by model inclusive learning of neural networks. The method is versatile in the sense that it can solve the problem in various circumstances. Almost all of the methods of recovering shape from shading proposed so far assume that illumination conditions are known a priori. It is, however, very difficult to identify them exactly. This paper discusses a method to solve the problem. We propose a model inclusive learning of neural networks which makes it possible to recover shape with simultaneously estimating illumination directions. The performance of the proposed method is demonstrated through some experiments.

Keywords: Model inclusive learning · Neural network · Shape from shading · Parameter estimation · Illumination direction

1 Introduction

The problem of surface-shape recovery of an object from a single intensity image is an important problem in computer vision and robotics and so on. The problem was first formulated in the general setting by B.K.P. Horn and several studies have been done based on the formulation [1,2]. The problem is essentially ill-posed and reduced to a nonlinear-function approximation problem.

In recent years, there have been increasing research interests of artificial neural networks and many efforts have been made on applications of neural networks to various fields. The most significant features of artificial neural networks are the extreme flexibility due to the learning capability of nonlinear function approximation and the generalization ability. It is expected, therefore, that neural networks make it possible to easily solve the ill-posed problem of shape from shading by their learning and generalization ability. G. Wei et al. presented

© Springer International Publishing Switzerland 2015
S. Arik et al. (Eds.): ICONIP 2015, Part I, LNCS 9489, pp. 501–511, 2015.
DOI: 10.1007/978-3-319-26532-2_55

a solution of the problem by using a multilayer feedforward network [3]. They proposed a method of recovering shape in which a feedforward neural network is trained so as to satisfy so called image irradiance equation which is a nonlinear partial differential equation. Motivated by the work [3] we already proposed a versatile method of solving the problem of recovering shape from shading by neural networks [4]. The proposed method is versatile in the sense that it can solve the problem in various circumstances. In order to realize the versatility, we introduced the concept of model inclusive learning of neural networks [5]. In the model inclusive learning a priori knowledge and inherent property of a target are incorporated into the formulation of learning problem, which could regularize an ill-posed problem and could improve learning and generalization ability of neural networks.

In almost all of the methods proposed so far for the problem of recovering shape from shading, the following two are assumed; (i) surface reflection properties of a target object are known a priori, and (ii) illumination conditions in which an image of a target object taken by a camera are identified. It is, however, very difficult to obtain surface reflection properties exactly and to identify illumination conditions exactly. We have already proposed a method which solves the first problem [6]. This paper discusses a method to solve the second problem that illumination conditions are not identified exactly. In general illumination conditions are identified through the calibration experiments. However it is very difficult to identify illumination conditions exactly. In addition to this the calibration experiments are complicated and troublesome and they cannot always be performed. We propose a method which makes it possible to recover shape with simultaneously estimating illumination directions by utilizing the versatileness of the model inclusive learning and present a practical procedure to realize it. The proposed method can reduce the burden of the calibration experiments. The performance of the proposed method is demonstrated through experiments.

2   Model Inclusive Learning for Shape from Shading-Simultaneous Estimation of Illumination Directions

In this section we explain the basic idea and concept of the proposed model inclusive learning for the problem of recovering shape from shading with simultaneously estimating illumination directions. An image of a three-dimensional object taken by a camera in an imaging condition depends on its geometric structure (shape), its reflectance properties and the imaging conditions (the distribution of light sources etc.). The image formation process can be illustrated as shown in the upper part of Fig. 1. The process can be regarded as a mapping from the geometric structure of the surface to the image. We call the mathematical model of the mapping 'image-formation model'. Note that the image-formation model, denoted by $\hat{F}$, depends on the reflectance properties and the imaging conditions. We assume that, in the image-formation model $\hat{F}$, the mathematical models of reflectance properties and the imaging conditions are known a priori except that the parameters of illumination directions in the the imaging conditions are unknown. This problem can be solved by the model inclusive learning of neural networks as follows. The schematic diagram of the proposed method is shown in Fig. 1. Suppose that an image of a three-dimensional target object is formed through an image formation process shown in the upper part of Fig. 1. Let $G(x, y)$ denote the brightness at a position (x, y) on the image. We formulate the learning problem of a neural network such that it recovers the geometric structure of the surface of the object as its input and output relation. In the formulation the neural network is trained with including the image-formation model $\hat{F}$ as follows. As shown in Fig. 1, we input a position (x, y) on the image to the neural network (NN), and we also input the corresponding output of the neural network to the image-formation model $\hat{F}$ together with the reflection properties and the imaging conditions. If the neural network is successfully trained so that the geometric structure of the surface of the object is realized as its input and output relation and if illumination directions are successfully estimated, the output of the image-formation model $\hat{F}$ becomes equal to the brightness data $G(x, y)$ which are taken from the object.

Therefore, training the neural network with simultaneously estimating the illumination directions so as to reduce the error between the output of the image-formation model $\hat{F}$ and the brightness $G(x, y)$ over all the data points to zero would make it possess the geometric structure of the surface as its input and output relation. Noting that we assume that the parameters of illumination directions in the imaging conditions are unknown, we adjust not only values of the neural network parameters but also those of illumination directions, as shown in Fig. 1, so as to minimize the error between the output of the image-formation model $\hat{F}$ and the brightness $G(x, y)$ over all the data points. If the error can be reduced enough small by the adjustments of the parameters, the surface recovery and the estimation of the illumination directions are achieved simultaneously.

3 Problem Formulation and Proposed Learning Method

3.1 Problem Formulation

Suppose that the surface of an object is represented by

$$z = f(x, y) \tag{1}$$

in a camera coordinate system x - y - z, with the x - y plane coinciding with the image plane, and z axis coinciding with the optical axis of the camera. It is known that, assuming that orthographic projection and uniform reflectance property of the object, the brightness at position (x, y) on the image plane can be described as

$$G(x, y) = R(p, q ; l), \qquad p = \frac{\partial f}{\partial x}, \quad q = \frac{\partial f}{\partial y} \tag{2}$$

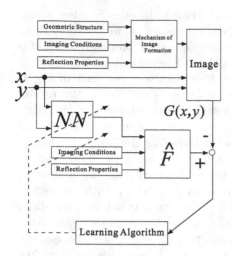

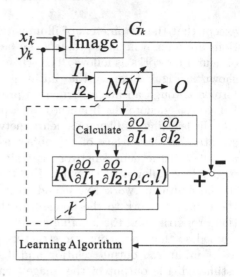

Fig. 1. Shape from shading by model inclusive learning of a neural network with simultaneous estimation of illumination directions.

Fig. 2. Proposed model inclusive learning method for shape from shading with simultaneous estimation of illumination directions.

where $l = (\ell_1, \ell_2, \ell_3)$ is the illuminant direction, and p and q are the surface gradient at (x, y). Equation (2) is called image irradiance equation. $R(p, q ; l)$ is called the reflectance map and represents reflection properties. Note that the image irradiance equation (2) is corresponding to the image-formation model $\hat{F}$. In general the image formation model (reflectance map) is modeled as being composed of the specular reflection $\Phi(\theta(p, q; l), c)$ and the diffuse reflection $\cos \phi(p, q; l)$ as follows:

$$R(p, q; \rho, c, l) = \rho \cdot \Phi(\theta(p, q; l), c) + (1 - \rho) \cdot \cos \phi(p, q; l) \tag{3}$$

where $\theta(p, q; l) = \cos^{-1} \frac{pl_1 + ql_2 - (l_3 - 1)}{\sqrt{p^2 + q^2 + 1}\sqrt{l_1^2 + l_2^2 + (l_3 - 1)^2}}$, $\cos \phi(p, q; l) = \frac{pl_1 + ql_2 - l_3}{\sqrt{p^2 + q^2 + 1}}$, ρ $(0 \leq \rho \leq 1)$ is the ratio parameter and c is the parameter describing extent of the specular reflection. Note that in (3) we use the expression $R(p, q; \rho, c, l)$ to represent the reflectance map in order to clarify that it depends on reflection parameters ρ and c. There have been several models proposed for the specular reflection $\Phi(\theta(p, q; l), c)$, a typical representative of which is the Torrance-Sparrow Model [8]:

$$\Phi(\theta(p, q; l), c) = \exp(-c^2\theta^2(p, q; l)) \tag{4}$$

The illumination direction $l = (\ell_1, \ell_2, \ell_3)$ is expressed as follows:

$$\ell_1 = \sin \theta \cos \varphi, \quad \ell_2 = \sin \theta \sin \varphi, \quad \ell_3 = \cos \theta \tag{5}$$

where θ is the polar angle and φ is the azimuth.

The objective here is to recover the geometric structure of the surface (1) from a single image. In this paper we propose a model inclusive learning method to solve the problem under the following conditions: (A1) the mathematical expression of the reflectance map $R(p, q; \rho, c, l)$ is known, (A2) the reflection parameters ρ and c are known, and (A3) the illuminant direction $l = (\ell_1, \ell_2, \ell_3)$ is not known, that is to say, the polar angle θ and the azimuth φ are not known. In the following we will discuss a method of recovering shape of an object and estimating the illuminant direction $l = (\ell_1, \ell_2, \ell_3)$ simultaneously.

3.2 Proposed Learning Method

Figure 2 shows the schematic diagram of the proposed model inclusive learning method of neural networks. Let G_k denote the brightness which is observed at a position (x_k, y_k) from an image taken from an object surface. We prepare a neural network (NN) with two inputs denoted by $I = [I_1, I_2]^T$ and one output denoted by O and consider that the input $I = [I_1, I_2]^T$ and the output O correspond to the position (x, y) on the images and the depth z of the surface, respectively. For an observed brightness G_k, we give its position (x_k, y_k) on the image to the input $I = [I_1, I_2]^T$ of the neural network and derive the derivatives of the output of the neural network with respect to the input, and obtain the values of the derivatives at $[I_1, I_2]^T = (x_k, y_k)^T$:

$$\left. \frac{\partial O}{\partial I} \right|_{I=(x_k, y_k)} = (\left. \frac{\partial O}{\partial I_1} \right|_{I=(x_k, y_k)}, \left. \frac{\partial O}{\partial I_2} \right|_{I=(x_k, y_k)})^T. \tag{6}$$

Note that those derivatives become equal to the surface gradients p and q at the position (x_k, y_k) if the input and output relation of the neural network exhibits the geometric structure (1) of the object. We substitute the values of the derivatives (6) into the surface gradient p and q of the image-formation model $R(\cdot, \cdot ; \rho, c, l)$. The obtained $R(\partial O/\partial I_1, \partial O/\partial I_2; \rho, c, l)$ corresponds to the outputs of the image formation model in Fig. 1 and is to be coincided with the brightness G_k. Accordingly, training the neural network so as to reduce the error between the brightness data G_k and $R(\partial O/\partial I_1, \partial O/\partial I_2; \rho, c, l)$ over all the data points to zero, we can obtain the geometric structure of the surface as the input and output relation of the neural network. Noting that we assume that the parameters of the illumination direction $l = (\ell_1, \ell_2, \ell_3)$ are not known, in the model inclusive learning we adjust not only values of the neural-network parameters but also those of the illumination direction $l = (\ell_1, \ell_2, \ell_3)$, as shown in Fig. 2, so as to minimize the error between the brightness data G_k and $R(\partial O/\partial I_1, \partial O/\partial I_2; \rho, c, l)$ over all the data points. Note also that the reflectance map $R(p, q; \rho, c, l)$ contains unknown parameters of the illumination direction l and in the calculation of $R(\partial O/\partial I_1, \partial O/\partial I_2; \rho, c, l)$ we use the current estimated values of the parameters of the illuminant direction $l = (\ell_1, \ell_2, \ell_3)$ (θ and φ).

Define the performance index by

$$J = \frac{1}{2} \sum_{(x_k, y_k) \in D_G} \left\{ R \left(\left. \frac{\partial O}{\partial I_1} \right|_{I=(x_k, y_k)}, \left. \frac{\partial O}{\partial I_2} \right|_{I=(x_k, y_k)} ; \rho, c, l \right) - G_k \right\}^2 \tag{7}$$

where D_G is a set of data points (x_k, y_k) at which the brightness data G_k are observed from the image. Note that J is the square error between G_k and $R(\partial O/\partial I_1, \partial O/\partial I_2; \rho, c, l)$ over the set of data points D_G. The problem is now reduced to finding values of parameters of the neural network and also the parameters of the illumination direction l (the polar angle θ and the azimuth φ) that minimize the performance index J, a solution of which could achieve simultaneous recovering shape and estimating the illumination direction.

In order to search values of the network parameters and the parameters of the illumination direction which minimize J, the gradient based methods can be used, in which several useful algorithms are available: the steepest descent algorithm, the conjugate gradient algorithm, the quasi-Newton algorithm and so on. The main problem associated with these algorithms is the computation of the gradients of J with respect to the parameter of the neural network and the parameters of the illumination direction l (θ and φ). Note that, as previously stated, the derivatives of the output with respect to the input of the neural network $\partial O/\partial I$ are also needed to be calculated. Efficient algorithms to calculate these gradients and the derivatives can be derived by introducing adjoint models of the neural network [7]. The derivation is omitted.

It is important to note that, in the minimization of the performance index J by the use of an appropriate gradient based algorithm, it may not succeed to converge if all the parameters, that is, the parameters of the neural network and the parameters of the illumination direction l (θ and φ) are simultaneously adjusted from the beginning of the optimization. In order to solve the problem we take two steps in the optimization procedure. In the first step we adjust only the parameters of the neural network, and in the second step we adjust all the parameters simultaneously as follows.

Step 1: Give adequate values to the parameters of the illumination direction l (θ and φ) and give random values to the parameters of the neural network as an initial guess. Adjust only the parameters of the neural network according to an appropriate gradient based algorithm with the parameters of the illumination direction being kept constant at the initial guess. The optimization iterations are continued until it converges and the tentative surface shape of the target object is obtained. We call it the initial surface shape.

Step 2: Starting from the initial surface shape, adjust all the parameters, that is, the parameters of the neural network and the parameters of the illumination direction l (θ and φ) simultaneously according to an appropriate gradient based algorithm. The optimization iterations are continued until it converges.

4   Experiments

In this section we show the results of the experiments in order to demonstrate the performance of the proposed method. In the following experiments a four-layer feedforward neural network with 15 hidden units is used. We utilize the quasi-Newton method with the Davidon-Fletcher-Powell algorithm as a gradient based method.

 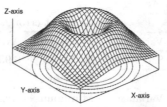 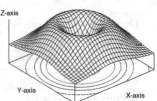

Fig. 3. Synthetic Image   **Fig. 4.** True surface.   **Fig. 5.** Recovered surface

4.1 Experiment 1 - Synthetic Image

In the first experiment we use a synthetic image in order to evaluate accuracy of the proposed method. We synthesize a target image from an artificial surface which is given by the following equation.

$$f(x,y) = 4[\exp\{-1.75(x^2 + y^2)\} - \exp\{-1.5(x^2 + y^2)\}] \qquad (8)$$

Figure 3 shows a synthetic image of size 32×32 generated by applying the image formation model (3) to the surface (8), where the Torrance-Sparrow Model given by (4) is used as the specular reflection model. In the model values of the reflection parameters ρ and c are given as $\rho = 0.7$, $c = 0.7$ and values of the parameters of the illumination direction θ and φ are given as $\theta = 2.922[rad](167.40°)$ and $\varphi = 4.249[rad](243.43°)$. Figure 4 shows the true three-dimensional surface given by (8). In the surface recovery experiment we take the two step procedure explained in the previous subsection. In Step 1 we give the initial guess of the parameters of the illumination direction θ and φ as $\theta = 2.526[rad](144.74°)$ and $\varphi = 3.927[rad](225.00°)$ and adjust (train) only the parameters of the neural network with those of the illumination direction being kept constant at the initial guess. The iterations are continued until the optimization converges and the initial surface shape is obtained. In Step 2 starting from the initial surface shape, we adjust the parameters of the neural network and those of the illumination direction l (θ and φ) simultaneously. Figure 5 shows an example of the recovered surface thus obtained.

In order to evaluate accuracy, for the synthesized image we performed surface recovery experiments 10 times by randomly changing values initial guess of the learning parameters of the neural networks in the range of [-0.1,0.1]. Table 1 shows the minimum, average and variance of the surface depth error and gradient error between the true surface and the recovered surface by the proposed method, which are calculated per one pixel. In the table the estimation errors of the parameters of the illumination direction θ and φ are also shown. Note that the minimum and maximum values of the depth of the target surface (8) are 0.227 and 3.0, and those values of its surface gradient are -0.461 and 0.461, respectively. It is observed that all the results are enough accurate for practically.

Table 1. Accuracy for the experiment using the synthetic image

Depth error			Gradient error		
Mimimun	Average	Variance	Mimimun	Average	Variance
3.45×10^{-3}	6.22×10^{-3}	8.51×10^{-6}	3.77×10^{-3}	6.90×10^{-3}	1.47×10^{-5}
Error of θ			Error of φ		
Mimimun	Average	Variance	Mimimun	Average	Variance
2.00×10^{-4}	2.02×10^{-3}	3.18×10^{-6}	2.81×10^{-2}	4.91×10^{-2}	8.58×10^{-4}

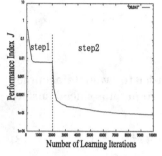 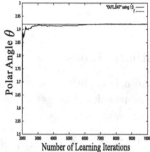 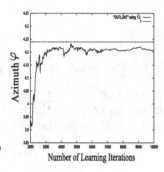

Fig. 6. An example of convergence behavior of J during the learning iteration.

Fig. 7. An example of convergence behavior of θ during the learning iteration.

Fig. 8. An example of convergence behavior of φ during the learning iteration.

Figure 6 shows an example of the convergence behavior of the proposed learning method. In the figure the variation of the performance index J versus the number of the learning iterations is plotted. It is observed that the values of J decreases and the learning converges. Note that at the beginning of Step 2 the values of J decreases rapidly, which reveal the effectiveness of the two step procedure. Figures 7 and 8 show examples of the convergence behavior of the estimation of the parameters of the illumination direction θ and φ. In the figures the variations of θ and φ versus the number of the learning iterations are plotted from the beginning of Step 2. It can be seen that θ converges to its true value $2.922[rad](167.40°)$ and φ converges approximately to its true value $4.249[rad](243.43°)$. Those results reveal the validity of the proposed method.

4.2 Experiment 2 - Real Image

We here present results of the experiment using a real image. The image we used is a Venus statue which is made up of curved surfaces shown in Fig. 9. The experiment is performed by using the image of size 71×51 shown in Fig. 10 which is the right eye of the Venus statue in Fig. 9. In the experiment we use the image formation model (3) with the specular reflection $\Phi(\theta(p,q;l),c)$ being Torrance-Sparrow Model (4) where the reflection parameters ρ and c are set be $\rho = 0.460$ and $c = 0.408$ which are obtained by the preparatory calibration experiment. In Step 1 of the two step procedure we give the initial guess of the illumination direction as $\theta = 3.000[rad]$ and $\varphi = 6.0828[rad]$ which are also obtained by the preparatory calibration experiment.

Fig. 9. Venus statue consisted of curved surfaces.

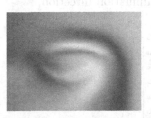

Fig. 10. Real images of the right eye of the Venus statue.

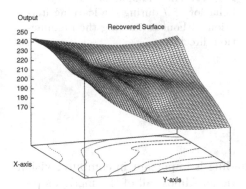

Fig. 11. Recovered eye-surface obtained by the proposed method.

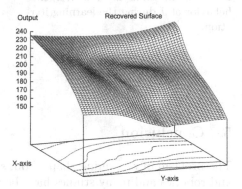

Fig. 12. Recovered eye-surface obtained by the method without estimating the illumination direction.

Figure 11 shows the recovered surface obtained by the proposed method. In the lower part of the figure the contour map of the surface is also shown. Figure 12 shows the recovered surface obtained by the method proposed in [4], that is, the model inclusive learning method without estimating the parameters of the illumination direction θ and φ. It is observed by comparing those figures that the result of Fig. 11 can captures the fine structure much more than that of Fig. 12, which reveals the effectiveness of the estimation of the parameters of the illumination direction. Figure 13 shows the convergence behavior of the proposed learning method in which the result in Fig. 11 is obtained, and Fig. 14 shows that in which the result in Fig. 12 is obtained. In those figures the variations of the performance index J versus the number of the learning iterations are plotted. It can be seen that the value of J in Fig. 13 converges to the value much smaller than that in Fig. 14 and the value of J in Fig. 13 decreases rapidly at the beginning of the Step 2, which also reveals the effectiveness of the estimation of the illumination direction.

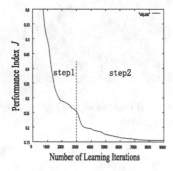 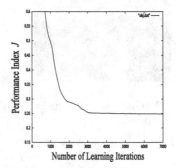

Fig. 13. An example of convergence behavior of J during the learning iteration.

Fig. 14. An example of convergence behavior of J during the learning iteration without estimating the illumination direction.

5   Conclusion

The problem of recovering shape from shading is important in computer vision and robotics and many studies have been done. Almost all of the methods proposed so far assume that illumination conditions in which an image of a target object taken by a camera are identified. It is, however, very difficult to identify them exactly. In this paper we have proposed a versatile method of recovering shape from shading with solving the above problem. The proposed method is a model inclusive learning of neural networks which makes it possible to recover shape with simultaneously estimating the parameters of the illumination direction.

References

1. Horn, B.K.P., Brooks, M.J. (eds.): Shape from Shading. The MIT Press, Cambridge (1989)
2. Klette, R., Koschan, A., Schlüns, K.: Computer Vision: Three-Dimensional Data from Images. Springer, Verlag (1996)
3. Wei, G.Q., Hirzinger, G.: Learning shape from shading by a multilayer network. IEEE Trans. Neural Netw. **7**(4), 985–995 (1996)
4. Kuroe, Y., Kawakami, H.: Versatile neural network method for recovering shape from shading by model inclusive learning. In: Proceedings of International Joint Conference on Neural Networks, pp. 3194–3199 (2011)
5. Kuroe, Y., Kawakami, H.: Estimation method of motion fields from images by model inclusive learning of neural networks. In: Alippi, C., Polycarpou, M., Panayiotou, C., Ellinas, G. (eds.) ICANN 2009, Part II. LNCS, vol. 5769, pp. 673–683. Springer, Heidelberg (2009)
6. Kuroe, Y., Kawakami, H.: Shape from shading by model inclusive learning with simultaneously estimating reflection parameters. In: Wermter, S., Weber, C., Duch, W., Honkela, T., Koprinkova-Hristova, P., Magg, S., Palm, G., Villa, A.E.P. (eds.) ICANN 2014. LNCS, vol. 8681, pp. 443–450. Springer, Heidelberg (2014)
7. Kuroe, Y., Nakai, Y., Mori, T.: A learning method of nonlinear mappings by neural networks with considering their derivatives, pp. 528–531. In: Proceedings of IJCNN, Nagoya, Japan (1993)
8. Torrance, K.E., Sparrow, E.M.: Theory for off-specular reflection from roughened surfaces. J. Opt. Soc. Am. **57**(9), 1105–1114 (1967)

A Computational Model of Match Decision-Making Problem Using Spiking SHESN with Reward-Modulated Reinforcement Learning

Zhidong Deng[✉] and Guorun Yang

Department of Computer Science, State Key Laboratory of Intelligent Technology and Systems, Tsinghua National Laboratory for Information Science and Technology, Tsinghua University, Beijing 100084, China
michael@tsinghua.edu.cn, ygr13@mails.tsinghua.edu.cn

Abstract. Match decision-making problem is one of the hot topics in the field of computational neuroscience. In this paper, we propose a spiking SHESN model with reward-modulated reinforcement learning so as to conduct computational modeling and prediction of such an open problem in a manner that has more neurophysiological characteristics. Neural coding of two sequentially-presented stimuli is read out from a collection of clustered neural populations in state reservoir through reward-modulated reinforcement learning. To evaluate match decision-making performance of our computational model, we set up three kinds of test datasets with different spike timing trains and present a criterion of maximum correlation coefficient for assessing whether match/nonmatch decision-making is successful or not. Finally, extensive experimental results show that the proposed model has strong robustness on interval of both spike timings and spike shift, which is consistent with monkey's behavior records exhibited in match decision-making experiment [1].

Keywords: Computational neural model · Match decision-making problem · Spiking SHESN · Reward-modulated reinforcement learning

1 Introduction

With advances in experimental techniques, a lot of secrets underlying biological brain are gradually unveiling. In biological nervous system, carriage of neural information is spike or action potential. Spike trains are generated due to change in neuronal membrane potential of being greater than a threshold and then transmitted along nerve fibers. Evidences indicate that any neural information can be represented in firing rate, spike timing, or assemblies [2]. Among them, a study of neuronal information in spike timing train is increasingly paid to attention in the past decade. A couple of spike timing-based learning algorithms were proposed such as spike-timing-dependent plasticity (STDP) [3].

Attractor is viewed as one of the most important ways for analysis of artificial recurrent neural network. In fact, the existence of attractor was neurophysiologically

© Springer International Publishing Switzerland 2015
S. Arik et al. (Eds.): ICONIP 2015, Part I, LNCS 9489, pp. 512–521, 2015.
DOI: 10.1007/978-3-319-26532-2_56

supported by experimental results on CA1 area of mouse brain that Wills et al. discovered in 2005 [4]. On the other hand, Jaeger et al. [5] proposed a new generation of recurrent neural network, called echo state network (or ESN), which has continuously been being improved. In general, ESN contains single hidden layer, i.e. state reservoir, where all internal/reservoir neurons are randomly connected. It enables state reservoir to have abundant state space and leads to ESN being able to store more complicated attractors. At the same time, a set of states that encode external stimuli can readily be read out from state reservoir by neurons in output layer. Synaptic weights between internal neurons and output ones can be shaped using supervised learning algorithm. Up to now, ESN has extensively been applied to many problems such as chaotic time sequence prediction [6].

In the study of decision-making task, the match decision-making problem is attracting attention in recent years [1]. Several computational neural models are proposed to investigate such a decision-making behavior [7, 8]. Decision-making often relies on abilities to compare memories with immediate sensations. In experimental neuroscience, the match decision-making problem is viewed as one of classical problems among decision-making tasks, where subjects are requested to classify, memorize, compare, and discriminate external consecutive stimuli that occur at slightly different times. Biophysically, neurons in lateral PFC are likely to get involved in simultaneously memory and comparison [1]. So far we have had, however, little knowledge of how both working memory and comparison are carried out neurobiologically in a single module without specific parameter's tuning. In existing computational models for such a problem, network structure is usually required to be specifically designed and does not reflect characteristics of biological nervous system in a sense [7, 8]. This paper attempts to develop a biologically inspired computational model to clarify the match decision-making problem. First, we give a description of match decision-making problem. Second, we propose a new spiking SHESN model with reward-modulated reinforcement learning. Third, neuronal spike timing trains are used to fulfill simulation on match/nonmatch decision-making problem. Furthermore, the experimental results are discussed. Finally, we draw conclusions.

2 Problem Description

Working memory is considered to play an important role in regular decision-making task of human beings. To have knowledge of functioning of working memory in decision-making process, two different types of classical experimental paradigms are generally designed by physiologists. One is to discriminate whether subsequent stimulus, which is produced after a short duration from the onset of the first stimulus, is identical to the preceding stimulus. Another is that given two consecutive stimuli, one needs to judge whether the two stimuli belong to the same class [1]. In the first type of decision-making problem, called a delayed discrimination task, it is only required to know if stimulus occurs repeatedly. But as for the second type, i.e. match/nonmatch decision-making problem, we must first make appropriate classification of stimulus and then decide whether the class of two consecutive stimuli is the same or different according to preceding classification results [1].

As shown in Fig. 1, an external stimulus is simply represented in a single neuronal spike in study of match/nonmatch decision-making problem. Suppose that there are two input stimuli of A and B presented. It is regarded as a match pattern if neuronal spike trains A (or B) contain two consecutive stimuli and B (or A) have nothing in stimuli. A nonmatch pattern is defined, if spike trains A (or B) consist of only one stimulus and B (or A) also include only one stimulus that triggers at different timings.

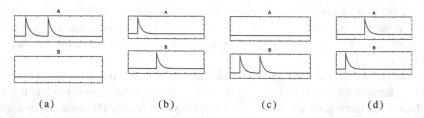

(a)   (b)   (c)   (d)

Fig. 1. Match and nonmatch stimuli expressed in neuronal spike trains. (a) and (c) Match patterns. (b) and (d) Nonmatch patterns.

3 Methods

3.1 Network Architecture

A novel spiking SHESN network is proposed to establish our computational neural model for match decision-making problem. The proposed clustered reservoir computing model consists of three layers: input, working memory/comparison/clustering (WCC), and output, as shown in Fig. 2. The input layer functions somatosensory neurons (S1) in brain. It is composed of two neurons, which receive two spike trains of external stimuli to be classified, respectively. The internal neurons in the hidden state reservoir layer, i.e. the WCC layer, are sparsely connected to form complex network according to naturally evolving rules presented in [9]. We unveiled that it is able to simultaneously perform working memory, comparison, and clustering for two consecutive stimuli in single WCC layer, which is significantly consistent with biological plausibility in brain cortex [1]. Additionally, there exhibit great heterogeneity over reservoir neurons in such a WCC layer. This is also observed in PFC neurons during recent monkey's experiments [2]. The output layer contains one neuron that is employed to read out neural coding of two

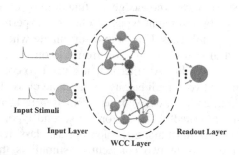

Input Stimuli

Input Layer   Readout Layer

WCC Layer

Fig. 2. The network structure of spiking SHESN model for match decision-making problem.

sequentially-presented stimuli from a collection of clustered neuronal populations in state reservoir through reward-modulated reinforcement learning. Using reinforcement learning for shaping plasticity of output synapses, activities of single neuron in the output/read-out layer are demonstrated to generate the same decision as that in match decision-making experiments. From this perspective of brain science, biological network is always sparsely connected and has very low connectivity. Similarly, the sparse connectivity from input neurons to any WCC neuron in our computational model is set to $c^{in} = 0.5$, whereas the WCC neurons themselves are also randomly connected with very low connectivity of $c^{res} = 0.01$. Interestingly, output synapses are found to be sparse after reinforcement learning completes. Totally, the whole spiking SHESN has very low connectivity. These data are in accord with recently reported experimental results on PFC activity [1].

The leaky integrator neuron is adopted to store the past state and improve learning and classification performance of network [10]. The leaky integrator neuron model of spiking SHESN is described below,

$$\tau \frac{dx(t)}{dt} = -\alpha x(t) + f(W^{res} x(t) + W^{fb} y(t) + W^{in} u(t) + v(t)), \tag{1}$$

$$y(t) = W^{out} x(t), \tag{2}$$

where $x(t)$ denotes the state vector of state reservoir, $y(t)$ the output vector of network, $u(t)$ the input vector, $v(t)$ the noise vector, W^{res} the reservoir synaptic matrix in the WCC layer, W^{fb} (W^{in}) the feedback (input) synaptic matrix, both of which take uniform distribution over $[-1, 1]$, W^{out} the output synaptic matrix, $f(\cdot) = \tanh(\cdot)$ the activation function of internal neurons, $\tau = 1.0$ the time constant, and $\alpha = 0.9$ the leaky decay rate. In our experiment, Eq. (1) was solved using Runge-Kutta-Fehlberg algorithm (or RKF45). The other parameters were selected below: the number of internal neurons $n = 1,000$, the number of backbone neurons $b = 2$, and the spectral radius $r = 0.5$.

3.2 Reward-Modulated Reinforcement Learning for Output Synapses

In our model, only output synaptic weights are updated based on reward-modulated reinforcement learning during training phase. Note that output weights remain unchanged in test phase. The exploratory Hebb rule (or EH-rule) [11] is given below,

$$w_{ji}^{out}(k+1) = w_{ji}^{out}(k) + \Delta w_{ji}^{out}(k), \tag{3a}$$

$$\Delta w_{ji}^{out}(k) = \eta(r(k) - E[r(k)]) x_i(k)(y_j(k) - E[y(k)]) / \parallel x(k) \parallel, \tag{3b}$$

and

$$y_j(k) = \sum_{i=1}^{n} w_{ji}^{out}(k) x_i(k), \tag{4}$$

where $\eta = 0.0001$ is the learning rate, $w_{ji}^{out}(k)$ represents the synaptic weights between neuron j in the readout layer and neuron i in the WCC layer, and $y_j(k)$ indicates responses of output neurons. The initial weights $w_{ji}^{out}(0)$ were found using the ridge regression.

The instantaneous reward signal $r(k)$ at the current trial k is defined as 1 if match/nonmatch is carried out and 0 otherwise. $E[r(k)]$ (or $E[y_j(k)]$) expresses the averaging of previously received rewards (or output responses), i.e.,

$$E[r(k)] = 0.8E[r(k-1)] + 0.2r(k), E[y_j(k)] = 0.8E[y_j(k-1)] + 0.2y_j(k). \quad (5)$$

Essentially, deviation of $r(k) - E[r(k)]$ denotes the difference between the actual and predicted rewards, namely, the reward prediction error, while deviation of $y_j(k) - E[y_j(k)]$ stands for exploratory signals for output neuron response [11].

4   Results and Discussion

4.1   Construction of Training Datasets and Test Datasets

We implemented a computational neural model of accomplishing match decision-making task (Fig. 1). In our experiment, the training and test duration all lasted for $T_f = 8s$ and the step-size for discretization was set to $\Delta t = 0.002$s (using RKF45). There were two spike train stimuli of $u_{t1}(t)$ and $u_{t2}(t)$ with a presentation of spike timing being assumed to be $t_1 = 1s$ and $t_2 = 3s$, respectively. Meanwhile, the amplitude of spike was set as $g_{in} = 0.675$ and the interval of spike timings as $\lambda = 1$. It had

$$u_{t1}(t) = \begin{cases} 0 & t \notin [t_1, t_1 + \lambda] \\ g_{in}e^{-\beta(t-t_1)} & t \in [t_1, t_1 + \lambda] \end{cases}, \quad (6a)$$

$$u_{t2}(t) = \begin{cases} 0 & t \notin [t_2, t_2 + \lambda] \\ g_{in}e^{-\beta(t-t_2)} & t \in [t_2, t_2 + \lambda] \end{cases}, \quad (6b)$$

where $\beta = 2$ was a time constant.

Suppose that the output spike train response y_{psp} was a sine signal with amplitude of spike $g_{out} = 1.2$, i.e.,

$$y_{psp}(t) = \begin{cases} 0 & t \notin [t_2, t_2 + \lambda] \\ g_{out}\sin\left(\frac{\pi}{\tau}(t - t_2)\right) & t \in [t_2, t_2 + \lambda] \end{cases}. \quad (7)$$

The four pairs of input-output spike trains are listed in Table 1, where M1 and M2 indicate a pair of spike trains that led to a Match response, and NM1 and NM2 to a Nonmatch behavior. The match/nonmatch response corresponds to excitatory/inhibitory post-synapse potential (EPSP/IPSP), respectively.

We used the above-mentioned four pairs of input-output spike trains for constructing the training and test datasets in our experiment, as listed in Table 2. The interval of spike timings was set to $\Delta = t_2 - t_1 = 2$. For each of the experiments, one

Table 1. Four pairs of input-output spike trains.

Pairs of spike trains	Input		Desired output
M1	$u_1 = u_1 + u_2$	$u_2 = 0$	$y_d = y_{psp}(t)$
M2	$u_1 = 0$	$u_1 = u_1 + u_2$	
NM1	$u_1 = u_1$	$u_2 = u_2$	$y_d = -y_{psp}(t)$
NM2	$u_1 = u_2$	$u_2 = u_1$	

Table 2. Training and test datasets used in the experiment.

Type	Dataset	Pairs of spike trains	Parameter
training	S1	M1, M2, NM1, NM2	$t_1 = 1s, \Delta = 2s$
test	S2	M1, M2, NM1, NM2	$t_1 = 1s, \Delta = 0.5s$ to 5s
test	S3	M1, M2, NM1, NM2	$t_1 = 0.5s$ to 4s, $\Delta = 2s$
test	S4	M1, M2, NM1, NM2	$t_1 = 0.5s$ to 3s, $\Delta = 0.5s$ to 3s

of the four pairs of input-output spike trains, i.e. M1, M2, NM1, or NM2, which constituted the training dataset S1, was selected for training. The test datasets of S2 to S4 had different parameters below. (a) The first spike timing $t_1 = 1$ was kept unchanged, but the interval of spike timings Δ ranged from 0.5 to 5. (b) The interval of

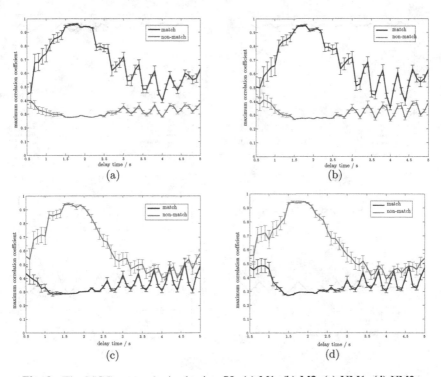

Fig. 3. The MCC curve obtained using S2. (a) M1. (b) M2. (c) NM1. (d) NM2.

spike timings $\Delta = 2$ was fixed, but the first spike timing t_1 had change from 0.5 to 4. (c) Both the first spike timing t_1 and the interval of spike timings Δ had variation.

In the test phase, we assessed actual response $y(k)$ of network according to the maximum correlation coefficient (MCC) ϑ_M (or ϑ_{NM}) evaluated for match (or non-match) response, which indicated the match level between $y(k)$ and the desired output $y_d(k)$ of match (or nonmatch) response, as described in (7). Specifically, it generated a match behavior if $\vartheta_M > 0.70$, while a nonmatch response if $\vartheta_{NM} < 0.35$. Similarly, a unmatch behavior is produced if $\vartheta_{NM} > 0.65$ and $\vartheta_M < 0.40$.

$$\vartheta_{M/NM} = \max_{t_2,\Delta} \text{corrcoef}\ (y(k), y_d(k)) = \max_{t_2,\Delta} \frac{\sum_{k=0}^{l}(y(k) - \bar{y})(y_d(k) - \bar{y}_d)}{\sqrt{\sum_{k=0}^{l}(y(k) - \bar{y})^2 \sum (y_d(k) - \bar{y}_d)^2}}.$$

$$(8)$$

4.2   The Behavioral Results of Model

The match and nonmatch MCC curves achieved using the test dataset S2 are shown in Fig. 3, which includes both the mean and variance over 50 independent runs. When the interval of spike timings in the input spike trains for a match stimulus $\Delta > 2.5$, the

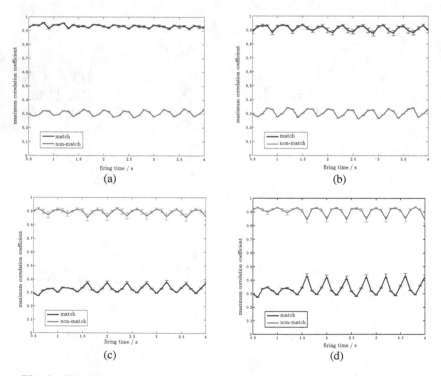

Fig. 4. The MCC curve achieved using S3. (a) M1. (b) M2. (c) NM1. (d) NM2.

match MCC of $\vartheta_M > 0.70$ decreased while the nonmatch MCC of $\vartheta_{NM} < 0.35$ went up, both of which led to the failure of decision-making that the network accomplished. Thus the spiking SHESN model was considerably sensitive to the interval of spike timings.

The match and nonmatch MCC curves yielded based on the test dataset S3 are shown in Fig. 4. Apparently, the curves of Fig. 4(a)–(c) manifested considerable smoothness, which meant that the network was not very sensitive to the spike timing.

According to the test dataset S4, Fig. 5 shows the match and nonmatch MCC curves. Different spike timings result in different MCC curves achieved for either match or nonmatch decision-making tasks, when the interval of spike timings was fixed. The match MCC variations got larger with the increase of the interval of spike timings and otherwise smaller for the nonmatch input spike trains. This suggested that sensitivity of network to the interval of spike timings was greater than to the spike timing.

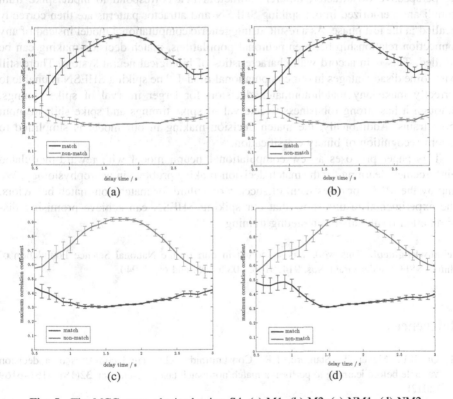

Fig. 5. The MCC curve obtained using S4. (a) M1. (b) M2. (c) NM1. (d) NM2.

5 Discussion and Conclusion

Many neurophysiological experimental results show that neurons in PFC is often persistently spiking so as to temporarily keep external stimuli, when subject performed match decision-making task. A couple of computational models have been proposed so far, in order to explore such a decision-making problem [1, 7, 8]. In these models, however, there exists common drawback that specific synaptic connection structures or supervised learning are required, although such specifically designed structures or learning mechanisms are not supported by any physiological evidence in biological brain. Thus resulting computational models seem to be not so convincing. Meanwhile, parameter settings in the models have to be provided empirically, which may have significant influence upon experimental results if there is any change in the initial values of parameters.

Our experimental results demonstrate that our spiking SHESN can classifying a couple of spike train stimuli that are used for match/nonmatch decision-making. It can learn from different input spike train samples through reward-modulated reinforcement learning, which implies judgments for success or failure of match/nonmatch decision-making. This seems to have direct physiological evidence [11, 12]. From the perspective of attractor, different attractors that correspond to input spike train stimuli are memorized in our spiking SHESN and attractor patterns are then correctly recalled in the test phase. As a result, using generic computational model instead of any connection relationship between neuronal populations, match decision-making can be fulfilled. This is in accord with characteristics of biological neural system. There still exist some disadvantages in our computational model. The spiking SHESN is unable to correctly make any match/nonmatch decision for larger interval of spike timings, although it has strong robustness to interval of spike timings and spike shift in input spike trains. Additionally, the match decision-making in our model is simplified to pattern recognition of binary classification.

This paper proposes a new computational neural model with reward-modulated reinforcement learning for the match decision-making problem in neurophysiology. We employ the MCC for assessment of success or failure for match/non-match behaviors. The experimental results show that our spiking SHESN can achieve promising discrimination results after completing training.

Acknowledgment. This work was supported in part by the National Science Foundation of China (NSFC) under Grant Nos. 91420106, 90820305, and 60775040.

References

1. Qi, X.L., Meyer, T., Stanford, T.R., Constantinidis, C.: Neural correlates of a decision variable before learning to perform a match/nonmatch task. J. Neurosci. **32**(18), 6161–6169 (2012)
2. Gerstner, W., Kistler, W.: Spiking Neuron Models: Single Neurons, Populations, Plasticity. Cambridge University Press, Cambridge (2002)
3. Caporale, N., Dan, Y.: Spike timing-dependent plasticity: a hebbian learning rule. Annu. Rev. Neurosci. **31**, 25–46 (2008)
4. Wills, T.J., Lever, C., Cacucci, F., Burgess, N., O'Keefe, J.: Attractor dynamics in the hippocampal representation of the local environment. Science **308**(5723), 873–876 (2005)

5. Jaeger, H.: The "echo state" approach to analysing and training recurrent neural networks - with an erratum note. German National Research Center for Information Technology GMD Technical report 48 (2001)
6. Jaeger, H., Haas, H.: Harnessing nonlinearity: predicting chaotic systems and saving energy in wireless telecommunication. Science **304**(5667), 78–80 (2004)
7. Smolinski, T.G., Prinz, A.A.: Multi-objective evolutionary algorithms for model neuron parameter value selection matching biological behavior under different simulation scenarios. BMC Neurosci. **10**(Suppl. 1), P260 (2009)
8. Krimm, R.F., Hill, D.L.: Neuron/target matching between chorda tympani neurons and taste buds during postnatal rat development. J. Neurobiol. **43**(1), 98–106 (2000)
9. Deng, Z.D., Zhang, Y.: Collective behavior of a small-world recurrent neural system with scale-free distribution. IEEE Trans. Neural Networks **18**(5), 1364–1375 (2007)
10. Jaeger, H., Lukoševičius, M., Popovici, D., Siewert, U.: Optimization and applications of echo state networks with leaky-integrator neurons. Neural Netw. **20**(3), 335–352 (2007)
11. Legenstein, R., Chase, S.M., Schwartz, A.B., Maass, W.: A reward-modulated hebbian learning rule can explain experimentally observed network reorganization in a brain control task. J. Neurosci. **30**(25), 8400–8410 (2010)
12. Dayan, P., Abbott, L.F.: Theoretical neuroscience: computational and mathematical modeling of neural systems. J. Cogn. Neurosci. **15**(1), 154–155 (2003)

Identify Website Personality by Using Unsupervised Learning Based on Quantitative Website Elements

Shafquat Chishti, Xiaosong Li, and Abdolhossein Sarrafzadeh[✉]

Department of Computing, Unitec Institute of Technology,
Auckland, New Zealand
shafquatac@yahoo.com, {xli,hsarrafzadeh}@unitec.ac.nz

Abstract. This paper reports a pilot study in identifying and ranking the personality of a website automatically and intelligently to help the users to find a more suitable website and to help the owners to improve the quality of their websites. The mapping between the selected items defined in WPS and the quantitative elements of a website was developed first. 240 valid websites were classified by using unsupervised clustering algorithm K-means. The classification was implemented for multiple times from $K = 2$ to $K = 15$. The average values for each attribute in each cluster were calculated, the standard deviation for all the clusters for a given K value was calculated to find out a suitable K value. A preliminary verification suggested that the attributes and the method used can properly identify the personality of a website. A software written in Java integrating other existing software packages was developed for the required experiments.

Keywords: Website · Unsupervised learning · K-means · Data extraction · Experiments · Ranking · Personality · Classification

1 Introduction

Websites become more and more important in people's life, from online shopping to routine business such as banking, booking and etc., websites play very important role. The business owners would like their websites to be more user-friendly and appealing to their users, so they can do their business easily and successfully. This becomes a challenge for the website developers.

A research showed that students with different personal attributes such as age, gender and academic achievement have different preferences on the website types [1]. Understanding the users' preferences can help the developers to make the website satisfying the users better. The developers could design and develop websites more effectively by introducing different attributes like colorful texts, search boxes and audio/video clips for the product information to make the website possess human-like characteristics. These characteristics altogether develop the website personality. The personality of the salesman plays a vital role in developing of long-lasting and trustworthy relationships between customers and business. Similarly the website of a

© Springer International Publishing Switzerland 2015
S. Arik et al. (Eds.): ICONIP 2015, Part I, LNCS 9489, pp. 522–530, 2015.
DOI: 10.1007/978-3-319-26532-2_57

business can be thought of as a salesman for the business, representing it to the potential customers. Therefore the website's personality is also vital, just like that of a salesman [2]. Research also showed that there is a relationship between the customer personalities and the website personalities for e-commerce websites. Website personalities, therefore, could help to maintain customer loyalty in e-commerce [3].

In most of the research work, website personality has been measured with the help of human interactions like surveys, interviews etc. [4], such as the work presented in [5]. This is influenced by the individual's own personality, way of thinking and likes and dislikes etc. The website personality scale (WPS) research [4] is a good effort for the diagnosis of a website personality. The researchers investigated the presence of human and brand personality attributes as well as information characteristics in over one hundred websites. They found that the human attributes correlated with overall attitude and liking of the websites [4]. Their research showed that website personality is based on different factors like intelligence, fun etc. Each of these factors further breaks down into different facets like proficiency, sophistication, engagement etc. and further depends on different factors such as whether it is searchable, satisfying, colorful etc.

This paper reports a pilot study in identifying and ranking the personality of a website automatically and intelligently to help the website users to find out which one is more suitable for them and to help the website owners to improve the quality of their websites such that they are more likely accepted by their target users.

The mapping between the selected items defined in WPS and the quantitative elements related to the website features such as search boxes, colorful texts etc. was developed first. The website features were then quantified by means of website content mining. The quantitative elements form the measurable framework for identifying and ranking the personality of a website. 250 website in five different website categories (Academic, Banks, E-commerce, News and Sports) were selected, where, there were 50 websites in each category. After the data extraction, the websites with invalid data were eliminated and 240 valid websites were obtained. These websites then were classified by using unsupervised clustering algorithm K-means aiming to construct a classifier, which can identify the personality characteristics for a new website. The classification were implemented for multiple times from $K = 2$ to $K = 15$. The results were analyzed to find a suitable K value. A preliminary verification was done by comparing the resulting clusters and their original categories. A software written in Java integrating other existing software packages such as Weka and Jsoup was developed to conduct the required experiments.

In the rest of the paper, the mapping from WPS to quantitative website elements is described first, the method used in this research is discussed after, then the data extraction is described, which is followed by a discussion of the experiments results, and a summary at last.

2 Mapping WPS Items to Quantitative Website Elements

WPS includes five factors (Intelligent, Fun, Organized, Candid, and Sincere), each of the factor includes a number of facets, and a facet includes a number of items. Due to the time and implementation limitations, five items from two facets were selected as the start point of our study. These include *Proficient* facet with *Informative*, *Satisfying* and *Searchable* items and *Systematic* facet with *Concise* and *Fast* items. The mapping between the WPS items and the quantitative website elements are based on the existing literature.

For Informative item, quantitative elements *Hyperlinks*, *Word Count* and *Video/Audio Clips* were chosen. The number and placement of links in a page provides valuable information about the broad category the page belongs to. A high ratio of the number of characters in links to the total number of characters in the page means the probability of the page being an information page is high [6]. The amount of text in a page gives an indication of the type of page, generally information and personal homepages are sparse in text compared to research pages [6]. People comprehend and grasp the content shown on a website much better when it is shown visually on a video, rather than just text. So a website that has videos and text, rather than just text will be more informative for the user since they will be able to understand more information [7].

For Satisfying item, quantitative elements *Search Boxes* and *Images* were chosen. Search boxes make it very easy for the user to use the website and the user is satisfied after the search for an item is ended. This makes the website itself satisfying and proficient to use. As the user came to purchase something, the user found the item on the website very easily without surfing through pages and content [8]. Images are easier for users to assess and help them to further broaden their horizon on the subject, by breaking the dullness and monotony of texts. This makes the website very interesting for the user and adds to the satisfying quality of the web pages [9].

For Searchable item, quantitative element *Search Boxes* was chosen. Most important characteristics to search a website is an implementation of a search box [10].

For Concise item, quantitative element *Density* was used, which is inversely proportional to the Item Concise. Webpage density is dependent on number of characters and number of tags. If the number of characters are very large when compare to number of tags, the density will be higher or vice versa [11].

For Fast item, quantitative elements *Webpage size*, *Duration* and *Graphic Files* were chosen. Large pages and applets slow down the page loading speed, to make a page fast loading avoid large pages and applets [12]. Page loading speed is the key factor in successful website designing [13]. Graphical files and animation makes websites so slow that the e-commerce business may lose their customers [14].

3   Method

250 website in five categories (Academic, Banks, E-commerce, News and Sports) were selected, where, there were 50 websites in each category. After the data extraction, the websites with invalid data were eliminated and 240 valid websites were obtained: Academic (50), Banks (49), E-commerce (49), News (45) and Sports (47).

The quantitative elements defined previously were used as attributes to model a website. Due the software performance issue, *Webpage size* and *Duration* elements were not implemented. So each website was modelled by seven attributes: *Hyperlinks*, *Word Count*, *Video/Audio Clips*, *Search Boxes*, *Images*, *Density* and *Graphic Files*. These websites then were classified into the clusters with different personalities based on these attributes aiming to construct a classifier, which can identify the personality characteristics for a new website.

Due to lack of the knowledge of existing websites' personality, unsupervised learning was employed [15, 16] and K-means clustering algorithm was used [17]. Other hierarchical clustering algorithms such as COBWEB [18] are available as well. While a hierarchical clustering can provide more accurate classification, it requires unification criteria and division criteria in the classification procedures. These criteria are hardly determined in this pilot study.

K-means is a simple and effective clustering algorithm, however, it requires pre-knowledge on the number of the clusters, i.e. the K value [17]. To gain knowledge on the suitable K value, the classification were implemented for multiple times from $K = 2$ to $K = 15$. The average values for each attribute in each clusters were calculated, the standard deviation for all the clusters for a given K value was calculated, assuming that the K value with the maximum standard deviation would provide maximum amount of information and should be a suitable K value.

It is fair to assume that the website in different categories (Academic, Banks, E-commerce, News and Sports) should have obvious difference in terms of personality. Based on this assumption, a preliminary verification was done by comparing the resulting clusters and their original categories to check that if this method is on the right track. If the system can correctly classify the selected websites in the right categories, the quantity elements defined in the proposed approach should be on the right track.

For a website, the data extraction starts from its home page and then propagates to the secondary pages linked to the home page. The home page and a few other pages get lots of views and most others get much less traffic, Chapter 5 of [19]. Most websites are organized in a multitier hierarchy from home pages to secondary pages. Home pages have four primary elements: Identity, Navigation, Timelines and Tools, Chapter 6 of [19]. These give us confidence to believe that the data from the home page and the secondary pages can capture the main characteristics of a website.

4 Data Extraction

The quantitative elements for a website defined previously can be extracted from the website by various techniques including structure mining, information entropy etc.

The *Images* element is achieved by detecting the "img" HTML tag in the source code of a web page. HTML tags are usually used for website user interface presentation, this element should reflect/contribute to the characteristics of visual appearance for the website. For *Graphic Files* element, a few well known media formats (.jpg,.tif,. gif,.png and.raw) are picked up from parsing the web page. This could include the images in the content and the images in the background, animations or slideshow. This element reflects the amount of the image files used in the website in general. For *Hyperlinks*, *Word Count*, *Video/Audio Clips*, and *Search Boxes* elements, each element is simply counted on a webpage. The *Density* element is calculated by using the method described in [11]. First the number of characters ('A', '3', '@' etc.), say C, on the web page is calculate, then the number of tags ('img', 'a' etc.), say n, on the web page is calculated, the web page *Density* is obtained by the following formula:

$$Density = C/n \qquad\qquad (1)$$

In order to rank the websites personality, a cluster is rated for each item respectively. A *Unit* is defined to allocate ratings to the quantitative elements of different clusters by the following formula, where D_n is the number of distinct values of a quantitative element in the cluster, 10 is the highest rating scale and 0 is the lowest:

$$Unit = 10/D_n - 1 \qquad\qquad (2)$$

1 is subtracted to assure that the smallest rating scale is 0.

For a given K value in the K-means, initially rating 0 is allocated to the cluster with the lowest quantity for a particular element, then this number is increased by the *Unit* defined in formula (2) which is allocated to the cluster with the second lowest quantity for the same element, this is repeated until the last cluster is reached, which will definitely get rating 10.

5 Experiment Results and Analysis

240 valid websites in five categories, Academic (50), Banks (49), E-commerce (49), News (45) and Sports (47), were used in the K-means algorithm, which was implemented for 14 times from K = 2 to K = 15. The average values for each attribute in each clusters were calculated, the standard deviation for all the clusters for a given K value was calculated. For each quantitative element in a website, two options were used: one is the sum of the values for all the pages reached in the data extraction; another is the average of all the pages reached in the data extraction. Table 1 shows the standard deviation obtained from the experiment when sum was used and Table 2 shows the standard deviation when average was used.

The above data are depicted in Fig. 1 to observe the trend of the standard deviation over the K values for both of the sum and average options.

For sum option, it is clear that when K = 8 all the attributes reach its maximum STDEV and that value remains after K > 9; on the other hand, no obvious trend is observed for average option.

The above observation can be supplemented by comparing the websites in each cluster with their original categories in Tables 3 and 4.

In Tables 3 and 4, short codes INFO, SAT, SEARCH, CON and FAST are used for items *Informative*, *Satisfying*, *Searchable*, *Concise* and *Fast* respectively; short codes C, A, B, E, N and S are used for Cluster, Academic, Banks, E-commerce, News and Sports respectively. Their ratings are displayed in the tables. The experiment results showed that when K = 7 in the sum option and K = 9 in the average option, the websites in each cluster only belong to one of the original categories: A, B, E, N and S, and that remains true for the rest of the K value with one exception: one news website stays with 8 e-commerce websites. This preliminarily verified that the attributes and the K-means algorithm used in this study can properly identify the personality of a website.

For sum option, when K = 9, the STDEV almost reached the lowest value, so K = 10 was selected for discussion in both of the options. An interesting fact is that all

Table 1. The STDEV when sum was used.

K	Density	Graphical files	Hyperlinks	Images	Search box	Video	Word count
2	315.970	846.190	5943.842	406.508	13.114	1803.504	47419.014
3	169.412	1174.103	2733.596	801.118	9.126	1492.724	23636.063
4	151.542	1803.013	7753.403	1424.544	21.097	1419.146	27885.943
5	356.866	3466.880	11627.801	2120.743	26.260	3969.308	70280.086
6	262.604	1893.566	7965.622	1304.320	25.456	1762.272	36290.168
7	469.261	7819.499	39431.165	6045.300	88.003	2439.092	122654.970
8	559.593	8306.351	36846.807	5394.063	72.096	5842.067	141229.674
9	270.878	2105.205	9796.252	1359.416	25.854	1157.128	44446.370
10	507.159	6703.533	36212.846	5280.065	66.473	6310.415	132807.482
11	484.832	6365.247	34464.775	5017.440	63.659	5989.316	126419.544
12	519.533	6604.232	36470.576	5179.422	117.374	6371.876	137997.621
13	504.686	6362.182	34873.606	4983.817	63.404	6118.294	131763.573
14	507.971	6662.538	34221.049	4947.990	63.760	5917.615	132308.078
15	490.344	5967.677	32945.650	4679.086	59.725	5736.350	125275.481

Table 2. The STDEV when average was used.

K	Density	Graphical files	Hyperlinks	Images	search box	Video	Word count
2	1.782	33.935	92.452	24.522	0.13991	13.031	379.765
3	1.470	22.945	66.548	17.874	0.09866	6.443	275.592
4	1.228	18.738	55.539	12.430	0.15074	11.127	502.627
5	1.037	16.626	50.675	11.228	0.14840	11.910	180.045
6	2.786	18.166	71.060	13.413	0.12887	96.923	302.599
7	2.366	18.284	75.553	14.674	0.14243	93.746	321.219
8	1.703	20.955	73.607	22.163	0.21546	90.786	426.141
9	1.921	22.863	58.204	21.549	0.21964	85.224	415.162
10	1.984	32.301	110.465	27.736	0.21207	80.721	444.059
11	4.126	29.508	107.103	24.907	0.22892	22.308	450.987
12	4.055	28.915	106.969	25.153	0.23361	74.529	431.453
13	4.153	49.044	131.274	33.398	0.20020	14.829	1099.768
14	4.505	47.961	129.298	32.481	0.20015	51.451	1070.459
15	7.173	17.434	91.608	14.086	0.78260	4.478	512.565

the bank websites stays together for most of the K values. By examining the ratings for each website category, it was found that in general, news websites are more informative, e-commerce websites are more satisfying, and sports websites are more concise and fast. Bank websites stay neutral in all aspects. In Table 3, the academic websites are classified into three clusters: 0, 4 and 6. The ratings for cluster 4 and 6 are quite similar. The ratings for cluster 0 suggest the websites in this cluster are quite informative. A close look at the actual websites in this cluster, it is found that these websites

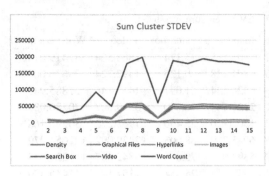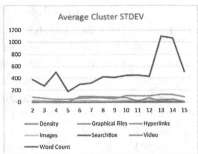

Fig. 1. The trend of the STDEV over the K values

Table 3. K = 10 (Sum)

C	A	B	E	N	S	INFO	SAT	SEARCH	CON	FAST
0	10	0	0	0	0	8.148148	5.555556	7.777778	2.222222	4.444444
1	0	0	33	0	0	2.592593	2.222222	2.222222	7.777778	7.777778
2	0	49	0	0	0	4.444444	3.333333	4.444444	4.444444	5.555556
3	0	0	0	0	7	8.148148	8.888889	8.888889	1.111111	1.111111
4	22	0	0	0	0	0.000000	0.000000	0.000000	10.000000	10.000000
5	0	0	8	0	0	5.925926	7.777778	6.666667	5.555556	2.222222
6	18	0	0	0	0	2.962963	4.444444	3.333333	6.666667	6.666667
7	0	0	0	0	40	1.111111	1.111111	1.111111	8.888889	8.888889
8	0	0	0	44	0	6.666667	6.666667	5.555556	3.333333	3.333333
9	0	0	8	1	0	10.000000	10.000000	10.000000	0.000000	0.000000

Table 4. K = 10 (Average)

C	A	B	E	N	S	INFO	SAT	SEARCH	CON	FAST
0	1	0	0	0	0	5.000000	1.428571	0.000000	0.000000	8.888889
1	0	0	36	0	0	1.944444	5.714286	0.000000	8.000000	5.555556
2	0	49	0	0	0	3.750000	4.285714	0.000000	6.000000	3.333333
3	0	0	0	0	33	0.370370	1.428571	0.000000	2.000000	6.666667
4	45	0	0	0	0	4.120370	1.428571	0.000000	4.000000	7.777778
5	0	0	3	0	0	7.314815	2.857143	0.000000	10.000000	4.444444
6	4	0	0	0	0	3.101852	0.000000	0.000000	0.000000	10.000000
7	0	0	0	0	14	7.731481	8.571429	0.000000	8.000000	1.111111
8	0	0	0	45	0	6.527778	7.142857	0.000000	4.000000	2.222222
9	0	0	10	0	0	8.888889	10.000000	0.000000	10.000000	0.000000

have more complex information structure and more images comparing with the other academic websites. In Table 4, the e-commerce websites are classified into three clusters: 1, 5 and 9. A close look at the actual websites in cluster 9, it is found that these websites have large amount of catalogues associated with nice images and rich options for each products. It is noticed that for the average option, the searchable rating is always 0 for all the clusters. This is reasonable as usually there is only one search box on each page in a website. This suggests that the formula of the rating could be improved.

6 Summary and Future Work

This paper reports a pilot study in identifying and ranking the personality of a website automatically and intelligently to help the website users to find out which one is more suitable for them and to help the website owners to improve the quality of their websites such that they are more likely accepted by their target users.

The mapping between the selected items defined in WPS and the quantitative elements related to the website features were developed first. The website features were then quantified by means of website content mining. The quantitative elements form the measurable framework for identifying and ranking the personality of a website. 240 valid websites were classified by using unsupervised clustering algorithm K-means. The classification were implemented for multiple times from K = 2 to K = 15. The average values for each attribute in each clusters were calculated, the standard deviation for all the clusters for a given K value was calculated to find a suitable K value. The results showed that K = 10 is a suitable value. A preliminary verification was done by comparing the resulting clusters and their original categories, the results suggested that the attributes and the K-means algorithm used in this study can properly identify the personality of a website. The personality of the websites in different clusters is discussed.

In the future study, more quantitative elements can be included in the experiments and more website categories can be included in the verification. The software will be improved accordingly and user evaluation will be conducted.

References

1. Li, X.: Mapping from student domain into website category. Neural Information Processing, pp. 11–17. Springer, Berlin (2011)
2. Smith, B.: Buyer-seller relationships: bonds, relationship management, and sex-type. Can. J. Adm. Sci./Revue Canadienne des Sciences de l'Administration 15(1), 76–92 (1998)
3. Poddar, A., Donthu, N., Wei, Y.: Web site customer orientations, web site quality, and purchase intentions: the role of web site personality. J. Bus. Res. 62(4), 441–450 (2009)
4. Chen, Q., Rodgers, S.: Development of an instrument to measure web site personality. J. Interact. Advertising 7(1), 4–46 (2006)
5. Katerattanakul, P., Siau, K.: Measuring information quality of web sites: development of an instrument. In: Proceedings of the 20th International Conference on Information Systems, Association for Information Systems pp. 279–285 (1999)
6. Asirvatham, A., Ravi K.: Web page categorization based on document structure. Centre for Visual Information Technology (2001)

7. Chtouki, Y., Harroud, H., Khalidi M., Bennani, S.: The impact of YouTube videos on the student's learning. In: 2012 International Conference on, Information Technology Based Higher Education and Training (ITHET),·pp. 1–4. IEEE (2012)
8. Thimthong, T., Chintakovid T., Krootjohn, S.: An empirical study of search box and autocomplete design patterns in online bookstore. In: 2012 IEEE Symposium on Humanities, Science and Engineering Research (SHUSER), pp. 1165–1170. IEEE (2012)
9. Jiang, N., Feng, X., Liu, H., Liu, J.: Emotional design of web page. In: 9th International Conference on, Computer-Aided Industrial Design and Conceptual Design, 2008. CAID/CD 2008, pp. 91–95. IEEE (2008)
10. Harpel-Burke, P.: Library homepage design at medium-sized universities: a comparison to commercial homepages via Nielsen and Tahir. OCLC Syst. Serv. Int. Digit. Libr. Perspect. 21(3), 193–208 (2005)
11. Sun, F., Song, D., Liao, L.: Dom based content extraction via text density. In: Proceedings of the 34th International ACM SIGIR Conference on Research and Development in Information Retrieval, pp. 245–254. ACM (2011)
12. Lau, T.: Revamping a corporate web site: strategies and implementation. In: International Professional Communication Conference, 1996. IPCC 1996 Proceedings. Communication on the Fast Track, pp. 104–107. IEEE (1996)
13. Gehrke, D., Turban, E.: Determinants of successful website design: relative importance and recommendations for effectiveness. In: HICSS-32. Proceedings of the 32nd Annual Hawaii International Conference on Systems Sciences, 1999, p. 8. IEEE (1999)
14. Busch, D.: Avoid the Five Cardinal Graphical Sins. Internet World, pp. 98–99 (1997)
15. Dayan, P.: Unsupervised learning. The MIT Encyclopedia of the Cognitive Sciences. The MIT Press, London (1999)
16. Iba, W., Langley, P.: Unsupervised Learning of Probabilistic Concept Hierarchies. In: Paliouras, G., Karkaletsis, V., Spyropoulos, C.D. (eds.) ACAI 1999. LNCS (LNAI), vol. 2049, pp. 72–112. Springer, Heidelberg (2001)
17. Jain, A.K.: Data clustering: 50 years beyond K-means. Pattern Recogn. Lett. 31(8), 651–666 (2010)
18. Theodorakis, M., Vlachos, A., Kalamboukis, T.: Using hierarchical clustering to enhance classification accuracy. In: Proceedings of 3rd Hellenic Conference in Artificial Intelligence, Samos (2004)
19. Lynch, P., Horton, S.: Web Style Guide, chapter 5, 6. Yale University Press, New Haven (2009)

Discriminative Dictionary Learning for Skeletal Action Recognition

Yang Xiang(✉) and Jinhua Xu(✉)

Shanghai Key Laboratory of Multidimensional Information Processing,
Department of Computer Science and Technology, East China Normal University,
500 Dongchuan Road, Shanghai, China
51131201034@ecnu.cn, jhxu@cs.ecnu.edu.cn

Abstract. Human action recognition is an important yet challenging task. With the introduction of RGB-D sensors, human body joints can be extracted with high accuracy, and skeleton-based action recognition has been investigated and gained some success. In this paper, we split an entire action trajectory into several segments and represent each segment using covariance descriptor of joints' coordinates. We further employ the projective dictionary pair learning (PDPL) and majority-voting for multi-class action classification. Experimental results on two benchmark datasets demonstrate the effectiveness of our approach.

Keywords: Action recognition · Covariance descriptor · Discriminative dictionary learning

1 Introduction

Action recognition is an active research field in computer vision. It has many applications, including video surveillance, human-computer interaction and health care. Traditional research mainly concentrates on action recognition from video sequence of 2D frames with RGB channels [1], which only capture projective information of the real world, and is sensitive to lighting conditions and occlusion. Recently, with the introduction of the low-cost RGB-D sensors such as Microsoft Kinect [7], depth information has been employed for action recognition [3,25].

There are two kinds of approaches for human action recognition based on the depth information, depth map-based approaches and skeleton-based approaches [25]. The depth map-based methods rely mainly on features that extracted from the space time volume (STV) [29]. In [11], Li *et al.* employed the concept of bag-of-points to construct the action graph to encode the human actions where a small number of 3D skeleton points were sampled from depth maps to describe the 3D shape of each salient posture. In [19], Vieira *et al.* presented a representation for 3D action recognition named Space-Time Occupancy Patterns (STOP). In order to address the noise and occlusion issues, Wang *et al.* proposed features named Random Occupancy Pattern (ROP) in [20].

© Springer International Publishing Switzerland 2015
S. Arik et al. (Eds.): ICONIP 2015, Part I, LNCS 9489, pp. 531–539, 2015.
DOI: 10.1007/978-3-319-26532-2_58

The study of skeleton-based activity recognition dates back to the early work by Johansson [9], which demonstrated that a large set of actions can be recognized solely from the joint positions. This concept has been extensively explored ever since. In [2], Bashir *et al.* segmented the trajectories into small units of perceptually similar pieces of motions and then built their models on Principal Component Analysis (PCA)-based representation of these trajectories. In [12], Lv *et al.* proposed a set of local features based on joints and then used Hidden Markov Models (HMMs) to model the temporal dynamics. In [21], differential invariants were proposed to describe trajectory features and then a nonlinear signature warping method was employed to recognize trajectories. In [22], Xia *et al.* proposed a feature named Histogram of 3D Joint Locations (HOJ3D) that encoded the distribution of joints around the skeleton root and employed a HMM to model the temporal changes of the features. In [26], Yuan *et al.* proposed an actionlet ensemble model to represent the interaction of a subset of human joints. In [23], EigenJoints were proposed to describe positional differences between joints, and then Naive Bayes nearest neighbour classifier was employed for action classification. Covariance descriptor captures the dependence of locations of different joints on one another during the performance of an action and has been used for action recognition [8,17]. In [17], joint trajectories were represented using vectorized log-covariances, then the sparse representation-based classifier was employed for action classification. In [8], a temporal hierarchy was used to encode the temporal information which was lost in the covariance descriptor, and SVM was used as the classifier.

One of the drawbacks of covariance descriptor for action recognition is the high dimensional problem. Different from the methods in [8,17] which used full covariance descriptor, we reduce the dimension of covariance descriptor by considering only the most correlated joints coordinates. In addition, we calculate the covariance matrix on multiple overlapped segments instead of the entire action sequences so that majority voting can be employed for classification. Finally, we employ the projective dictionary pair learning (PDPL) [6] for discriminative dictionary learning and feature encoding. This is different from the work in [8,17] where covariance descriptor was directly applied to sparse representation classifier [17] and Naive Bayes nearest neighbour classifier [8].

We make three contributions in this paper. First, we present the dimension-reduced log-covariance features. Second, we employ a discriminative dictionary learning method to encode the covariance features. The action label can be easily predicted based on the reconstruction error. Third, we split an entire action trajectory into some overlapped segments and predict the label for each segment, then the action label is predicted by majority voting. The remainder of this paper is organized as follows. In Sect. 2, we introduce log-covariances features for representation of trajectory and our dimension reduction method. In Sect. 3, we explain the projective dictionary pair learning (PDPL) for dictionary learning and multi-class action classifications. In Sect. 4, we report the experimental results of our method on two datasets and compare them to other methods. In Sect. 5, we briefly summarize this paper.

2 Representation of Trajectory Information

For the t-th frame, the 3D coordinates of J joints are available:

$$S^t = [x_1, ..., x_J, y_1, ..., y_J, z_1, ..., z_J]'.$$

A Joint trajectory information is specified in terms of positional information over time. The covariance matrix of the 3n joint coordinates over the T-length frames can be calculated as follows [8,17]:

$$cov(S) = \frac{1}{T-1} \sum_{t=1}^{T} (S^t - \bar{S})(S^t - \bar{S})' \tag{1}$$

where $\bar{S} = \sum_{t=1}^{T} S^t$. Covariance matrix lies on a Riemannian manifold. The log-covariance belongs to the tangent space of this Riemannian manifold. This space is Euclidean, giving rise to the Log-Euclidean distance metric, that is regarded as a good approximation of the geodesic distance on the manifold. Therefore, the matrix logarithm of the covariance descriptor is computed. Then the upper-triangular part of the resulting symmetric matrix is vectorized into an $\frac{3J \times (3J+1)}{2}$-dimensional feature vector. The covariance descriptor was considered as simple but robust feature for representing the human actions. However, the dimension can be very high, e.g. it is 1830 for $J = 20$ joints. For the temporal hierarchy in [8], multiple covariance descriptors were used, and the dimension increases linearly with the number of windows of each level and the number of levels.

The covariance in (1) calculates the correlations between each pair of elements. We assume among an action sequence the coordinate (x) of a joint is most correlated with this joint's other coordinates (y and z) and the same coordinate (x) of the other joints. Therefore, we calculate only partial elements (denoted as o as follows) of the covariance matrix in (1).

$$
\begin{array}{cccccccccc}
 & x_1 & \cdots & x_J & y_1 & \cdots & y_J & z_1 & \cdots & z_J \\
x_1 & o & o & o & o & & & o & & \\
\vdots & & o & o & & o & & & o & \\
x_J & & o & & & o & & & o & \\
y_1 & & & o & o & o & o & & & \\
\vdots & & & & o & o & & o & & \\
y_J & & & & & o & & o & & \\
z_1 & & & & & & o & o & o & \\
\vdots & & & & & & & o & o & \\
z_J & & & & & & & & o &
\end{array}
$$

As a result, the log-covariance is vectorized into a feature vector with $\frac{J \times (J+1)}{2} \times 3 + 3J$ dimension. For $J = 20$ joints, the dimension is 690, much lower than the original 1830.

As observed in [23], a short segment of the entire video sequence is sufficient to recognize the action. We split an entire trajectory into several segments, and calculate the covariance descriptor for each segment. Then we predict the label for each segment, and the label of the entire sequence can be predicted by majority voting.

3 Projective Dictionary Pair Learning (PDPL)

Supervised dictionary learning (SDL) has been widely employed in various classification problems. Most of the existing SDL methods aim to learn a synthesis dictionary to sparsely represent the input signal. We represent a set of p-dimensional training samples as $X = [X_1 \cdots X_k \cdots X_C]$, where C is the number of classes, $X_k \in \mathbb{R}^{p \times n}$ is the training samples of class k, and n is the number of samples of each class. The conventional supervised dictionary learning methods can be formulated as below:

$$\min_{D,\alpha} \|X - D\alpha\|_F^2 + \gamma\|\alpha\|_p + \lambda\Psi(D,\alpha,Y) \tag{2}$$

where D is the synthesis dictionary to be learned, α is the coefficient matrix, Y represents the class labels of samples. The third term $\Psi(D,\alpha,Y)$ is introduced to learn a discriminative dictionary [13]. γ and λ are parameters to balance the reconstruction error and regularization terms.

In [6], a projective dictionary pair learning (PDPL) method was proposed, in which a new analysis dictionary $P \in \mathbb{R}^{mk \times p}$ was introduced, such that the coefficient matrix was obtained as $\alpha = PX$. The PDPL model was formulated as below:

$$P^*, D^* = arg\min_{P,D} \|X - DPX\|_F^2 + \lambda\Psi(D,P,X,Y) \tag{3}$$

where $\Psi(D,P,X,Y)$ is some discrimination function. In [6], it was defined as $\|P_k\bar{X}_k\|_F$, and $\bar{X}_k$ denotes the complementary data matrix of X_k in the whole training set X. In the PDPL model, the analysis sub-dictionary P_k^* is trained to generate significant coding coefficients for sample from class k, and produce small coefficients for samples from classes other than k. And the synthesis sub-dictionary D_k^* is trained to reconstruct the samples of class k, that is the residual $\|X_k - D_k^*P_k^*X_k\|$ will be small, and the residual $\|X_i - D_k^*P_k^*X_i\|_F^2, i \neq k$ will be much larger than $\|X_k - D_k^*P_k^*X_k\|$.

The analysis dictionary $P = [P_1; \cdots ; P_k; \cdots ; P_C]$ and synthesis dictionary $D = [D_1, \cdots , D_k, \cdots , D_C]$, $D_k \in \mathbb{R}^{p \times m}, P_k \in \mathbb{R}^{m \times p}$ form a sub-dictionary pair corresponding to class k. Although this is a non-convex problem, it can be solved as introduced in [6].

In the test phase, the class l which gives the smallest reconstruction residual by this idea is assigned as the label of the test vector x:

$$label = arg\min_l \|x - D_l P_l x\|_2 \tag{4}$$

4   Experimental Results

We tested our method on two publicly available datasets: the Cornell Activity Dataset-60(CAD-60) [18] and the MSR Daily Activity3D dataset [26]. We constructed the first baseline in which the full covariance was used to demonstrate the partial covariance will capture most of the correlation information between joints coordinates. Also we constructed the second baseline in which the full covariance was calculated on the entire action sequence, rather than on multiple segments. We set the parameters empirically.

4.1   CAD-60 Dataset

The CAD-60 dataset [18] contains the RGB frames, depth information and the skeleton joint positions captured with Microsoft Kinect sensor. The actions in this dataset were performed in 5 different environment by 4 subjects (two males and two females). The frame rate is 30 frames per second with resolution of 640×480. The dataset contains twelve actions: *rinsing mouth, brushing teeth, wearing contact lens, talking on the phone, drinking water, opening pill container, cooking(chopping), cooking(strring), talking on couch, relaxing on couch, writing in whiteboard, working on computer.*

We followed the *"new person"* setting as [18]: the data of three people were used for training and the remaining one person for testing. We split each entire image sequence into 10 segments, and each segment contains 150 frames. The dictionary size for each category was 30, and λ in (3) was 0.002. Classification was performed on each of the segments from a test sequence, and class decision for the entire sequence was made by taking a majority-voting of the predicted labels from the individual segments. Our method achieved an accuracy of 85.3 %.

Table 1. Results of our method and other methods on the CAD-60 dataset.

Method	Accuracy(%)	Precision(%)	Recall(%)
Sung et al., AAAI PAIR 2011, ICRA 2012 [18]		67.9	55.5
Koppula, Gupta, Saxena, IJRR 2012 [10]		80.8	71.4
Zhang, Tian, NWPJ 2012 [28]		86	84
Yang, Tian, JVCIR 2013 [24]		71.9	66.6
Ni et al., Cybernetics 2013 [15]		75.9	69.5
Wang et al., PAMI 2014 [20]	74.70		
Gaglio, Lo Re, Morana, HMS 2014 [5]		77.3	76.7
Zhu, Chen, Guo, IVC 2014 [29]		93.2	84.6
Faria, Premebida, Nunes, RO-MAN 2014 [4]		91.1	91.9
Shan, Akella, ARSO 2014 [16]	91.9	93.8	94.5
Baseline 1 (Full Cov, multi Segments)	85.3		
Baseline 2 (Partial Cov, single segment)	79.4		
Ours (Partial Cov, multi segments)	85.3	87.73	84.82

We compared our results with baselines and other existing methods in Table 1. It can be seen that our result is better than some of existing methods, but not the best one. Compared with the two baselines, the result with partial covariance is similar to that with full covariance, and the result with multiple segments is significantly better than that with the entire sequence. This demonstrated our dimension reduction and majority voting methods are effective.

4.2 MSR Daily Activity3D Dataset

The MSR Daily Activity3D dataset is a daily activity dataset where each action was performed twice by 10 subjects. There are sixteen activities: *drink, eat, read book, call cellphone, write on a paper, use laptop, use vacuum cleaner, cheer up, sit still, toss paper, play game, lay down on sofa, walk, play guitar, stand up, sit down.*

The experimental setting was the same as in [26]: subjects 1,3,5,7,9 were used for training and 2,4,6,8,10 for testing. We split each entire image sequence into 5 segments, and each segment contains 70 frames. The dictionary size for each category was 50, and λ in (3) was 0.00004. We performed action classification at all segments and then made decisions by majority-voting. The accuracy of our algorithm was 73.1 %. The confusion matrices of our approach was shown in Fig. 1. All instances of "cheer up", "stand up" and "walk" were correctly recognized. However, for "call cellphone", only 2 out of 10 were correctly recognized, and four of them were recognized as "drink" by mistake. This is because the joint trajectories of "call cellphone" and "drink" are quite similar, and the main difference between them is the objects in hands.

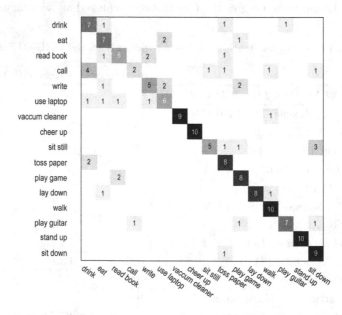

Fig. 1. The confusion matrix of our approach on the MSR Daily Activity3D dataset.

We compared our results with other existing skeleton-based methods in Table 2. Some best results [26] that simultaneously used the depth information on this dataset were not compared to our method. It can be seen that our result is comparable or better than other methods. Compared with the two baselines, the result of full covariance is slightly better than that of partial covariance, and the result with multiple segments is significantly better than that with the entire sequence.

Table 2. Results of our method and other methods on the MSR Daily Activity 3D dataset.

Method	Accuracy(%)
Dynamic Temporal Warping [14]	54
Only LOP features [26]	42.5
Only Joint Position features [26]	68
Moving Pose [27]	70.6
Baseline 1 (Full Cov, multi Segments)	73.7
Baseline 2 (Partial Cov, single segment)	65.6
Ours (Partial Cov, multi segments)	73.1

5 Conclusion

In this paper we proposed dimension-reduced log-covariance features to represent joints trajectories of human activity. Then we used Projective Dictionary Pair Learning to learn a set of discriminative dictionaries for feature encoding. Our method performed well on two benchmark datasets.

As discussed in [8], the covariance descriptor captures the dependence of locations of different joints on one another during the performance of an action. However, it does not capture the order of motion in time. Therefore, if the frames of a given sequence are randomly shuffled, the covariance matrix will not change. For skeleton-based action recognition, the temporal order in joints trajectories carry important information of action, therefore we will work on some more efficient features of actions which captures both joints dependence and temporal relation in the future work.

Acknowledgments. This work is supported by the National Natural Science Foundation of China under Project 61175116, and Shanghai Knowledge Service Platform for Trustworthy Internet of Things (No. ZF1213).

References

1. Aggarwal, J.K., Ryoo, M.S.: Human activity analysis: a review. ACM Comput. Surv. **43**(3), 16 (2011)
2. Bashir, F., Khokhar, A., Schonfeld, D.: Automatic object trajectory-based motion recognition using gaussian mixture models. In: 2012 IEEE International Conference on Multimedia and Expo, pp. 1532–1535 (2005)
3. Chen, L., Wei, H., Ferryman, J.M.: A survey of human motion analysis using depth imagery. Pattern Recogn. Lett. **34**, 1995–2006 (2013)
4. Faria, D.R., Premebida, C., Nunes, U.: A probabilistic approach for human everyday activities recognition using body motion from RGB-D images. In: IEEE International Symposium on Robot and Human Interactive Communication, pp. 732–737. IEEE (2014)
5. Gaglio, S., Re, G.L., Morana, M.: Human activity recognition process using 3-D posture data. IEEE Trans. Hum.-Mach. Syst. **45**(5), 586–597 (2014)
6. Gu, S., Zhang, L., Zuo, W., Feng, X.: Projective dictionary pair learning for pattern classification. In: Advances in Neural Information Processing Systems, pp. 793–801 (2014)
7. Han, J.: Enhanced computer vision with microsoft kinect sensor: a review. IEEE Trans. Cybern. **43**(5), 1318–1334 (2013)
8. Hussein, M.E., Torki, M., Gowayyed, M.A., El-Saban, M.: Human action recognition using a temporal hierarchy of covariance descriptors on 3D joint locations. In: The 23rd International Joint Conference on Artificial Intelligence (2013)
9. Johansson, G.: Visual motion perception. Sci. Am. **232**(6), 76–88 (1975)
10. Koppula, H.S., Gupta, R., Saxena, A.: Learning human activities and object affordances from RGB-D videos. Int. J. Robot. Res. **32**(8), 951–970 (2013)
11. Li, W., Zhang, Z., Liu, Z.: Action recognition based on a bag of 3D points. Workshop on Human Activity Understanding from 3D Data, pp. 9–14 (2010)
12. Lv, F., Nevatia, R.: Recognition and segmentation of 3-D human action using HMM and multi-class AdaBoost. In: Leonardis, A., Bischof, H., Pinz, A. (eds.) ECCV 2006. LNCS, vol. 3954, pp. 359–372. Springer, Heidelberg (2006)
13. Mairal, J., Bach, F., Ponce, J., Sapiro, G., Zisserman, A.: Supervised dictionary learning. In: arXiv preprint arXiv:0809.3083 (2008)
14. Muller, M., Roder, T.: Motion templates for automatic classification and retrieval of motion capture data. In: Proceedings of the ACM SIGGRAPH/Eurographics Symposium on Computer Animation, pp. 137–146. Eurographics Association Aire-la-Ville, Switzerland (2006)
15. Ni, B., Pei, Y., Moulin, P., Yan, S.: Multilevel depth and image fusion for human activity detection. IEEE Trans. Cybern. **43**(5), 1383–1394 (2013)
16. Shan, J., Akella, S.: 3D human action segmentation and recognition using pose kinetic energy. In: IEEE Workshop on Advanced Robotics and Its Social Impacts, pp. 69–75 (2014)
17. Sivalingam, R., Somasundaram, G., Bhatawadekar, V., Morellas, V., Papanikolopoulos, N.: Sparse representation of point trajectories for action classification. In: IEEE International Conference on Robotics and Automation, pp. 3601–3606 (2012)
18. Sung, J., Ponce, C., Selman, B., Saxena, A.: Unstructured human activity detection from RGBD images. In: IEEE International Conference on Robotics and Automation, pp. 842–849 (2012)

19. Vieira, A.W., Nascimento, E.R., Oliveira, G.L., Liu, Z., Campos, M.F.M.: STOP: space-time occupancy patterns for 3D action recognition from depth map sequences. In: Alvarez, L., Mejail, M., Gomez, L., Jacobo, J. (eds.) CIARP 2012. LNCS, vol. 7441, pp. 252–259. Springer, Heidelberg (2012)
20. Wang, J., Liu, Z., Wu, Y., Yuan, J.: Learning actionlet ensemble for 3D human action recognition. IEEE Trans. Pattern Anal. Mach. Intell. **36**(5), 914–927 (2014)
21. Wu, O., Li, Y.F., Zhang, J.: A hierarchical motion trajectory signature descriptor. In: IEEE International Conference on Robotics and Automation, pp. 3070–3075 (2008)
22. Xia, L., Chen, C.C., Aggarwal, J.K.: View invariant human action recognition using histograms of 3D joints. In: IEEE Conference on Computer Vision and Pattern Recognition, pp. 20–27 (2012)
23. Yang, X., Tian, Y.L.: Eigenjoints-based action recognition using Naive-Bayes-nearest-neighbor. In: IEEE Conference on Computer Vision and Pattern Recognition Workshops, pp. 14–19 (2012)
24. Yang, X., Tian, Y.L.: Effective 3D action recognition using eigenjoints. J. Visual Commun. Image Represent. **25**(1), 2–11 (2014)
25. Ye, M., Zhang, Q., Wang, L., Zhu, J., Yang, R., Gall, J.: A survey on human motion analysis from depth data. In: Grzegorzek, M., Theobalt, C., Koch, R., Kolb, A. (eds.) Time-of-Flight and Depth Imaging. LNCS, vol. 8200, pp. 149–187. Springer, Heidelberg (2013)
26. Yuan, J., Wu, Y., Liu, Z., Wang, J.: Mining actionlet ensemble for action recognition with depth cameras. In: IEEE Conference on Computer Vision and Pattern Recognition, pp. 1290–1297 (2012)
27. Zanfir, M., Leordeanu, M., Sminchisescu, C.: The moving pose: an efficient 3D kinematics descriptor for low-latency action recognition and detection. In: IEEE International Conference on Computer Vision, pp. 2752–2759 (2013)
28. Zhang, C., Tian, Y.: RGB-D camera-based daily living activity recognition. J. Comput. Vis. Image Process. **2**(4), 12 (2012)
29. Zhu, Y., Chen, W., Guo, G.: Evaluating spatiotemporal interest point features for depth-based action recognition. Image Vis. Comput. **32**(8), 453–464 (2014)

Single Face Image Super-Resolution via Multi-dictionary Bayesian Non-parametric Learning

Jingjing Wu[1,2], Hua Zhang[1,2], Yanbing Xue[1,2(✉)], Mian Zhou[1,2], Guangping Xu[1,2], and Zan Gao[1,2]

[1] Key Laboratory of Computer Vision and System, Tianjin University of Technology, Tianjin, China
yanbingxue@163.com
[2] Tianjin Key Laboratory of Intelligence Computing and Novel Software Technology, Tianjin University of Technology, Tianjin, China

Abstract. The face image super-resolution is a domain specific problem. Human face has complex, and fixed domain specific priors, which should be detail explored in super-resolution algorithm. This paper proposes an effective single image face super-resolution method by pre-clustering training data and Bayesian non-parametric learning. After pre-clustering, face patches from different clusters represent different areas in face, and also offer specific priors on these areas. Bayesian non-parametric learning captures consistent and accurate mapping between coupled spaces. Experimental results show that our method produces competitive results to other state-of-the-art methods, with much less computational time.

Keywords: Super-resolution · Multi-dictionary · Beta process · Pre-clustering

1 Introduction

The face image super-resolution algorithm aims to generate high-quality high-resolution (HR) face images from low-resolution (LR) inputs [1]. Image super-resolution is an ill-posed problem as some image details and texture characteristics in HR images are lost in LR images. Numerous methods have been proposed to address this problem, and sparse representation based single image super-resolution methods receive a lot of attention recently [2–4].

Yang et al. [2] first introduced the sparse representation method to super-resolution algorithms. It transfers HR-LR mapping into dictionary learning in sparse coding. Since the dictionaries are learned jointly, the computation cost is small. It only need to infer the sparse coefficient of LR, but it is not guaranteed to be consistent with the sparse representation of HR. Afterwards, [3] proposed Couple Dictionary Training to solve the problem using the stochastic gradient descent. However, in the coupled spaces, sparse coefficients are same and coefficients don't fit the coupled space at the same time.

In order to solve the problem of inconsistency between two coupled spaces sparse representations, [5] relaxes the coupled representation to a semi-coupled one and attempts to find the mapping relation between pairs of coefficients, by using a linear

© Springer International Publishing Switzerland 2015
S. Arik et al. (Eds.): ICONIP 2015, Part I, LNCS 9489, pp. 540–548, 2015.
DOI: 10.1007/978-3-319-26532-2_59

transformation. Yang [6] formulated the couple space sparse coding problem as a bilevel optimization problem and moves one of the optimization problem to the regularization term of the other problem. This method has less error than [3] but still required the same sparse coding coefficients for both feature spaces.

Recently, using non-parametric Bayesian method [7] to learn an over-complete dictionary offers several advantages and shows significant improvement in sparse representation based application, such as image denoising, image inpainting and compressive sensing [8]. Reference [9, 10] extended the non-parametric Bayesian approach to coupled feature spaces by using beta process model and get competitive result in general image super-resolution.

There are other factors which can enhance sparse representation based single image super-resolution result. Research in [11] shows that it is ideal to pre-cluster training examples by analyzing image contents to facilitate specific priors for different stuff in a single image simultaneously. Reference [12] has used this idea for textures super-resolution. Similar to training data pre-clustering, [13, 14] proposed position-patch based face hallucination methods. The method needs large database (the patch in the database and the test image should be aligned) and the coefficients are as same as in [2, 3]. A more complex face hallucination is proposed in [15]. The pose and landmark points in low resolution image should be extracted to locate facial components and contour, so as to capture the structural information.

This paper proposes an effective single image face super-resolution algorithm to capture the complex and nonlinear relationship between the coupled spaces. First, the face image patches are categorized to different clusters to represent different areas in face, which offers specific priors. Second, the number of dictionary elements is inferred non-parametrically by using beta process construction. Third, in coupled spaces for super-resolution, the sparse representation coefficients of face image patches can be decomposed to values and dictionary atom indicators. The coupled patches in HR and LR face image can be sparse represented with the same sparsity but different values, thus bringing consistent and accurate mapping.

2 Problem Formulation

In single face image super-resolution, there are two coupled spaces $X \in R^{B_x}$, $Y \in R^{B_y}$. The relationship between the dictionary and two spaces follow as:

$$x_i = D^{(x)}\alpha_i^{(x)} + \epsilon_i^{(x)}, y_i = D^{(y)}\alpha_i^{(y)} + \epsilon_i^{(y)}, M\,\alpha_i^{(y)} = \alpha_i^{(x)} \tag{1}$$

x_i, y_i denote training samples with dimensions B_x and B_y. $D^{(x)} \in R^{B_x*K}$, $D^{(y)} \in R^{B_y*K}$ are dictionaries of two coupled spaces. $\alpha_i^{(x)}$, $\alpha_i^{(y)}$ are sparse representation coefficients. K is the number of dictionary atoms. $\epsilon_i^{(x)}$, $\epsilon_i^{(y)}$ are the recovery errors. M is a mapping matrix from sparse coding of y_i to x_i.

We divide HR and LR face image patches into different clusters pairs. For each cluster pair, two coupled dictionaries (for low resolution patches and high resolution patches) and the mapping matrix M should be learned.

3 Proposed Approach

The proposed algorithm can be divided into three steps: First, training data are clustered by K-Means; Second, for each image patch pair, dictionary and mapping function are learned by beta process coupled dictionary learning; Finally, HR face images are synthesized. The schematic of the proposed method is shown in Fig. 1.

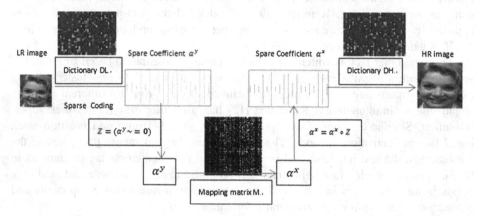

Fig. 1. Schematic of the proposed method. Dictionary atom indicators Z are the same in the coupled spaces to ensure coupled patches in HR and LR have the same sparse structure but different sparse representation value.

3.1 Training Set Clustering

HR and LR face image patches are extracted to make up training set. Due to the complex structures in face images, learning only one pair of dictionaries and an associated linear mapping function is often not enough to cover the relation of different resolution images. Therefore, the face image patches can be clustered into different categories couples and different categories image patches have different distribution. This paper adopts K-Means to divide the image patches of the training set into n classes. We can get n anchor points $v^c(c = 1,\ldots,n)$.

3.2 Dictionary Learning with Beta Process

For single face image super resolution, in coupled spaces of high resolution and low resolution, a pair of patches may have the same sparse representation structure but different values to capture the accuracy of the sparse representation. Beta process BP (a, b, H_0), one of the non-parametric Bayesian methods, can learn this kind of sparse representation with $a, b > 0$. The formula can be written as:

$$H = \sum_k^K \pi_k \delta_{d_k^{(y)}} \quad \pi_k \sim Beta(a/K, b(K-1)/K), d_k^{(x)} d_k^{(y)} \sim H_0 \tag{2}$$

where $\delta_{d_k^{(x)}}$ and $\delta_{d_k^{(y)}}$ are unit point mass at $d_k^{(x)}$ and $d_k^{(y)}$ which are dictionary atoms. π_k is the probability of the kth dictionary atom. H_0 is basic metric, and it is a fixed probability value in fact. In other words, H_0 is the initial value of π_k for all ($k = 1, \ldots, K$). In the dictionary learning stage, π_k is ever-changing and π_k is no less than H_0, otherwise the kth dictionary atom can't participate in the sparse representation, just as $d_k^{(x)}$, $d_k^{(y)} \sim H_0$. Equation (2) implies H is the sum of the probability of all dictionary atoms, namely $H = \sum_k^K \pi_k$. The beta process coupled dictionary learning model can be express as:

$$
\begin{aligned}
&x_i = D^{(x)}\alpha_i^{(x)} + \epsilon_i^{(x)}, y_i = D^{(y)}\alpha_i^{(y)} + y_i = D^{(y)}\alpha_i^{(y)}\epsilon_i^{(y)} \quad \alpha_i^{(x)} = z_i \circ s_i^{(x)}, \alpha_i^{(y)} = z_i \circ s_i^{(y)} \\
&d_k^{(x)} \sim N\left(0, B_x^{-1}I_{B_x}\right), d_k^{(y)} \sim N\left(0, B_y^{-1}I_{B_y}\right) \quad s_i^{(x)} \sim N(0, \gamma_{s^{(x)}}^{-1})I_K, s_i^{(y)} \sim N(0, \gamma_{s^{(y)}}^{-1})I_K \\
&\quad z_i \sim \prod_{k=1}^K Bernoulli(\pi_k), \pi_k \sim Beta(a/K, b(K-1)/K) \\
&\epsilon_i^{(x)} \sim N\left(0, \gamma_{\epsilon^{(x)}}^{-1}I_{B_x}\right), \epsilon_i^{(y)} \sim N\left(0, \gamma_{\epsilon^{(y)}}^{-1}I_{B_y}\right)\gamma_{s^{(x)}}, \gamma_{s^{(y)}} \sim \Gamma(c, d), \gamma_{\epsilon^{(x)}}, \gamma_{\epsilon^{(y)}} \sim \Gamma(e, f)
\end{aligned}
\tag{3}
$$

where $s_i^{(x)}$ and $s_i^{(y)}$ are the weight of sparse representation coefficients and z_i is dictionary atom indicator, namely a k-dimension binary vectors. The final coefficients are $\alpha_i^{(x)} = z_i \circ s_i^{(x)}$ and $\alpha_i^{(y)} = z_i \circ s_i^{(y)}$, where $\circ$ is an elementwise multiplication. $\gamma_{s^{(x)}}$ and $\gamma_{s^{(y)}}$ are the variance of $s_i^{(x)}$ and $s_i^{(y)}$. $\gamma_{\epsilon^{(x)}}$ and $\gamma_{\epsilon^{(y)}}$ are the variance of the recovery errors $\epsilon_i^{(x)}$ and $\epsilon_i^{(y)}$ respectively. I_{B_x}, I_{B_y} and I_K are identity matrix.

With the special characteristic of beta process, for $\alpha_i^{(x)}$ and $\alpha_i^{(y)}$, we use the same dictionary atom indicator z_i which have the same number of non-zero elements, following different distributions with $s_i^{x)}$ and $s_i^{(y)}$. It not only can ensure the consistency that a pair of patches have the same sparse representation that correspond to the same dictionary atoms but different values, but also ensure the sparsity employing $\alpha_i^{(x)} = z_i \circ s_i^{(x)}$. Thus, coupled patches have the same sparse structure but different sparse representation value. Besides, S^x and Z are made up of N weight vector s_i^x and N binary vector $z_i \epsilon \{0, 1\}^K$, $i = 1, \ldots, N$, respectively. Because *the Bernoulli distribution* is a discrete probability distribution which only has two values, 0 or 1. The value range is the same with z_i. And z_i is mainly decided by the probability π_k. When π_k reaches a certain value, $z_{ik} = 1$. So, we let $z_{ik} \sim Bernoulli(\pi_k)$. The *beta distribution* is often used to as the conjugate prior distribution density function of the *Bernoulli distribution*. But the *beta distribution* is a family of continuous probability distributions defined on the interval [0,1] parameterized by two positive shape parameters, denoted by a and b, that appear as exponents of the random variable and control the shape of the distribution. The interval is the same with the scope of the value of π_k. So, we use beta prior to set the value of π_k with $a = b = 1$ in Eq. (2).

Consecutive elements in the above hierarchical model are in the conjugate exponential family, and therefore inference may be implemented via a *variational Bayesian* or *Gibbs sampling analysis*. *Gibbs sampling* is adopted in this paper. In order to acquire the truth π_k, we do N independent repeated trials by the binomial distribution.

However, the binomial distribution consider as a normal distribution when $N \to \infty$. So, $d_k^{(x)}$, $s_i^{(x)}$ and $\epsilon_i^{(x)}$ follow the multivariate zero-mean Gaussian with variance $B_x^{-1} I_{B_x}$, $B_y^{-1} I_{B_y}$, $\gamma_{s^{(x)}}^{-1}) I_K$, $\gamma_{s^{(y)}}^{-1}) I_K$ and $\gamma_{\epsilon^{(x)}}^{-1} I_{B_x}$, $\gamma_{s^{(y)}}^{-1}) I_K$ respectively. In addition, because the inverse *Gamma distribution* is conjugate with the *Gaussian distribution*, $\gamma_{s^{(x)}}$, $\gamma_{s^{(y)}}$ and $\gamma_{\epsilon^{(x)}}$, $\gamma_{\epsilon^{(y)}}$ follow the *Gamma distribution*. In this paper, the parameter c and d initialize with 10^{-6}.

3.3   Synthesizing High-Resolution Face Image

In this stage, the input LR image Y is first magnified to the size of the desired HR image using Bi-cubic interpolation getting y. The synthesized high resolution image X_0 is initialized by y. X_0 and y are then partitioned into a set of overlapping 5 * 5 patches. We can select the anchor point v^c which is closest to x_{0i} by computing Euclidean distances for each input image patch x_{0i}. Then, we use the corresponding D and M to synthesize patches to reconstruct the high-resolution face image. In addition, because the recently introduced non-local redundancies in image have an advantage in image restoration [16, 17], we also incorporate the non-local self-similarities to our approach. The face image synthesis algorithm is summarized in Algorithm 1.

Algorithm 1. Face Image synthesis algorithm

Input: Test image Y, learned dictionary pair $D^{x,c}$, $D^{y,c}$ and the mapping M^c.

Initialize y and X_0 by bicubic interpolation. Initialize the number of iterations t. Conduct clustering according to K-means result.

For each iteration until convergence

For each cluster:

1. Update $\alpha^{y,c}$, get dictionary atom indicator Z^c

    If t=1 $D^{y,c} = D^{y,c}$, $y_c = y_c$

    else $D^{y,c} = [D^{y,c}; M^c]$, $y_c = [y_c; ah^{x,c}]$

    end

$$min_{D,S} \|y_c - D^{y,c} \alpha^{y,c}\|_F^2 + \lambda_y \|\alpha^{y,c}\|_1 \qquad Z^c = (\alpha^{y,c} \sim = 0)$$

2. Update $\alpha^{x,c}$

$$\alpha^{x,c} = M^c * \alpha^{y,c} \qquad ah^{x,c} = \alpha^{x,c} \qquad \alpha^{x,c} = \alpha^{x,c} \circ Z^c$$

3. Synthesize high-resolution patches x_c and get high-resolution image X_0

$$x_c = D^{x,c} * \alpha^{x,c}$$

Optimize

$$X^* = \arg min_X \|X - X_0\|^2$$

s.t. $\left\|x_i - \sum_{m=1}^M b^m x_{i(0)}^m\right\|_2^2 \le \epsilon$

where x_i and $x_{i(0)}$ are patches in X(the current high-resolution image) and X_0, respectively. $x_{i(0)}^m$ is the mth most similar patch to $x_{i(0)}$ and b^m is the non-local weight defined in [16].

Output: High-res image X^*

---

4   Experimental and Discussion

Three state-of-the-art dictionary learning based image super-resolution methods [4, 5, 8] were compared to our approach using the FERET and ORL datasets. SSIM, CSNR and recovery time are calculated for comparison. In the experiment, 24 images are used to train dictionary and 5 images to test. 40000 patches are randomly selected for each class at least. The LR image is directly down-sampled from the HR image. We use 5 * 5 patches, with overlap of 4 pixels between adjacent patches. And the training HR image size is 256 * 256. The size of test image is 86 * 86. Experiments were carried out on a Dell Precision workstation T7600 with 1.8G CPU*8 and 16 GB RAM.

4.1   Training Set Clustering

During the pre-cluster process, we have done a number of attempts, c are equal to 64, 40, 32, 30, 25, respectively. Only when c is equal to 32, the distribution of all clusters is the most homogeneous. And it produces the best super resolution results compared to other values.

4.2   Non-parametric Dictionary Learning

Because the proposed method has the non-parametric advantage, with different initial Ks, the dictionary size decreases rapidly and gradually converges to similar values, confirming that it can infer appropriate dictionary size no matter what the initial value is.

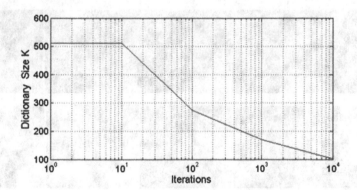

Fig. 2.   Our method infers dictionary size non-parametrically

Table 1.   Average results for the FERET dataset

Method	Time (s)	CNSR (dB)	SSIM
SCDL	41.478	43.288	0.988
ScSR	157.451	38.158	0.961
BPJDL	239.519	39.447	0.98
OURS	32.217	48.732	0.976

Table 2. Average results for the ORL dataset

Method	Time (s)	CNSR (dB)	SSIM
SCDL	42.5459	37.07	0.952
ScSR	162.147	36.003	0.906
BPJDL	301.124	37.806	0.959
OURS	32.725	38.178	0.954

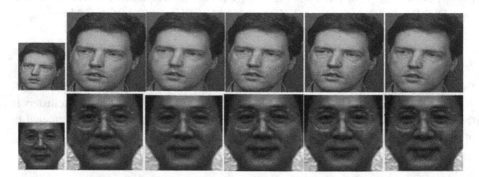

Fig. 3. The results on face image super-resolution both the FERET(bottom) and ORL (top) datasets (scaling factor: 3). From left to right: low resolution image, high resolution ground-truth reconstructed images by SCDL [5], ScSR [4], BPJDL [8] and our method.

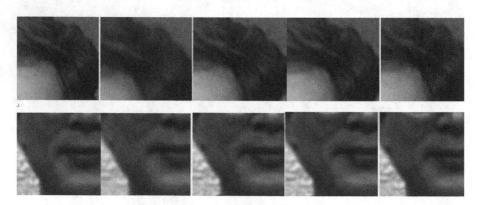

Fig. 4. The detail face images both the FERET(bottom) and ORL (top) datasets(scaling factor: 3). From left to right: high resolution, ground-truth, reconstructed images by SCDL [5], ScSR [4], BPJDL [8], and our method.

In terms of a single large dictionary, 771 is an appropriate dictionary size as in [8]. But our dictionaries are more accurate than others due to pre-cluster, dictionary size will shrink in some degree (See Fig. 2 for details). The initial value is 512.

4.3 Super Resolution Result

Tables 1 and 2 show the average results that our method is able to largely reduce the recovery time owe to pre-clustering and the smaller dictionary. And the CSNR of our method is the highest. Moreover, their structural similarities are similar.

Super resolution results on FERET and ORL are shown in Figs. 3 and 4. Our method is able to learn precious details competitive to previous methods.

5 Conclusions

In this paper, a novel method to generate HR face image by using domain specific priors and the non-parametric dictionary learning is proposed. In the algorithm, training data are clustered by K-Means to represent different areas in face due to different cluster has their specific distribution. Then, the dictionary atoms will more close to sample patches than the single big dictionary. At the same time, dictionary and mapping function are learned by beta process coupled dictionary learning to capture consistent and accurate mapping between the coupled spaces but they have different sparse representation. Of course, these will lead to better result. Experimental results show that the proposed method indeed generates high-quality images with favorable performance than state-of-the-arts methods with faster processing speed.

Acknowledgements. This research has been supported by funding from National Natural Science Foundation of China (61202168, and 61201234,61472278), Key project of Natural Science Foundation of Tianjin (14JCZDJC31700), and Tianjin Education Committee science and technology development Foundation (No. 20120802, 20130704).

References

1. Wang, N., Tao, D., Gao, X., Li, X., Li, J.: A comprehensive survey to face hallucination. Int. J. Comput. Vis. (IJCV) **106**(1), 9–30 (2014)
2. Yang, J., Wright, J., Huang, T., Ma, Y.: Image super-resolution as sparse representation of raw image patches. In: Proceedings of CVPR, pp. 1–8 (2008)
3. Yang, J., Wang, Z., Lin, Z., Huang, T.: Coupled dictionary training for image super-resolution. IEEE Trans. Image Process. (TIP) **21**(8), 3467–3478 (2012)
4. Yang, J., Wright, J., Huang, T., Ma, Y.: Image super-resolution via sparse representation. IEEE Trans. Image Process. (TIP) **19**(11), 2861–2873 (2010)
5. Wang, S., Zhang, L., Liang, Y., Pan, Q.: Semi-coupled dictionary learning with applications to image super-resolution and photo-sketch synthesis. In: Proceedings of CVPR, pp. 1–8 (2012)
6. Yang, J., Wang, Z., Lin, Z., Shu, X., Huang, T.: Bilevel sparse coding for coupled feature spaces. In: Proceedings of CVPR, pp. 2360–2367 (2012)
7. Paisley, J., Carin, L.: Nonparametric factor analysis with beta process priors. In: Proceedings of ICML, pp. 777–784, (2009)
8. Zhou, M., Chen, H., Paisley, J., Ren, L., Li, L., Xing, Z., Dunson, D., Sapiro, G., Carin, L.: Nonparametric bayesian dictionary learning for analysis of noisy and incomplete images. IEEE Trans. Image Process. (TIP) **21**(1), 130–144 (2012)

9. He, L., Qi, H., Zaretzki, R.: Beta process joint dictionary learning for coupled feature spaces with application to single image super-resolution. In Proceedings of CVPR, pp. 1–8 (2013)
10. Polatkan, G., Zhou, M., Carin, L., Blei, D., Daubechies, I.A.: Bayesian nonparametric approach to image super-resolution. IEEE Trans. Pattern Anal. Mach. Intell. (TPAMI) **37**, 346–358 (2015)
11. Yang, C.-Y., Ma, C., Yang, M.-H.: Single-image super-resolution: a benchmark. In: Fleet, D., Pajdla, T., Schiele, B., Tuytelaars, T. (eds.) ECCV 2014, Part IV. LNCS, vol. 8692, pp. 372–386. Springer, Heidelberg (2014)
12. Yang, C.-Y., Liu, S., Yang, M.-H.: Structured face hallucination. In: IEEE Conference on Computer Vision and Pattern Recognition (CVPR), pp. 1099–1106 (2013)
13. Ha Cohen, Y., Fattal, R., Lischinski, D.: Image up-sampling via texture hallucination. In: ICCP (2010)
14. Jiang, J., Hu, R., Han, Z., Lu, T., Huang, K.: Position-patch based face hallucination via locality-constrained representation. IEEE International Conference on Multimedia and Expo, pp. 212–217 (2012)
15. Jung, C., Jiao, L., Liu, B., Gong, M.: Position-patch based face hallucination using convex optimization. IEEE Sig. Process. Lett. **18**(6), 367–370 (2011)
16. Buades, A., Coll, B., Morel, J.: A non-local algorithm for Image denoising. In: CVPR, vol. 2, pp. 60–65. IEEE (2005)
17. Mairal, J., Bach, F., Ponce, J., Sapiro, G., Zisserman, A.: Non-local sparse models for image restoration. In: ICCV. IEEE (2009)

Sparse LS-SVM in the Sorted Empirical Feature Space for Pattern Classification

Takuya Kitamura[✉] and Kohei Asano

National Institute of Technology, Toyama College, 13 Hongo-cho, Toyama, Japan
kitamura@nc-toyama.ac.jp

Abstract. In this paper, we discuss an improved sparse least support vector training in the reduced empirical feature space which is generated by linearly independent training data. In this method, we select the linearly independent training data as the basis vectors of empirical feature space. Then, before we select these data, we sort training data in ascending order from the standpoint of classification with the values of objective function in training least squares support vector machines. Thus, good training data from the standpoint of classification can be selected in preference as the basis vectors of the empirical feature space. Next, we train least squares support vector machine in the empirical feature space. Then, the solution is sparse since the number of support vectors is equal to that of the basis vectors. Using two-class problems, we evaluate the effectiveness of the proposed method over the conventional methods.

Keywords: Classification · Empirical feature space · Least squares support vector machine · Sparse

1 Introduction

Since support vector machine (SVM) [1] is one of most powerful tool for pattern classification, SVM has been widely studied by many researchers. However, the computational of training the standard SVMs cost may be too large because it is necessary for standard SVM to solve a quadratic programming problem. To overcome this shortcoming, least squares SVM (LS-SVM) [2] was proposed. LS-SVM is trained by solving a set of linear equations. However, in LS-SVM, the solution is not sparse because all training data become support vectors (SVs) unlike standard SVM.

To overcome this problem, many types of sparse LS-SVMs (SLS-SVMs) [1,3–8] have been proposed by many researchers. As a type of SLS-SVMs, Abe proposed SLS-SVM in reduced empirical feature space [1,8]. Empirical feature space [9] is proposed by Xiong. The dimension of this space is the same as the number of training data at most. And the values of kernel function in this space are equivalent to those in feature space. Namely, the solution of training LS-SVM in empirical feature space is equivalent to that in feature space. Because

© Springer International Publishing Switzerland 2015
S. Arik et al. (Eds.): ICONIP 2015, Part I, LNCS 9489, pp. 549–556, 2015.
DOI: 10.1007/978-3-319-26532-2_60

empirical feature space is generated by eigenvalue problem with kernel matrix, the computational cost can be too large. To overcome this problem, Abe use the reduced empirical feature space, whose basis vectors are linearly independent training data instead of eigen vectors, for SLS-SVM. Namely, it is not necessary for generating this space to solve eigenvalue problem. However, the solution and the values of kernel functions in the reduced empirical feature space differ from those in feature space. And, in order of the given data, we determine whether the training data are linearly independent. Hence this selection of the linearly independent training data and the solution of Abe's SLS-SVM depend on this order. However, the order is not determined from standpoint of classification. Thus, the generalization capability of this method may be lower than that of LS-SVM.

In this paper, to overcome this problem, we propose SLS-SVM in the improved empirical feature space which is generated by the selected linearly independent training data from the standpoint of classification as the basis vectors. Namely, we sort the training data in ascending order from the standpoint of classification before we determine whether the training data are linearly independent. Then, we determine this order based on each value of the obtained objective function by training LS-SVM in one-dimensional space, whose basis vector is each training data. The smaller the value of the objective function is, the better the training data can be from the standpoint of classification as a basis vector of the empirical feature space. Thus, by the sorting, good training data from standpoint of classification can be selected in preference as the basis vectors of the empirical feature space. Finally, we train LS-SVM in the improved empirical feature space. Then, the solution is sparse since the number of support vectors is about equal to that of the selected linearly independent training data, like the conventional SLS-SVM. Here, we will call the proposed SLS-SVM "ISLS-SVM" in this paper.

This paper is organized as follows. In Sect. 2, we describe the conventional SLS-SVM in the reduced empirical feature space. In Sect. 3, we propose ISLS-SVM. In Sect. 4, we demonstrate the effectiveness of ISLS-SVM through computer experiments using bench mark data sets. And we conclude our work in Sect. 5.

2   Sparse Least Squares Support Vector Training in the Reduced Empirical Feature Space

2.1   Reduced Empirical Feature Space

Usually, the empirical feature space [9] is generated by using all the training data. Then, the kernel value in the empirical feature space is equivalent to that in the feature space. Hence, the obtained solution by training SVM or LS-SVM in the empirical feature space is equivalent to that in the feature space.

Let the number of training data, the m-dimensional training vectors, and the kernel be M, x_j $(j = 1, \ldots, M)$, and $H(x, x') = g^\top(x)g(x')$ where $g(x)$

is the mapping function into the l-dimensional feature space. It is necessary for generating the empirical feature space to solve the eigenvalue problem of kernel matrix K. However, it is time consuming to solve this problem and to transform input variables into variables in the empirical feature space. Hence, in Abe's SLS-SVM [1,8], the mapping function $h(x)$ is transformed into the empirical feature space as follows:

$$h(x) = (H(x, x_1^e), \ldots, H(x, x_{M'}^e))^\top, \tag{1}$$

where x_j^e ($i = j, \ldots, M'$) are the M' linearly independent data in the feature space. Namely, the basis vectors of the empirical feature space are these data. Here, the linearly independent training data are selected by Cholesky factorization. We assumed that the diagonal element in Cholesky factorization is zero if the argument of the square root in the diagonal element is less than or equal to a threshold value μ which is selected by cross-validation. We called the empirical feature space of Abe's SLS-SVM "reduced empirical feature space (REF)" in this paper.

2.2   Training SLS-SVM in REF

For two-class problem, let M training data pairs be $(x_1, y_1), \ldots, (x_M, y_M)$, where $y_j = 1$ or $y_j = -1$ if the j-th training data x_j belongs to class 1 or class 2. Then, we formulate SLS-SVM in REF as follows:

$$\min \qquad \frac{1}{2} w^\top w + \frac{C}{2} \sum_{j=1}^{M} \xi_j^2 \tag{2}$$

$$\text{s.t.} \qquad y_i(w^\top h(x_j) + b) = 1 - \xi_j \quad \text{for} \quad j = 1, \ldots, M, \tag{3}$$

where w, C, ξ_j, and b are M'-dimensional vector, the margin parameter which determines the trade-off between maximizing margins and minimizing misclassifications, the slack variable for x_j, and the bias term, respectively. Then, we solve the primal form of the optimization problem because the number of variables in the primal form is $M'(\leq M)$. The decision function is given by

$$D(x) = w^\top h(x) + b. \tag{4}$$

Here, (4) is obtained with the optimized w, b by solving (2), (3) and $h(x)$, which is given by the M' independent training data. Hence, the required training data for decision function are the M' linearly independent data and these data become SVs.

However, the solution in this method and the values of kernel functions in this space differ from those in feature space. REF is not generated from the standpoint of classification because the basis vectors of REF are selected by only determining whether each training data is linearly independent without the standpoint of classification ability. And, then, we determine in order of the given data. The selection of the linearly independent training data depend on this order, but this order is random. Hence, as the computer experiments in [8], this method often performs worse than the conventional LS-SVM.

3 Sparse Least Squares Support Vector Training in the Sorted Empirical Feature Space

In this section, we will describe SLS-SVM in the sorted empirical feature space, whose basis vectors are selected from the standpoint of linear independence and classification. We call this proposed method and the sorted empirical feature space "ISLS-SVM" (Improved SLS-SVM) and "SEF".

First, in this method, we obtain the mapping function $h_i(\boldsymbol{x})$ into one-dimensional feature space, whose basis vector is each training data, as follows:

$$h_i(\boldsymbol{x}) = \frac{H(\boldsymbol{x}_i, \boldsymbol{x})}{\|\boldsymbol{g}(\boldsymbol{x}_i)\|} \tag{5}$$

$$\text{for} \quad i = 1, \ldots, M.$$

With (5), we formulate optimization problem in primal form as follows:

$$\min \; Q_i = \frac{1}{2}w_i^2 + \frac{C}{2}\sum_{j=1}^{M}(y_j - w_i h_i(\boldsymbol{x}_j) - b_i)^2 \tag{6}$$

$$\text{for} \quad i = 1, \ldots, M,$$

where Q_i, $\boldsymbol{w}_i$ and b_i are the objective function, the weight value and the bias term. $\boldsymbol{w}_i$ and b_i are obtained by solving (6) as follows:

$$w_i = (\frac{1}{C} + \sum_{j=1}^{M} h_i^2(\boldsymbol{x}_j) - \frac{1}{M}\sum_{j,k=1}^{M} h_i(\boldsymbol{x}_j)h_i(\boldsymbol{x}_k))^{-1} \tag{7}$$

$$(\sum_{j=1}^{M} y_j h_i(\boldsymbol{x}_j) - \frac{1}{M}\sum_{j,k=1}^{M} y_j h_i(\boldsymbol{x}_k)),$$

$$b_i = \frac{1}{M}\sum_{j=1}^{M}(y_j - w_i h_i(\boldsymbol{x}_j) \tag{8}$$

$$\text{for} \quad i = 1, \ldots, M.$$

And, we obtain the values of objective function Q_i $(i = 1, \ldots, M)$ with (6)–(8). If the value of Q_i is small, $\boldsymbol{x}_i$ is good from the standpoint of classification as a basis vector of SEF because this optimization problem (6) is a minimization problem. Hence, we sort the training data in the order of the associated values from smallest to largest. Thus, in the order of classification ability from best to worst as the basis vectors of SEF, we can determine whether the training data are linearly independent. Namely, in preference, we can select the good linearly independent training data as the basis vectors. After sorting, we select the linearly independent training data by Cholesky factorization and generate SEF and train LS-SVM in primal form like the conventional SLS-SVM.

In our following study, we use linear kernels: $\boldsymbol{x}^T\boldsymbol{x}'$, polynomial kernels: $(\boldsymbol{x}^T\boldsymbol{x}' + 1)^d$, where d is a positive integer, and radial basis function (RBF) kernels: $\exp(-\gamma\|\boldsymbol{x} - \boldsymbol{x}'\|^2)$, where γ is the width of the radius. Then, d and γ are kernel parameters while linear kernels do not have those. In training ISLS-SVM, we need to determine a kernel type, an associated kernel parameter, a margin parameter C, and a threshold value μ in Cholesky factorization. In the following, we show the training algorithm using these selected kernel and parameters by cross-validation.

Algorithm of ISLS-SVM

Step 1: Set $i = 1$.

Step 2: Calculate $h_i(x_j)$ $(j = 1, \ldots, M)$ with (5).

Step 3: With $h_i(x_j)$ $(j = 1, \ldots, M)$, calculate Q_i by (6)–(8).

Step 4: If $i \neq M$, we set $i = i + 1$ and go to Step 2, otherwise go to Step 5.

Step 5: Sort the training data in the order of the associated Q_i from smallest to largest.

Step 6: Select the linearly independent training data by Cholesky factorization in the sorted order and those selected training data become the basis vector of SEF.

Step 7: Using the selected linearly independent training data x_k^e ($k = 1, \ldots, M'$) in Step 6, calculate the mapping function $h(x)$ into SEF with (1).

Step 8: Using $h(x)$ determined in Step 7, calculate w and b by solving the optimization problem (2), (3) in primal form.

Step 9: Using w and b determined in Step 8, calculate $D(x)$ by (4).

4 Experimental Results

We compared the generalization ability and the number of SVs of ISLS-SVM, SLS-SVM, and LS-SVM, using two-class benchmark data sets [1,8,10–13] listed in the Table 1 that shows the number of inputs, training data, test data, and data sets. We measured training time using a personal computer (3.10 GHz, 2.0 GB memory, Windows 7 operating system).

4.1 Parameter Setting

In the following study, we normalized the input ranges into $[0, 1]$. For ISLS-SVM, SLS-SVM, LS-SVM, we determined a kernel type, a kernel parameter of the selected kernels, and a margin parameter C by five-fold cross-validation for each problem, where training SLS-SVM in REF is called "SLS-SVM" simply. For ISLS-SVM and SLS-SVM, we determined the threshold value μ of Cholesky factorization by five-fold cross-validation. We selected a kernel type from linear, polynomial, and RBF kernels. If we selected polynomial or RBF kernels, we selected d or γ from $\{2, 3, 4, 5\}$ or $\{0.1, 0.5, 1, 1.5, 3, 5, 10, 15, 20\}$. For multi-class problem, if we selected RBF kernels, we selected γ from $\{0.1, 0.5, 1, 1.5, 3, 5, 10, 15, 20, 30, 50, 100, 200\}$. And we selected C from $\{0.1, 1, 5, 10, 50, 100, 500, 10^3, 5 \times 10^3, 10^4\}$ and μ from $\{10^{-2}, 10^{-3}, 10^{-4}, 10^{-5}\}$. Table 2 shows the selected type of kernels and parameters by the above procedure. In this table, "Pol." denote polynomial kernels.

Table 1. Two class Benchmark data sets

Data	Inputs	Training	Test	Sets
Banana	2	400	4900	100
B. cancer	9	200	77	100
Diabetes	8	468	300	100
German	20	700	300	100
Heart	13	170	100	100
Image	18	1300	1010	20
Ringnorm	20	400	7000	100
F. solar	9	666	400	100
Splice	60	1000	2175	20
Thyroid	5	140	75	100
Titanic	3	150	2051	100
Twonorm	20	400	7000	100
Waveform	21	400	4600	100

Table 2. Determined kernels and parameters values by five-fold cross-validation

Data	LS-SVM			SLS-SVM				ISLS-SVM			
	Kernels	d or γ	C	Kernels	d or γ	C	μ	Kernels	d or γ	C	μ
Banana	RBF	$\gamma = 20$	10^3	RBF	$\gamma = 20$	5×10^3	10^{-3}	RBF	$\gamma = 20$	5×10^3	10^{-4}
B. cancer	RBF	$\gamma = 10$	1	Pol.	$d = 2$	100	10^{-3}	Pol.	$d = 2$	500	10^{-5}
Diabetes	RBF	$\gamma = 10$	5	Pol.	$d = 3$	5	10^{-5}	Pol.	$d = 2$	10^3	10^{-2}
German	RBF	$\gamma = 3$	50	RBF	$\gamma = 5$	50	10^{-3}	RBF	$\gamma = 10$	10	10^{-3}
Heart	RBF	$\gamma = 1.5$	10	RBF	$\gamma = 15$	0.1	10^{-2}	RBF	$\gamma = 0.5$	50	10^{-5}
Image	Pol.	$d=5$	10^4	Pol.	$d = 5$	10^4	10^{-5}	Pol.	$d = 5$	10^4	10^{-6}
Ringnorm	RBF	$\gamma = 20$	5	RBF	$\gamma = 0.1$	50	10^{-4}	RBF	$\gamma = 0.5$	1	10^{-3}
F. solar	Pol.	$d = 3$	1	Pol.	$d = 3$	10	10^{-2}	Pol.	$d = 3$	5	10^{-2}
Splice	RBF	$\gamma = 10$	50	RBF	$\gamma = 10$	100	10^{-2}	RBF	$\gamma = 15$	5×10^3	10^{-2}
Thyroid	RBF	$\gamma = 20$	10^3	RBF	$\gamma = 20$	5×10^3	10^{-2}	RBF	$\gamma = 20$	10^4	10^{-2}
Titanic	Linear	–	5	RBF	$\gamma = 10$	10	10^{-2}	Linear	–	50	10^{-2}
Twonorm	Pol.	$d = 2$	0.1	RBF	$\gamma = 1.5$	5	10^{-4}	RBF	$\gamma = 1$	10	10^{-6}
Waveform	RBF	$\gamma = 15$	1	RBF	$\gamma = 3$	100	10^{-4}	RBF	$\gamma = 3$	10^3	10^{-4}

4.2 Performance Comparison

Table 3 shows the average recognition rates of the test data sets, their standard deviations, which are denoted in columns of "Rec.", and the average number of support vectors which are denoted in columns of "SVs". In this table, the best results of the average recognition rates and the fewest number of SVs in each row of the data sets are shown in boldface. From the standpoint of classification ability, ISLS-SVM performs the best among all the methods for five problems and better than SLS-SVM for four problems. But, for other problems, CSLS-SVM performed about the same as SLS-SVM. Additionally, for the most problems, the average number of SVs for ISLS-SVM is about the same as that for

Table 3. Comparison of the average recognition rates in percent, standard deviations of the rates, and the average number of support vectors

Data	LS-SVM		SLS-SVM		ISLS-SVM	
	Rec.	SVs	Rec.	SVs	Rec.	SVs
Banana	**89.5 ± 0.5**	400	89.2 ± 0.5	44	89.2 ± 0.5	63
B. cancer	73.6 ± 4.5	200	**74.1 ± 4.5**	52	**74.1 ± 4.6**	52
Diabetes	**77.0 ± 1.6**	468	**77.0 ± 1.7**	165	**77.0 ± 1.6**	39
German	**76.2 ± 2.1**	700	75.9 ± 2.1	189	76.0 ± 2.2	321
Heart	**84.2 ± 3.1**	170	**84.2 ± 3.3**	126	84.0 ± 3.6	53
Image	**95.5 ± 0.7**	1300	91.7 ± 1.2	279	92.1 ± 1.1	451
Ringnorm	**96.3 ± 0.5**	400	94.2 ± 3.0	22	93.9 ± 3.7	33
F. solar	66.6 ± 1.6	666	66.6 ± 1.6	32	**66.8 ± 1.9**	36
Splice	**89.4 ± 0.7**	1000	89.3 ± 0.7	977	89.1 ± 0.8	977
Thyroid	**93.8 ± 2.8**	140	92.7 ± 2.8	29	92.7 ± 2.7	32
Titanic	**77.3 ± 1.2**	150	77.2 ± 0.8	10	77.1 ± 1.3	3
Twonorm	97.4 ± 0.2	400	**97.5 ± 0.2**	306	**97.5 ± 0.2**	239
Waveform	**90.3 ± 0.4**	400	89.6 ± 0.6	393	90.2 ± 0.5	392

SLS-SVM and much fewer than that for LS-SVM. Therefore, from Table 3, we can conclude that ICSLS-SVM performed better than SLS-SVM from the standpoint of classification ability, and the solution of ISLS-SVM is sparse as compared with that of LS-SVM like that of SLS-SVM.

5   Conclusions

In this paper, we proposed ISLS-SVM that is trained in the improved empirical feature space. Because, unlike SLS-SVM, the improved empirical feature space is generated by the selected linearly independent training data from the standpoint of classification with the values of the objective functions as basis vectors, the generalization ability of ISLS-SVM can be higher than that of SLS-SVM. And, like SLS-SVM, the solution of ISLS-SVM is sparse.

According to the computer experiments using two-class benchmark data sets, ISLS-SVM performed better than SLS-SVM from the standpoint of classification. And ISLS-SVM performed about the same as SLS-SVM from the standpoint of sparsity.

References

1. Abe, S.: Support Vector Machines for Pattern Classification. Advances in Pattern Recognition. Springer, London (2010)
2. Suykens, J.A.K., Vandewalle, J.: Least squares support vector machine classifiers. Neural Process. Lett. **9**(3), 293–300 (1999)

3. Cawley, G.C., Talbot, N.L.C.: Improved sparse least squares support vector machines. Neurocomputing **48**, 1025–1031 (2002)
4. Jiao, L., Bo, L., Wang, L.: Fast sparse approximation for least squares support vector machine. Neural Netw. **18**(3), 1025–1031 (2007)
5. Suykens, J.A.K., Lukas, L., Vandewalle, J.: Sparse least squares support vector machine classifiers. In: European Symposium on Artificial Neural Networks (ESANN 2000), pp. 37–42 (2000)
6. Liu, J., Li, J., Xu, W., Shi, Y.: A weighted Lq adaptive least squares support vector machine classifiers - robust and sparse approximation. Expert Syst. Appl. **38**(3), 2253–2259 (2011)
7. Kitamura, T., Sekine, T.: A novel method of sparse least squares support vector machines in class empirical feature space. In: Huang, T., Zeng, Z., Li, C., Leung, C.S. (eds.) ICONIP 2012, Part II. LNCS, vol. 7664, pp. 475–482. Springer, Heidelberg (2012)
8. Abe, S.: Sparse least squares support vector training in the reduced empirical feature space. Pattern Anal. Appl. **10**(3), 203–214 (2007)
9. Xiong, H., Swamy, M.N.S., Ahmad, M.O.: Optimizing the kernel in the empirical feature space. IEEE Trans. Neural Netw. **16**(2), 460–474 (2005)
10. Kitamura, T., Takeuchi, S., Abe, S., Fukui, K.: Subspace-based support vector machines for pattern classification. Neural Netw. **22**, 558–567 (2009)
11. Kitamura, T., Takeuchi, S., Abe, S.: Feature selection and fast training of subspace based support vector machines. In: International Joint Conference on Neural Networks (IJCNN 2010), pp. 1967–1972 (2010)
12. Rätsch, G., Onda, T., Müller, K.R.: Soft margins for AdaBoost. Mach. Learn. **42**(3), 287–320 (2001)
13. http://archive.ics.uci.edu/ml

A Cost Sensitive Minimal Learning Machine for Pattern Classification

João Paulo P. Gomes[1] (✉), Amauri H. Souza Jr.[2], Francesco Corona[3],
and Ajalmar R. Rocha Neto[2]

[1] Department of Computer Science, Federal University of Ceará,
Fortaleza, Ceará, Brazil
jpaulo@lia.ufc.br
[2] Department of Computer Science, Federal Institute of Ceará,
Maracanaú, Ceará, Brazil
{amauriholanda,ajalmar}@ifce.edu.br
[3] Department of Teleinformatics Engineering, Federal University of Ceará,
Fortaleza, Ceará, Brazil
francesco.corona@deti.ufc.br

Abstract. The present work proposes a variant of the Minimal Learning Machine (MLM) in a cost sensitiVe framework for classification. MLM is a recently proposed supervised learning algorithm with a simple formulation and few hyperparameters. The proposed method is tested under two classification problems: imbalanced classification and classification with reject option. The results are comparable to other to state of the art classifiers.

Keywords: Supervised learning · Minimal Learning Machine · Cost sensitive classification

1 Introduction

Standard supervised classification techniques have been successfully applied in many real world. However, in some specific domains, the classifier needs to incorporate in its formulation different misclassification costs. A typical example can be found in medical applications. Misclassifying a patient with cancer may lead to death and on the other hand, misclassifying a patient without cancer may only lead to extra spendings in further examinations.

To overcome this problem, many researchers have proposed cost sensitive variants of standard classifiers, such as ELM [1] and SVM [2]. Loosely speaking, the proposed methods add extra terms to its cost function. The terms refer to the cost of misclassifying each of the examples.

Recently, a new supervised learning algorithm called Minimal Learning Machine (MLM) was proposed in [3]. MLM is based on the idea of the existence of a mapping between the geometric configurations of points in the input and output space. The main advantages of MLM are its easy understanding, simple implementation and the use of only one hyperparameter.

© Springer International Publishing Switzerland 2015
S. Arik et al. (Eds.): ICONIP 2015, Part I, LNCS 9489, pp. 557–564, 2015.
DOI: 10.1007/978-3-319-26532-2_61

In this paper, a variant of the original Minimal Learning Machine called Weighted Minimal Learning Machine (wMLM) is proposed. The main idea of wMLM is to weight the instances of the training set and modify the contribution of each sample on the definition of the final MLM model. This novel method adds flexibility to the original MLM, since it is able to deal with a variety of classification problems while maintaining MLM advantages. In order to show the wMLM effectiveness, we propose two ways of configuring its weights to deal with the problems of imbalanced classification and classification with reject option.

The remainder of the paper is organized as follows. Section 2 introduces the Minimal Learning Machine. Section 3 presents the proposed method and its applications. The experiments are reported in Sect. 4. Conclusions are given in Sect. 5.

2   Minimal Learning Machine

We are given a set of N input points $X = \{\mathbf{x}_i\}_{i=1}^{N}$, with $\mathbf{x}_i \in \mathbb{R}^D$, and the set of corresponding outputs $Y = \{\mathbf{y}_i\}_{i=1}^{N}$, with $\mathbf{y}_i \in \mathbb{R}^S$. Assuming the existence of a continuous mapping $f : \mathcal{X} \to \mathcal{Y}$ between the input and the output space, we want to estimate f from data with the multiresponse model $\mathbf{Y} = f(\mathbf{X}) + \mathbf{R}$. The columns of the matrices $\mathbf{X}$ and $\mathbf{Y}$ correspond to the D inputs and S outputs respectively, and the rows to the N observations. The columns of the $N \times S$ matrix $\mathbf{R}$ correspond to the residuals. The MLM design can be divided in two steps: distance regression and output estimation.

2.1   Distance Regression

For a selection of reference input points $R = \{\mathbf{m}_k\}_{k=1}^{K}$ with $R \subseteq X$ and corresponding outputs $T = \{\mathbf{t}_k\}_{k=1}^{K}$ with $T \subseteq Y$, define $\mathbf{D}_x \in \mathbb{R}^{N \times K}$ in such a way that its kth column $\mathbf{d}(X, \mathbf{m}_k)$ contains the distances $d(\mathbf{x}_i, \mathbf{m}_k)$ between the N input points $\mathbf{x}_i$ and the kth reference point $\mathbf{m}_k$. Analogously, define $\boldsymbol{\Delta}_y \in \mathbb{R}^{N \times K}$ in such a way that its kth column $\boldsymbol{\delta}(Y, \mathbf{t}_k)$ contains the distances $\delta(\mathbf{y}_i, \mathbf{t}_k)$ between the N output points $\mathbf{y}_i$ and the output $\mathbf{t}_k$ of the kth reference point.

The mapping g between the input distance matrix $\mathbf{D}_x$ and the corresponding output distance matrix $\boldsymbol{\Delta}_y$ can be reconstructed using the multiresponse regression model

$$\boldsymbol{\Delta}_y = g(\mathbf{D}_x) + \mathbf{E}.$$

The columns of the matrix $\mathbf{D}_x$ correspond to the K input vectors and columns of the matrix $\boldsymbol{\Delta}_y$ correspond to the K response vectors, the N rows correspond to the observations. The columns of matrix $\mathbf{E} \in \mathbb{R}^{N \times K}$ correspond to the K residuals.

Assuming that mapping g between input and output distance matrices has a linear structure for each response, the regression model has the form

$$\boldsymbol{\Delta}_y = \mathbf{D}_x \mathbf{B} + \mathbf{E}. \tag{1}$$

The column $\mathbf{b}_k$ of the $K \times K$ regression matrix $\mathbf{B}$ corresponds to the coefficients for the kth response and it can be solved from data through a minimization of the corresponding residual sum of squares as loss function:

$$\text{RSS}(\mathbf{b}_k) = (\boldsymbol{\delta}(Y, \mathbf{t}_k) - \mathbf{D}_x \mathbf{b}_k)'(\boldsymbol{\delta}(Y, \mathbf{t}_k) - \mathbf{D}_x \mathbf{b}_k). \tag{2}$$

When the number of selected reference points is smaller than the number of available points available (i.e., $K < N$), the regression vectors $\mathbf{b}_k$ can be approximated by the usual least squares estimate. Due to the independence of the K least-squares estimates, the estimated regression matrix $\hat{\mathbf{B}}$ can be written in compact matrix notation

$$\hat{\mathbf{B}} = (\mathbf{D}'_x \mathbf{D}_x)^{-1} \mathbf{D}'_x \boldsymbol{\Delta}_y. \tag{3}$$

For an input test point $\mathbf{x} \in \mathbb{R}^D$ whose distances from the K reference input points $\{\mathbf{m}_k\}_{k=1}^K$ are collected in the vector $\mathbf{d}(\mathbf{x}, R) = [d(\mathbf{x}, \mathbf{m}_1) \ldots d(\mathbf{x}, \mathbf{m}_K)]$, the corresponding distances between its unknown output $\mathbf{y}$ and the known outputs $\{\mathbf{t}_k\}_{k=1}^K$ of the reference points is

$$\hat{\boldsymbol{\delta}}(\mathbf{y}, T) = \mathbf{d}(\mathbf{x}, R)\hat{\mathbf{B}}. \tag{4}$$

The vector $\hat{\boldsymbol{\delta}}(\mathbf{y}, T) = [\hat{\delta}(\mathbf{y}, \mathbf{t}_1) \ldots \hat{\delta}(\mathbf{y}, \mathbf{t}_K)]$ provides an estimate of the geometrical configuration of $\mathbf{y}$ and the reference set T, in the $\mathcal{Y}$-space.

2.2 Output Estimation

The problem of estimating the output $\mathbf{y}$, given the outputs $\{\mathbf{t}_k\}_{k=1}^K$ of all the reference points and estimates $\hat{\boldsymbol{\delta}}(\mathbf{y}, T)$ of their mutual distances, can be understood as a multilateration problem to estimate its location in $\mathcal{Y}$.

The location of $\mathbf{y}$ is estimated from the minimization of the objective function

$$J(\mathbf{y}) = \sum_{k=1}^K \left((\mathbf{y} - \mathbf{t}_k)'(\mathbf{y} - \mathbf{t}_k) - \hat{\delta}^2(\mathbf{y}, \mathbf{t}_k) \right)^2. \tag{5}$$

An optimal solution to (5) can be achieved by any minimizer $\hat{\mathbf{y}} = \arg\min_{\mathbf{y}} J(\mathbf{y})$ like the nonlinear least square estimates from standard gradient descent methods.

3 Weighted Minimal Learning Machine

The Weighted MLM (wMLM) consists of a generalized least-squares fit between distances in the input and output spaces. The idea is to bring in a differential weighting of residuals in the regression step of the MLM to account for errors that are not independently and identically distributed with zero mean and constant variance.

This is practically achieved by extending the least-squares criterion to the generalized form

$$\text{RSS}(\mathbf{b}_k) = (\boldsymbol{\delta}(Y, \mathbf{t}_k) - \mathbf{D}_x \mathbf{b}_k)' \mathbf{W}_k (\boldsymbol{\delta}(Y, \mathbf{t}_k) - \mathbf{D}_x \mathbf{b}_k), \tag{6}$$

where $\mathbf{W}_k$ is a symmetric positive definite matrix that allow for an unequal weighting of squares and products of residuals. Considering that the matrix $\mathbf{D_x}$ comprise the training examples non-linearly projected on a space defined by the reference points, one can weight the columns of $\mathbf{D_x}$ instead of the vectors $\mathbf{x}$ directly. Under this assumption, the distance regression problem can be seen as a weighted linear regression problem. Defining a weight matrix $\mathbf{W}$, the MLM loss function can be rewritten as:

$$RSS(\mathbf{B}) = \text{tr}\left(\mathbf{W}(\boldsymbol{\Delta}_y - \mathbf{D}_x \mathbf{B})' \mathbf{W}(\boldsymbol{\Delta}_y - \mathbf{D}_x \mathbf{B})\right). \tag{7}$$

where $\mathbf{W}$ is a diagonal matrix and each element w_{ii} of its diagonal represents the weight of each training sample $\mathbf{x}_i$.

The optimization can now be defined as a weighted least squares problem and the matrix $\mathbf{B}$ can be estimated by:

$$\hat{\mathbf{B}} = (\mathbf{D}_x' \mathbf{W} \mathbf{D}_x)^{-1} \mathbf{D}_x' \mathbf{W} \boldsymbol{\Delta}_y. \tag{8}$$

The modified calculation of $\hat{\mathbf{B}}$ defines the wMLM since all other step can be performed according to the original formulation presented in section 2.The possibility to choose different values for $\mathbf{W}$ allows the application of wMLM in different classification problems.

3.1   Imbalanced Data Classification

Among the strategies for imbalanced learning, one can cite the use of a cost sensitive learning framework. In this framework, the error cost of each sample is weighted according to the sample class. The main idea is to provide a higher weight to those examples that belong to the class that is less represented on the training set. Choosing the weights for each sample determines the degree of re-balance towards the minority class. A possible choice for the weights is given by:

$$w_{ii} = \frac{1}{C_{y_i}} \tag{9}$$

where C_{y_i} is the number of elements in the training set that belong to class y_i.

3.2   Classification with Reject Option

Classification with reject option consists in withholding the automatic classification of an item, if the decision is considered not sufficiently reliable. Rejected patterns can then be handled by a different classifier, or manually by a human. Implementation of reject option strategies requires finding a trade-off between the achievable reduction of the cost due to classification errors, and the cost of handling rejections (which are application-dependent).

The wMLM-based classifier with reject option is built upon a two classifiers framework [5]. The first classifier (classifier 1) is designed with a bias for the class C_1, as well as the second classifier (classifier 2) is designed with a bias for the class C_2. This can be performed by tuning the weight matrix so that the examples of class C_1 are more weighted for the classifier 1 and examples of class C_2 are more weighted for the classifier 2. An example is rejected whenever the classifiers disagree.

In this work, the weighting schemes for the classifier 1 and 2 are given respectively by:

$$w_{ii} = \begin{cases} 1 - R_s & \text{if } \mathbf{x}_i \text{ is from class } C_1; \\ R_s & \text{otherwise.} \end{cases} \quad \text{and} \quad w_{ii} = \begin{cases} R_s & \text{if } \mathbf{x}_i \text{ is from class } C_1; \\ 1 - R_s & \text{otherwise.} \end{cases}$$

where R_s is an input parameter in $[0, 0.5]$ that defines the degree of bias towards class s. The choice of R_s impacts on the number of rejected samples.

4 Results

4.1 Imbalanced Classification

The performance of wMLM on imbalanced classification was assessed using 4 real world binary classification datasets. Table 1 present the characteristics of the datasets used in this work.

Table 1. Datasets characteristics

Dataset	Attributes	Instances	IR
Pima	8	768	0.536
Haberman	3	306	0.360
Vertebral	6	310	0.476
Yale	105	164	0.071

The first three datasets are available on UCI [6]. The last one is based on the Yale face recognition database. This database is composed of 164 images of 15 subjects. For this classification task, the feature vectors were generated using PCA and preserving 99 % of the variance. This resulted on feature vectors of dimension 105.

The last column of Table 1 presents the imbalance ratio (IR). IR measures the imbalance degree of each dataset and is defined by the number of examples in the minority class divided by the number of examples of the majority class.

The performance of the methods were analyzed using accuracy and F-measure. All datasets were normalized within the range of $[0, 1]$. 50 % of the data were used for training and 50 % for testing.

Table 2. F-measure

Dataset	MLM	wMLM	WELM
Pima	0.624	0.670	0.675
Haberman	0.316	0.468	0.461
Vertebral	0.700	0.731	0.742
Yale	0.407	0.444	0.448

Table 3. Accuracy

Dataset	MLM	wMLM	WELM
Pima	0.765	0.741	0.735
Haberman	0.737	0.699	0.687
Vertebral	0.816	0.807	0.812
Yale	0.939	0.928	0.923

wMLM was compared to MLM and the recently proposed Weighted Extreme Learning Machine (WELM, [1]). For MLM and ELM, all hyperparameters (number of reference point for MLM and number of hidden neurons for ELM) were chosen using grid search and 10 fold cross validation. ELM's hidden layer weights and MLM's reference points were randomly assigned.

Tables 2 and 3 present the performance comparison for all datasets.

As expected, the wMLM achieved a better F-measure performance for all datasets when compared to MLM. wMLM and WELM achieved similar results. As a consequence of the F-measure increase, the total accuracy is reduced for all classifiers.

4.2   Classification with Reject Option

The performance of the wMLM was assessed for the same UCI datasets used for imbalanced learning. wMLM was compared to a Gaussian Quadratic Discriminant classifier (QDC) and an ELM. The ELM was designed as a regression algorithm and then a rejection threshold could be used. Similarly, for the QDC, a rejection threshold could be used since the QDC provides a probabilistic output.

Comparisons of classifiers with reject option are usually performed using Accuracy Rejection (A-R) curves. This curve shows the relation between classification accuracy and rejection rate.

The curves were built under the framework of reject option, proposed by Chow [4]. In this work, the empirical risk is given by:

$$E_R = w_R R + E \tag{10}$$

where w_R is the rejection cost, R is the rejection rate and E is the misclassification rate (error rate). A-R curves are generated by finding the rejection threshold

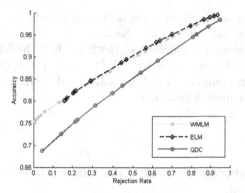

Fig. 1. A-R curve for Pima dataset

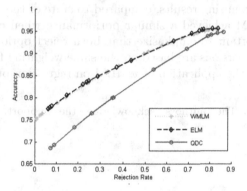

Fig. 2. A-R curve for Haberman dataset

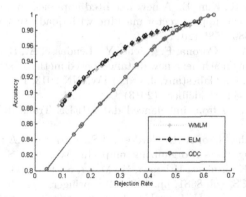

Fig. 3. A-R curve for Vertebral column dataset

that minimize 10 for a given rejection cost. Figures 1, 2 and 3 present the A-R curves for all datasets.

In A-R curves, a method performs better as it achieves a higher accuracy for a given rejection rate. Based on that, it can be seen that ELM and the wMLM achieved a similar performance. Both methods had better results than the QDC. This may be explained by the fact that the QDC can only design a quadratic decision boundary and both ELM and wMLM can generate other nonlinear decision boundary functions.

5 Conclusion

The current work proposed a variant of the Minimal Learning Machine for cost sensitive classification.

Two possible applications of the Weighted MLM were shown: imbalanced data classification and classification with reject option. For both applications wMLM achieved promising results, compared to state of the art classifiers.

Although wMLM achieved a similar performance when compared to ELM versions, it is important to emphasize that both reject option and imbalanced classification MLM versions are based on the same weighting framework. Future works may address the application of wMLM on regression problems.

Acknowledgments. The authors acknowledge the support of CNPq (Grant 456837/2014-0).

References

1. Zonga, W., Huang, G.-B., Chenb, Y.: Weighted extreme learning machine for imbalance learning. Neurocomputing **101**, 229–242 (2013)
2. Hwang, J.P., Park, S., Kim, E.: A new weighted approach to imbalanced data classification problem via support vector machine with quadratic cost function. Expert Syst. Appl. **38**, 8580–8585 (2011)
3. de Souza Junior, A.H., Corona, F., Miche, Y., Lendasse, A., Barreto, G.A., Simula, O.: Minimal learning machine: a new distance-based method for supervised learning. In: Rojas, I., Joya, G., Gabestany, J. (eds.) IWANN 2013, Part I. LNCS, vol. 7902, pp. 408–416. Springer, Heidelberg (2013)
4. Chow, C.K.: Learning from imbalanced data. IEEE Trans. Inf. Theory **1**, 41–46 (1970)
5. Sousa, R., da Rocha Neto, A.R., Cardoso, J.S., Barreto, G.A.: Classification with reject option using the self-organizing map. In: Wermter, S., Weber, C., Duch, W., Honkela, T., Koprinkova-Hristova, P., Magg, S., Palm, G., Villa, A.E.P. (eds.) ICANN 2014. LNCS, vol. 8681, pp. 105–112. Springer, Heidelberg (2014)
6. Frank, A., Asuncion, A.: UCI machine learning repository. School of Information and Computer Sciences, University of California, Irvine (2010)

A Minimal Learning Machine for Datasets with Missing Values

Diego P. Paiva Mesquita[1], João Paulo P. Gomes[1]([✉]),
and Amauri H. Souza Jr.[2]

[1] Department of Computer Science, Federal University of Ceará,
Fortaleza, Ceará, Brazil
{diegoparente,jpaulo}@lia.ufc.br
[2] Department of Computer Science, Federal Institute of Ceará,
Maracanaú, Ceará, Brazil
amauriholanda@ifce.edu.br

Abstract. Minimal Learning Machine (MLM) is a recently proposed supervised learning algorithm with simple implementation and few hyperparameters. Learning MLM model consists on building a linear mapping between input and output distance matrices. In this work, the standard MLM is modified to deal with missing data. For that, the expected squared distance approach is used to compute the input space distance matrix. The proposed approach showed promising results when compared to standard strategies that deal with missing data.

1 Introduction

Missing attributes in feature vectors are present in many real world machine learning applications. Despite that, traditional machine learning models rely on the assumption that feature vectors have a fixed dimension and none of its features are missing. To overcome this issue, many strategies have been proposed. Some naive approaches include eliminating examples that have missing attributes and filling the missing feature with the mean value for this feature, estimated using all other examples.

Eliminating feature vectors with missing values can degrade the model performance, specially when the dataset comprises a small number of samples. Filling the missing entries with the mean value of the correspondent features may also result in a poor performance when features exhibit high variance.

Another simple strategy of imputation consist on finding the nearest neighbor among the fully known feature vectors and filling the missing feature with the feature value of this nearest neighbor. This approach becomes ineffective as the number of examples with missing feature increase [1].

More sophisticated approaches include the imputation of missing values based on the knowledge of the other features that belong to the same vector. Using a probabilistic model, the missing value can be estimated by using, for instance, the Expectation-Maximization (EM) algorithm [2].

© Springer International Publishing Switzerland 2015
S. Arik et al. (Eds.): ICONIP 2015, Part I, LNCS 9489, pp. 565–572, 2015.
DOI: 10.1007/978-3-319-26532-2_62

Recently, the authors in [3] proposed an approach for estimating the pairwise distance between vectors with missing values. Assuming that data is Missing-at-Random (MAR) [4] and that the samples originate from some probability distribution, statistical techniques were applied to find an expression for the expectation of the squared Euclidean distance (ESD) between samples.

In the present work, we propose an adaptation of the Minimal Learning Machine (MLM, [5]) capable of handling missing vales, based in the ESD calculation. MLM is a supervised learning method that builds a linear mapping between distance matrices on the input and output space. The ESD is used to as part of the input space matrix estimation. The proposed approach is compared to other imputation methods under real world datasets and achieved the best overall performance.

The remainder of the paper is organized as follows. Section 2 introduces the Minimal Learning Machine. Section 3 presents the method for missing values imputation on MLM. The experiments are described in Sect. 4 and the results are shown in Sect. 6. Conclusions are given in Sect. 6.

2 Minimal Learning Machine

We are given a set of N input points $X = \{\mathbf{x}_i\}_{i=1}^N$, with $\mathbf{x}_i \in \mathbb{R}^D$, and the set of corresponding outputs $Y = \{\mathbf{y}_i\}_{i=1}^N$, with $\mathbf{y}_i \in \mathbb{R}^S$. Assuming the existence of a continuous mapping $f : \mathcal{X} \to \mathcal{Y}$ between the input and the output space, we want to estimate f from data with the multiresponse model

$$\mathbf{Y} = f(\mathbf{X}) + \mathbf{R}.$$

The columns of the matrices $\mathbf{X}$ and $\mathbf{Y}$ correspond to the D inputs and S outputs respectively, and the rows to the N observations. The columns of the $N \times S$ matrix $\mathbf{R}$ correspond to the residuals.

The MLM is a two-step method designed to

1. reconstruct the mapping existing between input and output distances;
2. estimating the response from the configuration of the output points.

In the following, the two steps are discussed.

2.1 Distance Regression

For a selection of reference input points $R = \{\mathbf{m}_k\}_{k=1}^K$ with $R \subseteq X$ and corresponding outputs $T = \{\mathbf{t}_k\}_{k=1}^K$ with $T \subseteq Y$, define $\mathbf{D}_x \in \mathbb{R}^{N \times K}$ such that its kth column $\mathbf{d}(X, \mathbf{m}_k)$ contains the distances $d(\mathbf{x}_i, \mathbf{m}_k)$ between the N input points $\mathbf{x}_i$ and the kth reference point $\mathbf{m}_k$. Analogously, define $\mathbf{\Delta}_y \in \mathbb{R}^{N \times K}$ in such a way that its kth column $\delta(Y, \mathbf{t}_k)$ contains the distances $\delta(\mathbf{y}_i, \mathbf{t}_k)$ between the N output points $\mathbf{y}_i$ and the output $\mathbf{t}_k$ of the kth reference point.

We assume that there exists a mapping g between the input distance matrix $\mathbf{D}_x$ and the corresponding output distance matrix $\boldsymbol{\Delta}_y$ that can be reconstructed using the multiresponse regression model

$$\boldsymbol{\Delta}_y = g(\mathbf{D}_x) + \mathbf{E}.$$

The columns of the matrix $\mathbf{D}_x$ correspond to the K input vectors and columns of the matrix $\boldsymbol{\Delta}_y$ correspond to the K response vectors, the N rows correspond to the observations. The columns of matrix $\mathbf{E} \in \mathbb{R}^{N \times K}$ correspond to the K residuals.

Assuming that mapping g between input and output distance matrices has a linear structure for each response, the regression model has the form

$$\boldsymbol{\Delta}_y = \mathbf{D}_x \mathbf{B} + \mathbf{E}. \tag{1}$$

The columns of the $K \times K$ regression matrix $\mathbf{B}$ correspond to the coefficients for the K responses. Under the normal conditions where the number of selected reference points is smaller than the number of available points available (i.e., $K < N$), the matrix $\mathbf{B}$ can be approximated by the usual least squares estimate:

$$\hat{\mathbf{B}} = (\mathbf{D}_x' \mathbf{D}_x)^{-1} \mathbf{D}_x' \boldsymbol{\Delta}_y. \tag{2}$$

For an input test point $\mathbf{x} \in \mathbb{R}^D$ whose distances from the K reference input points $\{\mathbf{m}_k\}_{k=1}^K$ are collected in the vector $\mathbf{d}(\mathbf{x}, R) = [d(\mathbf{x}, \mathbf{m}_1) \ldots d(\mathbf{x}, \mathbf{m}_K)]$, the corresponding estimated distances between its unknown output $\mathbf{y}$ and the known outputs $\{\mathbf{t}_k\}_{k=1}^K$ of the reference points are

$$\hat{\boldsymbol{\delta}}(\mathbf{y}, T) = \mathbf{d}(\mathbf{x}, R)\hat{\mathbf{B}}. \tag{3}$$

The vector $\hat{\boldsymbol{\delta}}(\mathbf{y}, T) = [\hat{\delta}(\mathbf{y}, \mathbf{t}_1) \ldots \hat{\delta}(\mathbf{y}, \mathbf{t}_K)]$ provides an estimate of the geometrical configuration of $\mathbf{y}$ and the reference set T, in the $\mathcal{Y}$-space.

2.2 Output Estimation

The problem of estimating the output $\mathbf{y}$, given the outputs $\{\mathbf{t}_k\}_{k=1}^K$ of all the reference points and estimates $\hat{\boldsymbol{\delta}}(\mathbf{y}, T)$ of their mutual distances, can be understood as a multilateration problem [6] to estimate its location in $\mathcal{Y}$.

Numerous strategies can be used to solve a multilateration problem [9]. From a geometric point of view, locating $\mathbf{y} \in \mathbb{R}^S$ is equivalent to solve the overdetermined set of K nonlinear equations corresponding to S-dimensional hyperspheres centered in $\mathbf{t}_k$ and passing through $\mathbf{y}$. Figure 1 graphically depicts the problem for $S = 2$.

Given the set of $k = 1, \ldots, K$ spheres each with radius equal to $\hat{\delta}(\mathbf{y}, \mathbf{t}_k)$

$$(\mathbf{y} - \mathbf{t}_k)'(\mathbf{y} - \mathbf{t}_k) = \hat{\delta}^2(\mathbf{y}, \mathbf{t}_k), \tag{4}$$

the location of $\mathbf{y}$ is estimated from the minimization of the objective function

$$J(\mathbf{y}) = \sum_{k=1}^K \left((\mathbf{y} - \mathbf{t}_k)'(\mathbf{y} - \mathbf{t}_k) - \hat{\delta}^2(\mathbf{y}, \mathbf{t}_k) \right)^2. \tag{5}$$

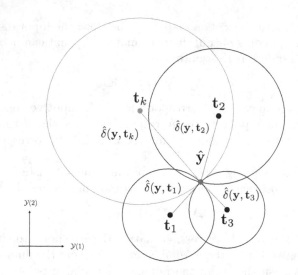

Fig. 1. Output estimation.

The cost function has a minimum equal to 0 that can be achieved if and only if $\mathbf{y}$ is the solution of (4). If it exists, such a solution is thus global and unique. Due to the uncertainty introduced by the estimates $\hat{\delta}(\mathbf{y}, \mathbf{t}_k)$, an optimal solution to (5) can be achieved by any minimizer $\hat{\mathbf{y}} = \underset{\mathbf{y}}{\operatorname{argmin}}\, J(\mathbf{y})$ like the nonlinear least square estimates from standard gradient descent methods. The original MLM proposal applies the Levenberg-Marquardt (LM) method [7] to solve the output estimation step.

3   Computing $\mathbf{D}_x$ in the Presence of Missing Data

The training phase of the MLM model consists on fitting a linear model of the form $\boldsymbol{\Delta}_y = \mathbf{D}_x \mathbf{B} + \mathbf{E}$. That being said, given a pair of labeled examples feature vectors X_i and X_j, instead of estimating what values to impute in the missing entries of these feature vectors, we want to estimate its pairwise distance $\mathbf{D}_x^{(i,j)}$. For that purpose, we use the Expected Squared Distance method, which was introduced by Eirola et al. in [3]. In the subsequent section, we describe the ESD method and how to proceed with its computation.

3.1   Expected Squared Distance (ESD) Calculation

Consider $\alpha, \beta \in \mathbb{R}^D$ drawn from a same multivariate probability distribution, but possibly with deleted entries. It is reasonable to estimate the squared euclidean norm between α and β as

$$\mathrm{E}[\|\alpha - \beta\|_2^2] = \sum_{i=1}^{D} \mathrm{E}[(\alpha_i - \beta_i)^2] \tag{6}$$

Let M_α and M_β be sets containing the index of the missing entries of α and β, respectively. We can further expand (6) into

$$
\begin{aligned}
\mathrm{E}[\|\alpha - \beta\|_2^2] = \sum_{i \notin M_\alpha \cup M_\beta} (\alpha_i - \beta_i)^2 + \sum_{i \in M_\alpha \setminus M_\beta} \mathrm{E}[(\alpha_i - \beta_i)^2] \\
+ \sum_{i \in M_\beta \setminus M_\alpha} \mathrm{E}[(\alpha_i - \beta_i)^2] + \sum_{i \in M_\alpha \cap M_\beta} \mathrm{E}[(\alpha_i - \beta_i)^2]
\end{aligned}
\tag{7}
$$

By performing simple expected value calculations, we can develop the terms comprising the entries $i \in M_\alpha \setminus M_\beta$ into.

$$
\begin{aligned}
\mathrm{E}[(\alpha_i - \beta_i)^2] &= \mathrm{E}[\alpha_i^2 + \beta_i^2 - 2\alpha_i\beta_i] = \mathrm{E}[\alpha_i^2] + \beta_i^2 - 2\mathrm{E}[\alpha_i]\beta_i \\
&= \mathrm{E}[\alpha_i^2] - \mathrm{E}[\alpha_i]^2 + \mathrm{E}[\alpha_i]^2 + \beta_i^2 - 2\mathrm{E}[\alpha_i]\beta_i \\
&= (\mathrm{E}[\alpha_i] - \beta_i)^2 + \mathrm{Var}[\alpha_i]
\end{aligned}
$$

By symmetry the terms involving entries $i \in M_\beta \setminus M_\alpha$ resume to

$$
\mathrm{E}[(\alpha_i - \beta_i)^2] = (\alpha_i - \mathrm{E}[\beta_i])^2 + \mathrm{Var}[\beta_i]
$$

In case both α_i and β_i are missing, assuming these are uncorrelated ($\mathrm{E}[\alpha_i\beta_i] = \mathrm{E}[\alpha_i]\mathrm{E}[\beta_i]$) we have

$$
\begin{aligned}
\mathrm{E}[(\alpha_i - \beta_i)^2] &= \mathrm{E}[\alpha_i^2] + \mathrm{E}[\beta_i]^2 - 2\mathrm{E}[\alpha_i]\mathrm{E}[\beta_i] \\
&= \mathrm{E}[\alpha_i^2] - \mathrm{E}[\alpha_i]^2 + \mathrm{E}[\beta_i^2] - \mathrm{E}[\beta_i]^2 + \mathrm{E}[\alpha_i]^2 + \mathrm{E}[\beta_i]^2 - 2\mathrm{E}[\alpha_i]\mathrm{E}[\beta_i] \\
&= (\mathrm{E}[\alpha_i] - \mathrm{E}[\beta_i])^2 + \mathrm{Var}[\alpha_i] + \mathrm{Var}[\beta_i]
\end{aligned}
$$

If we assume that the distribution from which both α_i and β_i were drawn is a multivariate normal distribution, we can use the Expectation Maximization (EM) algorithm to estimate a mean vector μ and a covariance matrix Σ based on training data. We can then use μ and Σ to substitute each expected value by its conditional mean and substitute the variance of an entry by its conditional variance.

It is interesting to notice that the only real difference between the estimate of the ESD and the result obtained by the direct substitution of a missing value by its EM estimate is the presence of the variance terms, which manifest some uncertainty on the operation being performed using the imputed values.

4   Performance Evaluation

Standard techniques to deal with missing data include imputing the missing entries with the sample mean or EM estimates. A more naive approach is to simply drop the examples that contain missing data. To evaluate the performance

Table 1. Datasets characteristics

Dataset	Attributes	Instances
Concrete compression	8	1030
Boston Housing	13	506
Servo	4	167
Stocks	9	950
Breast cancer	30	569
Wine	13	178

of the proposed algorithm, we compare it with MLMs trained on datasets treated with these techniques.

To carry the experiments, 6 different regression datasets from the UCI machine learning repository [8] were chosen. Details about each dataset are presented in Table 1.

To simulate the impact of the number of missing values for each method, we gradually increase the number of examples with missing attributes in each dataset. The artificial missing values were generated in a manner to guarantee that they are Missing at Random [4]. The experiments were repeated 10 times. At each trail, the data was split between training and test sets (70 % and 30 %, respectively).

For all experiments, the MLMs used all the examples with no missing entries as reference points. The adopted performance measure was the Root Mean Square Error (RMSE).

5 Results

Figure 2 presents the RMSE for all datasets as the number of examples with missing data increase. Each point consists on the mean value for all 10 runs.

As expected, more naive strategies lead to the worst results. Using the mean value for each missing feature (input sample mean) achieved the worst overall result in 3 of the 6 datasets. Discarding examples (Drop entries) achieved the worst results in the other 3 datasets.

It is interesting to notice that the Drop entries approach obtained a similar result to EM and ESD in 2 of the 6 dataset. This can be explained by the fact both Stocks and Concrete Compression datasets have a higher number of examples when compared to the other datasets. This can be observed in Table 1. In such cases, discarding some examples my not have a significant impact on the resulting model.

Analyzing the performance of ESD it can be noticed that the proposed method achieved the lowest RMSE in 5 datasets. For the Stocks dataset, ESD achieved the best results for datasets with up to 30 % of missing data. For values above 30 %, ESD obtained the second best overall result.

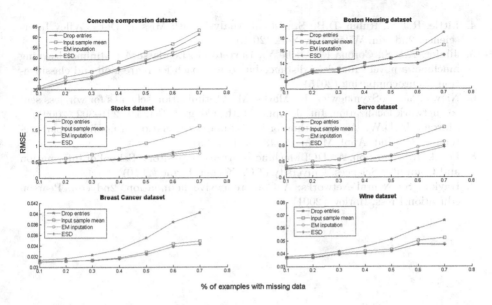

Fig. 2. RMSE for all datasets

6   Conclusion

This work presented a variation of the Minimal Learning Machine (MLM) with a built-in mechanism to deal with missing attributes. The proposed approach uses the Expected Squared Distance (ESD) calculation to estimate the pairwise distance between input point, potentially with missing attributes, and the reference points.

It is worth noting that the proposed algorithm can handle not only missing data on the training set but also capable of making decision under new data with missing attributes.

Results showed that MLM with ESD achieved the best overall results when compared to different standard strategies to deal with missing data such as sample mean imputation and EM imputation. Future works shall include a variant of MLM with ESD to handle missing values in the output space.

References

1. Hruschka, E.R., Jr., E.R.H., Ebecken, N.F.F.: Evaluating a nearest-neighbor method to substitute continuous missing values. In: Gedeon, T.T.D., Fung, L.C.C. (eds.) AI 2003. LNCS (LNAI), vol. 2903, pp. 723–734. Springer, Heidelberg (2003)
2. Lin, T.H.: A comparison of multiple imputation with EM algorithm and MCMC method for quality of life missing data. Qual. Quant. **44**, 277–287 (2010)
3. Eirola, E., Doquire, G., Verleysen, M., Lendasse, A.: Distance estimation in numerical data sets with missing values. Inf. Sci. **240**, 115–128 (2013)

4. Little, R.J.A., Rubin, D.B.: Statistical Analysis with Missing Data. Wiley Interscience, 2nd edn. Wiley, New York (2002)
5. Júnior, A.H.S., Corona F., Miché Y., Barreto G., Lendasse A.: Minimal learning machine: a novel supervised distance-based approach for regression and classification. Neurocomputing (2015)
6. Niewiadomska-Szynkiewicz, E., Marks, M.: Optimization schemes for wireless sensor network localization. Int. J. Appl. Math. Comput. Sci. **19**, 291–302 (2009)
7. Marquardt, D.W.: An algorithm for least-squares estimation of nonlinear parameters. J. Soc. Ind. Appl. Math. **11**, 431–441 (1963)
8. Frank, A., Asuncion, A.: UCI Machine Learning Repository. School of Information and Computer Sciences, University of California, Irvine (2010)
9. Haykin, S.: Neural Networks: A Comprehensive Foundation, 2nd edn. Pearson education Press, Harlow (2001)

Calibrated k-labelsets for Ensemble Multi-label Classification

Ouadie Gharroudi[✉], Haytham Elghazel, and Alex Aussem

Université Lyon 1, LIRIS, UMR 5205, 69622 Villeurbanne, France
{ouadie.gharroudi,haytham.elghazel,aaussem}@univ-lyon1.fr

Abstract. RAndom k-labELsets (RAkEL) is an effective ensemble multi-label classification (MLC) model where each base-classifier is trained on a small random subset of k labels. However, the model construction does not fully benefit from the diversity of the ensemble and the label probability estimates obtained with RAkEL are usually badly calibrated due to the problems raised by the imbalanced label representation. In this paper, we propose three practical solutions to overcome these drawbacks. One is to increase the diversity of the base classifiers in the ensemble. The second to smooth the label powerset probability estimates during the ensemble aggregation process, and the third to calibrate the label decision thresholds. Experimental results on various benchmark data sets indicate that the proposed approach outperforms significantly recent state-of-the-art MLC algorithms, including RAkEL and its variants.

Keywords: Multi-label classification · Ensemble learning

1 Introduction

In contrast to traditional single-label classification, instances in Multi-label classification (MLC) can be associated simultaneously with more than one label. Given a training sample (x, y) such as, $x \in \mathcal{X}$ is a feature vector and $y \in Y = \{0, 1\}^Q$ indicates its affected labels in $\mathcal{L} = \{\lambda_1, \ldots, \lambda_Q\}$, MLC task learns the decision function $h : \mathcal{X} \to \mathcal{Y}$. There are mainly two categories of MLC methods [6]: (a) *algorithm adaptation methods* and (b) *problem transformation methods*. Algorithm adaptation methods extend specific learning algorithms to handle multi-label (ML) data directly. Problem transformation methods, on the other hand, transform the ML learning problem into either several binary classification problems, such as the Binary Relevance (BR) approach, or one multiclass classification problem, such as the Label Powerset (LP) approach. The BR method ignores the probabilistic dependency structure of the labels (given the inputs) as each label is predicted independently of the other. In contrast, by combining all the labels into one multi-class label, the LP method takes into account the label correlation (to some extent). Unfortunately, the multi-class learning task in LP is known to be difficult as the effective number of classes

© Springer International Publishing Switzerland 2015
S. Arik et al. (Eds.): ICONIP 2015, Part I, LNCS 9489, pp. 573–582, 2015.
DOI: 10.1007/978-3-319-26532-2_63

grows swiftly with the number of labels, especially when most label subsets are associated with very few examples only.

With a view to trade off the BR label independence assumption with LP complexity, Tsoumakas *et al.* [11] proposed an effective ensemble method named RAndom k-labELsets (RAkEL). Each base-classifier in RAkEL is a LP based on a random small subset of k labels (k-labelsets). By construction, RAkEL takes into account the correlation structure between the labels, and at the same time, reduces the number of labels handled by each LP. Prediction of new instances is achieved by combining the ensemble binary outputs through a label majority voting process. The aggregation step helps correct potential uncorrelated errors and improves the overall performance. RAkEL is now considered as a state-of-the-art MLC algorithm due to its simplicity and practical efficiency. However, as all heuristic methods, RAkEL has several shortcomings: First; considered as an ensemble approach, RAkEL does not fully take advantage of the best part of the diversity concept when constructing its base-classifiers since that each label combination is allowed to appear at most once. Second, the labels that are drawn more often than the others during the random label sampling should not be overly weighted in the aggregation process. Third, the use of a unique threshold to select the final predicted labelset is not well adapted for the data sets having many labels associated with few training examples. In this paper, we discuss a method to improve the overall performance of k-labelsets based ensemble models. Our contribution is three-fold: First, we use Bagging in tandem with random k-labelsets to increase the diversifity of the base classifiers and thus the robustness of the ensemble. Second, the label set probabilities are calibrated to account for the effective label occurrence rate in the random labelsets sampling. Third, a finely-tuned threshold is associated to each label instead of using a single threshold for all the labels [11].

The paper is organized as follows: Sect. 2 presents an overview of the related work. Our contribution is discussed in Sect. 3. Extensive experiments on *Mulan* [12] benchmark data sets are reported in Sect. 4 to demonstrate the usefulness of the method.

2   Related Work

In this Section, we provide an overview of the recent attempts to improve the RAkEL approach and discuss some state-of-the art thresholding schemes in MLC.

As a first extension of RAkEL, Kouzani et al. [7] combine a random selection of labels, feature subsets and instance subsets, to build a Triple-Random Ensemble Multi-Label Classification (TREMLC). Each base-classifier in TREMLC is trained using a portion of data drawn randomly without replacement and trained to predict k-labelsets using only a subset of features. They reported that the model performance was especially susceptible to the percentage of instance selection and the random subspace size. In fact, such diversity is hard to manage and requires a large ensemble size. The ensemble size depends on the number of labels since the k-labelsets selection is carried by a random selection without replacement from all possible k-labelsets in $\mathcal{L}$. In [10], an improved version of RAkEL [11] named RAkEL++ is presented. The idea is to, (i) aggregate the probabilities provided by the base-classifiers rather than 0/1 votes in the original RAkEL, and (ii) use in the labelling step a single threshold for all labels, calibrated by optimizing a performance measure of interest via a cross validation (CV) procedure.

It should also be emphasized that RAkEL - as many ML classifier methods - is trained to output a score for each label, and thus requires a thresholding strategy to implement a decision function. Works on thresholding MLC outputs reveals that thresholding can improve dramatically the performance of most MLC algorithm [3–5,8]. In [4], Ioannou *et al.* conduct a comparative study on thresholding strategies based on multi-label output scores which conclude that calibrating one single threshold for all labels remain the most promising strategy. Authors attribute its, compared to a strategy using distinct thresholds for each label $T = \{t_1, \ldots, t_Q\}$, to the limited number of calibrated thresholds which attenuate risks of overfitting. However, the study was only carried on *HammingLoss* metric [6]. A distinct label thresholds calibration strategy, not included in the study of [4], was proposed in [3]. The threshold calibration problem was handled via a greedy cyclic optimization algorithm named FBR. The heuristic aims to update iteratively each label threshold t_k to optimize either $MicroF_1$ or $MacroF_1$ for BR models. In [8], the authors analyzed the optimization strategies proposed in [3] and concluded that Maximizing $MicroF_1$ on a given data set can easily lead to overfitting. Beside Authors derived $MicroF_1$ and $MacroF_1$ properties to reduce computational cost of such optimization strategy. Recently, a theoretical study on threshold optimization for F_1 measures has been proposed in [5]. This study confirms previous experimental finding and demonstrate that $MicroF_1$ could be optimized by predicting all instances to be negative for high imbalance labels. Thereby calibrating threshold via CV can be too intensive and may lead to overfitting. An alternative was proposed in [1] for Ensemble Classifier Chain with Random Forest (ECC-RF) in which out-of-bag instances are used for threshold calibration. Nevertheless, only the decomposable ML metric $HammingLoss$ [6] was the only loss function considered.

3   Calibrated k-labelsets Ensemble Method

In this section, we discuss our Calibrated k-labelsets for Ensemble Multi-Label Classification method (termed CkMLC as a shorthand).

Committee Construction: In k-labelsets ensemble models, diversity is carried in the output space by a random selection of m k-labelsets without replacement from the set of all possible label sets of size k in $\mathcal{L}$. However nothing hinders two base-classifiers to share the same output space if diversity is maintained in the input space. On the contrary this will improve the predictive performance of the ensemble since more votes for each k-labelsets lead to more accurate estimates of the true value of this k-labelsets. In this work we propose to induce diversity in the input space by a bootstrapping strategy. In contrast to [7] we enforce diversity only in instance space via random sampling with replacement from the instance set. Combining input and output strategies has two advantages: The k-labelsets strategy provides a specific output view to the base-classifier. Meanwhile, the latter strategy enforces diversity by allocating distinct samples to the classifiers. Last but not least, in each bootstrap, almost 33 % are left out-of-bag (oob), *i.e.*, they are not used for the construction of their corresponding model. These samples can be used as an unbiased validation set of the thresholding strategy in the sequel.

Aggregating Base-Classifier Predictions: The overlapping property of k-labelsets ensembles induces different numbers of predictions for each label. Thus, for a test instance x_t, prediction for λ_j is the average of all base-classifiers predictions. Therefore, the aggregation involves that each label prediction has it own significance since it depends on the label occurrence in the ensemble. Obviously the probability predicted for a label appearing in several k-labelsets is more accurate than the probability predicted for label appearing in one k-labelsets. The following example illustrates the potential problem with such ensemble aggregation. Assume that λ_1 and λ_2 appear respectively in 10 and 3 k-labelsets. If for a test sample (x_t), 9 base-classifiers predict λ_1 and 3 classifiers predict λ_2, the probabilities to assign λ_1 and λ_2 to x_t given by the original RAkEL are respectively $\hat{P}_{\lambda_1} = (9/10) = 0.9$ and $\hat{P}_{\lambda_2} = (3/3) = 1$. Our confidence in the probably prediction based on 3 classifiers is not as good as for 10 classifiers of course. Therefore, we propose to smooth the ensemble probabilities for each label using the Laplace estimate as: $\hat{P}_{\lambda_i} = \frac{h(x_t,\lambda_i)+1}{n+C}$ (C is the number of classes per label, here $C = 2$). In our example, the Laplace estimate yields a probability of $\frac{9+1}{10+2} = 0.83$ for λ_1 and $\frac{3+1}{3+2} = 0.8$ for λ_2. This smoothing strategy flattens the label probability distribution and improve the MLC performance in terms of probability-based ranking measure.

Threshold Calibration: In this work we propose a simple forward algorithm easy to implement with a low computational cost for calibrating label decision thresholds. The algorithm benefits from oob instances and does not need to carry CV procedure to create a validation data set. The proposed optimization algorithm is valid for both, decomposable and non-decomposable performance measure. To calibrate decision threshold for a specific performance measure. First the best thresholds are selected independently for each label $\lambda \in \mathcal{L}$, then the label achieving the best performance λ^* is selected as well as its optimal threshold t_{λ^*}. Then, λ^* is removed from the search space $\mathcal{L}$ and added to $\mathcal{L}^*$. Afterwards, for each label in $\mathcal{L}$ the best thresholds are selected as having the best performance jointly with labels in $\mathcal{L}^*$ associated with their calibrated thresholds. The process is repeated until calibrating all thresholds. Algorithm 1 gives a formal description of the procedure. To the best of our knowledge, this is the first attempt to propose an algorithm for selecting a distinct threshold per label by optimizing any ML performance measure of interest for any ML classifier.

Algorithm 1 *Forward Multi-label Thresholds Calibration*

Require: *Oob* predictions probabilities $(\hat{Y})$, *Oob* real labels (Y), label set $\mathcal{L}$, multi-label performance measure to optimize $(MLmeasure)$. $=0$

$\mathcal{L}^* \leftarrow \emptyset;\ T \leftarrow \emptyset$

while $\mathcal{L} \neq \emptyset$ **do**

$\quad \lambda^*, t_{\lambda^*} \leftarrow \underset{\lambda \in \mathcal{L}, t \in [0,1]}{\mathrm{argmax}}\ MLmeasure([\hat{Y}_{\mathcal{L}^*}/T \cup \{\hat{Y}_\lambda/t\}], [Y_{\mathcal{L}^*} \cup \{Y\}])$

$\quad \mathcal{L}^* \leftarrow \mathcal{L}^* \cup \lambda^*$

$\quad T \leftarrow T \cup t_{\lambda^*}$

$\quad \mathcal{L} \leftarrow \mathcal{L} \backslash \lambda^*$

end while

return T

---

4 Experimental Evaluation

In this section, we investigate the effectiveness of the proposed approach and compare its performances against several state-of-art MLC methods.

4.1 Data Sets and Experimental Setup

To thoroughly evaluate the performance of our approach, a total of twelve real-word ML data sets given from the *Mulan's repository* [12] are employed in this paper. Table 1 summarizes their basic statistics: **M** number of features, **Q** number of labels; Label Cardinality **LC**= $\frac{1}{N} \sum_{i=1}^{N} |Y_i|$ and Label Density **LD**= $\frac{1}{N} \sum_{i=1}^{N} \frac{|Y_i|}{Q}$.

Table 1. Description of the multi-label data sets used in the experiments.

Data	Domain	N	M	Q	LC	LD
Arts	Text	5000	462	26	1.636	0.063
Bird	Audio	645	260	19	1.014	0.053
Business	Text	5000	438	30	1.588	0.053
Education	Text	5000	550	33	1.460	0.044
Emotions	Music	593	72	6	1.869	0.311
Enron	Text	1702	1001	53	3.378	0.064
Flag	Images	194	19	7	3.392	0.485
Health	Text	5000	612	32	1.662	0.052
Image	Image	2000	249	5	1.236	0.247
Medical	Text	978	1449	45	1.245	0.028
Scene	Image	2407	294	6	1.074	0.179
Slashdot	Text	3782	1079	22	1.180	0.041

In the experiments, both the new ensemble construction and the threshold calibration strategies combined together in our CkMLC approach are studied and compared against several state-of-the-art MLC methods, namely RAkEL taken as our gold standard ML ensemble approach, RAkEL++ [10] and TREMLC [7] that should be viewed as another variants of RAkEL, two recently proposed MLC techniques named ECC-RF [1] and FBR [3] which implement (as for CkMLC) respectively two different thresholding strategies for the prediction step.

To make fair comparisons, the same experimental setting in [7] was adopted here for the RAkEL approach [11] and its variants (RAkEL++ and TREMLC), *i.e.*, the number of models was set to $m = min(2Q, 100)$ and a size of labelsets k of 3. These values were found to yield the most satisfactory performances in [7,11]. The remaining parameters of TREMLC are tuned as suggested by the authors in [7]. In our CkMLC approach, the number of label per bag k was set to 3 as for RAkEL and the committee size m was computed using the following formula: $m = 10 \times ceil(\log(\alpha)/\log(1 - 1/k))$. This formula ensures that each label is drawn 10 times at a confidence level of $\alpha = 1\,\%$. To ensure that the total number of classifiers which cast a vote in every prediction is the same between CkMLC and ECC-RF, the number of iterations done for ECC and RF within ECC-RF were both taken to be $\sqrt{m}$ [1]. The *classregtree* Matlab implementation of decision tree was used as the base learner in all compared algorithms. Finally, instead of manually setting up the single threshold for all labels to 0.5 to output the final decision as in RAkEL and TREMLC, this threshold was tailored to each dataset in RAkEL++ using a 5-fold CV procedure [10]. On the other hand, FBR, ECC-RF and CkMLC select a separate threshold for each label, calibrated using 5-fold CV for FBR and oob estimation for both ECC-RF and CkMLC. We tested 9 different threshold values ranging from 0.1 to 0.9 in 0.1 steps.

The algorithm's performances were analyzed according to three standard ML measure: *RankingLoss*, *MicroF*$_1$ and *MacroF*$_1$ [6] using a 2-fold CV [9]. To get reliable statistics over the performance metrics, experiments were repeated 5 times. So the results obtained were averaged over 10 iterations. Significant differences among the methods were established using statistical tests.

4.2   Results

Performances are tabulated in terms of averaged values as well as standard deviations on each data set. The symbol '↓/↑' indicates the smaller/larger the better. To examine whether the results are statistically significant, paired t-tests were carried out at 5 % significance level. The marker '•/∘' suggests that our approach is statistically superior/inferior to others. Otherwise, a tie is counted and no marker is placed. Furthermore, following [2], if two compared algorithms are, as assumed under the null-hypothesis, equivalent, each should win on approximately $n/2$ out of n data sets. The number of wins is distributed according to the binomial distribution and the critical number of wins at $\alpha = 5\,\%$ is equal to 10 in our case. Since tied matches support the null-hypothesis we should not discount them but split them evenly between the two classifiers when counting the number of wins; if there is an odd number of them, we again ignore one. The

Table 2. Predictive performances in terms of *RankingLoss*.

↓	CkMLC	RAkEL	RAkEL++	TREMLC	FBR	ECC-RF
Arts	.080±.029	.100±.034●	.082±.029●	.096±.033●	.235±.082●	.124±.043●
Birds	.169±.026	.315±.057●	.176±.037	.279±.044●	.235±.040●	.271±.047●
Business	.029±.009	.052±.017●	.031±.010	.044±.015●	.122±.042●	.052±.018●
Education	.058±.019	.089±.031●	.060±.019	.078±.026●	.283±.101●	.088±.029●
Emotions	.099±.034	.164±.063●	.123±.043●	.158±.049●	.230±.084●	.158±.045●
Enron	.057±.018	.089±.030●	.065±.020●	.094±.030●	.155±.050●	.072±.025●
Flags	.138±.038	.204±.045●	.153±.041●	.202±.041●	.202±.052●	.208±.041●
Health	.034±.012	.056±.019●	.033±.011	.045±.015●	.176±.061●	.055±.019●
Image	.088±.031	.169±.066●	.129±.050●	.150±.041●	.270±.093●	.161±.040●
Medical	.020±.008	.064±.021●	.024±.011●	.033±.013●	.070±.026●	.037±.015●
Scene	.045±.015	.104±.034●	.077±.028●	.098±.025●	.193±.065●	.096±.027●
Slashdot	.048±.018	.118±.034●	.044±.018	.079±.025●	.085±.028●	.085±.028●
(win/tie/loss)		(12/0/0)	(8/4/0)	(12/0/0)	(12/0/0)	(12/0/0)

●/○ CkMLC is significantly better/worse, at a level of significance of 5 %.

obtained (*win/tie/loss*) counts for CkMLC against the compared algorithms are reported in the bottom row of each Table. Note that calibrating the thresholds should not affect the *RankingLoss* as the latter is a probability-based ranking metric [6]. However, calibration can greatly help to reduce example-based metrics like the $MicroF_1$ and $MacroF_1$ measures.

As may be observed in Table 2, the performances of CkMLC in terms of *RankingLoss* are statistically distinguishable from the performance of all other algorithms. CkMLC outperforms the other methods by generally achieving the smallest *RankingLoss* values. This firstly validates the motivation behind our method CkMLC that encouraging diversity in the committee construction achieves more robust votes per label and thus more accurate probability estimates for each label. Moreover, results in Table 2 also confirms the effectiveness of the smoothing strategy in CkMLC to rank the labels properly. Compared to TREMLC for which the idea is to mainly encourage the diversity in RAkEL using a triple randomization, the combination of our diverse committee construction and probability smoothing strategy in CkMLC shows promise for obtaining a ML ensemble classification framework that enjoys significant improvements in terms of *RankingLoss* metric.

Tables 3 and 4 depict the performances in terms of $MicroF_1$ and $MacroF_1$ measures respectively. In the sequel, the thresholding strategies proposed respectively in CkMLC, FBR and RAkEL++ are implemented twice as we tune the decision thresholds for optimizing either the $MicroF_1$ or the $MacroF_1$. So each approach has two variants: one devoted to optimizing the $MicroF_1$ measure ' superscript 'm', the other devoted to optimizing the $MacroF_1$ denoted with the superscript 'M'.

Table 3. Predictive performances in terms of $MicroF_1$.

↑	$CkMLC^m$	RAkEL	$RAkEL++^m$	$CkMLC^{0.5}$	TREMLC	FBR^m	ECC-RF
Arts	.650±.123	.620±.144•	.599±.149•	.539±.106•	.528±.107•	.508±.107•	.521±.091•
Birds	.560±.080	.498±.079•	.418±.110•	.476±.068•	.420±.056•	.470±.074•	.472±.057•
Business	.807±.057	.799±.044•	.796±.061•	.801±.043	.796±.045•	.757±.055•	.779±.040•
Education	.655±.119	.637±.144•	.585±.153•	.585±.127•	.568±.130•	.534±.106•	.545±.098•
Emotions	.798±.062	.728±.081•	.742±.080•	.778±.071•	.699±.078•	.727±.085•	.726±.063•
Enron	.716±.059	.698±.075•	.720±.083	.670±.060•	.654±.059•	.631±.071•	.678±.060•
Flags	.809±.043	.787±.039•	.804±.037	.806±.027	.764±.036•	.778±.044•	.773±.039•
Health	.769±.072	.755±.080•	.751±.087•	.736±.070•	.732±.068•	.674±.077•	.715±.057•
Image	.789±.070	.707±.095•	.707±.105•	.745±.091•	.681±.081•	.673±.090•	.685±.070•
Medical	.871±.031	.857±.034•	.862±.035•	.859±.032•	.860±.035•	.853±.032•	.853±.025•
Scene	.840±.059	.769±.077•	.773±.084•	.803±.072•	.746±.070•	.725±.077•	.752±.058•
Slashdot	.818±.021	.816±.030	.817±.029	.815±.019•	.816±.017	.814±.030	.807±.016•
(win/tie/loss)		(11/1/0)	(9/3/0)	(10/2/0)	(11/1/0)	(11/1/0)	(12/0/0)

•/∘ $CkMLC^m$ is significantly better/worse, at level of significance of 5 %.

Table 4. Predictive performances in terms of $MacroF_1$.

↑	$CkMLC^M$	RAkEL	$RAkEL++^M$	$CkMLC^{0.5}$	TREMLC	FBR^M	ECC-RF
Arts	.405±.131	.429±.136	.402±.140	.299±.075•	.282±.073•	.398±.114	.326±.079•
Birds	.384±.097	.324±.082•	.253±.098•	.276±.056•	.220±.035•	.336±.067•	.294±.049•
Business	.385±.125	.295±.098•	.329±.126•	.271±.079•	.245±.075•	.375±.110	.278±.073•
Education	.349±.117	.360±.123	.294±.109•	.256±.081•	.226±.068•	.316±.105•	.249±.071•
Emotions	.784±.078	.721±.083•	.736±.082•	.767±.078•	.689±.078•	.721±.085•	.715±.069•
Enron	.382±.135	.279±.076•	.386±.133	.204±.044•	.184±.032•	.320±.087•	.257±.066•
Flags	.738±.054	.724±.050	.757±.046	.742±.058	.664±.048•	.734±.052	.713±.052
Health	.440±.092	.439±.099	.445±.109	.352±.060•	.333±.048•	.416±.091•	.351±.059•
Image	.786±.074	.707±.093•	.707±.104•	.746±.090•	.682±.081•	.674±.089•	.687±.069•
Medical	.541±.091	.524±.083•	.525±.091•	.511±.086•	.481±.090•	.524±.083•	.430±.070•
Scene	.842±.058	.773±.076•	.778±.084•	.805±.074•	.748±.070•	.730±.075•	.756±.058•
Slashdot	.294±.093	.205±.063•	.263±.081•	.151±.029•	.127±.018•	.284±.088	.145±.026•
(win/tie/loss)		(8/4/0)	(9/3/0)	(11/1/0)	(12/0/0)	(8/4/0)	(11/1/0)

•/∘ $CkMLC^M$ is significantly better/worse, at level of significance of 5 %.

In order to better assess the effectiveness of our thresholding strategies, Tables 3 and 4 report also the results of our algorithm using a 0.5 single threshold for all labels. This approach without threshold selection is denoted with the superscript '0.5'. There are several remarks we can draw for these observations:

- CkMLC exhibits the highest performances in terms of $MicroF_1$ ($CkMLC^m$) and the $MacroF_1$ ($CkMLC^M$) measures than the original RAkEL and TREMLC.
- $CkMLC^m$ (respectively $CkMLC^M$) significantly outperforms $CkMLC^{0.5}$ (without threshold calibration) by a noticeable margin in terms of $MicroF_1$ (respectively $MacroF_1$). This confirms the ability of the proposed greedy thresholding algorithm to optimize the performance measure of interest (here $MicroF_1$ and $MacroF_1$).
- The strategy proposed in CkMLC to calibrate a separate threshold per label seems to perform better than selecting one single threshold for all labels in RAkEL++.

- CkMLC is shown to reduce significantly the micro-average (respectively macro-average) F-measure when optimizing $MicroF_1$ (respectively $MacroF_1$) compared to ECC-RF. It is interesting to notice that ECC-RF computes a separate threshold for each label by optimizing the $HammingLoss$ performance measure.
- FBR is worse than CkMLC in all comparisons. Even if our thresholding greedy algorithm has no optimality guarantee (as for FBR), the results in Tables 3 and 4 confirm its ability, compared to FBR, to select the relevant thresholds accurately by optimizing the performance measure of interest.

5   Conclusion

In this paper, we discussed a novel strategy to build and aggregate k-labelsets in the context of ensemble multi-label classification. The proposed strategy extends and improves upon the original RAkEL algorithm in three ways: (i) new randomization strategy using bagging in tandem with random k-labelsets; (ii) accounting for the imbalanced label representation when aggregating the base-classifiers predictions; and (iii), a specific label threshold calibration procedure on out-of-bag instances. Experimental results on twelve benchmark data sets indicate that the proposed model outperforms the RAkEL algorithm and other recent state-of-the-art MLC algorithms. Future works will be conducted to analyze the thresholding strategy on different ensemble MLC approaches and to adapt, in a more principled way, the aggregation procedure to the specific loss function.

References

1. Briggs, F., Fern, X.Z., Irvine, J.: Multi-label classifier chains for bird sound. In: Workshop on Machine Learning for Bioacoustics, ICML 2013 (2013)
2. Demsar, J.: Statistical comparisons of classifiers over multiple data sets. J. Mach. Learn. Res. **7**, 1–30 (2006)
3. Fan, R., Lin, C.: A study on threshold selection for multi-label classification. Department of Computer Science, National Taiwan University (2007)
4. Marios, I., George S., Tsoumakas, G., Vlahavas, I.: Obtaining bipartitions from score vectors for multi-label classification. In: ICTAI (2010)
5. Lipton, Z.C., Elkan, C., Naryanaswamy, B.: Optimal thresholding of classifiers to maximize F1 measure. In: Calders, T., Esposito, F., Hüllermeier, E., Meo, R. (eds.) ECML PKDD 2014, Part II. LNCS, vol. 8725, pp. 225–239. Springer, Heidelberg (2014)
6. Madjarov, G., Kocev, D., Gjorgjevikj, D., Džeroski, S.: An extensive experimental comparison of methods for multi-label learning. Pattern Recogn. **45**, 3084–3104 (2012)
7. Kouzani, A.Z., Nasierding, G., Tsoumakas, G.: A triple-random ensemble classification method for mining multi-label data. In: Data Mining Workshops, ICDMW 2010 (2010)
8. Pillai, I., Fumera, G., Roli, F.: Threshold optimisation for multi-label classifiers. Pattern Recogn. **46**, 2055–2065 (2013)

9. Payam, R., Lei, T., Huan, L.: Cross-validation. In: Encyclopedia of Database Systems. Springer, New York (2009)
10. Rokach, L., Schclar, A., Itach, E.: Ensemble methods for multi-label classification. Expert Syst. Appl. **41**, 7507–7523 (2014)
11. Tsoumakas, G., Katakis, I., Vlahavas, I.: Random k-labelsets for multilabel classification. IEEE Transa. Knowl. Data Eng. **23**, 1079–1089 (2011)
12. Tsoumakas, G., Xioufis, E.S., Vilcek, J., Vlahavas, I.P.: Mulan: a java library for multi-label learning. J. Mach. Learn. Res. **12**, 2411–2414 (2011)

EMG Signal Based Knee Joint Angle Estimation of Flexion and Extension with Extreme Learning Machine (ELM) for Enhancement of Patient-Robotic Exoskeleton Interaction

Tanvir Anwar[✉], Khairul Anam, and Adel Al Jumaily

School of Electrical, Mechanical and Mechatronic Engineering,
University of Technology, Sydney, Australia
Tanvir.anwar@student.uts.edu.au, Adel.Ali-Jumaily@uts.edu.au

Abstract. To capture the intended action of the patient and provide assistance as needed, the robotic rehabilitation device controller needs the intended posture, intended joint angle, intended torque and intended desired impedance of the patient. These parameters can be extracted from sEMG signal that are associated with knee joint. Thus an exoskeleton device requires a multilayer control mechanism to achieve a smooth Human Machine Interaction force. This paper proposes a method to estimate the required knee joint angles and associate parameters. The paper has investigated the feasibility of Extreme Learning Machine (ELM) as a estimator of the operation range of extension $(0° - 90°)$ and The performance is compared with Generalized Regression Neural Network (GRNN) and Neural Network (NN). ELM has performed relatively better than GRNN and NN.

Keywords: EMG · Flexion · Extension · Interaction force · Neural network · ELM

1 Introduction

The conventional Robotic Rehabilitation Device (RRD) is in the pattern of industrial robot which still behaves like master-slave manner (MIT-MANUS). One of the main objectives of a RRD is to obtain a smooth human machine interaction in different phases of gait cycle at the interaction point by considering patient-exoskeleton interaction is bidirectional rather than unidirectional. To achieve bidirectional interaction, a design of an effective wearable exoskeleton is possible where minimum interaction force is experienced, since patient now become an active element of the conventional closed loop control system of a RRD. In human body, brain is the controller that generates necessary signal for muscle which is the actuator. In the exoskeleton device, Robot is the controller and generates necessary signal for the actuator. When patient's brain is affected due to stroke or any other injury then brain cannot generate necessary signal for limb movement. Then Robotic Rehabilitation device shares most of the joint activities and help patient perform the required movements. As a result of robotic therapy, the brain will be able to generate necessary EMG signal responsible for lower limb movement in the long run. This process is called plasticity. But for plasticity to occur, prolonged and

© Springer International Publishing Switzerland 2015
S. Arik et al. (Eds.): ICONIP 2015, Part I, LNCS 9489, pp. 583–590, 2015.
DOI: 10.1007/978-3-319-26532-2_64

intensive therapy is required. Whole process of brain plasticity has to undergo different phases of rehabilitation. For lower limb robotic rehabilitation, knee and hip joint requires certain set of joint kinematic and dynamic properties in each phase [1]. Dynamic parameters like muscle sEMG and interaction force are always changing with respect to time and it is very important that these parameters are captured to produce suitable control signal to generate desired knee joint kinematics and dynamics.

2 Proposed Model

Figure 1 shows the complete proposed closed loop adaptive control schematics. It consists of task level, Low Level and High Level controller.

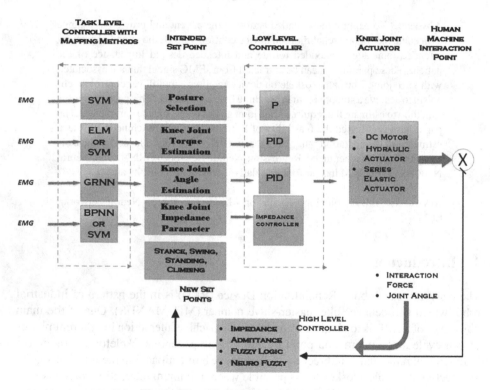

Fig. 1. Wearable bidirectional exoskeleton.

In the manipulation of interaction force of exoskeleton which is interacting with environment, it is very important that enough effort has been made to learn the environment. If the environment is static, then a fixed exoskeleton will interact with the environment with a fixed interaction force. But if the environment (muscle activity) is changing with respect to time then exoskeleton will interact with the environment with a manipulated interaction force which is also changing adaptively [2]. So an extracted EMG signal can be a function of stiffness, damping, mass, force parameters for knee joint that behaves like impedance (spring, mass and damper) like knee joint rather than rigid joint. Extreme Learning Machine (ELM) has been used here to represent the function that maps EMG signal to above mentioned parameters so that knee joint properties can be changed dynamically. So far researchers are able to extract muscle force, fatigue, muscle activation time, muscle load, joint angle and torque based on EMG signal. EMG is a bio- signal that can be analyzed in time domains and frequency domain and use features that describes the behavior of EMG. On set and off set time of EMG signal best describes the transient behavior of EMG signal and it is triggered when knee joint flexes or extends to the operating range of angle. Transient of EMG also describes as to how knee joint torque is changing [3]. To boost the system accuracy, different approach like number of electrode per muscle, the electrode configuration, selection of good filters for noise reduction and various features (application dependent) are used so far. The signals that are very dynamic in nature and their composites are changing with respect of time, for such signal frequency domain features are very useful. For dynamic signal, features that contains derivative $\frac{dx(t)}{dt}$ from $x(t)$ yields higher accuracy than without $\frac{dx(t)}{dt}$. Root mean square (RMS), Integral of absolute value (IAV), Auto Regression Coefficients, Slope Sign Changes, Zero Crossing, Fourier Coefficients, Wavelets are most popular features used in research [4]. The EMG signal can also be mapped to swing, stance, support, standing, sitting and climbing phases of lower limb gait cycle [5]. Movement of any limb of the body is a synchronized co-ordination of various muscle activities with different degree of intensity. In standing position, estimated knee joint torque is observed to be decreasing and torque of hip is increasing. To lift upper body, a large hip torque is observed. Similarly, in sitting and climbing also such variations in EMG signal intensity at various muscle involved in knee and hip are observed.

3   Data Acquisition

In order to obtain the convincing experimental data, six able bodied subjects participated in leg extension exercise. The subjects were asked to do the leg extension exercise movement completed in 2 s, 3 s, 4 s, and 5 s. The knee joint flexion has an operational range of 50° and extension has an operational range of 125°. But operational range of extension has been limited to 90° due to sitting position. Two sets of muscles are involved. 1st set is called Quardriceps which consists of Rectus Femoris, Vastus medialis and vastus lateralis. This set is responsible for extension. The 2nd set of muscle is called Hamstrings and it consists of bicep femoris, semitendinosus and semimembranosus. This set is responsible for flexor movement. To extract information about information of flexor and extensor, one channel has been used for each posture. So Rectus Femoris is used for extensor and Bicep Femoris is used for flexor. The muscles are selection for this specific application of posture selection based on their relatively better intensity over other muscle and they can serve the purpose better. For angle estimation, Rectus femoris and Vastus medialis are selected for extension angle estimation. Bicep femoris and semitendinosus are selected for flexion angle estimation. sEMG measuring device recorded sEMG data of these muscles simultaneously. So each subject has been made to repeat a complete flexion and extension for six consecutive trials of EMG data. sEMG signal acquisition equipment used in the experiment named "FlexiCom" which is a product of Thought Technology Ltd, from Canada. The device can simultaneously capture 10 channels of sEMG data with sampling rate at 2048 Hz for each channel. The joint angle measurement device MPU6050, 6 axis gyroscope is used to measure knee joint angles. The MPU6050 is incorporated with Arduino nano microcontroller mounted the lower limb that is used to extend or flex about knee joint. The sensor MPU6050 communicate with Arduino through I2C protocol. Then collected angle data is then transferred to Matlab in PC from Arduino through serial port. Recording of EMG and recording of knee joint angle have been done simultaneously so that for each EMG data there is a corresponding angle data. The sampling rate of the joint angles is selected as 100 Hz. Shaving and cleaning of the skin surface is desired of all the muscles in order to reduce input resistance and the external disturbance. Ag/AgCl electrodes with glue solution were used for measuring the analog sEMG signal. Each of the electrodes in a pair was separated from each other by 2 cm. The tissue underlying the sEMG electrodes on the skin filter the muscle action potentials. The filtering characteristics of this tissue depend on day to day variation in the position of sEMG electrodes, skin preparation, ambient temperature and electrical impedance. The tissue filtering characteristics are implicitly accounted for by the sEMG to activation filter [6].

4 Signal Processing

It is prominent that original sEMG signals are contaminated during signal acquisition. These noise signals may come from inherent noise in electronic equipment such as industrial frequency interference, DC bias and baseline noise. Motion artifact which is mainly caused by electrode interface and electrode cable will also cause irregularities in sEMG data. Firing rate of the motor units and the firing frequency region 0–20 Hz also affect the sEMG signals. So the removal of the noise required. The power density spectra of the EMG contains most of its power in the frequency range of 5–500 Hz at the extremes, so the signal over the high cut off frequency 500 Hz should be eliminated. In the present work, a notch filter with 50 Hz and a band pass filter with low cut off frequency 500 Hz should be applied to the raw sEMG signals to remove the noise signal. The sEMG data is high pass filtered between 20 Hz to 450 Hz with a forth order recursive Butterworth filter (30 Hz) to remove the movement artifact. Then EMG is filtered again with Butterworth low-pass filter with a 6 Hz low-pass cut-off frequency.

5 Proposed Knee Model

Due to non-linear relationship of EMG and angle, it is very difficult to model such system. So ELM, GRNN and NN are used as model to establish EMG-angle relationship. Figure 2 show the steps required to prepare the data set for training the ELM model. Most of the blocks have been described in the preceding sections. The data have been rectified to avoid negative values of input vector. The filtered data have been resampled to reduce the dimension of the input vector of ELM. The input vector dimension has to match target vector dimension. So it is desired that input vector is reduced by resampling it with sampling rate that ensure a dimension that match target vector. The data is filtered with a 2nd order digital filter so that there is not too much variation in the data and it becomes relatively smoother. A recursive filter has been used which is a second order discrete linear mode to model muscle excitation from the rectified and the low-pass filtered EMG data. The filter used is as follows [7],

$$u_j(t) = \alpha e_j(t-d) - \beta_1 u_j(t-1) - \beta_2 u_j(t-2) \tag{1}$$

Where $e_j(t)$ the high-pass filtered, full wave is rectified and low-pas filtered EMG of muscle j at time t, $u_j(t)$ the post-processed EMG of muscle j at time t, α the gain coefficient for muscle j, β_1, β_2 the recursive coefficients for muscle j, d the electromechanical delay. A set of constraints are employed. Normalized data is quite an obvious issue to be addressed for ELM due to too much variation in the signal. Normalization improves the signal to noise ration. No feature has been taken of the data.

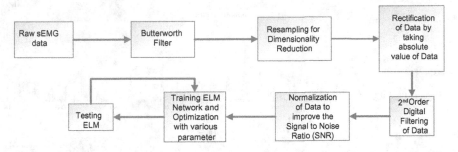

Fig. 2. The steps required to prepare the data set for training the ELM model

Next normalized filtered data is used to train ELM model. To optimize the perform-ance of ELM as joint angle estimator the node activation function, number of hidden layer neuron are changed to improve accuracy.

6 Data Analysis

For estimation of flexion and extension angle, The Generalized Regression Neural Network (GRNN) has been used with radial basis activation function in neurons, Feed forward Neural Network (NN) is used with 150 hidden layer neurons and 'tansig' as activation function in the neurons [8]. GRNN has Mean Square Error (MSE) of 14.9058, NN has MSE of 13.9623. ELM exhibits an MSE of 9.8746; ELM is trained with 300 hidden layer neurons and node activation function as 'Sig'. Now the ELM model is optimized with parameter tuning the testing of model with testing data set. ELM is relatively more consistent along the entire range of extension angles. The estimated angle decently follows the knee joint target angles and does not change so abruptly as it is very conspicuous in Figs. 4 and 5. The ELM regression model has been optimized with 200 hidden layer neurons and activation function 'Triangular Basis Function' at the hidden layer.

MSE	**ELM activation Function**				
	'Sig'	**'Sin'**	**'hardlim'**	**'tribas'**	**'radbas'**
	11.9248	*11.6478*	*10.3921*	*9.8746*	*11.08178*

MSE	**ELM Number of Hidden Nodes**				
	50	**100**	**150**	**200**	**300**
	12.3659	*11.9831*	*11.8889*	*11.8889*	*11.9303*

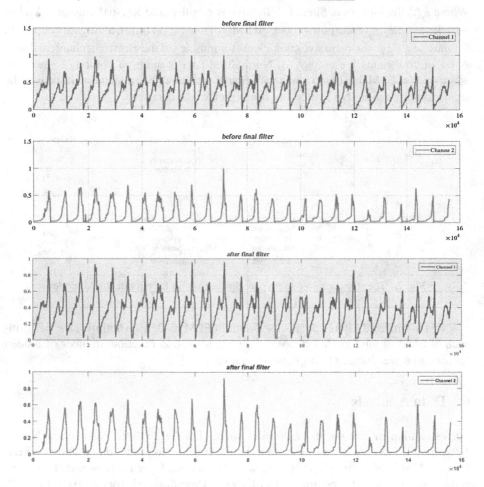

Fig. 3. EMG data before filtering and after filtering

The changes in number of hidden layers nodes do not make any significant changes in the system accuracy. But accuracy does improve with the change in node activation function. Figure 3 shows the Channel 1 and Channel 2 data before filtering and after filtering. There are 26 trials of extension data in each channel. The 26 available data has been divided into 5 equal test data sets and each one of them has been tested on the ELM Regression Model. The average Mean Square Error of five test sets has been shown in the table above. For training and testing both are done with ELM type as regression. Regression is required when we are trying to fit a function (activation function or kernel function) into our EMG data that best estimates knee joint angle.

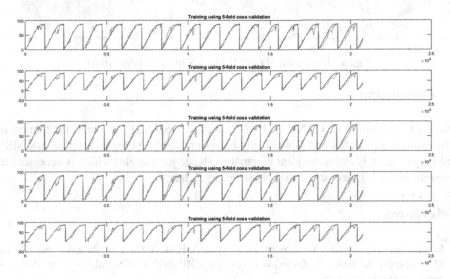

Fig. 4. Training the ELM with 5 different sets of training data.

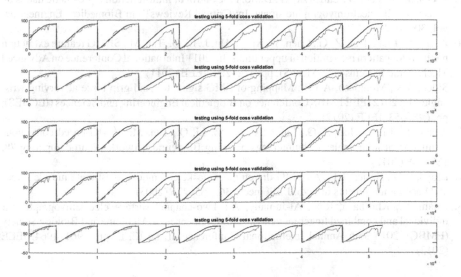

Fig. 5. Cross validation of ELM with 5 sets of testing data.

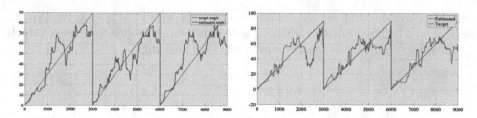

Fig. 6. Estimation of GRNN and NN of Knee Joint Angle

7 Conclusion

The angle estimation of GRNN and NN has high MSE and too much fluctuation in estimation make GRNN and NN not suitable estimator. ELM regression has low MSE and very efficiently estimates joint kinematics (Knee joint angle) which is necessary to generate actuator trajectory of the robotic exoskeleton.

References

1. Kawamoto, H., Sankai, Y.: Power assist system HAL-3 for gait disorder person. In: Miesenberger, K., Klaus, J., Zagler, W.L. (eds.) ICCHP 2002. LNCS, vol. 2398, pp. 196–203. Springer, Heidelberg (2002)
2. Oskoei, M.A., Hu, H.: Myoelectric control systems—A survey. Biomed. Signal Process. Control **2**, 275–294 (2007)
3. Zecca, M., Micera, S., Carrozza, M., Dario, P.: Control of multifunctional prosthetic hands by processing the electromyographic signal. In: Critical Reviews™ in Biomedical Engineering, vol. 30 (2002)
4. Zhang, F., Li, P., Hou, Z.-G., Chen, Y., Xu, F., Hu, J., Li, Q., Tan, M.: SEMG feature extraction methods for pattern recognition of upper limbs. In: 2011 International Conference on Advanced Mechatronic Systems (ICAMechS), pp. 222–227. IEEE (2011)
5. Sidek, S.N., Mohideen, A.J.H.: Mapping of EMG signal to hand grip force at varying wrist angles. In: 2012 IEEE EMBS Conference on Biomedical Engineering and Sciences (IECBES), pp. 648–653. IEEE (2012)
6. Zhang, F., Li, P., Hou, Z.-G., Lu, Z., Chen, Y., Li, Q., Tan, M.: sEMG-based continuous estimation of joint angles of human legs by using BP neural network. Neurocomputing **78**, 139–148 (2012)
7. Lloyd, D.G., Besier, T.F.: An EMG-driven musculoskeletal model to estimate muscle forces and knee joint moments in vivo. J. Biomech. **36**, 765–776 (2003)
8. Anam, K., Khushaba, R.N., Al-Jumaily, A.: Two-channel surface electromyography for individual and combined finger movements. In: Engineering in Medicine and Biology Society (EMBC), 2013 35th Annual International Conference of the IEEE, pp. 4961–4964. IEEE (2013)

Continuous User Authentication Using Machine Learning on Touch Dynamics

Ştefania Budulan[1,2]($\boxtimes$), Elena Burceanu[1], Traian Rebedea[2], and Costin Chiru[2]

[1] Bitdefender, Bucharest, Romania
{sbudulan,eburceanu}@bitdefender.com
[2] Faculty of Automatic Control and Computers, University Politehnica of Bucharest,
Bucharest, Romania
{traian.rebedea,costin.chiru}@cs.pub.ro

Abstract. In the context of constantly evolving carry-on technology and its increasing accessibility, namely smart-phones and tablets, a greater need for reliable authentication means comes into sight. The current study offers an alternative solution of uninterrupted testing towards verifying user legitimacy. A continuously collected dataset of 41 users' touch-screen inputs provides a good starting point into modeling each user's behavior and later differentiate among users. We introduce a system capable of processing features based on raw data extracted from user-screen interactions and attempting to assign each gesture to its originator. Achieving an accuracy of over 83 %, we prove that this type of authentication system is feasible and that it can be further integrated as a continuous way of disclosing intruders within given mobile applications.

1 Introduction

The need for technology increases every day, from using basic electronic devices, even up to efficiently organizing our daily activities. The mobile market embraces new consumers steadily, facilitating access to thousands of applications. The fact that mobile phones can successfully replicate most of the computers' features leads to the imperative upgrade towards a lower-risk security. In terms of rejecting access to illegitimate users, most smart-phone security is limited to setting a Personal Identification Number (PIN) (usually) or a graphical pattern for screen unlock. Additionally, each application may require a user-name and password in order to allow access, but many users choose to opt for permanently "signed in", increasing the risk of unauthorized access. Considering all that, an attacker, if having the means of entering someone else's phone (e.g. shoulder surfing or guessing a common password), he/she might gain permanent access to that phone or have at his/her disposal a sufficiently amount of time to retrieve the owner's personal information.

An authentication system which can perform seamless licit user acknowledgment without disrupting in any way the user's activity is needed. We propose a model that can discriminate among users based on data taken from user-screen interactions (e.g. x,y coordinates, time). Our project stands out from previous

S. Arik et al. (Eds.): ICONIP 2015, Part I, LNCS 9489, pp. 591–598, 2015.
DOI: 10.1007/978-3-319-26532-2_65

similar work in the field in terms of feature exploitation engaged in different types of classification and its results. Moreover, our aim is to achieve a high accuracy in the smallest amount of time possible while distinguishing gestures for a relatively large number of users (41 users, leading to a total of 912,133 records), using different types of classifiers. The results are conclusive enough to infer that this type of authentication is relevant, its error rate can be lowered and even if it does not perform at its best as a standalone authorization system, it represents a valid and suitable complementary security solution for mobile devices.

2   Previous Work

The research in this field has become more prominent in the past couple of years, but there are discussions related to this topic at least since 2011 (mouse dynamics [6], biometric gestures [4]). One of the former experiments involved a Sensor Glove [2] for recording data such as screen coordinates, finger direction, speed, etc., but became obsolete due to the usage difficulty. Also, it is uncertain if wearing the glove determined an unnatural behavior of the handler, entailing artificial dissimilarities among users. However, they proved that, when having recorded the features previously mentioned from 40 users, a system can be designed to lead to a False Acceptance Rate (FAR) of 4.66 % and a False Rejection Rate (FRR) of 0.13 % for user authentication (with Random Forest, Decision Tree and Bayes Net).

Nowadays, the Android API has built-in functions allowing an application to acquire basic data about touch-screen interactions. Touchalytics [3], a project gathering data through 2 types of applications - a document reader and an image comparison game, collected data from 41 participants across multiple sessions during a day, but also at one week distance. Their approach consists in applying two classifiers: support-vector machine (SVM) and k-nearest-neighbors (kNN) for each user, in a one-vs-rest classification, reaching a 0 % Equal Error Rate (EER) during the same authentication session, 2 %−3 % in-between sessions and almost 4 % on data recorded the following week. Using this dataset, we decided to tackle the problem differently: given all users' input, what is the best accuracy obtained with one-vs-one classifiers, instead of one-vs-rest, that way solving a much more difficult problem: dissociating the gestures of all users.

More recent studies, such as TIPS [1], came to the conclusion that the context of each application plays a critical role when discriminating among users' inputs. Therefore, they modified the Android kernel so that interactions were recorded from all applications, not just one. They used One Nearest Neighbor (1NN) and Dynamic Time Warping (DTW) in order to capture variances among users' touch screen data by saving several gesture templates per application and using them for comparing data over time and across users. Their system reached 90 % accuracy within what they reported as "real-life naturalistic conditions". However, in order to benefit from that kind of security system, the user needs to install a different version of Android OS and that constitutes not only a privacy issue, but can also lead to losing the device's warranty in most cases.

3 Proposed Approach

3.1 Dataset

The dataset we employed was gathered for extracting biometrics also for authentication[1]. We created several features using the given raw data for extracting our own sets of touch gestures and characteristics needed for best shaping each user's input. In Fig. 1, all data recorded for the first 8 users is plotted. The fact that even the human eye can differentiate among these is conspicuous.

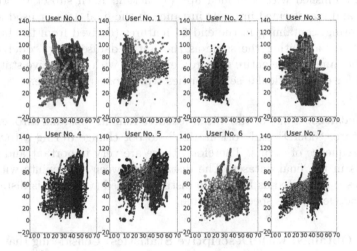

Fig. 1. Input data for the first 8 users. Every point marks a touch with its true coordinates on screen; pressure is represented by color" the higher the pressure, the darker the color; the dimension of bullets on screen is proportional to the area recorded for a particular touch.

For each user-screen interaction, the dataset contains: phone ID, User ID, document ID, action, phone orientation, Ox-coordinate, Oy-coordinate, pressure, area covered and finger orientation. The "Phone ID" column identifies the phone and the experimenter who conducted data recording. "Doc ID" is a number directly related to the ongoing session. There were 4 Wikipedia articles (IDs 0–2 and 5) and 3 comparison games (IDs 3, 4 and 6). The "Action" column may take one of the following values: 0 - touch down, 1 - touch up or 2 - displacement. Timestamp measures the absolute time (milliseconds since 1970).

This data was collected from 41 users operating on 5 Android phones; a distinct stroke is considered an input sequence starting with a touch-down action and ending with a touch-up action (e.g. records for clicks span from action 0 to action 1, without any coordinate displacement in-between). Since the participants received a similar device to their own and therefore the time allocated

---

[1] http://www.mariofrank.net/touchalytics/.

usually for getting familiar with the phone could be neglected, most of data was recorded during the same day. The rest of it was logged a week after the initial experiment took place, permitting an inter-week analysis.

3.2 Feature Processing

As previously shown, the way data is stored makes it easier to divide it into smaller subsets, depending on the current action. Every gesture starts with a "touch down" (0) action, followed by several "move on screen" (2) actions and usually finished with a "touch up" (1) action. Each subset (gesture) was processed and transformed into one instance in the final dataset used for training and testing, containing in the end 64 features (derived from the initial 10). Those features are either the same as the initial dataset (e.g. "docID" which remains the same for the entire gesture), extracted with descriptive statistics, or computed for each particular set of entries which forms a gesture.

Initial Features. Some of the initial features remain constant throughout the entire gesture. "docID" and "orientation" do not change during a touch-down touch-up sequence of events. Therefore, these are very important and are kept as features in the final dataset. The distance between two points with screen coordinates is computed as Euclidean distance, speed (velocity) is distance over time and acceleration is speed over time.

Features Obtained with Descriptive Statistics. Considering that multiple touch events constitute a given gesture for a certain user, it is very important to memorize the particularities of the gesture with respect to finger orientation, pressure, area covered, distance on screen, end-to-end distance. In terms of pressure and area, it was found to be very important to keep the mean, the minimum, the maximum, the mid one and the standard deviation of the set of pressures/areas for each gesture.

We extract (for each gesture) the mean, maximum, mid and standard deviation, having distance(s), speed(s) and acceleration(s) for each two consecutive points (that make up a segment). Screen coordinates proved to be very important in the final modeling of data. We stored the start point, the stop point, the minimum, the maximum, the mid, the mean and the standard deviation for both x and y coordinates. A user tends to be almost repetitively precise when touching the screen targeting a certain movement (e.g. click, scroll, swipe). Also, we compute the mean distance, speed and acceleration for the initial 3 and final 3 segments as they tend to capture additional information.

Computed Features. On the column "click" is stored 0/1 if the user had/hadn't moved his finger on the screen in between touching and releasing the finger of the screen.

We consider a trajectory as being a trail of points characterized by coordinates on screen, speed and acceleration, contained by a single gesture. Considering this, we determined as being particularly important to model a trajectory for each gesture of each user. After some research and several attempts, it resulted that the best way of doing this is by modeling a trajectory as a 3rd degree polynomial function of time, with 4 coefficients - matrices of shape $(2, 1)$.

A deviation of a certain (intermediate) point in relation to the line given by initial and final points of a trajectory can show the curvature of a complete gesture and it was computed as the distance from a point to a line (given by two points). Then we computed maximum, median, mean and standard deviation of all deviations of a gesture and selected them as features for the final set.

3.3 Models and Tools

After each step of processing we applied several models, usually including AdaBoostClassifier - an ensemble method implementing a boosting technique in which many classifiers with weaker performances are built into one stronger classifier, for instance over DecisionTreeClassifier. For each of the trained classifiers we used GridSearchCV - a tool found in Scikit-learn [5] for searching the best parameters for a specific estimator, applied over CrossValidation - a tool for computing a more accurate score than the one obtained for only one train-test cycle. It resides in splitting the data (train, validation) for a number of times (default 3) and applying the same estimator with the same parameters each time, obtaining an array containing the scores for each run. The final score was computed as a mean among those contained by the array.

Other models (from Scikit-learn [5] or XGBoost[2]) can be found in the final results, where all 64 features were present. Their descriptions are available usually in Scikit-learn, by following the links. Most of them are ensemble methods applied over much simple estimators such as Decision Tree.

4 Results

Before any further modification on raw data, we computed GridSeachCV over AdaBoostClassifier with DecisionTreesClassifier, in order to obtain an initial cross validation score for further references and comparisons. The running time was approximately 7 min and the accuracy for the untouched dataset, having only the relevant features (e.g. time stamp is not a relevant feature) was 35.90 %. It is not very small, but certainly not enough to discriminate against similar users. Here, even the same user would not manage to authenticate on his/her own phone.

The next steps were based on extracting new features, from every complete set of actions that define a gesture. Some features increased the accuracy very little on their own, but were very powerful combined with similar statistics for the

---

[2] https://github.com/dmlc/xgboost.

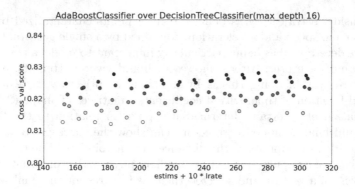

Fig. 2. AdaBoost Classifier over Decision Trees learning curve with respect to parameters for AdaBoost number of estimators between 150−300 and learning rates between 0.2−0.6. darker dots represent better results.

Table 1. Importance of features extracted after fitting ExtraTreesClassifier

Feature name	Feature importance (out of 100 %)
maxPressure	3.7839
meanPressure	3.5867
midPressure	3.3212
docID	2.8676
initPressure	2.8258
meanArea	2.7452
minX	2.5355
maxX	2.5074
startX	2.3952
stopX	2.3818

same initial feature. For instance, pressure was a strong feature from the beginning, but transformed only as the mean per gesture, it loses power. Therefore, we produced as final features along with the mean pressure, also the minimum, the maximum, the mid one and the standard deviation pressure. This ensures a score increase by almost 5 %.

AdaBoostClassifier was one of the best estimators on our data. Figure 2 shows learning curves for AdaBoost over DecisionTreeClassifier, in order to find best suited parameters which obtain the greatest score using Cross-Validation. The best score, achieved for a depth of 16, for the Decision Tree, and number of estimators equal to 270 and a learning rate of 0.4, for AdaBoost Classifier, was 82.83 %.

In Table 1 are showed the first 10 features, according to their importance when training the ExtraTrees classifier. Pressure alone represents over 13 % of the total feature importance and also some of the most relevant features were

Table 2. Accuracy and duration for several estimators

Estimator	Accuracy(%)	Running time (s)
XGBoost	83.60	774
AdaBoost over DecisionTree	82.73	367
AdaBoost over ExtraTrees	82.05	639
ExtraTrees	81.85	19
RandomForest	80.24	551
GradientBoosting	77.10	2201
Bagging over DecisionTree	77.04	327

given by user's coordinates on screen. Features with (almost) 0 % importance were removed from dataset.

After all features were extracted, the next steps include running different estimators, finding the best set of parameters for each of them, preferable using Grid Search, and comparing the results (depicted in Table 2).

5   Conclusions

The dataset provided by Touchalytics team was really appreciated and demanding into conceiving new and relevant features. The fact that it contained mainly raw data (e.g. pressure, timestamp) was both challenging and convenient. It is hard to imagine a set of features that are important to our case, but, once you have an idea, it is easier to compute them out of an essential given dataset.

The best score obtained of 83.60 % demonstrates that an authentication of users can be made, using solely their interactions with smart-phone screens. Even several minutes can be a little too long only for training, but the model can be trained previously and tested with a smaller set of gestures at a time. Therefore, it does not lose the feeling of real-time authentication. Even if we intend to retrain the model after continuously recording data, the faster model obtains an accuracy of 81.85 % in just a few seconds.

Considering the outcomes of the current state-of-the-art research, it is safe to assume that the user's touch-screen input can be used as a continuous form of authentication within the context of a specific application. Furthermore, the accuracy can be improved when combined with other types of authentication. Moreover, it will be a binary classification task, where the legitimate user will have a complete recorded gesture-profile and new input data has to be compared only against the owner's data, not against other 40 users, as in our case.

6   Future Work

There are several main improvements with respect to our work. The first is to implement our system on a phone, gather real-time data, and re-analyze its performance. In addition, users often give more feedback than the accuracy score itself, if it is a matter of security on their phones. There are still some unanswered questions, raised during the developing and also collecting results stages of our work. For instance, it is important to know if there are people with similar touch-patterns and they are the ones who always get mis-classified data. Also, it is important to know how hard it is for a human or agent to emulate other user's behavior when interacting with the screen.

Another direction would be to develop a model that analyses time-series data (e.g. use Recurrent Neural Networks: having a set of consecutive gestures, find an appraisal for the next one). This matters because, usually, the user's behavior changes in time and, at a certain point, he/she would not be able to authenticate, and the model will become purposeless. One other important betterment resides in training various Neural Networks on the data. Auto-encoders are suitable for obtaining new features, with a non-linear method, which can be furthermore trained and improve overall accuracy.

References

1. Feng, T., Yang, J., Yan, Z., Tapia, E.M., Shi, W.: Tips: context-aware implicit user identification using touch screen in uncontrolled environments (2014)
2. Feng, T., Liu, Z., Kwon, K.A., Shi, W., Carbunary, B., Jiangz, Y., Nguyen, N.: Continuous mobile authentication using touchscreen gestures (2012)
3. Frank, M., Biedert, R., Ma, E., Martinovic, I., Song, D.: Touchalytics: On the applicability of touchscreen input as a behavioral biometric for continuous authentication (2012)
4. Meng, Y., Duncan, S.W., Schlegel, R., Kwok, L.f.: Touch gestures based biometric authentication scheme for touchscreen mobile phones (2012)
5. Pedregosa, F., Varoquaux, G., Gramfort, A., Michel, V., Thirion, B., Grisel, O., Blondel, M., Prettenhofer, P., Weiss, R., Dubourg, V., Vanderplas, J., Passos, A., Cournapeau, D., Brucher, M., Perrot, M., Duchesnay, E.: Scikit-learn: machine learning in python. J. Mach. Learn. Res. **12**, 2825–2830 (2011)
6. Zheng, N., Paloski, A., Wang, H.: An efficient user verification system via mouse movements (2011)

Information Theoretical Analysis of Deep Learning Representations

Yasutaka Furusho, Takatomi Kubo, and Kazushi Ikeda[✉]

Nara Institute of Science and Technology, Ikoma, Nara 630-0192, Japan
{furusho.yasutaka.fm1,takatomi-k,kazushi}@is.naist.jp
http://mi.naist.jp/

Abstract. Although deep learning shows high performance in pattern recognition and machine learning, the reasons are little clarified. To tackle this problem, we calculated the information theoretical variables of representations in hidden layers and analyzed their relationship to the performance. We found that the entropy and the mutual information decrease in a different way as the layer gets deeper. This suggests that the information theoretical variables may become a criterion to determine the number of layers in deep learning.

Keywords: Entropy · Conditional entropy · Mutual information

1 Introduction

Deep learning, a multi-layered neural network, has been changing the history of pattern recognition and machine learning in performance [1] and is applied to computer vision, automatic speech recognition and translation, and so on [2,3]. However, the reasons of its high performance are little clarified since layered models have singular points that are difficult to treat statistically [4,5]. In addition, the performance has improved by combining several heuristics such as pre-training [1,6] and drop-out [7].

We tackled the problem of why deep learning works well by analyzing its representations in hidden layers from the information theoretical point of view. Let us consider the simplest case where the network has the input, hidden and output layers and the activation functions are linear (Fig. 1):

$$h_j = \sum_{i=1}^{I} w_{ji} x_i, \qquad\qquad j = 1,\ldots,J \qquad (1)$$

$$y_k = \sum_{j=1}^{J} v_{kj} h_j, \qquad\qquad k = 1,\ldots,K \qquad (2)$$

where I, J and K are the number of nodes in the input, hidden and output layers. Suppose that we set $I = K$ and train the network so that it becomes an encoder, that is, minimizes the squared errors, $\sum_i |x_i - y_i|^2$, for training samples. Then, it

© Springer International Publishing Switzerland 2015
S. Arik et al. (Eds.): ICONIP 2015, Part I, LNCS 9489, pp. 599–605, 2015.
DOI: 10.1007/978-3-319-26532-2_66

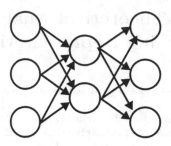

Fig. 1. Linear neural network.

is easily shown that the three-layered network works as the principle component analysis (PCA) and projects a vector $x = \{x_i\}$ onto the space spanned by the first and second principal components of training samples. Since the PCA is equivalent to the maximum entropy method, autoencoder may be trained to maximize the entropy of the data. In fact, the denoising autoencoder maximizes a lower bound of the cross entropy (or the mutual information) between a layer and the succeeding layer for each local connection [8].

Another key is the discrepancy between the classes. Since deep neural networks are mainly used as classifiers, the labels should be taken into account. The key is considered as the variance between the classes and the variance within the classes in the Fisher discriminant analysis (FDA), assuming that all the distributions are Gaussian. From the information theoretical point of view, the variance within the classes is the conditional entropy of data given labels to be minimized and the variance between the classes is the conditional entropy of labels given data to be maximized.

In this paper, to see how deep learning "encodes" input data in the hidden layers, we trained a deep neural network with major (pre)training methods using real data (MNIST database) and investigated the entropy and the conditional entropies in the hidden layers. As results, we found that the entropies of the representations of data and their labels remain high in early layers and decrease in higher layers and that the mutual information between the representations and the labels as well as the prediction performances decrease in earlier layers. This suggests that the information theoretical variables may become a criterion to determine the number of layers in deep learning.

2   Materials and Methods

2.1   MNIST Database

The MNIST database was used for the experiments [9]. Each image was decimated to one fourth of the original (196 pixels) by merging four pixels to one to reduce the complexity for brevity. 70,000 images in 10 categories are divided to three sets for training (50,000 images), validation for early-stopping (10,000 images) and test for evaluation (10,000 images).

2.2   Deep Neural Network

Our deep neural network has one input layer (196 nodes), six hidden layers (150, 120, 90, 60, 30, 10 nodes, respectively) and one output layer (10 nodes), corresponding to the number of pixels and that of categories (Fig. 2). The activation function of each node in the hidden layers was the sigmoid while the category was determined by winner-take-all in the output layer.

2.3   Learning Algorithms

Pretraining is an essence of deep learning. Among the restricted Boltzmann machine [1], the autoencoder [8] and their variants, we chose four representative methods below:

1. Stacked Autoencoder (SAE) [10]
2. Stacked Denoising Autoencoder (SDAE) [8]
3. Stacked Contractive Autoencoder (SCAE) [11]
4. Deep Belief Network (DBN) [1]

After pretraining with one of the algorithms above, the network was fine-tuned using the stochastic gradient method with early-stopping using the validation data.

2.4   Evaluation

The entropy and the conditional entropies,

$$H(h^i, t) = \mathrm{E}\left[-\log_2 P(h^i, t)\right], \tag{3}$$

$$H(h^i|t) = \mathrm{E}\left[-\log_2 P(h^i|t)\right], \tag{4}$$

$$H(t|h^i) = \mathrm{E}\left[-\log_2 P(t|h^i)\right], \tag{5}$$

were calculated for the deep neural network with each algorithm, where $h^i = \{h^i_j\}$ is the activation of the jth node in the ith layer and is binarized to 0 or 1 using the threshold 0.5. In our analysis below, the mutual information $I(h^i, t)$ was considered instead of $H(t|h^i)$ since they are essentially equivalent and the former seems easier to understand the meaning due to

$$I(h^i, t) = H(t) - H(t|h^i), \tag{6}$$

$$H(h^i) = H(h^i|t) + I(h^i, t). \tag{7}$$

The prediction error of the deep neural network as a classifier was also evaluated to see the relationship between the information theoretical variables and the performance. Here, the prediction error of each layer was calculated using a linear classifier that outputs the signature of a weighted sum of the hidden nodes.

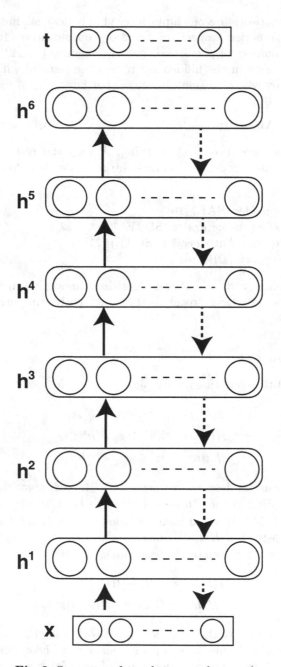

Fig. 2. Structure of our deep neural network.

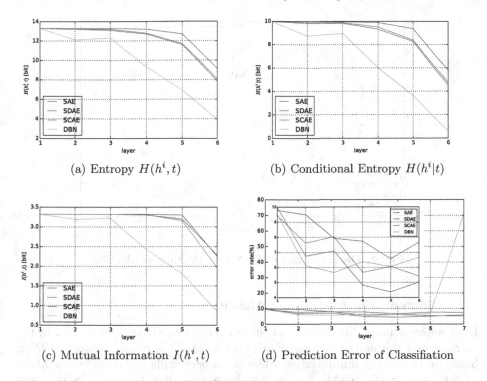

(a) Entropy $H(h^i, t)$

(b) Conditional Entropy $H(h^i|t)$

(c) Mutual Information $I(h^i, t)$

(d) Prediction Error of Classifiation

Fig. 3. Entropies of hidden nodes trained by the four methods and their performance.

3   Result

As the index of a hidden layer i increased, the entropy $H(h^i, t)$, the conditional entropy $H(h^i|t)$, and the mutual information $I(h^i, t)$ decreased for any method, without a small number of exception (Fig. 3a,b,c). However, the prediction error was not monotonically decreasing but took a minimum when $i = 3, 4$ or 5, depending on the pretraining algorithms (Fig. 3d).

4   Discussion

4.1   Entropy, Mutual Information and Performance

The conditional entropy $H(h^i|t)$ is a kind of variances within the classes and the mutual information $I(h^i, t)$ corresponds to the variance between the classes. As the FDA maximizes the ratio of the latter to the former as a separation criterion, we plotted the performance vs. the ratio $I(h^i, t)/H(h^i|t)$ and found that the curves have an L-shape (Fig. 4), indicating that the ratio may be a criterion to stop increase of layers.

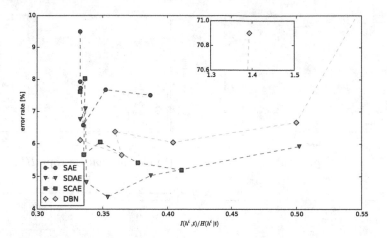

Fig. 4. The performance of DNN vs. the ratio of MI/CE.

4.2 Differences in Pretraining Algorithms

DBN shows a large difference from the others, SAE, SDAE and SCAE, in the entropy and the mutual information (Figs. 3, 4). This seems because DBN is based on the energy minimization while the others explicitly consider encoding.

Although SAE, SDAE and SCAE have a similar tendency, SDAE and SCAE perform better than SAE. This will be an effect of regularization in the algorithms [8, 11].

5   Conclusion

We analyzed the representations in hidden layers of deep neural networks from the information theoretical point of view. We calculated the entropy and the conditional entropies in each of the hidden layers when the deep neural network was trained with one of the four major pretraining algorithms and found that the entropies of the representations of data and their labels remain high in early layers and decrease in higher layers while the mutual information between the representations and the labels decrease in earlier layers. The ratio of the mutual information to the conditional entropy is related to the prediction performances, suggesting that the information theoretical variables may work as a criterion for model selection.

References

1. Hinton, G.E., Osindero, S., Teh, Y.: A fast learning algorithm for deep belief nets. Neural Comp. **12**, 531–545 (2006)
2. Seide, F., Li, G., Yu, D.: Conversational speech transcription using context-dependent deep neural networks. In: Proceedings Interspeech, pp. 437–440 (2011)

3. Zeiler, M.D., Fergus, R.: Visualizing and understanding convolutional networks. In: Fleet, D., Pajdla, T., Schiele, B., Tuytelaars, T. (eds.) ECCV 2014, Part I. LNCS, vol. 8689, pp. 818–833. Springer, Heidelberg (2014)
4. Watanabe, S.: Algebraic analysis for nonidentifiable learning machines. Neural Comp. **13**, 899–933 (2001)
5. Fukumizu, K., Akaho, S., Amari, S.: Critical lines in symmetry of mixture models and its application to component splitting. NIPS **15**, 857–864 (2003)
6. Erhan, D., Bengio, Y., Courville, A., Manzagol, P.A., Vincent, P., Bengio, S.: Why does unsupervised pre-training help deep learning? JMLR **11**, 625–660 (2010)
7. Hinton, G.E., Srivastava, N., Krizhevsky, A., Sutskever, I., Salakhutdinov, R.R.: Improving neural networks by preventing co-adaptation of feature detectors. arXiv:1207.0580 (2012)
8. Vincent, P., Larochella, H., Lajoie, I., Bengio, Y., Manzagol, P.-A.: Stacked demonising autoencoders: learning useful representations in a deep network with a local denoising criterion. JMLR **11**, 3371–3408 (2010)
9. LeCun, Y., Bottou, L., Bengio, Y., Haffner, P.: Gradient-based learning applied to document recognition. Proc. IEEE **86**, 2278–2324 (1998)
10. Bengio, Y., Lamblin, P., Popovici, D., Larochelle, H.: Greedy layer-wise training of deep networks. NIPS **19**, 153–160 (2007)
11. Rifai, S., Vincent, P., Muller, X., Glorot, X., Bengio, Y.: Contracting auto-encoders: explicit invariance during feature extraction. In: Proceedings ICML, pp. 833–840 (2011)

Hybedrized NSGA-II and MOEA/D with Harmony Search Algorithm to Solve Multi-objective Optimization Problems

Iyad Abu Doush$^{(\boxtimes)}$ and Mohammad Qasem Bataineh

Computer Science Department, Yarmouk University, Irbid, Jordan
iyad.doush@yu.edu.jo

Abstract. A multi-objective optimization problem is an area concerned an optimization problem involving more than one objective function to be optimized simultaneously. Several techniques have been proposed to solve Multi-Objective Optimization Problems. The two most famous algorithms are: NSGA-II and MOEA/D. Harmony Search is relatively a new heuristic evolutionary algorithm that has successfully proven to solve single objective optimization problems. In this paper, we hybridized two well-known multi-objective optimization evolutionary algorithms: NSGA-II and MOEA/D with Harmony Search. We studied the efficiency of the proposed novel algorithms to solve multi-objective optimization problems. To evaluate our work, we used well-known datasets: ZDT, DTLZ and CEC2009. We evaluate the algorithm performance using Inverted Generational Distance (IGD). The results showed that the proposed algorithms outperform in solving problems with multiple local fronts in terms of IGD as compared to the original ones (i.e., NSGA-II and MOEA/D).

Keywords: Multi-objective optimization problems · Harmony search algorithm · Multi-objective optimization evolutionary algorithms

1 Introduction

Multi-objective Optimization Problems (MOPs) is an area concerned with optimization problems that have two objective functions or more to be optimized simultaneously [4]. It can be found everywhere in nature, and we deal with such problems on a daily basis. From a person who tries to optimize a budget on a supermarket, trying to get better quality products for less amounts of money; industries trying to optimize their production, reducing their production costs and increasing their quality. The objectives in most of engineering problems are often conflicted. (i.e., maximize performance function, minimize cost function, maximize reliability function, etc.). In this case, one solution would not satisfy both objective functions and the optimal solution of one objective will not

Dr. Iyad Abu Doush, Department of Computer Sciences, Yarmouk University, Zip Code 21163, Irbid, Jordan. Phone: 00962-2-7211111 ext: 3858, Fax: 00962-2-7211128.

S. Arik et al. (Eds.): ICONIP 2015, Part I, LNCS 9489, pp. 606–614, 2015.
DOI: 10.1007/978-3-319-26532-2_67

necessary be the best solution for other objective(s). Therefore, different solutions will produce trade-offs between different objectives and a set of solutions is required to represent the optimal solutions of all objectives. One of the effective approaches that solve MOPs is Multi-Objective Evolutionary Algorithm.

EA was inspired by natural evolution, and relies on the concept of survival of the fittest. Evolutionary method imitating the evolution principle of nature to reach best solutions. In general, all evolutionary algorithms start with an initial population of random solutions, then the population is updated in each generation based on fitness function by using three operators: selection, reproduction and mutation. EA runs until reaches the wanted convergence. One of the important feature of evolutionary algorithm is that it have a population nature, which is very useful to solve MOPs. Since when we tend to solve MOPs, we seek to find a set of non-dominated solutions. EA has the ability to get a set of non-dominated solutions in each generation. EA characterized by less allergies to shape of PF and continuity, robustness and the ability to be implemented in parallel mode [1]. The previous reasons make EA suitable for solving MOPs.

Harmony Search (HS) algorithm, an Evolutionary Algorithm (EA), was proposed by Geem et al. [6]. It is inspired by musical improvisation process, where a group of musicians improvise the pitches of their musical instruments seeking for a fantastic melody on determined aesthetic estimation, practice after practice. Similarly, in optimization, iteration after iteration, HS seeks a good enough solution by evaluating a set of decision variables using fitness function. HS proved to be successful in many single objective optimization problems such as: multi-buyer multi-vendor supply chain problem, timetabling, and flow shop scheduling [7,9,12]. HS can be extended to solve MOPs in collaboration with Multi-Objective Evolutionary Algorithm (MOEA) frameworks, which have the ability to solve this kind of problems.

There are many frameworks for solving MOP. The following are the frameworks we used to develop our novel algorithms based on HSA. This work summarizes part of the results from the authors' master thesis [3]. **Non-domination Sorting Genetic Algorithm (NSGA-II)** [5] is based on the genetic operators, Pareto optimality and density estimation. **Multi-objective Evolutionary Algorithm Based on Decomposition (MOEAD)** [10,13] framework breaks a Multi-objective Optimization Problem (MOP) into several number of scalar optimization sub problems and starts to optimize them simultaneously by taking information from its neighbor sub problems.

2   Methodology

2.1   The Proposed Hybridized Frameworks

Harmony Search (HS) Algorithm: HS has been proved its success in many optimization problems such as multi-buyer multi-vendor supply chain problem, timetabling, flow shop scheduling [7,9,12], and others as recorded in [8]. HS starts with a set of solutions stored in Harmony Memory (HM). At each generation, a new harmony is generated as a new solution using three operators: (a) memory consideration, to select the variable values from HM; (b) random consideration, to maintain the diversity of the new solution, and (c) pitch adjustment to do local enhancements.

HS need a set of parameter to be tuned:

1. HMCR: The Harmony Memory Consideration Rate, to determine the number of decision variable that will be selected in New Harmony from HM.
2. HMS: The Harmony Memory Size, maximum number of solutions that will be stored in HM.
3. PAR: The Pitch Adjustment Rate, to determine the rate of decision variables will be changed to their neighboring values.
4. BW: The bandwidth, to determine the allowed amount of change in the pitch adjustment operator.
5. NI: The number of iterations.

Harmony Search has been used in large number of applications and problems in single objective optimization problems. HS can be extended to solve multi-objective optimization problems by using multi-objective evolutionary algorithm (MOEA) frameworks. We propose two hybrid multi-objective evolutionary algorithm frameworks based on Harmony Search. These modified frameworks are based on previous frameworks [5,10].

Harmony NSGA-II: Initially, Harmony NSGA-II generates an initial HM of size HMS randomly. Then non-dominated sorting procedure is used to sort the population based on Pareto optimality, best solution is given rank 1 and the second best rank 2 and so on. After that Harmony NSGA-II use Harmony operators to create a new Harmony from the original HM and insert it into a new empty population with size HMS (HM2), creation process continues until it is filled HM2. The two populations (HM and HM2) are combined to form a large population of size 2HMS. Again the sorting procedure runs to sort the two populations to several ranks rank 1,..., rank m. Figure 1 summarizes the steps of Harmony NSGA-II.

Harmony MOEA/D: MOEA/D was proposed by [10], this algorithm break a Multi-objective Optimization Problem (MOP) down into several number of scalar optimization sub-problems and starts to optimize them simultaneously by taking information from its neighbor sub-problems. To decompose a MOP into several number of scalar optimization sub-problems there are many number of approaches (e.g., weighted sum). MOEA/D uses the Tchebycheff approach to decompose a MOP.

Harmony MOEA/D need to tune a set of parameters: N: the number of sub-problem of MOP (represent population size), N weight vectors (are generated by uniform distribution) and T: the number of the weight vectors in the neighborhood of each weight vector (as desired). Then Harmony MOEA/D creates an empty External Population (EP) to be used as the set of non-domination solutions, next Harmony MOEA/D finds the indexes of closest weight vectors of size T formulating B sets. Next Harmony MOEA/D generate an initial population HM randomly. For all solutions in HM Harmony MOEA/D computes

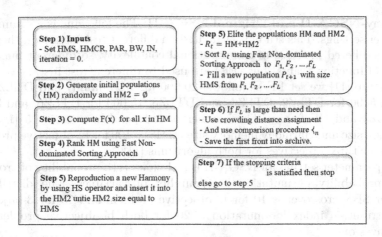

Step 1) Inputs
- Set HMS, HMCR, PAR, BW, IN, iteration = 0.

Step 2) Generate initial populations (HM) randomly and HM2 = ∅

Step 3) Compute F(x) for all x in HM

Step 4) Rank HM using Fast Non-dominated Sorting Approach

Step 5) Reproduction a new Harmony by using HS operator and insert it into the HM2 untie HM2 size equal to HMS

Step 5) Elite the populations HM and HM2
- R_t = HM+HM2
- Sort R_t using Fast Non-dominated Sorting Approach to $F_1, F_2, ..., F_L$
- Fill a new population P_{t+1} with size HMS from $F_1, F_2, ..., F_L$

Step 6) If F_L is large than need then
- Use crowding distance assignment
- And use comparison procedure $\langle_n$
- Save the first front into archive.

Step 7) If the stopping criteria is satisfied then stop else go to step 5

Fig. 1. Harmony NSGA-II steps.

F(x) which consist of a set of f objective functions with size m. Lastly, in initialization process Harmony MOEA/D assigns a target point in objective space Z from problem-specific method to guide the solutions towered Pareto front and in each iterations Z vector is modified. Figure 2 summarizes the steps of Harmony MOEA/D.

3 Experimental Results

In this section, we first present a numerical experiment studies for our four algorithms. In addition we present parameter settings.

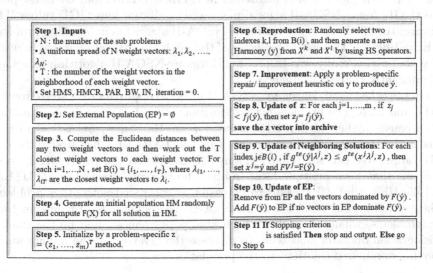

Step 1. Inputs
• N : the number of the sub problems
• A uniform spread of N weight vectors: $\lambda_1, \lambda_2, ...,$ λ_N;
• T : the number of the weight vectors in the neighborhood of each weight vector.
• Set HMS, HMCR, PAR, BW, IN, iteration = 0.

Step 2. Set External Population (EP) = ∅

Step 3. Compute the Euclidean distances between any two weight vectors and then work out the T closest weight vectors to each weight vector. For each i=1,....,N , set B(i) = {$i_1, ..., i_T$}, where $\lambda_{i1}, ...,$ λ_{iT} are the closest weight vectors to λ_i.

Step 4. Generate an initial population HM randomly and compute F(X) for all solution in HM.

Step 5. Initialize by a problem-specific z = $(z_1, ..., z_m)^T$ method.

Step 6. Reproduction: Randomly select two indexes k,l from B(i) , and then generate a new Harmony (y) from X^k and X^l by using HS operators.

Step 7. Improvement: Apply a problem-specific repair/ improvement heuristic on y to produce ŷ.

Step 8. Update of z: For each j=1,....,m , if z_j < $f_j(ŷ)$, then set z_j= $f_j(ŷ)$. **save the z vector into archive**

Step 9. Update of Neighboring Solutions: For each index j∈B(i) , if $g^{te}(ŷ|\lambda^j, z) \leq g^{te}(x^j\lambda^j, z)$, then set x^j=ŷ and FV^j=F(ŷ) .

Step 10. Update of EP:
Remove from EP all the vectors dominated by F(ŷ) . Add F(ŷ) to EP if no vectors in EP dominate F(ŷ) .

Step 11 If Stopping criterion is satisfied **Then** stop and output. **Else** go to Step 6

Fig. 2. Harmony MOEA/D steps.

Harmony-NSGA-II and Original NSGA-II: The following is a summary of parameter settings. This study is evaluated on different test instances, these test instances differed in terms of computational complexity, so different maximum number of function evaluations (FE) are used to different test instances. Based on literature [11] we set FE as follows: 20,000 for ZDT family and DTLZ1 and DTLZ7, 5000 for DTLZ2, DTLZ4, and DTLZ5, 15,000 for DTLZ6, and 50,000 for DTLZ3 and UF family. The number of independent runs is 15 [11]. The population size and Harmony memory size (HMS)is 100 for bi-objective problems and 200 for three objectives for both algorithms [11].

The parameter setting for NSGA-II is as follows [11]: probability of crossover is 0.9, probability of mutation is 1/number of variables. The distribution index for SBX crossover is 10 for bi-objective problem and 15 for 3-objectives one, distribution index for mutation is 20 for both bi-objective problem and 3-objectives one.

The harmony parameters are set from literature and empirically as follows: memory consideration rate (MCR)= 0.95 empirically, some of MCR values are tested like 0.90, 0.94, 0.98. The pitch adjustment rate (PAR)= 0.4 empirically, some of PAR values are tested like 0.1, 0.2, 0.5. The bandwidth (bw)= 0.01 [2,7].

We compare Harmony-NSGA-II and NSGA-II on the three aforementioned datasets (22 test instances). Table 1 shows the minimum, median, and maximum of the IGD-metric values of the 15 independent run. Harmony-NSGA-II outperformed in 13 problems out of 22 problems in total. Meantime in non-convex problem tests (ZDT2, ZDT6 and UF4) showed a rapprochement in IGD values in both algorithms. The nondominated solutions of the Harmony-NSGA-II algorithm showed slightly lower IGD values when compared to the NSGA-II algorithm in this type of problems. Test problems ZDT4, DTLZ1, DTLZ3, UF1 and UF3 have multiple local fronts. When the comparing Harmony-NSGA-II algorithm with the NSGA-II algorithm in DTLZ1, DTLZ3 and UF1 test problems the Harmony-NSGA-II algorithm showed significant lower IGD values. This indicates the efficiency of the Harmony-NSGA-II algorithm in handling problems with multiple local fronts. The success of the pitch adjustment operator in reaching global region really was helpful for Harmony-NSGA-II algorithm to achieve a significantly faster convergence.

On the other hand, Harmony-NSGA-II failed to approximate Pareto fronts of ZDT4 and UF3. The problem is that Harmony-NSGA-II stuck in local minima which causes higher median IGD values for the Harmony-NSGA-II algorithm. This is due to the unfortunate tuning of the PAR. UF3 has a slightly high IGD values. We can see that Harmony- NSGA-II has the preference to solve problems that have multiple local fronts. The test problems DTLZ2, DTLZ4, DTLZ5, UF2, UF4, UF7 seems to be easier as compared with other problems. Relatively, Harmony-NSGA-II and NSGA-II have similar IGD values in these test problems, what causes this comparable performance could be the random consideration operator that seeks to increase the diversity of the population by inserting dominated solutions to the population.

Table 1. The minimum, median (best is in bold), and maximum of the IGD-metric values for Harmony-NSGA-II, NSGA-II, Harmony-MOEAD and MOEAD.

Test problem	Type	HarmonyNSGA-II	NSGA-II	HarmonyMOEAD	MOEAD
ZDT1	Best	6.54e-04	7.21e-04	1.37e-03	2.55e-04
	Median	**8.03e-04**	2.52e-03	1.86e-03	**3.14e-04**
	Worst	1.14e-03	8.13e-03	2.18e-03	4.54e-04
ZDT2	Best	9.98e-04	8.20e-04	2.26e-03	1.82e-04
	Median	**1.12e-03**	2.50e-03	3.01e-03	**2.80e-04**
	Worst	1.32e-03	5.01e-03	4.01e-03	4.16e-04
ZDT3	Best	6.63e-04	5.18e-04	9.15e-04	3.13e-04
	Median	**5.01e-04**	2.66e-03	1.19e-03	**3.53e-04**
	Worst	7.67e-04	3.23e-03	1.76e-03	5.09e-04
ZDT4	Best	1.68e-02	7.09e-04	1.76e-04	6.75e-02
	Median	8.33e-02	**4.22e-03**	**1.64e-04**	2.08e-01
	Worst	2.19e-01	1.04e-02	4.23e-03	4.45e-01
ZDT6	Best	1.97e-04	6.71e-04	1.59e-04	1.39e-04
	Median	**2.11e-04**	7.73e-04	1.91e-04	**1.42e-04**
	Worst	2.41e-04	8.89e-04	2.42e-04	1.83e-04
DTLZ1	Best	7.40e-04	1.38e-02	4.69e-03	5.23e-02
	Median	**6.28e-03**	2.63e-02	**1.28e-02**	3.05e-02
	Worst	1.22e-02	5.01e-02	4.16e-02	1.07e-01
DTLZ2	Best	6.08e-04	6.49e-04	3.23e-03	2.73e-03
	Median	7.55e-04	**5.25e-04**	4.08e-03	**3.92e-03**
	Worst	1.04e-03	8.05e-04	4.71e-03	4.57e-03
DTLZ3	Best	1.75e-02	2.11e-01	8.62e-02	1.59e-01
	Median	**5.20e-02**	3.01e-01	**1.33e-01**	4.86e-01
	Worst	8.19e-02	5.31e-01	2.00e-01	1.22e+00
DTLZ4	Best	8.45e-04	8.14e-04	8.43e-03	8.78e-03
	Median	**9.24e-04**	9.64e-04	**9.76e-03**	1.09e-02
	Worst	7.59e-03	1.31e-03	1.06e-02	1.27e-02
DTLZ5	Best	8.65e-05	4.04e-05	1.34e-03	1.40e-03
	Median	1.54e-04	**4.83e-05**	**1.40e-03**	1.44e-03
	Worst	2.29e-04	9.23e-05	1.45e-03	1.45e-03
DTLZ6	Best	4.62e-04	2.06e-02	1.21e-02	6.40e-03
	Median	**5.83e-04**	2.20e-02	2.21e-02	**1.74e-02**
	Worst	1.02e-03	2.37e-02	3.35e-02	3.51e-02
DTLZ7	Best	2.84e-03	1.73e-03	2.89e-02	2.29e-02
	Median	3.32e-03	**1.90e-03**	3.00e-02	**2.71e-02**
	Worst	3.55e-03	1.99e-03	3.07e-02	2.97e-02

(Continued)

Table 1. *(Continued)*

Test problem	Type	HarmonyNSGA-II	NSGA-II	HarmonyMOEAD	MOEAD
UF1	Best	2.98e-03	4.04e-03	1.68e-03	1.68e-03
	Median	**3.69e-03**	6.86e-03	2.56e-03	**2.08e-03**
	Worst	4.64e-03	2.23e-02	3.79e-03	7.32e-03
UF2	Best	3.51e-03	1.04e-03	1.49e-03	9.16e-04
	Median	**3.45e-02**	3.45e-02	**1.75e-03**	1.88e-03
	Worst	3.86e-02	1.50e-02	2.68e-03	6.15e-03
UF3	Best	7.72e-03	4.07e-03	5.04e-03	4.78e-03
	Median	8.54e-03	**6.83e-03**	**6.69e-03**	8.37e-03
	Worst	1.00e-02	9.56e-03	8.90e-03	1.07e-02
UF4	Best	3.02e-03	3.41e-03	1.93e-03	1.93e-03
	Median	**3.29e-03**	9.26e-03	**2.07e-03**	2.15e-03
	Worst	3.46e-03	1.32e-02	2.23e-03	2.66e-03
UF5	Best	4.32e-02	4.61e-02	3.94e-02	9.31e-02
	Median	**4.57e-02**	8.14e-02	**4.88e-02**	1.28e-01
	Worst	4.78e-02	1.56e-01	6.29e-02	1.74e-01
UF6	Best	5.88e-03	6.21e-03	6.94e-03	3.36e-03
	Median	8.91e-03	**5.86e-03**	9.22e-03	**5.82e-03**
	Worst	1.24e-02	1.99e-02	1.46e-02	3.00e-02
UF7	Best	2.76e-03	9.40e-02	1.30e-03	7.14e-04
	Median	**1.13e-02**	1.33e-01	1.18e-02	**1.18e-03**
	Worst	1.70e-02	1.65e-02	1.64e-02	2.20e-02
UF8	Best	2.71e-03	1.97e-03	4.98e-03	5.14e-03
	Median	2.78e-03	2.78e-03	**5.38e-03**	6.42e-03
	Worst	2.89e-03	2.88e-03	5.72e-03	1.00e-02
UF9	Best	3.78e-03	1.30e-03	4.94e-03	5.17e-03
	Median	4.39e-03	**2.51e-03**	**6.04e-03**	6.18e-03
	Worst	4.63e-03	3.82e-03	6.78e-03	7.03e-03
UF10	Best	2.89e-03	2.89e-03	7.12e-04	1.06e-02
	Median	3.63e-03	**3.52e-03**	5.91e-03	9.67e-02
	Worst	4.78e-03	6.65e-03	9.67e-03	1.60e-01

Harmony-MOEAD and Original MOEAD: Here, we present a study for Harmony-MOEAD and original MOEAD. The parameters setting is as follows: MOEAD parameters: $T = 20$, delta $= 0.9$ and $nr = 2$ [10]. For all other parameters we used values similar to the previous experiment. A comparison between Harmony-MOEAD and MOEAD is presented in this section. Table 1 shows the minimum, median, and maximum of the IGD-metric values of the 15 independent run. Harmony-MOEAD outperformed in 11 problems out of 22 problems in total. The test problems DTLZ7 and DTLZ6 are formulated to be more complicated when compared to DTLZ5 by adjusting the g function. Harmony-MOEAD algorithm assigns a higher IGD value in DTLZ6 and DTLZ7 as compared to the MOEAD algorithm. The reason for this result is due to Harmony-MOEAD attempts to increase the diversity by inserting dominated solutions. However, Harmony-MOEAD and MOEAD have similar IGD median values in DTLZ6 and DTLZ7.

Test problems ZDT4, DTLZ1, DTLZ3, UF1 and UF3 have multiple local fronts. When comparing Harmony-MOEAD algorithm with the MOEAD algorithm in this type of problem, we found that Harmony-MOEAD algorithm showed lower median IGD values except UF1 test problem. This give an indication that Harmony-MOEAD algorithm has the efficiency in handling problems with multiple local fronts. The success of the pitch adjustment operator in reaching a relatively global region really was helpful for Harmony-MOEAD algorithm to achieve a faster Convergence.

Three objectives space problems UF8 UF9 and UF10 were approximated by Harmony-MOEAD and MOEAD algorithms. Harmony-MOEAD got lower mean IGD values as compare to MOEAD. Generally the Harmony-MOEAD outperformed on MOEAD in terms of median IGD values. MOEAD algorithm showed relatively lower median IGD values as compared to the Harmony-MOEAD algorithm in convex test problems ZDT1 and ZDT3, but in UF2 Harmony-MOEAD outperformed MOEAD. Meantime in non-convex problem tests ZDT2 and ZDT6, but in UF4 Harmony-MOEAD outperformed MOEAD. In both type of problems (convex and non-convex) Harmony-MOEAD failed to outperform on MOEAD, this due to unfortunate tuning parameters for this kind of problems. The test problems UF4, DTLZ2, UF7, DTLZ4, DTLZ5 and UF2 seems to be easier as compared with other problems. Relatively, Harmony-MOEAD and MOEAD have similar IGD values in these test problems, what causes this comparable performance could be the random consideration operator that seeks to increase the diversity of the population by inserting a random values to the solutions.

4   Conclusion

In this paper, new algorithms for multi-objective optimization field are developed. A Comprehensive study is conducted to study the performance of the proposed algorithms in terms of IGD. This research is done by using three problem families: ZDT, DTLZ and CEC2009 as datasets. Our work has investigated solving multi-objective optimization problem by using various parameter settings such as HMCR and PAR that are being used in enhancing solutions. Also, both the experimental results discussion and the compared results from both sides our proposed experiments and the state-of-the-art algorithms were clearly stated.

In general, all proposed algorithms which hybridized with Harmony Search are superior in solving multiple local front problems and 3-objectives problems in terms of IGD. Harmony NSGAII presented a better solutions in solving non-convex test problems as compare to the original algorithms. Harmony MOEAD algorithm showed a better IGD values as compared to the original ones when solving discontinuous problems. However, Harmony MOEAD algorithm suffered in solving ZDT family test instances except ZDT4 which has multiple fronts. The results shows that Harmony MOEAD algorithm did not give good results when solving ZDT family test instances, so we suggest to enhance these algorithms to be able to deal with this kind of problems as a future work.

References

1. Abraham, A., Jain, L.: Evolutionary Multiobjective Optimization. Advanced Information and Knowledge Processing. Springer, London (2005)
2. Al-Betar, M.A., Doush, I.A., Khader, A.T., Awadallah, M.A.: Novel selection schemes for harmony search. Appl. Math. Comput. **218**(10), 6095–6117 (2012)
3. Bataineh, M.Q.: Hybridizing evolutionary multi-objective optimization algorithms with the harmony search algorithm. Master's thesis, Yarmouk University, Irbid, Jordan, January 2015
4. Deb, K.: Multi-objective Optimization Using Evolutionary Algorithms. Wiley, Chichester (2001)
5. Deb, K., Pratap, A., Agarwal, S., Meyarivan, T.A.M.T.: A fast and elitist multi-objective genetic algorithm: NSGA-II. IEEE Trans. Evol. Comput. **6**(2), 182–197 (2002)
6. Geem, Z.W., Kim, J.H., Loganathan, G.V.: A new heuristic optimization algorithm: harmony search. Simulation **76**(2), 60–68 (2001)
7. Hasan, B.H.F., Doush, I.A., Maghayreh, E.A., Alkhateeb, F., Hamdan, M.: Hybridizing harmony search algorithm with different mutation operators for continuous problems. Appl. Math. Comput. **232**, 1166–1182 (2014)
8. Ingram, G., Zhang, T.: Overview of applications and developments in the harmony search algorithm. In: Geem, Z.W. (ed.) Music-Inspired Harmony Search Algorithm. Studies in Computational Intelligence, vol. 191, pp. 15–37. Springer, Heidelberg (2009)
9. Lee, K.S., Geem, Z.W.: A new meta-heuristic algorithm for continuous engineering optimization: harmony search theory and practice. Comput. Meth. Appl. Mech. Eng. **194**(36), 3902–3933 (2005)
10. Li, H., Zhang, Q.: Multiobjective optimization problems with complicated pareto sets, MOEA/D and NSGA-II. IEEE Trans. Evol. Comput. **13**(2), 284–302 (2009)
11. Sindhya, K., Miettinen, K., Deb, K.: A hybrid framework for evolutionary multi-objective optimization. IEEE Trans. Evol. Comput. **17**(4), 495–511 (2013)
12. Wang, L., Pan, Q.-K., Tasgetiren, M.F.: A hybrid harmony search algorithm for the blocking permutation flow shop scheduling problem. Comput. Ind. Eng. **61**(1), 76–83 (2011)
13. Zhang, Q., Li, H.: MOEA/D: a multiobjective evolutionary algorithm based on decomposition. IEEE Trans. Evol. Comput. **11**(6), 712–731 (2007)

A Complex Network-Based Anytime Data Stream Clustering Algorithm

Jean Paul Barddal$^{(\boxtimes)}$, Heitor Murilo Gomes, and Fabrício Enembreck

Graduate Program in Informatics (PPGIa),
Pontifícia Universidade Católica do Paraná,
R. Imaculada Conceição, 1155, Curitiba, Brazil
jean.barddal@ppgia.pucpr.br

Abstract. Data stream mining is an active area of research that poses challenging research problems. In the latter years, a variety of data stream clustering algorithms have been proposed to perform unsupervised learning using a two-step framework. Additionally, dealing with non-stationary, unbounded data streams requires the development of algorithms capable of performing fast and incremental clustering addressing time and memory limitations without jeopardizing clustering quality. In this paper we present CNDenStream, a one-step data stream clustering algorithm capable of finding non-hyper-spherical clusters which, in opposition to other data stream clustering algorithms, is able to maintain updated clusters after the arrival of each instance by using a complex network construction and evolution model based on homophily. Empirical studies show that CNDenStream is able to surpass other algorithms in clustering quality and requires a feasible amount of resources when compared to other algorithms presented in the literature.

1 Introduction

Data stream clustering can be described as the act of grouping streaming data in meaningful classes [4]. Data stream clustering is subject to acting within limited time, memory and treating data incrementally with single pass processing. Apart from the time and memory space constraints, two requirements are long-awaited for data stream clustering algorithms. Data stream clustering algorithms must not make assumptions about the number of clusters, since it is not often known in advance and due to the temporal aspect, the number of ground-truth clusters may change regularly [13] and they must be capable of discovering clusters with arbitrary shapes, since most of data streams are not Gaussian distributed.

In this paper we present an extension of the DenStream algorithm: CNDenStream. CNDenStream, in opposition to other data stream clustering algorithms, performs a single step processing to find clusters. CNDenStream does so by using a complex network construction and evolution model based on homophily. Moreover, CNDenStream is not bounded to the k-means algorithm [11] to find clusters, therefore it is able to discover clusters with arbitrary shapes and not only hyper-spherical clusters.

© Springer International Publishing Switzerland 2015
S. Arik et al. (Eds.): ICONIP 2015, Part I, LNCS 9489, pp. 615–622, 2015.
DOI: 10.1007/978-3-319-26532-2_68

The remainder of this work is organized as follows: Section 2 surveys related work for data stream clustering. Section 3 introduces basic concepts of complex networks. In Sect. 4 we present our proposal: CNDenStream. In Sect. 5 we present a performance study and discuss about parametrization sensitivity. Finally, Sect. 6 concludes this paper and presents future work.

2   Related Work

A variety of data stream clustering algorithms were developed throughout the last decade. Generally, these algorithms are divided in online and offline steps.

During the online step, algorithms incrementally update specific data structures aiming at dealing with the evolving nature of data streams and time-space constraints. One widely used data structure is the feature vector, a triplet $CF = \langle LS, SS, N \rangle$, where LS stands for the sum of the objects x_i summarized, SS is the squared sum of these objects and N is the amount of objects [13]. Feature vectors are able to represent hyper-spherical clusters incrementally due to its incremental and additive properties. Basically, an instance x_i can increment an feature vector CF_j as follows: $LS_j \leftarrow LS_j + x_i$, $SS_j \leftarrow SS_j + (x_i)^2$ and $N_j \leftarrow N_j + 1$. As for the additive property, two feature vectors CF_i and CF_j can be merged into a third CF_l as follows: $LS_l \leftarrow LS_i + LS_j$, $SS_l \leftarrow SS_i + SS_j$, and $N_l \leftarrow N_i + N_j$. Also, in order to assign more importance to recently retrieved instances in clustering, various window models featuring sliding, damped and landmark were developed [13]. During the offline step, conventional batch clustering algorithms are used to form final clusters using the CFs.

In the following sections we describe other data stream clustering algorithms.

2.1   CluStream

CluStream adopts the landmark windowing technique, treating the stream based on data chunks of size $\mathcal{H}$ [1]. CluStream assumes a number q of CFs that are maintained at any instant of the stream. Initial CFs are computed with an amount of instances $\mathcal{N}$, also determined by the user. CluStream computes an Euclidean distance for each instance x_i to each CF, then, determines whether the distance to the closest CF_j is less or equal to its radius. Positively, x_i is merged within CF_j. Conversely, x_i starts a new CF_k. If the amount of CFs is above q, the two closest CFs are merged. When $\mathcal{H}$ is reached, all q CFs are recomputed with the next N instances obtained from the stream.

On the offline step, CluStream uses a modification of the k-means or DBSCAN algorithms to obtain clusters based on the q CFs computed during the online step. In this paper, we compare the DBSCAN version, since k-means is highly dependent of the user-given parameter of ground-truth clusters K.

2.2   ClusTree

ClusTree [9] maintains CFs in a hierarchy with different granularity levels. Depending on how much time is available to process each instance, ClusTree performs a search in the R-Tree in order to find the most similar CF. Accordingly to user-given thresholds, it is determined whether this instance should or not be merged. In the negative case, a new CF is then created and added to the R-Tree. ClusTree also copes with noisy data by using outlier-buffers.

In order to assign more importance to recent data, ClusTree assigns a exponentially decaying weight for all CFs' components.

On the offline step, algorithms such as k-means and DBSCAN are used in order to obtain clusters, where CFs centers are treated as centroids.

2.3   DenStream

DenStream is based on the DBSCAN algorithm, which guarantees the union of the ϵ-neighborhood of clusters which covers all dense areas of the attribute space. A core object is an object which ϵ-neighborhood has at least ψ neighbors and a dense area is the union of all ϵ-neighborhoods of all core objects. DenStream defines the concept of a core-micro-cluster in a time instant t, which is a temporal extension to a CF, as $CMC(w, c, r)$ to a group of near instances $x_i, x_{i+1}, \ldots, x_n$ where w is its weight, c its center and r its radius. DenStream assumes two types of micro-clusters: potential and outlier micro-clusters. A micro-cluster is said potential or outlier based on its weight restriction, where $w \geq \beta\psi$ implies in a potential micro-cluster and outlier otherwise and $0 \leq \beta \leq 1$.

The online step of DenStream has the objective of maintaining a group of potential and outlier micro-clusters. At the arrival of each instance x_i, DenStream tries to aggregate x_i to the closest potential micro-cluster accordingly to the weight restrictions. In the negative case, the same occurs for outlier micro-clusters. If x_i was aggregated in an outlier micro-cluster, the weight restriction is checked to determine whether this micro-cluster should be promoted to a potential micro-cluster. Conversely, if x_i was not merged with any micro-cluster at all, it starts a new outlier micro-cluster.

The offline step of DenStream uses the DBSCAN algorithm to find clusters based on current potential micro-clusters.

3   Complex Networks

Complex Network Theory has been applied in many research fields, from computer science to sociology, mainly due to its formal description of structural variables. Although complex networks analysis are mixed with social network analysis for subjective topics, such as an individual behavior in society, both of its building blocks can be represented computationally as a graph.

Different complex network models were developed over the years, aiming at representing the evolutionary aspects of real networks [3,7,12,14]. The first complex network model is denominated random [7]. This model is based on the hypothesis that the existence of a connection between any pair of nodes is given by a probability p. The Small-world model, based on the studies of "Small-World" conducted in [12], incorporates attributes of both random and regular (lattices) networks and presents high clustering coefficient and a small average path length. Finally, the Scale–free model aims on modeling networks presented in real–world situations with higher accuracy than both random and small–world networks by performing nodes additions and edges rewirings throughout time. In this paper we adapt the rewiring component for the clustering task so they occur accordingly to homophily. Homophily is a characteristic of real networks where nodes tend to eradicate connections with dissimilar nodes and replace these by new connections with more similar ones.

4   CNDenStream

Complex Network-based DenStream (CNDenStream) is based on the hypothesis that intra-cluster data are related due to high similarity and inter-cluster data are not related, due to high dissimilarity. CNDenStream generates a complex network $G = (V, E, W)$, where the set of nodes V are micro-clusters, edges E represent connections between these nodes, W is a set of weights (Euclidian distances) w_i associated to each edge $e_i \in E$ where subgroups in this network represent clusters and an outlier micro-cluster buffer $\mathcal{B}$. In order to keep track of clusters during the stream without the need of batch processing during the offline step, CNDenStream uses an homophily-based insertion and rewiring procedures inspired in complex networks theory.

Initially, CNDenStream stores the first N instances retrieved from $\mathcal{S}$ in a buffer to an initial DBSCAN run, thus finding initial potential and outlier micro-clusters. While outlier micro-clusters are stored in an outlier buffer $\mathcal{B}$, potential micro-clusters PMC_i are added to the network G, where each potential micro-cluster establishes with the k closest possible neighbors (considering Euclidian distances) currently in V.

Afterwards, PMC_i is added to V and edges and corresponding weights are added to its correspondent sets E and W. Figure 1 presents the insertion of 4 potential micro-clusters, namely PMC_1 to PMC_4. The insertion procedure is able to connect the last added node to the k-most similar nodes in G, nevertheless, the same can not be said for the other nodes currently in G. In Fig. 1(d) one can see that after the addition of PMC_4, PMC_1 should be connected to PMC_4 instead of PMC_3, since $d(PMC_1, PMC_4) < d(PMC_1, PMC_3)$.

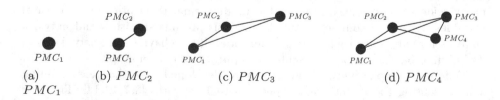

(a)
PMC_1 (b) PMC_2 (c) PMC_3 (d) PMC_4

Fig. 1. Insertion example using $k = 2$.

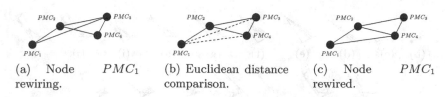

(a) Node PMC_1 rewiring.

(b) Euclidean distance comparison.

(c) Node PMC_1 rewired.

Fig. 2. Example of node PMC_1 rewiring.

After the addition of each PMC_i obtained from the DBSCAN initial run to the network, all nodes $PMC_i \in V$ perform rewirings based on homophily, such that each PMC_i replaces its edges with higher dissimilarities w by edges to its closest neighbors, i.e., edges with lower dissimilarity. For every PMC_i the Euclidean distances for all of its 2-hop neighbors are then computed. A 2-hop neighborhood is assumed since potential closest nodes are likely to be neighbors (2-hop) of the current neighbors (1-hop). This 2-hop neighborhood is an approximation in order to prevent distance computation between all nodes, which would be computationally costly. With the results of these Euclidean distances, PMC_i replaces edges by the most dissimilar instances with some similar ones, yet, maintaining its degree d_i.

In order to present how the rewiring procedure works, we refer back to the addition of nodes presented in Fig. 1, where one could see that PMC_1 is capable of connecting itself with higher similar nodes. Therefore, Fig. 2 presents the rewiring of node PMC_1. Firstly, Euclidean distances between PMC_1 and its 2-hop neighborhood are computed, and compared to its current neighbors (1-hop). In Fig. 2(b) one can see that $d(PMC_1, PMC_4) < d(PMC_1, PMC_3)$. Consequently, PMC_1, in order to maintain its degree $d_1 = 2$, must eliminate the its current most dissimilar edge to replace it with a similar one. Figure 2(c) the edge between PMC_1 and PMC_3 is removed from G and a new one connecting PMC_1 and PMC_4 and its corresponding weight $d(\cdot, \cdot)$ are added to E and W.

Due to the rewiring process, communities of potential micro-clusters tend to appear naturally since the amount of intra-clusters edges between similar micro-clusters grows, while those of dissimilar micro-clusters shrinks. Figure 3 presents the evolution of a network as instances arrive, where one can see that the rewiring procedure enlarges the amount of intra-cluster edges and diminishes the amount of inter-clusters connections. This procedure is repeated until, in Fig. 3(r), two clusters emerge.

After the DBSCAN execution and the initial network is build, all arriving instances x_i are processed according to an adaptation of the DenStream algorithm. Firstly, CNDenStream finds the potential micro-cluster in V which minimizes the dissimilarity with x_i: PMC_i. Afterwards, CNDenStream verifies whether the addition of x_i with PMC_i results in a micro-cluster with a radius below ϵ, if true, then x_i is added to PMC_i. Otherwise, this process is repeated within the outlier micro-clusters: the most similar outlier micro-cluster OMC_j to x_i is found and if the addition of x_i results in a micro-cluster with radius below ϵ, x_i is then added to OMC_j.

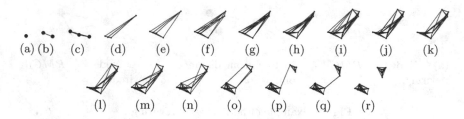

(a) (b)   (c)   (d)   (e)   (f)   (g)   (h)   (i)   (j)   (k)

(l)   (m)   (n)   (o)   (p)   (q)   (r)

Fig. 3. Insertion of potential micro-clusters obtained from the insertion of micro-clusters during the RBF$_2$ experiment.

When an outlier micro-cluster OMC_j is promoted to a potential micro-cluster, i.e., $w(OMC_j) \geq \beta\psi$, it is removed from the outlier buffer $\mathcal{B}$, and thus it is inserted in the network G.

As in DenStream, micro-clusters weights' decay exponentially with time. When the weight $w(PMC_i)$ of an micro-cluster PMC_i is below $\beta\psi$, it is removed from the network, or from $\mathcal{B}$. In the first case, all neighbors PMC_j of PMC_i are allowed to rewire in order to maintain their degree d_j after the PMC_i's removal. When the micro-cluster is an outlier, it is simply removed from $\mathcal{B}$.

CNDenStream's power resides in the rewiring process which aims to enable each micro-cluster to establish connections with the most similar micro-clusters. As presented in Fig. 1, the rewiring procedure finds clusters without using any batch clustering algorithm at the offline step such as k-means or DBSCAN.

One could argue about the effects of the parameter k on the construction and evolution of the network, therefore, in Sect. 5, we discuss about the parameter sensitivity and show that $k = 4$ is a good choice for many data streams domains.

5   Experimental Evaluation

Our proposal is evaluated in several experiments with different types of data domains. Synthetic data streams were generated using the Radial Basis Function (RBF) generator, which creates a user-given number of drifting centroids, each defined by a class label, position, weight and standard deviation accordingly to a Gaussian distribution. In our experiments, the RBF generator is used for modeling concept drifts every 500 instances. Three data streams using the RBF generator were created changing the dimension of the instances $d = \{2, 5, 10\}$.

Additionally, we evaluated algorithms in two massive datasets, namely Forest Covertype [8] and KDD'99 [2], where clusters are non-hyper-spherical.

Algorithms parameters were set accordingly to its original papers. CluStream parameters are: a horizon $\mathcal{H} = 1000$ and $q = 1000$ [1]. ClusTree parameters are: a horizon $\mathcal{H} = 1000$ and a maximum tree height $= 8$ [9]. DenStream parameters are: $\psi = 1$, $N = 1000$, $\lambda = 0.25$, $\epsilon = 0.02$, $\beta = 0.2$ and an offline step multiplier $\eta = 2$ [6]. Finally, CNDenStream parameters are: $\psi = 1$, $N = 1000$, $\lambda = 0.25$, $\epsilon = 0.02$ and $\beta = 0.2$ and $k = 4$. All experiments were performed on a Intel

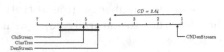

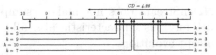

(a) CMM comparison with other algorithms.

(b) CMM comparison when varying the parameter k.

Fig. 4. Critical distances CMM comparison for results obtained in experiments.

Xeon CPU E5649 @ 2.53 GHz ×8 based computer running CentOS with 16 GB of memory at MOA framework [5].

5.1 Discussion

In order to evaluate algorithms in terms of clustering quality, we have adopted the Cluster Mapping Measure (CMM). CMM is an external clustering evaluation metric that accounts for non-associated and misassociated instances and noisy data inclusion [10]. Also, CMM considers recently retrieved instances with more weight than older ones by using an exponential decay function inside evaluation windows. In Fig. 4(a) we summarize the results obtained by algorithms after applying Friedman's and Nemenyi's tests, where one can see that CNDenStream is superior when compared to others with a 95 % confidence level.

Besides CMM, we evaluated both CPU Time and RAM-Hours, however, Friedman test pointed out that there is no significant difference between algorithms in these two dimensions.

5.2 Parameter Sensitivity

In opposition to pure density-based algorithms, CNDenStream relies on the amount of connections k established at the arrival of each instance parameter to find and keep track of clusters. Therefore, to determine whether different values of k affect results directly, we ran all experiments varying it in the $[1; 10]$ interval.

In Fig. 4(b) we summarize the results obtained by applying Friedman and Nemenyi's tests, where one can see that $k \in [2; 10] \succ k = 1$. Also, one can see that $k = 4$ presents the best averaged rank, therefore, this value is adopted as a default value for CNDenStream.

6 Conclusion

In this paper CNDenStream algorithm was presented. CNDenStream is a one-step incremental complex network-based data stream clustering algorithm. It was empirically evaluated in both real and synthetic datasets where one can see that it achieves significant superior CMM when compared to others algorithms, while demanding similar resources (CPU Time and RAM-Hours). Additionally,

CNDenStream does not make assumptions about the number of ground-truth clusters. This characteristic also allows the algorithm to naturally cope with concept evolutions.

In future works we expect to use archive programming techniques to optimize distance computation and develop a specific graph implementation to reduce memory usage. Besides, we envision experiments with other evaluation metrics, algorithms and datasets.

References

1. Aggarwal, C.C.: A framework for diagnosing changes in evolving data streams. In: Proceedings of the 2003 ACM SIGMOD International Conference on Management of Data, SIGMOD 2003, pp. 575–586. ACM, New York, NY, USA (2003)
2. Aggarwal, C.C., Han, J., Wang, J., Yu, P.S.: A framework for clustering evolving data streams. In: Proceedings of the 29th International Conference on Very Large Data Bases - Volume 29, VLDB 2003, pp. 81–92. VLDB Endowment (2003)
3. Albert, R., Barabási, A.L.: Statistical mechanics of complex networks. In: Reviews of Modern Physics, pp. 139–148. The American Physical Society, January 2002
4. Amini, A., Wah, T.Y.: On density-based data streams clustering algorithms: a survey. J. Comput. Sci. Technol. 29(1), 116–141 (2014)
5. Bifet, A., Holmes, G., Kirkby, R., Pfahringer, B.: MOA: massive online analysis. J. Mach. Learn. Res. 11, 1601–1604 (2010)
6. Cao, F., Ester, M., Qian, W., Zhou, A.: Density-based clustering over an evolving data stream with noise. In: SDM, pp. 328–339 (2006)
7. Erdos, P., Rényim, A.: On the evolution of random graphs. In: Publication of the Mathematical Institute of the Hungarian Academy of Sciences, pp. 17–61 (1960)
8. Kosina, P., Gama, J.: Very fast decision rules for multi-class problems. In: Proceedings of the 27th Annual ACM Symposium on Applied Computing, SAC 2012, pp. 795–800. ACM, New York, NY, USA (2012)
9. Kranen, P., Assent, I., Baldauf, C., Seidl, T.: The clustree: indexing micro-clusters for anytime stream mining. Knowl. Inf. Syst. 29(2), 249–272 (2011)
10. Kremer, H., Kranen, P., Jansen, T., Seidl, T., Bifet, A., Holmes, G., Pfahringer, B.: An effective evaluation measure for clustering on evolving data streams. In: Proceedings of the 17th ACM Conference on Knowledge Discovery and Data Mining (SIGKDD 2011), San Diego, CA, USA, pp. 868–876. ACM, New York, NY, USA (2011)
11. Lloyd, S.: Least squares quantization in pcm. IEEE Trans. Inf. Theor. 28(2), 129–137 (1982)
12. Milgram, S.: The small world problem. Psychol. Today 1(1), 61–67 (1967)
13. Silva, J.A., Faria, E.R., Barros, R.C., Hruschka, E.R., de Carvalho, A.C.P.L.F., Gama, J.: Data stream clustering: a survey. ACM Comput. Surv. 46(1), 13:1–13:31 (2013)
14. Watts, D.J., Strogatz, S.H.: Collective dynamics of small-world networks. Nature 393(6684), 440–442 (1998)

Robust Online Multi-object Tracking by Maximum a Posteriori Estimation with Sequential Trajectory Prior

Min Yang$^{(\boxtimes)}$, Mingtao Pei, Jiajun Shen, and Yunde Jia

Beijing Laboratory of Intelligent Information Technology, School of Computer Science, Beijing Institute of Technology, Beijing 100081, People's Republic of China
{yangminbit,peimt,shenjiajun,jiayunde}@bit.edu.cn

Abstract. This paper address the problem of online multi-object tracking by using the Maximum a Posteriori (MAP) framework. Given the observations up to the current frame, we estimate the optimal object trajectories by solving two MAP estimation problems: object detection and trajectory-detection association. By introducing the sequential trajectory prior, *i.e.*, the prior information from previous frames about "good" trajectories, into MAP estimation, the output of the pre-trained object detector is refined and the correctness of the association between trajectories and detections is enhanced. In addition, the sequential trajectory prior allows the two MAP stages interact with each other in a sequential manner, which facilitates online multi-object tracking. Our experiments on publicly available challenging datasets demonstrate that the proposed algorithm provides superior performance in various complex scenes.

Keywords: Online multi-object tracking · Data association · Maximum a posteriori estimation · Sequential trajectory prior

1 Introduction

Multi-object tracking is a very challenging problem, especially in complex scenes, due to frequent occlusions and interactions among similar-looking objects. Driven by the recent development of object detectors [1–3], tracking-by-detection has become a popular technique for multi-object tracking. With the detection responses provided by detectors, tracking-by-detection approaches associate these detections across frames to form the trajectories of objects.

Many tracking methods [4–6] address the association problem in a large temporal window, which seek for the optimum detection assignments by considering a batch of frames at a time. Due to the utilization of future information, they can handle detection errors and tracking failures caused by occlusions. However, it is difficult to apply the batch methods to time-critical applications, since they provide tracking results with a significant temporal delay.

Our work focuses on online multi-object tracking which only considers observations up to the current frame and sequentially builds trajectories via frame-by-frame association with online provided detections. Compared with the batch

© Springer International Publishing Switzerland 2015
S. Arik et al. (Eds.): ICONIP 2015, Part I, LNCS 9489, pp. 623–633, 2015.
DOI: 10.1007/978-3-319-26532-2_69

methods, online tracking systems [7–10] can be applied to real-time applications, but suffer from performance degradation in complex scenes. We aim to overcome the limitations for online multi-object tracking and to achieve high quality tracking results in complex scenes.

In this paper, we formulate the online multi-object tracking problem under a Beyesian framework, and treat detection and association as two collaborative maximum a posteriori (MAP) estimation problems by introducing the *sequential trajectory prior*. The basic idea is that the observations from previous frames contain useful prior information to assist the estimation of object trajectories in the current frame. Intuitively, it is better to allow the high-confidence trajectories to guide the current estimation of hard-to-see detections. And, for trajectory-detection association, more reliable detections are likely linked to high-confidence trajectories. We thus model such cues as the sequential trajectory prior, and use MAP estimation to simultaneously refine the detector output and enhance the trajectory-detection association correctness. We show that the two MAP stages interact with each other via the sequential trajectory prior: high-confidence trajectories from previous frame provide reliable prior information to refine the detections in the detection stage, and accurate detections facilitate the association stage to generate more confident trajectories. Our experiments demonstrate that the resulting algorithm provides superior tracking performance in various complex scenes.

Previous methods [10–12] exploit the prior information from previous frames for online multi-object tracking. Luo *et al.* [11] introduced a spatio-temporal consistency constraint to their online detector learning stage. Bae and Yoon [10] used trajectory confidence to assist their local and global association approach. Their work is extended in [12] by introducing a track existence probability into data association. However, these methods utilize the prior information only in the detection or association task. In contrast, we explicitly introduce the sequential trajectory prior into both the detection and association stages by using a unified MAP framework. As a result, the online multi-object tracking performance is significantly improved especially in complex scenes.

2   Our Approach

2.1   Problem Formulation

Let $\mathbb{X}_{1:t}$, $\mathbb{Y}_{1:t}$ and $\mathbb{Z}_{1:t}$ be the trajectories, detections and observed images up to frame t, respectively. We adopt a Bayesian approach to formulate the online multi-object tracking problem, where trajectories $\mathbb{X}_{1:t}$ and detections $\mathbb{Y}_{1:t}$ are random variables and the goal is to maximize the joint posterior distribution over $\mathbb{X}_{1:t}$ and $\mathbb{Y}_{1:t}$ given observed images $\mathbb{Z}_{1:t}$. Formally,

$$
\begin{aligned}
(\mathbb{X}_{1:t}^*, \mathbb{Y}_{1:t}^*) &= \underset{\mathbb{X}_{1:t}, \mathbb{Y}_{1:t}}{\arg\max} \, P\left(\mathbb{X}_{1:t}, \mathbb{Y}_{1:t} | \mathbb{Z}_{1:t}\right) \\
&= \underset{\mathbb{X}_{1:t}, \mathbb{Y}_{1:t}}{\arg\max} \, P\left(\mathbb{X}_{1:t} | \mathbb{Y}_{1:t}, \mathbb{Z}_{1:t}\right) P\left(\mathbb{Y}_{1:t} | \mathbb{Z}_{1:t}\right),
\end{aligned}
\tag{1}
$$

where the second equation used the definition of conditional probability. Since it is impossible to globally optimize Eq. (1) using brute force search, we expand the original formulation by sequentially estimating the current trajectories $\mathbb{X}_t$ and detections $\mathbb{Y}_t$ conditional on the previous results using the tracking-by-detection strategy. The problem is then decomposed into two MAP estimation stages:

$$(\text{detection}) \quad \mathbb{Y}_t^* = \arg\max_{\mathbb{Y}_t} P\left(\mathbb{Y}_t | \mathbb{Z}_{1:t}\right), \tag{2}$$

$$(\text{association}) \quad \mathbb{X}_t^* = \arg\max_{\mathbb{X}_t} P\left(\mathbb{X}_t | \mathbb{Y}_t^*, \mathbb{X}_{t-1}\right). \tag{3}$$

Specifically, in the detection stage, we obtain a MAP estimation of the detections $\mathbb{Y}_t^*$ by considering the observed images up to the current frame $\mathbb{Z}_{1:t}$. The trajectory estimation problem is then reformulated as a MAP estimation of pairwise associations between $\mathbb{X}_{t-1}$ and $\mathbb{Y}_t^*$ in the association stage.

2.2   Detection Refinement with MAP Estimation

Based on the Bayesian rule, the MAP estimation of the detections $\mathbb{Y}_t^*$ defined in Eq. (2) can be represented as

$$\mathbb{Y}_t^* = \arg\max_{\mathbb{Y}_t} \frac{P\left(\mathbb{Z}_t | \mathbb{Y}_t, \mathbb{Z}_{1:t-1}\right) P\left(\mathbb{Y}_t | \mathbb{Z}_{1:t-1}\right)}{P\left(\mathbb{Z}_t | \mathbb{Z}_{1:t-1}\right)}, \tag{4}$$

where $P\left(\mathbb{Z}_t | \mathbb{Y}_t, \mathbb{Z}_{1:t-1}\right)$ models the observation likelihood function which measures how well the hypothetical detections explain the observed image, and $P\left(\mathbb{Y}_t | \mathbb{Z}_{1:t-1}\right)$ is a prior detection probability which represents the prior information collected from the previous observations.

Prior Detection Probability. We approximately compute the prior detection probability based on the spatio-temporal consistency assumption during tracking. That is, the object states in two subsequent frames should not change drastically. Intuitively, the detections in frame t are much likely to appear around the trajectories from frame $(t-1)$. To utilize such prior, we predict the object states of high-confidence trajectories through Kalman filters, and use the predicted states to produce a density map to represent the prior detection probability. The trajectory confidence is defined by Eq. (8) in Sect. 2.3. Formally, we compute a density map D_t^k for a specific confident object k at frame t as

$$D_t^k(\mathbf{p}) = \exp(-\frac{\|\mathbf{p} - \mathbf{p}_k\|^2}{2\sigma_k^2}), \tag{5}$$

where $\mathbf{p}$ is the image position, $\mathbf{p}_k$ is the predicted position of object k, and σ_k is the scale parameter which is proportional to the scale of object k (set to 5 times the object scale in our implementation). Suppose that we have c confident objects from high-confidence trajectories in frame $(t-1)$, the density map D_t corresponding to $P\left(\mathbb{Y}_t | \mathbb{Z}_{1:t-1}\right)$ is generated by combining the density maps of all confident objects, expressed as $D_t = \max(D_t^0, D_t^1, \ldots, D_t^c)$. Note that D_t^0 is a const density map where the prior detection probability for each position is equal to 0.5, which is used to prevent the suppression of newly appeared objects.

Observation Likelihood Function. We revisit the detection confidence map produced by the pre-trained object detector to represent the observation likelihood function $P(\mathbb{Z}_t|\mathbb{Y}_t, \mathbb{Z}_{1:t-1})$. Following the general object detection strategy, we generate the hypothetical detections $\mathbb{Y}_t$ in multiple scales. Hence, $P(\mathbb{Z}_t|\mathbb{Y}_t, \mathbb{Z}_{1:t-1})$ is expressed as multiple confidence maps by applying the object detector to the observed image $\mathbb{Z}_t$ in multiple scales.

Posterior Detection Probability. Combining the observation likelihood function and the prior detection probability mentioned above, we can estimate the posterior detection probability as indicated in Eq. (4). Since the normalized term $P(\mathbb{Z}_t|\mathbb{Z}_{1:t-1})$ is constant, we simply use the density map D_t to refine the multiple confidence maps produced by the detector. Then the optimal detections $\mathbb{Y}_t^*$ is obtained by applying non-maximum suppression to the refined confidence maps. Most existing methods use the observation likelihood $P(\mathbb{Z}_t|\mathbb{Y}_t, \mathbb{Z}_{1:t-1})$ to approximate the posterior $P(\mathbb{Y}_t|\mathbb{Z}_{1:t})$, which actually ignores the useful prior information. In this paper, we employ the prior information from previous frames to model a prior detection probability $P(\mathbb{Y}_t|\mathbb{Z}_{1:t-1})$ which actually refines the detector output in a principle manner.

2.3   Data Association with MAP Estimation

Since the number of all possible enumerations of $\mathbb{X}_t$ given the existing trajectories $\mathbb{X}_{t-1}$ and the refined detections $\mathbb{Y}_t^*$ is huge, directly solving Eq. (3) is intractable. We turn to solve a data association problem and then obtain the optimal trajectories $\mathbb{X}_t^*$ by updating $\mathbb{X}_{t-1}$ with the associated detections.

Suppose that we have m trajectories $\mathbb{X}_{t-1} = \{X^i\}_{i=1}^m$ at frame $t-1$ and n refined detections $\mathbb{Y}_t^* = \{\mathbf{y}^j\}_{j=1}^n$ at frame t, where X^i is the trajectory of the i-th object and $\mathbf{y}^j$ is the j-th refined detection. Note that we drop the time index for simplicity since the association is exactly between $\mathbb{X}_{t-1}$ and $\mathbb{Y}_t^*$. We define an event $\Psi_{i,j}$ to represent that the j-th refined detection is associated with the i-th trajectory. Then, the pairwise association problem between $\mathbb{X}_{t-1}$ and $\mathbb{Y}_t^*$ can be expressed as a MAP estimation formulation,

$$\Psi_{i,j}^* = \arg\max_{\Psi_{i,j}} P(\Psi_{i,j}|\mathbb{Y}_t^*, \mathbb{X}_{t-1}), \tag{6}$$

where $P(\Psi_{i,j}|\mathbb{Y}_t^*, \mathbb{X}_{t-1})$ is the the posterior association probability. It can be computed by applying the Bayesian rule,

$$P(\Psi_{i,j}|\mathbb{Y}_t^*, \mathbb{X}_{t-1}) = \frac{P(\mathbb{Y}_t^*|\Psi_{i,j}, \mathbb{X}_{t-1}) P(\Psi_{i,j}|\mathbb{X}_{t-1})}{P(\mathbb{Y}_t^*|\mathbb{X}_{t-1})}, \tag{7}$$

where $P(\mathbb{Y}_t^*|\Psi_{i,j}, \mathbb{X}_{t-1})$ is the likelihood that indicates the possibility of observing the detections $\mathbb{Y}_t^*$ given the existing trajectories $\mathbb{X}_{t-1}$ and the association

$\Psi_{i,j}$, and $P\left(\Psi_{i,j}|\mathbb{X}_{t-1}\right)$ is the prior association probability that measures the possibility of the association $\Psi_{i,j}$ before data association.

Prior Association Probability. To compute the prior association probability $P\left(\Psi_{i,j}|\mathbb{X}_{t-1}\right)$, we exploit two kinds of prior information before performing data association: the trajectory confidence and the detection reliability.

Similar to [10], we use a trajectory confidence score function $\Delta(X^i)$ to measure the reliability of an existing trajectory X^i,

$$\Delta(X^i) = \exp\left(-\beta \cdot \frac{M}{L}\right) \times \left(\frac{1}{L}\sum_{k \in \Omega^i} \Phi_k^i\right), \tag{8}$$

where L is the number of frames in which the trajectory has associated detections, $M = |X^i| - L$ is the number of frames in which the object is missing, Ω^i indicates the set of frames in which the trajectory X^i has associated detections, Φ_k^i is the posterior association probability between X^i and the associated detection at frame k, and β is a control parameter depending on the detection performance. Since the trajectory confidence lies in $[0,1]$, we consider a trajectory as a high-confidence when $\Delta(X^i) > 0.5$.

The reliability of a detection $\mathbf{y}^j$ can be directly represented as the posterior defined in Sect. 2.2, simply denoted as $\delta(\mathbf{y}^j)$. Then the prior $P\left(\Psi_{i,j}|\mathbb{X}_{t-1}\right)$ can be intuitively approximated as

$$P\left(\Psi_{i,j}|\mathbb{X}_{t-1}\right) \approx \frac{\delta(\mathbf{y}^j)}{\sum_{v=1}^n \delta(\mathbf{y}^v)} \cdot \Delta(X^i), \tag{9}$$

where we impose the constraint that the association events for a trajectory X^i are mutually exclusive.

Observation Likelihood Function. We assume that the detections in $\mathbb{Y}_t^*$ are conditionally independent given the existing trajectories $\mathbb{X}_{t-1}$ and the association $\Psi_{i,j}$. Then the likelihood $P\left(\mathbb{Y}_t^*|\Psi_{i,j}, \mathbb{X}_{t-1}\right)$ can be computed as

$$P\left(\mathbb{Y}_t^*|\Psi_{i,j}, \mathbb{X}_{t-1}\right) = \prod_{v=1}^n P\left(\mathbf{y}^v|\Psi_{i,j}, \mathbb{X}_{t-1}\right). \tag{10}$$

Note that $P\left(\mathbf{y}^j|\Psi_{i,j}, \mathbb{X}_{t-1}\right) = P\left(\mathbf{y}^j|X^i\right)$ is the association likelihood between $\mathbf{y}^j$ and X^i. We compute the the association likelihood by using the appearance, shape, and motion cues, similar to [7]. The remaining task is to estimate the likelihood $P\left(\mathbf{y}^v|\Psi_{i,j}, \mathbb{X}_{t-1}\right)$ with $v \neq j$ which can be explained as the probability that the detection $\mathbf{y}^v$ is not originated from the trajectory X^i.

We consider two situations where the detection $\mathbf{y}^v$ can be observed: $\mathbf{y}^v$ is originated from other trajectories except X^i, or $\mathbf{y}^v$ is a false positive detection. Using the definition of marginal probability, the likelihood $P\left(\mathbf{y}^v|\Psi_{i,j}, \mathbb{X}_{t-1}\right)$ with $v \neq j$ can be computed by

$$P\left(\mathbf{y}^v|\Psi_{i,j},\mathbb{X}_{t-1}\right) = P\left(\mathbf{y}^v,\Psi_{0,v}|\Psi_{i,j},\mathbb{X}_{t-1}\right) + \sum_{u\neq i}P\left(\mathbf{y}^v,\Psi_{u,v}|\Psi_{i,j},\mathbb{X}_{t-1}\right)$$

$$= P\left(\mathbf{y}^v,\Psi_{0,v}|\mathbb{X}_{t-1}\right) + \sum_{u\neq i}P\left(\mathbf{y}^v,\Psi_{u,v}|\mathbb{X}_{t-1}\right), \tag{11}$$

where $\Psi_{0,v}$ means that the detection $\mathbf{y}^v$ is not associated with any trajectory. Denote $P_{u,v} = P\left(\Psi_{u,v}|\mathbb{X}_{t-1}\right)$ as the prior association probability defined in Eq. (9), and $\rho = P\left(\mathbf{y}^v|\Psi_{0,v},\mathbb{X}_{t-1}\right)$ as the const probability that a detection becomes false positive, we have

$$P\left(\mathbf{y}^v,\Psi_{0,v}|\mathbb{X}_{t-1}\right) = P\left(\mathbf{y}^v|\Psi_{0,v},\mathbb{X}_{t-1}\right)P\left(\Psi_{0,v}|\mathbb{X}_{t-1}\right) = \rho\cdot\prod_{u=1}^{m}\left(1-P_{u,v}\right), \tag{12}$$

$$P\left(\mathbf{y}^v,\Psi_{u,v}|\mathbb{X}_{t-1}\right) = P\left(\mathbf{y}^v|\Psi_{u,v},\mathbb{X}_{t-1}\right)P\left(\Psi_{u,v}|\mathbb{X}_{t-1}\right) = P\left(\mathbf{y}^v|X^u\right)P_{u,v}, \tag{13}$$

and thus

$$P\left(\mathbf{y}^v|\Psi_{i,j},\mathbb{X}_{t-1}\right) = \rho\cdot\prod_{u=1}^{m}\left(1-P_{u,v}\right) + \sum_{u\neq i}P\left(\mathbf{y}^v|X^u\right)P_{u,v}. \tag{14}$$

Then the observation likelihood function $P\left(\mathbb{Y}_t^*|\Psi_{i,j},\mathbb{X}_{t-1}\right)$ can be obtained by substituting Eq. (14) into Eq. (10),

$$P\left(\mathbb{Y}_t^*|\Psi_{i,j},\mathbb{X}_{t-1}\right) = P\left(\mathbf{y}^j|X^i\right)\prod_{v\neq j}\Theta_{i,j}^v, \tag{15}$$

where we denote $\Theta_{i,j}^v = P\left(\mathbf{y}^v|\Psi_{i,j},\mathbb{X}_{t-1}\right)$ with $v\neq j$ for simplicity.

Posterior Association Probability. Denote the normalization term in Eq. (7) as $\gamma = P\left(\mathbb{Y}_t^*|\mathbb{X}_{t-1}\right)$, we can derive the posterior as

$$P\left(\Psi_{i,j}|\mathbb{Y}_t^*,\mathbb{X}_{t-1}\right) = \gamma^{-1}P_{i,j}P\left(\mathbf{y}^j|X^i\right)\prod_{v\neq j}\Theta_{i,j}^v. \tag{16}$$

In a similar manner, the posterior association probability for the non association event $\Psi_{i,0}$ of the trajectory X^i can be acquired by

$$P(\Psi_{i,0}|\mathbb{Y}_t^*,\mathbb{X}_{t-1}) = \frac{P\left(\mathbb{Y}_t^*|\Psi_{i,0},\mathbb{X}_{t-1}\right)P\left(\Psi_{i,0}|\mathbb{X}_{t-1}\right)}{P\left(\mathbb{Y}_t^*|\mathbb{X}_{t-1}\right)}$$

$$= \gamma^{-1}\left(1-\sum_{j=1}^{n}P_{i,j}\right)\prod_{v=1}^{n}\Theta_{i,0}^v. \tag{17}$$

Using the fact that $\sum_{j=1}^{n}P\left(\Psi_{i,j}|\mathbb{Y}_t^*,\mathbb{X}_{t-1}\right) + P\left(\Psi_{i,0}|\mathbb{Y}_t^*,\mathbb{X}_{t-1}\right) = 1$, the normalization term γ can be computed as

$$\gamma = \sum_{j=1}^{n}\left(P_{i,j}P\left(\mathbf{y}^j|X^i\right)\prod_{v\neq j}\Theta_{i,j}^v\right) + \left(1-\sum_{j=1}^{n}P_{i,j}\right)\prod_{v=1}^{n}\Theta_{i,0}^v$$

$$= \left(1-\sum_{j=1}^{n}P_{i,j}+\sum_{j=1}^{n}Q_{i,j}\right)\prod_{v=1}^{n}\Theta_{i,0}^v, \tag{18}$$

where $Q_{i,j} = P_{i,j} P\left(\mathbf{y}^j | X^i\right) / \Theta_{i,0}^j$, and the second equation uses the fact $\Theta_{i,j}^v = \Theta_{i,0}^v$ when $v \neq j$.

Data Association. With the posterior probabilities given by Eqs. (16) and (17), the data association problem of Eq. (6) can be solved by the Hungarian algorithm [13]. Specifically, a association cost matrix $S = [s_{ij}]_{m \times n}$ is constructed with each entry $s_{ij} = -\log(P(\Psi_{i,j} | \mathbb{Y}_t^*, \mathbb{X}_{t-1}))$ to indicate the cost when j-th refined detection is associated with the i-th trajectory. Then the optimal trajectory-detection pairs are determined by minimizing the total cost in $S_{m \times n}$. When the association cost of a trajectory-detection pair is less than the cost of non association $-\log(P(\Psi_{i,0} | \mathbb{Y}_t^*, \mathbb{X}_{t-1}))$, the detection $\mathbf{y}^j$ is associated with X^i. A Kalman filter is used to refine the object states for a trajectory, with the associated detections as the measurement data. Then the confidence $\Delta(X^i)$ is updated using Eq. (8). The detections that are not associated with any existing trajectories are used to initialize a new potential trajectory. Once the length of a potential trajectory grows over a threshold (set to 5 frames in our implementation), it gets formally initialized.

3   Experiments

In this section, we give a detailed analysis of our approach compared to the state-of-the-art in multi-object tracking. The state-of-the-art trackers include DP [4], TBD [6], CEM [5] and CMOT [10], in which the CMOT tracker is online algorithms while the other trackers perform multi-object tracking in a batch mode. We report the results by using the source codes publicly provided by the authors with the same object detector and their default parameters.

3.1   Implementation Details

Our online multi-object tracking algorithm is implemented in MATLAB, and operates entirely in the image coordinate without camera or ground plane calibration. Without code optimization and parallel programming, our algorithm runs at about 10 fps on an Intel Core i7 3.5 GHz PC with 16 GB memory. The system parameters that need to be set beforehand include the control factor β in Eq. (8), and the const probability ρ in Eq. (14). In our implementation, we empirically set $\beta = 2$ and $\rho = 0.1$ for all experiments.

3.2   Datasets and Object Detector

We use the following datasets for performance evaluation: *PETS2009* dataset [14], *TUD* dataset [15], and *ETH Mobile Scene (ETHMS)* [16]. The *PETS2009* dataset shows an out door survivance scene where large amount of pedestrians enter and exit the filed-of-view. We adopt the widely used *S2L1* and *S2L2* sequences for evaluation. In the *TUD* dataset, the sequences *Campus*, *Crossing* and *Stadtmitte* are used, where the challenges include severe occlusions between

Table 1. Quantitative comparison results. Batch methods are marked with an asterisk. Bold scores highlight the best results.

Method	MOTA↑	MOTP↑	FP↓	FN↓	MT↑	ML↓	IDS↓	FG↓
⋆DP [4]	31.1%	**71.6%**	3,695	11,890	19.6%	33.2%	3,177	1,277
⋆CEM [5]	39.7%	70.7%	4,656	11,411	24.5%	34.0%	349	**640**
⋆TBD [6]	35.4%	71.4%	6,267	9,995	27.5%	**31.3%**	1,329	1,025
CMOT [10]	21.7%	69.9%	7,912	11,354	20.1%	33.4%	1,998	1,139
Ours (w/o all)	27.3%	70.5%	4,855	13,293	21.5%	41.3%	679	990
Ours (w/o MAP assoc.)	43.5%	71.2%	3,982	10,764	24.7%	37.0%	634	931
Ours (w/o MAP det.)	40.8%	70.9%	4,292	11,521	28.1%	34.0%	312	790
Ours (with all)	**49.0%**	71.2%	**3,603**	**9,942**	**31.0%**	32.8%	**235**	754

objects and low viewpoint. In the *ETHMS* dataset, we evaluate our algorithm on the sequences *Bahnhof*, *Jelmoli* and *SunnyDay*, which are taken by a moving camera in crowded street scenes. In total, the test datasets contain over 3500 frames and 368 annotated trajectories (27240 bounding boxes). For fair comparison, we use the ground truth publicly provided by Milan *et al.* [5].

To efficiently acquire online detections, we use the aggregate channel features object detector [3] which can be operated in almost real time. The detector is trained on the INRIA dataset [1] with default parameters.

3.3 Evaluation Metrics

We use the widely accepted CLEAR performance metrics [17] for quantitative evaluation: the multiple object tracking precision (MOTP↑) that evaluates average overlap rate between true positive tracking results and the ground truth, and the multiple object tracking accuracy (MOTA↑) which indicates the accuracy composed of false positives (FP↓), false negatives (FN↓) and identity switches (IDS↓). Additionally, we report measures defined by Li *et al.* [18], including the percentage of mostly tracked (MT↑) and mostly lost (ML↓) ground truth trajectories, as well as the number of times that a ground truth trajectory is interrupted (FG↓). Here, ↑ means that higher scores indicate better results, and ↓ represents that lower is better.

3.4 Results and Discussion

Quantitative results of our algorithm compared with the state-of-the-art tracking methods on the datasets are listed in Table 1, and sample results are shown in Fig. 1. Overall, our algorithm outperforms the competing online tracker CMOT, and achieves competitive results compared to the state-of-the-art batch methods (*i.e.*, DP, TBD and CEM). It owes to the proposed two collaborative MAP estimation stages which simultaneously incorporate the sequential trajectory prior into both the detection and association procedures during tracking. As can be observed from the quantitative evaluation results, our algorithm achieves far

Fig. 1. Sample tracking results of our method on three representative test video sequences (*PETS2009-S2L2, TUD-Stadtmitte* and *ETHMS-Jelmoli*). At each frame, objects with different IDs are indicated by bounding boxes with different colors.

superior performance in terms of MOTA, FP and FN, which indicates that the detection refinement stage integrating with the sequential trajectory prior significantly facilitates the tracking process. In addition, we achieve excellent results in terms of MT, ML, IDS and FG, demonstrating that the combination of association likelihood and sequential prior benefits the correct association between trajectories and detections. As shown in the qualitative examples of tracking results in Fig. 1, our method is able to accurately track the target persons under various challenging conditions.

To demonstrate the effectiveness of the proposed two MAP estimation stages with the sequential trajectory prior, we build three baseline algorithms to do validation and analyze various aspects of our approach. The comparison results between our approach and three baseline algorithms are also listed in Table 1, where removal of the MAP estimation stage means removal the prior and only using the likelihood as most tracking methods do. As can be seen from the comparison results, the baseline algorithm without both of the two MAP estimation stages shows severe performance degradation. Using sequential trajectory prior to refine the detections results in significant improvement on MOTA and FN, which validates that the sequential trajectory prior indeed assists the detector to recall more accuracy detections. In addition, incorporating sequential trajectory prior to trajectory-detection association apparently improves the accuracy in terms of MT, ML, IDS and FG, which demonstrated that the association correctness is improved by using the MAP estimation of the posterior association probability. The proposed algorithm considers the sequential trajectory prior in both the detection and association stages, and thus shows the best performance.

4   Conclusion

We have proposed an online multi-object tracking-by-detection algorithm by using the Maximum a Posteriori (MAP) framework. To account for noisy detections and improve trajectory-detection association correctness, we exploit the prior information contained in previous frames, such as the positions of objects that most likely to appear, the adaptive confidences of trajectories and the detection reliability, to guide the detection and association stages in the current frame. By using these sequential trajectory priors in MAP, the tracker is able to recall more reliable detections and alleviate the ambiguity of trajectory-detection association, and achieves great improvement on tracking performance.

References

1. Dalal, N., Triggs, B.: Histograms of oriented gradients for human detection. In: CVPR, pp. 886–893 (2005)
2. Felzenszwalb, P.F., Girshick, R.B., McAllester, D., Ramanan, D.: Object detection with discriminatively trained part-based models. IEEE TPAMI **32**(9), 1627–1645 (2010)
3. Dollár, P., Appel, R., Belongie, S., Perona, P.: Fast feature pyramids for object detection. IEEE TPAMI **36**(8), 1532–1545 (2014)
4. Pirsiavash, H., Ramanan, D., Fowlkes, C.C.: Globally-optimal greedy algorithms for tracking a variable number of objects. In: CVPR, pp. 1201–1208 (2011)
5. Milan, A., Roth, S., Schindler, K.: Continuous energy minimization for multitarget tracking. IEEE TPAMI **36**(1), 58–72 (2014)
6. Geiger, A., Lauer, M., Wojek, C., Stiller, C., Urtasun, R.: 3D traffic scene understanding from movable platforms. IEEE TPAMI **36**(5), 1012–1025 (2014)
7. Wu, B., Nevatia, R.: Detection and tracking of multiple, partially occluded humans by bayesian combination of edgelet based part detectors. IJCV **75**(2), 247–266 (2007)
8. Breitenstein, M.D., Reichlin, F., Leibe, B., Koller-Meier, E., Van Gool, L.: Online multiperson tracking-by-detection from a single, uncalibrated camera. IEEE TPAMI **33**(9), 1820–1833 (2011)
9. Shu, G., Dehghan, A., Oreifej, O., Hand, E., Shah, M.: Part-based multiple-person tracking with partial occlusion handling. In: CVPR, pp. 1815–1821 (2012)
10. Bae, S.H., Yoon, K.J.: Robust online multi-object tracking based on tracklet confidence and online discriminative appearance learning. In: CVPR, pp. 1218–1225 (2014)
11. Luo, W., Kim, T.K., Stenger, B., Zhao, X., Cipolla, R.: Bi-label propagation for generic multiple object tracking. In: CVPR, pp. 1290–1297 (2014)
12. Bae, S.H., Yoon, K.J.: Robust online multi-object tracking with data association and track management. IEEE TIP **23**(7), 2820–2833 (2014)
13. Kuhn, H.W.: The Hungarian method for the assignment problem. Nav. Res. Logistics Q. **2**(1–2), 83–97 (1995)
14. Ellis, A., Shahrokni, A., Ferryman, J.M.: PETS2009 and Winter-PETS 2009 results: a combined evaluation. In: IEEE International Workshop on Performance Evaluation of Tracking and Surveillance (PETS-Winter), pp. 1–8 (2009)
15. Andriluka, M., Roth, S., Schiele, B.: People-tracking-by-detection and people-detection-by-tracking. In: CVPR, pp. 1–8 (2008)

16. Ess, A., Leibe, B., Schindler, K., Van Gool, L.: A mobile vision system for robust multi-person tracking. In: CVPR, pp. 1–8 (2008)
17. Keni, B., Rainer, S.: Evaluating multiple object tracking performance: the CLEAR MOT metrics. EURASIP J. Image Video Process. (2008)
18. Li, Y., Huang, C., Nevatia, R.: Learning to associate: hybridboosted multi-target tracker for crowded scene. In: CVPR, pp. 2953–2960 (2009)

Enhance Differential Evolution Algorithm Based on Novel Mutation Strategy and Parameter Control Method

Laizhong Cui[✉], Genghui Li, Li Li, Qiuzhen Lin,
Jianyong Chen, and Nan Lu

College of Computer Science and Software Engineering, Shenzhen University,
Shenzhen 518060, Guangdong, China
{cuilz,qiuzhlin,jychen,lunan}@szu.edu.cn,
{ligenghuigm,chariotlily}@gmail.com

Abstract. Differential evolution (DE) algorithm is a very effective and efficient approach for solving global numerical optimization problems. However, DE still suffers from some limitations. Moreover, the performance of DE is sensitive to its mutation strategy and associated parameters. In this paper, an enhanced differential evolution algorithm called EDE is proposed, which including a new mutation strategy and a new control method of parameters. Compared with other DE algorithms including four classical DE and two state-of-the-art DE variants on ten numerical benchmarks, the experiment results indicate that the performance of EDE is better than those of the other algorithms.

Keywords: Differential evolution · Mutation strategy · Parameter control method · Exploration and exploitation

1 Introduction

Differential evolution algorithm (DE) is a simple yet effective heuristic algorithm firstly proposed by Storn and Price [1] for dealing with global optimization over continuous space. Due to its outstanding characteristics, such as compact structure, ease to use, speediness and robustness, it has become more and more popular and been employed to handle many optimization problem in real-world applications [2,3]. However, like other evolutionary algorithms, DE does not ensure to find the global optimum, especially for the complicated multimodal functions. Many approaches have been proposed to improve the optimal performance of DE, which can be mainly divided into three categories:

1. New mutation strategies. There have been some classical DE mutation strategies, such as DE/rand/1, DE/rand/2, DE/best/1, DE/best/2, DE/current-to-rand/1. Different DE mutation strategies have distinct characteristics, that means they may bring different effects when solving various global optimization problems. For further improving the performance of DE, many new DE mutation strategies are put forward, such as [4,5].

© Springer International Publishing Switzerland 2015
S. Arik et al. (Eds.): ICONIP 2015, Part I, LNCS 9489, pp. 634–643, 2015.
DOI: 10.1007/978-3-319-26532-2_70

2. New parameter control methods. The classical DE algorithm has three control parameters, which significantly affect the performance of DE. Therefore, many different parameter control methods are proposed to enhance the performance of DE, such as [4–8].

3. New universal DE operators. In recent years, more and more universal operators have been proposed, which can be applied to different DE algorithms for effectively improving the performance of DE, such as [9,10].

This paper pays attention to the first and the second category for improving the performance of differential evolution algorithm. An enhanced DE called EDE with a new mutation strategy and a new parameter control method is proposed, which can effectively enhance the performance of the classical differential evolution algorithm.

The remainder of this paper is organized as follows. In Sect. 2, we introduce the classical DE algorithm briefly. The details of our proposed algorithm are described in Sect. 3, which includes the new mutation strategy and the new parameter adjustment method. The experiment and evaluation of our proposed algorithm are presented in Sect. 4. Finally, Sect. 5 concludes this paper.

2 Classical Differential Evolution Algorithm

Differential evolution algorithm (DE) is a branch of Evolution Algorithm (EA) that also follows the general procedures of EA. More specifically, three are three basic operators of DE, including mutation, crossover and selection operator, and the algorithm flowchart of DE is illustrated in Fig. 1.

As described in Fig. 1, after the initialization process, DE turns into the loop including the process of mutation, crossover, and selection, until the termination condition is satisfied. The processes of these operators are described as follows.

2.1 Initialization

In the initialization process, the control parameters and the initial population are produced. The initial population includes NP solutions (vectors), each of which

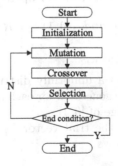

Fig. 1. The flowchart of differential evolution algorithm (DE)

contains D variables. NP is the size of the population and D is the dimension of the search space. A solution (individual) of the population at the generation G is defined as follows:

$$X_i^G = (X_{i,1}^G, X_{i,2}^G, \ldots, X_{i,D}^G), \qquad i = 1, 2, \ldots, NP \tag{1}$$

Generally speaking, the initialization population should be evenly distributed throughout the whole search space [4,5]. The commonly used initialization method for solutions (individuals) is:

$$X_{i,j}^0 = X_j^{min} + rand(0,1) \cdot (X_j^{max} - X_j^{min}) \tag{2}$$

where $rand(0,1)$ is a uniformly distributed random real number in the range of 0 to 1, X_j^{max} and X_j^{min} are the maximum and minimum bound of the jth dimension of the search space respectively.

2.2 Mutation Operator

DE adopts the mutation strategy to create a mutant vector for each individual (also called a target vector) at each generation G. Some of the most widely used DE mutation strategies are shown as follows [7],

$$\text{DE/rand/1:} \qquad V_i^G = X_{r1}^G + F \cdot (X_{r2}^G - X_{r3}^G) \tag{3}$$

$$\text{DE/rand/2:} \quad V_i^G = X_{r1}^G + F \cdot (X_{r2}^G - X_{r3}^G) + F \cdot (X_{r4}^G - X_{r5}^G) \tag{4}$$

$$\text{DE/best/1:} \qquad V_i^G = X_{best}^G + F \cdot (X_{r1}^G - X_{r2}^G) \tag{5}$$

$$\text{DE/current-to-rand/1:} \quad V_i^G = X_i^G + F \cdot (X_{r1}^G - X_i^G) + F \cdot (X_{r2}^G - X_{r3}^G) \tag{6}$$

where $r1, r2, r3, r4, r5$ are the distinct integers randomly generated from the range of $[1, NP]$, and they are not equal to i. X_{best}^G is the best individual with the best fitness value at generation G. The parameter F is the scaling factor to control the mutation scale, which is generally restricted in the range of $(0, 1]$.

2.3 Crossover Operator

After mutation, a trial vector $U_i^G = (U_{i,1}^G, U_{i,2}^G, \ldots, U_{i,D}^G)$ is generated for each individual according to a binomial crossover operator on X_i^G and V_i^G as follows,

$$U_{i,j}^G = \begin{cases} V_{i,j}^G & \text{if } (rand_{i,j}(0,1) \leq CR \text{ or } j == j_{rand}) \\ X_{i,j}^G & \text{otherwise} \end{cases} \tag{7}$$

In the above equation, j_{rand} is a uniformly distributed random integer in the range of $[1, D]$, which should be generated for each individual. CR is the crossover rate, which is restricted in range of $[0, 1]$.

If the jth variable $U_{i,j}^G$ of the trial vector U_i^G violates the boundary constraints, it will be reset as follows

$$U_{i,j}^G = X_j^{min} + rand(0,1) \cdot (X_j^{max} - X_j^{min}) \tag{8}$$

2.4   Selection Operator

Selection operator determines whether the target or the trial vector survives and goes into the next generation based on their fitness values. For a minimization problem, the decision vector with the lower fitness value (objective value) could enter the next generation, which can be defined as follows.

$$X_i^{G+1} = \begin{cases} U_i^G & \text{if } (fit(U_i^G) \leq fit(X_i^G)) \\ \\ X_i^G & \text{otherwise} \end{cases} \tag{9}$$

3   The Enhanced Differential Evolution Algorithm

In this section, a new mutation strategy and a new parameter control method are proposed to enhance the performance of DE.

3.1   The New Mutation Strategy

As described in the previous section, there are some mutation strategies in DE, which have different characteristics and are suitable for solving different optimization problems. For example, DE/rand/1 and DE/rand/2 pay more attention to exploration and are suitable for solving multimodal problem, while DE/best/1 and DE/best/2 pay more attention to exploitation and are suitable for solving unimodal problems [4,6]. So, in this paper, we design a new mutation strategy to combine the merits of DE/rand/1 and DE/best/1 mutation strategy, which is called DE/rand-superior/1.

DE/rand-superior/1:

$$X_{base} = \lambda \cdot X_{r1}^G + (1 - \lambda) \cdot X_{superior}^G \tag{10}$$

$$V_i^G = X_{base} + F \cdot (X_{r2}^G - X_{r3}^G) \tag{11}$$

where $r1, r2, r3$ are the distinct integers randomly generated in the range of $[1, NP]$, and they are not equal to i. $X_{superior}^G$ is randomly selected from the superior individuals, which contains the top $floor(p \cdot NP)$ individuals in the current population. $floor(x)$ is a rounding function, returning the largest integer that is smaller than its parameter x. The parameter λ controls the base vector X_{base} to make it close to a randomly selected individual or a randomly selected superior individual. Obviously, when $\lambda = 1$, DE/rand-superior/1 degenerates to DE/rand/1, and when $\lambda = 0$, DE/rand-superior/1 degenerates to DE/superior/1 [5]. Therefore, the parameter λ can adjust the exploration and exploitation ability of DE/rand-superior/1. The parameter p and λ are set respectively as follows:

$$p = rand(0.05, 0.15) \tag{12}$$

$$\lambda = (\frac{G_{max} - G}{G_{max}})^4 \tag{13}$$

where $rand(0.05, 0.15)$ is a uniformly distributed random real number in the range of $[0.05, 0.15]$. G is the current generation count, and G_{max} is the maximal generation count. The changing curve of parameter λ is shown in Fig. 2. Obviously, at the early stage of the evolution, DE/rand-superior/1 pays more attention to exploration for locating the promising area, while at the late stage of the evolution, it pays more attention to exploitation for finding out the global optimal solution.

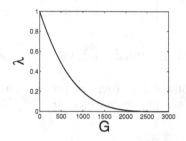

Fig. 2. The changing curve of λ

3.2   The New Parameter Control Method

The performance of DE is significantly affected by its parameters (scaling factor and crossover rate) [4,8]. Generally speaking, at the early stage of evolution, DE need to locate an area that contains the global optimal solution, and at the late stage of evolution, DE should search at a fine-grained level. According to this principle, we propose a new control method for scaling factor F and crossover CR as follows.

Scaling Factor F. A population level scaling F_P is assigned to the whole population, which linearly decreases from F_P^{max} to F_P^{min} as follows

$$F_P = F_P^{max} - (G/G_{max}) \cdot (F_P^{max} - F_P^{min}) \tag{14}$$

The individual level parameter F_i for each individual in population is generated by Gaussian distribution [4,6] based on the population level parameter F_P as follows

$$F_i = randn(F_P, 0.1) \tag{15}$$

F_i is truncated to be 1 when $F_i > 1$, and F_i is regenerated when $F_i < 0$.

This control method can make sure that scaling factor F gradually decreases as the evolution process from the overall level. Moreover, the Gaussian distribution can provide the flexibility for F from the individual level.

Crossover Rate CR. A CR candidate pool is established which includes some typical values, such as 0.1, 0.5 and 0.9. At each generation, each individual randomly picks up a value from CR candidate pool as its crossover rate. The CR candidate pool is set as follows

$$CR_{candidate} = \{0.1, 0.5, 0.9\} \tag{16}$$

The rationality of this setting is that the values of 0.1, 0.5 and 0.9 are widely assigned to CR in other literatures [11,12]. Besides, $CR = 0.1$ emphasizes on exploitation, and $CR = 0.9$ emphasizes on exploration and $CR = 0.5$ can balance between exploration and exploitation.

3.3   The Enhance Differential Evolution Algorithm (EDE)

Our new mutation strategy and new parameter control method are integrated with the framework of classical DE to form the EDE algorithm. The pseudo-code of the complete EDE is demonstrated in Algorithm 1.

Algorithm 1. The procedure of the EDE algorithm

1: **Initialization**: Generate a uniformly distributed random initial population
2: **while** termination condition is not satisfied **do**
3:     $F_P = F_P^{max} - (G/G_{max}) \cdot (F_P^{max} - F_P^{min})$
4:     $\lambda = (G/G_{max})^4$
5:     $p = rand(0.0.5, 0.15)$
6:     **for** $i = 1$ to NP **do**
7:         $F_i = randn(F_P, 0.1)$
8:         Select random indexes $r1, r2$ and $r3$, $r1 \neq r2 \neq r3 \neq i$ //mutation
9:         Select $X_{superior}^G$ randomly from top $floor(p \cdot NP)$ individuals
10:        $V_i^G = \lambda \cdot X_{r1}^G + (1 - \lambda) \cdot X_{superior}^G + F \cdot (X_{r2}^G - X_{r3}^G)$ // end mutation
11:        $j_{rand} = randint(1, D)$ //crossover
12:        Select a value for CR from candidate pool randomly
13:        **for** $i = 1$ to D **do**
14:            **if** $rand(0, 1) \leq CR$ or $j == j_{rand}$ **then**
15:                $U_{i,j}^G = V_{i,j}^G$
16:            **else**
17:                $U_{i,j}^G = X_{i,j}^G$
18:            **end if**
19:        **end for**//end crossover
20:        **if** $f(U_i^G) \leq f(X_i^G)$ **then**
21:            $X_i^{G+1} = U_i^G$ //selection
22:        **else**
23:            $X_i^{G+1} = X_i^G$
24:        **end if**//end selection
25:    **end for**
26: **end while**

---

4   Experiments and Results

EDE will be tested by ten benchmark numerical functions with 30D to evaluate the performance, which are proposed in the special session on real-parameter
optimization of CEC 2014 [13] (the first ten test functions). The detailed description of these benchmark functions can be found in [13].

Table 1. The parameter settings for the compare algorithms

Algorithm	Parameter values
DE	$NP = 100, F = 0.5, CR = 0.9$
ODE	$NP = 100, F = 0.5, CR = 0.9, J_r = 0.3$
SaDE	$NP = 100, k = 4, \epsilon = 0.01, L = 50$
EDE	$NP = 100, F_P^{max} = 0.9, F_P^{min} = 0.1, CR \in \{0.1, 0.5, 0.9\}$

In our experimental studies, the average and the standard deviation of the
function error value $f(X_{best}) - f(X^*)$ are adopted to evaluate the optimization
performance, where X_{best} is the best solution found by the algorithm in each run
and X^* is the actual global optimal solution of the test function. The maximal
function evaluation (max_FES) is adopted as the termination condition, which
is set to $10000 \cdot D$. For all experiments, 20 independently runs are conducted for
each test function. Wilcoxon's rank-sum test is conducted on the experimental
results at the 5 % significant level to obtain the reliable statistic conclusion. For
clarity, the best results for each test problem are marked in boldface.

EDE is compared with other six DE algorithms, which are DE/rand/1/bin,
DE/rand/2/bin, DE/current-to-rand/1/bin, DE/best/1/bin, ODE [12] and
SaDE [6]. The parameter settings for these seven algorithms are given in Table 1
and the experimental results are shown in Table 2.

Table 2 clearly indicates that EDE has the best performance among the
seven algorithms on unimodal functions F1-F3, and DE/rand/1/bin has the second best performance. This phenomenon effectively demonstrates that EDE not
only retains the merits of DE/rand/1/bin, but also improves the performance
of DE. On multi-modal functions F4-F10, EDE outperforms DE/rand/1/bin,
DE/rand/2/bin, DE/current-to-rand/1/bin, DE/best/1/bin, ODE and SaDE
on most test functions and EDE is only beaten by DE/rand/1/bin and ODE
on F6, and SaDE on F10. Overall, EDE is significantly better than DE/rand/1,
DE/rand/2, DE/current-to-rand/1/bin, DE/best/1/bin, ODE and SaDE on 5,
10, 10, 10, 5 and 6 test functions respectively, while DE/rand/1/bin, ODE
and SaDE are better than EDE only on one of the test functions respectively.
Moreover, DE/rand/2/bin, DE/current-to-rand/1/bin and DE/best/1/bin are
unable to outperform EDE on any of 10 test functions. The box plot of 20 function error values of each algorithm on each test function are plotted in Fig. 3,
where 1, 2, 3, 4, 5, 6, and 7 denote EDE, DE/rand/1/bin, DE/rand/2/bin,

Table 2. The parameter settings for the compare algorithms

Func \ Alg	rand/1/bin mean error (std dev)	rand/2/bin mean error (std dev)	cur-to-rand/bin mean error (std dev)	best/1/bin mean error (std dev)	ODE mean error (std dev)	SaDE mean error (std dev)	EDE mean error (std dev)
F1	7.36e+04 4.55e+04−	2.71e+07 8.09e+06−	9.19e+06 4.99e+06−	6.42e+08 2.48e+08−	1.11e+05 7.64e+04−	3.70e+05 2.31e+05−	**4.86e+04** **2.92e+04**
F2	0.00e+00 0.00e+00=	4.58e+06 1.66e+06−	2.44e+09 1.23e+09−	5.84e+10 1.50e+10−	7.80e-13 8.01e-13−	2.65e-14 8.31e-14−	0.00e+00 0.00e+00
F3	0.00e+00 0.00e+00=	2.79e+01 6.93e+00−	4.08e+03 2.02e+03−	1.27e+05 3.06e+04−	1.14e-14 2.31e-14=	9.25e+00 2.52e+01−	0.00e+00 0.00e+00
F4	4.93e-01 1.31e-01=	1.76e+02 1.50e+00−	2.63e+02 9.92e+01−	9.27e+03 3.38e+03−	**3.50e-01** **2.07e-01**−	3.39e+01 3.47e+01−	6.88e-01 1.54e-01
F5	2.09e+01 5.63e-02−	2.09e+01 5.12e-02−	2.09e+01 6.66e-02−	2.09e+01 5.12e-02−	2.09e+01 4.75e-02−	2.06e+01 6.15e-02=	**2.05e+01** **3.98e-02**
F6	**2.54e-04** **4.16e-04**+	3.46e+01 1.14e+00−	1.27e+00 7.52e-01−	3.61e+01 1.78e+00−	6.00e-02 1.73e-01−	4.79e+00 1.45e+00−	8.43e-01 9.34e-01
F7	0.00e+00 0.00e+00=	9.79e-00 3.77e-02−	2.54e+01 8.95e+00−	5.30e+02 7.87e+01−	3.79e-15 2.08e-14=	9.84e-03 1.39e-02−	0.00e+00 0.00e+00
F8	1.30e+02 2.37e+00−	2.18e+02 6.46e+00−	1.48e+02 9.62e+00−	3.04e+02 3.34e+01−	1.53e+02 2.08e+01−	3.32e-02 1.82e-01−	0.00e+00 0.00e+00
F9	1.82e+02 1.08e+00−	2.35e+02 9.15e+00−	1.59e+02 9.00e+00−	3.56e+02 3.88e+01−	1.83e+02 1.25e+01−	**3.50e+01** **7.79e+00**=	3.78e+01 1.88e+01
F10	5.76e+03 4.85e+00−	6.34e+03 1.84e+02−	5.60e+03 2.91e+02−	5.92e+03 5.52e+02−	6.00e+03 5.01e+02−	**2.33e+01** **4.76e-01**+	7.33e+00 4.13e+00
+/ = /−	1/4/5	0/0/10	0/0/10	0/0/10	1/4/5	1/3/6	———

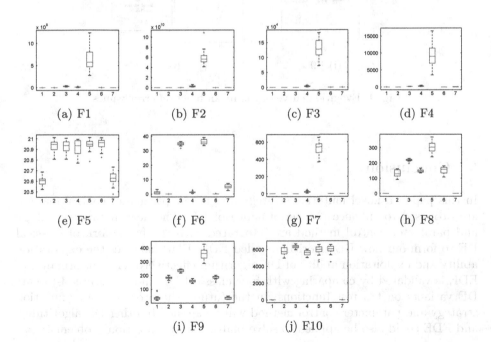

(a) F1   (b) F2   (c) F3   (d) F4

(e) F5   (f) F6   (g) F7   (h) F8

(i) F9   (j) F10

Fig. 3. Box plot of 20 function error values

DE/current-to-rand/1/bin, DE/best/1/bin, ODE and SaDE respectively. Obviously, EDE has better robustness and accuracy than other algorithms. The evolution curves of the *mean function error values* derived from EDE, DE/rand/1/bin, DE/rand/2/bin, DE/current-to-rand/1/bin, DE/best/1/bin, ODE and SaDE versus the number of FES are plotted in Fig. 4.

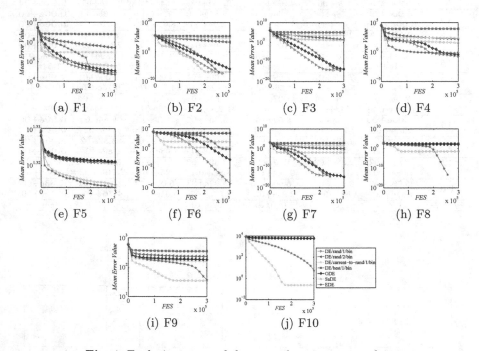

Fig. 4. Evolution curve of the mean function error values

5   Conclusion

In this paper, a novel mutation strategy and a new parameter control method are proposed to enhance the performance of DE. The new mutation strategy and parameter control method are integrated with the framework of classical DE to form our new DE algorithm, called EDE. EDE balances the exploration ability and exploitation ability of DE algorithm effectively. The performance of EDE is validated by comparing with four classical DE and two state-of-the-art DE variants on ten test functions. In the future, the proposed new mutation strategy and parameter control method will be applied to other DE algorithms, and EDE could also be applied to solve real-world optimization problems.

Acknowledgments. This work was supported in part by National Natural Science Foundation of China (Grant no. 61402294, 61170283 and 61402291), National

High-Technology Research and Development Program (863 Program) of China (Grant no. 2013AA01A212), Ministry of Education in the New Century Excellent Talents Support Program (Grant no. NCET-12-0649), Guangdong Natural Science Foundation (Grant no. S2013040012895), Foundation for Distinguished Young Talents in Higher Education of Guangdong, China (Grant no. 2013LYM_0076 and 2014KQNCX129), Major Fundamental Research Project in the Science and Technology Plan of Shenzhen (Grant no. JCYJ20130329102017840, JCYJ20130329102032059, JCYJ20140418095735608 and JCYJ20140828163633977).

References

1. Storn, R., Price, K.V.: Differential evolution: a simple and efficient heuristic for global optimization over continuous space. J. Glob. Optim. **11**(4), 341–359 (1997)
2. Gao, Z., Pan, Z., Gao, J.: A new highly efficient differential evolution scheme and its application to waveform inversion. IEEE Geosci. Remote Sens. Lett. **11**(10), 1702–1706 (2014)
3. Tenaglia, G.C., Lebensztajn, L.: A multiobjective approach of differential evolution optimization applied to electromagnetic problems. IEEE Trans. Magn. **50**(2), 625–628 (2014)
4. Zhang, J., Sanderson, A.C.: JADE: adaptive differential evolution with optional external archive. IEEE Trans. Evol. Comput. **13**(5), 945–958 (2009)
5. Yu, W.J., Shen, M., Chen, W.N., Zhan, Z.H., Gong, Y.J., Lin, Y.: Differential evolution with two-level parameter adaption. IEEE Trans. Cybern. **44**(7), 1080–1099 (2014)
6. Qin, A.K., Huang, V.L., Suganthan, P.N.: Differential evolution algorithm with strategy adaptation for global numerical optimization. IEEE Trans. Evol. Comput. **12**(1), 64–79 (2008)
7. Liu, J., Lampinen, J.: A fuzzy adaptive differential evolution algorithm. Soft Comput. **9**(6), 448–462 (2005)
8. Brest, J., Greiner, S., Boskovic, B., Mernik, M., Zumer, V.: Self-adapting control parameters in differential evolution: a comparative study on numerical benchmark problems. IEEE Trans. Evol. Comput. **10**(6), 646–657 (2006)
9. Gong, W., Cai, Z.: Differential evolution with ranking-based mutation operators. IEEE Trans. Cybern. **43**(6), 2066–2081 (2013)
10. Cai, Y., Wang, J.: Differential evolution with neighborhood and direction information for numerical optimization. IEEE Trans. Cybern. **43**(6), 2202–2215 (2013)
11. Wang, Y., Cai, Z., Zhang, Q.: Differential evolution with composite trial vector generation strategies and control parameters. IEEE Trans. Evol. Comput. **15**(1), 55–66 (2011)
12. Rahnamayan, S., Tizhoosh, H.R., Salama, M.M.A.: Opposition-based differential evolution. IEEE Trans. Evol. Comput. **12**(1), 64–79 (2008)
13. Liang, J.J., Qu, B.Y., Suganthan, P.N.: Problem definitions and evaluation criteria for CEC 2014 special session and competition on single objective real-parameter numerical optimiation. Technical Report, Nanyang Technological University, Singapore, Zhenzhou University, China, December 2013. http://www.ntu.edu.sg/home/epnsugan/

Hybrid Model for the Training of Interval Type-2 Fuzzy Logic System

Saima Hassan[1,2]([✉]), Abbas Khosravi[3], Jafreezal Jaafar[1],
and Mojtaba Ahmadieh Khanesar[4]

[1] Department of CIS, Universiti Teknologi PETRONAS, Tronoh, Malaysia
saimahassan@kust.edu.pk
[2] Kohat University of Science and Technology, Kohat, Pakistan
[3] Centre for Intelligent Systems Research, Deakin University, Dandenong, Australia
[4] Faculty of Electrical and Computer Engineering, Semnan University, Semnan, Iran

Abstract. In this paper, a hybrid training model for interval type-2 fuzzy logic system is proposed. The hybrid training model uses extreme learning machine to tune the consequent part parameters and genetic algorithm to optimize the antecedent part parameters. The proposed hybrid learning model of interval type-2 fuzzy logic system is tested on the prediction of Mackey-Glass time series data sets with different levels of noise. The results are compared with the existing models in literature; extreme learning machine and Kalman filter based learning of consequent part parameters with randomly generated antecedent part parameters. It is observed that the interval type-2 fuzzy logic system provides improved performance with the proposed hybrid learning model.

Keywords: Hybrid learning model · Extreme learning machine · Genetic algorithm · Interval type-2 fuzzy logic system · Prediction

1 Introduction

Information deficiencies such as incomplete, fragmentary, not fully reliable, vague and contradictory information [1] results in uncertainties in data and a process. Type-1 fuzzy logic system (T1FLS) can only handle the uncertainties about the meaning of the words by using precise membership functions. The choice of T1FLS is not appropriate in the presence of other sources of uncertainties in the real world data as it may cause problem in determining the exact and precise parameters of both the antecedents and consequents [2]. However, T2FLS can handle all type of uncertainties with their fuzzy grades [3].

T2FLS is computationally demanding because of the extra dimension. Interval T2FLS (IT2FLS) is the simplest form of T2FLS as all points in the third dimension are at unity and can be ignored for modelling purposes [4]. Though improvements of IT2FLS to its earlier version have been evidenced, yet it still lacks a systematic and coherent design. Different learning algorithms proposed for parameters optimization of IT2FLS include back propagation based learning

© Springer International Publishing Switzerland 2015
S. Arik et al. (Eds.): ICONIP 2015, Part I, LNCS 9489, pp. 644–653, 2015.
DOI: 10.1007/978-3-319-26532-2_71

method [5], genetic and other bio-inspired algorithms [6–9], ant colony optimization [10], and extended Kalman filter based learning algorithm [11]. A hybrid model for IT2FLS was also proposed by [12] using orthogonal least-squares and back-propagation methods. The distressing issues of learning algorithms i.e. stopping criteria, learning rate, learning epochs and local minima may not be handled by the conventional learning algorithms. Huang et al. [13] introduced extreme learning machine (ELM) that can solve the stated issues of conventional training methods. Jang et al. established a functional equivalent between fuzzy and single hidden layer feed-forward neural network (SLFN) [14], that made it possible to hybridize fuzzy and ELM. Different hybrid models of fuzzy and ELM reported in literature include an evolutionary fuzzy extreme learning machine analyzed for mammographic risk [15], a hybrid model of fuzzy and ELM for fault detection method in power generation plant [16], ELM based fuzzy inference system [17] and an online sequential fuzzy ELM for function approximation and classification problems [18]. The antecedent parameters in the above mentioned models were randomly assigned and the consequent parameters were determined analytically. However, there are chances that the randomly assigned parameters might not create suitable membership function in fuzzy model. As it is noted that the randomly generated parameters may not be effective for network output [19] and can cause high learning risks [20] due to overfitting. Soon after the realization of this issue, optimal parameters (hidden node) are reported for ELM [19–22]. However the hybrid model of fuzzy and ELM have not yet been reported with optimal parameters and are generated randomly.

ELM is an efficient learning algorithm for T2FLSs [23], however, the antecedent part parameters were generated randomly. Inspired by the competitive performance of the T2FLS with ELM and motivated from the issue to find the optimal parameters of fuzzy and ELM hybrid model, this paper proposes a hybrid training model for IT2FLS, where the antecedent parameters are optimized using genetic algorithm (GA) and consequent parameters are determined analytically through ELM. GA proposes multiple solutions which evolve to find best point, so it is less probable that it fells in a local minima than other optimization methods. Moreover, this optimization method is suitable for the nonlinear optimization problems. These are the reasons why GA is proposed to be used to optimize the parameters of antecedent part. The proposed hybrid learning model for IT2FLS is described in Sect. 3. The parameters of optimization, such as length of chromosome, fitness function and simulation results are discussed in Sect. 4. Section 5 concludes the paper with some remarks and guidelines for future work.

2 Structure of the Interval Type-2 FLS Used in This Paper

An IT2FS $\tilde{A}$ can be defined as follows:

$$\tilde{A} = \int_{x \in X} \int_{u \in J_x} 1/(x, u) \qquad J_x \subseteq [0, 1] \tag{1}$$

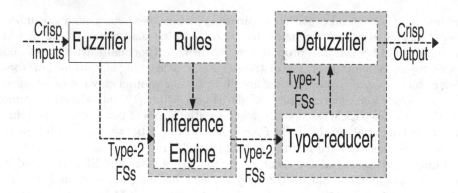

Fig. 1. Block diagram of type-2 FLS.

A type-2 FLS (see Fig. 1) maps crisp input into type-2 fuzzy sets by assigning membership grade to each fuzzy set in the Fuzzifier block. There are various type-2 fuzzy membership functions (MFs), however the Gaussian MF is utilized here because of less parameters. The Gaussian MF with fixed mean m_i^n and uncertain deviation $[\sigma_i^{n1}, \sigma_i^{n2}]$ can be represented as:

$$\mu_{\tilde{A}_i^n}(x_i) = exp[-\frac{1}{2}(\frac{x_i - m_i^n}{\sigma_i^n})^2], \quad \sigma_i^n \in [\sigma_i^{n1}, \sigma_i^{n2}] \tag{2}$$

where $\mu_{\tilde{A}_i^n}$ is a Gaussian MF that has upper and lower MFs $[\overline{\mu}_{\tilde{A}_i^n}(x_i) \; \underline{\mu}_{\tilde{A}_i^n}(x_i)]$ as:

$$\overline{\mu}_{\tilde{A}_i^n}(x_i) = N(m_i^n, \sigma_i^{n2}; x_i) \tag{3}$$

$$\underline{\mu}_{\tilde{A}_i^n}(x_i) = N(m_i^n, \sigma_i^{n1}; x_i) \tag{4}$$

A fuzzy Rule Base is a set of linguistic rules in the form of a two parts IF-THEN conditional statements. The IF part (known as antecedents) need to be satisfied to inferred the THEN part (known as consequents). The interval type-2 FLS's with a rule-base of N_{th} rules (R^n) are taken as:

R^n: if x_1 is $\tilde{A}_1^n \wedge x_2$ is $\tilde{A}_2^n \wedge \cdots \wedge x_d$ is $\tilde{A}_d^n$

Then $w^n(\mathbf{x}) = p_0^n + p_1^n x_1 + \cdots + p_d^n x_d, \quad n = 1, \cdots, N$

In the Inference Engine, each fuzzy rule is premised on the input vector $\mathbf{x} = [x_1, x_2, \cdots, x_n]^T$ as a varying singleton w^n. $\tilde{A}_i^n$ is the ith IT2 fuzzy subset generated from the input variable x_i in the nth rule domain. N and $\wedge$ represent the number of fuzzy rules and conjunction operator respectively. $p_n = [p_0^n, p_1^n, \cdots, p_d^n]^T$ denotes the nth fuzzy rule parameters of the consequent.

The Output Processing block in type-2 FLS comprises of an additional component called the Type reducer followed by a Defuzzifier block. Because of the distinct nature of type-2 fuzzy membership functions, the output from the inference engine is type-2 FS. Since the defuzzifier block can only input the type-1 FSs to produce crisp output therefor, a type reducer is needed after the inference engine to produce a type-reduced set using a centroid calculation. This type-reduced set can be then defuzzified to crisp output.

3   Structure of the Hybrid Learning Model for Interval Type-2 FLS (IT2FELM-GA)

The major task in the design process of a type-2 FLS involves the selection of optimal parameters. In this paper, a hybrid learning model for IT2FLS is proposed based on ELM and GA. The proposed hybrid learning model tune the consequent part parameters using ELM with randomly generated antecedent part parameters initially. The antecedent part parameters are then encoded as chromosome and optimize using GA in the direction of having better performance. Figure 2 shows the flowchart of the hybrid learning model of IT2FLS using ELM and GA.

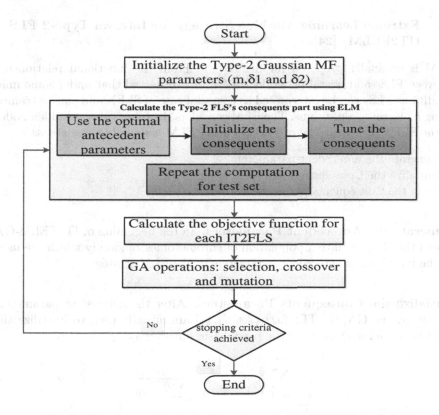

Fig. 2. Flowchart of the hybrid learning model of IT2FLS model.

3.1   Optimal Parameters Using GA

GA, an optimization tool, is based on a formalization of natural selection and genetics. A population of chromosomes, objective function and stopping criteria are defined in GA. The population then undergoes genetic operation to evolve and the best population is selected based on the objective function. An IT2FS described by a Gaussian membership function with fixed mean and uncertain standard deviation is encoded into a population of chromosomes. Root means square error (RMSE) is defined as the fitness function for the determination of the best chromosomes. Maximum number of iterations and relatively small changes in the value of RMSE are the stopping criteria. The GA runs for each iteration and calculated the RMSE for the IT2FLS with the consequents parameters are learnt through ELM. Learning of consequents through ELM is described in the next section. The optimal parameters are achieved once GA stops with the minimum RMSE. These optimal parameters are then used in the ELM strategy to develop a hybrid learning model for IT2FLS.

3.2 Extreme Learning Machine Strategy for Interval Type-2 FLS (IT2FELM) [24]

ELM is originally proposed for SLFN [13]. From the functional relationship between FLSs and neural networks [14,25], it is observed that under some mild conditions FLSs can be interpreted as a special case of SLFN and can be trained using its learning algorithms. The ELM considers the fuzzy rules as hidden nodes of the SLFN [18]. Learning of IT2FLS using ELM is done in three steps.

– Generate the antecedent parameters.
– Initialize the Consequents parameters.
– Tune the Consequents parameters.

Generate the Antecedent Parameters. In the beginning of IT2FELM-GA model the GA initialize a population of chromosomes randomly which are used as the initial set of antecedent parameters in the ELM strategy.

Initialize the Consequents Parameters. After the antecedent parameters are set by the GA, the IT1 fuzzy set $[y_l, y_r]$ are initially used to initialize the consequent parameters using the fuzzy basic function as:

$$y_l = \sum_{n=1}^{N} \acute{f}_n w^n, \quad \acute{f}_n = \frac{f_n}{\sum_{\acute{n}=1}^{N} \bar{f}_{\acute{n}}} \tag{5}$$

$$y_r = \sum_{n=1}^{N} \underline{f}_n w^n, \quad \underline{f}_n = \frac{\underline{f}_n}{\sum_{\acute{n}=1}^{N} f_{\acute{n}}} \tag{6}$$

The initialized consequent parameters in IT2FELM-GA are then expressed as a function of linear system using the ELM strategy. Under the constraints of minimum least square, the linear system is optimized by ELM. Huang et al. [13] observed that such an optimal solution has the smallest least-squares norm and has a unique solution.

Tune the Consequents Parameters. Having the optimized initial consequent parameters of the IT2FLS in Sect. 3.2, the K-M algorithm [26] is utilized to obtain the final consequent parameters. The obtained final parameters are once again expressed as a function of linear system and is optimized using ELM. This step gives an optimized output of the IT2FLS.

Repeat the Computation for Test Data Set. The above parts of subsections 3.2 described training of IT2FLS with ELM strategy. Since the antecedents are computed once, the test data set is utilized with the last two parts for the prediction purposes.

3.3 Objective Function Evaluation

Once the ELM strategy is finished, the chromosome in each iteration is evaluated using the objective function of RMSE. The chromosomes' population having lowest RMSE represents the best population of the solution. The chromosome having best value of the objective function is saved in each iteration.

3.4 GA Operations

The current population of chromosome in IT2FELM-GA is updated to generate the new set of chromosomes for the next iteration using genetic operations of selection, crossover and mutation. Parents are selected using the tournament selection mechanism. Crossover operation creates new chromosomes inherit information (genes) from parents. The mutation operation introduce new genetic information hence promote diversity in population. These genetic operation are performed to evolve and optimize the encoded antecedent part parameters (chromosomes). These chromosomes are iteratively utilized in the ELM strategy of IT2FLS for several generations until the optimum solution is achieved.

4 Simulation Results

In this paper, an IT2FS described by a Gaussian membership function with 5 number of MFs (nMF) and 4 inputs is optimized using GA. This means that each parameter of the Gaussian MF requires 20 chromosomes. Thus, a total of 60 chromosomes of population 10 are generated randomly in the range [0,1]. One-point crossover with a probability of 0.8 is utilized.

The proposed hybrid learning model of interval type-2 fuzzy logic system is tested on the prediction of Mackey-Glass time series data sets with different levels of noise. The results are compared with the existing model in literature; extreme learning machine and Kalman filter based learning of consequent parameters with randomly generated antecedent parameters are used.

The IT2FEKM-GA is tested on the prediction of Mackey-Glass time series data sets with different levels of noise. The noise-free Mackey-Glass time series

Table 1. Numerical values of Mackey-Glass time series data

Parameter	Value
a	0.2
b	0.1
τ	17
x0	1.2
ts	0.1
nData	12000

data is generated using a nonlinear time-delay differential equation as expressed as follows:

$$\frac{dx(t)}{dt} = \frac{ax(t-\tau)}{1+x^n(t-\tau)} - bx(t) \tag{7}$$

where $x(t)$ is the time series data at time t, a, b and n are constants and τ is the delay parameter used to produce chaotic behavior in the data. The discretisized data is obtained for simulation using the Fourth-Order Runge-Kutta method with an initial condition x_0 and a time step ts. Table 1 shows the numerical values to generate Mackey-Glass time series data. The dataset with 4 inputs and one output is extracted in the form of $x(t-18), x(t-12), x(t-6), x(t)$ and $x(t+6)$. By adding different levels of noise to the Mackey-Glass time series data, five noisy data sets are generated. The training and testing data sets are obtained with a ratio of (70/30).

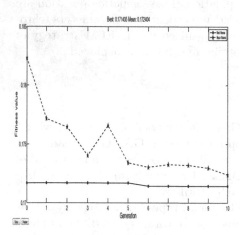

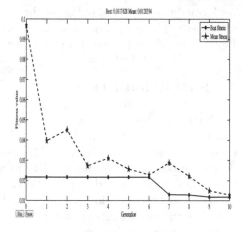

Fig. 3. The convergence of IT2FELM-GA for 0db Mackey Glass time series data.

Fig. 4. The convergence of IT2FELM-GA for 40db Mackey Glass time series data.

Table 2. RMSE of IT2FELM-GA, IT2FELM and IT2FKF obtained with noisy Mackey-Glass data sets.

	IT2FELM-GA	IT2FELM	IT2FKF
0db	0.171	0.198	0.192
10db	0.075	0.125	0.106
20db	0.030	0.105	0.073
30db	0.014	0.095	0.112
40db	0.011	0.060	0.110

The best and average trend of convergence of the IT2FELM-GA can be seen in Figs. 3 and 4. Continuous reduction of the best and average values of the fitness function is observed. The average value of the fitness function drops from 0.182 to 0.173 for the most noisy data (0db) and from 0.1 to 0.02 for 40db.

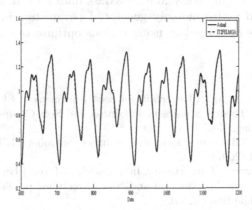

Fig. 5. Actual and forecasted times series data using IT2FELM-GA.

In order to show the effectiveness of optimal parameters, the IT2FELM-GA is compared with IT2FELM [23,24] and Kalman filter based IT2FLS (IT2FKF) [11], where the antecedent parts are generated randomly and consequent parts are learnt using ELM and KF respectively. The results of the last two models are taken after 10 runs. The minimum among the 10 RMSE is selected as the best performance of these models. Table 2 shows the results of IT2FELM-GA over IT2ELM and IT2FKF in terms of RMSE. The prediction results of IT2FEL-GA gradually decrease with decrease in the level of noise in data. It is also observed that the prediction results of IT2FELM-GA is very stable even with higher level of noise. The IT2FELM produces higher errors with higher level of noise. Whereas the IT2FKF produces good results in the presence of high level of noise as compared to IT2FELM. However, increase in RMSE value is observed with lower level of noise. It may be due to the fact that the KF algorithm are designed to perform well with noisy data [11].

Figure 5 shows the actual data and the forecasts obtained with the hybrid learning algorithm of IT2FLS. As can be seen from the figure, the results of IT2FELM-GA is quite satisfactory.

5   Conclusion

In this paper, effectiveness of the optimal parameters in ELM based fuzzy model is demonstrated with a hybrid learning model for IT2FLS using GA and ELM. The consequent parameters are tuned using ELM whereas the antecedent parameters are encoded as a chromosome and are optimized using GA. The proposed hybrid learning model is compared with IT2FELM and IT2FKF where the antecedent parameters are generated randomly. Competitive performance of the proposed hybrid learning model with optimal parameters is observed as compared to IT2FELM and IT2FKF in the presence of uncertainty. Uncertainty in models is introduced using noisy Mackey-Glass data sets. It is concluded that, the results achieved with randomly generated parameters in the ELM based fuzzy models can be optimized by using various optimization algorithms.

References

1. Klir, G.J., Wierman, M.J.: Uncertainty-Based Information: Elements of Generalized Information Theory. Studies in Fuzziness and Soft Computing, vol. 15, 2nd edn. Physica-Verlag, Heidelberg (1999)
2. Hagras, H.: Type-2 flcs: a new generation of fuzzy controllers. IEEE Comput. Intell. Mag. **2**(1), 30–43 (2007)
3. Mendel, J.M.: Sources of uncertainty. In: Mendel, J.M. (ed.) Uncertain Rule-Based Fuzzy Logic Systems: Introduction and New Directions, pp. 66–78. Prentice-Hall PTR, Upper Saddle River (2001)
4. Mendel, J., John, R., Liu, F.: Interval type-2 fuzzy logic systems made simple. IEEE Trans. Fuzzy Syst. **14**(6), 808–821 (2006)
5. Mendel, J.M.: Computing derivatives in interval type-2 fuzzy logic systems. IEEE Trans. Fuzzy Syst. **12**(1), 84–98 (2004)
6. Wu, D., Tan, W.W.: Genetic learning and performance evaluation of interval type-2 fuzzy logic controllers. Eng. Appl. Artif. Intell. **19**(8), 829–841 (2006)
7. Castillo, O., Melin, P., Alanis, A., Montiel, O., Sepulveda, R.: Optimization of interval type-2 fuzzy logic controllers using evolutionary algorithms. Soft Comput. **15**(6), 1145–1160 (2011)
8. Khosravi, A., Nahavandi, S., Creighton, D.: Short term load forecasting using interval type-2 fuzzy logic systems. In: 2011 IEEE International Conference on Fuzzy Systems (FUZZ), pp. 502–508, June 2011
9. Maldonado, Y., Castillo, O., Melin, P.: Particle swarm optimization of interval type-2 fuzzy systems for FPGA applications. Appl. Soft Comput. **13**(1), 496–508 (2013)
10. Juang, C.F., Hsu, C.H., Chuang, C.F.: Reinforcement self-organizing interval type-2 fuzzy system with ant colony optimization. In: IEEE International Conference on Systems, Man and Cybernetics, SMC 2009, pp. 771–776, October 2009

11. Khanesar, M., Kayacan, E., Teshnehlab, M., Kaynak, O.: Extended kalman filter based learning algorithm for type-2 fuzzy logic systems and its experimental evaluation. IEEE Trans. Ind. Electron. **59**(11), 4443–4455 (2012)
12. Mendez, G.M., de los Angeles Hernandez, M.: Hybrid learning for interval type-2 fuzzy logic systems based on orthogonal least-squares and back-propagation methods. Inf. Sci. **179**(13), 2146–2157 (2009)
13. Huang, G.B., Zhu, Q.Y., Siew, C.K.: Extreme learning machine: theory and applications. Neurocomputing **70**(1–3), 489–501 (2006)
14. Jang, J.S., Sun, C.T.: Functional equivalence between radial basis function networks and fuzzy inference systems. IEEE Trans. Neural Netw. **4**(1), 156–159 (1993)
15. Qu, Y., Shang, C., Wu, W., Shen, Q.: Evolutionary fuzzy extreme learning machine for mammographic risk analysis. Int. J. Fuzzy Syst. **13**(4), 282–291 (2011)
16. Aziz, N.L.A., Yap, S., Bunyamin, M.A.: A hybrid fuzzy logic and extreme learning machine for improving efficiency of circulating water systems in power generation plant. In: IOP Conference Series: Earth and Environmental Science, vol. 16 (2013)
17. Sun, Z.L., Au, K.-F., Choi, T.M.: Neuro-fuzzy inference system through integration of fuzzy logic and extreme learning machines. IEEE Trans. Syst. Man Cybern. B Cybern. **37**(5), 1321–1352 (2007)
18. Rong, H.J., Huang, G.B., Sundararajan, N., Saratchandran, P.: Online sequential fuzzy extreme learning machine for function approximation and classification problems. IEEE Trans. Syst. Man Cybern. B Cybern. **39**(4), 1067–1072 (2009)
19. Huang, G.B., Chen, L.: Enhanced random search based incremental extreme learning machine. Neurocomput. **71**(16–18), 3460–3468 (2008)
20. Zhang, Y., Cai, Z., Gong, W., Wang, X.: Self-adaptive differential evolution extreme learning machine and its application in water quality eva. Comput. Inf. Syst. **11**(4), 1443–1451 (2015)
21. Feng, G., Huang, G., Lin, Q., Gay, R.: Error minimized extreme learning machine with growth of hidden nodes and incremental learning. IEEE Trans. Neural Netw. **20**(8), 1352–1359 (2009)
22. Zhang, R., Lan, Y., Huang, G.-B., Soh, Y.C.: Extreme learning machine with adaptive growth of hidden nodes and incremental updating of output weights. In: Kamel, M., Karray, F., Gueaieb, W., Khamis, A. (eds.) AIS 2011. LNCS, vol. 6752, pp. 253–262. Springer, Heidelberg (2011)
23. Deng, Z., Choi, K.S., Cao, L., Wang, S.: T2fela: Type-2 fuzzy extreme learning algorithm for fast training of interval type-2 tsk fuzzy logic system. IEEE Trans. Neural Netw. Learn. Syst. **25**(4), 664–676 (2014)
24. Hassan, S., Khosravi, A., Jaafar, J.: Training of interval type-2 fuzzy logic system using extreme learning machine for load forecasting. In: Proceedings of the 9th International Conference on Ubiquitous Information Management and Communication, IMCOM 2015, pp. 87–91 (2015)
25. Castro, J., Mantas, C., Benitez, J.: Interpretation of artificial neural networks by means of fuzzy rules. IEEE Trans. Neural Netw. **13**(1), 101–117 (2002)
26. Wu, D., Mendel, J.: Enhanced karnik-mendel algorithms. IEEE Trans. Fuzzy Syst. **17**(4), 923–934 (2009)

A Numerical Optimization Algorithm Based on Bacterial Reproduction

Peng Shao[1,2], Zhijian Wu[1(✉)], Xuanyu Zhou[2], Xinyu Zhou[1,4],
Zelin Wang[1,2], and Dang Cong Tran[3]

[1] State Key Lab of Software Engineering, Wuhan University,
Wuhan 430072, Hubei, China
sp198310@163.com, zhijianwu@whu.edu.cn
[2] Computer School, Wuhan University, Wuhan 430072, Hubei, China
[3] Vietnam Academy of Science and Technology, Hanoi, Vietnam
[4] School of Computer and Information Engineering, Jiangxi Normal University,
Nanchang 330022, China

Abstract. According to characteristics of rapid speed and large quantity in the process of bacterial reproduction, and natural selection, survival of the fittest in the process of evolution, the framework of bacterial reproduction optimization(BRO) algorithm is proposed from a macro perspective of bacteria reproduction. The process of bacteria reproduction is divided to four periods with lag period, logarithmic period, stable period and decline period. Likewise, the process of optimization algorithm proposed by this paper is segmented into four periods with initial period, iteration period, stable period and decline period. Based on the framework, strategies are introduced to design BRO more efficiently. Experimental results and theoretical analysis show that BRO has faster convergence speed and higher accuracy for high-dimensional problems.

Keywords: Intelligent algorithms · Reproduction · Bacterial swarm · Particle swarm optimization

1 Introduction

Intelligent algorithms [1] have aroused attention of researchers in recent years because of their excellent optimization performance, a part of which mimic biological social behaviors of foraging such as particle swarm optimization (PSO) [2], artificial bee colony (ABC) [3], bacterial foraging optimization(BFO) [4] and so on. Besides these algorithms mentioned above, there are other algorithms such as genetic algorithm (GA) [5], cultural quantum-inspired shuffled frog leaping algorithm [6], group search optimization (GSO) [7], and social emotional optimization algorithm [8] and so forth. However, these algorithms have better effect on optimization for low-dimensional problems, but for high-dimensional problems they have some disadvantages over optimization such as curse of dimensionality easily, lower accuracy and slower convergence speed and higher time complexity

© Springer International Publishing Switzerland 2015
S. Arik et al. (Eds.): ICONIP 2015, Part I, LNCS 9489, pp. 654–661, 2015.
DOI: 10.1007/978-3-319-26532-2_72

and so on [9, 10]. This paper, from the macro perspective of biological individuals reproduction, proposes bacterial reproduction optimization(BRO)according to reproduction, another social behavior different from foraging.

2   Design of BRO Algorithm

2.1   The Frame of BRO

The reproduction, a phenomenon that biological individuals reproduce their offsprings, is a basic characteristic of biological individuals. For bacterial swarm, reproductive process undergoes four periods with lag period, logarithmic period, stable period and decline period [11, 12]. Likewise, the process of BRO is also divided into four periods and described as follows. The process of BRO initialization is called initial period; the process of BRO iterating evolution is called iteration period; the process that BRO is trapped into the local optima is called stable period; the dead process of the Mothers is called decline period.

Definition 1. *The best individual reproduced on some condition is called the Mother in the first generation, and in the later generations the best individuals are called the Sub Mothers at the same condition.*

In the four periods of BRO, the function of the first period is to implement initialization of the Mother, the Sub Mother and parameters of BRO. BRO achieves reproducing offsprings in the second period. BRO, in the third period, will fall into local optima, and the last period makes BRO out of local optimal values. The basic idea of BRO is described in Fig. 1.

2.2   BRO Algorithm Design

The minimization principle is adopted to choose the Mother and the sub Mother from bacterial swam, and its basic idea is described as follows. n bacterial individuals are generated randomly regarded as x_i, $i = 1, 2, \cdots, n$ and then fitness values $f(x_i)$are calculated. The fitness values, next, are compared with each other to find out the bacterial individual of minimal fitness value. The method that generates the sub Mother is the same as the Mother except for one different point that the individual selected as the Mother reproduces n-1 bacterial individuals, and the sub Mother is selected from n-1 individuals which are born by reproduction and inherit their Mother genetic information. Therefore, strategies that generate them are shown as follows.

$$Mother = \min(f(x_i)), i = 0, 1, \cdots, n-1 \tag{1}$$

$$subMother = \left\{ \begin{array}{c} f(x_0) = Mother \\ \min(f(x_i)), i = 0, 1, \cdots, n-1 \end{array} \right\}. \tag{2}$$

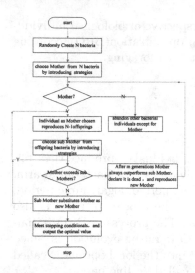

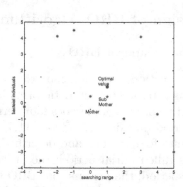

Fig. 1. The flowchart of BRO frame

Fig. 2. The graph of optimizing process

where $f(\bullet)$ is the fitness function and x_i is the ith bacterial individual.

A bacterial individual is regarded as a possible solution x_i, in D dimension space, which is represented as $x_i^D = (x_{i1}, x_{i2}, \cdots, x_{iD})$. Hypothesis that the searching area of the Mother x_0 is $[a, b]$ and its offspring bacterial searching area is $[ave\_a, ave\_b]$. Hence we can get the formula as $x_0^D = (x_{01}, x_{02}, \cdots, x_{0D}), x_0 \in [a, b]$ and ave_a and ave_b is shown as

$$ave\_a = - \left| \frac{x_{01} + x_{02} + \cdots + x_{0D}}{D} \right| \tag{3}$$

$$ave\_b = \left| \frac{x_{01} + x_{02} + \cdots + x_{0D}}{D} \right| \tag{4}$$

According to Eqs. 3 and 4, ave_a and ave_b are the lower limit and upper limit of offspring bacterial searching area respectively. By this way, it can implement information exchange among individuals. Since offsprings inherit their Mother information, this paper adopts the searching area designed above and takes into consideration the following equation to reproduce offsprings.

$$x_i^D = k_1 x_0^D + k_2, i = 1, 2, \cdots, n - 1 \tag{5}$$

where x_i^D is the offspring bacterial individual the Mother reproduced which locates at the searching area $[ave\_a, ave\_b]$; k_1 and k_2 are the random numbers uniformly distributed between $(-1, 1)$.

Definition 2. *Suppose that x is a real number that defines in the range $[a, b]$, that is, $x \in [a, b]$, the definition of the escaping value x_1 of x is as follows.*

$$x_1 = x - h \tag{6}$$

where h is a random number that locates at [-0.1, 0.1] uniformly distributed.

The advantage of Eq. 6 is that the Mothers which have been dead are adjusted slightly to escape the local optima. Meanwhile, fitness values of x_1 and the Mother are calculated. If the fitness of the Mother is worse than x_1, it is regards as the new Mother. Otherwise, x_1 insteads of x to continue calculating the escaping value until the new Mother appears or the algorithm meets the stop condition. Optimization process of BRO with 10 offsprings is shown in Fig. 2. In Fig. 2, the sub Mother closed to the optimal solution is better than the Mother obviously and then the sub Mother substitutes for it to continue reproducing and other individuals are abandoned except for the Mother.

3   Experimental Results and Analysis

3.1   Comparative Algorithms Parameters Setting

Eight benchmark functions are chosen and divided into two categories, which are all high and changeable dimension functions [13,14]. The first category functions with no peaks or simply multi peaks and the second with no rotated multi peaks, all minimization problems, are (1) Sphere(F_1), Rosenbrock(F_2); (2) Ackley(F_3), Griewanks(F_4), Weierstrass(F_5), Rastrigin(F_6), Noncontinuous Rastrigin(F_7), Schwefel(F_8), which are described in literatures [15,16].

For testing the performance of BRO, PSO, GA and BFO are chosen as comparative algorithms with 500 dimension. The parameters initial values of PSO and BFO are as follows: acceleration factors $c_1 = c_2 = 1.49$, inertia weight $\omega = 0.729$, maximum speed $= 2.0$, population size $= 30$, iteration numbers $= 20000$, run times $= 30$, chemotactic steps $= 20$, swimming length $= 5$, reproduction steps $= 8$. r_1 and r_2 are random numbers uniformly distributed between$(0, 1)$.

3.2   Experimental Results and Analysis

The Performance Evaluation of PSO and GA. PSO, because of few complicate operating factors, has a better effect on some typical optimizing problems. The research of literature [16] shows that PSO ameliorates the speed and accuracy of convergence compared with GA with selection, crossover and mutation for most non-linear and complex problems, which is hard to meet the needs of practical applications. In other word, the performance of PSO exceeds GA for problems above. Therefore, GA is no longer to be tested.

The Performance Evaluation of BRO, PSO and BFO. Eight fitness functions are tested for 30 times by PSO and BRO separately in experiments with 500 dimension. The comparison results are shown in Table 1. In Table 1, 'mean' represents average values of 30 times and 'std' represents standard deviation. 'RA' is the ratio of convergence accuracy of PSO and BRO. 'RT' is the ratio of average running time of 30 times of BRO and PSO (seconds).

Table 1. The experimental results with 500 dimension

Functions	Indexes	PSO	BRO	RA
F_1	Mean/Std	5.61e+02/4.93e+01	5.26e-06/1.35e-05	$10^8/10^6$
F_2	Mean/Std	1.31e+04/1.18e+03	7.92e-04/1.57e-03	$10^8/10^6$
F_3	Mean/Std	1.75e+01/2.92e-01	2.38e-04/2.14e-04	$10^5/10^3$
F_4	Mean/Std	1.92e+03/1.36e+02	1.62e-08/3.09e-08	$10^{11}/10^{10}$
F_5	Mean/Std	5.02e+02/2.18e+01	3.73e+00/2.42e+00	$10^2/10^1$
F_6	Mean/Std	3.49e+03/1.11e+02	5.98e-04/1.04e-03	$10^7/10^5$
F_7	Mean/Std	1.30e+04/9.72e+02	7.71e-04/1.47e-03	$10^8/10^5$
F_8	Mean/Std	1.64e+05/3.72e+03	2.01e+05/1.92e+03	$10^0/10^0$

As it can be seen from the Table 1, BRO outperforms PSO on seven out of eight functions except for F_8 (equivalent performance both PSO and BRO), especially for F_4, the ratio of accuracy is about 10^{11}. Table 1 and following Table 3 show that, for F_1, F_2, F_3, F_4, F_6, F_7, BRO has better optimizing performance. Table 3, meanwhile, shows that, the highest time ratio of the seven functions above is only 17.8 and the lowest is only 4.3. The time ratio of BRO is more than PSO but the lowest accuracy ratio of BRO is 10^6 times and the highest is 10^{16} times compared with PSO. For F_5, PSO consumes about 18 s and BRO improves the accuracy but consumes more than 1000s. BRO and PSO both trap into the local optimum and running time of BRO is 20 times more than PSO for F_8. Hence the two algorithms both have worse effect on the function in that its local extreme point is hidden near the optimal solution. In addition, it is obvious from Fig. 3 that convergence speed of BRO is far faster than PSO for the top 7 functions but slightly slower than PSO for F_8. In sum, the performance of BRO is far better than PSO.

BFO [4] is proposed according to the social behavior of bacterial foraging but BRO is proposed according to bacterial reproduction. The experimental results are shown in Table 2 when the dimension is 500. As it can be seen from Table 2, BRO outperforms BFO obviously for the top 7 functions except for F_8. Therefore, BRO has better performance over high-dimensional problems.

4   The Theoretical Analysis of BRO

4.1   The Analysis of Convergence

Theorem 1. *The ith and (i+1)th optimum are regarded as the Mother(i) and the subMother(i) respectively, y(i) represents the ith optimal value and the highest accuracy of convergence M will be produced when meets the stop condition.*

Proof. The Eq. 7 is obtained according to the idea of BRO.

$$y(i) = \min(Mother(i), subMother(i + 1)) \tag{7}$$

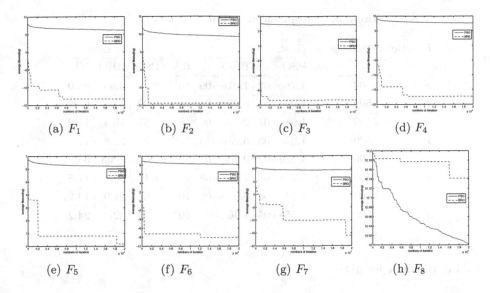

Fig. 3. The convergence curve of BRO

Table 2. The experimental results of BFO and BRO

Functions	Dimensions	Optimum values		
		BFO	BRO	RA
F_1	500	3.90e+03	1.07e−09	10^{12}
F_2	500	1.90e+05	8.95e−08	10^{13}
F_3	500	2.10e+01	2.22e−05	10^{6}
F_4	500	1.35e+04	5.98e−13	10^{17}
F_5	500	9.34e+02	2.74e−01	10^{3}
F_6	500	8.41e+03	7.75e−08	10^{11}
F_7	500	6.27e+04	7.94e−08	10^{12}
F_8	500	2.01e+05	1.96e+05	10^{0}

If $Mother(i) \geq subMother(i+1)$

$$y(i) = subMother(i+1) \tag{8}$$

Conversely, the following equation can be obtained

$$y(i) = Mother(i) \tag{9}$$

With incensement of iterating numbers, the following relationship is obtained

$$y(i) \geq y(i+1) \tag{10}$$

When BRO meets the stop condition or accuracy, the optimal M is obtained

$$y(0) \geq y(1) \geq y(2) \geq \cdots \geq y(n) = M \tag{11}$$

Table 3. The optimal values of BRO and PSO

Functions	Dimension	Best_value			Cost_time		
		PSO	BRO	RA	PSO	BRO	RT
F_1	500	4.78e+02	1.07e−09	10^{11}	0.9	3.95	4.30
F_2	500	1.09e+04	8.95e−08	10^{12}	0.96	6.04	6.29
F_3	500	1.68e+01	2.22e−05	10^6	1.37	15.4	11.2
F_4	500	1.63e+03	5.98e−13	10^{16}	1.34	22.8	17.0
F_5	500	4.57e+02	2.74e−01	10^3	18.3	1003	54.8
F_6	500	3.25e+03	7.75e−08	10^{11}	1.11	19.8	17.8
F_7	500	1.14e+04	7.94e−08	10^{12}	1.65	21.9	13.2
F_8	500	1.55e+05	1.96e+05	10^0	1.33	32.3	24.2

Taking the limit for $y(i)$

$$\lim_{i \to n} y(i) = y(n) = M \tag{12}$$

$\square$

It is obvious from Eq. 12 that BRO converges to higher accuracy M.

4.2 The Analysis of Time Complexity

According to *Cost_time* in Table 3, compared with PSO, excluding F_5, the highest time ratio of BRO is 24.2, for the other 7 functions with 500 dimension. The consuming time of BRO is slightly higher than PSO, but except for F_5, F_8, from the ratio of accuracy, it is improved greatly for the other 6 functions. The lowest accuracy ratio of BRO is 10^6, and the highest is 10^{16}. For F_5, the time is more than 50 times but the accuracy is only 1000 times, not by much but at least the accuracy increases slightly. For F_8, whether time or accuracy BRO is not better than PSO and the optimization performance is poor.

5   Conclusion

This paper, from the macro perspective of biological reproduction, proposes the frame of BRO according to characteristics of biological reproduction and introduces some strategies to design BRO based on the frame. A large number of experimental results and the analysis of time complexity and accuracy, show that BRO is an optimization algorithm, simply, few parameters and easy to implement. What more important is that BRO has faster convergence speed and higher accuracy dealing with high-dimensional problems.

Acknowledgments. This work was supported by the National Natural Science Foundation of China (No.61070008 and 70971043), the Science and Technology Foundation of Jiangxi Province(No.20151BAB217007), the Foundation of State Key Laboratory of Software Engineering(No.SKLSE2014-10-04) and Application research project of Nantong science and Technology Bureau(No.BK2014057).

References

1. Christian, B., Xiaodong, L.: Swarm intelligence in optimization. In: Blum, C., Merkle, D. (eds.) Swarm Intelligence: Introduction and Applications. Natural Computing Series, pp. 43–85. Springer, Heidelberg (2008)
2. Kennedy, J., Eberhart, R.C.: Particle swarm optimization. In: Proceedings of IEEE International Conference on Neural Networks, pp. 1942–1948 (1995)
3. Dervis, K.: An idea based on honey bee swarm for numerical optimization. Erciyes University, Turkey (2005)
4. Passino, K.M.: Bacterial foraging optimization. Int. J. Swarm Intell. Res. (IJSIR) **1**(1), 1–16 (2010)
5. Holland, J.H.: Adaptation in Natural and Artificial Systems. University of Michigan Press, Ann Arbor (1975)
6. Hongyuan, G., Congqiang, X.: Cultural quantum-inspired shuffled frog leaping algorithm for direction finding of non-circular signals. Int. J. Comput. Sci. Math. **4**(4), 321–331 (2013)
7. Changcheng, W., Juanyan, F.: Group search optimiser: a brief survey. Int. J. Comput. Sci. Math. **4**(1), 42–50 (2013)
8. Zhihua, C.: Social Emotional Optimization Algorithm. Publishing House of Electronics Industry Press, Beijing (2011)
9. Yang, Z., Tang, K., Yao, X.: Large scale evolutionary optimization using cooperative coevolution. Inf. Sci. **178**(15), 2985–2999 (2008)
10. van den Frans, B., Engelbrecht, A.P.: A cooperative approach to particle swarm optimization. IEEE Trans. Evol. Comput. **8**(3), 225–239 (2004)
11. Narra, H.P., Ochman, H.: Of what use is sex to bacteria? Curr. Biol. **16**(17), 705–710 (2006)
12. Willey, J., Sherwood, L., Woolverton, C.: Prescott's Microbiology. McGraw-Hill Ryerson, New York, NY (2010)
13. Xin, Y., Yong, L., Guangming, L.: Evolutionary programming made faster. IEEE Trans. Evol. Comput. **3**(2), 82–102 (1999)
14. Wenying, G.: Differential Evolution Algorithm and Its Application in Clustering Analysis, China University of Geosciences, 5 (2010). In Chinese
15. Liang, J.J., Qin, A.K., Suganthan, P.N.: Comprehensive learning particle swarm optimizer for global optimization of multimodal functions. IEEE Trans. Evol. Comput. **10**(3), 281–295 (2006)
16. Shi, Y., Eberhart, R.C.: Experimental study of particle swarm optimization. In: Proceedings of Fourth World Conference on Systems, Cybernetics and Informatics (2000)

Visual-Textual Late Semantic Fusion Using Deep Neural Network for Document Categorization

Cheng Wang$^{(\boxtimes)}$, Haojin Yang, and Christoph Meinel

Hasso Plattner Institute, University of Potsdam, Prof.-Dr.-Helmert-Str. 2-3,
14482 Potsdam, Germany
{cheng.wang,haojin.yang,christoph.meinel}@hpi.de

Abstract. Multi-modality fusion has recently drawn much attention due to the fast increasing of multimedia data. Document that consists of multiple modalities i.e. image, text and video, can be better understood by machines if information from different modalities semantically combined. In this paper, we propose to fuse image and text information with deep neural network (DNN) based approach. By jointly fusing visual-textual feature and taking the correlation between image and text into account, fusion features can be learned for representing document. We investigated the fusion features on document categorization, found that DNN-based fusion outperforms mainstream algorithms include K-Nearest Neighbor(KNN), Support Vector Machine (SVM) and Naive Bayes (NB) and 3-layer Neural Network (3L-NN) in both early and late fusion strategies.

Keywords: Categorization · Semantic feature · Late fusion · Deep neural network

1 Introduction

Over the past decade, the tremendous increasing of multimedia data (e.g.image, audio and video) brings difficulties to information processing. Traditional approach for representation learning, classification and retrieval tasks usually focus on singe modality. However, in reality, we receive data from different information channels, one of the most common scenarios is image-text paired document. It is worth to note that different data modalities actually carry different information at different semantic levels. As shown in Fig. 1, an example document which contains an image and a loosely related descriptive text. If image and text information can be semantically fused, the more expressive and representative features can be learned for representing this document, and further improve multimodal document classification accuracy. Realizing the importance of multimodal information, in this work, we propose to address this problem by fusing visual and textual information with deep neural network. Multi-modality joint modeling is an open problem in bridging "semantic gap" across modalities. The procedure for

© Springer International Publishing Switzerland 2015
S. Arik et al. (Eds.): ICONIP 2015, Part I, LNCS 9489, pp. 662–670, 2015.
DOI: 10.1007/978-3-319-26532-2_73

 "Though adult lions have no natural predators, evidence suggests that the majority die violently from humans or other lions.Schaller, p. 183 This is particularly true of male lions, who, as the main defenders of the pride, are more likely to come into aggressive contact with rival males..." (—-from Wikipedia)

Fig. 1. An example document that paired with an image and a descriptive text

multimodal data modeling generally falls into two stages: (1) modality representation and (2) correlation learning. In modality representation, one popular approach for image representation is to represent images as "bag-of-visual-words" (BOVW) using scale-invariant feature transform (SIFT) [10] or Dense SIFT [16] descriptor. In text representation, text is represented as topic feature that derived from Latent Dirichet Allocation [2]. Recently, many approaches have been proposed to explore the correlation between different modalities, including Canonical Correlation Analysis (CCA) [14], Semantic Correlation Match (SCM)[12], Cross-Modal Topic Correlations (CMTC) [19].

Unfortunately, the problem of fusing and combining different modalities was rarely discussed for multimedia data classification. In this paper, from different perspective, we focus on multimodal fusion problem [1]. In [3] St. Clinchant et al. proposed semantic combination approach for late fusion and image re-ranking in multimedia retrieval. D.Liu et al. [9] proposed Sample Specific Late Fusion (SSLF) method, which learns the optimal sample-specific fusion weights and enforces the positive sample have the highest fusion scores. In [17] deep neural network has been proved effective at fusion video keyframe and audio information for video classification. Considering the powerful capability of late fusion in areas such as video analysis [18], image retrieval [5] and object recognition [13].

Our work is distinguished from previous works in two aspects. First, we investigated deep convolutional neural network(CNN) features as image feature, this is motivated by recent success of deep CNN feature in addressing various research questions such as speech recognition [6], image classification [8] and multimodal learning [11]. Compare to commonly used SIFT feature, we prove that deep CNN features are more robust and representative in multi-modality fusion. Second, we propose to use deep neural network (DNN) to capture the highly non-linear dependency between different modalities, besides, late fusion with linear interpolation rule is adopted to capture the semantic contribution of image and text. Our contributions can be summarized as follows:

1. We propose to represent image and text to higher level feature using deep CNN feature and topic feature respectively.
2. We propose a novel approach to learn visual-textual fusion feature, which is seen as a unified representation for document categorization.
3. Extensive experiments and discussion were provided to show the effectiveness of DNN based late semantic fusion.

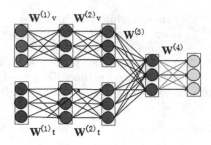

Fig. 2. DNN late fusion framework. Red nodes are visual (image) feature inputs and blue nodes are textual (text) feature inputs. The visual-textual fusion feature can be extracted from the output of 4th layer (Color figure online).

The remainder of this paper is organized as follows. Section 2 states proposed approach for late semantic fusion and then we describe our implementation details in Sect. 3. Section 4 presents the experimental evaluation which illustrates the effectiveness of DNN-based late semantic fusion. Section 5 concludes this work.

2   DNN Late Semantic Fusion

This section introduces proposed DNN late semantic fusion. Given a set of N documents $S = \{D_n\}, \forall n = 1, 2...N$, where D_i is image-text paired document. We extracted the deep CNN feature and topic feature for each document D_n, then document D_n can be represented as $D_n = \{I_n, T_n\}, I_n \in \mathbb{R}^{d_i}, T_n \in \mathbb{R}^{d_t}$ at feature level, where d_i and d_t are the dimensionality of visual and textual feature respectively. Traditionally, if we combine visual feature I_n and textual feature T_n at feature level, called early fusion, formulated as

$$F_{(n)} = \alpha_v f_r(I_n) + (1 - \alpha_v) f_r(T_n), \ \forall n = 1, 2, .., N, r \in [0, 1] \tag{1}$$

where $F_{(n)}$ is early fused feature which is used to represent given document D_n and $f_r(\cdot)$ is normalization operator. $\alpha_v(0 \le \alpha_v \le 1)$ and 1-α_v denotes the fusion weight of visual and textual feature respectively. Another common approach is depicted in Eq. (2) called late fusion that performs fusion at decision level by combining the prediction scores of M pre-trained classifiers C_m.

$$P_{(n)} = \alpha_v C_m(I_n) + (1 - \alpha_v) C_m(T_n), \forall n = 1, .., N, \forall m = 1, .., M \tag{2}$$

In both approaches, α_v is usually assigned according to empirical experiments for demonstrating the importance of individual feature or classifier. Unfortunately, both fusion strategies do not take the correlations between visual and textual feature into account. A good fusion approach should consider the underlying shared semantic correlation between different modalities and take the advantage of the complementarity of modalities. To address the problem, besides heuristically assign α_v from 0 to 1 to capture the semantic contribution of each modality, we also learn latent fusion weights using deep neural network to capture the relationships across image and text. To achieve this goal, we propose a DNN fusion architecture which is shown as in Fig. 2. For a given single training sample $\{I_n, T_n, Y_n\}$, where I_n and T_n are input image and text feature respectively, Y_n is ground truth category label. The final output the global network can be represented as

$$\begin{cases} \hat{Y}^{(5)} = g^{(5)}(\hat{Y}^{(4)}W^{(4)} + b^{(4)}) \\ \hat{Y}^{(4)} = f^{(4)}((\alpha_v P_v + (1 - \alpha_v)P_t)W^{(3)} + b^{(3)}) \end{cases} \tag{3}$$

where $\hat{Y}^{(l)}$ is the output of lth layer and $W^{(l)}$ denotes the weights that connect to $(l-1)^{th}$ layer(also see from Fig. 2). $g(\cdot)$ and $f(\cdot)$ are activation functions and $b^{(l)}$ is bias item corresponding to l^{th} layer. P_v and P_t are prediction scores that computed from input feature I_n and T_n by

$$\begin{cases} P_v = f^{(3)}[f^{(2)}(I_n W_v^{(1)} + b_v^{(1)})W_v^{(2)} + b_v^{(2)}] \\ P_t = f^{(3)}[f^{(2)}(T_n W_t^{(1)} + b_t^{(1)})W_t^{(2)} + b_t^{(2)}] \end{cases} \tag{4}$$

We unitized sigmoid function $f^{(2)}(x) = f^{(3)}(x) = f^{(4)}(x) = \frac{1}{1+e^{-x}}$ and softmax function $g^{(5)}(x) = e^{(x-\varepsilon)} / \sum_{k=1}^{K} e^{(x_k-\varepsilon)}$ where $\varepsilon = max(x_k)$. To learn optimal weight set $\mathbf{W} = \{W^{(l)}\}$ and $\mathbf{b} = \{b^{(l)}\}$, $\forall l = 1, 2, 3, 4$, with all training samples, the objective is to minimize following loss function

$$\underset{\mathbf{W}, \mathbf{b}}{\operatorname{argmin}} \quad \frac{1}{2N} \sum_{n=1}^{N} \| \hat{Y}_n^{(5)} - Y_n \|^2 + \frac{\lambda}{2} \sum_{l=1}^{L-1} \| W_l \|^2 \tag{5}$$

Where the second part is weight decay item for preventing overfitting. In learning procedures, the weights $W_m^{(1)}$ and $W_m^{(2)}$, m={v,t} are first learned by intra-modality training. Those weights can be regarded as local weights for achieving better prediction results. $W^{(3)}$ and $W^{(4)}$ are learned globally by fusing the scores of predicting image and text feature. The output of the 4th layer are fusion features which combines visual and textual predictions.

3   Implementation

Our experimental configuration are Ubuntu 12.04, Nvidia GTX 780 GPU with 3G memory for image feature extraction. Ubuntu 12.04, Intel 3.20GHz×4 CPU, 8G RAM for text model training and feature extraction. And Window 8, Intel 3.20GHz×4 CPU, 8G RAM for training visual-textual joint model on Matlab.

Dataset: Our experiments were conducted on open benchmark Wikipedia dataset[1], which contains 2886 documents (2173 for training and 693 for test).

---

[1] http://www.svcl.ucsd.edu/projects/crossmodal/.

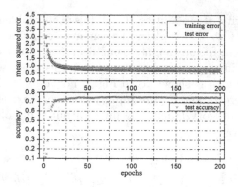

Fig. 3. Top: The mean squared error of training and test against epochs. Bottom: Test accuracy against epochs

This dataset has 10 semantic categories such as "biology","geography". Each document is comprised of an image and a short descriptive text as the example we given in Fig. 1.

Image Representation: We use deep convolutional neural network (deep CNN) [8] that has been proved its effectiveness in image representation in recent years. Based on Caffe framework [7] we extracted the image feature with Caffe model that on ImageNet [4] ILSVRC2012 dataset (more than 1.2M training images). By extracting the output of the 7th layer(F7), each image can be represented as a 4096-dimension vector, that is, visual feature $I \in \mathbb{R}^{4069}$. Due to image features are highly learned by deep CNN, it can be considered as kind of high level semantic feature.

Text Representation: To represent text as semantic feature, Latent Dirichlet Allocation(LDA) [2] was used to generate 20 topics. We compute the topic distribution of given text document d over 20 topics and finally obtain a 20-dimension vector, that is, textual feature $T \in \mathbb{R}^{20}$.

Training: In DNN learning, the first three layers are designed for intra-modal regularization which optimizes the weights within each modality to improve performance firstly. Thus we named our fusion framework as RE-DNN. The networks are designed as [4096/100/10] and [20/100/10] for image and text receptively, and the last three layers is set as [20/100/10]. In our experiments, learning rate α=0.001, momentum=0.9 achieved the best performance. According the scale of our training data (2173 training samples), we adopted the mini batch gradient descent with batch size 41. The epoch number fixed at K=200. Figure 3 shows the change of mean squared error of training and test during training procedure as well as the increasing of test accuracy against training epochs. We obtained the final test accuracy is 74.6 %.

Table 1. Comparison between unimodal and multimodal fusion feature. Top: visualization of visual feature(a(1)), textual feature(a(2)) and fusion features (a(3)) from test examples. Bottom: classification results include precision (P), recall (R), F1-score (F1) and Accuracy (A).

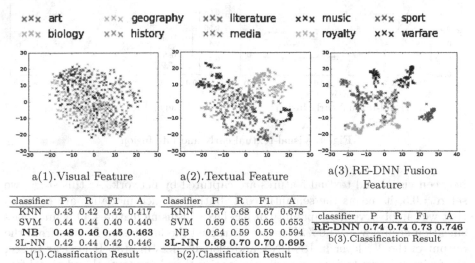

a(1).Visual Feature   a(2).Textual Feature   a(3).RE-DNN Fusion Feature

classifier	P	R	F1	A
KNN	0.43	0.42	0.42	0.417
SVM	0.44	0.44	0.40	0.440
NB	**0.48**	**0.46**	**0.45**	**0.463**
3L-NN	0.42	0.44	0.42	0.446

b(1).Classification Result

classifier	P	R	F1	A
KNN	0.67	0.68	0.67	0.678
SVM	0.69	0.65	0.66	0.653
NB	0.64	0.59	0.59	0.594
3L-NN	**0.69**	**0.70**	**0.70**	**0.695**

b(2).Classification Result

classifier	P	R	F1	A
RE-DNN	**0.74**	**0.74**	**0.73**	**0.746**

b(3).Classification Result

4   Experimental Evaluation

To validate the effectiveness of proposed RE-DNN approach for multimodal feature fusion. Our experiment first consider unimodal (visual or textual feature separately) to perform document categorization task and then compared with RE-DNN approach. In this work, we also explored early fusion and late fusion on some mainstream classifiers such as K-Nearest Neighbor(KNN), Support Vector Machine (SVM), Naive Bayes (NB) and Neural Network(NN).

Table 1 shows the comparison between unimodal feature and multimodal fusion feature based classification. We visualized visual feature $I \in \mathbb{R}^{4096}$ and textual feature $T \in \mathbb{R}^{20}$ to 2D by using t-SNE [15] as shown in a(1) and a(2) respectively. By visually comparing visual and textual feature from a(1) and a(2) find that the margin of textual feature tend to be clearer. Meanwhile, we applied those features to classification. We note that text-based classification outperforms image-based classification for all employed classifiers. This confirms previous research that text information is easier to be perceived and recognized by machines compare to image information. The best performed classification accuracy of visual feature is achieved by NB with 0.463, and a 3L-NN achieved the best classification accuracy 0.695 for textual feature. The configuration of 3L-NN are {4096/100/10} for visual feature and {20/100/10} for textual feature. The learning rate is adjusted as 0.001 and momentum=0.9. However, the further improvements are made by fusing visual and textual feature with deep neural network. This relies on the fact that paired image and text are perceived by machines that they belong to same semantic and the latent relationships

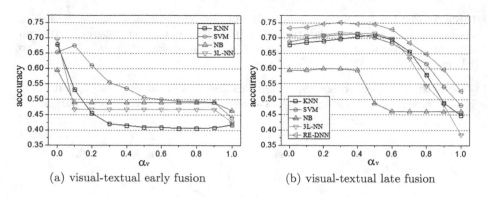

(a) visual-textual early fusion   (b) visual-textual late fusion

Fig. 4. Visual-textual early and late fusion

between visual and textual features are captured by network. At this stage, we set $\alpha_v = 0.5$, it means the semantic contribution of each modality are equal so that we can observe the capability of RE-DNN in fusing features. Our final classification accuracy is 74.6 %. Here we extracted the late fusion feature from the output of the 4th layer in RE-DNN and visualized as in a(3). It is clear to see, the fusion features tend to more discriminative than both textual and visual features. Compare Table 1 b(1)–b(3) we see that the overall performance including precision, recall, F1 and accuracy of RE-DNN approach are higher than unimodal based classification. The result shows that late fusion based RE-DNN improves on the approaches "3L-NN for textual" and "NB for visual" by 5.1 % and 28.3 % respectively.

Further experiments were conducted to explore visual-textual early fusion and late fusion by taking the semantic contribution of each modality into consideration. In both fusion strategies, according to Eqs. (1) and (2) we heuristically assign α_v(image modality weight) from 0 to 1. For early fusion, the inputs are raw image and text features. For late fusion, the inputs are prediction scores of different classifiers. Figure 4(a) shows the accuracy changes in early fusion and Fig. 4(b) describes late fusion results. It is observed that late fusion outperforms early fusion at most of levels of α_v. In early fusion approach, almost the accuracy for all classifiers decreasing along with the increasing of α_v. When we impose linear interpolation on RE-DNN, we note that for all levels of α_v, RE-DNN late fusion with linear interpolation further improved the classification accuracy to 75.3 % at $\alpha_v = 0.3$. It proves the effectiveness of our approach.

5   Conclusions

In this paper, we have proposed a DNN framework for fusing visual and textual features. By imposing linear interpolation on DNN, more discriminative and representative fusion feature can be extracted. Our experiments on document categorization show that our proposed approach outperforms mainstream classifiers in both early fusion and late fusion.

References

1. Atrey, P.K., Hossain, M.A., El-Saddik, A., Kankanhalli, M.S.: Multimodal fusion for multimedia analysis: a survey. Multimedia Syst. **16**(6), 345–379 (2010)
2. Blei, D.M., Ng, A.Y., Jordan, M.I.: Latent dirichlet allocation. J. Mach. Learn. Res. **3**, 993–1022 (2003)
3. Clinchant, S., Ah-Pine, J., Csurka, G.: Semantic combination of textual and visual information in multimedia retrieval. In: Proceedings of the 1st ACM International Conference on Multimedia Retrieval, p. 44. ACM (2011)
4. Deng, J., Dong, W., Socher, R., Li, L.-J., Li, K., Fei-Fei, L.: Imagenet: a large-scale hierarchical image database. In: IEEE Conference on Computer Vision and Pattern Recognition, 2009 (CVPR 2009), pp. 248–255. IEEE (2009)
5. Escalante, H.J.: Late fusion of heterogeneous methods for multimedia image retrieval (2008)
6. Hinton, G., Deng, L., Dong, Y., Dahl, G.E., Mohamed, A., Jaitly, N., Senior, A., Vanhoucke, V., Nguyen, P., Sainath, T.N., et al.: Deep neural networks for acoustic modeling in speech recognition: the shared views of four research groups. IEEE Sig. Process. Mag. **29**(6), 82–97 (2012)
7. Jia, Y., Shelhamer, E., Donahue, J., Karayev, S., Long, J., Girshick, R., Guadarrama, S., Darrell, T.: Caffe: Convolutional architecture for fast feature embedding. arXiv preprint arXiv:1408.5093 (2014)
8. Krizhevsky, A., Sutskever, I., Hinton, G.E.: Imagenet classification with deep convolutional neural networks. In: Advances in Neural Information Processing Systems, pp. 1097–1105 (2012)
9. Liu, D., Lai, K.-T., Ye, G., Chen, M.-S., Chang, S.-F.: Sample-specific late fusion for visual category recognition. In: IEEE Conference on Computer Vision and Pattern Recognition (CVPR), 2013, pp. 803–810. IEEE (2013)
10. Lowe, D.G.: Distinctive image features from scale-invariant keypoints. Int. J. Comput. Vis. **60**(2), 91–110 (2004)
11. Ngiam, J., Khosla, A., Kim, M., Nam, J., Lee, H., Ng, A.Y.: Multimodal deep learning. In: Proceedings of the 28th International Conference on Machine Learning (ICML 2011), pp. 689–696 (2011)
12. Rasiwasia, N., Pereira, J.C., Coviello, E., Doyle, G., Lanckriet, G.R.G., Levy, R., Vasconcelos, N.: A new approach to cross-modal multimedia retrieval. In: Proceedings of the International Conference on Multimedia, pp. 251–260. ACM (2010)
13. Terrades, O.R., Valveny, E., Tabbone, S.: Optimal classifier fusion in a non-bayesian probabilistic framework. IEEE Trans. Pattern Anal. Mach. Intell. **31**(9), 1630–1644 (2009)
14. Thompson, B.: Canonical correlation analysis. In: Everitt, B., Howell, D. (eds.) Encyclopedia of Statistics in Behavioral Science. Wiley, New York (2005)
15. Van der Maaten, L., Hinton, G.: Visualizing data using t-sne. J. Mach. Learn. Res. **9**(2579–2605), 85 (2008)
16. Vedaldi, A., Fulkerson, B.: Vlfeat: An open and portable library of computer vision algorithms. In: Proceedings of the International Conference on Multimedia, pp. 1469–1472. ACM (2010)
17. Wu, Z., Jiang, Y.-G., Wang, J., Pu, J., Xue, X.: Exploring inter-feature and inter-class relationships with deep neural networks for video classification. In: Proceedings of the ACM International Conference on Multimedia, pp. 167–176. ACM (2014)

18. Ye, G., Liu, D., Jhuo, I.-H., Chang, S.-F.: Robust late fusion with rank minimization. In: 2012 IEEE Conference on Computer Vision and Pattern Recognition (CVPR), pp. 3021–3028. IEEE (2012)
19. Yu, J., Cong, Y., Qin, Z., Wan, T.: Cross-modal topic correlations for multimedia retrieval. In: 2012 21st International Conference on Pattern Recognition (ICPR), pp. 246–249. IEEE (2012)

Prototype Selection on Large and Streaming Data

Lakhpat Meena[✉] and V. Susheela Devi

Department of Computer Science and Automation, Indian Institute of Science,
Bangalore 560 012, India
lakhpatmeena01@gmail.com, susheela@csa.iisc.ernet.in

Abstract. Since streaming data keeps coming continuously as an ordered sequence, massive amounts of data is created. A big challenge in handling data streams is the limitation of time and space. Prototype selection on streaming data requires the prototypes to be updated in an incremental manner as new data comes in. We propose an incremental algorithm for prototype selection. This algorithm can also be used to handle very large datasets. Results have been presented on a number of large datasets and our method is compared to an existing algorithm for streaming data. Our algorithm saves time and the prototypes selected gives good classification accuracy.

Keywords: Prototype selection · One-pass algorithm · Streaming data

1   Introduction

Streaming algorithms are one-pass algorithms, used for processing data streams which can be examined in only one-pass. Typical prototype selection methods work with static data sets which means we have the whole data set on disk and we can access the data any number of times to run the algorithm on the whole data set. In case of stream data and very large data sets, it may be impossible to store the entire data set on disk.In order to reduce the space and time complexity for classification, it is better to work on the condensed set instead of the large training set. Prototype selection refers to the process of reducing the size of the training set to get a training set consistent subset. Let,

$$T = \{(x_1, c_1), (x_2, c_2), \ldots, (x_n, c_n)\} \tag{1}$$

be the given labelled training set of n patterns. Prototype selection gives reduced condensed set

$$S = \{(x^1, c^1), (x^2, c^2), \ldots, (x^k, c^k)\}. \tag{2}$$

S is the obtained condensed set, where k <n and S is a subset of T and each $x^i, 1 \leq i \leq k$ with class label c^i is obtained from patterns in T.

The first and well known algorithm for prototype selection is Condensed nearest-neighbour rule (CNN rule) introduced by P.E. Hart [2]. CNN rule uses

© Springer International Publishing Switzerland 2015
S. Arik et al. (Eds.): ICONIP 2015, Part I, LNCS 9489, pp. 671–679, 2015.
DOI: 10.1007/978-3-319-26532-2_74

nearest neighbour rule to compute misclassified data points. Cover and Hart [1] proposed that single-NN rule has lower probability of error than other k-NN rules and hence it is admissible among all k-NN rules. Initially it takes a random data point from the training set and adds it to the condensed set. After initialization of condensed set, it checks the next selected point from training set T using nearest-neighbour algorithm. If it is misclassified, then it is included in condensed set S. The algorithm terminates when all data points are correctly classified using the condensed set. CNN is *order-dependent* as the points of condensed set S depends on selection of initial sample point and order in which data is presented to the algorithm. The Fast CNN rule [5] uses the voronoi set to compute representative data points. The Voronoi cell of point $p \in S$ set is the set of all points of T that are closer to p than any other point p' in set S. Voronoi enemies of $p \in S$ is set of data points that have different class label. During each iteration, for each point $p \in S$, nearest neighbour of point p in Voren(p,S,T) is selected and inserted to condensed set S. Karacali and Krim [6] proposed Structural Risk Minimization using the NN rule (NNSRM). The complexity of NNSRM algorithm is $O(n^3)$. The Reduced nearest neighbour rule [3] reduces the size of the final condensed set. In Bien and Tibshirani [12], the problem of finding prototypes is translated into a set cover optimization problem. Verbiest et al. [14], use a fuzzy rough approach to carry out prototype selection. In Li and Wang [15], multi-objective optimization and partitioning is used to carry out prototype selection. Garcia and Derrac [10] give a taxonomy of methods for prototype selection. In Gadodiya and Chandak [13], a survey is made of prototype selection algorithms for kNN classification.

All these prototype selection algorithms are for static data sets. Prototype selection algorithms for streaming data should run in linear time. In Czarnowski and Jedrzejowicz [11], ensemble classifier is used for classification of data streams, where each ensemble is induced from the incoming chunks of data in the data stream. This is the algorithm WECU which we have compared with our method. In Law et al. [7], they proposed an incremental classification algorithm based on multi-resolution data representation to compute nearest neighbors of a point. Beringer and Hüllermeier [8] have proposed instance-based learning on data streams. The subset D is considered as training set for operations on given query. Tabata et al. [9] proposed a volume prototype selection algorithm for streaming data. The algorithm uses the acceptance region to update prototypes. If the new sample data point falls into the acceptance region, then this point is added to update the prototypes.

Prototype selection algorithms compute condensed set, that provides faster classification without sacrificing the classification accuracy. In this paper, prototype selection is carried out to get condensed set for streaming data sets and large data sets. As the data keeps coming, the condensed set has to be modified to reflect the data coming in. This should be done in an incremental manner without having to go through the entire prototype selection algorithm again. Also, as the data stream comes in, stays for some time and goes away, we can sample the data only once and we need a one-pass algorithm to incrementally

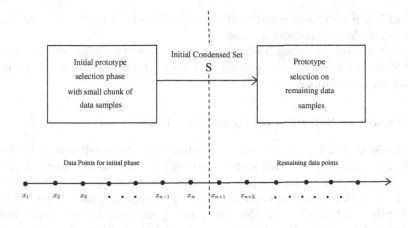

Fig. 1. Algorithm flow

obtain the condensed set. Our algorithm is a one-pass algorithm and computation time is linear. This algorithm is used with big data sets, where each instance comes in one by one.

2   Methodology

The algorithm deals with data streams, which have infinite data points. To deal with these infinite and large data sets, the algorithm executes in two phases. The initial phase deals with small amount of data points and computes initial prototypes. Second phase deals with remaining incoming streaming data points to compute next prototypes for condensed set. In the initial prototype selection phase of the algorithm, it considers the data set

$$D = \{(x_1, c_1), (x_2, c_2), \dots, (x_n, c_m)\} \tag{3}$$

here c_i is the class label of x_i, as initial training set, where (i) m $<$n and (ii) D is a subset of T. We can choose initial data size σ according to our memory space. To compute initial condensed set, we apply MCNN rule [4] on the initial set D (Fig. 1).

2.1   MCNN Rule

Initial condensed set is computed by MCNN rule [4]. MCNN rule does partitioning of class regions to non-overlapping regions. It computes prototype samples in an incremental manner. MCNN rule starts with taking centroid points of each class of training set into condensed set. In each iteration, it applies this method on misclassified data points. In each iteration, MCNN adds at most c prototypes to condensed set S , where c is the number of classes in the training set T. Unlike

CNN, MCNN is *order-independent*. We use MCNN with a small subset of stream points to get the initial condensed set.

The MCNN rule terminates when all data samples of training set are correctly classified using condensed set *cond* and therefore the condensed set gives 100 % accuracy on the training set.

2.2 Prototype Selection on Remaining Data Points

Initial phase is used to compute initial condensed set for small subset D of data stream. The second phase deals with remaining data points. It requires only one-time scan of incoming data samples. In second phase, it takes initial condensed set D from initial phase as condensed set. Two methods have been used for updating the condensed set. In Method 1, data points which are misclassified by condensed set are added directly to the condensed set. In Method 2, misclassified samples are collected together and every now and then MCNN is run on these samples to get a condensed set which is added on to the original condensed set.

2.3 Algorithm

Method 1: In phase, the algorithm considers each incoming data sample and incrementally adds misclassified data samples in to the current condensed set.

Algorithm 1. Prototype selection on streaming data set

Input: $T=$ training set, $cond = \phi$.
1. Initialize set $D = \{(x_1, c_1), (x_2, c_2), \ldots, (x_m, c_m)\}$ dataset for first phase.
2. Initialize set $R = T \setminus D$ remaining data set.
3. Set $cond = \text{mcnn}(D)$.
4. Classify remaining data samples.
for all x_i in R **do**
    find neigh = nearest-neighbour(x_i , *cond*) .
    if $label(x_i) \neq label(neigh)$ **then**
       update *cond* by adding x_i sample into *cond*.
    end if
end for
5. Stop and *cond* is final condensed set.

It requires only one-pass to get remaining prototypes of condensed set from remaining data stream. It will terminate after full scan of training data set T. The algorithm of Method 1 is shown in Algorithm 1. In the Algorithm 1, We used set D as input to MCNN rule to compute initial condensed set. All remaining data samples x_i, where $i > m$, are data samples belonging to R.

In step 2–3, it creates initial subset D of size m and set R is the remaining data set. The algorithm then computes initial condensed set using MCNN rule. MCNN rule is applied on set D. Step 4 is used to select prototypes in remaining data samples of data stream i.e. from set R. When new data sample comes, it computes its nearest-neighbour in current *cond* set. For all data samples, If class label of data sample is different than class label of computed neighbour data point, then add this data sample into *cond* (condensed set). For streaming data set, when we want to have final condensed set, we use the current condensed set as final condensed set. With big data sets, it will terminate after full scan of training data set T.

The algorithm also computes condensed set for big data sets in only one-pass scan. The algorithm in this paper, computes condensed set without the availability of complete data set at same time. To use this algorithm on big data sets, we can handle big data set as data streams. In this approach, data samples of big data set will come as sequence of data samples. So with big data sets, it handles both memory management and execution time requirements.

Method 2: This is a variation of method 1. The algorithm described in this section results in a smaller condensed set than method 1. On the other hand, it takes more time to compute the final condensed set than method 1. In method 2, prototype set is built incrementally with different small subsets of data points. The algorithm of Method 2 is shown in Algorithm 2.

Algorithm 2. Method 2 for Prototype selection on Streaming data set

---

Input: $T=$ training set, $cond = \phi$, $counter = 0$.
1. Initialize set $D = \{(x_1, c_1), (x_2, c_2), \ldots, (x_m, c_m)\}$ dataset for first phase.
2. Initialize set $R = T \setminus D$ remaining data set.
3. Set $cond = \text{mcnn}(D)$.
4. Classify remaining data samples.
for all x_i in R **do**
  Increment *counter* by one.
  Find neigh $=$ nearest-neighbour$(x_i , cond)$.
  if $label(x_i) \neq label(neigh)$ **then**
    Update *typical* by adding x_i sample into *typical*.
  end if
  if $counter == L$ **then**
    $cond = cond \cup \text{mcnn}(typical)$.
    Set $counter = 0$ and $typical = \phi$.
  end if
end for
5. Stop and *cond* is final condensed set.

The main difference between method 1 and method 2 occurs in step 4. In this algorithm L is the number of analyzed data points after which the algorithm carries out prototype set selection on the misclassified data points from these L data points using MCNN. In this step, it classifies each incoming data sample using current condensed set *cond* as in method 1. For each data sample, if it is misclassified then add this data sample into *typical* set and increase the *counter* value by one. When *counter* value is equal to L, it applies MCNN rule on *typical* set and sets *counter* value to zero. It adds this condensed set computed from *typical* to the current condensed set and sets *typical* $= \phi$. As in Method 1, it computes the final condensed set after full scan of data set.

Table 1. Datasets used in experiments

Dataset	No. of samples	No. of features	No. of classes
Optical dig. rec	5,620	64	10
Pen dig. rec	10,992	16	10
Letter image rec	20,000	16	26
Gisette	15,300	5000	2
Forrest cover type	581,012	54	7

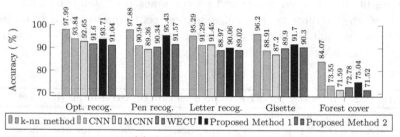

(a) Accuracy comparison

	Algorithms	optical recog.	pen digit recog.	letter recog.	gisette	forest cover
				Datasets		
Total Execution time (in seconds)	K-NN	219.367	367.252	1881.049	8592.908	67485.967
	CNN	108.543	113.988	1088.749	3261.942	39481.851
	MCNN	123.475	151.387	1258.749	5136.131	45541.165
	WECU	103.162	109.370	835.046	2409.370	23109.370
	Method 1	51.404	84.035	587.914	1732.530	17489.347
	Method 2	57.017	96.460	746.046	2318.855	20942.107
Size of dataset for initial phase	K-NN	-	-	-	-	-
	CNN	-	-	-	-	-
	MCNN	-	-	-	-	-
	WECU	-	-	-	-	-
	Method 1	260	230	702	978	12922
	Method 2	260	230	702	978	12922
Size of condensed set	K-NN	-	-	-	-	-
	CNN	306	313	1532	195	31204
	MCNN	68	45	189	71	18809
	WECU	-	-	-	-	-
	Method 1	217	256	708	152	24763
	Method 2	143	189	504	108	19045
Time taken for condensation process (in seconds)	K-NN	-	-	-	-	-
	CNN	74.531	78.875	781.824	2065.741	23849.168
	MCNN	107.562	121.387	1142.120	3036.458	32513.984
	WECU	-	-	-	-	-
	Method 1	31.348	47.374	410.568	1426.945	12943.985
	Method 2	38.407	58.975	574.349	1823.367	17063.451
Accuracy (%)	K-NN	97.997 (k=1)	97.885 (k=3)	95.292 (k=1)	96.200 (k=1)	84.074 (k=1)
	CNN	93.843	90.939	91.292	88.913	73.549
	MCNN	92.654	89.365	91.456	87.200	71.592
	WECU	91.597	90.337	88.969	89.900	72.783
	Method 1	93.712	95.426	90.061	91.700	75.042
	Method 2	91.041	91.567	89.021	90.300	71.523

(b) Experimental results for all datasets used

Fig. 2. Comparison of all six algorithms for five different data sets

In streaming data set, whenever we want to have a condensed set for classification, we use current condensed set as the final condensed set. During classification of testing data set, algorithm will use this condensed set instead of complete training data set. It takes some time for condensation but after condensation, classification requires less time using the condensed set.

3   Results

We used five data sets for performance analysis. The data set details are given in Table 1. The algorithm is also applied on large data set. The last data set, Forest cover type data set is a large data set having 581,012 data points. Experimental results of both the proposed methods for streaming data sets are compared with previous algorithms.

Figure 2(a) gives the comparison of classification accuracy on different datasets for all methods. Figure 2(b) shows experimental results of all the data sets used for performance analysis. From Fig. 2(b), In each part, the first row of table shows the result obtained by using k-nn rule on complete dataset. The given result is the highest accuracy for a specific k value in k-nn rule. No condensation approach is applied in this classification. The execution time which is given in the table for this approach, is the overall time it takes to classify testing data set using whole training data set. As expected this takes the most time.

The condensed set produced by proposed methods give better classification accuracy on most datasets than other approaches since it includes all important required data points in condensed set with respect to their distribution. From the experimental results, we can see that proposed methods of our paper give better

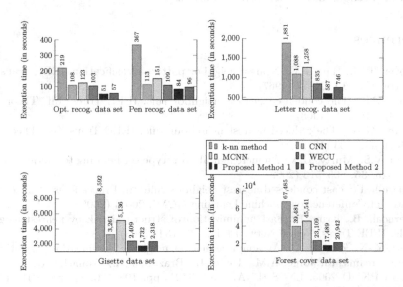

Fig. 3. Execution time comparison of all six algorithms for five different data sets

results than the result obtained by ANNCAD approach [7] for both letter recognition and forest cover datasets. Experimental results show that the proposed methods gives good accuracy with less execution time. On every data set, proposed methods compute very small condensed set. We can observe from results that both proposed methods give better results than existing WECU approach for data streams with less execution time and memory space. In all five data sets, there is a huge saving in the time required. Method 1 gives higher no. of prototypes than method 2 but gives better results. Also, the entire data set need not be stored. Only the initial training set and the condensed set formed needs to be stored. We get performance comparable to CNN and MCNN and often gives better result than CNN and MCNN. To give an idea of the huge saving in times Fig. 3, gives the time taken by the various algorithms for all the datasets.

4   Conclusion

In this paper, we proposed two prototype selection methods for data streams. The algorithms give good results for both streaming data set and large data sets. Streaming data comes as a sequence of data points with infinite length. The proposed methods do not require complete data set at same time, therefore they handle memory requirement and strict time constraints for streams. It has also been shown how for large data by dividing it into chunks or using one pattern at a time, this algorithm can be used leading to considerable saving in time and space complexity. We have discussed both proposed methods with regard to accuracy and execution time and also comparison with existing approaches. Method 1 requires less execution time and Method 2 computes smaller condensed set. Both methods have their own advantages.

References

1. Cover, T.M., Hart, P.E.: Nearest neighbor pattern classification. IEEE Trans. Inf. Theor. (IT) **13**, 21–27 (1967)
2. Hart, P.E.: The condensed nearest neighbor rule. IEEE Trans. Inf. Theor. (IT) **14**(3), 515–516 (1968)
3. Gates, G.W.: The reduced nearest neighbour rule. IEEE Trans. Inf. Theor. (IT) **18**(3), 431–433 (1972)
4. Devi, V.S., Murty, M.N.: An incremental prototype set building technique. Pattern Recogn. **35**, 505–513 (2002)
5. Angiulli, F.: Fast condensed nearest neighbor rule. In: Proceedings of 22nd International Conference on Machine Learning (ICML 2005) (2005)
6. Karacali, B., Krim, H.: Fast minimization of structural risk by nearest neighbor rule. IEEE Trans. Neural Netw. **14**(1), 127–134 (2003)
7. Law, Y.-N., Zaniolo, C.: An adaptive nearest neighbor classification algorithm for data streams. In: Jorge, A.M., Torgo, L., Brazdil, P.B., Camacho, R., Gama, J. (eds.) PKDD 2005. LNCS (LNAI), vol. 3721, pp. 108–120. Springer, Heidelberg (2005)

8. Beringer, J., Hüllermeier, E.: Efficient instance-based learning on data streams. Intell. Data Anal. **11**(6), 627–650 (2007)

9. Tabata, K., Sato, M., Kudo, M.: Data compression by volume prototypes for streaming data. Pattern Recogn. **43**, 3162–3176 (2010)

10. Garcia, S., Derrac, J.: Prototype selection for nearest neighbor classification : taxonomy and empirical study. IEEE Trans. PAMI **34**, 417–435 (2012)

11. Czarnowski, I., Jedrzejowicz, P.: Ensemble classifier for mining data streams. In: 18th International Conference on Knowledge-Based and Intelligent Information and Engineering Systems(KES 2014), Procedia Computer Science, vol. 35, pp. 397–406 (2014)

12. Bien, J., Tibshirani, R.: Prototype selection for interpretable classification. Ann. Appl. Stat. **5**(4), 2403–2424 (2011)

13. Gadodiya, S.V., Chandak, M.B.: Prototype selection algorithms for kNN classifier: a survey. Int. J. Adv. Res. Comput. Commun. Eng. (IJARCCE) **2**(12) (2013)

14. Verbiest, N., Cornelis, C., Herrera, F.: FRPS : a fuzzy rough prototype selection method. Pattern Recogn. **46**(10), 2770–2782 (2013)

15. Li, J., Wang, Y.: A nearest prototype selection algorithm using multi-objective optimization and partition. In: 9th International Conference on Computational Intelligence and Security, pp. 264–268, December 2013

GOS-IL: A Generalized Over-Sampling Based Online Imbalanced Learning Framework

Sukarna Barua[1]([✉]), Md. Monirul Islam[1], and Kazuyuki Murase[2]

[1] Bangladesh University of Engineering and Technology (BUET), Dhaka, Bangladesh
sukarna.barua@gmail.com
[2] University of Fukui, Fukui, Japan

Abstract. Online imbalanced learning has two important characteristics: samples of one class (minority class) are under-represented in the data set and samples come to the learner online incrementally. Such a data set may pose several problems to the learner. First, it is impossible to determine the minority class beforehand as the learner has no complete view of the whole data. Second, the status of imbalance may change over time. To handle such a data set efficiently, we present here a dynamic and adaptive algorithm called Generalized Over-Sampling based Online Imbalanced Learning (GOS-IL) framework. The proposed algorithm works by updating a base learner incrementally. This update is triggered when number of errors made by the learner crosses a threshold value. This deferred update helps the learner to avoid instantaneous harms of noisy samples and to achieve better generalization ability in the long run. In addition, correctly classified samples are not used by the algorithm to update the learner for avoiding over-fitting. Simulation results on some artificial and real world datasets show the effectiveness of the proposed method on two performance metrics: recall and g-mean.

Keywords: Imbalanced learning · Online learning · Oversampling

1 Introduction

A class imbalance problem deals with data sets where number of samples of one class is very few compared to the number of samples of other classes. Whenever a classifier is presented with such a dataset, it usually performs badly. Specifically, its performance is worse on the minority class, the class having under-representation in the data set. Online learning adds an additional dimension to the problem in the sense that data samples come to the classifier one by one in online fashion. Online imbalance learning has practical applications such as web click data [1], spam filtering [2], and credit card transactions [3].

The online nature of data creates some additional problems for a classifier. In offline version, the classifier can detect the minority class and majority class before learning begins. So, the majority or minority status of a class is static and it does not change over time. However, in online learning, it is not possible to determine in advance which class will be minority or majority as the classifier has

© Springer International Publishing Switzerland 2015
S. Arik et al. (Eds.): ICONIP 2015, Part I, LNCS 9489, pp. 680–687, 2015.
DOI: 10.1007/978-3-319-26532-2_75

no complete view of the data. In addition, classes can even change their minority or majority status over time. So, classifiers have to be adaptive to detect minority and majority classes dynamically and to cope with the changing nature of the data. Although, imbalance learning problem has been studied extensively in literature [4–8], few studies have been done to tackle the online version [9–11]. The authors in [11] proposed two algorithms based on online bagging: Over-sampling based Online Bagging (OOB) and Under-sampling based Online Bagging (UOB) algorithm. Two perceptron based methods were proposed in [9,10] that adjusts misclassification costs to update the weights of underlying learning model.

In this paper, we propose a general framework to handle online imbalanced learning problems. We named our framework as Generalized Over-Sampling based Online Imbalanced Learning (GOS-IL) algorithm. The proposed algorithm has the following key characteristics that make it different from existing approaches:

- GOS-IL works in online fashion and keeps a base classifier model that is updated dynamically. Unlike existing approaches, GOS-IL does not update the learner by samples that are correctly classified by the learner. It is done to avoid over-fitting to the majority class samples. Because, the number of majority samples is far greater than the minority, and therefore updating learner by correctly classified samples may over-fit it towards majority class.
- GOS-IL uses a deferred update mechanism in which it does not update the learner after receiving each sample. Errors are inherent to any learning model. GOS-IL allows its underlying classifier to mis-classify samples up-to an acceptable amount that may help the learner to gain more generalization ability and to avoid any erroneous update by noisy samples. To the best of our knowledge, no other works has such characteristics.

The rest of the works are described in four sections. In Sect. 2, we describe the proposed framework GOS-IL and its features. Section 3 presents experimental designs and results. In Sect. 4, we look into some future works and conclude the paper.

2 Proposed Framework: GOS-IL

In this section, we present GOS-IL algorithm. The complete algorithm is shown in **Algorithm 1: GOS-IL**. For each class w_i, GOS-IL maintains three dynamic parameters e_{w_i}, n_{w_i}, and S_{w_i} where e_{w_i} is the number of samples of class w_i that are mis-classified by the model so far, n_{w_i} is the number of samples of class w_i encountered by the model so far, and S_{w_i} stores the samples of class w_i encountered by the model so far.

Let, (x_t, w_t) be the data sample comes at time t to the algorithm where x_t is the feature vector and w_t is the class identifier. GOS-IL updates the parameters e_{w_i}, n_{w_i}, and S_{w_i} at each time step t depending on the target class w_t and learner's current prediction y_t (Lines 6–10). These parameters are also reset to

0 after certain number of samples (T_n) of a class are processed or after certain amount of misclassification (T_e) of a class occurs.

Algorithm 1. GOS-IL

---

1. $L \leftarrow$ Initialize learning model
2.   for each class w_i
3. $e_{w_i} \leftarrow 0, n_{w_i} \leftarrow 0, S_{w_i} \leftarrow \phi, P_{w_i} \leftarrow \phi$
4.   end for
5.   for each sample (x_t, w_t) received at time step t
6. $n_{w_t} \leftarrow n_{w_t} + 1, S_{w_t} \leftarrow S_{w_t} \cup \{x_t\},$
7. $y_t \leftarrow$ Prediction of x_t by the current model L
8.   if $y_t \neq w_t$
9. $e_{w_t} \leftarrow e_{w_t} + 1, P_{w_t} \leftarrow P_{w_t} \cup \{x_t\}$
10.   end if
11.   if $e_{w_t} > T_e$
12. $r_t \leftarrow \frac{n_{w_t}}{\sum_i n_{w_i}}$
13.   if $r_t < \delta_r$
14. $O_{w_t} \leftarrow$ Apply oversampling on P_{w_t} using S_{w_t}
15.   else
16. $O_{w_t} \leftarrow P_{w_t}$
17.   end if
18. $L \leftarrow$ Update model L using samples in O_{w_t}
19. $e_{w_t} \leftarrow 0, n_{w_t} \leftarrow 0, S_{w_t} \leftarrow \phi, P_{w_t} \leftarrow \phi$
20.   else if $n_{w_t} > T_n$
21. $e_{w_t} \leftarrow 0, n_{w_t} \leftarrow 0, S_{w_t} \leftarrow \phi, P_{w_t} \leftarrow \phi$
22.   end if
23.   end for

GOS-IL uses the dynamic parameters e_{w_i} and n_{w_i} to determine the following: (i) when to update the underlying learner (Line 11 of Algorithm 1), and (ii) whether current data stream is imbalanced (Line 12). Although, GOS-IL works in an online fashion, it stores the last T_n data samples (at a maximum) of each class w_i in the set S_{w_i} in memory. These T_n samples include both misclassified samples (to be used for future update) and correctly classified samples (used in over-sampling).

The key characteristics of the proposed GOS-IL algorithm are described below.

Deffered Update and Permissible Error Level (PEL). GOS-IL uses a differed update technique. Rather than updating the learner after receiving each sample, GOS-IL allows the underlying classifier to make errors up to a certain amount during the online learning phase. This amount is denoted as permissible

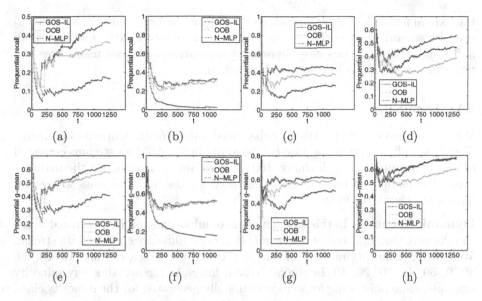

Fig. 1. Prequential minority recall values for (a) dynamic, (b) static high, (c) static moderate, and (d) static low data stream. Prequential g-mean values for (a) dynamic, (b) static high, (c) static moderate, and (d) static low data stream.

error level(PEL). The algorithm waits to find whether the cumulative number of errors is within PEL (Line 11). PEL can be calculated as $PEL = \frac{T_e}{T_n}$. This property can help the learner to achieve more generalization capability. A PEL value of 0.05 means the learner is allowed to make at most 5 % errors on each class individually. Typical values of T_e and T_n can be set so that PEL is within $0.01 - 0.05$.

No Update for Correct Classifications. Unlike existing works, GOS-IL does not update the model with samples that are correctly classified by the learner. This is done to increase the generalization capability of the online learner and to avoid any over-fitting issues. Specifically the learner will be over-fitted to majority class as the number of samples of majority class will be abundant in the stream. So, it will result in severe mis-classifications in future minority instances and poor performance on the minority class.

Detection of Imbalance. To detect imbalance, GOS-IL uses the class fraction of samples r_t of a class w_t to detect imbalance in the online stream. If r_t is less than the threshold value δ_r (Line 12), GOS-IL marks that class as a minority. The value of r_t is also used in over-sampling phase to determine the amount of over-sampling required for the current misclassified samples.

3   Experimental Study

In this section, we show the effectiveness of the proposed algorithm GOS-IL. We use some artificially generated data sets as well as some real world data sets in the experiments. As for performance criteria, we use prequential measures [12] used by several other earlier works [11]. For imbalance study, minority recall and G-mean are the two most important measures [4]. So we used these two performance measures to compare the methods.

3.1   Data Sets

We created two types of artificial imbalanced data stream: static and dynamic. The details of data set generation can be found in [5]. All data streams consisted of ten data chunks. In each chunk, 100 samples were generated for the majority class. The number of data samples generated for the minority class was varied among the chunks as follows.

Dynamic Scenario. In this case, numbers of minority samples generated in ten chunks were varied to reflect a dynamic change of imbalance in the data stream. We created following number of minority examples in ten chunks: 10, 20, 30, 40, 0, 50, 40, 30, 20, 10. In the 5th chunk instead of generating any minority example, some noisy samples were intentionally generated for the minority class.

Static Scenario. In this case, we created three different data sets of three different imbalance ratios. The number of minority examples generated were kept same for all chunks and it was set to 10 (high imbalance), 20 (moderate imbalance), and 30 (low imbalance) respectively in three imbalance scenarios.

So, we had four artificial data sets that were termed as dynamic, static high, static moderate, and static low. As for real world data sets, we collected following four real world data sets from UCI machine learning repository [13]: Glass, Pima, Semeion, and Yeast. Some of these original data sets had more than two classes and we converted them to a two-class data set according to [5]. The details of these data sets can be found in [5].

3.2   Methods and Learner Settings

We compared the performance of GOS-IL with the following two methods:

(1) A naive MLP, denoted as N-MLP, works in online fashion. After each data sample is encountered, the sample is simply fed to the N-MLP and its weights are updated incrementally. No over-sampling and imbalance detection methods are used for the N-MLP.
(2) The algorithm proposed in [11] is one of the most recent and preliminary works in online imbalanced learning. However, as GOS-IL is a over-sampling based approach, we chose the Over-sampling based Online Bagging (OOB) method of [11].

For the GOS-IL, we used a multi-layer perceptron (MLP) as the base classifier model that works in online mode. Similar to GOS-IL, one MLP was used as the base classifier model for OOB. In all experiments, the number of hidden layer for the MLP was set to one and ten (10) nodes were used for the hidden layer. For GOS-IL, different parameters were set as follows: T_e was set to 2 and T_n was set to 40. This implies that permissible error level was set to 5 %. ($PEL = 2/40 = 0.05$). For imbalance detection, the threshold δ_r was set to 0.5 and random over-sampling was used. For the OOB algorithm, parameters were set according to [11].

3.3 Experimental Results

The simulation results obtained are shown in Table 1. The prequential values of minority recall and g-mean are shown at four different time steps: at the time when processing of 25 %, 50 %, 75 %, and 100 % of the total number of samples completes. The best performance values are shown in bold-faced type. The mean of all performance values over the eight data sets are also shown in Table 1.

From Table 1, we see that GOS-IL outperforms both N-MLP and OOB comprehensively in most of the experiments. For the static low data stream, we see that the g-mean values of N-MLP are almost near to values of OOB and GOS-IL.

Table 1. Prequential minority recall and g-mean values of N-MLP, OOB [11], and GOS-IL on four artificial data sets and four real world data sets.

Dataset	Method	Prequential minority recall				Prequential G-mean			
		t = 25 %	t = 50 %	t = 75 %	t = 100 %	t = 25 %	t = 50 %	t = 75 %	t = 100 %
Dynamic	N-MLP	0.0980	0.1000	0.1442	0.1654	0.3125	0.3156	0.3769	0.4044
	OOB	**0.2549**	0.2909	0.3269	0.3615	**0.4630**	0.5122	0.5472	0.5789
	GOS-IL	**0.2549**	**0.3273**	**0.4231**	**0.4654**	0.4503	**0.5202**	**0.5893**	**0.6282**
Static high	N-MLP	0.0667	0.0333	0.0222	0.0250	0.2561	0.1818	0.1487	0.1578
	OOB	0.2333	**0.2667**	0.2889	0.3083	0.4593	**0.4888**	0.5072	0.5200
	GOS-IL	**0.3000**	**0.2667**	**0.3111**	**0.3250**	**0.4911**	0.4727	**0.5116**	**0.5241**
Static moderate	N-MLP	0.1429	0.2000	0.2416	0.2500	0.3696	0.4414	0.4869	0.4962
	OOB	0.2857	0.3500	0.3691	0.3700	0.5061	0.5587	0.5828	0.5888
	GOS-IL	**0.4082**	**0.4400**	**0.4564**	**0.4450**	**0.5760**	**0.6029**	**0.6163**	**0.6059**
Static low	N-MLP	0.3250	0.4000	0.4515	0.4656	0.5527	0.6216	0.6625	0.6738
	OOB	0.2750	0.2875	0.3291	0.3844	0.4986	0.5221	0.5590	0.6014
	GOS-IL	**0.4375**	**0.4750**	**0.5316**	**0.5563**	**0.5931**	**0.6218**	**0.6675**	**0.6872**
Glass	N-MLP	0.0000	0.0000	0.0000	0.7451	0.0000	0.0000	0.0000	0.8605
	OOB	0.0000	0.0000	0.0000	0.8235	0.0000	0.0000	0.0000	0.8906
	GOS-IL	0.0000	0.0000	0.0000	**0.9216**	0.0000	0.0000	0.0000	**0.9391**
Pima	N-MLP	0.2286	0.3446	0.3413	0.3657	0.4435	0.5431	0.5457	0.5660
	OOB	0.3286	0.4189	0.4038	0.4254	0.4923	**0.5589**	0.5553	**0.5775**
	GOS-IL	**0.3857**	**0.4392**	**0.4519**	**0.4478**	**0.5029**	0.5474	**0.5618**	0.5670
Semion	N-MLP	0.4250	0.5000	0.5932	0.6329	0.6474	0.7017	0.7659	0.7911
	OOB	0.3500	0.5128	0.5424	0.5633	0.5833	0.7081	0.7292	0.7440
	GOS-IL	**0.6750**	**0.6923**	**0.7203**	**0.6962**	**0.7960**	**0.8068**	**0.8268**	**0.8126**
Yeast	N-MLP	0.1154	0.2796	0.2712	0.3026	0.3315	0.5074	0.5051	0.5343
	OOB	**0.4231**	**0.4355**	**0.4491**	**0.4671**	**0.5893**	**0.5989**	0.6035	**0.6348**
	GOS-IL	0.3974	0.4086	0.4110	0.4507	0.5875	0.5924	**0.6119**	0.6292
Mean	N-MLP	0.1752	0.2321	0.2581	0.3690	0.3641	0.4140	0.4364	0.5605
	OOB	0.2688	0.3202	0.3386	0.4629	0.4489	0.4934	0.5105	0.6420
	GOS-IL	**0.3573**	**0.3811**	**0.4131**	**0.5385**	**0.4996**	**0.5205**	**0.5481**	**0.6741**

However, N-MLP performs very poorly for dynamic and static high imbalanced data streams. This observation implies that treatment of imbalance is very necessary for data streams that have high imbalance and that changes imbalance over time.

In some cases, e.g., Yeast data set, GOS-IL is beaten by OOB method in both recall and g-mean values. However, on average GOS-IL achieves the best value of prequential recall and g-mean among the three methods as can be seen from the mean values of the three methods (Table 1). These results prove the superiority of the GOS-IL algorithm over the other two methods.

We also show the graphs of prequential values vs. t for the four artificial data sets in Fig. 1. Due to space constraints, we do not show the similar graphs for the other data sets. From Fig. 1(a)–(h), we see that GOS-IL achieves better minority recall and g-mean values than the other two methods in most portion of the time domain. For the dynamic data set, the performance difference between GOS-IL and OOB is not significant when time steps t is less than 500 (Fig. 1(a) and (e)). Because, learner is not stable during the initial phases as it requires a sufficient amount of samples to learn the underlying concepts. However, after that point, the performance difference between GOS-IL and OOB is significant and GOS-IL clearly beats OOB with very good margin values (Fig. 1(a) and (e) after $t > 500$). This is due to the fact that GOS-IL's deferred update method and permissible error level helps it to gain better generalization ability in the long run. Similarly, for the other three data sets, GOS-IL achieves better values specially in prediction of minority samples (Fig. 1(b)–(d)).

4   Conclusion

In this paper, we present a new framework for handling online class imbalance problem. The proposed framework GOS-IL uses a base learner model that is incrementally updated by data samples. However, unlike other works, GOS-IL updates its learner only after it makes a certain number of mis-classifications. Moreover, GOS-IL does not update the learner by samples that are correctly classified by the learner. This is done to avoid over-fitting to the majority class samples as they are abundant in imbalanced data. The prequential recall and g-mean performances of GOS-IL method are measured by simulation experiments over artificial and real world data sets. The results show that GOS-IL achieves better performance values than other methods. Several other research issues can be investigated such as how GOS-IL performs for multi-class problems, whether GOS-IL's performance remain same when some other learning model (e.g., decision tree, support vector machine) are used, and whether use of some other state of the art over-sampling methods (e.g., synthetic over-sampling) can bring better performance values.

Acknowledgments. This research work has been done in the Department of Computer Science & Engineering of Bangladesh University of Engineering and Technology (BUET). The authors would like to acknowledge BUET for its generous support.

References

1. Ciaramita, M., Murdock, V., Plachouras, V.: Online learning from click data for sponsored search. In: International World Wide Web Conference, pp. 227–236 (2008)
2. Wang, H., Fan, W., Yu, P.S., Han, J.: Mining concept-drifting data streams using ensemble classifiers. In: 9th ACM SIGKDD International Conference on Knowledge Discovery and Data Mining, pp. 226–235 (2003)
3. Nishida, K., Shimada, S., Ishikawa, S., Yamauchi, K.: Detecting sudden concept drift with knowledge of human behavior. In: IEEE International Conference on Systems, Man and Cybernetics, pp. 3261–3267 (2008)
4. He, H., Garcia, E.A.: Learning from imbalanced data. IEEE Trans. Knowl. Data Eng. **21**(10), 1263–1284 (2009)
5. Barua, S., Islam, M.M., Yao, X., Murase, K.: MWMOTE-majority weighted minority oversampling technique for imbalanced data set learning. IEEE Trans. Knowl. Data Eng. **26**(2), 405–425 (2014)
6. Chawla, N.V., Bowyer, K.W., Hall, L.O., Kegelmeyer, W.P.: SMOTE: synthetic minority over-sampling technique. J. Artif. Intell. Res. **16**, 321–357 (2002)
7. He, H., Bai, Y., Garcia, E.A., Li, S.: ADASYN: adaptive synthetic sampling approach for imbalanced learning. In: IEEE International Joint Conference on Neural Networks, pp. 1322–1328. IEEE, Hong Kong (2008)
8. Barua, S., Islam, M.M., Murase, K.: ProWSyn: proximity weighted synthetic over-sampling technique for imbalanced data set learning. In: Pei, J., Tseng, V.S., Cao, L., Motoda, H., Xu, G. (eds.) PAKDD 2013, Part II. LNCS, vol. 7819, pp. 317–328. Springer, Heidelberg (2013)
9. Ghazikhani, A., Monsefi, R., Yazdi, H.S.: Recursive least square perceptron model for non-stationary and imbalanced data stream classification. Evol. Syst. **4**(2), 119–131 (2013)
10. Mirza, B., Lin, Z., Toh, K.A.: Weighted online sequential extreme learning machine for class imbalance learning. Neural Process. Lett. **38**(3), 465–486 (2013)
11. Wang, S., Minku, L.L., Yao, X.: A learning framework for online class imbalance learning. In: Computational Intelligence and Ensemble Learning (CIEL), pp. 36–45 (2013)
12. Dawid, A.P., Vovk, V.G.: Prequential probability: principles and properties. Bernoulli **5**(1), 125–162 (1999)
13. UCI Machine Learning Repository. http://archive.ics.uci.edu/ml/

A New Version of the Dendritic Cell Immune Algorithm Based on the K-Nearest Neighbors

Kaouther Ben Ali$^{(\boxtimes)}$, Zeineb Chelly, and Zied Elouedi

LARODEC, Institut Supérieur de Gestion de Tunis, Tunis, Tunisia
kaoutherbenali17@gmail.com, zeinebchelly@yahoo.fr, zied.elouedi@gmx.fr

Abstract. In this paper, we propose a new approach of classification based on the artificial immune Dendritic Cell Algorithm (DCA). Many researches have demonstrated the promising DCA classification results in many real world applications. Despite of that, it was shown that the DCA has a main limitation while performing its classification task. To classify a new data item, the expert knowledge is required to calculate a set of signal values. Indeed, to achieve this, the expert has to provide some specific formula capable of generating these values. Yet, the expert mandatory presence has received criticism from researchers. Therefore, in order to overcome this restriction, we have proposed a new version of the DCA combined with the K-Nearest Neighbors (KNN). KNN is used to provide a new way to calculate the signal values independently from the expert knowledge. Experimental results demonstrate the significant performance of our proposed solution in terms of classification accuracy, in comparison to several state-of-the-art classifiers, while avoiding the mandatory presence of the expert.

Keywords: Artificial immune systems · K-Nearest Neighbors · Classification

1 Introduction

The Dendritic Cell Algorithm (DCA) is an immune inspired classification algorithm based on the abstract model of immune Dendritic Cells (DCs) [1]. It was applied to a wide range of applications, precisely in data classification such as in [2,3]. The DCA performance relies on its data-preprocessing phase where a signal dataset is generated for classification and which is based on three input signals pre-categorized as Pathogenic Associated Molecular Patterns (PAMPs), Danger Signals (DS) and Safe Signals (SS). All input signals cooperate with each other to give a final decision; i.e., the class of the data item. To achieve this classificaion task, the expert knowledge is required to calculate the signal values of the new data item to classify. More precisely, the expert has to give specific formula to calculate the signal values and specifically the DS values that play the leading role in assigning the class of each data instance.

Our aim, in this paper, is to propose a new DCA version capable of overcoming the mentioned DCA restriction which is based on the need of the expert

© Springer International Publishing Switzerland 2015
S. Arik et al. (Eds.): ICONIP 2015, Part I, LNCS 9489, pp. 688–695, 2015.
DOI: 10.1007/978-3-319-26532-2_76

knowledge to provide a complete signal data set for classification. Our proposed method, named KNN-DCA, is a new version of the DCA hybridized with the K-Nearest Neighbors (KNN) machine learning technique [4]. KNN is used as a technique to automatically calculate signals, precisely, the danger signal values that strongly depend on expert knowledge. We will show that our KNN-DCA is capable of processing the danger signal values and finding out the class of a new item without the need of an expert knowledge.

To guarantee the effectiveness and the efficiency of our proposed method, the material in this paper is organized as follows: In the remainder of this introduction, we specify our issue. In Sect. 2, the problem statement is highlighted. In Sect. 3, a detailed description of our proposed approach KNN-DCA is presented. This is followed by Sect. 4, where the results obtained from a set of experiments are discussed. Finally, Sect. 5 concludes the paper and presents some future directions.

2 Problem Statement

In this section, we will mainly clarify the main DCA limitation while performing its classification task. Yet, first, we have to elucidate one important characteristic of the DCA. Actually the algorithm, in literature, was applied in two different manners to machine learning datasets depending on the presence or absence of the expert. The first application manner of the DCA is its application as an unsupervised algorithm. In this case, no information about the previous data item classes are needed while classifying a new data instance. Technically and in this case, the presence of the expert is mandatory where he/she will provide a specific formula showing how to calculate the signal values of that new instance. Once the signal values are calculated, the DCA will generate the MCAVs and classify the new antigen. The second case is where the DCA is applied as a semi-supervised algorithm and in this case some information is required to acquire from the initial training dataset that includes the classes of all antigens. Here, the expert knowledge is not needed. Meanwhile, the classes of the data items have to be known and based on that, the algorithm applies a formula to classify the new data item. More precisely, the needed information from the initial training data set is the number of data items belonging to the normal class (class 1). Based on this information, a formula is applied to calculate only the danger signal values of the new data instance. However, to calculate the values of PAMPs and SS of the same new instance, a second formula is applied which does not depend on the class type; either class 1 or class 2. As discussed, the manner of how to apply the DCA strongly depends on the presence or absence of the expert knowledge. Yet, in most cases no information is afforded about the classes of the data items belonging to the training dataset and at the same time we want to avoid the expert knowledge. In this case, DCA is not capable of performing its classification task nor able it is to classify a new data instance. Thus, it would be very interesting to propose a new DCA version capable of performing its classification task in an autonomous way; i.e., independently from the need to the expert knowledge.

3 The Proposed Approach: A New Dendritic Cell Algorithm Based on the K-Nearest Neighbors

In this section, we will give a detailed description of our proposed new DCA version; the Dendritic Cell Algorithm based on the K-Nearest Neighbors (KNN-DCA). First, we will highlight the KNN-DCA architecture and then we will explain how our KNN-DCA is capable of performing its classification task without the mandatory presence of the expert nor the need of his/her guidelines on how to generate the signal values of the new data item to classify.

3.1 The KNN-DCA Architecture

The contribution of our work is to present a new DCA version capable of surmounting the mentioned DCA restriction which is based on the need of the expert knowledge to provide a signal dataset for classification. Our proposed method is a new version of the DCA hybridized with the K-Nearest Neighbors (KNN) machine learning technique. KNN is used as a technique to automatically calculate signals, precisely, the danger signal values that strongly depend on expert knowledge. The algorithmic steps of our KNN-DCA are as follows:

1. Preprocessing and Initialization phase.
2. Detection phase.
3. Context Assessment phase.
4. Classification phase based on KNN.

As presented in the itemized list, our proposed KNN-DCA is based on the same DCA steps [1,5,6] except for the classification phase which is based on the KNN concept. That is why in this section, we will focus mainly on the KNN-DCA classification step where we will explain in details how our KNN-DCA is capable of classifying a new data item via a new calculation process showing how to generate the signal values, without calling the expert knowledge.

3.2 The KNN-DCA Classification Phase

To classify a new data instance, KNN-DCA has to calculate a set of signals which are the PAMP signals, the safe signals and the danger signals without referring to the expert guidelines to do so. In what follows, we will give the algorithmic steps showing how to perform the signals calculation process.

Calculating the SS and the PAMP Signal Values. Based on immunological concepts, both PAMP and SS are considered as positive indicators of an anomalous and normal signal. This is because the PAMP signals are essential molecules produced by microbes but not produced by the host. They are definite indicators of abnormality indicating the presence of a non-host entity. However, the SSs are released as a result of a normal programmed cell death. They are indicators of normality which means that the antigen collected by the DC was found in a normal context. Hence, tolerance is generated to that antigen.

Mapping the immunological semantics of these two signals to the algorithmic KNN-DCA signal calculation process, one attribute is used to form both PAMP and SS values. The selected attribute is the one having the highest standard deviation among the feature set presented in the input training dataset. Using one attribute to derive the signal values of the new data item X to classify requires a threshold level to be set: values greater than this can be classified as a safe signal with a specific value, while values under this level would be used as a PAMP signal with another specific value. The process of calculating PAMP and SS values is itemized as follows:

1. Recall the most interesting feature which was selected to represent both SS and PAMPs, i.e., the selected one having the highest standard deviation among the feature set presented in the input training dataset; during the data pre-processing phase.
2. Calculate the median (M) of that attribute for all data instances in the training data set.
3. For the data item to classify determine its PAMP and safe signal values based on its attribute value (Val_X) which is the same used to calculate the median in Step 2. If the attribute value is greater than the median then this value is used to form the safe signal of the new data item. The absolute distance from the median is calculated and attached to the safe signal value and the PAMP signal value takes 0 (and vise versa). This can be seen as follows:

$$\text{If } (Val_X > M) \text{ then } SS_X = |M - Val_X| \text{ and } PAMP_X = 0; \qquad (1)$$

As noticed, the process of calculating SS_X and $PAMP_X$ is independent from the need to the expert knowledge. This process is the same one performed by the standard DCA to calculate the values of PAMP and SS for all data items in the training dataset [1].

Calculating the DS Values. Let us remind that the standard DCA version, to calculate the DS value of a new data item X, has either to call the expert or to use the information related to the number of data items having the label "normal" in the training dataset. Yet, we aim to avoid the expert mandatory presence and to avoid using the mentioned needed information from the training dataset. This is because this information, in most cases, is not accessible. Thus, we propose to apply the KNN machine learning technique in order to calculate automatically the DS value of X without referring to the two main restrictions; i.e., expert knowledge and the number of class 1 data items.

Our proposed KNN-DCA is based on the idea of applying KNN in order to select the nearest neighbors to the new data item X that will be classified and to find out an adequate formula mapping the nearest neighbors danger signal values to the DS_X value. This is how the DS_X value will be automatically calculated. Yet, while applying the KNN machine learning technique we may face two possible cases; either we may select the first nearest neighbor to X (k=1) and define the DS_X value based on that or to select a set of the nearest neighbors to X (k>1) and define the DS_X value. Based on these two possibilities, in what follows, we will propose two KNN-DCA methodologies to calculate the DS_X value.

First Case: DS Calculation Process based on the 1-Nearest Neighbor (K=1). At this stage the $PAMP_X$ and the SS_X values are calculated and what is missing is the DS_X value. Based on the KNN-$DCA_{k=1}$, we will search the input signal dataset which includes all the signal values of the antigens of the training dataset (PAMP, SS and DS) and from there we will select the nearest antigen (K=1). We will apply the Euclidian distance to calculate the similarities between all antigens and the new data item X. The similarity is calculated between both $PAMP_X$ and SS_X and between $PAMP_{y_i}$ and the SS_{y_i}; where y is referring to a data instance belonging to the training data set and $i \in \{1, n\}$ where n refers to the length of the input signal dataset; i.e., the number of all antigens in the training data base. The calculation of the DS_X value is given by Eq. 2.

$$DS_X = \frac{DS_Y * Signal_X}{Signal_Y} \tag{2}$$

In Eq. 2, Y refers to the nearest object to the X data item to be classified and DS_Y is the value of the Y danger signal. Let us remind that while calculating the PAMP and SS values, if the PAMP has a value different than zero then the SS value equals zero and vise versa. Thus, if both SS_Y and SS_X are null then $Signal_X$ equals the value of $PAMP_X$ and $Signal_Y$ equals the value of $PAMP_Y$. In the opposite case, where both SS_Y and SS_X are different from null and where $PAMP_X$ and $PAMP_Y$ are null then $Signal_X$ equals the value of SS_X and $Signal_Y$ equals the value of SS_Y.

Second Case: DS Calculation Process based on the K-Nearest Neighbors (K>1). The second case is focused on calculating the DS_X value based on the KNN-$DCA_{k>1}$. In this case, we will search the same input signal dataset used before and from there we will select the K nearest antigens (K>1). Just like the first case, we will apply the Euclidian distance to calculate the similarities between all antigens and the new data item X but instead of selecting one nearest neighbor we will select a set of nearest neighbors. The calculation of the DS_X value is given by Eq. 3.

$$DS_X = \frac{mean(DS_{Y_k}) * mean(Signal_{X_k})}{mean(Signal_{Y_k})} \tag{3}$$

In Eq. 3, Y_k refers to the set of the k nearest objects to X and DS_{Y_k} is the set of the Y_k danger signal values. The semantics of both $Signal_{X_k}$ and $Signal_{Y_k}$ hold as in Eq. 2. So, we will calculate the mean of all DS_{Y_k} values and the mean of both $Signal_{X_k}$ and $Signal_{Y_k}$ values and apply Eq. 3 to calculate the DS_X value.

Based on the set of these three calculated signal values, $PAMP_X$, SS_X and DS_X, and just like the standard DCA process, KNN-DCA can generate the

MCAV of the new data item and then compare the later value to the anomaly threshold which is generated automatically from the data at hand. So, if the MCAV of the new instance is greater than the anomaly threshold; the data item will be classified as a dangerous one (class 2) else it will be tolerated and assigned a normal label (class 1).

4   Experimental Setup and Results

In this section, we try to show the effectiveness of our KNN-DCA as well as its performance. The aim of our method is to show that our KNN-DCA is capable of performing well its classification task, in an autonomous way, in comparison to a set of well known state-of-the-art classifiers. We will, also, test the performance of our proposed KNN-DCA under a variation of the k parameter which is referring to the number of the nearest neighbors to the object to be classified. We have developed our program in Eclipse V 4.2.2 for the evaluation of our KNN-DCA. Different experiments are performed using two-class data sets from the UCI Machine Learning Repository [7]. The used datasets are described in Table 1.

Table 1. Details about the used Datasets

Data set	Ref	Instances	Attributes
Wisconsin Breast Cancer	WBC	699	10
SPECTF Heart	SH	267	45
Pima Indians	PI	768	6
Blood Transfusion	BT	748	5
Haberman's Survival	HS	306	4

In all experiments, each data item is mapped as an antigen, with the value of the antigen equals to the data ID of the item. A population of 100 cells is used. The DC migration threshold is set to 10. To perform anomaly detection, a threshold is applied to the MCAVs. The threshold is calculated by dividing the number of anomalous data items presented in the used data set by the total number of data items. So, if the MCAV is greater than the anomaly threshold then the antigen is classified as anomalous else it is classified as normal. For each experiment, the results presented are based on mean MCAVs generated across a 10-folds cross validation. We evaluate the performance of our KNN-DCA in terms of classification accuracy where we compare it with a set of well known classifiers, namely, the Decision Tree (C4.5), the Support Vector Machine (LibSVM), BayesNet, NaiveBayes, Hoeffding Tree (HT), Wrapper Classifier and the K star classifier. All the parameters used for these classifiers are set to the most adequate values to perform the classification results based on the Weka Software. We will divide our comparison methodology into two main phases.

Table 2. Comparison of Classifiers in terms of Classification Accuracy (%)

DataSets	KNN-DCA			DT	HT	Wrapper	BayesNet	NaiveBayes	SVM	K Star
	k=1	k=3	k=5							
WBC	99.71	99.57	99.86	94.70	96.20	42.90	96.20	97.20	96.70	95.60
SH	99.63	99.25	98.87	72.10	69.30	30.30	69.30	69.30	66.80	65.10
PI	99.61	99.35	99.22	74.40	75.70	42.40	79.80	79.80	77.60	70.90
BT	99.73	99.47	99.33	58.10	58.10	58.10	71.60	70.80	87.17	82.10
HS	87.26	86.60	84.64	68.60	71.50	54.10	69.10	73.80	63.80	71.00

First, we will test the influence of the variation of the k parameter on the KNN-DCA classification performance. We have chosen three different values of k; $k \in \{1, 3, 5\}$. The objective is to select the most convenient k value where KNN-DCA gives the most interesting results. Second, we will compare the KNN-DCA classification results with the already mentioned state-of-the-art classifiers.

Studying the influence of the k parameter on our KNN-DCA classification results and from Table 2, we notice that the three KNN-DCA PCCs are close to each other. For example, applying the KNN-DCA to the PI dataset, the classification accuracies are set to 99.61 %, 99.35 % and 99.22 % for KNN-$DCA_{k=1}$, KNN-$DCA_{k=3}$, KNN-$DCA_{k=5}$, respectively. The same remark is noted for the rest of the used datasets. Yet, we can notice that even if the classification accuracies are roughly the same, the PCC of KNN-$DCA_{k=1}$ is better than KNN-$DCA_{k=3}$ and KNN-$DCA_{k=5}$; in most datasets. Thus, we can conclude that the variation of k keeps the good performance of our proposed KNN-DCA and, as a consequence, this shows that our proposed danger signal calculation methodologies are valid as good classification results are obtained. Now, from Table 2, we can notice that in most cases our KNN-DCA is outperforming the used state-of-the-art classifiers in terms of classification accuracy. For instance, while applying the algorithms to the HS database and for the different values of k, the lowest PCC value of our KNN-DCA is set to 84.64 % with k=5. Yet, this value is greater than 68.60 %, 71.50 %, 54.10 %, 69.10 %, 73.80 %, 63.80 % and 71 % which are given by C4.5, HT, Wrapper, BayesNet, NaiveBayes, LibSVM and K Star; respectively.

To summarize, with the variation of the parameter k on the used datasets, our KNN-DCA gives better classification results when compared to other classifiers. Moreover, we have shown that despite of varying the k value, the KNN-DCAs are producing close PCCs which endorses the validity of our mathematical DS calculation processes which were detailed in Sect. 4.

5   Conclusion and Future Directions

This work extends the original DCA to be more applicable to the problem of interest without the need of expert knowledge. KNN-DCA has provided good classification results on a number of datasets. Indeed, our proposed algorithm is based on robust mathematical formula based on the KNN machine learning technique allowing the DCA classification task in an autonomous way. As future work, we intend to further explore the new instantiation of our KNN-DCA by extending the applicability of the KNN algorithm within a fuzzy context.

References

1. Greensmith, J., Aickelin, U., Cayzer, S.: Introducing dendritic cells as a novel immune-inspired algorithm for anomaly detection. In: Jacob, C., Pilat, M.L., Bentley, P.J., Timmis, J.I. (eds.) ICARIS 2005. LNCS, vol. 3627, pp. 153–167. Springer, Heidelberg (2005)
2. Greensmith, J., Aickelin, U.: Dendritic cells for syn scan detection. In: GECCO, pp. 49–56 (2007)
3. Chelly, Z., Elouedi, Z.: Hybridization schemes of the fuzzy dendritic cell immune binary classifier based on different fuzzy clustering techniques. In: New Generation Computation, vol. 33(1), pp. 1–31. Ohmsha, Chiyoda-ku (2015)
4. Ghosh, A.: On optimum choice of k in nearest neighbor classification. Comput. Stat. Data Anal. 50(11), 3113–3123 (2006)
5. Chelly, Z., Elouedi, Z.: Supporting fuzzy-rough sets in the dendritic cell algorithm data pre-processing phase. In: Lee, M., Hirose, A., Hou, Z.-G., Kil, R.M. (eds.) ICONIP 2013, Part II. LNCS, vol. 8227, pp. 164–171. Springer, Heidelberg (2013)
6. Chelly, Z., Elouedi, Z.: RST-DCA: a dendritic cell algorithm based on rough set theory. In: Huang, T., Zeng, Z., Li, C., Leung, C.S. (eds.) ICONIP 2012, Part III. LNCS, vol. 7665, pp. 480–487. Springer, Heidelberg (2012)
7. Asuncion, A., Newman, D.J.: UCI machine learning repository, (2007). http://mlearn.ics.uci.edu/mlrepository.html

Impact of Base Partitions on Multi-objective and Traditional Ensemble Clustering Algorithms

Jane Piantoni[1], Katti Faceli[1(✉)], Tiemi C. Sakata[1], Julio C. Pereira[1], and Marcílio C.P. de Souto[2]

[1] Universidade Federal de São Carlos, Sorocaba, Brazil
jpiantoni@gmail.com, {katti,tiemi,julio-pereira}@ufscar.br
[2] University of Orleans, INSA Centre Val de Loire, LIFO EA 4022, Orléans, France
marcilio.desouto@univ-orleans.fr

Abstract. This paper presents a comparative study of cluster ensemble and multi-objective cluster ensemble algorithms. Our aim is to evaluate the extent to which such methods are able to identify the underlying structure hidden in a data set, given different levels of information they receive as input in the set of base partitions (BP). To do so, given a gold/reference partition, we produced nine sets of BP containing properties of interest for our analysis, such as large number of subdivisions of true clusters. We aim at answering questions such as: are the methods able to generate new and more robust partitions than those in the set of BP? are the techniques influenced by poor quality partitions presented in the set of BP?

Keywords: Clustering · Cluster ensemble · Multi-objective clustering · Performance evaluation

1 Introduction

Ensemble approaches for clustering are aimed at dealing with many drawbacks of traditional clustering methods [2,3]. The goal of cluster ensemble methods is to obtain a partition that represents the consensus among a set of base partitions (BPs) [6,9]. For this, it first produces a set of BPs and, then, applies a consensus function to combine the members in the set into a single partition. Several strategies for generating BPs and different types of consensus function have been described in the literature. In this paper, we focus on heterogeneous cluster ensembles, where the set of BPs is assumed to have been generated by clustering algorithms with different biases/criteria. Without loss of generality to our experimental setting, we assume that the set of BPs is provided.

In traditional cluster ensembles, the consensus function combines all partition in a single step. As a consequence, high quality clusters with respect to one criterion can be diluted by weak clusters when they are combined. This can lead to an overall poor quality of the consensus partition, even if good clusters (or even if the whole true partition) is presented in the set of BPs [2]. Another

© Springer International Publishing Switzerland 2015
S. Arik et al. (Eds.): ICONIP 2015, Part I, LNCS 9489, pp. 696–704, 2015.
DOI: 10.1007/978-3-319-26532-2_77

drawback of traditional cluster ensembles is the fact that they are not intended for generating multiple partitions as solution. Multiple partition solutions is a relevant issue in many real scenarios.

Multi-objective clustering approaches are based on the simultaneous optimization of two or more complementary clustering criteria [2,3]. By doing so, they are able to produce as solution a set of diverse partitions. Such properties tend to lead the algorithm to produce high quality solutions. However, as a disadvantage, the algorithm can yield a set of solutions with a large number of partitions. This can make the step of analyzing/interpreting the solutions very time consuming. In the previous context, a hybrid approach of multi-objective ensemble can minimize the problem of the possible large number of solutions, while maintaining the advantage of producing diverse partitions as results [2]. In this paper, we will focus on this multi-objective ensemble approach. Other multi-objective clustering frameworks, such as the algorithm presented in [3], has a built in heuristic to generate the set of BPs. Thus, in such a context, the analysis proposed in this paper does not apply.

The purpose of this paper is to present a comparative study of the traditional and multi-objective ensemble approaches analyzing: (1) capability of them in terms of generating solutions containing partitions not presented in the set of BPs (novelty); and (2) how the quality of final solution is influenced by the different levels of information provided in the BPs. To do so, taking as reference a given gold/reference partition (GD), we produced scenarios where the set of BPs were composed by (i) partitions whose clusters contained only partial clusters in GD, (ii) partitions whose clusters contained partial clusters in GD, as well as complete clusters, (iii) partitions containing clusters as (i), (ii), as well as GD itself, and (iv) partitions containing (i)–(iii), as well as random partitions. This way, we aim at answering the following questions: are the techniques able to generate new and more robust partitions than those in the set of base partitions? are the techniques influenced by extremely poor quality partitions?

2   Related Work

Several papers have been published comparing ensembles and/or multi-objective approaches for clustering. In [6], the authors compared the quality of 24 alternatives for building an ensemble. The study taked into account different strategies to generate the base partitions, as well as the use of different consensus functions. In [9], the authors presented a theoretical comparative study of several methods. Their analysis considered six properties relevant to an ensemble. They also discussed the advantages and disadvantages of each method.

In [4,7], the authors presented detailed reviews on multi-objective evolutionary algorithms applied to clustering. The analysis presented in these two works are more oriented to intrinsic issues of evolutionary algorithms, such as chromosomes representation, objective functions and evolutionary operators, rather than to the characteristics and/or performance of the techniques.

Comparison among ensembles and multi-objective approaches are presented in [2, 3]. These works introduced, respectively, the multi-objective clustering algorithm MOCK (Multi-Objective Clustering with automatic K-determination) and the multi-objective ensemble MOCLE (Multi Objective Clustering Ensemble). In order to evaluate their framework, the authors compared them with other ensembles and multi-objective algorithms, pointing out the advantages of using multi-objective approaches over traditional ensembles.

All the studies previously described present very general comparisons regarding the characteristics of the techniques or a performance evaluation. Indeed, differently from the work we present in this paper, none of them explores the behavior of the algorithms with respect to the quality of the set of base partitions used as input.

3   Materials and Methods

In this section, we briefly describe the clustering algorithms employed in our study. More specifically, we introduce the traditional cluster ensembles CSPA (Cluster-based Similarity Partitioning Algorithm), HGPA (HyperGraph-Partitioning Algorithm), MCLA (Meta-CLustering Algorithm) [8] and BCE (Bayesian Cluster Ensembles) [10] and the multi-objective ensemble MOCLE [2]. Figure 1 summarizes the main similarities and differences between traditional and multi-objective ensembles. In this figure, X is a data set and Π_I and Π_F are collections of partitions with n_I and n_F partitions, respectively.

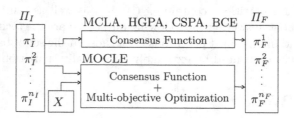

Fig. 1. Overview of the approaches investigated

Regarding the input, traditional cluster ensembles depend on a set of base partitions, which can be externally provided, not requiring access to the original data. MOCLE, on its turn, has as input the data set together with a set of base partitions (initial population). To generate the solution, traditional ensembles make use of consensus functions while MOCLE combines consensus function together with multi-objective optimization. Finally, regarding the output, for traditional ensembles, $n_F = 1$, that is, for each run, a single partition is generated while for MOCLE a higher number of partitions are produced as solution.

Further details about the algorithms we employ in this paper is provided next. The ensembles CSPA, HGPA and MCLA are based on graph and hipergraph [8], whereas BCE is based on probability. **CSPA** generates a co-association matrix from the base partitions and uses this matrix as input to a graph partitioning algorithm to produce the consensus partition. **MCLA** constructs a meta-graph where each cluster of the base partitions is considered as a vertex and the edges link only clusters from different partitions, with weights proportional to the similarity between the clusters. This meta-graph is partitioned producing meta-clusters which encompasses the corresponding clusters. Finally, the consensus partition is produced by assigning each object to the meta-cluster it is more strongly associated with. **HGPA** produces a hypergraph where the clusters of base partitions are hyperedges. A hypergraph partitioning algorithm is used to partition the hypergraph into k unconnected components of approximately the same size, where k is a predetermined number of clusters for the consensus partition. **BCE** treats all base clustering results for each data point as a vector with a discrete value on each dimension, and learns a mixed-membership model from such a representation [10].

MOCLE is a multi-objective clustering ensemble. Starting with a diverse set of base partitions, it employs the multi-objective evolutionary algorithm NSGA-II to generate an approximation of the Pareto optimal set. It simultaneously optimizes the compactness and the connectivity criteria and uses a special crossover operator, which combines pairs of partition using an ensemble method. No mutation is employed. By iteratively combining pairs of partitions, MOCLE can avoid the negative influence of low quality partitions presented among the base partitions.

4 Experimental Protocol

In this study, experiments were performed with ensemble clustering methods applied to different sets of base partitions (BPs), in order to evaluate them with respect to their ability to identify high quality partitions given different initial scenarios. For this, we used an artificial data set called 2sp2glob — Fig. 2a. This data set contains 2000 objects, grouped in four clusters with 500 objects each: two spirals (C_1 and C_2) and two globular shaped clusters (C_3 and C_4). Based on this data set, we artificially generated nine sets of BPs with the properties of interest for analysis. These sets are described in Sect. 4.3. The data set and BPs are available at http://lasid.sor.ufscar.br/2sp2globBPCollection/.

4.1 Performance Evaluation

We used the AR index (Adjusted Rand Index) [5] to measure similarity between a partition generated with the algorithm with the true structure (gold/reference partition) presented in the data set. The better the partition, the closest to 1 the value of AR is. To provide a sound and reliable analysis, we made a statistical analysis. For this, we performed the Friedman test, together with Nemenyi post test when appropriate [1]. More specifically, we performed two types of analysis. For all cases, the significance level was set to 5 %.

In one type of analysis, we considered the algorithms as the treatments being compared and the set of BPs (the initial conditions) as the blocks. In this way, we tested the null hypothesis H_0: there is no difference among the algorithms, against H_1: there is significant statistical difference between at least two algorithms. When the null hypothesis was rejected, we performed the Nemenyi post test with the hypothesis H_0: algorithms i and j present the same performance, against H_1: performance of algorithms i and j differ.

In the other type of analysis, we compared the influence of different initial conditions (given by BPs) on the performance of the algorithms. For this, we tested the null hypothesis H_0: the initial conditions does not influence the results of the algorithms, against H_1: at least two different initial conditions influenced the performance of algorithms. When the null hypothesis was rejected, we proceeded the Nemenyi post test with the hypothesis H_0: BPs i and j lead to the same performance of the algorithms, against H_1: BPs i and j lead to different performances.

4.2   Parameters' Settings

The only parameter of BCE, CSPA, MCLA and HGPA is k, the number of clusters for the consensus partition. We run experiments with $k = [4, 8]$, obtaining a set of five partitions as the result of each algorithm. Then, for each algorithm, we calculated the AR for the respective partitions generated and the best one was selected for the analysis.

The current version of MOCLE allows the variation of two parameters, L and G, respectively related to the connectivity and to the number of generations of the genetic algorithm. In addition, two crossover operators are available, being the MCLA ensemble considered more effective in analysis reported by the authors [2]. Thus, we employed MCLA. For determining values for L and G, we run some preliminary tests with various combinations of these values. Applying Friedman test [1], there was no difference in the result's quality at a significance level of 0.5. Thus, we decided to employ the default values: $L = 5$ and $G = 100$.

Since MOCLE is non-deterministic, it was run 30 times, resulting in 30 sets of partitions. For each of these 30 sets, we calculated the AR for its partitions and selected the best one. Then, we calculated the mean of the 30 solutions selected. This is the value that we will use in the analysis.

4.3   Sets of Base Partitions

We produced nine sets of BPs (**BP1–BP9**). They were generated so as to contain partial information concerning the gold/reference partition, with properties of interest for our analysis. Figure 2 illustrates examples of partitions in BPs. More specifically, each partition in **BP1**, **BP2** and **BP3** represents partial information regarding the true partition. Each of these three sets contains 12 partitions. More specifically, each cluster in these partitions represents one true cluster or part of a true cluster. The difference among them is the degree to which the true clusters are subdivided. These BPs are used to analyze the ability of the methods in

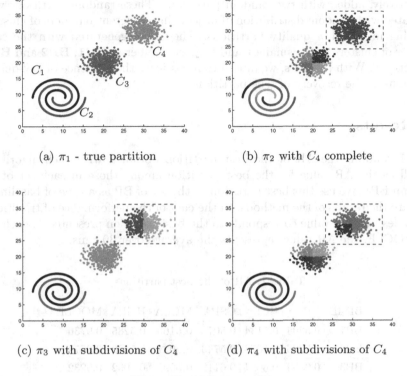

(a) π_1 - true partition

(b) π_2 with C_4 complete

(c) π_3 with subdivisions of C_4

(d) π_4 with subdivisions of C_4

Fig. 2. Partitions examples

reconstructing the true partition, given only partial information. That is, if a new and higher quality partition with respect to BPs can be obtained. The differences among these sets allow to analyze whether the level of details of partial information influences the final solution outputted by algorithms.

BP1 contains partitions with 15 to 24 clusters, all presenting a large number of subdivisions of all true clusters. The subdivisions in one partition complement the subdivisions in other partitions, meaning that two partitions contain part of clusters that overlap. For example, Fig. 2c and d illustrates overlapping subdivisions of cluster C_4. **BP2** contains partitions with 10 to 22 clusters. Each partition has one of the true cluster together with subdivisions of the others, as the partition in Fig. 2b, where C_4 is present together with subdivisions of C_1, C_2 and C_3. **BP3** contains one or more of the true clusters together with subdivisions of the others. In this case, a smaller number of subdivisions than in **BP2** is provided. Thus, the partitions contain between 5 and 8 clusters.

BP4, **BP5** and **BP6** contain 13 partitions each, corresponding respectively to **BP1**, **BP2** and **BP3** added with the true partition. With such BPs, we aim at investigating if the presence of true partition helps to improve the quality of consensus or if this information is lost during the processing. **BP7**, **BP8** and **BP9** contain 14 partitions each, also corresponding to **BP1**, **BP2** and **BP3**, respectively, added with two random partitions. These random partitions were generated with random distribution of objects into different numbers of clusters, resulting in very low quality partitions. They were generated with the same number of clusters as the smallest and biggest number in **BP1**, **BP2** and **BP3**, respectively. With this BPs, we intend to investigate the influence of low-quality partitions in the recovery of true partition.

5   Results

Table 1 contains the AR of the best partition obtained with each algorithm, as well as the AR value for the best partition among those in each set of BP (column BP). We use this best partition in the set of BP as a type of baseline to compare the results of the methods. In the case of the performance of traditional ensembles, the AR value corresponds to the best partition presented in each set. For MOCLE, the AR value represents the average over 30 runs.

Table 1. AR of the best partitions

BP ID	BP	BCE	CSPA	MCLA	HGPA	MOCLE
BP1	0.5204	0.5134	0.5078	0.5165	0.5285	0.9736
BP2	0.7226	0.5678	0.5743	0.6021	0.5832	0.9678
BP3	0.9354	0.6084	0.6112	0.5991	0.6002	0.9232
BP4	1	0.5208	0.5122	0.5321	0.5397	0.9743
BP5	1	0.6032	0.5832	0.5973	0.5899	0.9433
BP6	1	0.6008	0.6218	0.6199	0.6078	0.9671
BP7	0.5204	0.3832	0.3944	0.4231	0.3755	0.9429
BP8	0.7226	0.3784	0.3983	0.3678	0.3813	0.9534
BP9	0.9354	0.3988	0.3786	0.4009	0.4023	0.9734

From Table 1, we can observe a clear distinction among the results of traditional ensembles and MOCLE, which presented an overall superior performance. Moreover, compared to the baseline (**BP**), that is, the best partition in the set of BP, we can see that MOCLE provided solutions with similar or better quality. In contrast, with respect to the baseline, traditional cluster ensembles consistently generated partitions of inferior quality.

To check if the differences in performance observed was not due to chance, we applied the Friedman test [1] as described in Sect. 4. We performed the statistical analysis considering the algorithms as treatments and BPs as blocks. Applying the Friedman test, we obtained $\chi^2 = 30.27$, with p-value $= 1.305 \times 10^{-5} < 0.05$. We thus rejected H_0 and can conclude that there is a difference between at least two algorithms or between an algorithm and the baseline (**BP**). In order to verify which algorithms are different, we proceeded with Nemenyi post test.

The results of the test confirmed that the results of MOCLE is superior to those of the traditional cluster ensemble methods (BCE, CSPA, MCLA and HGPA). Moreover, if we consider the traditional methods only (BCE, CSPA, MCLA and HGPA), there was no statistical evidence of difference among them.

With respect to the comparison to the baseline (**BP**), there was not enough statistical evidence to state the results obtained by MOCLE were different from those of the baseline. That is, MOCLE in the worst case, at least, managed to keep the good information already presented in the set of base partitions. Indeed, when one observes the results in Table 1, one can verify that for several configurations of base partitions (e.g., **BP1**, **BP2**, **BP7** and **BP8**), MOCLE generated solutions with AR much larger than those of the baseline (best partition of the set of base partition). This means it was able to discover (novelty) high quality partition not initially presented in the base partitions.

Differently from the case of baseline versus MOCLE, for the traditional ensembles the results of the statistical test showed that the AR of the cluster ensembles are significantly inferior to those of the baseline (**BP**). In this context, we performed a statistical analysis, now considering the BPs as treatments and the traditional ensembles as blocks. Applying the Friedman test, we obtained $\chi^2 = 29.667$ with a p-value of $0.0002 < 0.05$. We thus rejected H_0, concluding that at least two BPs lead to different performance of the traditional ensembles. Proceeding with the Nemenyi post test, we identified that **BP6** leads to different performance than **BP7**, **BP8** and **BP9**; and **BP3** leads to different performance than **BP8**. That is, the BPs containing random partitions led to significant differences in performances of the traditional ensembles, specifically when compared to the BPs designed with the smallest amounts of subdivisions (**BP3** and **BP6**). This means that most of the initial conditions did not influence the results of the traditional ensembles. On the other hand, differently from what happened with MOCLE, the best consensus partitions found presented AR values smaller than those of the baseline.

6 Final Remarks

In this paper, our aim was to evaluate the extension to which some advanced clustering ensemble strategies are able to identify the underlying structure hidden in a data set, given different levels of information they receive as input. Specifically, we presented a comparative study of four traditional cluster ensemble and a multi-objective ensemble.

Taking into account the results presented in previous sections, we observed that traditional cluster ensembles presented a poor overall performance for all levels of information provided in the set of base partitions (BPs). For example, the results of our experiments showed that the presence of random partitions among the set of BPs had a great negative influence in the consensus. The results of the experiments also showed that the presence of true clusters, or even whole true partition (gold/reference partition), in the set of BP was not enough to provide information to ensembles recover the true structure.

In contrast, our experimental results, showed that multi-objective ensemble could produce new high quality solutions, even when fed only with partial information. Moreover, they were not negatively influenced by extremely poor quality partitions presented in the set of BPs. In other words, according to our experimental results, when compared to traditional cluster ensembles, for data with a heterogeneous underlying structure (clusters with different shapes and sizes) the multi-objective approach showed to be more adequate.

Finally, in terms of limitation and further work, we could point out that, besides the AR, other evaluation indices could have been used to calculate the performance of the algorithm (e.g., normalized mutual information index). As further work, we can also increase the number of methods used in the experiments — the experiments in this paper were conducted only with the algorithms whose code was made available by the authors.

Acknowledgments. This work was partially funded by FAPESP and a CAPES/ COFECUB project.

References

1. Demšar, J.: Statistical comparisons of classifiers over multiple data sets. J. Mach. Learn. Res. **7**, 1–30 (2006)
2. Faceli, K., de Souto, M.C.P., de Araújo, D.S.A., Carvalho, A.C.P.L.F.: Multi-objective clustering ensemble for gene expression data analysis. Neurocomputing **72**(13–15), 2763–2774 (2009)
3. Handl, J., Knowles, J.: An evolutionary approach to multiobjective clustering. IEEE Trans. Evol. Comput. **11**(1), 56–76 (2007)
4. Hruschka, E., Campello, R., Freitas, A., Carvalho, A.: A survey of evolutionary algorithms for clustering. IEEE Trans. Syst. Man Cybern. Part C Appl. Rev. **39**(2), 133–155 (2009)
5. Hubert, L., Arabie, P.: Comparing partitions. J. Classif. **2**(1), 193–218 (1985)
6. Kuncheva, L.I., Hadjitodorov, S.T., Todorova, L.P.: Experimental comparison of cluster ensemble methods. In: Proceedings of 9th International Conference on Information Fusion, pp. 1–7 (2006)
7. Mukhopadhyay, A., Maulik, U., Bandyopadhyay, S., Coello, C.A.C.: Survey of multiobjective evolutionary algorithms for data mining: part II. IEEE Trans. Evol. Comput. **18**(1), 20–35 (2014)
8. Strehl, A., Ghosh, J.: Cluster ensembles - a knowledge reuse framework for combining multiple partitions. J. Mach. Learn. Res. **3**, 583–617 (2002)
9. Vega-Pons, S., Ruiz-Shulcloper, J.: A survey of clustering ensemble algorithms. Int. J. Pattern Recogn. Artif. Intell. **25**(3), 337–372 (2011)
10. Wang, H., Shan, H., Banerjee, A.: Bayesian cluster ensembles. Statistical Anal. Data Min. **4**(1), 54–70 (2011)

Multi-Manifold Matrix Tri-Factorization
for Text Data Clustering

Kais Allab$^{(\boxtimes)}$, Lazhar Labiod, and Mohamed Nadif

LIPADE, University of Paris Descartes, 45, rue des Saints-Peres, Paris, France
{kais.allab,lazhar.labiod,mohamed.nadif}@parisdescartes.fr

Abstract. We propose a novel algorithm that we called Multi-Manifold Co-clustering (MMC). This algorithm considers the geometric structures of both the sample manifold and the feature manifold simultaneously. Specifically, multiple Laplacian graph regularization terms are constructed separately to take local invariance into account; the optimal intrinsic manifold is constructed by linearly combining multiple manifolds. We employ multi-manifold learning to approximate the intrinsic manifold using a subset of candidate manifolds, which better reflects the local geometrical structure by graph Laplacian. The candidate manifolds are obtained using various representative manifold-based dimensionality reduction methods. These selected methods are based on different rationales and use different metrics for data distances. Experimental results on several real world text data sets demonstrate the effectiveness of MMC.

Keywords: Multi-manifold · Matrix tri-factorization · Co-clustering

1 Introduction

Clustering methods based on matrix factorization have recently been emerging as beneficial tools, mainly because of the simplicity of the formalization and their close relationships to other well-studied problems, such as spectral clustering or matrix decomposition. Recently, Non-negative Matrix Factorization (NMF) [12] has become one of the most frequently used matrix factorization tool. NMF was proposed to learn a parts-based representation, but it focuses on unilateral clustering, i.e., on only one of the two sets of samples or features of a data matrix. Largely because of this, Non-negative Matrix Tri-Factorization (NMTF) [12] was presented for co-clustering dyadic data (on both sets of samples and features) whose interest is well established (see for instance, [2,3,7,8]). NMTF is 3-factor decomposition, seeking an approximation of the data matrix via the product of a row-coefficient matrix, a block value matrix and a column-coefficient matrix, with the restriction that these three matrices are all non-negative [4,14]. Regardless of them merits, one drawback of factorization-based co-clustering methods is that they see only the global Euclidean geometry, and the local manifold geometry is not fully considered. To address this major limitation, some researchers have sought to take into account the local geometrical structure in

© Springer International Publishing Switzerland 2015
S. Arik et al. (Eds.): ICONIP 2015, Part I, LNCS 9489, pp. 705–715, 2015.
DOI: 10.1007/978-3-319-26532-2_78

matrix-factorization-based co-clustering. It has been shown that just as sample data lie on a nonlinear low-dimensional manifold, namely the data manifold, features too lie on a manifold, namely the feature manifold [9,17].

2   Related Works and Aims

The Multi-Manifold Learning was proposed to approximate the intrinsic manifold using a subset of candidate manifolds, which can better reflect the local geometrical structure of data. Kmanifolds [18], which starts by estimating geodesic distances between points, is the first method to classify unorganized data nearly lying on multiple intersecting nonlinear manifolds. Unfortunately, this method is limited to deal with intersecting manifolds since the estimation of geodesic distances will fail when there are widely separated clusters. On the contrary, the Spectral Multi-Manifold Clustering (SMMC) [19], which is able to handle intersections, is well suited to group samples generated from separated manifolds. In SMMC, the data are assumed to lie on or close to multiple smooth low-dimensional manifolds, where some data manifolds are separated but some are intersected. Then, local geometric information of the sampled data is incorporated to construct a suitable affinity matrix. Finally, spectral method is applied to this affinity matrix to group the data.

Recently, Relational Multi-manifold Co-clustering (RMC) [13] has been proposed for co-clustering relational data via ensemble manifold learning. RMC estimate the data geometric structure through a K nearest neighbour (KNN) graph on a scatter of objects. The geometric structure modelled by the KNN graph learns incomplete and inaccurate intra-type relationships, i.e., only finding favourable neighbours that are close in Euclidean space instead of finding distant objects that are within manifold neighbours. In addition, in real world applications, there exists no unique (global) manifold but a number of manifolds with possible intersections [11]. KNN graph fails to distinguish the manifolds that are intersecting due to that objects located at manifold intersections share almost the same K nearest neighbours. Moreover, RMC construct the manifolds not only from the informative part, but also from the noisy part of the different candidate manifolds. However, it is difficult to learn accurate intra-type relationships in the presence of noise and outliers. To address the above problems, the Robust High-order Co-clustering via Heterogeneous Manifold Ensemble (RHCHME) [11] method has been proposed. RHCHME incorporates multiple subspace learning with a heterogeneous manifold ensemble to learn complete and accurate intra-type relationships.

Furthermore, RMC and RHCHME are based on a very sparse matrix and are applied to modified block matrices instead of the original data and Laplacian matrices. The RMC and RHCHME Algorithms employ the alternately iterative method, and involve intensive matrix multiplication at each iteration step. The high computational cost of such algorithms makes them unsuitable for large-scale real-world data. To help overcome such problems that are encountered when processing large-scale and noisy data, we propose a novel algorithm that we call Multi-Manifold Co-clustering (MMC). We constrain the factor matrices of MMC to be cluster indicator matrices, which dramatically reduces the computational complexity of co-clustering. We also attempt to consider simultaneously the diversity of geometric structures in the sample manifold and the feature manifold, with the aim of discarding the noisy part in each candidate manifold. Specifically, this involves constructing multiple low-dimensional manifold regularization terms separately, using state of the art dimensionality reduction methods to take account of local invariance; the optimal intrinsic manifold is constructed by linearly combining multiple manifolds.

The selected dimensionality reduction methods includes Canonical Discriminant Analysis (CDA), Multi-Dimensional Scaling (MDS), Isometric Feature Mapping (ISO), Locally Linear Embedding (LLE), Locally Preserving Projections (LPP) and Stochastic Neighbor Embedding (SNE) (for details see for instance [5,6,15]). These methods include different techniques for capturing the non-linearity of the underlying manifold, and they incorporate local distance information in different ways. This is the idea behind our work. Furthermore, the effectiveness of different methods varies, and it has been shown that no single method constantly outperforms the others. Rather than choosing a single method, therefore, we seek to apply a set of dimensionality reduction methods and to merge the output of the different methods. Our multi-manifold learning algorithm aims to overcome the drawbacks of single manifold learning methods and to combine the different data structures to which they give rise.

This paper is organized as follows. In Sect. 3 we introduce our new approach that we called Multi-Manifold Co-clustering algorithm (MMC) and we represent the simplified process to resolve the Matrix Tri-Factorization based Co-clustering optimization problem. In Sect. 4 the proposed algorithm is evaluated and compared against other algorithms designed to solve the same tasks, on both single manifold and multiple manifolds cases. Finally, a conclusion summarizes the main points of our contribution.

3 Multi-Manifold Matrix Tri-Factorization Based Co-clustering

3.1 Problem Formalization

Given a data set $X \in \mathbb{R}^{d \times n}$ and defined by $X := \{x_{ji}; j = 1, \ldots, d; i = 1, \ldots, n\}$, the co-clustering considers simultaneously the set of samples $\{\mathbf{x}_{.1}, \ldots, \mathbf{x}_{.n}\}$ and the set of features $\{\mathbf{x}_{1.}, \ldots, \mathbf{x}_{d.}\}$ in order to organize data matrix X into homogeneous blocks. This block structure can be obtained by a couple of partitions $\mathcal{P} = \{\mathcal{P}_1, \ldots, \mathcal{P}_k\}$ of columns into k clusters and $\mathcal{Q} = \{\mathcal{Q}_1, \ldots, \mathcal{Q}_\ell\}$ of rows into ℓ clusters. Then a summary defined by a matrix $S := \{(s_{qp}; q = 1, \ldots, \ell; p = 1, \ldots, k\}$ of size $\ell \times k$ can be computed. Each summary s_{qp} corresponding to block (q, p) is a real number and the row and column vectors of S are noted $\mathbf{s}_{.q}$ and $\mathbf{s}_{p.}$. The partitions $\mathcal{P}$ and $\mathcal{Q}$ can be respectively expressed as binary matrices $G := \{g_{ip}; i = 1, \ldots, n; p = 1, \ldots, k\}$ with $g_{ip} = 1$ if $i \in \mathcal{P}_p$ and $g_{ip} = 0$

Table 1. Notation used in this paper

Notation	Description
X	Data matrix of size $(d \times n)$
n, d	Number of data samples and data features
k, ℓ	Number of sample clusters and feature clusters
$\mathcal{P}, \mathcal{Q}$	Data samples and data features partitions
G	Sample partition matrix of size $(n \times k)$ $G \in \{0,1\}^{n \times k}$
F	Feature partition matrix of size $(d \times \ell)$; $F \in \{0,1\}^{d \times \ell}$
S	Block value matrix of size $(\ell \times k)$
L_g	Multi-manifold sample graph Laplacian of size $(n \times n)$
L_f	Multi-manifold feature graph Laplacian of size $(d \times d)$

otherwise, and $F := \{f_{jq}; j = 1, \ldots, d; q = 1, \ldots, \ell\}$ with $f_{jq} = 1$ if $j \in \mathcal{Q}_q$ and $f_{jq} = 0$ otherwise. For convenience, Table 1 gives the notation used throughout this paper.

The co-clustering can be formulated as a matrix approximation problem that consists in minimizing the approximation error between the original data matrix X and the reconstructed matrix based on $\mathcal{P}$, $\mathcal{Q}$ and S, defined by

$$\min_{G,F,S} \left\| X - FSG^T \right\|^2. \quad G \in \{0,1\}^{n \times k}, \ F \in \{0,1\}^{d \times l}. \tag{1}$$

where $\|.\|$ denotes the the Frobenius norm.

3.2   Multi-Manifold Co-clustering Algorithm (MMC)

To consider different data manifolds, a set of C candidate graph Laplacians are defined. The intrinsic manifold of the sample or feature space lies in the convex hull of these pre-given candidate manifolds. Sample multi-manifold learning means that the manifold ensemble L_g is represented as a linear combination of the predefined sample candidate manifolds $\{L_g^1, \ldots, L_g^C\}$. Each candidate L_g^c is linked to a coefficient γ_g^c. Similarly, the manifold ensemble L_f is represented as a linear combination of the predefined feature candidate manifolds $\{L_f^1, \ldots, L_f^C\}$ and each candidate L_f^c is linked to a coefficient γ_f^c.

$$L_g = \sum_{c=1}^{C} \gamma_g^c L_g^c, \quad s.t. \quad \sum_{c=1}^{C} \gamma_g^c = 1, \ \gamma_g^c \geq 0. \tag{2}$$

$$L_f = \sum_{c=1}^{C} \gamma_f^c L_f^c, \quad s.t. \quad \sum_{c=1}^{C} \gamma_f^c = 1, \ \gamma_f^c \geq 0. \tag{3}$$

In Eqs. 2 and 3, if we assume the candidate graph Laplacian are obtained directly from the original data matrix X, this is why each of them contains an informative part and a noisy part. Consequently, we may consider that the learned compromise L is made on both the informative and the noisy parts.

In order to discard the noisy part in each of the C candidate manifolds, we propose using a set of C low-dimensional data representations $\{B^1, ..., B^C\}$ instead the data matrix X in the candidate graph Laplacians construction. These low-dimensional data representations $\{B^c\}_{c=1..C}$ are obtained using C selected dimensionality reduction methods.

Since G is a binary matrix, the following loss function is used as a measure of disagreement between each low-rank manifold representation B_g^c and the clustering matrix G with respect to Q_g:

$$\sum_{c=1}^{C} \gamma_g^c \|G - B_g^c Q_g\|^2 \quad \text{s.t. } \{Q_g^T Q_g = I\}. \tag{4}$$

where each candidate distance $\|G - B_g^c Q_g\|^2$ has a corresponding coefficient γ_g^c. In the same way, for the feature space, we consider a set of C feature candidate low-dimensional data representations $\{B_f^1, ..., B_f^C\}$. Multiple manifolds are integrated using a similar loss function:

$$\sum_{c=1}^{C} \gamma_f^c \|F - B_f^c Q_f\|^2 \quad \text{s.t. } \{Q_f^T Q_f = I\}. \tag{5}$$

To preserve the local geometrical structure of data samples and data features spaces, we integrate the two multi-manifold regularizing terms defined in Eqs. 4 and 5.

$$\min_{G,F,S} \|X - FSG^T\|^2 + \alpha \sum_{c=1}^{C} \gamma_g^c \|G - B_g^c Q_g\|^2 + \beta \sum_{c=1}^{C} \gamma_f^c \|F - B_f^c Q_f\|^2 \tag{6}$$

$$\text{s.t.}, Q_g^T Q_g = I, Q_f^T Q_f = I.$$

where the parameters α and β are used to trade-off the contribution of the multi-manifold regularizing. We also introduce the l_2 norm of the variable γ (i.e., $\|\gamma\|^2$) to avoid over-fitting on only one manifold. After some simple algebraic manipulations, the MMC objective function is formulated as:

$$\min_{G,F,S} \|X - FSG^T\|^2 - 2\alpha Tr[G^T(\sum_{i=c}^{C} \gamma_g^c B_g^c) Q_g] + \theta_g \|\gamma_g\|^2$$

$$-2\beta Tr[F^T(\sum_{c=1}^{C} \gamma_f^c B_f^c) Q_f] + \theta_f \|\gamma_f\|^2 \tag{7}$$

$$\text{s.t.}, Q_g^T Q_g = I, Q_f^T Q_f = I.$$

where θ_g and θ_f controls the regularization terms $\|\gamma_g\|^2$ and $\|\gamma_f\|^2$, respectively.

3.3   Optimization

To solve (7), we use an alternated iterative method. The problem is simplified using the following theorem.

Theorem 1. *Let $G_{n \times k}$ and $B_{n \times k}$ be two matrices. Consider the constrained minimization problem*

$$Q^* = \arg \min_Q \left\| G - BQ^T \right\|^2 \text{ subject to. } Q^T Q = I. \tag{8}$$

Let $U \Lambda V^T$ be the SVD for $G^T B$, then $Q^* = UV^T$.

Proof. Expanding the matrix norm $\left\| G - BQ^T \right\|^2$ leads to

$$Tr(G^T G) - 2Tr(G^T BQ^T) + Tr(QB^T BQ^T).$$

Since $Tr(G^T G) = n$ and $Q^T Q = I$, the last term is equal to $Tr(B^T B)$ and the optimization problem (8) is equivalent to

$$\arg \max_Q Tr(G^T BQ^T) \text{ subject to. } Q^T Q = I.$$

Let $G^T B = U \Lambda V^T$ be the SVD for $G^T B$, the $Tr(G^T BQ^T)$ term becomes

$$\text{Tr}(U \Lambda V^T Q^T) \quad = \quad \text{Tr}((U \Lambda^{0.5})(\Lambda^{0.5} V^T Q^T)) \quad = \quad \langle U \Lambda^{0.5}, \Lambda^{0.5} V^T Q^T \rangle.$$

By the Cauchy-Schwartz inequality, we get

$$\langle U \Lambda^{0.5}, \Lambda^{0.5} V^T Q^T \rangle \leq \|(U \Lambda^{0.5})\| \|(\Lambda^{0.5} V^T Q^T))\| \quad = \quad \|\Lambda^{0.5}\| \|\Lambda^{0.5}\| \quad = \quad \text{Tr}(\Lambda)$$

due to the invariance of $\|\cdot\|$ under orthogonal transformations. Hence, the sum in (9) is maximized if $U^T QV = I$ and the solution Q^* to (9) is given by $Q^* = UV^T$.

Hereafter we present the computation of all matrices and parameters.

* **Computation of S:** Fixing G and F, by setting the derivative of $W(G, F, S)$ with respect to S as 0, we obtain:

$$S = (F^T F)^{-1} F^T XG(G^T G)^{-1} \tag{9}$$

* **Computation of Q_g and Q_f:** Fixing G, F and S, we can separate (7) into two sub-problems:

$$\max_{Q_g^T Q_g = I} Tr[G^T (\sum_{c=1}^C \gamma_g^c B_g^c) Q_g] \quad \text{and} \quad \max_{Q_f^T Q_f = I} Tr[F^T (\sum_{c=1}^C \gamma_f^c B_f^c) Q_f].$$

Based on Theorem 1, by applying SVD on $G^T (\sum_{c=1}^C \gamma_g^c B_g^c)$, we obtain $Q_g = U_g V_g^T$. Similarly, applying SVD on $F^T (\sum_{c=1}^C \gamma_f^c B_f^c)$ yields $Q_f = U_f V_f^T$.

* **Computation of G:** We fix S, F and Q_g, and let be $\widetilde{B}_g = (\sum_{c=1}^C \gamma_g^c B_g^c) Q_g$.

$$g_{ip}^{(t+1)} = \begin{cases} 1 & p = \arg \min_{p'} \|(\mathbf{x}_{i.})^{(t)} - s_{p'.}^{(t)}\|^2 - 2\alpha(\widetilde{B}_g)_{ip'} \\ 0 & \text{otherwise.} \end{cases}$$

* **Computation of F:** We fix S, G and Q_f, and let be $\widetilde{B}_f = (\sum_{c=1}^{C} \gamma_f^c B_f^c) Q_f$.

$$f_{jq}^{(t+1)} = \begin{cases} 1 & q = \arg\min_{q'} \|(\mathbf{x}_{.j})^{(t)} - s_{q'.}^{(t)})\|^2 - 2\beta(\widetilde{B}_f)_{jq'} \\ 0 & \text{otherwise.} \end{cases}$$

* **Computation of γ_g and γ_f:** Fixing α, β, G and F, the objective function in Eq. 7 reduces to two subproblems:

$$1: \quad \max_{\gamma_g} \; Tr[G^T(\textstyle\sum_{c=1}^{C} \gamma_g^c B_g^c)Q_g] + \theta_g \|\gamma_g\|^2, \quad \text{s.t.,} \sum_{c=1}^{C} \gamma_g^c = 1, \; \gamma_g^c \geq 0.$$

$$2: \quad \max_{\gamma_f} \; Tr[F^T(\textstyle\sum_{c=1}^{C} \gamma_f^c B_f^c)Q_f] + \theta_f \|\gamma_f\|^2, \quad \text{s.t.,} \sum_{c=1}^{C} \gamma_f^c = 1, \; \gamma_f^c \geq 0.$$

To optimize the multi-manifold coefficients γ_g and γ_f, we can use the entropic mirror descent algorithm (EMDA) [1], which is especially well suited for dealing with convex problems. In the interests of simplicity, we present the EMDA process for sub-problem 1 only. If θ_g equals 0, then γ_g will have the trivial solutions 0 and 1. If θ_g approaches infinity, the manifolds L_g^c will be treated equally. Hence, we need to assign a proper value to θ_g to guarantee the effectiveness of multi-manifold learning. EMDA can use a general distance-like function rather than Euclidean squared distance. Since the constraints imposed on γ_g is a unit simplex: $\Delta_g = \left\{ \gamma_g \in \mathbb{R}^c, \; \sum_{c=1}^{C} \gamma_g^c = 1, \; \gamma_g \geq 0 \right\}$. EMDA requires the objective function Φ to be a convex Lipschitz continuous function with Lipschitz constant Z_Φ w.r.t. a fixed norm. In our approach, this Lipschitz constant is computed for data samples by $\|\nabla \Phi(\gamma_g)\|_1 \leq 2\theta_g + s_g = Z_\Phi$ where $s_g = \text{tr}(G^T(\sum_{c=1}^{C} \gamma_g^c B_g^c)Q_g)$. The pseudocode of EMDA is given in Algorithm 1, and the steps of MMC are shown in Algorithm 2.

Algorithm 1. Entropic Mirror Descent Algorithm.

Input : Lipschitz constant Z_Φ, θ, L, G;
Output : Multi-manifold ensemble coefficient γ;
Initialize : γ_i with identical weights $\frac{1}{C}$; m =1 (number of iterations);
for $c = 1$ **to** C **do**
  repeat
    (a) - $t_m = \sqrt{\dfrac{2 \ln C}{m Z_\Phi^2}}$

    (b) - $\gamma_c^{m+1} \leftarrow \dfrac{\gamma_c^m \exp\left[-t_m \Phi'(\gamma_c^m)\right]}{\sum_{c=1}^{C} \gamma_c^m \exp\left[-t_m \Phi'(\gamma_c^m)\right]}$,   where $\Phi'(\gamma_c^m) = 2\theta \gamma_c^m + s_c^m$
  until *Convergence*;

---

4 Numerical Experiments

In this section we investigate the use of our proposed MMC algorithm for data co-clustering. First, we present the performance of MMC on single manifold. The single candidate manifold is constructed using each of the selected dimensionality reduction methods. Second, we evaluate the impact that combining all the manifolds has on the quality of the co-clustering.

Algorithm 2. MMC algorithm

Input: - Data matrix X, the trade-off parameters α and β,
- C sample candidate manifolds $\{B_g^1, .., B_g^C\}$ and C feature candidate manifolds
$\{B_f^1, .., B_f^C\}$.
Output: Partition matrices G and F.
Initialize: G and F using Kmeans ; S by (9).
repeat
    (a) - Compute $Q_g^{(t)}$ and $Q_f^{(t)}$.
    (b) - Compute $\gamma_g^{(t)}$ and $\gamma_f^{(t)}$ using the EMDA algorithm.
    (c) - Update $G^{(t+1)}$ by (10).
    (d) - Update $F^{(t+1)}$ by (10).
    (e) - Update $S^{(t+1)}$ by (9).
until *convergence*;

The selected dimensionality reduction methods that we compared and combined are CDA, LPP, LLE, MDS, ISO and SNE. Note that CDA is a supervised method, which is why its candidate manifold is computed using the partitions obtained by *Spherical Kmeans* rather than the correct data set partitions. These numerical experiments were performed using some benchmark text data sets from the clustering and co-clustering literature. Table 2 summarizes the characteristics of these data sets.

4.1 Parameter Settings

To measure the clustering performance of the proposed algorithm, we use the commonly adopted metrics, the Accuracy (Acc), the Normalize Mutual Information (NMI) [16] and the Adjusted Rand Index (ARI) [10]. We focus only on the quality of row clustering. We run each method under different parameter settings 50 times, and the average result is computed. We report the best average result for each method. We set the number of sample clusters equal to the true number of classes in data sets (k).

For each of the compared approaches: Kmanifolds, RHCHME, SMMC and RMC, the best parameters are used, as suggested in each of the reference articles (see for details [9,11,13,17,19]). For MMC, the graph Laplacian is constructed using the Cosine-distance-based K-Nearest Neighbors in which the neighborhood size is fixed to 5. The regularization parameter α is searched from the grid (0.01,

Table 2. Data set characteristics.

Data set	Type	Samples	Features	Classes
CSTR	Text	1428	1024	4
WebKB4	Text	4199	1000	4
WebACE	Text	2340	1000	20
RCV1	Text	9625	29992	4
Ng20	Text	19949	43586	20

0.1, 1, 10, 100, 500, 1000). We set $\beta = \alpha$ for both the sample and feature graphs. Moreover, the spherical Kmeans (*SKmeans*) algorithm [2] has computational advantages for sparse high-dimensional data vectors. For this reason, we use *SKmeans* to initialize the factor matrices.

Note that for all the compared methods, it was suggested that the number of feature clusters is equal to the number of sample clusters $\ell = k$. However, in our approach, the candidate manifolds are generated by using some reduction dimension methods. Taking a too small value of ℓ for determining the number of components (dimensions), may cause a loss of the information provided by the initial features. Contrariwise, a too large number increases the computational complexity. Then in order to assess the number of feature clusters, we varied ℓ between 2 and $10\,k$, and retained the one that optimizes the criterion.

4.2   Results

First, we present the effectiveness of MMC on single manifold. The single candidate manifold is constructed using each of the dimensionality reduction methods, i.e., CDA, LPP , LLE, MDS, ISO or SNE. Next, we report the performance of MMC when the dimensionality reduction algorithms are combined. MMC is compared against the multi-manifold approaches Kmanifolds, RHCHME, SMMC and RMC. The main comments arising from our experiments are the following.

– In Table 3, we observe that RMC outperforms Kmanifolds, RHCHME and SMMC for all data sets. Note that RMC is a KNN based method and uses the squared loss function to measure the quality of the matrix decomposition, which is unstable with respect to noise and outliers. This is why MMC which exploits only the informative part of the data and removes the noisy part, is clearly more efficient than RMC; we confirm this thanks to t-tests on 50 random initialisations. We show that the improvement is statistically significant; all p-values are less than $10^{-8}\%$.

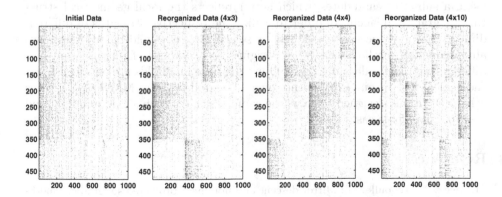

Fig. 1. CSTR: reorganized data visualisation according row and column clusters $(k \times \ell)$.

Table 3. MMC results on the manifold-based dimensionality reduction methods.

Data set	Metric	SKmeans	Single manifold ($C = 1$)						Multi-manifolds				
			CDA	LLE	LPP	MDS	SNE	ISO	Kmanifolds	RHCHME	SMMC	RMC	MMC
CSTR	Acc	0.903	0.909	0.929	0.920	0.939	0.922	0.903	0.745	0.806	0.512	0.898	0.956
	NMI	0.780	0.783	0.915	0.903	0.902	0.814	0.833	0.689	0.751	0.461	0.766	0.906
	ARI	0.818	0.822	0.852	0.824	0.829	0.795	0.754	0.616	0.721	0.355	0.733	0.860
WebKB4	Acc	0.778	0.783	0.853	0.893	0.892	0.807	0.897	0.730	0.775	0.652	0.835	0.928
	NMI	0.551	0.557	0.691	0.778	0.717	0.476	0.873	0.666	0.683	0.637	0.573	0.843
	ARI	0.563	0.566	0.558	0.735	0.628	0.457	0.682	0.493	0.536	0.419	0.587	0.772
WebACE	Acc	0.653	0.686	0.722	0.776	0.802	0.760	0.735	0.602	0.713	0.597	0.718	0.888
	NMI	0.698	0.704	0.836	0.818	0.880	0.833	0.790	0.640	0.727	0.625	0.766	0.916
	ARI	0.586	0.603	0.538	0.616	0.659	0.600	0.578	0.399	0.485	0.368	0.530	0.766
RCV1	Acc	0.681	0.715	0.734	0.784	0.759	0.764	0.739	0.721	0.730	0.556	0.777	0.807
	NMI	0.523	0.512	0.602	0.641	0.615	0.629	0.622	0.469	0.495	0.287	0.608	0.651
	ARI	0.485	0.565	0.509	0.566	0.523	0.533	0.522	0.465	0.523	0.230	0.551	0.618
NG20	Acc	0.388	0.406	0.437	0.517	0.488	0.492	0.456	0.414	0.431	0.288	0.504	0.533
	NMI	0.365	0.397	0.416	0.491	0.485	0.485	0.450	0.392	0.422	0.307	0.493	0.526
	ARI	0.148	0.155	0.227	0.286	0.265	0.271	0.243	0.162	0.223	0.113	0.266	0.311

- In the multi-manifold case, the candidate manifolds are weighted according to them quality in reflecting the local geometrical structure of data. These coefficients are an additional indicator of the effectiveness of each method.
- MMC is a co-clustering method revealing a reorganization into homogeneous blocks of data. In Fig. 1, we illustrate visually the obtained co-clusters according to several numbers of the feature clusters $\ell = 3, 4$ and 10.

5   Conclusion

We propose a novel algorithm, MMC, which simultaneously considers the geometric structures of both the sample manifold and the feature manifold. Specifically, we employ multi-manifold learning to approximate the intrinsic manifold using a subset of candidates, which better reflects the local geometrical structure by graph Laplacian. In order to utilize the respective strengths of different dimensionality reduction techniques, we selected six manifold-based dimensionality reduction methods that were designed for a variety of purposes and use different metrics for data distances. Our candidate manifolds are obtained using these methods. In our experiments on real text data sets, MMC outperforms other algorithms designed to solve the same tasks, on both single manifold and multiple manifolds cases.

References

1. Beck, A., Teboulle, M.: Mirror descent and nonlinear projected subgradient methods for convex optimization. Oper. Res. Lett. **31**(3), 167–175 (2003)
2. Dhillon, I.S.: Co-clustering documents and words using bipartite spectral graph partitioning. In: SIGKDD, pp. 269–274 (2001)

3. Dhillon, I.S., Mallela, S., Kumar, R.: A divisive infomation-theoretic feature clustering algorithm for text classification. Mach. Learn. Res. **3**, 1265–1287 (2003)
4. Ding, C., Li, T., Peng, W., Park, H.: Orthogonal nonnegative matrix trifactorizations for clustering. In: ACM SIGKDD, pp. 126–135 (2006)
5. Engel, D., Hüttenberger, L., Hamann, B.: A survey of dimension reduction methods for high-dimensional data analysis and visualization. In: IRTG 1131 Workshop, vol. 27, pp. 135–149 (2012)
6. Gittins, R.: Canonical Analysis: A Review with Applications in Ecology. Biomathematics, vol. 12. Springer, Heidelberg (1985)
7. Govaert, G., Nadif, M.: Clustering with block mixture models. Pattern Recogn. **36**(2), 463–473 (2003)
8. Govaert, G., Nadif, M.: Co-Clustering: Models, Algorithms and Applications. Wiley, London (2013)
9. Gu, Q., Zhou, J.: Co-clustering on manifolds. In: ACM SIGKDD (2009)
10. Hubert, L., Arabie, P.: Comparing partitions. J. Classif. **2**, 193–218 (1985)
11. Jun, H., Richi, N.: Robust clustering of multi-type relational data via a heterogeneous manifold ensemble. In: The 31st International Conference on Data Engineering, ICDE 2015 (2015)
12. Lee, D., Seung, H.: Learning the parts of objects by non-negative matrix factorization. Nature **401**(6755), 788–791 (1999)
13. Li, P., Bu, J., Chen, C., He, Z.: Relational co-clustering via manifold ensemble learning. In: Proceedings of the 21st ACM International Conference on Information and Knowledge Management, CIKM 2012, pp. 1687–1691, ACM (2012)
14. Long, B., Zhang, Z., Yu, P.S.: Unsupervised learning on k-partite graphs. In: ACM SIGKDD, pp. 317–326 (2005)
15. van der Maaten, L.J.P., Postma, E.O., van den Herik, H.J.: Dimensionality Reduction: A Comparative Review. Tilburg University Technical Report, TiCC-TR 2009-005 (2009)
16. Strehl, A., Ghosh, J.: Cluster ensembles:a knowledge reuse framework for combining multiple partitions. Mach. Learn. Res. **3**, 583–617 (2002)
17. Wang, H., Nie, F., Huang, H., Makedon, F.: Fast nonnegative matrix trifactorization for large-scale data co-clustering. In: IJCAI (2011)
18. Wang, Y., Jiang, Y., Wu, Y., Zhou, Z.-H.: Multi-manifold clustering. In: Zhang, B.-T., Orgun, M.A. (eds.) PRICAI 2010. LNCS, vol. 6230, pp. 280–291. Springer, Heidelberg (2010)
19. Wang, Y., Jiang, Y., Wu, Y., Zhou, Z.: Spectral clustering on multiple manifolds. IEEE Trans. Neural Netw. Learn. Syst. **22**(7), 1149–1161 (2011)

Clustering of Binary Data Sets
Using Artificial Ants Algorithm

Nesrine Masmoudi[1,3,4][✉], Hanane Azzag[2], Mustapha Lebbah[2],
Cyrille Bertelle[1], and Maher Ben Jemaa[3,4]

[1] Normandie University, ULH, LITIS, ISCN, FR-CNRS-3638,
25 Rue Ph. Lebon, 76600 Le Havre, France
Nesrine.masmoudi@redcad.org, Cyrille.bertelle@gmail.com
[2] LIPN, University Paris 13, Av. J.-B. Clnt, 93430 Villetaneuse, France
{Hanane.Azzag,Mustapha.Lebbah}@lipn.univ-paris13.fr
[3] University of Sfax, Sfax, Tunisia
[4] ReDCAD, National School of Engineers of Sfax, 3038 Sfax, Tunisia
Maher.benjemaa@gmail.com

Abstract. As an important technique for data mining, clustering often consists in forming a set of groups according to a similarity measure such as hamming distance. In this paper, we present a new bio-inspired model based on artificial ants over a dynamical graph of clusters using colonial odors and pheromone-based reinforcement process. Results analysis are provided and based on the impact of parameter values on purity index which is a measure of clustering quality. Dynamic evolution of cluster graph topologies are presented on two databases from Machine Learning Repository.

Keywords: Swarm intelligence · Data clustering · Binary data · Artificial ants model

1   Introduction

Clustering in data mining is a discovery process that groups each set of similar data in clusters. Clustering consists in either constructing a hierarchical structure, or forming a set of groups. It is a useful technique for knowledge discovery from a data set. However, when the amount of data is huge, it is necessary to form homogeneous groups to allow a better understanding and operational reasoning. Swarm intelligence is a broader issue that suggests a new approach to the clustering of individuals in groups: this method draws upon the behavior of ants as a source of inspiration for the concept of clustering. Despite the lack of cognition in individual ants, they can instinctively group themselves with similar individuals and groups to appear as a distinct and homogeneous group of individuals. Swarm intelligence is a relevant technique in dynamical situations. In this paper, we will focus on qualitative variables with several modalities. This type of data is used for example in surveys and polls whose answer to a question must be unique by choosing one method among the methods proposed. Thus, we propose in this work a new algorithm named CL-Ant dealing with this new challenging type of data.

© Springer International Publishing Switzerland 2015
S. Arik et al. (Eds.): ICONIP 2015, Part I, LNCS 9489, pp. 716–723, 2015.
DOI: 10.1007/978-3-319-26532-2_79

2   Related Work

Some clustering models are based on neighboring graphs requiring distances or similarities to define between couples of data. Using this distance, appropriate heuristics can discover underlying topological information about the dataset. This learned topology can be used in various ways: for instance, starting from a selected datum, one may explore the content of the dataset by following edges of the graph to find interesting neighbors (i.e., similar data). Another possibility is to use the topology to define clusters. Building a proximity graph consists in using an existing distance (or similarity) between data in order to establish binary relations between nodes. Several standard methods exist [1]. Building a neighborhood or proximity graph over a dataset is an interesting process which is used in many domains [2]. The graphs of neighborhoods involved in many areas such data mining [3], the pattern recognition [4] and the spatial data mining [5] due to their intrinsic qualities [6–8]. In [9], they are interested in the complex structures that are built by real ants and studied the self-assembly behavior of real ants, in order to define a hierarchical clustering algorithm called AntTree where each tree node represents one data object. It is a new algorithm to perform a hierarchical clustering inspired from the ants self-assembly behavior. In this paper, we want to use such a graph to help a domain expert to discover knowledge about a large dataset.

3   Swarm Intelligence and Clustering

We describe the artificial ants approaches for clustering as a rich source of inspiration. The initial and pioneering work in this area is due to [10] where the way real ants sort objects in their nest is modeled. A complete study of approaches using Ant-based and Swarm-based clustering is presented in [11]. Ant Colony System (ACS) and Ant Colony Optimization (ACO) were developed from the foraging behavior of real ants [12] based on the pheromone where the basic entities are virtual ants which cooperate to find the solution of graph-based problems, like network routing problems, for example. Authors [12] simulated the way ants work collaboratively in the task of grouping dead bodies. Monmarche et al. [13] combined the stochastic principles of an ant colony in conjunction with the deterministic principles of the K-means algorithm. Labroche et al. [14] presented a new model called AntClust based on an ant clustering system using the colonial odors. In this new version, the same principles of real ants behavior and the chemical odor in ant species proposed in [16] are applied. We focus an important real ant collective behavior, namely the construction of a colonial odor and its use for determining the ant nest membership. It involves modeling the way ants recognize the odor of their nest that allows them to protect it by recognizing, rejecting intruders and sharing a common colonial odor to all ants of the same nest. In [17] authors describe the mechanisms of recognition in most ant species, colonies are made up of individuals from one or more queens. Between individuals of the same colony, the atmosphere is generally peaceful as opposed to individuals from different colonies. All work on the subject confirms that the discrimination between individuals of different colonies is based on chemical recognition. This notion of development by allowing colonial odor exchanges more or less important chemicals was expressed in [14] by a similarity measure and a set of behavioral rules for performing clustering. Also our method CL-Ant simulates how ants move following volatile substances called pheromones on a graph of cluster [15].

4   The CL-Ant Clustering Algorithm

For all of the algorithm that we propose, we used the same similarity measure $Sim(X, Y)$ between two data X and Y, belonging to a metric space:

$$Sim(X, Y) = 1 - \parallel X - Y \parallel^2 \tag{1}$$

The data are aggregated in clusters, the jth is noted CL_j, located by its centroid W_j.

We therefore use the following parameters:

– a_i, i = 1,...,n are the set of ants. Each ant is associated to a data.
– CL_j, j = 1,...,k are the set of clusters.
– W_j is the centroid of the cluster CL_j.
– Ph_{ij} is the pheromone rate of each edge.
– Th_{CL_j}: each cluster CL_j has a threshold of affiliation defined as follows:

$$Th_{CL_j} = \frac{\sum_{i=1}^{n} \sum_{j=i+1}^{n} \frac{Sim(x_{a_i}, x_{a_j})}{N} + \min(Sim(x_{a_i}, x_{a_j}))}{2} \tag{2}$$

Where a_i and $a_j \in CL_j$, $i \in \{1, ..., n\}$, $N = \frac{n(n-1)}{2}$, x_{a_i} is the data vector associated to ant a_i.

Our model lies within an environment represented by a graph (see Fig. 1). The CL-Ant algorithm is divided into two main steps:

1. **Initialization step using K-means algorithm:** The K-means algorithm [18] chooses inadvertently K first points representing the centers of K classes. The initial graph is a complete graph whose nodes are groups of data known a priori and whose edges represent the neighborhood relations between clusters. From there, a first partition is formed by allotting each data to the class K to which the center is closest in terms of hamming distance.
2. **The first step of clustering:** In this step, the CL-Ant algorithm [16] is applied to build a dynamic graph that represents topology preservation of data clusters. The main idea is to associate data to be classified to an artificial ant. Each cluster represents a node on the dynamic graph. Each edge connecting two neighboring clusters is weighted by hamming distance between the centers of two clusters. We aim to improve the partitioning of K-means using

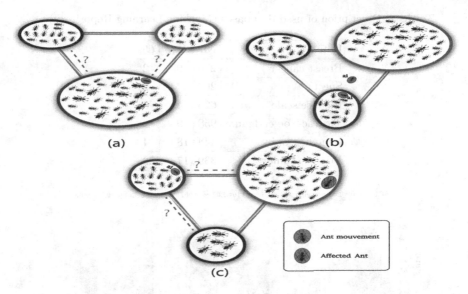

Fig. 1. General principles of graph building with artificial ants, and the computation of an ant's neighborhood.

artificial ant rules to produce dynamic graph after a fixed number of iterations. Once the ant has left its cluster, it moves towards another cluster by choosing a path which has a stronger concentration of pheromone (see Fig. 1 (a, b)). If an ant cannot integrate another nest (cluster), it moves towards another one taking the densest way. The ant is accepted by a cluster $CL_{j'}$ if $(\text{Sim}(a_i, W_{j'}) >= Th_{CL_{j'}})$. After each assignment of data to a new cluster, an update of the pheromone rate on edge is carried out. The weight of edge is then increased by a pheromone rate with α value $(Ph_{jj'} = Ph_{jj'} + \alpha)$, corresponding to the path crossing by ants when they find a better cluster to belong. This process corresponds to a reinforcement process (see Fig. 1 (c)). If $\text{Sim}(a_i, W_{j'}) < Th_{CL_{j'}}$ all edges weight decreased with γ value $(Ph_{hl} = Ph_{hl}(1 - \gamma); 1 \leq h, l \leq k)$, corresponding to clusters of height dissimilarity. When the weight of some edge becomes under a threshold the edge is removed.

5   Experimental evaluation

5.1   Experiments with Binary Data Sets

To validate our approch. We used different binary data sets extracted from the Machine Learning Repository [19] whose general characteristics are summarized in Table 1 (a): Nb is the total number of data forming the data bases, N_{Att} of the number of attribute, K the number of theoretical cluster fixed by k-means algorithm in initial step, C_R the number of real classes.

Table 1. Description of used databases (Machine Learning Repository).

Datasets	Nb	N_{Att}	C_R
Breast cancer	286	9	2
Spect	267	22	2
Balance scale	625	4	3
Tic-Tac-Toe endgame	958	9	2
Lymphography	148	18	4
Primary tumor	339	17	22

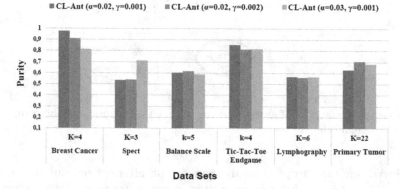

Fig. 2. Results of purity index obtained with CL-Ant (Test 1, Test 2, Test 3) on binary data sets after 1000 iterations (Color figure online).

There are many suggestions for a quality measure [21]. Such a purity measure [20] can be used to compute the quality of a clustering. It takes its value in $[0, 1]$; 1 indicates whether all clusters are pure. Clustering methods group these objects into K clusters, thus two partitions to compare are defined:

$$Purity(P_1, P_2) = \frac{1}{N} \sum_i^k \operatorname*{argmax}_j | W_i \bigcap C_j | \qquad (3)$$

Where $P_1 = W_1, W_2, ..., W_k$ is the set of clusters result, $P_2 = C_1, C_2, ..., C_j$ is the set of reference cluster and i, j $\in \{1, ...k\}$.

We tested our method with different values of α and γ in Fig. 2. For all databases and as far as the purity values in Fig. 2 are concerned, the results obtained by CL-Ant (Test 1) with $\alpha = 0.02$ and $\gamma = 0.001$ are globally similar to those obtained by CL-Ant (Test 2) and CL-Ant (Test 3) on balance scale and lymphography databases. CL-Ant (Test 1) with $\alpha = 0.02$ and $\gamma = 0.001$ yielded the best results compared to CL-Ant (Test 1) and CL-Ant (Test 2) on breast cancer dataset. We obtained a higher purity value equals to 0.9755 for breast cancer dataset with K = 4 after 1000 iterations. Results of CL-Ant with three

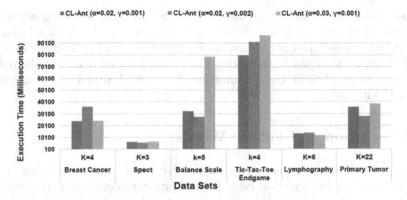

Fig. 3. Processing time in millisecondes obtained with CL-Ant (Test 1, Test 2, Test 3) after 1000 iterations (Color figure online).

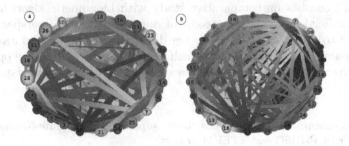

Fig. 4. An example of primary tumor dataset with k = 23, $\alpha = 0.02$ and $\gamma = 0.001$; (a) Complete graph before clustering, (b) Dynamic graph after CL-Ant.

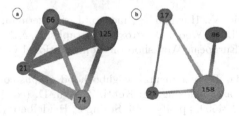

Fig. 5. An example of breast cancer dataset with k = 4, $\alpha = 0.02$ and $\gamma = 0.001$; (a) Complete graph before clustering, (b) Dynamic graph after CL-Ant.

tests were comparable: equal for different databases, and one test better than the other for the other databases. We give the computation time of CL-Ant with different values of α and γ (see Fig. 3). All three CL-Ant tests are outperformed in processing time. It is noteworthy that the CL-Ant (Test 1) with $\alpha = 0.02$ and $\gamma = 0.001$ is faster than CL-Ant with other values of α and γ (see Fig. 3). Experimental results are given by Figs. 4 and 5. Figure 4 shows different graphs tested on the primary Tumor dataset with k = 23, $\alpha = 0.02$ and $\gamma = 0.001$. The purpose of our algorithm is to reduce the initial graph by removing edges with a low rate of pheromone and maintaining the strong neighborhood relationship represented by a high rate of pheromone. Figures 4 and 5 (a) present the complete graph obtained after the initialization step (execution of K-means algorithm). Figures 4 and 5 (b) show the dynamic graph obtained after CL-Ant algorithm during 100 iterations.

6   Conclusion and Future Works

We presented in this paper a new model for binary data clustering named CL-Ant based on the chemical odor in real ants considered as a source of bio-mimetic inspiration. CL-Ant algorithm introduced new heuristics for unsupervised clustering techniques and also the field of artificial ants simulation methods. The obtained results are encouraging in terms of the quality of data clustering. Future work consists on comparative study with biomimetic algorithms could be very beneficial in terms of outcome of our work. Another perspective work is that how to apply our models on big data? To handle large data sets, one of newest paradigm is MapReduce which performs map and reduce operations. Further investigation shall focus on the study of another type of datasets such as text, images.

Acknowledgments. This work has been supported by Haute-Normandie and European FEDER-RISC and XTERM project.

References

1. Bose, P., Dujmovic, V., Hurtado, F., Iacono, J., Langerman, S., Meijer, H., Sacristan, V., Saumell, M., Wood, D., Proximity graphs: E, δ, Δ, $\chi and \omega$. In: Proceedings of the 28th European Workshop on Computational Geometry (EuroCG12) (2012)
2. Hacid, H., Yoshida, T.: Incremental neighborhood graphs construction for multidimensional databases indexing. In: Kobti, Z., Wu, D. (eds.) Canadian AI 2007. LNCS (LNAI), vol. 4509, pp. 405–416. Springer, Heidelberg (2007)
3. Zighed, D.A., Lallich, S., Muhlenbach, F.: Separability index in supervised learning. In: Elomaa, T., Mannila, H., Toivonen, H. (eds.) PKDD 2002. LNCS (LNAI), vol. 2431, pp. 475–487. Springer, Heidelberg (2002)
4. Angot, F., Revenu, M., Elmoataz, A., Clouard, R.: Les graphes de voisinage comme outil de mise en oeuvre de méthodes de segmentation hiérarchique d'images. 16e Colloque GRETSI, pp. 399–402 (1997)
5. Aufaure, M.A., Yeh, L., Zeitouni, K.: Le temps, L'espace et l'évolutif en sciences du traitement de l'information. In: Fouille de données spatiales, pp. 319–328 (2000)
6. Edelsbrunner, H.: Algorithms in Combinatorial Geometry. Springer, New York (1987)
7. Huy, T-D.: Sectorisation contrainte de l'espace aérien (2004)
8. Gabriel, K.R., Sokal, R.R.: A new statistical approach to geographic variation analysis. Syste Zool. **18**(3), 259–278 (1969)

9. Azzag, H., Venturini, G., Oliver, A., Guinot, C.: A hierarchical ant based clustering algorithm and its use in three real-world applications. Eur. J. Oper. Res. **179**, 906–922 (2007)

10. Deneubourg, J.L., Goss, S., Franks, N.R., Franks, S.A., Detrain, C., Chretien, L.: The dynamics of collective sorting: robot-like ant and ant-like robots. In: Proceedings of the First International Conference on Simulation of Adaptive Behavior, pp. 356–365 (1990)

11. Handl, M., Bernd, J.: Ant-based and swarm-based clustering. Swarm Intel. **1**(2), 95–113 (2007)

12. Bonabeau, E., Dorigo, M., Theraulaz, G.: Swarm Intelligence: From Natural to Artificial Systems. Oxford University Press, New York (1999)

13. Monmarché, N., Venturini, G., Slimane, M.: On how Pachycondyla apicalis ants suggest a new search algorithm. Future Gener. Comp. Syst. **16**(8), 937–946 (2000)

14. Labroche, N.: Méthodes d'apprentissage automatique pour l'analyse des interactions utilisateurs. HDR, Université Pierre et Marie Curie (2012)

15. Dorigo, M., Caro, G.D., Gambarella, L.M.: Ant algorithm for discrete optimization. Artif. life **5**(3), 137–172 (1994)

16. Masmoudi, N., Azzag H., Lebbah M., Bertelle C.: Clustering using chemical and colonial odors of real ants. In: 5th World Congress on Nature and Biologically Inspired Computing. NaBIC, 12–14 August, Fargo, North Dakota, USA, IEEE (2013)

17. Calin, N.F., Holldober, B.: The kin recognition system of carpenter ants(camponot us spp).ii. larger colonies. Behav. Ecool Sociobiol. **20**, 209–217 (1987)

18. Macqueen, J.B.: Some methods for classification and analysis of multivariate observations. In: Proceedings of 5th Berkeley Symposium on Mathematical Statistics and Probability. University of Calfornia, Berkeley (1967)

19. Blake, C.L., Merz C.L.: Uci repository of machine learning databases. Technical report, University of California, Department of information and Computer science, Irvine, CA (1998)

20. Jain, A.K., Dubes, R.C.: Algorithms for Clustering Data. Prentice Hall Advanced Reference Series (1988)

21. Everitt, B., Landau, S., Leese, M., Stahl, D.: Cluster Analysis, 5th edn. Wiley, Chichester (2011). pp. 11–45

Inverse Reinforcement Learning Based on Behaviors of a Learning Agent

Shunsuke Sakurai[✉], Shigeyuki Oba[✉], and Shin Ishii[✉]

Department of Systems Science, Graduate School of Informatics,
Kyoto University, Kyoto, Japan
sakurai.shunsuke.35r@st.kyoto-u.ac.jp, {oba,ishii}@i.kyoto-u.ac.jp

Abstract. Reinforcement learning agents can acquire the optimal policy to achieve their objectives based on trials and errors. An appropriate design of reward function is essential, because there are variety of reward functions for the same objective whereas different reward functions would give rise to different learning processes. There is no systematic way to determine a good reward function for a given environment and objective. One possible way is finding a reward function to imitate the learning strategy of a reference agent which is intelligent enough to efficiently adapt even variable environments. In this study, we extended the apprenticeship learning framework in order to imitate a learning reference agent, whose policy may change on the process of optimization. For the imitation above, we propose a new inverse reinforcement learning based on that agent's history of states and actions. When mimicking a reference agent that was trained with a simple 2-state Markov decision process, the proposed method showed better performance than that by the apprenticeship learning.

Keywords: Reinforcement learning · Inverse reinforcement learning · Apprenticeship learning

1 Introduction

Reinforcement learning (RL) is a framework to achieve the optimal action policy that maximizes expected cumulative reward [1]. When applying the RL to an optimal control problem, the reward function is arbitrarily set such to allow the learning agent to achieve an intended objective in a given environment. An appropriate design of the reward function is important to efficiently learn the control policy that realizes the objective, because there are variety of reward functions for the same objective whereas different reward functions would give rise to different RL processes. A simple reward function which applies a large positive reward at a goal state and uniform, negative reward at the other states leads to a policy to track the shortest path to the goal state. However, the RL process with this simple reward function is usually very slow because the agent with an initial policy barely reaches the goal state. On the other hand, a gradually increasing reward function as approaching the goal state would accelerate

© Springer International Publishing Switzerland 2015
S. Arik et al. (Eds.): ICONIP 2015, Part I, LNCS 9489, pp. 724–732, 2015.
DOI: 10.1007/978-3-319-26532-2_80

the RL, but it can lead to local optimal solutions. There is no systematic way to determine a good reward function that accelerates RL while stably achieving the optimal policy in a given environment and objective.

Conversely to the RL, inverse reinforcement learning (IRL) estimates the reward function when an agent, called a reference agent, optimizes the expected cumulative reward, based on observation of the agent's behaviors. Especially when the reference agent achieves the objective in our optimal control problem, one possible way to set the reward function is to imitate this agent's one. In this study, we are interested in a case where a reference agent is adaptive to even variable environments; such a reference agent is called a learning agent. Estimating the reward function that a learning reference agent employed, we can expect that our RL agent acquires the optimal policy and even adapts variety of environments appropriately. The IRL was first applied to modeling of animal and human learning processes [2]. There are many studies to improve the estimation accuracy of reward functions in the framework of IRL [3–5]. IRL has also been introduced to apprenticeship learning which allows an agent to mimic the expert's policy [4,6,7]. In these studies, the reference agent was regarded as an expert which had realized the optimal policy based on its reward function.

In this study, we propose the new IRL which allows our agent to mimic a learning reference agent whose policy may not be optimal because of the reference agent still being in a learning process. The proposed method is preferable because of the following three advantages. First, learning behaviors of the learning reference agent would provide additional information to improve estimation of the reward function. Second, our method focuses on imitating learning process rather than on copying the optimal policy, which means that our agent can mimic the reference agent's way to adapt to even changing environments. Third, our agent does not need to know the state transition probability of the environment, because it also follows the model-free RL scheme of its own.

2   Background

2.1   Markov Decision Process (MDP)

A Markov decision process (MDP) is characterized by a state transition probability $p(s'|s,a)$ denoting the probability from state s to state s' by taking an action a, where $s, s' \in S$ and $a \in A$, and S and A are finite sets of possible states and actions, respectively. We assume action a is probabilistically determined at state s, according to a stochastic policy $\pi(a|s)$ that satisfies $\sum_{a \in A} \pi(a|s) = 1$. For a fixed policy π, action-value function Q^π is defined as

$$Q^\pi(s,a) = E_\pi \left[\sum_{t=0}^{\infty} \gamma^t R(s_t, a_t) | s_0 = s, a_0 = a, a_t \sim \pi(a_t|s_t) \right], \quad (1)$$

respectively, where $R(s,a)$ and $\gamma \in [0,1)$ are the reward function and the discount factor, respectively. From the definition of action-value function (1), the following Bellman equation for a policy π is derived,

$$Q^\pi(s,a) = R(s,a) + \gamma \sum_{\substack{s' \in S \\ a' \in A}} p(s'|s,a)\pi(a'|s')Q^\pi(s,a). \tag{2}$$

2.2   Reinforcement Learning

Reinforcement learning (RL) is a machine learning framework to obtain the optimal policy π^* that maximizes the expected cumulative reward, which is estimated by the summation of instantaneous rewards obtained in a course of trials and errors [1]. In this study, we assume each agent, either of a reference agent or our agent in hand, produces its actions based on its own policy and also updates its policy according to an RL algorithm. In particular, we used the following SARSA algorithm, which is a typical on-policy temporal difference learning, to update the action-value function,

$$Q_{t+1}(s_t,a_t) = (1-\alpha)Q_t(s_t,a_t) + \alpha\{R(s_t,a_t)+\gamma Q_t(s_{t+1},a_{t+1}))\}. \tag{3}$$

As a stochastic policy, we used the following soft-max (Boltzmann) policy

$$\pi_t(a_t|s_t) = \frac{\exp(\frac{1}{\tau}Q_t(s_t,a_t))}{\sum_a \exp(\frac{1}{\tau}Q_t(s_t,a))}, \tag{4}$$

where $\alpha \in [0,1)$ and τ are parameters that signify the learning rate and the greediness of policy, respectively. In this study, we assume the stochastic policy (4) and the value updating rule (3) of the reference agent are known for our agent which attempts to mimic the reference agent. On the other hand, the reward function $R(s,a)$ and hence the action-value function $Q_t(s,a)$ are unknown for our agent. Then, in order for our agent to mimic the reference agent, our agent should estimate the reward function and then the action-value function of the reference agent from its behaviors. In this estimation, we assume that the time-course of policy update $\pi_t = \pi_t(a|s; O_{t-1}, R)$ is characterized by the past observation of behaviors O_{t-1}, the unknown reward function R, the known action-value update rule (SARSA), and the known stochastic policy (soft-max).

2.3   Related Works

Inverse reinforcement learning (IRL) is a framework to determine the reward function that an agent is optimizing under the following conditions: (1) the behaviors of the reference agent are accessible in variety of circumstances, (2) the sensory inputs to the reference agent are accessible, and (3) the dynamic character of the environment (including the agent's dynamics) is known [2]. These conditions are translated into MDP framework by determining the unknown reward function R from the observed behaviors $O_T = \{(s_0,a_0),(s_1,a_1),\cdots,(s_T,a_T)\}$ with the known (stochastic) dynamics of the environment $p(s'|s,a)$.

   The first IRL was implemented as an iterative algorithm [2]. The IRL was used for apprenticeship learning [6], which tried to mimic experts' behaviors and a gradient-based optimization method for IRL was introduced [7]. In many IRL studies including those, the reward function of the reference agent was estimated based on the known state transition probability of the environment and the Bellman equation assumed to be used by the reference agent. Such information, however, may not be sufficient to estimate the unique reward function, then additional restrictions can be necessary , such as assuming that the difference in the value between optimal actions and non-optimal ones is large [3] and introducing a prior distribution of rewards [4]. When the action-value update rule is assumed to be known, a reward function can be estimated on situations where the state transition probability is unknown [8]. In the IRL studies above, the observation of the reference agent's behaviors O_T was implicitly assumed to be generated by the optimal policy. Because of this assumption, two issues arise; first, if O_T was generated by a premature policy, we cannot estimate the accurate reward function. Especially when the reference agent is adaptively behaving in, for example, a changing environment, it would be difficult for us to assume the agent employs the optimal policy in the present environment. Such situations are typical when mimicking behaviors of animals including humans. Second, even if the observation of behaviors has been generated by the optimal policy of the reference agent, the reward function estimated by the existing IRL methods does not incorporate the information observed when the reference agent is in the learning process. Hence, existing IRL methods have not provided any way to allow us to imitate learning process of the reference agent.

To overcome these two issues, we propose a new IRL method to estimate a learning reference agent's reward function based on the history of its behaviors on the process of optimization, and hence to allow a new agent to mimic the RL process of the reference agent.

3   Inverse Reinforcement Learning with a Developing Agent

3.1   Likelihood of Reward Function

Here, we present an idea that the stochastic policy $\pi_t(a_t|s_t; R, O_{t-1}, Q_0)$ is regarded as a likelihood of reward R and initial action-value Q_0, given the observations, s_t, a_t, and O_{t-1}. Note here that although this stochastic policy itself is non-stationary, our unknown reward function is assumed to be stationary. Then, our problem is to estimate the unknowns, R and Q_0, based on the available observations. This estimation can be performed in the framework of the maximum likelihood estimation. Given the observed history of the reference agent's states and actions, the likelihood becomes

$$\prod_{t=0}^{T} \pi_t(a_t|s_t) = \prod_{t=0}^{T} \frac{\exp(\frac{1}{\tau}Q_t(s_t, a_t))}{\sum_a \exp(\frac{1}{\tau}Q_t(s_t, a))}. \tag{5}$$

Note that we used the policy's non-stationarity here. The log likelihood becomes

$$L(R, Q_0) = \sum_{t=0}^{T} \left\{ \frac{1}{\tau} Q_t(s_t, a_t) - \ln \left(\sum_a \exp(\frac{1}{\tau} Q_t(s_t, a)) \right) \right\}. \tag{6}$$

Note that Q_t is an implicit function of R. There RL settings of the reference agent are arbitrary, but they are assumed to be known.

3.2   Matrix Representation of SARSA

Vector forms of the reward $\mathbf{R}$ and the action-value $\mathbf{Q}_t$ are defined as vectors whose each element corresponds to each pair of state s and action a. Using those vector notations, the SARSA update rule (3) is rewritten as

$$\mathbf{Q}_{t+1} = (\mathbf{I} - \mathbf{A}_{t,t})\mathbf{Q}_t + \mathbf{A}_{t,t+1}\{\mathbf{R} + \gamma \mathbf{Q}_t\}$$
$$= (\mathbf{I} - \mathbf{A}_{t,t} + \gamma \mathbf{A}_{t,t+1})\mathbf{Q}_t + \mathbf{A}_{t,t+1}\mathbf{R}, \tag{7}$$

where the $(i_{t_1} \times m + j_{t_1}, i_{t_2} \times m + j_{t_2})$-th element of $\mathbf{A}_{t_1,t_2}$ is α and all the other elements are 0, and $(s_t, a_t) = (s^{i_t}, a^{j_t})$. m is the size of the finite action set. Equation (7) is further transformed into

$$\mathbf{Q}_t = \mathbf{C}_t \mathbf{Q}_0 + \mathbf{D}_t \mathbf{R}, \tag{8}$$

where we defined $\mathbf{B}_t = \mathbf{I} - \mathbf{A}_{t,t} + \gamma \mathbf{A}_{t,t+1}$, $\mathbf{C}_{t+1} = \mathbf{B}_t \mathbf{C}_t$, $\mathbf{D}_{t+1} = \mathbf{A}_{t,t} + \mathbf{B}_t \mathbf{D}_t$, $\mathbf{C}_0 = \mathbf{I}$, and $\mathbf{D}_0 = \mathbf{0}$. Then, we can optimize the log likelihood with respect to the reward and the initial action-value according to the gradient-based optimization, leading to the maximum likelihood estimates. In actual implementation, we used BFGS method for the optimization [9].

4   Numerical Experiment

Here, we evaluated how well our new IRL method worked for mimicking a learning reference agent; the evaluation criteria are: how well the reward function of the learning reference agent was estimated, and how well the RL behaviors of the learning reference agent could be mimicked by a new RL agent employing the reward function estimated above. We prioritized the latter rather than the former. In this evaluation, we used a simple 2-state MDP task (Fig. 1), whose optimal policy is to move to state S2 when it is at an initial state S1 and to stay on S2 after having come to S2. The evaluation procedure consisted of the following four steps: (1) The reference agent performed the RL to obtain its optimal policy in the MDP above during which its history of states and actions was recorded. (2) The reward function of the reference agent was estimated by our IRL and other existing IRL methods, based on the behavioral history recorded in (1). (3) A new agent, called the target agent, was prepared and performed the RL with the reward function estimated in (2). (4) Compare the RL processes between the reference agent and the target agent. The reference agent performed the RL

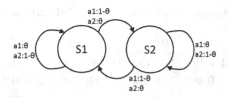

Fig. 1. A simple 2-state MDP task. When an agent takes an action $a1$ at state $S1$, it continues to stay at $S1$ with probability θ or moves to $S2$ with probability $1 - \theta$. θ is the parameter that characterizes this MDP. Since the reward at $S2$ is higher than that at $S1$, the agent should move to and stay at $S2$ to behave optimally.

with SARSA with a soft-max policy, whose RL settings were shared by the target agent. In steps (1) and (3) above, each history consisted of 200 episodes each of which contained 50 steps in the 2-state MDP environment. The true reward function $R^{\dagger}$ of the reference agent was $R^{\dagger}(S1) = -1.0$ and $R^{\dagger}(S2) = 0.1$. The learning parameters of the reference agent were set at $\alpha = 0.001$, $\gamma = 0.9$, and $\tau = 0.1$, which were also shared by the target agent.

We compared our IRL method (learning agent: LA) with three existing IRL methods, gradient method: GM, maximal posterior: MP, and dynamic programming: DP. The GM [7] minimizes the squared error between the reference agent's policy and the target agent's policy by a gradient-based optimization method. The MP [4] maximizes the posterior of reward functions by setting a prior on reward functions. These two methods, GM and MP, need the true state transition of the environment. The DP [8] assumes that the reference agent employs dynamic programming to obtain its own policy, but the target agent does not need to know the true transition probability.

Because the reference agent was continuously learning in its history of 200 RL episodes and hence changing by itself, the reward function estimated by the IRL could be different by using different periods in the history. To examine such dependency on the behavioral history, we prepared three kinds of sub-history by the reference agent: (1) an earlier part of the history (history 1), from the 1st to the 100th episodes, (2) a middle part of the history (history 2), from the 26th to the 125th episodes, and (3) a later part of the history (history 3), from the 101st to the 200th episodes. For each IRL algorithm, LA, GM, MP, or DP, we trained 100 target agents based on the reward function estimated by the IRL algorithm.

Table 1 shows the reward functions estimated by the four IRL methods. When estimated from earlier part of the history (history 1), our method (LA) exhibited the most accurate reward function. When estimated from middle or later part of the history, on the other hand, the DP was the best. Since our method (LA) assumed that the reference agent is on the course of learning, it showed the best estimation performance especially when the reference agent was in a premature stage, as expected. The other IRL methods performed better when they used observations after the reference agent was matured enough.

Table 1. Reward functions estimated by GM, MP, DP, and LA (proposed).

	GM	MP	DP	LA (proposed)
History 1	[−0.0384, 0.192]	[−0.134, 0.0936]	[−0.104, 0.104]	[−0.955, 0.104]
History 2	[−0.0365, 0.282]	[−0.0621, 0.106]	[−0.152, 0.152]	[−1.86, 0.105]
History 3	[−0.0305, 0.468]	[−0.118, 0.108]	[−0.238, 0.238]	[−4.25, 0.253]

Next, we examined the RL process of the target agent employing the reward function estimated by the four IRL algorithms, GM, MP, DP, and LA (Fig. 2). When evaluating a target agent's RL process, we defined a policy imitation error between the reference and target agents, as

$$J_e = \sum_s \mu'(s) \sum_a \left(\frac{\mu(s,a)}{\mu(s)} - \frac{\mu'(s,a)}{\mu'(s)} \right)^2, \qquad (9)$$

where $\mu(s)$ $(\mu'(s))$ and $\mu(s,a)$ $(\mu'(s,a))$ count the numbers of visits to state s and selecting action a at state s by the reference (target) agent, respectively.

Figure 2(a), (b) and (c) show profiles of the average policy imitation error by the four IRL methods, each of which used the earlier part of the reference agent's history (history 1), the middle part (history 2), and the later part (history 3), respectively. When the target agent performed the RL from the earlier or middle history part, our IRL (LA) showed best mimicking performance of the RL process by the target agent. When learning from the middle part, the GM showed performance being comparable to or even better than our LA after the RL process approached convergence. When learning from the later part, our LA showed a little worse mimicking performance than that by GM or DP when the target agent was premature, but became comparable after the target agent's RL approached convergence.

When comparing summation of policy imitation error over the 200 episodes of the target agent (Table 2), our LA showed the apparent minimum error with the earlier (history 1) or middle (history 2) part of the reference agent's behavioral history. When estimated from the latter part (history 3), on the other hand, our method (LA) was slightly inferior to the GM. When the information of the RL processes of the reference agent is available, like in history 1 or 2, our LA agent could fully utilize the information and then mimic the RL process of the reference agent. The GM was superior to other IRL methods when mimicking the reference agent's RL from its later learning stage (history 3), because it knows the dynamics of the MDP environment.

In total, our new IRL outperformed the existing IRL methods especially when reference agent's behaviors are available on its learning process, while the other IRL methods performed well only when the behaviors after the reference agent got matured were available.

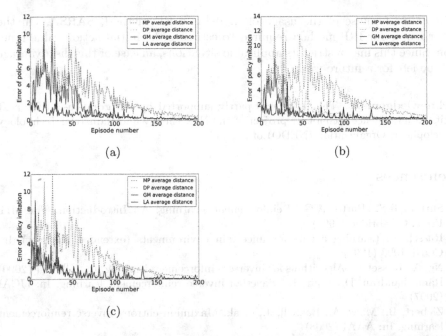

Fig. 2. Time-course of the average policy imitation error by the four IRL methods, each of which used the earlier part of the reference agent's history (history 1, panel (a)), the middle part (history 2, panel (b)), and the later part (history 3, panel (c)). The horizontal and vertical axes denote the episode number experienced by both of the reference and target agents, and the average policy imitation error, respectively.

Table 2. Summation of policy imitation errors along over the episodes.

	GM	MP	DP	LA (proposed)
History 1	293.6	552.9	559.3	115.9
History 2	172.3	698.9	272.5	106.5
History 3	114.7	492.6	135.6	132.4

5   Conclusion

We proposed an new IRL method that estimates the reward function of a reference agent based on its behaviors along its learning process and then allows our agent to mimic the learning reference agent. According to the numerical experiment using a simple 2-state Markov decision process, the proposed method showed better estimation performance of the reward function than that by existing IRL methods. Moreover, by letting our new agent do the RL based on the estimated reward function, we found our agent exhibited similar RL profiles with the reference agent, which suggests preferable performance of our new method when mimicking a reference agent being adaptive to even changing environments. In addition, our method did not need to know the true state transition probability because of the usage of model-free RL method, SARSA. On the other hand, our IRL method required to estimate the initial action-value function. Since this may restrict its applicable situations, an ease of this disadvantage will be left for a future work.

Acknowledgement. This work was partly supported by the grant-in-aid for next artificial intelligence technology project of the New Energy and Industrial Technology Development Organization (NEDO) of Japan.

References

1. Sutton, R.S., Barto, A.G.: Reinforcement Learning: An Introduction. The MIT Press, Cambridge (1998)
2. Russel, S.: Learning agents for uncertain environments (extended abstract). In: COLT 1998 (1998)
3. Ng, A., Russel, S.: Algorithms for inverse reinforcement learning. In: ICML17 (2000)
4. Ramachandran, D., Amir, E.: Bayesian inverse reinforcement learning. In: IJCAI (2007)
5. Ziebert, B., Maas, A., Bagnell, J., et al.: Maximum entropy inverse reinforcement learning. In: AAAI (2008)
6. Abbeel, P., Ng, A.: Apprenticeship learning via inverse reinforcement learning. In: ICML21 (2004)
7. Neu, G., Szepesvari, C.: Apprenticeship learning using inverse reinforcement learning and gradient methods. In: UAI 2007 (2007)
8. Tossou, A.C.Y., Dimitrakakis, C.: Probabilistic inverse reinforcement learning in unknown environments, arXiv (2013)
9. Shanno, D., Kettler, P.: Optimal conditioning of quasi-Newton methods. Math. Comput. **24**, 657–664 (1970)

Author Index

Printed in the United States
By Bookmasters